THE BERBERIS
OF CHINA
AND VIETNAM

A Revision

Xinjiang

Nei Mongol

Heilongjiang

Jilin

Liaoning

Shanxi

Hebei

1

4

Qinghai

Ningxia

Shandong

Gansu

Shaanxi

Henan

1 Beijing
2 Chongqing
3 Shanghai
4 Tianjin

Xizang

Sichuan

Hubei

Jiangsu

Anhui

2

Hunan

Jiangxi

Zhejiang

Guizhou

Yunnan

Guangxi

Guangdong

Fujian

Taiwan

Hainan

THE BERBERIS OF CHINA AND VIETNAM
A Revision

Julian Harber

MISSOURI BOTANICAL GARDEN PRESS

ISBN 978-1-935641-18-6
Library of Congress Control Number 2019957442
Monographs in Systematic Botany from the Missouri Botanical Garden, Volume 136
ISSN 0161-1542

Publisher: Liz Fathman
Managing Editor: Allison M. Brock
Editor: Lisa J. Pepper
Press Coordinator: Amanda Koehler
Cover design: Sam Balmer

This monograph was printed on 7 March 2020.

Front cover photo: *Berberis verruculosa*, photo by Dave Boufford. Back cover photos (clockwise): *Berberis nantoensis*, photo by Chih-Chieh Yu; *Berberis pratensis*, photo by author; *Berberis wilsoniae*, photo by David Boufford; *Berberis abbreviata*, photo by David Boufford.

Frontispiece: map of Chinese provinces, produced by Bruce Bartholomew.

This study is dedicated to the memory of my mother,
Mary N. Harber (née Whittaker), 1915–1967,
who taught me a love of plants.

Contents

Preface

This study originated in February 2006 when I was invited to be the non-Chinese co-author of the entry on *Berberis* in the English-language *Flora of China*, the Chinese author being Ying Junsheng. My starting point was an English translation of Ying's treatment of *Berberis* in the Chinese-language *Flora Reipublicae Popularis Sinicae* (2001: 54–214). I duly completed and submitted a revision of this treatment, but my changes and additions were not accepted by Ying. At this stage, I could have exercised my right under protocols of the Flora of China project to have my disagreements appended to Ying's text, but the changes were so numerous that the result would have been unreadable and would have generated confusion. Hence, it was agreed that Ying's original 2001 text should be published largely as it stood and I should publish my treatment separately. My treatment has been substantially revised since my original *Flora of China* submission and now includes many more new species. It was my original intention to publish two additional articles, the first providing a checklist of the locations of all relevant type specimens, lectotypifying where appropriate, and the second a revision of the *Berberis* of Vietnam. Both of these exercises are now incorporated into the current study. In addition, I have attempted to map the distribution of each species: in China at the level of Xian (often translated as county) and in Vietnam at the level of province.

Acknowledgments

This study would not have been possible without the help of a multitude of people and institutions.

For funding, I am grateful to the Royal Horticultural Society to enable me to visit herbaria in Beijing and Kunming; to the Missouri Botanical Garden who supported my visit to the St. Petersburg herbarium; and to the National Geographic Society who funded a collecting expedition in Yunnan and Sichuan in 2013 and in Taiwan in 2014. I am also grateful to the Vienna and Geneva herbaria for funding my visits there in return for identifying various *Berberis* specimens in their collections.

In the course of this study, I have visited 32 different herbaria in China (including Taiwan), the U.K., Europe, and the United States. These are listed in my introduction, and I am grateful for all the help I have received from their curators and staff. Many other herbaria have kindly sent me images of specimens that are not currently available on their websites or have helped in other ways. Thanks to the Edinburgh herbarium I have been able to receive specimens on loan from various non-U.K. herbaria. The Chinese Virtual Herbarium (CVH) and the Chinese National Specimen Information Infrastructure (NSII) websites have proved invaluable resources.

This study has benefitted from my being able to examine living *Berberis* in cultivation, and I owe a particularly big thanks to the curators and staff of the Royal Botanic Gardens at Kew, the Royal Botanic Garden Edinburgh (both at Inverleith and Dawyck), and the Royal Horticultural Society's garden at Wisley. Lord Charles Howick of Howick Hall, Northumberland, and the late Sir John Quick of Sherwood Gardens, Exeter, and their respective head gardeners Robert Jamieson and Vaughan Gallavan, made me welcome and allowed me to take specimens for study, as did the staff of the Sir Harold Hillier Gardens, Hampshire. As I note in my introduction, I have also benefitted from visits to the Exbury Gardens, Hampshire; St. Andrews Botanic Garden, Fife; Trewithen Gardens, Cornwall; and Quarryhill Botanical Garden, California, United States. I am also grateful to Bleddyn Wyn-Jones of Crûg Farm Plants, Caernarfon, Wales, and the late Michael Wickenden of Cally Gardens, Castle Douglas, Scotland, for (legally collected) *Berberis* seeds and plants.

For access to the manuscript material cited here, I am grateful to the Lindley Library of the Royal Horticultural Society, the British Library, and the libraries of the Royal Botanic Garden Edinburgh, the Royal Botanic Gardens, Kew, and the botanic library of the Natural History Museum, London.

I am grateful to Tatyana Shulkina for translations from Russian, to Gunhild Woodcock for help with difficult bits of German, and to Libing Zhang for help with difficult bits of Chinese and much else.

I owe particular thanks to the following people: Anthony Brach for searching out and scanning rare botany articles for me; Nicholas Turland for guiding me patiently through the intricacies of typification; Mike Gilbert who parsed my descriptions thus making the construction of my keys possible; and Bruce Bartholomew who produced my distribution maps. David Boufford has been a constant source of advice, information, and encouragement and a meticulous reader of various drafts of this study. Chih-Chieh Yu and Bhaskar Adhikari, with whom I have exchanged hundreds of emails over the years, have carried out numerous investigations on my behalf. My editor Lisa Pepper and copyeditor Mary Ann Schmidt subjected drafts of my text to forensic interrogation for which I am very grateful.

The help of other individuals is acknowledged in the discussions under various of the species' entries.

Finally, I need to thank Peter Raven who suggested this study, Roy Lancaster for his enthusiastic support over the years, and Jill Liddington for tolerating a garden full of spiky plants.

INTRODUCTION

Berberis is a member of the Berberidaceae. Historically, there has been disagreement as to which genera the family should include. For example, Hutchinson (1959), Airy Shaw (1965), and Takhtajan (1969) recognized Nandinaceae (*Nandina* Thunb.) and Podophyllaceae (*Podophyllum* L.) as separate families while APG II (2003) and APG III (2009) placed both of these in the Berberidaceae. Mabberley (2008) included 14 genera in the family. Of these, 11 are in China (Ying et al., 2011). Yu and Chung (2017) published two further genera, *Alloberberis* C. C. Yu & K. F. Chung and *Moranothamnus* C. C. Yu & K. F. Chung. While *Berberis* is woody, all other genera except *Mahonia* Nutt., *Nandina*, and these two newly described genera are herbaceous.

Berberis has two main centers of distribution—first, across the Eurasian landmass stretching from the Sierra Nevada of southern Spain to the Russian Far East and Japan and as far south as Sri Lanka and second, Latin America with a concentration in the Andean spine. There are outposts from both of these areas: *B. hispanica* Boiss. & Reut. is found in North Africa and *B. holstii* Engl. intermittently in East Africa all the way down to Malawi. Madeira hosts one species as do the Philippines and Indonesia. There are two species in the United States and two more in the remote Juan Fernández Islands off the coast of Chile. The total number of *Berberis* species is unknown. The last worldwide monograph of the genus (Ahrendt, 1961) listed 496, but a much lower number is suggested by the fact that Landrum (1999) recognized only 20 of the 60 species reported by Ahrendt for Chile, or that Adhikari et al. (2012) recognized only 19 of the 27 species Ahrendt reported for Nepal. This might appear to be reinforced here, given that of the 192 species Ahrendt reported for China only 125 are accepted in this revision (with an additional 12 he reported from outside China now being reported as also being found in China). However, the situation is somewhat more complicated than this in that 140 species not to be found in Ahrendt are recognized here. Given that this study is not exhaustive and I am confident that more new species from China will be published in the future, the total number of species worldwide may prove to be nearer to that given by Ahrendt than has been recently thought.

TAXONOMIC HISTORY OF *BERBERIS* IN CHINA AND VIETNAM

CHINA

Given the part it plays in Chinese traditional medicine, *Berberis* was almost certainly known to Chinese herbalists in pre-Linnaean times, but, as was the case with many other genera in China, its delineation as *Berberis* was late and slow. The first Chinese species to be validly published was *B. chinensis* (Poiret, 1808), based on cultivated material dating from 1789 that was grown from seeds from China; the first wild-collected herbarium specimens of the species were made in 1793 by Sir George Staunton, the secretary to the first British embassy to China on a journey to meet the Qian Long Emperor. But the long-standing closure of the interior of China to foreigners meant that, with three exceptions, no further species were added until after the second Opium War (1856–1860), when the victorious powers of Britain and France forced the Qing dynasty to concede the right of foreigners to travel beyond the coast. The exceptions to this came from the very edge of the Chinese empire where Russian incursions into Songaria in the far west of Xinjiang led to the publication of *B. heteropoda* (Schrenk, 1841) and *B. integerrima* (Bunge, 1843) and the discovery of *B. sibirica* (Pallas, 1773, 1789) in the same area.

Little progress had been made by the time Regel (1873: 407–421) attempted a treatment of deciduous *Berberis* in the Northern Hemisphere. His study can now be seen as largely misguided since it treated a large number of taxa as varieties of two species, *B. vulgaris* and *B. sinensis* (itself an illegitimate name). If anything, Regel regressed matters in that he treated *B. chinensis* as a synonym of a variety, *B. sinensis* var. *angustifolia* (for the multiple confusions this caused subsequently, see the discussion of *B. chinensis* below) and mistakenly reduced *B. amurensis* (Ruprecht, 1857), whose range includes northeastern China, to a variety of *B. vulgaris* while not noticing *B. kaschgarica* (Ruprecht, 1869), treated here as a synonym of *B. ulicina*, also from west Xinjiang.

Shortly after Regel's 1873 publication, large numbers of specimens of many genera began to arrive in European herbaria, particularly those of St. Petersburg and Paris. In the former, these specimens were largely from further Russian incursions into China as part of the "Great Game." In the latter, they were mostly from collections made by French Catholic missionaries (Kilpatrick, 2014; C. M. Hu & Watson, 2015). Both the Russian and French collections included *Berberis* and led Maximowicz (1877, 1889a, 1889b, 1891) and Franchet (1883, 1885, 1887, 1889) to describe and publish new species from north and west China respectively. However, no species from Xinjiang and none of Franchet were among the eight *Berberis* wild species that were listed by Forbes and Hemsley (1886–1888) in their attempt to enumerate all the plants in China then known in the West.

In the early 20th century, the key centers for the study of the flora of China shifted from Russia and France to Germany and Britain and a little later to America (the emergence of Germany as a major center for the study of Chinese plants is illustrated by the fact that the entire Chinese botanic collections of the Italian missionary Giraldi and the Austro-Hungarian diplomat Von Rosthorn were sent from Florence and Oslo respectively to Berlin for identification). The British and American interest in Chinese flora led to and was stimulated by the botanic expeditions of George Forrest, Frank Kingdon-Ward, Joseph Rock, Augustine Henry, Ernest Wilson, and others.

But the actual study of *Berberis* in China only really took off with the publication in 1904–1905 by a German horticulturist turned botanist, Camillo Schneider (1876–1951), of the first attempt to delineate the genus worldwide. This initial study (which reported 18 species from China) was based on specimens in five European herbaria (B, G, M, W, and WU), but his knowledge was subsequently increased through an expedition to Yunnan and Sichuan with Heinrich Handel-Mazzetti and by being employed both before he left for China and in 1914–1915 by the Arnold Arboretum in the United States to catalog and identify the numerous *Berberis* specimens collected in China by Ernest Wilson, as well as by visits to or the loan of specimens from at least 10 other herbaria in Europe and the United States. This led to a long series of publications that established him as the world's leading expert on the genus, especially the *Berberis* of China.

In November 1923, Schneider approached the British Royal Horticultural Society (RHS) with a proposal for a definitive study of the genus. In 1929, after some considerable negotiation terms were agreed. It is clear from the minutes of the RHS that both parties expected this study to be completed within a few years. However, the task proved much more difficult than Schneider anticipated (though it is apparent from Vierle [1998] that Schneider's *Berberis* project was only one among his numerous other activities in the 1930s), and it was only in late 1938 that he sent the RHS a manuscript of the evergreen section *Wallichianae*, promising a further three parts, "the Asiatic deciduous group, the South Americas Group [and] the garden hybrids." In January 1939, the RHS Publications Committee debated whether this manuscript should be published separately but decided to await the further parts, a decision which Schneider welcomed. Eight months later, war was declared and contact was lost, and subsequently Schneider (1942) published his study of section *Wallichianae* in German. On 27 January 1947, contact was restored when Schneider wrote to the RHS stating,

"When the Botanical Museum in Berlin-Dahlem was entirely burnt out in March 1943 I lost all that I had there. Grievous above all was the loss of the manuscripts, plates, figures and other material I had brought together in the course of 35 years for my *Berberis* monograph . . . There is not the slightest possibility of resuming this work" (reported in minutes of RHS Publications Committee, 4 March 1947). Despite this tragedy, Schneider's contribution to the study of *Berberis* was outstanding. Of the 105 species he identified as being in China, 88 are accepted here (but none of his 13 varieties), and 52 of these were published by him. From the evidence provided, by his immaculate line drawings of his dissections of flowers and fruit found on various specimen sheets of deciduous *Berberis* in A, C, E, K, S, US, and elsewhere, there is little doubt that a number of the new species published here would have been published by Schneider had he been able to complete his monograph (for further details of Schneider's life, see Kreichbaum [1951], Stearn [1951], and Vierle [1998]). All in all, he was a remarkable taxonomist, especially given the relatively limited amount of material he had to work on. One of the reasons he has perhaps not been given the full due he deserves is that the majority of his publications are in German.

The torch was then taken up by an Englishman, the Reverend Leslie Ahrendt (1903–1969), rector of Broughton near Banbury in Oxfordshire and another horticulturist turned botanist. He had already started publishing articles in 1939 (it is possible that one of the reasons for the publication by Schneider in the same year of extracts from his *Wallichianae* treatment detailing new species was to forestall Ahrendt from doing so). In 1941, Ahrendt also approached the RHS Publications Committee with a proposal for a definitive study of *Berberis*, but presumably because they were still hopeful that the contract with Schneider would eventually be fulfilled, the proposal was merely noted and it was not until 1961 that such a study by Ahrendt was published by the Linnean Society of London (Chamberlain & Hu's statement [1985: 529] that this is "an extensively edited version of notes left by Ahrendt on his death" is mistaken).

Although 28 Chinese *Berberis* species published by Ahrendt are accepted here, Ahrendt was not as accomplished a botanist as Schneider. This can be perhaps partially explained by his limited experience. On his own account (1961: 2), Ahrendt examined specimens in only four herbaria, BM, E, K, and OXF. And it appears the only specimens he saw from non-British herbaria were a few Nepalese ones sent to him for identification by Geneva (see Ahrendt, 1956). Nor, it seems, did he ever see any *Berberis* in the wild, not even the only possibly native British species *B. vulgaris*. From contextual evidence, he also appears not to have been able to read German. But one of his other key limitations has long been known: his publication of new varieties based on minute variations of such characteristics as leaf size, shape, and marginal spines, characteristics which both Chamberlain and Hu (1985: 530) and Landrum (1999: 795) note are so variable that they can sometimes occur on the same plant or on plants of the same species growing together. Given that Ahrendt had ready access to his own large collection of living *Berberis* growing in his parents' garden outside Oxford, it is somewhat of a mystery why he failed to or chose not to notice this variability. The only explanation seems to be an insatiable appetite for novelties. It is perhaps this appetite for the new that explains another and, in this case, little noticed of Ahrendt's limitations and that is his lack of rigor in relation to the provenance of cultivated plants and specimens. On a number of occasions, which are recorded here, he published new species on the basis of cultivated plants reputedly grown from wild-collected seed, although there are wild-collected specimens (mostly in the British herbaria with which he was familiar) with the same collector's number that are unmistakably different. *Berberis* is notoriously difficult to propagate vegetatively, and species brought together in cultivation are often subject to hybridization (see Schneider, 1923; Ahrendt, 1961 passim; Harber, 2010); and it is likely that plants supplied to Ahrendt were on occasion hybrids grown from cultivated rather than wild-collected seed (for further details, see under Excluded Taxa).

The charges against Ahrendt unfortunately do not end there. As noted by Boufford (2013) in relation to *Mahonia*, when faced with protologues of *Berberis* in which no type was indicated and when Ahrendt had not seen any of the specimens cited (and in some cases, as with *B. zanlancianensis*, did not even know where they were to be found), he simply chose the first one listed as the type.

He also treated as types all cultivated specimens grown from seeds of type collections. His reporting (Ahrendt, 1961) of types was also at times inaccurate (see, e.g., under *Berberis feddeana* and for non-Chinese species *B. sieboldii* Miq. and *B. amurensis* var. *latifolia* Nakai). On one occasion, having designated a holotype of *B. humidoumbrosa* var. *inornata* in one publication (1945a: 116), a quite different type was cited in a subsequent publication (Ahrendt, 1961: 160).

Further charges include making radical changes to protologue descriptions based on specimens clearly of quite different species (see, e.g., under *Berberis dawoensis*); providing keys that on occasion contradict information given in the species descriptions (see, e.g., under *B. minutiflora*); and citing cultivated specimens as grown from particular wild-collected numbers, even though these collections were gathered well outside the fruiting season and have only floral material (see, e.g., under *B. microtricha*). For some particular confusions caused by Ahrendt that are unscrambled here for the first time, see under *B. sanguinea* and under *B. veitchii* in relation to *B. gagnepainii* var. *lanceifolia*.

But perhaps the best evidence of Ahrendt's limitations can be seen in his approach to various specimens that arrived in British herbaria after Schneider's last publication covering deciduous species (Schneider, 1916, 1918) or in the case of specimens from Bhutan, Myanmar, and Tibet collected in the period 1936–1947, ones that Schneider never saw. Here, with no previous studies to build on, he appears often to be at sea and made identifications that can be at best described as speculative or even eccentric. This included publishing new varieties of existing species from exceptionally poor specimens (see, e.g., under *Berberis franchetiana* var. *glabripes*), ones whose characteristics bear little relationship to the taxon from which they are proposed to differ varietally (see, e.g., under *B. minutiflora* var. *glabramea* and *B. amoena* var. *moloensis*) and including in descriptions features taken from specimens that are clearly different species from those that he designated as being the type (see, e.g., under *B. ludlowii*).

In the 1920s and 1930s, important botanical institutions were established in China (Z. G. Hu et al., 2015), and foreign herbaria came increasingly dependent for Chinese specimens on local collectors, including K. L. Chu, W. P. Fang, C. Y. Chiao, R. C. Ching, H. T. Tsai, and T. T. Yu. However, the only *Berberis* species found in China authored by a Chinese botanist that Schneider and Ahrendt could report in their various publications was *B. chingii* (Cheng, 1934). *Berberis* did not appear at all in *Icones Plantarum Sinicarum* (H. H. Hu & Chien, 1927–1937).

In the post-war period, the first new *Berberis* taxa for China authored by Chinese botanists were published by P. Y. Li (1965) and Hsiao and Sung (1974), with further new species in Ying's (1985) contribution to the *Flora Xizangica* and in C. M. Hu (1986) and Zhang (1991). But the most significant publications of new species were by Bao (1985) and Ying (1999), the former drawing largely on specimens at KUN identified as types of new species by C. Y. Wu but not published by him. Chamberlain and Hu (1985) published a synopsis of section *Wallichianae*. They recognized only two new taxa for China but many names of Chinese species in synonymy, a majority of which are accepted here. In 2001, Ying published a full-length study of Chinese *Berberis* for the *Flora Reipublicae Popularis Sinicae*, which as noted in the preface appeared largely unchanged in the English-language *Flora of China* (Ying, 2011). Of the 215 species recognized by Ying, 151 are accepted here.

Ying's (1985) contribution to *Flora Xizangica* was only one among the various provincial floras issued from 1981 onward with entries for *Berberis* (the process is still incomplete; at the time of writing, that for Sichuan has not yet reached Berberidaceae). The published provincial floras (e.g., Institute of Botany in Jiangsu Province & Chinese Academy of Sciences, 1982; Liu, 1988; He, 1992; Zhu, 1997; Liu & the Editorial Committee of Shanxi Flora, 1998, etc.) are all listed in the bibliography, but it needs to be reported that, with one exception (C. M. Hu [1995] on Guangdong), they all contain mistakes. For some from northeast provinces, though, the mistakes are minor, amounting to no more than reporting *B. chinensis* by its synonym *B. poiretii*. In contrast, only 49 of the 89 species claimed for Yunnan by Bao (1997) are accepted here.

Taiwan

The taxonomic history of *Berberis* in Taiwan differs considerably from that of mainland China. The island was occupied by Japan from 1895 to 1945, and the first five Taiwanese *Berberis* species were published by the Japanese botanist Hayata (1911, 1913, 1915). They were based on specimens made by Japanese collectors, as were two further species published by Japanese botanists in the post war period: Mizushima (1954) and T. Shimizu (1963, 1964). Two other Taiwanese species were published by Schneider (1939: 252–253) and one, *B. formosana*, by Ahrendt (1941b: 24), later (1961: 65) reduced by him to a variety of *B. kawakamii*. Unfortunately, a number of Taiwanese authors subsequently muddied the situation by mistakenly synonymizing many of these species. Mistaken synonyms of Taiwanese *Berberis* were also made by Chamberlain and Hu (1985: 539), and their mistakes were repeated by Ying (2001, 2011), the latter also failing to notice *B. tarokoensis* (S. Y. Lu & Yang, 1996). A full account of this history is in Yu and Chung (2014), their table 2 being particularly useful. I have drawn heavily on their article, in which they also published three new species.

VIETNAM

None of the major works concerning Asian *Berberis* (Schneider, 1905, 1942; Ahrendt, 1961; Chamberlain & Hu, 1985) described the genus as being found in Vietnam. This was perhaps not surprising since the only references to *Berberis* in Vietnam at the time they were writing were easily overlooked (I only researched Indochina flora after finding *Berberis* from Vietnam in the Paris herbarium). Thus, Gagnepain (1938: 144), citing Petélot as the collector, reported *B. wallichiana* from Chapa (Sapa) in Lao Cai Province in northern Vietnam, the report being repeated by Petélot himself

(1952: 45). Given that *B. wallichiana* is otherwise only found in Nepal, this identification was always implausible but appears never to have been challenged. Indeed, it has been reproduced by successive authors, e.g., Nguyên (1998: 45), Pham (1999: 326), Võ (2007: 120), and Hien et al. (2018). Under this name it was listed in the *Red Data Book of Vietnam* in 1996 and was subsequently awarded protected status by the Government of Vietnam (Government of the Socialist Republic of Vietnam, 2006). The species is identified here as *B. ferdinandi-coburgii*, widely distributed in south Yunnan in China.

The *Red Data Book of Vietnam* listed a second species from Sapa, *Berberis julianae*, which was also awarded protective status in 2006 and is included in Nguyên (1998: 45), Võ (2007: 120), and Hien et al. (2018). However, *B. julianae* is endemic to Hubei and neighboring areas of Chongqing, Guizhou, Hunan, and Shanxi in China. This second species is identified here as *B. subacuminata*, also widely distributed in south Yunnan.

Two other *Berberis* species from Vietnam are reported here, the first, *B. hypoxantha* (Bao, 1985), again from the north of the country, is also found in China; the second, *B. vietnamensis*, is new and is from southern Vietnam.

Scope, Materials, and Methods

This study covers Vietnam and the whole of China (including Taiwan) with the exception of that part of southeast Xizang/Arunachal Pradesh where China and India have conflicting territorial claims. This exclusion should not be taken as expressing any view on the issue but simply reflects the fact that I do not know enough about *Berberis* in this particular area to be able to give any comprehensive account. There are a small number of specimens of evergreen species from the area in U.K. herbaria (principally at BM), and various plants (both evergreen and deciduous) grown from wild-collected seed have appeared in cultivation in the U.K. in recent years (some of the deciduous plants being almost certainly of unrecognized species); but together they provide only very limited evidence. It is likely that further evidence can be obtained from additional specimens in Indian herbaria, but I have not investigated this. It should be noted, however, that the maps in this study use software developed for the Flora of China project and, as such, show Xizang as including this disputed territory.

The study is based mainly on specimens in the following herbaria listed in Index Herbariorum (Thiers, 2017): A/GH, AU, B, BISH, BJ, BKL, BM, BNU, BR, C, CAF, CAL, CAS, CCAU, CDBI, CGE, CM, CNU, CQNM, CSFI, CSH, CZH, DB, DD, E, F, FAN, FI,

FJSI, FNU, FUS, G, GB, GF, GFS, GXMG, GXMI, GZAC, GZTM, GZU, H, HAST, HBG, HEAC, HENU, HGAS, HHBG, HHE, HHNNR, HIB, HILL, HIMC, HITBC, HK, HN, HNWP, HNU, HSIB, HTC, IBK, IBSC, IFP, IGA, IMC, IMDY, IMM, JIU, JJF, JMSMC, K, KH, KUN, KYO, L, LBG, LD, LE, LINN, LIV, LZU, M, MANCH, MBK, MEXU, MO, MPU, MSB, MW, N, NA, NAS, NAU, NENU, NF, NIMM, NKU, NMAC, NS, NTUF, NY, O, OXF, P, PE, PEM, PEY, PH, PHN, PR, PRG, S, SHM, SING, SITC, SKK, SM, SNU, SNUA, SRP, SWCTU, SWFC, SYS, SZ, SZG, TAI, TAIF, TCD, TCF, TEX, TI, TIE, TNM, TUS, U, UBC, UPS, URV, US, VNM, W, WCSBG, WH, WNU, WU, WUK, WYS, XBGH, XJA, XJBI, XJU, YAK, YUKU, ZM, ZY. Articles published by Chinese botanists between 1965 and 1986 sometimes cited herbarium acronyms at variance with those found in Index Herbariorum; these have been deciphered by reference to Anonymous (1982).

The study has also drawn on specimens from various Chinese herbaria not currently listed in the Index, those of: Hunan Food and Drug Vocational College (HUFD); North East Agricultural University, Heilongjiang (NEAU); Guizhou University of Engineering Science (GUES); Baxian Shan Nature Reserve, Tianjin; Dapan Shan Nature Reserve, Zhejiang; Ganzi Institute of Forestry, Sichuan; Mayang He, Guizhou; Shennongjia Forest, Hubei; Southwest Jiaotong University, Sichuan; Wuling Shan Research Institute, Hubei; Wutai Shan National Nature Reserve, Hebei; and Yaoluoping National Nature Reserve, Anhui. Simple logistics has meant that of these I have been able to visit only A/GH, B, BKL, BM, BR, CAS, CGE, E, G, GZTM, HGAS, HILL, K, KUN, LE, LIV, MANCH, MO, NY, OXF, P, PE, PR, PRG, SWFC, TAI, TAIF, TNM, UC, W, WU, and WYS. For the remainder, I have relied on loans of specimens to E or much more frequently on images either from herbarium websites or from those sent to me.

As a result, I have seen either in person or via an image one or more specimens of the type collection of every taxon of *Berberis* recorded for China, with only two exceptions: *B. feddeana* and *B. tischleri*. The only specimen of the type collection of *B. feddeana* was lost with the destruction of the Berlin herbarium. The type specimen of *B. tischleri* may have suffered the same fate in that it was loaned to Schneider from LE in 1905 and may never have been returned (although all the other accompanying loans appear to have been). Although many hitherto unrecorded duplicates of type specimens are recorded here, it is likely that others await discovery.

It is necessary to report here that no fewer than 20 type specimens of *Berberis* from China (and one from India relevant to this study) that Ahrendt (1961) cited as being at Kew are missing (there may be others missing from India and from elsewhere but I have not investigated this). All of these are of species published in 1939 or before (mostly before and mostly by Schneider and only one, *B. panlanensis*, by Ahrendt himself). Kew's loans books confirm that on various occasions numerous unspecified specimens of *Berberis* were sent to Ahrendt's home in Oxfordshire. For the post-war period, the entries appear to confirm their return. The pre-war loan books, however, are more difficult to interpret, and my provisional conclusion is that these specimens were sent to Ahrendt at the start of his *Berberis* career just before WWII and were never returned. I have lodged an account of the missing specimens with Kew. Fortunately, there are duplicates of all of them elsewhere.

There is also a need to report on another tranche of *Berberis* type specimens that have been lost. These are those described in Schneider (1939) as being in "Herb. Dendrol. C. Schneider" and are ones he collected in China in 1914. He almost certainly kept these in the Berlin herbarium, and, as such, they will have been lost with its destruction. But again, fortunately, duplicates of all of these are elsewhere. (Confusingly, there is a second Schneider herbarium with the same name at BKL, but this consists of cultivated specimens from the Arnold Arboretum.)

It should be noted that whenever possible barcode (or sheet number) references are given for all specimens cited. This is for three reasons. First, this is usually the easiest way to locate a specimen in a herbarium database and often the only way if it does not have a collector's number. Secondly, a very high percentage of specimens in herbaria (particularly Chinese ones) are either misidentified or unidentified, in the latter case sometimes only as Berberidaceae rather than *Berberis*. Thirdly, because so many specimens are misidentified trying to locate them in situ (especially in herbaria with large *Berberis* collections) is likely to prove very time-consuming, but with a barcode it is possible to enter them in the relevant database and find out what they have been misidentified as. When feasible, I have added my own determination slips to specimens, but to do so for all the herbaria listed above would be an impossible task. Even were this to be possible, few herbaria have systems for regularly updating their databases of existing specimens. It should also be borne in mind that (a) for various herbaria the process of adding barcodes to sheets is ongoing and may not extend beyond type specimens and (b) Chinese websites do not distinguish between barcodes and accession numbers; the barcode field 条形码 (tiaoxingma) being used for both.

In addition to herbarium specimens, I have drawn on living *Berberis* seen in the field and collected in Qinghai, Sichuan, Yunnan, and Taiwan, as well as living plants in cultivation in the U.K. in my own garden in Mytholmroyd in West Yorkshire and at the Royal Bo-

tanic Gardens, Kew; the Royal Botanic Garden Edin-
burgh; Exbury Gardens, Hampshire; Howick Hall,
Northumberland; Ness Botanic Garden, Cheshire; the
Royal Horticultural Society Garden, Wisley, Surrey;
Sherwood Gardens, Devon (now closed); the Sir Harold
Hillier Gardens, Hampshire; St. Andrews Botanic Gar-
den, Fife; Trewithen Gardens, Cornwall; and Quarry-
hill Botanical Garden, California, United States.

BERBERIS AND *MAHONIA*

The close relationship between *Berberis* and *Mahonia* (Nuttall, 1818: 212) was recognized early, and indeed many species of *Mahonia*, particularly those from
the United States and Mexico, were named as *Berberis*
before Fedde (1901) published the first account of the
genus *Mahonia* as a whole. This led to widespread
(though not universal) acceptance that *Berberis* and *Mahonia* were separate genera, so much so that Ahrendt
(1961: 296) could refer to *Mahonia* as a name "which
has now become thoroughly established," giving as the
key difference that "The leaves of *Berberis* are simple,
of *Mahonia* pinnate. This is the one universal distinction to which there is no exception."

Debate as to whether *Mahonia* was separate from
Berberis was recommenced by Moran (1982), who in
publishing *B. claireae* Moran, a species with both simple and pinnate leaves endemic to a few canyons in
Baja California, Mexico, asked, "Is it certain that *Berberis*, separated only by simple leaves is monophyletic?" It is worth quoting the key paragraph of his answer (Moran, 1982: 222).

> The primary leaves in most species of *Berberis* s.s. are re-
> duced to simple or branching spines and in a few others are
> transitional between leaves and spines; and the foliage
> leaves are crowded on axillary short shoots. Thus the char-
> acter of simple leaves in *Berberis* s.s. usually is supported
> by the presence of spiniform primary leaves and of short
> shoots. However, at least *B. insignis* Hook. f. & Thomps.
> and relatives, of the eastern Himalaya, have simple primary
> leaves and no spines or short shoots. The leaves are pin-
> nately veined as in *Berberis* s.s. and like the leaflets of most
> American species of *Mahonia* but unlike those of (most?)
> Asian species. The question is whether these plants can be
> simple-leaved members of *Mahonia*, spoiling the one key
> character of a one-character genus? Or are they best called
> a third group more or less coordinate with the other two?
> Thus the distinction of *Mahonia* from *Berberis* remains un-
> clear. Shifting species from one group to the other may per-
> haps make the groups more natural if less easily defined.
> For the present, however, *Mahonia* is hard to defend as a
> distinct natural group, and I prefer to keep it in *Berberis*.

The argument of this paragraph (where the concept
of "*Berberis* s.s." seems to have made its first appearance) rests on a false premise concerning *B. insignis*
Hook. f. & Thompson and "relatives of the eastern Himalaya," a reference to the four species in Ahrendt's

subsection *Insignis* of section *Wallichianae* (1961: 36–
39), the other three being *B. dasyclada* Ahrendt, *B.
incrassata*, and *B. hypokerina* Airy Shaw. Here, contrary to both Ahrendt and Moran, only *B. hypokerina*
appears to be completely spineless (and even may not
be, the evidence here is extremely limited). The other
three species can have spines; for *B. insignis*, see
Adhikari et al. (2012: 516); for *B. incrassata*, see the
discussion under that species below; for *B. dasyclada*,
the evidence comes from plants at Sherwood Gardens,
Devon, U.K., and in my own collection, both grown
from *A. Clark 5384*, West Kameng, Arunachal Pradesh,
India, autumn 2004. In all three of these species, the
spines are small, weak, and infrequent, and an obvious
explanation is that, as with *B. hypokerina*, these species are found in areas of dense vegetation and hence
in evolutionary terms have less or no need for spines as
a defense against animal depredations (for similar reduced spines, see also *B. hypoxantha* below from the
warm temperate forests of southern Yunnan and northern Vietnam).

It is unclear what significance Moran attributed to
his assertion that the leaves of all *Berberis* and American *Mahonia* are pinnately veined, whereas those of all
or most Asian *Mahonia* are not (the actual situation is
somewhat more complex than this). Laferrière (1997),
however, misread Moran to be saying that whereas *B.
insignis* etc. were pinnately veined the majority of Asian
Berberis were not. And this misreading was part of his
justification for transferring all *Mahonia* to *Berberis*,
asserting that "*Mahonia* is probably paraphyletic, while
Berberis sensu stricto is quite possibly polyphyletic,"
giving only as evidence of the latter that "Differences
in venation in various simple-leaved species suggest
that they may be more closely related to certain com-
pound-leaved taxa than they are to each other (Moran,
1982). Hence, reduction in leaflet number may have
occurred more than once, making *Berberis* sensu stricto
an artificial group" (Laferrière, 1997: 96).

Subsequent molecular analyses, Kim et al. (2004),
Adhikari (2010), Roy et al. (2010), and Adhikari et al.
(2012) (and which in the case of both Kim et al. and
Roy et al. included *Berberis insignis*), however, confirmed *Berberis* to be a monophyletic taxon. This both
Kim et al. and Adhikari et al. reported was not the case
with *Mahonia*, which they confirmed to be paraphyletic
due to species within "Gruppe" *Horridae* Fedde (1901),
section *Horridae* (Fedde) Ahrendt (Ahrendt, 1961).
Given this, within their conceptual framework, their
conclusion was that the only way that all *Mahonia* species could be grouped under a single genus was through
the concept of "*Berberis* s.l.," a name which of course
included "*Berberis* s. str."

The way out of this taxonomic impasse was finally
provided by Yu and Chung (2017), who on the basis of

an extensive phylogenetic analysis proposed that *Mahonia* sect. *Horridae* should be regarded as constituting a separate genus, *Alloberberis*, and that *Berberis claireae* should be reclassified as a species of a further new genus, *Moranothamnus*.

Subgeneric Classifications of *Berberis* and Molecular Analysis

Both Schneider (1905, 1908, 1916, 1918, 1942) and Ahrendt (1941b, 1942, 1944a, 1945a, 1961) classified *Berberis* into various sections and subsections based on apparent morphological similarities (but in some cases clearly also taking into account geographic considerations), there being significant revisions over time in both of their cases as well as differences between them.

Ahrendt's classifications were first called into doubt by Landrum (1999), who synonymized three species from Chile and Argentina, each of which Ahrendt had placed in a different section. Some of the synonyms made in this study also bring into question Ahrendt's classifications. For example, he placed *Berberis solutiflora* and *B. spraguei* var. *pedunculata* in his section *Polyantha* and section *Franchetiana* respectively. But both are treated here as synonyms of *B. concolor*, which Ahrendt placed in his section *Vulgaris*.

With the development of molecular studies, the subgeneric classification of *Berberis* has largely fallen into disuse, though it is still followed by some Indian botanists, see, e.g., Rao et al. (1998a, 1998b) or Tiwari and Adhikari (2011). In China, the last authors to use infrageneric classification were Hsiao and Sung (1974). It was not used in Ying (2001, 2011) or in any of the Chinese provincial floras. I have followed this practice but with one exception: section *Wallichianae* (Schneider, 1905). This, as Chamberlain and Hu (1985: 529) noted, "is a well-defined natural group of species, the circumspection of which is not disputed." I have not, however, followed their division of section *Wallichianae* into subsections and series.

This study is by necessity purely a morphological one, given that I do not have access to the facilities that would enable me to carry out any molecular investigations. But, since enumerating the number of species in a particular genus needs to be more than a bean count, I have tried to keep abreast of the various developments in the molecular study of *Berberis*. And here I hope two comments are in order.

The first is that a number of researchers have not taken enough care in ensuring that the material they are analyzing is what they state it to be. Without such assurances, molecular analysis is pointless and the results worthless (except possibly when investigating the relationship between *Berberis* from widely separated regions, e.g., the Himalayas and the Andes, where individual species determinations may not be important). There is certainly a challenge here since, as Landrum (1999: 794) noted, large groups such as *Berberis* "are taxonomic 'black holes' because no one can understand them in a reasonable number of years, or even a lifetime." The one obvious way to ensure accuracy in identification would be to use material from type specimens. But even if allowed (and many herbaria would not give permission) such material is unlikely to be of the necessary quality. Therefore, researchers usually have to rely on material from living plants either in cultivation or in the wild, often relying on the identification of others, identifications which may not be accurate. And here I must declare myself an aider and abettor of misidentification in that Adhikari et al. (2015) include in their molecular analysis material from a dwarf *Berberis* in my living collection. It was grown from seed from *Alpine Garden Society Exped. to China (ACE) 1847*, collected on Daxue Shan on the Yunnan–Sichuan border. From a wild-collected specimen of *ACE 1847* in the Edinburgh herbarium, which was single-fruited and which had leaves resembling those of the type of *B. minutiflora*, I assumed this must be that species and as such supplied the authors with material from it. The plant subsequently died, but a further plant from the same collection finally produced flowers in 2015, revealing it to be a new species that appears here as *B. scrithalis*.

Given this, I am therefore not in a strong position to criticize others. Nevertheless, here are examples from two other studies.

The first are from Roy et al. (2010). This includes as evidence specimens of *Berberis chitria* Buch.-Ham. ex Ker Gawl. from the Indian west Himalayas and ones of *B. replicata* from northwest Arunachal Pradesh. But, as was noted long ago (Stapf, 1926), *B. chitria* is an illegitimate name for *B. aristata* (treated separately in the authors' specimen evidence, which implies their *B. chitria* is something different) while *B. replicata* is not found in India, being endemic to a small area of volcanic lava beds in western Yunnan. (It should be noted that a subsequent molecular study of three Guizhou *Berberis* species by H. N. Li et al. [2011] uncritically accepted the identifications made by Roy et al. with the consequence that their placement of Guizhou species in a clade with other species almost certainly has no validity.)

The second is from Kim et al. (2004), whose study includes cultivated material from living plants of Chinese *Berberis* at the Royal Botanic Garden Edinburgh. Among them were a plant dating from 1939 identified as *B. sanguinea*, a plant of *B. lecomtei*, one of *B. candidula*, and one of *B. verruculosa*. The first is a cultivar, *B. sanguinea* var. *microphylla* (apud Creasey, 1935), of unknown origin often mistakenly referred to as *B. pan-*

lanensis and which bears little resemblance to *B. sanguinea* or any other known *Berberis* species, while the second (grown from *ACE 2072*, collected in northwest Yunnan) is actually *B. dictyophylla*. From the evidence of the RBGE accession book, the provenance of the particular plant of *B. candidula* they selected is also unknown beyond being donated to the garden in 1990 by the then Superintendent of Grounds of the University of Edinburgh. The provenance of the *B. verruculosa* planted in 1968 is not recorded at all. Seeds of *B. candidula* have only been collected once in the wild, by Farges, either in 1900 or sometime before, while up until 1988, when Charles Howick et al. collected seed in Baoxing Xian in Sichuan, all plants in the U.K. named as *B. verruculosa* were descendants of plants grown from seed collected by Wilson before WWI. In cultivation, *B. candidula* has frequently hybridized with *B. verruculosa*, producing various cultivars (see Hillier Nurseries, 2014: 39).

The conclusion I draw from all this is that, however time-consuming it may be, those engaged in constructing the phylogeny of *Berberis* should (a) be familiar with the type specimens of all the species they include in their molecular analysis and (b) subject non-type material and especially cultivated material to detailed interrogation, particularly to its provenance, as well as to how it was identified and by whom. This interrogation is particularly necessary for Chinese *Berberis* grown outside of China, where many plants such as RBGE's plants of *B. candidula* and *B. verruculosa* are descendants of ancient ones dating from the period between 1900 and 1938 (for further details of this in relation to the U.K., see Harber [2010]). It should be noted even for the RBGE, whose accession records go back to the early 19th century, the prescriptive recording of all new accessions to their living collections dates only from 1969 (P. Brownless of RBGE, pers. comm. 27 Nov. 2017). The challenge will, of course, vary from species to species. Thus, for example in relation to China, on current evidence, any wild-collected material from Jilin in northeast China or from any cultivated plant from wild-collected seed from there can only be either of *B. amurensis* or *B. chinensis*. At the other end of the spectrum, the utmost caution may be needed in relation to material originating from Sichuan, where 88 different species are reported here.

The second comment relates to my acceptance of section *Wallichianae*. With even a minimum of experience, identifying herbarium specimens or living plants as belonging to this section is straightforward. It is of course possible that the section is not monophyletic but polyphyletic, but no one has yet directly suggested this. Such polyphyly is however depicted in the Bayesian tree in Kim et al. (2004: 180) based on ITS sequences of 79 taxa (13 of which are from section *Wallichianae*),

but the authors note of the tree "divergence of the ITS sequence in *Berberis* is somewhat too low to provide a solid relationship at the species level" and only go as far as concluding that most *Berberis* sections or subsections are questionable. Polyphyly could be implied by figures 1 and 2 in Adhikari et al. (2015), which show four species of section *Wallichianae* in a monophyletic group but two, *B. hookeri* and *B. coxii*, in different groups along with various non-section *Wallichianae* species. However, as one of the authors, Bhaskar Adhikari (pers. comm. 25 June 2015), notes, "if you look at fig. 1, its position is unresolved, and in fig 2. the support value is very low so we can't make any conclusions about its position based on those phylogenies."

The cohesion of section *Wallichianae* was also brought into question by the phylogenetic tree in Yu and Chung (2014: 71), in which the non-section *Wallichianae* species *Berberis asiatica* was placed in a monophyletic group along with 34 species of section *Wallichianae*, but for *B. asiatica* the support value was again very low. I conclude from all of this that until any conclusive evidence to the contrary is produced the very distinct morphological characteristics of *Wallichianae* species justify its continued acceptance as a section.

MORPHOLOGY

Habit and reproduction

Berberis species in China range from semi-prostrate shrubs (*B. erythroclada*, *B. tsangpoensis*) through dwarf ones no more than 1 m tall (e.g., *B. minutiflora*, *B. qiaojiaensis*, *B. mianningensis*, *B. tsarica*, *B. ulicina*) to treelike shrubs (e.g., *B. brachypoda*, *B. dasystachya*, *B. polyantha*) recorded as being up to 4.5 m tall.

Most species are erect, at least two, *Berberis taliensis* and *B. grodtmanniana*, rigidly so. Two species, *B. mingetsensis* and *B. ravenii*, are recorded as being decumbent. One species, *B. hypoxantha*, is recorded in Vietnam as being a climbing or pendant shrub. Most species branch profusely from the base.

Berberis reproduces itself through seed, but some species can also spread by suckering. How common this is is unknown since it is not something that general collectors are likely to record. In 2013, I observed suckering with *B. phanera* in Muli Xian in Sichuan and in 2014 with *B. brevisepala* and *B. morii* in Taiwan. I have not observed this in cultivated plants except for *B. calliantha*, *B. cavaleriei*, *B. griffithiana*, *B. lijiangensis*, and *B. wui*. These eight species are all in section *Wallichianae*.

Seedlings

The first leaves of *Berberis* seedlings always have a long petiole and spinose margins, but in most species the former feature is only temporary and soon disappears with age, whereas the latter feature persists in many species albeit usually with differently shaped leaves. For species of section *Wallichianae*, the first leaves of seedlings are mostly or perhaps always abaxially white or glaucous pruinose. These leaves may persist in young plants beyond the seedling stage but in most species soon wither. Abaxially pruinose leaves in very young plants in section *Wallichianae* are therefore not reliable characteristics for distinguishing species.

Roots

The roots of *Berberis* are a distinctive vivid yellow.

Stem

Stems are terete, angled, or sulcate. They are sometimes categorized as being of two types: primary stems (long shoots) and short axillary stems (short shoots); but in many species, though the longest stems grow from or near the base, short stems often produce their own axillary stems as the shrub ages. Young shoots are usually green but may become partially or wholly reddish, reddish brown, or purplish with exposure to sunlight. In at least two species (*Berberis dictyophylla*, *B. lijiangensis*), young shoots are densely white pruinose, and in *B. tianchiensis* they are pinkish and partially pruinose. In the majority of species, toward the end of the first year or in the second year, the stems turn pale yellow, grayish yellow, pale brownish yellow, or pale brown, but in a significant number of species the stems turn dark brown, purple reddish or brownish purple, or even in the case of *B. aemulans* and *B. ziyunensis* blackish purple. Following both Schneider (1916) and Ahrendt (1961), I describe stems at this stage as mature. In a few species (e.g., *B. calcipratorum*, *B. dictyophylla*, *B. leptoclada*, *B. pruinosifolia*), the mature stems are partially pruinose. In some species, this initial color is short-lived, changing within a few months; in others, particularly dark-stemmed species (e.g., *B. purpureocaulis*, *B. umbratica*), the color may persist for several years. Ultimately, the stems become ash gray. The wood of freshly cut old stems is yellow.

Spines

All Chinese and Vietnamese *Berberis* have spines on their branches, but, as noted above, in the two species *B. hypoxantha* and *B. incrassata* these are infrequent and insignificant. Spines are usually 3-fid, but 5-fid in *B. tsarica*, 5- to 9-fid in *B. erythroclada*, and up to 9-fid in *B. sibirica* (where they are also partially foliaceous). In many species, particularly deciduous ones, the spines are solitary or absent toward the ends of twigs. Spines are immediately below the leaves in recent growth and are usually terete or abaxially sulcate or subsulcate. The distance between spines on branches is variable and usually increases as the plant ages. In four species, *B. markamensis*, *B. potaninii*, *B. saxatilis*, and *B. ulicina*, the spines are often longer than the internodes, resulting in particularly ferocious looking shrubs.

Leaves

Leaves are always simple and are always pinnately veined with two exceptions, *Berberis bowashanensis* and *B. tengchongensis*, where they are semi-camptodromous. Leaves are arranged in whorls or (in species with long petioles) in fascicles, with substantial variation in both shape and size. Leaves are elliptic, oblong, ovate, obovate, lanceolate, or oblanceolate; in three species, *B. insolita*, *B. ninglangensis*, and *B. wuchuanensis*, they are sublinear. In many species, variation in leaf shape is substantial, even on the same plant. Length varies from 0.5 cm in *B. tsarica* to 20 cm in *B. acuminata*. Very rarely a plant will produce a trifurcate leaf (e.g., two such leaves can be found on the isotype of *B. delavayi* var. *wachinensis* at A; for an example from a non-section *Wallichianae* species see *B. wilsoniae*, E E00612661).

Leaves of most Chinese and all Vietnamese *Berberis* are subsessile or very shortly petiolate. A notable exception is *B. dasystachya*, where petioles are up to 5 cm long. The leaf apex is obtuse to acute with the exception of *B. retusa*, where the apex is often retuse or truncate (an occasional retuse apex sometimes occurs on other species). In many species, particularly those in section *Wallichianae*, the apex is mucronate. The base is cuneate or attenuate, sometimes narrowly so, the most extreme example being in *B. ninglangensis*, where the elongated base is up to 2.2 cm long with a width of only 0.1–0.2 cm. The margins are entire, spinulose, spinose, or dentate. The number of teeth is variable within species, and a few teeth may occasionally be found on the margins in species whose leaves are otherwise entire. In a few section *Wallichianae* species (e.g., *B. griffithiana*, *B. nantoensis*, *B. replicata*, *B. taliensis*), margins exposed to sunlight are conspicuously revolute. Venation is consistent within species but is variable both in density and prominence between species. Venation is frequently more prominent on herbarium specimens than on living plants (sometimes venation which is more or less invisible to the naked eye on living plants can be seen in high resolution images of well-preserved specimens). Leaf texture varies from papery to rigidly leathery. Leaves are adaxially glabrous except in *B. brachypoda*, where they are pilose. The leaves of most non-section *Wallichianae* species are abaxially papillose, an exception being *B. gilgiana*, where they are pubescent. The leaves of some species are abaxially white or glaucous pruinose, though this is often variable within species (e.g., *B. pruinosa*, *B. wilsoniae*). Distinguishing between abaxially papillose and abaxially pruinose leaves in herbarium specimens is not always easy, and on occasion both Schneider and Ahrendt described as pruinose leaves that are actually densely papillose.

Inflorescences

All Vietnamese species have flowers in fascicles, while in China most of the basic forms of inflorescences, solitary flowers, fascicles, spikes, racemes, and panicles, are present. However, in many deciduous species and in some evergreen non-section *Wallichianae* species, the inflorescence is variable within species not just between plants, but often on the same plant and even on the same branch and may consist of some or all of fascicles, sub-fascicles, umbels, sub-umbels, racemes, and sub-racemes.

Flowers

Flowers range from 0.3 to ca. 2 cm in diameter. With five exceptions from section *Wallichianae*, they are all yellow, greenish yellow, or very rarely orange-yellow (e.g., *Berberis amoena*, *B. calcipratorum*). The exceptions are *B. bicolor*, where they are red and white; *B. sanguinea*, where they are partially purple; and *B. incrassata*, *B. triacanthophora*, and *B. veitchii*, where they are partially pink or pinkish red. Reddish stripes have been recorded on the outer sepals of other species, particularly those from Taiwan, though these stripes sometimes disappear as the flower matures.

The flowers have two or three whorls (rarely four) of three sepals. Below the lowest whorl of sepals are bracteoles (called prophylls by Ahrendt and bractlets by Ying [2011]). These bracteoles are sometimes treated as sepals (e.g., Adhikari et al., 2012: 453), but as Ahrendt (1961: 14) noted, whereas sepals are in whorls of three, a flower rarely has more than one or two bracteoles. Although sepals vary in shape between species, bracteoles are always triangular or lanceolate with an acute or acuminate apex. It should also be added that in cultivated plants at least bracteoles can be caducous. Petals are in two whorls of three and in most species are obovate and usually smaller than the inner sepals. The base of the petals has two glands, the exact position of which varies between species but appears to be largely consistent within species. The apex of the petals range from entire or emarginate to distinctly notched. Each petal has a stamen attached at the base. The anther connective of the stamen in some species is distinctly extended beyond the anthers, and the shape of the apex of the connective appears to vary little within species, thus sometimes making it an important character for species identification. The pistil is simple, and the ovary contains one to 15 ovules. Ovaries with large numbers of ovules are confined to single-flowered or few-flowered fascicled species mostly from Tibet.

Fruits

The fruits are ellipsoid, oblong, ovoid, obovoid, or globose berries. In a few species (e.g., *Berberis circumserrata*, *B. longipedicellata*), the apex is bent. The color ranges from pink or red to black. In many section *Wallichianae* species, the immature berries are partially or wholly pruinose. This usually disappears with ripening but persists in some species, most notably *B. pruinosa*. The fruits are stylose or estylose.

Phenology

The flowering period of *Berberis* varies between species. Although most species are recorded as flowering in April and May, *B. ferdinandi-coburgii* and *B. replicata* from Yunnan and *B. hayatana* from Taiwan are recorded as flowering as early as February, while *B. wilsoniae* is not recorded as flowering before June and *B. gyalaica* not before July. The fruits take many weeks and often months to mature. When not eaten by birds and other animals, the fruit of section *Wallichianae* species can persist into the following year, though due to the paucity of specimens from the winter months, it is unknown how prevalent this is.

Species Delimitation

From the discussion of morphology above it should be obvious why species delimitation in *Berberis* can be particularly challenging. The only attempt at defining explicit criteria for *Berberis* with universal application appears to have been that of Schneider (1916, 1939, 1942), and it is worth mapping the development of his criteria.

In his first thoughts, Schneider (1916: 314) laid out six criteria as follows:

(a) The shape and color of ripe fruit particularly in relation to the presence or absence of a distinct style.

(b) The number and configuration of the ovules in the ovary.

(c) The form and in particular the length of bracts in relation to the length of the flower or fruit stalk.

(d) The color and other characteristics of mature shoots (defined as those at the end of the first year or in the second year), especially those of long shoots.

(e) The character, venation, and serrations of the leaves of evergreen species, in that the shape of the blade and the spines of the margin are usually reliable evidence when combined with the texture of the leaf and absence or presence of reticulation.

(f) For deciduous species, the diversity of leaves on fruiting branches and young long shoots.

Subsequently, Schneider (1939: 245) wrote, "a detailed examination of the characteristics of the flowers of *Berberis* is far more significant than has been hitherto accepted," later (1942: 1) expanding on this by noting:

> I have more and more come to a realization that a very important criterion for the identification of species and their place in the Sections can be found in the flowers. In particular it is the shape and venation of the petals and sepals that is very significant for all groups. From the beginning I have highlighted the importance of the gynoecium with respect to the number of ovules and the absence or presence of a style. But hitherto I did not consider details of the structure of the perianth to be needed in order to draw firm conclusions. I now do so.

Of Schneider's initial six criteria I have found all but (c) and (f) of use. In relation to (c), I agree with Ahrendt (1961: 13) that bract length is only of diagnostic value in a small number of species, e.g., *Berberis aggregata*, and I do not really understand Schneider's point (f). His point (e) is important in relation to venation, which as noted above appears to be consistent within evergreen species (but also within deciduous ones), though given the variations noted under leaf morphology (before leaf shape and margin spines above) care needs to be taken in relation to their diagnostic value in many species. For an interesting study of venation in *Berberis* in Guizhou, see Wang and He (2015), unfortunately marred by many misidentifications of specimens.

Schneider's later comments on flower structure are of particular importance. For although they all have the same basic structure, the variation between species is quite remarkable and the variation within species minimal. This is not to say there is no variation within species in that there is some variation in the size of sepals and petals and sometimes in the number of ovules, but with sufficient examples the limits to this variation can be established.

For the three most important recent revisions of *Berberis* (Landrum, 1999; Adhikari et al., 2012; Yu & Chung, 2014), the issue of flower structure was for the most part not a problem. From their own collections in the field, they were able to verify descriptions of flowers of published species, add descriptions where these were lacking, and give descriptions of the flowers of new species. But they were describing only 20, 21, and 11

species respectively, whereas this study describes 277 species and collecting flowering specimens in the wild of all of them would be an impossible task for one person or even a team, always assuming all of the species could be found. Hence, the limitations of this study are at least four. First, some species are known only from the type collection, and in most cases I have relied on the account of the flower structure given in the protologue (although where the account is rudimentary or non-existent, I have on occasion been given permission to dissect flowers). Secondly, some species are known only from specimens with fruit, and in these cases the number of seeds and unfertilized ovules has to substitute for an account of the flowers. Thirdly, even when there are specimens with floral material beyond the type collections, the task of checking flower structure for each of them would have meant that this study would never have been completed. This is not to say that I have never checked the accounts given in protologues and elsewhere. I have where accounts are incomplete or dubious, and I have had access to accessible herbarium specimens or living plants of proven wild-collected origin either in my own collection or elsewhere. Fourthly, I have on occasion had to rely on accounts given by Ahrendt from cultivated material that I have been unable to trace, although in the relevant discussion I have indicated such as the description source, sometimes accompanied by a cautionary note as to possible inaccuracy.

What the importance of flower structure reinforces is the lesson I learned from my National Geographical Society-funded field studies to northwest and northeastern Yunnan and southwestern Sichuan in August and September 2013. There, I collected a significant number of deciduous species that were previously undescribed. Of those I have judged that there is enough evidence to publish 13 new species (mostly where there were occasional late flowers on the plants or where there were other distinctive morphological features). However, for some 15 other collections, all with fruit and all apparently different both from each other and from any published species, I have found it impossible to give watertight descriptions sufficient to enable subsequent collections to be identified as the same species. In all these cases, I have judged that the only way these might be published as new species (if that is what they are) would be with a description of flowers. This experience is not unique. Over the years, David Boufford (who was a fellow collector on the NGS expedition) has kindly sent me large numbers of Chinese *Berberis* specimens from expeditions he has participated in mostly from western and northern Sichuan and Tibet. They have resulted in 12 new species. But there is at least the same number of species again that appear to be undescribed, all from Sichuan and all with only fruit, where it has proved impossible to describe them in a way that would unambiguously differentiate them from already published species.

To further illustrate some of the difficulties of species identification and the importance of flower structures, it is worth drawing attention here to a particular group of very similar species where flower structure is absolutely crucial for distinguishing between them. These are from northwest Yunnan and neighboring areas of Sichuan with mostly narrowly obovate leaves and largely fascicled, sub-fascicled, and subumbellate inflorescences and flowers with two ovules.

From the evidence of both Chinese and non-Chinese herbaria, specimens from this group have mostly been identified as *Berberis lecomtei* or *B. microtricha* or more rarely as *B. papillifera*. But detailed investigations show that where floral material is available most of these can be shown to be none of these. One of these, *B. bowashanensis* (Harber, 2017), has already been published, and a further four new species from this first group, *B. difficilis*, *B. yingii*, *B. yulongshanensis*, and *B. zhongdianensis*, are published here, as well as recognizing *B. tischleri* var. *abbreviata* as a distinct species, *B. abbreviata*. The similarities and differences between these nine species are detailed in a multi-access key (Key 9).

It is unlikely that this is an exhaustive list, and it is likely that further species will be identified in the future.

What this shows is the need for expeditions specifically aimed at collecting flowers of *Berberis*. This is not without problems in that, as noted above, different species flower at different times and flowering seasons are often short (and the time when flowers are at their best even shorter, sometimes no more than a few days). However, fruiting seasons extend over much longer periods, and an alternative approach would be to collect seeds and use cultivated plants as evidence once they flower. (Seeds collected on the NGS 2013 expedition may eventually produce further new species.)

NATURAL HYBRIDS

As noted earlier, *Berberis* species brought together artificially in cultivation are subject to hybridization. The extent to which hybridization also occurs in the wild is little explored. Landrum (1999) reported a number of hybrids from Chile and Argentina, while Adhikari et al. (2012: 476) reported only one possible such from Nepal. The only published hybrid from China is that of *B.* ×*baoxingensis* X. H. Li (*B. sanguinea* × *B. verruculosa*) from Baoxing (X. H. Li et al., 2015a). In "*Berberis* sect. *Wallichianae* from the Kunming Area" below, I discuss the likely hybridization between *B. ferdinandi-coburgii* and *B. pruinosa* on Xi Shan (Western Hills) just south of Kunming in Yunnan. But be-

yond this, from my experiences in the field, I can report only two possible occurrences of ?*B. derongensis* × *B. dictyophylla* in Derong Xian in Sichuan (for which, see under *B. dictyophylla*) and one of possibly *B. aristatoserrulata* × *B. kawakamii* in Huili Hsien in Taiwan (for which, see under *B. aristatoserrulata*). In all these cases the plants were growing among or very near to both parents. But if hybridization is common in cultivation but apparently rare in nature the question is, why? It is unlikely that it is simply because such hybrids are sterile (though the plants of *B. aristatoserrulata* × *B. kawakamii* were) because hybrids in cultivation can produce viable seed. One explanation could be that *Berberis* species in any particular locality grow at different elevations and/or flower at different times so the opportunities for cross-fertilization between species are rare, but from personal observation this could only be a partial explanation. To give one example, in August 2013 on a fairly level 200 m or so stretch of road between Gongxia and Bigu Tianchi Hu in Zhongdian (Xianggelila) Xian in northwestern Yunnan, I made six collections of *Berberis*, all with fruit and all of apparently different species, only one of which, *B. wilsoniae*, I could identify; the other five (only represented by one plant each) being among the 15 unidentified ones cited above. *Berberis wilsoniae* certainly flowers significantly later than most other species in Yunnan, but it seems quite unlikely that the other five flowered in neat succession with no overlaps. Perhaps, therefore, the question posed at the beginning of this paragraph really should be rephrased to ask, why *Berberis* hybridizes rarely in the wild but does so frequently in cultivation? Whichever way the question is posed, much more research is needed to answer it.

ECOLOGY

In China, *Berberis* occurs in 23 of the 24 floristic regions detailed by Sun and Wu (2015), the exception being region IVG South China Sea. *Berberis* occurs at elevations as low as 100 m (*B. virgetorum*) and up to 4700 m (*B. tsarica*). The lowest elevations are mostly in the east, northeast, and southeast of the country, the highest in Tibet, Yunnan, and Sichuan. No evergreen species occur north of southern Gansu and southern Shaanxi. In south China, no deciduous species except *B. virgetorum* occur in Guangxi, Guangdong, or Fujian. All species in Taiwan are evergreen with the exception of *B. morrisonensis*. In Vietnam, *Berberis* is found only in forested mountainous areas bordering China, with the exception of *B. vietnamensis*, which occurs over 1000 km further south. All species in Vietnam are evergreen.

Berberis occupies a range of habitats but most commonly forest margins, sparse forests, and open pastures. Semi-prostrate and dwarf species are found in open often rocky areas mostly at high elevations. Some section *Wallichianae* species (e.g., *B. impedita*, *B. photiniifolia*, *B. jinfoshanensis*, *B. xanthoclada*) are found only or mainly on mountain summits at ca. 1000–2500 m. *Berberis* appears to be a great survivor of human disturbance being found on road and field sides and can be particularly prevalent in cut-over areas. In Yunnan and Sichuan, at least, it is often one of the first shrubs to appear in regenerating areas after logging and fires.

DISTRIBUTION

CHINA

As Table 1 and Figure 1 show, *Berberis* is found in all but three of China's 34 provinces (although only one specimen from Jiangsu has been located). The high number of both species and endemism in Sichuan, Yunnan, and Xizang is not surprising since these have long been recognized as centers of high biodiversity for many genera. Most of the endemic species are found in the "Hengduan Mountains and South-Central China hotspot" as defined by Boufford and van Dijk (2000), which includes most of western and southwestern Sichuan, northwestern and part of northeastern Yunnan, and eastern and southeastern Xizang. Only Taiwan is host to wholly endemic species.

Distribution within provinces is also important. As noted in the preface, this study presents evidence for distribution by Xian (or its equivalent) plotted on maps for each species. This information should not be treated as definitive in that there are now over 300 herbaria in China and this study draws on evidence from only a fraction of them, albeit a fraction that includes all the largest and most important ones. One example of the necessary incompleteness of distribution information is that no *Berberis* is reported here from three of the eight Xian bordering Nepal or from Nagarzê (Langkazi) Xian bordering Bhutan. It may be because none occur there, but given these areas are remote and that the leading herbaria hold few specimens of any genera from them this currently remains an open question. Conversely, because of the pace of destruction of China's natural environment, particularly since the time of the "Great Leap Forward" of 1958–1961, it should not be assumed that *Berberis* species collected in a particular area decades ago are still there or even in some cases that the species is still extant. No collection of *B. arguta* has been made in Shuifu Xian in northeastern Yunnan since the type was collected there in 1882, and only one other collection has been made elsewhere, in 1975. The only collection of *B. lubrica* was made in 1914. No collection of *B. pallens* would seem to have been made since 1886 and none of *B. candidula* since 1907. Only three collections of *B. asmyana* have been made, the last in 1940. But I take heart from the fact that, as noted above, *Berberis* appears to be a great survivor.

Even if they are not definitive, what the distribution maps show is that there is high endemism in the Hengduan Mountains and south-central China hotspot and that within this hotspot the overwhelming majority of species have very limited ranges. Indeed, within this area, only a handful of species are reported from 10 or more Xian or Shi, with only *Berberis dictyophylla*, *B. pruinosa*, and *B. wilsoniae* being reported from 15 or more. As such, this pattern in *Berberis* seems to mirror that of *Rhododendron* L. as reported by the various publications from the Royal Botanic Garden Edinburgh (Cullen, 1980; Chamberlain, 1982; Philipson & Philipson, 1986; Chamberlain & Rae, 1990; Kron, 1993; Judd & Kron, 1995).

VIETNAM

Berberis in Vietnam occurs only in the north on the border with China in Hà Giang and Lào Cai Provinces and in the south in Lâm Đồng Province.

CHINESE PLACE NAMES

Chinese place names, particularly historic ones, are problematic and not just for foreigners. Some of the difficulties can be illustrated by considering the example of what is currently known in Mandarin Chinese as 康定 (Kangding) and in Tibetan as རྡར་མདོ (Dardo) or རྡར་རྩེ་མདོ (Darzêdo) and which is in the Ganzi (Mandarin) or Garzê (Tibetan) Prefecture in western Sichuan. Kangding can refer to an administrative area (Kangding Xian) or the main town within that area. Its current placement in Sichuan is relatively recent. The first western botanical collections were made when the area was in disputed territory. Specimens collected by A. E. Pratt in the late 1880s locate it as being on the "West Szechuan and Tibetan Frontier," while those of J. A. Soulié in the 1890s refer to "Principauté de Kiala" in "Thibet Oriental," Kiala being reference to Chakla the capital of an ancient kingdom within the Tibetan region of Kham. Neither Pratt nor Soulié used the name Kangding but referred respectively to "Tachienlu" and "Ta-tsien-lou," these being transliterations of the Chinese transliteration 打箭炉 (Dajianlu) of the Tibetan Darzêdo. Later, in 1928 when W. P. Fang collected in the area he referred to it as "Kangtin Hsien" in "Szechuan" while H. Smith in 1932 referred to it as "Kangting" in "Sikang," a short-lived province that covered part of what is now eastern Tibet and western Sichuan and which was abolished in 1955. Dajianlu and Kangding Xian are transliterations from Mandarin using the Pinyin system adopted by China in 1958. Ta-tsien-lou and Kantin(g) Hsien are transliterations using the earlier Wade-Giles system.

Table 1. Distribution of *Berberis* species by province (mainland China) and Taiwan. A = number of species; B = number of species endemic to China; C = number of species endemic to that province.

Province	A	B	C	Province	A	B	C	Province	A	B	C	Province	A	B	C
Sichuan	88	86	54	Qinghai	12	12	3	Guangxi	5	5	0	Heilongjiang	2	0	0
Yunnan	87	77	51	Hunan	9	9	0	Guangdong	4	4	0	Jiangsu	1	1	0
Xizang	65	49	38	Ningxia	7	6	0	Henan	3 (?4)	3 (?4)	0	Tianjin	1	0	0
Guizhou	22	21	7	Jiangxi	6	6	0	Hebei	3	0	0	Hainan	0	0	0
Gansu	18	18	3	Shanxi	6	3	0	Liaoning	3	0	0	Shanghai	0	0	0
Hubei	17	17	1	Nei Mongol	5	2	0	Beijing	3	0	0	Hong Kong / Macau	0	0	0
Chongqing	17	17	1	Fujian	5	5	0	Anhui	2	2	0				
Shaanxi	16	16	2	Xinjiang	5	0	0	Shandong	2	1	0				
Taiwan	14	14	14	Zhejiang	4	4	1	Jilin	2	0	0	ALL CHINA	276	246	

Figure 1.   Distribution of *Berberis* in China by Xian.

In attempting to deal with such complexities, my aim has been wherever possible to assist the reader in locating a collection area on a current map. This is no easy task. Early foreign collectors often did not use the Wade-Giles system (or its modification by the Chinese Postal Map Romanization system) and produced their own transliterations of what they heard (or thought they heard), which in some cases was a dialect word with only local usage, a word in a major non-Chinese language such as Tibetan, Uyghur, or Mongolian, or a word in one of the more localized languages.

In locating historic place names I have drawn in particular on the Gazetteer on the website http://hengduan .huh.harvard.edu/fieldnotes/gazetteer, which includes many of the place names recorded for the collections of E. H. Wilson and J. F. Rock. For collections made by J. M. Delavay, the little-known map of Bretschneider (1898a) has proved indispensable, as has Stearn (1976)

for the collections of F. Ludlow and G. Sherriff in Tibet (this article being the source of the coordinates given here for their collections). For collections made by C. K Schneider and H. Handel-Mazzetti, I have drawn on the glossary in Handel-Mazzetti (1996), though it is not always reliable. Herner (1988) is useful for place names recorded by H. Smith. For collections made in Tibet by F. Kingdon-Ward, I have found useful his maps in Kingdon-Ward (1926a, 1934, 1936).

The historic place names that I have found for the most part impossible to correlate with current ones are those recorded on the sheets of *Berberis* specimens collected by the Italian missionaries G. Giraldi (between 1895 and 1900) and C. Silvestri (between 1907 and 1913). The exact role of Giraldi and Silvestri in relation to these names appears to be uncertain, and it is possible that the Italian botanist A. Biondi, who was the initial recipient of the specimens (see Kilpatrick,

2014: 165–173), had a hand in it. But whoever the author was, the place names given are in an eclectic mix of Italian and a highly idiosyncratic system of transliteration from Chinese. They (and the place names recorded for Giraldi and Silvestri specimens of many other genera) await their decipherer. In the meantime, it is important to note that, with one exception (*Giraldi 49*, a syntype of *B. dolichobotrys*), Giraldi *Berberis* specimens are recorded as being from "Shen-si Settentr.," i.e., north Shaanxi, whereas they clearly are from the middle or south of that province, a mistake that must have been made by somebody other than Giraldi.

For mainland China, wherever possible, I have given Chinese place names in Pinyin, these being in parentheses when different from the name found on the specimen sheet. The exceptions are the names of Xian and Shi and their equivalent, Qi (旗) in Nei Mongol (often misleadingly translated as Banner, another of its meanings in Mandarin), in what are known as autonomous prefectures (自治州 zizhizhou), which are those where the majority of the population is of an ethnic minority. Here, names are mostly given in the language of that minority (the exception being some Mongolian place names) with the Mandarin name (if different) given in parentheses. Place names in square brackets are those that have no equivalent on the specimen label but which can be deduced from other evidence.

For the most part and for reasons of clarity, I have not translated the place names into English for specimens whose collection details are recorded in Chinese. For example, Zhen (镇), Cun (村), and Xiang (乡) are all often translated simply as village, but within the hierarchy of administrative divisions in China each has different functions. Moreover, sometimes two settlements with the same name but each with a different one of these epithets can be found in the same area. Places with these three epithets can be found on current maps, but two others often cited on the collection details of specimens collected between 1958 and the mid 1980s cannot. These are Gongshe (公社) Commune and Dadui (大队) Production Brigade or Team, the latter applying not just to a division within a commune, but the geographical area for which they were responsible. Many of these commune names (but not usually those of dadui) are on current maps but now with one of the three epithets listed above. But this is not true of all; some communes had revolutionary names, e.g., 红卫公社 (Hongwei Gongshe, Red Guard Commune) and 前锋公社 (Qianfeng Gongshe, Vanguard Commune), and these have disappeared from view. In the 1950s (and sometimes into the 1970s) some Xian had subdivisions—Qu (区) Districts—with numerical epithets, e.g., 二区 (Er Qu, Number 2 District) or 第四区 (Disi Qu, Fourth District), and these too are not on current maps. I have not translated any of these.

Other geographical words I have not translated include Gou (沟) (sometimes mistranslated as ditch), gorge or ravine, but also used for side or minor valleys; Gu (谷), valley; He (河), river; Hu (湖), lake; Jiang (江), river; Linchang (林场), sometimes rendered in English as "forest farm," a term which is either meaningless or misleadingly implies some sort of commercial plantation whereas actually the nearest English equivalent would be something like "managed forest"; Qiao (桥), bridge, where it is an integral part of a settlement name (cf U.K. names such a Stamford Bridge), and Shan (山), hill(s) or mountain(s), but which can also be an integral part of a settlement name.

It should be noted that Chinese place names are still subject to change. Of particular importance are changes in relation to Xian (县) and Shi (市), sometimes misleadingly translated as "county" and "city" respectively; misleading because Xian do not have the same functions as counties in the English-speaking world and especially misleading in relation to Shi, which can apply to an administrative area that is partially or even wholly rural rather than urban. Because Shi have more powers than Xian, there are regular applications for upgrading, which, when successful, can cause confusion when trying to locate where historic specimens were collected, especially when boundary changes are involved. Thus, many specimens are recorded as being from Lijiang Xian, which included Yulong Shan. But this no longer exists, and its replacement Lijiang Shi covers only the urban area while Yulong Shan is now in Yulong Xian. Just to add to these complexities, Diqu (地区) the administrative level above Xian and Shi, can also be promoted to Dijishi (地级市) with consequent name changes lower down. Thus, when this happened to Qamdo (Chamdo) Diqu in northeast Tibet in 2014, the name of its capital Qamdo Xian was changed to Karub (Karuo) Qu. Wherever possible in such cases, I have tried to take these changes into account.

Chinese handwriting is notoriously difficult to decipher. Therefore, for specimens whose collection details are wholly in Chinese, I have relied on the transcriptions given on the website of the herbarium where the specimen is to be found and have only checked this against the details on the actual sheet where the transcription is obviously wrong or where two or more herbaria possess specimens of the same collection but give different transcriptions. Some Chinese characters have two or more pronunciations. Though some unusual pronunciations for major place names are well known. For instance, 铅山县 in Guangxi is pronounced Yanshan Xian not Qianshan Xian, but for smaller settlements, the correct pronunciation may only be known by those familiar with the area. So in some instances, I may have chosen the wrong one.

Finally, it should be noted that while I use Tibet and Xizang interchangeably in this introduction depending on context, in the main text I usually use Xizang (Tibet).

Place names in Taiwan are reported differently because specimens in Taiwanese herbaria (including many collected in the time of the Japanese occupation) usually have collection details recorded in English with places names given in Wade-Giles. Rather than change these into Pinyin, I have simply reported them as so recorded.

CHINESE COLLECTORS' AND AUTHORS' NAMES

In Chinese, the position of surnames and given names is the reverse of their positioning in English. However, I have followed the usual practice of Chinese authors publishing in English of putting the initials of the given names first. These are given in Pinyin.

In relation to specimens, the exceptions relate to various Chinese collectors in the period 1920–1940 who collected partially or primarily for foreign herbaria, e.g., R. C. Ching, K. L. Chu, H. T. Tsai, and T. T. Yu, and to Taiwanese collectors. In such cases, their names appear on specimen labels using Wade-Giles transcriptions. Rather than transcribe these into Pinyin, I have followed convention and report them as recorded.

Until the late 1980s, Chinese authors publishing in English usually used Wade-Giles transcriptions for their own names, only changing over to Pinyin thereafter. Thus, the given names for 俊生 are in Wade-Giles—Tsun-Shen—in Ying (1985, 1999) but in Pinyin—Junsheng (but with Tsun-Shen in parentheses)—in Ying (2011). To avoid confusion in such cases, works in the bibliography are all under the name used in their author's first listed publication. For Ying's own *Berberis* collections (recorded on the sheets only in Chinese), I have used T. S. rather than J. S. Ying.

CHINESE NAMES FOR *BERBERIS*

Chinese publications (including all provincial floras) invariably give Chinese names in addition to Latin ones for *Berberis*. These appear rarely if ever to be vernacular names (though unsurprisingly these certainly exist; see, e.g., Tibetan names recorded on *Berberis* specimens at MO collected in 2003 by D. M. Anderson et al. in Dêqên Xian in northwest Yunnan) but instead are wholly invented ones. Since there is no equivalent to the Shenzhen Code (Turland et al., 2018) and its predecessor Codes for names in Chinese, authors are free to follow previous usage or to ignore it. The results can only be described as chaotic with the same species being given different names by different authors and reversely the same name being applied by different authors (or even on occasion the same author) to differ-

ent species. There is not even any consistent name for the genus, in that, though most authors stick to 小檗 (xiaobo), the word for *Berberis* in Chinese dictionaries, not all species are given this epithet. Ying (2001, 2011), for instance, gives 黄芦木 (huang lu mu) for *B. amurensis*, 刺红珠 (ci hong zhu) for *B. dictyophylla*, and 豪猪刺 (hao zhu ci) for *B. julianae*. Chinese names for varieties of *Berberis* sometimes have no connection to the name of the species they are varieties of.

Given all this, I have been faced with three choices for previously published species: (a) choose one Chinese name per species; (b) list all Chinese names for a particular species; or (c) not give any Chinese names. The first approach was that adopted for the Flora of China project by Ying (2001, 2011) but where there were two or more names to choose from his choices appear arbitrary. They were not based, e.g., on any genealogy to establish which was published first. The second option seemed initially attractive, but I soon realized that constructing a comprehensive list of previously used names would be near impossible (though necessary were I to decide to give unique Chinese names to the new species published here). I have therefore opted not to give any Chinese names whether for existing species or for novelties.

IDENTIFYING HERBARIUM SPECIMENS AND LIVING PLANTS

The number of herbarium specimens of *Berberis*, misidentified, unidentified, or even just recorded as Berberidaceae, is testimony to the difficulties in assigning them to the correct species. The best single tool for identification would be a computer-based interactive key. This would be better than any paper-based tool at taking into account the often variable nature of the leaves and inflorescences of many species as well as the incomplete information in a number of cases as to flower structure or the nature of the berry. But the task of constructing such a computer-based key would be substantial. In its absence, the following strategies are recommended for using the information presented here.

The most important starting point is the provenance of the material being examined, i.e., where does it come from? For living plants in the field, this, of course, presents no problem, and the more recently herbarium specimens were collected the more accurate and detailed the collection data are likely to be, sometimes including exact coordinate information. For cultivated plants, for the reasons given in the introduction, provenance is all important. Trying to identify most cultivated plants whose provenance is unknown is largely an unproductive exercise, given the high probability that they may be hybrids (see Harber, 2010).

Once the provenance is established, the following steps are recommended. For Taiwan and Vietnam, go to the keys, then to the descriptions to verify the determination. For mainland China, first go to Table 2 or 3 below (species of section *Wallichianae* and non-section *Wallichianae* species) to see the species that are recorded for the particular province. Except for Sichuan, Yunnan, and for non-section *Wallichianae* plants from Xizang, this will give a manageable list of the likely possibilities. For most provinces, the list will be so small that the descriptions alone should be sufficient to arrive at an identification (e.g., for Anhui there are only two possible species, for Liaoning only three, and for Xinjiang only five).

For Sichuan, Yunnan, and for non-section *Wallichianae* plants from Xizang, the best starting point is probably the keys (augmented where appropriate by the distribution maps, which give a quick visual guide to the locations in the province where a particular species has been found). The keys have been constructed to highlight the most obvious and easily discernible characteristics, but on occasion the keys can only distinguish between very similar species by reference to the structure of the flowers, which makes dissection necessary; this is particularly the case with a number of deciduous species with two or three ovules from northwestern Yunnan and neighboring areas of Sichuan. In these cases, it may not be possible to identify specimens (or in some cases even living plants) with only fruiting material.

Finally, it should be noted that keys were developed as an aid in an age when researchers had to rely almost exclusively on written descriptions, augmented on occasion by line drawings. In an era of digitization and internet access this is no longer the case. Already for the majority of species listed here, it is possible to sit at a computer and download often high-quality images not just of type specimens but of a large percentage of other specimens cited here, thus providing an additional tool for confirming an identification reached through the keys and descriptions. As more and more herbaria put images online, this facility can only increase.

It is of course possible that a species has not previously been recorded for a particular province (this being most likely when the plant is from the border region of a neighboring province where it has been reported) or that it is an unrecognized species. This latter is most likely if it is from Sichuan, Yunnan, and Xizang, but from my list of specimens that I have been unable to identify, southern Qinghai, Chongqing, and possibly Guizhou and Hubei should be included as well.

It should be emphasized that identification is not always possible. This is often the case with sterile specimens unless they have particularly distinctive vegetative features (e.g., as with the dense ferocious spines of *Berberis ulicina* or the large abaxially pruinose leaves of *B. hsuyunensis*).

KEY TO SECTIONS

1a. Plants evergreen; flowers solitary or in fascicles (except *B. centiflora* which has partially racemose inflorescences); leaves thinly or thickly leathery . . . . *Berberis* sect. *Wallichianae* C. K. Schneid. (spp. 1–90).

1b. Plants deciduous, evergreen, or semi-evergreen; inflorescences solitary flowers, fascicles, racemes, umbels, panicles, or a mixture of inflorescence types; leaves of semi-evergreen and most deciduous species papery, leaves of evergreen plants leathery and inflorescences other than solitary flowers or fascicles (except in *B. asiatica*, which in China has mostly fascicled flowers). . . . . species not in *Berberis* sect. *Wallichianae* (spp. 91–277).

Table 2.   Species of *Berberis* sect. *Wallichianae*.

PROVINCE	Species number
Chongqing	12, 26, 43, 44, 68, 70, 76, 78
Fujian	16, 29, 60, 84
Gansu	8, 70, 79
Guangdong	16, 39, 60
Guangxi	14, 16, 59, 77
Guizhou	1, 9, 14, 16, 21, 33, 44, 62, 67, 75, 76, 77, 81, 85, 88, 90
Hubei	12, 16, 26, 44, 46, 66, 68, 70, 76, 78
Hunan	16, 39, 42, 44, 66, 75, 76
Jiangxi	16, 39, 42, 84
Shaanxi	44, 70, 76
Sichuan	1, 4, 6, 7, 8, 21, 25, 30, 32, 37, 41, 51, 52, 59, 61, 62, 65, 69, 70, 76, 79, 87
Taiwan	2, 5, 10, 17, 34, 45, 53, 54, 55, 57, 61, 63, 67, 74
Xizang	11, 18, 23, 31, 36, 48, 56, 62
Yunnan	1, 3, 4, 13, 15, 19, 20, 21, 22, 23, 24, 27, 28, 32, 35, 37, 38, 40, 49, 50, 51, 58, 62, 64, 71, 72, 73, 82, 83, 86, 88, 89
Zhejiang	29, 47

Table 3.   Species not in section *Wallichianae*.

PROVINCE	Species number
Anhui	99, 257
Beijing	97, 123, 233
Chongqing	93, 130, 132, 154, 162, 211, 236, 261
Fujian	257
Gansu	93, 113, 122, 132, 136, 137, 142, 152, 168, 213, 217, 227, 235, 238, 246
Guangdong	257
Guangxi	257
Guizhou	162, 208, 253, 257, 259, 261
Hebei	97, 123, 233
Heilongjiang	97, 233
Henan	124, 132?, 154, 236
Hubei	116, 130, 132, 154, 162, 236, 257
Hunan	116, 257
Jiangsu	257
Jiangxi	116, 257
Jilin	97, 123
Liaoning	97, 123, 233
Nei Mongol	97, 122, 123, 142, 233
Ningxia	113, 122, 132, 136, 142, 227, 233
Qinghai	111, 113, 122, 132, 133, 136, 142, 144, 223, 233, 273
Shaanxi	93, 113, 124, 132, 154, 162, 213, 217, 219, 224, 227, 236, 258
Shandong	97, 257
Shanxi	97, 113, 123, 132, 154?, 219, 233
Sichuan	91, 92, 93, 95, 96, 103, 104, 106, 110, 112, 113, 122, 131, 132, 133, 135, 136, 137, 138, 139, 142, 144, 145, 146, 152. 153, 160, 162, 170, 171, 173, 174, 178, 184, 185, 187, 191, 192, 193, 195, 197, 198, 202, 211, 213, 214, 215, 216, 220, 222, 226, 228, 234, 235, 237, 241, 247, 260, 261, 264, 266, 267, 268, 269, 270, 271
Taiwan	196
Tianjin	123
Xinjiang	119, 163, 167, 233, 254
Xizang	94, 98, 100, 101, 102, 105, 107, 114, 120, 121, 125, 126, 127, 138, 140, 144, 147, 148, 155, 156, 159, 161, 165, 166, 169, 170, 172. 175, 176, 177, 182, 183, 185, 186, 189. 194, 199, 200, 204, 205, 209, 210, 211, 221, 225, 231, 239, 243, 249, 250, 251, 252, 255, 256, 261, 263, 275
Yunnan	55, 96, 108, 109, 115, 117, 118, 121, 125, 128, 129, 134, 135, 138, 141, 143, 149, 150, 151, 157, 158. 164, 170, 178, 179, 180, 181, 188, 190, 193, 195, 201, 203, 206, 207, 209, 212, 218, 222, 229, 230, 232, 240, 242, 244, 245, 248, 252, 253, 261, 262, 265, 272, 273, 276, 277
Zhejiang	99, 257

A Note on "Dustbin Species"

The difficulty of identifying *Berberis* in China has led over the years to the emergence of various "dustbin species." In the 19th and early 20th century, the two such dustbin species were *B. wallichiana* (a native of Nepal) and *B. sinensis* (for which, see under *B. chinensis*). The identifications under these names undoubtedly had their origins in the species list in Candolle (1821), reinforced in the case of *B. sinensis* by the treatment in Regel (1873). Thus, *B. arguta*, *B. candidula*, *B. davidii*, *B. griffithiana*, and *B. sublevis* were all first published as varieties or forms of *B. wallichiana*, while *B. photiniifolia* in Guangdong was identified as *B. wallichiana* by Dunn and Tutcher (1912: 32). Similarly

B. amoena was first published by Delavay (1889) as *B. sinensis* var. *elegans*, while the type specimens of *B. aggregata* had previously been identified by Maximowicz (1891: 40) as examples of a *B. sinensis* var. *crataegina*. When I first visited the Paris herbarium in 2009 before its reorganization, I found the ghostly remnants of these classifications in the way *Berberis* was filed.

These two particular misidentifications have long since ceased being made (though, as noted in the introduction, not in relation to Vietnam in the case of *Berberis wallichiana*) but have been replaced by others. In the case of section *Wallichianae*, for reasons I am unable to explain, the favored species has been *B. julianae*, which is endemic to Hubei and surrounding areas, but as noted in the discussion under that species has been misreported in the literature as occurring over a wide area of southern and eastern China as well as (again) in Vietnam. For deciduous species, the favored misidentifications have been *B. dictyophylla* (or *B. dictyophylla* var. *epruinosa*) and *B. muliensis* for single-flowered specimens (even though the latter is not in fact 1-flowered), whereas as noted in the introduction the favored misidentifications for fascicled and semi-fascicled species have been *B. lecomtei*, *B. papillifera*, and *B. microtricha*. For racemose species, it has been *B. henryana*, which has been confused inter alia with *B. anhweiensis*, *B. dasystachya*, *B. elliptifolia*, and *B. emeishanensis*. For species with mixed inflorescences, *B. silva-taroucana* has been a favored misidentification, probably the result of the confused account of it given in its protologue (for which, see under that species).

TAXONOMIC TREATMENT AND KEYS

Berberis L., Sp. Pl. 1: 330. 1753. TYPE: *Berberis vulgaris* L. (lectotype, designated by Hitchcock [1929: 147]).

Shrubs or sometimes small trees, evergreen or deciduous, wood and roots yellow. Young shoots green, sometimes partially or wholly shades of pink or red. Stems and branches terete, angled, or sulcate, initially shades of yellow, brown, red, or purple, sometimes pruinose, turning into ash gray when older, shoots and mature stems usually verruculose. Spines 1- to 9-fid, rarely absent. Leaves simple, alternate, in fascicles or in whorls; petiole usually almost absent, rarely to 50 mm; leaf blade margin entire, serrate, to spinose-toothed, venation usually pinnate, prominent to obscure. Inflorescences of solitary flowers, fascicles, sub-fascicles, racemes, sub-racemes, umbels, sub-umbels, or panicles. Bracts (when present) usually 1 or 2, smaller in size than sepals and petals; bracteoles (when present) usually 1 or 2, smaller in size than sepals and petals. Flowers perfect, shades of yellow or orange, rarely red and white, partially red, pinkish, or purple. Sepals 3- or 4-merous in 2 to 4 whorls; petals 6 in 2 whorls, with a pair of nectariferous glands at the base of adaxial surface; stamens opposite to petals; pistil usually barrel-shaped; ovary superior; stigma peltate; ovules 1 to 15. Fruit a berry, ellipsoid, globose, ovoid, ovoid to oblong, or obovoid to oblong, red, pink, orange, black, dark blue, or dark purple, sometimes partially or wholly white or glaucous pruinose. Style present or absent.

Landrum (1999), Adhikari et al. (2012), and Yu and Chung (2014) all cite Britton and Brown (1913: 127) as designating *Berberis vulgaris* as the type species of *Berberis*. However Art. 10.5 of the Shenzhen Code (Turland et al., 2018) states a previous designation can be superseded if "it was based on a largely mechanical method of selection," and Art. 10.6, example 6, specifically notes Britton and Brown as adopting such a method.

Keys

As should be apparent from previous remarks, constructing keys to identify the species in this study has been exceptionally difficult given the variation in leaf size and shape and more particularly the variation in inflorescence within many species, this latter being particularly the case for many non-section *Wallichianae* species. Nevertheless, for these latter species it appears useful to give separate keys for different types of inflorescence. These categories, however, should be regarded for the most part as only guidelines in that for a number of species they describe the dominant form of the peduncle and/or pedicel based on the evidence available; thus, Key 5 includes racemes that sometimes have some flowers in fascicles at the base. Where appropriate, some species appear in more than one key. Anything stricter than this would result in a considerable expansion of the mixed inflorescence key at the expense of Keys 3–6 and one whose consequent length would likely deter its actual use.

As noted in the introduction, there are nine species recognized here from northwest Yunnan and neighboring areas of southwest and west Sichuan with mostly narrowly obovate leaves and largely fascicled, sub-fascicled, and subumbellate inflorescences with flowers with two ovules. Distinguishing between them is exceptionally difficult. These species are in Keys 4 and 7, but are also in the multi-access Key 9.

Evergreen species not in section *Wallichianae* and semi-evergreen species are few, hence the separate Key 8. However, since it may be difficult to determine the persistence of leaves on specimens collected outside the winter months, those species are also to be found in Keys 3, 4, 6, and 7.

Some readers may be unhappy with the inclusion in the keys of geographical locations, whether they are particular provinces or on occasion specific Xian within a province. I have done this only where the current evidence indicates a species is limited to a specific area. Such inclusion, it is readily conceded, is of no use when considering specimens whose provenance is recorded simply as "China" or, e.g., "Western China," but having looked at perhaps 20,000 or more herbarium specimens of *Berberis* from China, either in situ or as images, I can report that the percentage of those that have this bare minimum of provenance information is tiny. It should be added that because Taiwan has been politically separate from mainland China, whether under the Japanese occupation or subsequently, the chances of specimens from Taiwan being simply recorded as being from China would seem remote. Hence, no problems should arise from species from Taiwan being recorded only in Key 10.

List of Keys

Keys 3–7. Species not in section *Wallichianae*; plants mostly deciduous

Key 3. Flowers solitary

Key 4. Flowers in fascicles, sub-fascicles, umbels, or sub-umbels

Key 5. Flowers in racemes (including racemes sometimes with fascicles at base)

Key 6. Flowers in panicles

Key 7. Mixed inflorescences (one or more of fascicles, sub-fascicles, umbels, and sub-umbels plus one or both of racemes and sub-racemes)

Key 8. Species not in section *Wallichianae*; plants semi-evergreen and evergreen

Key 9 (multi-access key). Deciduous species from NW Yunnan and SW and W Sichuan with mostly narrowly obovate leaves and largely fascicled, sub-fascicled, and subumbellate inflorescences and flowers with 2 ovules

Key 10. Plants of Taiwan

Key 11. Plants of Vietnam

KEY 1. PLANTS OF CHINESE MAINLAND

1a. Plants evergreen; flowers solitary or in fascicles (except *B. centiflora*, which has partially racemose inflorescences); leaves thinly or thickly leathery . . . . Key 2 (section *Wallichianae*)

1b. Plants deciduous, evergreen, or semi-evergreen; inflorescences solitary flowers, fascicles, racemes, umbels, panicles, or a mixture of inflorescence types; leaves of evergreen plants leathery and inflorescences other than solitary flowers or fascicles (except in *B. asiatica*, which in China has mostly fascicled flowers). . . . .
. . . . .Keys 3–8 (species not in section *Wallichianae*)

KEY 2. SPECIES IN SECTION *WALLICHIANAE*

1a. Spines on stems mostly absent 2

1b. Spines on stems present, (1- to)3-fid. 5

2a. Mature stems pale brownish yellow or brownish gray; ovules 1 to 6 3

2b. Mature stems purplish black or purple; ovules 2 or 3 4

3a. Leaf blade elliptic-lanceolate, sometimes oblong-elliptic, 9–15 × 2–6 cm, lateral veins and reticulation conspicuous adaxially; flowers partially pinkish red; ovules 4 to 6 . . . . 40. *B. incrassata* Ahrendt

3b. Leaf blade elliptic, oblong-elliptic, or elliptic-obovate, 3–7 × 2–2.5 cm, lateral veins inconspicuous or obscure adaxially; flowers yellow; ovules 1 to 3 . . . . 38. *B. hypoxantha* C. Y. Wu ex S. Y. Bao

4a. Mature stems purplish black; leaf blade narrowly elliptic, oblanceolate, or narrowly obovate, 4–10 × 1–3 cm; flowers per inflorescence 4 to 10; sepals in 2 whorls; ovules 3 . . . . 90. *B. ziyunensis* P. K. Hsiao & Z. Yu Li

4b. Mature stems purple turning reddish brown; leaf blade narrowly elliptic, oblanceolate, obovate, or narrowly obovate, 3–7 × 1–2 cm; flowers per inflorescence 4 to 30; sepals in 3 whorls; ovules 2 . . . . . 77. *B. uniflora* F. N. Wei & Y. G. Wei

5a. Inflorescence always or mostly 1-flowered 6

5b. Inflorescence in fascicles 9

6a. Leaf blade elliptic, obovate-elliptic, ovate-elliptic, or rhomboid-elliptic, 1–3 × 0.6–1.1 cm, abaxial leaf surface pruinose or epruinose. 7

6b. Leaf blade narrowly elliptic, narrowly obovate-elliptic, or oblong-lanceolate to lanceolate, 1.5–4 × 0.4–1 cm, abaxial leaf surface pruinose. 8

7a. Shrubs to 1.5 m tall; leaf blade elliptic or obovate-elliptic, abaxial leaf surface epruinose; pedicel 20–40 mm; ovules 4 or 5 . . . . 6. *B. asmyana* C. K. Schneid.

7b. Shrubs to 1 m tall; leaf blade elliptic, ovate-elliptic, or rhomboid-elliptic, abaxial leaf surface pruinose (but sometimes epruinose on older leaves); pedicel 4–10 mm; ovules 4 to 6 . . . . 79. *B. verruculosa* Hemsl. & E. H. Wilson

8a. Shrubs to 60 cm tall; mature stems reddish brown; leaf blade narrowly elliptic or narrowly obovate-elliptic, 1.5–4 × 0.4–1 cm; ovules 9 to 12. . . . . 18. *B. chrysosphaera* Mulligan

8b. Shrubs to 45 cm tall; mature stems yellow; leaf blade oblong-lanceolate to lanceolate, 1.2–2.6(–3) × 0.4–1 cm; ovules 4 or 5 . . . . 12. *B. candidula* (C. K. Schneid.) C. K. Schneid.

9a. Fascicles often mixed with 2- or 3-flowered racemes; mature stems pale yellow. . . . . 15. *B. centiflora* Diels

9b. Fascicles not mixed with racemes; mature stems shades of purple, brown, or yellow 10

10a. Ovules 1 11

10b. Ovules more than 1. 18

11a. Mature stems very dark purplish brown . . . . 83. *B. wuliangshanensis* C. Y. Wu ex S. Y. Bao

11b. Mature stems yellow, very pale yellow, yellowish gray, brownish yellow, yellowish brown, or pale reddish brown 12

12a. Leaf blade thickly leathery. 13

12b. Leaf blade thinly leathery. 15

13a. Mature stems yellow; leaf blade elliptic, obovate-elliptic, obovate, or oblanceolate, 4–10 × 1–3 cm . . . .
. . . . .44. *B. julianae* C. K. Schneid.

13b. Mature stems yellowish brown or brownish yellow; leaf blade elliptic, elliptic-oblanceolate, elliptic-lanceolate, oblanceolate, obovate, or rarely linear-lanceolate, 2.5–10(–16) × 1–2.5 cm. 14

14a. Leaf blade elliptic, oblanceolate, or obovate, 2.5–6 × 1–1.5 cm, reticulation inconspicuous adaxially, margin spinulose with 6 to 15(to 18) teeth on each side; outer sepals ovate or narrowly ovate, inner sepals obovate or narrowly obovate . . . . 14. *B. cavaleriei* H. Lév.

14b. Leaf blade narrowly elliptic, elliptic-oblanceolate, elliptic-lanceolate, rarely linear-lanceolate, 4–10(–16) × 1.5–2.5 cm, reticulation conspicuous and dense adaxially, margin spinulose with 35 to 60 teeth on each side; outer sepals lanceolate, inner sepals ovate . . . . .28. *B. ferdinandi-coburgii* C. K. Schneid.

15a. Mature stems pale reddish or yellowish brown (sometimes purplish when dry); leaf blade oblong-elliptic, elliptic, elliptic-lanceolate, or lanceolate, 3–13 × 1.5–3.5 cm, reticulation conspicuous adaxially . . . . .21. *B. deinacantha* C. K. Schneid.

15b. Mature stems very pale yellow, yellowish gray, or yellow; leaf blade narrowly lanceolate, elliptic, narrowly elliptic, elliptic-lanceolate, or oblanceolate, 2–12.5 × 0.7–3.5 cm, reticulation inconspicuous adaxially 16

16a. Mature stems very pale yellow or yellowish gray; leaf blade elliptic, narrowly elliptic, or lanceolate, 3–8(–12.5) × 0.7–2(–3.5) cm . . . . .49. *B. levis* Franch.

16b. Mature stems yellow; leaf blade narrowly lanceolate, oblanceolate, or elliptic-lanceolate, 2–8 × 0.7–2 cm 17

17a. Leaf blade narrowly lanceolate, 4–12 × 1–1.5 cm; species of W Yunnan . . . . .72. *B. sublevis* W. W. Sm.

17b. Leaf blade oblanceolate, elliptic-lanceolate, or elliptic, 2–6 × 0.7–2 cm; species of S Sichuan and NE Yunnan . . . . .51. *B. liophylla* C. K. Schneid.

18a. [10b] Mature stems dark reddish purple, brown, purplish brown, reddish brown, or pale reddish brown 19

18b. Mature stems pale yellowish brown, yellow, pale brownish yellow, or grayish yellow 30

19a. Mature stems dark reddish purple; leaf blade lanceolate, rarely elliptic-lanceolate, 3–11 cm, abaxial leaf surface epruinose, margin spinose with 6 to 10 widely spaced coarse teeth on each side . . . . .25. *B. ebianensis* Harber

19b. Mature stems brown, purplish brown, reddish brown, or pale reddish brown; leaf blade linear-lanceolate, narrowly lanceolate, oblong-lanceolate, oblong-elliptic, elliptic-lanceolate, oblanceolate, or narrowly or broadly elliptic, 1.2–17 × 0.2–7 cm, abaxial leaf surface pruinose or epruinose, margin spinose or spinulose with 1 to 25 teeth on each side or entire 20

20a. Abaxial leaf surface pruinose 21

20b. Abaxial leaf surface epruinose 23

21a. Ovules 10 to 15 . . . . 11. *B. calliantha* Mulligan

21b. Ovules 4 to 6 22

22a. Mature stems pale reddish brown turning yellow; leaf blade narrowly elliptic or oblanceolate, 1.2–3 × 0.3–0.5 cm; ovules 4 . . . . .73. *B. taliensis* C. K. Schneid.

22b. Mature stems purplish brown; leaf blade oblong-elliptic, elliptic, or broadly elliptic, 6–9 × 3–5 cm; ovules 5 or 6 . . . . 37. *B. hsuyunensis* P. K. Hsiao & W. C. Sung

23a. Leaf blade always or mostly linear-lanceolate 24

23b. Leaf blade lanceolate, narrowly lanceolate, elliptic-lanceolate, broadly or narrowly elliptic, oblong-elliptic, obovate-elliptic, oblanceolate, or ovate 25

24a. Leaf bade linear-lanceolate, (3.5–)6–10.5 × (0.2–)0.4–0.8 cm; flowers pale yellow . . . . 81. *B. wuchuanensis* Harber & S. Z. He

24b. Leaf blade linear-lanceolate, oblong-lanceolate, or narrowly elliptic, 2–4.5(–6) × 0.25–0.5 cm; flowers salmon pink and pale whitish yellow . . . . .76. *B. triacanthophora* Fedde

25a. Leaf blade more than 9 cm 26

25b. Leaf blade less than 9 cm, very rarely longer 27

26a. Mature stems purplish brown; leaf blade elliptic, 8–17 × 3.5–7 cm, margin entire . . . . .61. *B. pingshanensis* W. C. Sung & P. K. Hsiao

26b. Mature stems pale reddish brown; leaf blade oblong-elliptic, 4–15 × 1.5–6.5 cm, margin spinose with 10 to 25 teeth on each side . . . . .66. *B. sargentiana* C. K. Schneid.

27a. Leaf blade elliptic or broadly elliptic or occasionally ovate, 4–8 × 1.5–3 cm; pedicel 7–13 mm . . . . 85. *B. xanthoclada* C. K. Schneid.

27b. Leaf blade broadly or narrowly lanceolate, narrowly elliptic, or obovate-elliptic, 1–8.5(–14) × 0.4–2.8 cm; pedicel 1–15 mm 28

28a. Leaf blade broadly lanceolate, 4–8.5 × 1.5–2.8 cm; pedicel 10–15 mm . . . . .60. *B. photiniifolia* C. M. Hu

28b. Leaf blade narrowly lanceolate, narrowly elliptic, or obovate-elliptic, 1–7(–14) × 0.4–1.2(–2.4) cm; pedicel 1–10 mm 29

29a. Leaf blade narrowly lanceolate, sometimes narrowly elliptic, 3–7(–14) × 0.5–1.2(–2.4) cm, margin conspicuously revolute especially on upper leaves; inflorescence 5- to 10-flowered; pedicel 1–4 mm . . . . .32. *B. grodtmanniana* C. K. Schneid.

29b. Leaf blade narrowly elliptic or obovate-elliptic, 1–4.5(–5.5) × 0.4–0.8(–1.4) cm, margin flat; inflorescence 10- to 15-flowered; pedicel 4–10 mm . . . . .88. *B. zhaotongensis* Harber

30a. [18b] Leaf blade abaxially pruinose 31

30b. Leaf blade abaxially epruinose 43

31a. Leaf blade thickly leathery 32

31b. Leaf blade thinly leathery 33

32a. Leaf blade oblong-oblanceolate, elliptic-oblanceolate, or narrowly oblong-elliptic, 2–8 × 0.8–2.5 cm; sepals in 2 whorls; ovules 2 or 3 . . . . .62. *B. pruinosa* Franch.

32b. Leaf blade oblanceolate or narrowly elliptic, 3.5–8 × 1–2 cm; sepals in 3 whorls; ovules 4 . . . . 23. *B. dongchuanensis* T. S. Ying

33a. Leaf blade to 5 cm 34

33b. Leaf blade more than 5 cm 38

34a. Margin conspicuously revolute 35

34b. Margin flat or inconspicuously revolute 37

35a. Leaf blade adaxially dull pale green, oblong-elliptic or oblong-lanceolate, 1.5–3.5(–4.5) × 0.3–0.5(–0.8) cm . . . . .64. *B. replicata* W. W. Sm.
35b. Leaf blade adaxially shiny green, narrowly elliptic, oblanceolate, or narrowly elliptic-lanceolate, 1.2–5 × 0.3–0.9 cm 36
36a. Leaf blade narrowly elliptic or oblanceolate, 1.2–3 × 0.3–0.5 cm; species of NW Yunnan. . . . . .73. *B. taliensis* C. K. Schneid.
36b. Leaf blade narrowly elliptic-lanceolate, 1.2–5 × 0.4–0.9 cm; species of Cona (Cuona) Xian in SE Xizang (Tibet). . . . . 31. *B. griffithiana* C. K. Schneid.
37a. Shrubs to 1 m tall; leaf blade elliptic or narrowly elliptic, 1–2.3 × 0.4–0.7 cm. . . . . .46. *B. laojunshanensis* T. S. Ying
37b. Shrubs to 2 m tall; leaf blade oblong-elliptic or narrowly elliptic, (1.5–)3–5 × (0.8–)1.4–1.8 cm . . . . .50. *B. lijiangensis* C. Y. Wu ex S. Y. Bao
38a. Leaf blade elliptic or ovate-elliptic. 39
38b. Leaf blade lanceolate, lanceolate-elliptic, oblong-lanceolate, oblong-elliptic, oblong-oblanceolate, elliptic-oblanceolate, or obovate. 40
39a. Leaf blade elliptic to ovate-elliptic, (3–)4–6(–7) × (1.2–)1.6–2.6(–3) cm, abaxial surface densely pruinose; pedicel 10–15(–20) mm; sepals in 3 whorls; species of Lushi Xian in NW Yunnan . . . . 19. *B. coxii* C. K. Schneid.
39b. Leaf blade elliptic, 3–7 × 1.3–2.5 cm, abaxial surface lightly pruinose; pedicel 15–22 mm; sepals in 2 whorls; species of Jinfo Shan, Nanchuan Xian, Chongqing (SE Sichuan) . . . . 43. *B. jinfoshanensis* T. S. Ying
40a. Leaf blade to 10 cm, margin entire, rarely spinose with 1 to 5(to 8) teeth on each side. . . . .35. *B. holocraspedon* Ahrendt
40b. Leaf blade to 8 cm, margin spinose or spinulose with 2 to 10(to 20) teeth on each side 41
41a. Leaf blade lanceolate-elliptic or oblong-elliptic or sometimes obovate, 3–7 × 0.8–3 cm; ovules 3 to 6 . . . . 36. *B. hookeri* Lem
41b. Leaf blade oblong-oblanceolate, elliptic-oblanceolate, or narrowly oblong-elliptic, 2–8 × 0.6–2.5 cm; ovules 2 or 3. . . . 42
42a. Leaf margin revolute, particularly in winter; species of Gongshan Xian in NW Yunnan. . . . 3. *B. amabilis* C. K. Schneid.
42b. Leaf margin flat; species of Fujian, Guangdong, Guangxi, Guizhou, Hubei, and Hunan. . . . . . 16. *B. chingii* W. C. Cheng
43a. [30b] Leaf blade thickly leathery 44
43b. Leaf blade thinly leathery 45
44a. Leaf blade oblong-oblanceolate, elliptic-oblanceolate, or narrowly oblong-elliptic, 2–8 × 0.8–2.5 cm, spinose with 1 to 6 coarse teeth on each side or occasionally entire or spinulose with 8 or 9 teeth on each side. . . . . .62. *B. pruinosa* Franch.
44b. Leaf blade oblong-elliptic or elliptic, (3–)5–8 × 1.6–2.5 cm, margin spinose with 6 to 10 often coarse teeth on each side. . . . . 8. *B. bergmanniae* C. K. Schneid.
45a. Leaf blade to 5 cm 46
45b. Leaf blade more than 5 cm 53
46a. Pedicel more than 10 mm 47
46b. Pedicel 10 mm or less. 50
47a. Leaf blade lanceolate, 2.5–4 × 0.6–0.8 cm, apex narrowly acuminate; pedicel 10–20 mm . . . . 56. *B. nujiangensis* Harber
47b. Leaf blade narrowly to broadly elliptic or lanceolate-elliptic, rarely lanceolate, 1–4.5 × 0.2–1.5 cm, apex acute or obtuse; pedicel 5–32 mm 48
48a. Mature stems pale yellow; leaf blade narrowly elliptic or elliptic-lanceolate, rarely lanceolate, 1–3.5 × 0.2–0.6 mm; pedicel 5–12 mm . . . . 20. *B. davidii* Ahrendt
48b. Mature stems pale yellowish brown; leaf blade narrowly to broadly elliptic, or lanceolate-elliptic, 3–5 × 0.8–1.5 cm; pedicel 6–32 mm 49
49a. Leaf blade narrowly to broadly elliptic, 2–5 × 0.8–1.5 cm; pedicel 6–15 mm; outer sepals broadly obovate, 6–7 × 5.5–6.5 mm; inner sepals obovate-orbicular, 6.5–8 × 6.5–7.5 mm . . . . .59. *B. phanera* C. K. Schneid.
49b. Leaf blade narrowly elliptic to lanceolate-elliptic, 3–4.5 × 1–1.5 cm; pedicel 18–32 mm; outer sepals oblong-ovate, 4.5–5 × 3–4 mm; inner sepals broadly obovate, 7–8 × 5.5–6 mm. . . . . 82. *B. wui* Harber
50a. Species of Jiangxi, Hunan, Fujian, and Zhejiang. 51
50b. Species of NW Yunnan, W Hubei, and Chongqing 52
51a. Mature stems pale yellow; leaf blade elliptic or elliptic-oblanceolate, 3–5(–7) × 1–2(–2.5) cm; pedicel 4–7 mm; species of Fujian and Zhejiang. . . . . 29. *B. fujianensis* C. M. Hu
51b. Mature stems yellowish brown; leaf blade elliptic, elliptic-oblanceolate, or oblong, 1.5–4 × 0.5–1.2 cm; pedicel 6–10 mm; species of Jiangxi and Hunan. . . . . 42. *B. jiangxiensis* C. M. Hu
52a. Mature stems yellow; leaf blade elliptic or oblong, 2–5 × 1–2 cm; pedicel 5–8 mm; species of W Hubei and Chongqing . . . . .68. *B. silvicola* C. K. Schneid.
52b. Leaf blade ovate to elliptic, 1.1–2.2 × 0.6–1.2(–1.5) cm; pedicel 7–10 mm; species of NW Yunnan . . . . 58. *B. petrogena* C. K. Schneid.
53a. Leaf blade more than 10 cm, narrowly lanceolate, lanceolate, elliptic-lanceolate, or narrowly elliptic. 54
53b. Leaf blade 10 cm or less, elliptic, linear, oblong, ovate, obovate, oblanceolate, or various combinations of these shapes. 58
54a. Mature stems yellow 55
54b. Mature stems yellowish brown or brownish yellow 56
55a. Leaf blade lanceolate, or sometimes narrowly elliptic, 5–15 × 1–2 cm, margin spinose with 10 to 24 coarse teeth on each side; flowers pale pinkish and white. . . . . 78. *B. veitchii* C. K. Schneid.
55b. Leaf blade narrowly elliptic, elliptic-lanceolate, or lanceolate, 3.5–11 × 0.4–2.2 cm, margin spinose with 6 to 20 teeth on each side; flowers greenish yellow. . . . . .30. *B. gagnepainii* C. K. Schneid.
56a. Leaf blade very narrowly lanceolate, 4.5–12(–18) mm; pedicel 7–13 mm . . . . .87. *B. yingjingensis* D. F. Chamb. & Harber

56b. Leaf blade lanceolate or narrowly lanceolate, 7–20 × 0.6–3.5 mm; pedicel 14–15 mm 57
57a. Leaf blade lanceolate, 7–20 × 1–3.5 cm; pedicel 15–25 mm . . . . 1. *B. acuminata* Franch.
57b. Leaf blade narrowly lanceolate, 7–14 × 0.6–2.5 cm; pedicel 14–23 mm . . . . 13. *B. caudatifolia* S. Y. Bao
58a. Leaf blade 0.6 cm wide or less 59
58b. Leaf blade 0.7–4 cm wide 60
59a. Leaf blade linear-oblong, 3–8 × 0.3–0.4 cm; pedicel 2–5 mm; flowers yellow . . . . 41. *B. insolita* C. K. Schneid.
59b. Leaf blade linear-lanceolate, 1.6–6 × 0.3–0.6 cm, pedicel 4–12(–15) mm; flowers partially purple . . . .
     . . . . 65. *B. sanguinea* Franch.
60a. Reticulation of adaxial leaf surface conspicuous and dense 61
60b. Reticulation of adaxial leaf surface inconspicuous or conspicuous but not dense 62
61a. Spines 0.4–0.8 cm; leaf blade elliptic, elliptic-lanceolate, or broadly lanceolate, 3–6.5 × 1.2–2.2 cm, margin spinu-
     lose with 12 to 50 teeth on each side; ovules (seeds) 2 or 3 . . . . 4. *B. arguta* (Franch.) C. K. Schneid.
61b. Spines 1–2 cm; leaf blade elliptic, sometimes narrowly elliptic, 5–9.5 × 1–1.8 cm, margin spinulose with 20 to 40
     teeth on each side; ovules 1 or 2 . . . . 24. *B. dumicola* C. K. Schneid.
62a. Ovules (seeds) 3 to 6 63
62b. Ovules no more than 3 66
63a. Leaf blade narrowly lanceolate, margin undulate, revolute; ovules 4 . . . . 52. *B. lubrica* C. K. Schneid.
63b. Leaf blade elliptic, oblong, oblong-lanceolate, oblong-elliptic, lanceolate-elliptic, or obovate, margin flat; ovules/
     seeds 3 to 6 64
64a. Leaf blade lanceolate-elliptic or oblong-elliptic, sometimes obovate; pedicel 15–25(30) mm; ovules 3 to 6 . . . .
     . . . . 36. *B. hookeri* Lem
64b. Leaf blade elliptic, oblong, oblong-lanceolate, or oblong-elliptic; pedicel 8–16 mm; ovules 4 to 6 65
65a. Leaf blade oblong-lanceolate or oblong-elliptic, 2–6 × 0.8–1.6 cm; seeds 4 or 5; species of NW Yunnan . . . .
     . . . . 27. *B. fallax* C. K. Schneid.
65b. Leaf blade elliptic or oblong, 4–6.5 × 1–2.5 cm; ovules 4 to 6; species of NE Guangxi, N Guangdong, S Hunan, and
     W Jiangxi . . . . 39. *B. impedita* C. K. Schneid.
66a. Leaf blade dull gray-green or bluish green; species of Zhejiang . . . . 47. *B. lempergiana* Ahrendt
66b. Leaf blade green or dark green, ± shiny; species not found in Zhejiang 67
67a. Leaf blade margin entire, spinose, or spinulose with no more than 6 teeth on each side 68
67b. Leaf blade margin spinose or spinulose with more than 6 teeth on each side 70
68a. Leaf blade oblanceolate or elliptic-obovate; pedicel 8–10 mm . . . . 84. *B. wuyiensis* C. M. Hu
68b. Leaf blade oblong-lanceolate, oblong-elliptic, narrowly elliptic, ovate-elliptic, or lanceolate-elliptic; pedicel 10–25
     mm 69
69a. Leaf blade oblong-lanceolate or oblong-elliptic, 4–10 × 1.5–2.5 cm; species of Yunnan . . . . . 35. *B. holocraspedon* Ahrendt
69b. Leaf blade narrowly elliptic, oblong-elliptic, ovate-elliptic, or lanceolate-elliptic, 3.3–6.5(–9) × 0.8–1.8 cm; species
     of Mêdog (Motuo) Xian in southeast Xizang (Tibet) . . . . 48. *B. leptopoda* Ahrendt
70a. Species endemic to Yunnan 71
70b. Species not found in Yunnan 74
71a. Leaf blade ovate, oblong-ovate, or oblong-elliptic, 2.5–5.5 × 0.8–1.3(–1.7) cm . . . . 22. *B. delavayi* C. K. Schneid.
71b. Leaf blade lanceolate, elliptic-lanceolate, elliptic, or narrowly elliptic, rarely obovate, (2.5–)3–9 × 0.8–2.5 cm 72
72a. Spines 0.4–1.5 cm; flowers per inflorescence 2 to 6(to 17); berry subglobose, style not persistent . . . .
     . . . . 71. *B. subacuminata* C. K. Schneid.
72b. Spines (0.8–)1.2–3.2 cm; flowers per inflorescence 1 to 9; berry oblong or oblong-elliptic, style persistent 73
73a. Leaf blade lanceolate or very narrowly elliptic, (2.5–)4–9 × 0.8–1.4 cm; flowers per inflorescence 1 to 5; pedicel
     10–15 mm . . . . 89. *B. zhenxiongensis* Harber
73b. Leaf blade narrowly elliptic or elliptic, rarely obovate-elliptic, (3–)4.5–8.5 × (0.8–)1.5–2(–2.5) cm; flowers per inflo-
     rescence 2 to 9; pedicel 12–26 mm . . . . 86. *B. yiliangensis* Harber
74a. Leaf blade broadly or narrowly ovate-elliptic, 6–10 × 2–3(–4) cm; berry 10–12 × 6–8 mm . . . . . 69. *B. simulans* C. K. Schneid.
74b. Leaf blade elliptic, elliptic-lanceolate, elliptic-oblanceolate, obovate-elliptic, oblong, oblong-elliptic, oblong-oblan-
     ceolate, oblong-obovate, oblong-ovate, or lanceolate, 2–10 × 0.5–2.5 cm; berry 5–9 × 3–6 mm 75
75a. Leaf blade oblong-oblanceolate, elliptic-oblanceolate, or narrowly obovate-elliptic, abaxially often white pruinose,
     margin teeth sometimes confined to upper half of leaf . . . . 16. *B. chingii* W. C. Cheng
75b. Leaf blade elliptic, elliptic-lanceolate, oblong, oblong-elliptic, oblong-obovate, oblong-ovate, or lanceolate, abaxially
     epruinose, margin teeth along whole length of leaf 76
76a. Flowers per inflorescence (2 to)5 to 20 77
76b. Flowers per inflorescence (1 to)2 to 7(to 8) 78
77a. Leaf blade lanceolate or narrowly oblong-elliptic, 3–6(–8) × 0.7–2 cm; outer sepals oblong-obovate, apex of petals
     deeply narrowly incised . . . . 7. *B. atrocarpa* C. K. Schneid.
77b. Leaf blade narrowly oblong, oblong-ovate, narrowly oblong-elliptic, or narrowly oblong-obovate, 3.5–10 × 1–1.5 cm;
     outer sepals oblong-elliptic, apex of petals emarginate . . . . 70. *B. soulieana* C. K. Schneid.
78a. Leaf blade lanceolate or narrowly lanceolate; pedicel 20–25(–32) mm . . . . 75. *B. tengii* Harber
78b. Leaf blade elliptic, narrowly elliptic, oblong, narrowly elliptic-lanceolate, or lanceolate; pedicel 10–25 mm 79
79a. Leaf blade elliptic, narrowly elliptic, or oblong, 2–6.5 × 0.8–1.6 cm, margin conspicuously revolute, undulate . . . . .
     . . . . 33. *B. guizhouensis* T. S. Ying
79b. Leaf blade lanceolate or narrowly elliptic-lanceolate or elliptic, margin flat 80
80a. Leaf blade lanceolate or narrowly elliptic-lanceolate, 3–7 × 1–1.8 cm; flowers yellow; berry obovoid, 6–9 × 5–6 mm
     . . . . 26. *B. fallaciosa* C. K. Schneid.

80b. Leaf blade narrowly elliptic or elliptic-lanceolate, 7–10 × 2–2.5 cm; flowers pale pinkish red outside, white within; berry ellipsoid, 7 × 5 mm . . . . 9. *B. bicolor* H. Lév.

SPECIES NOT IN SECTION *WALLICHIANAE*

KEY 3. FLOWERS SOLITARY

1a. Shrubs less than 1 m tall, semi-evergreen or deciduous 2
1b. Shrubs more than 1 m tall, deciduous. 17
2a. Shrubs semi-evergreen, sub-prostrate, to 20 cm tall; mature stems pale yellow; leaf blade obovate; pedicel 25–30 mm; ovules 12 to 15 . . . . 250. *B. tsangpoensis* Ahrendt
2b. Shrubs deciduous, erect; mature stems shades of yellow, brown, red, or purple; leaf blade elliptic, obovate, or oblanceolate; pedicel 2–30 mm; ovules 2 to 15. 3
3a. Mature stems pale yellow, pale brownish yellow, pale yellowish brown, or pale reddish brown 4
3b. Mature stems dark brown, dark red, reddish brown, reddish purple, or purple. 9
4a. Spines 3- to 9-fid, sometimes spreading at base or foliaceous; ovules 5 to 8. 233. *B. sibirica* Pall.
4b. Spines 3-fid or 3- to 5-fid, not spreading at base, not foliaceous; ovules 1 to 11 5
5a. Spines 3(to 5)-fid; ovules 11; pedicel 15–25 mm . . . . 130. *B. daiana* T. S. Ying
5b. Spines 3-fid; ovules 1 to 7; pedicel 2–14 mm 6
6a. Ovules 1; pedicel 10–14 mm . . . . .191. *B. mianningensis* T. S. Ying
6b. Ovules 3 to 7; pedicel 2–12 mm 7
7a. Ovules 5 to 7; pedicel 4–5 mm . . . . .173. *B. kangdingensis* T. S. Ying
7b. Ovules 3 or 4; pedicel 2–12 mm 8
8a. Mature stems pale yellow; pedicel 2–3 mm; species of W Himalayas. . . . . 148. *B. everestiana* Ahrendt
8b. Mature stems pale reddish brown; pedicel 10–12 mm; species of NW Yunnan. . . . . .230. *B. scrithalis* Harber
9a. [3b] Spines more than 3-fid. 10
9b. Spines 1- to 3-fid. 11
10a. Spines (3- to)5-fid; leaf blade entire; ovules 3 to 5 . . . . .251. *B. tsarica* Ahrendt
10b. Spines 5- to 7(to 9)-fid; leaf blade spinose with (5 to)7 to 15 teeth on each side; ovules 6 to 9 . . . .
. . . . 147. *B. erythroclada* Ahrendt
11a. Seeds 6 or 7; berry ca. 20 mm. . . . . .150. *B. fengii* S. Y. Bao
11b. Seeds fewer than 7; berry 10 mm or less. 12
12a. Seeds 1 to 5; leaf blade obovate or obovate-elliptic. . . . . .128. *B. crassilimba* C. Y. Wu ex S. Y. Bao
12b. Seeds fewer than 5; leaf blade obovate, narrowly obovate, oblanceolate, or oblong-oblanceolate 13
13a. Seeds 1 or 2 14
13b. Seeds 3 or 4 15
14a. Leaf blade obovate, margin spinulose with 5 to 14 prominent teeth on each side, sometimes entire. . . . .
. . . . .267. *B. yalongensis* Harber
14b. Leaf blade narrowly obovate or oblanceolate, margin entire or sometimes spinulose with 1 to 3 teeth on each side . .
. . . . 193. *B. minutiflora* C. K. Schneid.
15a. Leaf blade oblong-oblanceolate; mature stems dark brown; seeds 4 . . . . 222. *B. qiaojiaensis* S. Y. Bao
15b. Leaf blade narrowly obovate or oblanceolate; mature stems reddish purple or purple; seeds 3 or 4 16
16a. Mature stems reddish purple; berry narrowly ovoid, 12 × 6 mm; style not persistent; seeds 3 . . . . .149. *B. exigua* Harber
16b. Mature stems purple; berry subglobose, 8–9 × 5–6 mm; style persistent, short; seeds 4. . . . . . . .201. *B. nanifolia* Harber
17a. [1b] Mature stems yellow, yellowish brown, or yellowish gray. 18
17b. Mature stems red, reddish brown, reddish purple, or purple. 24
18a. Leaf blade entire. 19
18b. Leaf blade spinose or spinulose 20
19a. Shrubs to 1.2 m tall; leaf blade obovate, 2–2.2 × ca. 0.5–0.8 cm; spines 0.3–0.5 cm; pedicel 7–14 mm; sepals in 2 whorls. . . . . .200. *B. nambuensis* (Ahrendt) Harber
19b. Shrubs to 3 m tall; leaf blade narrowly obovate or obovate-elliptic, (1.2–)2–2.8 × 0.6–0.9 cm; spines 0.6–1.6 cm; pedicel 18–28 mm, sepals in 3 whorls . . . . 127. *B. cornuta* Harber
20a. Leaf blade margin spinose or spinulose with 12 to 40 teeth on each side 21
20b. Leaf blade margin spinose or spinulose with fewer than 12 teeth on each side. 22
21a. Leaf margin spinose with 5 to 40 closely spaced teeth on each side, leaf blade obovate-orbicular or obovate-elliptic, rarely elliptic, 1.4–2.8(–3.4) × 0.5–2.5 cm; pedicel 15–45 mm; ovules 4. . . . .
. . . . 124. *B. circumserrata* (C. K. Schneid.) C. K. Schneid.
21b. Leaf margin minutely spinulose with 12 to 20 teeth on each side or entire, leaf blade obovate, 2.5–3.5(–4) × 1.2–1.6 cm; pedicel (15–)25–30(–35) mm; seeds 2 . . . . 185. *B. longipedicellata* Harber
22a. Stems pale yellowish gray; leaf blade up to 4 cm; pedicel 10–16 mm; sepals in 3 whorls; ovules 5 or 6. . . . .
. . . . 106. *B. bawangshanensis* (Ahrendt) Harber
22b. Mature stems pale yellowish brown; leaf blade less than 2 cm; pedicel 5–12 mm; sepals in 2 or 3 whorls, ovules 3 or 4 23
23a. Leaf blade narrowly obovate or narrowly elliptic, 1.3–1.7 × 0.4–0.6 cm; pedicel 5–10 mm; sepals in 2 whorls; ovules 3 or 4 . . . . 223. *B. qinghaiensis* Harber
23b. Leaf blade oblanceolate or narrowly elliptic, (0.7–)1.4–1.5 × 0.2–0.3 cm; pedicel 12 mm; sepals in 3 whorls; ovules 4
. . . . 111. *B. bouffordii* Harber

24a. [17b] Pedicel absent or to 4 mm 25
24b. Pedicel longer than 4 mm 30
25a. Leaf blade oblanceolate, narrowly obovate, obovate, or narrowly elliptic, 2 cm or longer 26
25b. Leaf blade obovate, narrowly obovate, or oblong-obovate, less than 2 cm 27
26a. Leaf blade oblanceolate or narrowly obovate, 2–4(–5.2) × 0.4–1(–1.3) cm; mature stems pale reddish brown ......
.... 117. *B. brevipedicellata* Harber
26b. Leaf blade obovate or narrowly elliptic, 2–2.5 × 1–2 cm; mature stems dark reddish purple ....
.... 104. *B. barkamensis* Harber
27a. Ovules or seeds 1 or 2 .... 189. *B. markamensis* Harber
27b. Ovules or seeds 3 to 5 28
28a. Seeds (ovules) 5; leaf blade narrowly obovate or oblong-obovate, 0.8–1.2 × 0.15–0.4 cm, reticulation inconspicuous
or obscure adaxially; mature stems purplish red .... 95. *B. ambrozyana* C. K. Schneid.
28b. Seeds/ovules 3 or 4; leaf blade obovate or obovate-elliptic, (0.7–)1–1.7 × 0.3–1.2(–2) cm, reticulation dense and
conspicuous or inconspicuous or indistinct adaxially; mature stems reddish brown, not pruinose, or pale or dark
purple and mostly pruinose 29
29a. Leaf blade obovate (0.7–)1–1.7 × 0.3–0.7 cm, abaxially not pruinose; reticulation dense and conspicuous adaxially;
mature stems reddish brown, not pruinose .... 103. *B. baiyuensis* Harber
29b. Leaf blade obovate or obovate-elliptic, 1–1.6 × 0.8–1.2(–2) cm, abaxially pruinose; reticulation inconspicuous or
indistinct adaxially; mature stems pale or dark purple, mostly pruinose .... 146. *B. epedicellata* Harber
30a. [24b] Abaxial leaf surface white or glaucous pruinose 31
30b. Abaxial leaf surface not pruinose 33
31a. Pedicel 15–25 mm; young shoots bright pink, sometimes pruinose; seeds 5 to 8 .... 245. *B. tianchiensis* Harber
31b. Pedicel 6–15 mm; young shoots green, mostly white pruinose; ovules 3 to 11 32
32a. Ovules 3 to 5; leaf blade obovate or obovate-elliptic, very rarely obovate-orbicular, 1–2.5(–5) × 0.4–1(–2) cm, margin
entire, very rarely spinulose .... 138. *B. dictyophylla* Franch.
32b. Ovules 6 to 11; leaf blade oblong-obovate, 2–7 × 1.2–3 cm, margin dentate with 3 to 9 teeth on each side or sometimes entire .... 239. *B. temolaica* Ahrendt
33a. Pedicel to 12 mm 34
33b. Pedicel more than 12 mm 36
34a. Leaf blade (2–)3–4(–5.5) cm, narrowly elliptic, obovate, or oblanceolate; seeds 4 to 6 .... 228. *B. saltuensis* Harber
34b. Leaf blade less than 3 cm, oblanceolate, obovate, narrowly obovate, narrowly elliptic, or elliptic-obovate, ovules 4 or
5 35
35a. Mature stems reddish brown; leaf blade oblanceolate, narrowly obovate, or narrowly elliptic; pedicel 3–5(–10) mm;
anther connective extended, ovules 4; species of W Sichuan .... 153. *B. gaoshanensis* Harber
35b. Mature stems purple or reddish purple; leaf blade obovate or elliptic-obovate; pedicel 6–12 mm; anther connective
not extended; ovules 5 to 7; species of C Xizang (Tibet) .... 183. *B. lhunzhubensis* Harber
36a. Pedicel 25 mm or less 37
36b. Pedicel usually more than 25 mm 40
37a. Species of SE Gansu and SW Shaanxi (inflorescences sometimes also loose 2- to 6-flowered racemes or sub-umbels)
.... 217. *B. pseudothunbergii* P. Y. Li
37b. Species of Xizang (Tibet) and NW Yunnan; inflorescences consistently 1-flowered 38
38a. Leaf blade deep shiny green; berry subglobose; species of E Himalayas .... 98. *B. angulosa* Wall. ex Hook. f. & Thomson
38b. Leaf blade mid-green or dull green; berry ovoid, oblong-ovoid, or ellipsoid; species of NW Yunnan 39
39a. Berry ovoid or oblong-ovoid; leaf blade narrowly oblong-obovate or narrowly elliptic; outer sepals elliptic; inner sepals oblong; ovules 4 or 5 .... 244. *B. tianbaoshanensis* S. Y. Bao
39b. Berry ovoid or ellipsoid; leaf blade oblong-oblanceolate or narrowly obovate; outer sepals broadly ovate; inner sepals
broadly elliptic; ovules 5 to 7 .... 108. *B. beimanica* (Ahrendt) Harber
40a. Shrubs to 4 m tall; leaf blade (2.5–)3.3–7 cm; pedicel 25–50(–55) mm .... 186. *B. ludlowii* Ahrendt
40b. Shrubs to 2 m tall; leaf blade 1.5–3(–5) cm; pedicel (8–)13–35 mm 41
41a. Mature stems pale reddish brown; leaf blade narrowly obovate, often more broadly obovate on plants in more exposed
conditions; pedicel (8–)13–30 mm; seeds 5 or 6 .... 195. *B. monticola* Harber
41b. Mature stems reddish purple; leaf blade obovate, obovate-elliptic, or oblong-obovate; pedicel (15–)25–35 mm;
ovules (5 to)8 or 9(or 10) .... 121. *B. capillaris* Cox

Key 4. Flowers in Fascicles, Sub-fascicles, Umbels, or Sub-umbels

1a. Leaves abaxially pruinose 2
1b. Leaves not abaxially pruinose 5
2a. Mature stems partially glaucous; inflorescences fascicles; flowers per inflorescence 1 to 3 ....
.... 216. *B. pruinosifolia* Harber
2b. Mature stems not glaucous; inflorescences fascicles or very short umbels or sub-umbels; flowers per inflorescence 1
to 5 3
3a. Mature stems pale reddish brown; leaf blade narrowly obovate, (0.9–)1.7–3.2 × 0.5–0.8 cm; inflorescences umbels;
flowers per inflorescence 3 to 5 .... 268. *B. yanyuanensis* Harber
3b. Mature stems pale yellowish brown or reddish purple; leaf blade narrowly obovate or broadly obovate, narrowly
elliptic-obovate, narrowly elliptic, or oblanceolate, 0.8–2.8 × 0.3–1 cm; inflorescences fascicles or sub-umbels;
flowers per inflorescence 1 to 5 4

4a. Mature stems purple; leaf blade narrowly obovate or narrowly elliptic-obovate, 1.5–2.8 × 0.6–1 cm, margin spinulose with 4 to 8 coarse teeth on each side; inflorescences fascicles or sub-umbels. . . . . .129. *B. dahaiensis* Harber
4b. Mature stems reddish purple; leaf blade narrowly elliptic, oblanceolate, or narrowly or broadly obovate, 0.8–1.6(–2) × 0.3–0.7(–1) cm; margin entire or spinulose with 1 to 5 teeth on each side; inflorescences fascicles . . . . 120. *B. campylotropa* T. S. Ying
5a. Inflorescences solely subumbellate. 6
5b. Inflorescences partially fascicles and partially short umbels or sub-umbels or fascicles and sub-fascicles, or inflorescences solely fascicles 7
6a. Leaf blade obovate or obovate-lanceolate, 1.5–5 × 0.6–2.1 cm; berry black; species of Bomê (Bomi) Xian, SE Xizang (Tibet). . . . . .255. *B. umbratica* T. S. Ying
6b. Leaf blade obovate to obovate-elliptic, 1.5–5 × 1–2.2 cm; berry red; species of Gyirong (Jilong) Xian, S Xizang (Tibet). . . . . .155. *B. gilungensis* T. S. Ying
7a. Inflorescences partially fascicles and partially short umbels or sub-umbels. 8
7b. Inflorescences fascicles and sub-fascicles, or inflorescences solely fascicles 9
8a. Mature stems purple; leaf blade narrowly obovate or narrowly elliptic-obovate, 1.5–2.8 × 0.6–1 cm, margin spinulose with 4 to 8 coarse teeth on each side; pedicel 5–8 mm; berry oblong or ellipsoid, 4–5 × 8 mm . . . . .129. *B. dahaiensis* Harber
8b. Mature stems pale brownish yellow; leaf blade narrowly oblong-obovate or spatulate, 1–3 × 0.3–1 cm, margin entire; pedicel 10–15 mm; berry ellipsoid, 9 × 6 mm . . . . .207. *B. papillifera* (Franch.) Koehne
9a. Inflorescences fascicles and sub-fascicles 10
9b. Inflorescences solely fascicles 15
10a. Ovules/seeds 2 [see also multi-access key]. 11
10b. Ovules/seeds 3 to 8. 13
11a. Whorls of sepals 3. . . . . .91. *B. abbreviata* (Ahrendt) Harber
11b. Whorls of sepals 2. 12
12a. Flowers per inflorescence 1 to 6; pedicel 5–15(–22) mm; outer sepals elliptic-ovate, inner sepals oblong-elliptic; anther connective not or slightly extended, truncate. . . . . .139. *B. difficilis* Harber
12b. Flowers per inflorescence 3 to 9; pedicel 9–14 mm; outer sepals obovate-oblong, inner sepals broadly obovate; anther connective distinctly extended, obtuse . . . . 112. *B. bowashanensis* Harber
13a. Mature stems pale yellow; leaf blade adaxially shiny; pedicel 6–12(–16) mm; ovules 4 to 8 . . . . 126. *B. cooperi* Ahrendt
13b. Mature stems reddish or brownish purple; leaf blade adaxially dull; pedicel (10–)15–25(–40) mm; ovules/seeds 3 or 4 14
14a. Apex of leaf blade acute or subobtuse; flowers per inflorescence 1 to 3; pedicel (12–)20–30 mm; berry sometimes slightly bent at tip. . . . . .198. *B. muliensis* Ahrendt
14b. Apex of leaf blade rounded, obtuse, or subacute; flowers per inflorescence 3 to 5(to 7); pedicel (10–)15–25(–40) mm; berry not bent at tip . . . . .273. *B. yunnanensis* Franch.
15a. [9b] Leaf blade oblong, obovate-oblong, obovate, narrowly or broadly elliptic, obovate, or obovate-elliptic, (0.8–)1.2–6 × 0.5–2.5 cm; ovules 4 to 7. 16
15b. Leaf blade obovate, narrowly obovate, obovate-elliptic, narrowly elliptic, suborbicular, or oblanceolate, 0.7–2.5(–3) × 0.3–1.7 cm; ovules 1 to 4. 20
16a. Mature stems grayish yellow, pale reddish, or yellowish brown. 17
16b. Mature stems dark purple. 19
17a. Mature stems grayish yellow; leaf blade thickly and rigidly leathery, margin spinose or dentate with 1 to 4 teeth on each side, rarely entire; flowers (8 to)15 to 25(to 35). . . . . .101. *B. asiatica* Roxb. ex DC.
17b. Mature stems pale reddish or yellowish brown; leaf blade papery, margin entire or spinulose with 1 to 14 teeth on each side; flowers 1 to 5 18
18a. Mature stems pale reddish brown; leaf margin entire, rarely spinulose with 1 to 4 teeth on each side; pedicel 4–8 mm; flowers per inflorescence 4 to 6, fascicles for entire length of branches . . . . .182. *B. lhunzensis* Harber
18b. Mature stems pale yellowish brown; leaf margin spinulose with 2 to 14 teeth on each side, or entire; pedicel 12–22 mm; flowers per inflorescence 1 to 5, frequently solitary or paired toward the end of branches . . . . 136. *B. diaphana* Maxim.
19a. Leaf blade broadly obovate or elliptic, 2–5 × 1–1.4 cm, margin entire, very rarely spinose-serrulate with 1 to 6 teeth on each side; flowers per inflorescence (1 or)2 to 4. . . . . 209. *B. platyphylla* (Ahrendt) Ahrendt
19b. Leaf blade obovate, (2–)2.7–6 × (1.2–)1.5–2.5 cm, margin dentate with 2 to 10 widely spaced coarse teeth on each side; flowers per inflorescence 4 to 9. . . . . .157. *B. gongshanensis* Harber
20a. [15b] Species of Xizang (Tibet). 21
20b. Species of Guizhou, Sichuan, and Yunnan 24
21a. Leaf blade obovate, occasionally suborbicular, 0.7–2.5 × 0.4–1.7 cm; species of S Xizang (Tibet). . . . . .166. *B. hypericifolia* T. S. Ying
21b. Leaf blade narrowly elliptic, obovate-elliptic, oblanceolate, or narrowly or broadly obovate, 0.7–2.1 × 0.3–0.8(–1) cm; species of E and NE Xizang (Tibet) 22
22a. Mature stems pale orange-brown; spines 1–2.5 cm; leaf blade oblanceolate or narrowly elliptic, 0.7–2.1 × 0.3–0.8 cm . . . . .225. *B. reticulinervis* T. S. Ying
22b. Mature stems reddish purple or reddish brown; spines 0.4–1.4 cm; leaf blade narrowly elliptic, obovate-elliptic, oblanceolate, or narrowly or broadly obovate, 0.8–1.6(–2) × 0.3–0.7(–1) cm 23
23a. Mature stems reddish purple; spines 0.4–0.8 cm; pedicel 6–12 mm; berry oblong, 10–12 × 4–5 mm . . . . 120. *B. campylotropa* T. S. Ying

23b. Mature stems reddish brown; spines 0.7–1.4 cm; pedicel 3–4 mm; berry ellipsoid, 6–7 × 5–6 mm . . . .
. . . . 221. *B. qamdoensis* Harber
24a. Spines 0.2–0.5 cm; leaf blade elliptic or obovate-elliptic, sometimes obovate, 0.7–1.5 × 0.3–0.6 cm, adaxial surface shiny; fascicles 3–4 mm; seeds 1 . . . . 253. *B. tsienii* T. S. Ying
24b. Spines 0.5–2.3 cm; leaf blade narrowly obovate, oblong-obovate, narrowly elliptic, or oblanceolate, (0.8–)1.2–2.5 × 0.3–0.6 cm, adaxial surface dull; fascicles 4–10 mm; ovules/seeds 2 to 4 25
25a. Spines 1.2–2.3 cm; flowers per inflorescence 2 to 7; ovules 2 . . . . 242. *B. tenuispina* Harber
25b. Spines 0.5–1.5 cm; flowers per inflorescence 1 to 4; ovules 3 or 5 26
26a. Shrubs to 50 cm tall; leaf blade oblong-obovate or oblanceolate, 1–2 × 0.5–0.7 cm; pedicel 5–10 mm . . . .
. . . . 262. *B. woomungensis* C. Y. Wu ex S. Y. Bao
26b. Shrubs to 2 m tall; leaf blade narrowly obovate or narrowly elliptic, (0.8–)1.2–1.6 × 0.4–0.6 cm; pedicel 5–6 mm . .
. . . . 214. *B. pratensis* Harber

KEY 5. FLOWERS IN RACEMES (INCLUDING RACEMES SOMETIMES WITH FASCICLES AT BASE)

1a. Raceme spikelike, flowers dense 2
1b. Raceme open, flowers distinctly separate from each other 8
2a. Leaf blade abaxially pubescent or villous 3
2b. Leaf blade abaxially glabrous 4
3a. Leaf blade lanceolate, obovate-lanceolate, or narrowly elliptic, (1.5–)2.5–4(–6) × 0.3–1.5 cm, abaxially pubescent, adaxially glabrous . . . . 154. *B. gilgiana* Fedde
3b. Leaf blade elliptic, obovate, oblong-elliptic, or lanceolate, 3–8(–14) × 1.5–3.5(–5) cm, abaxial veins villous, adaxially rugose . . . . 113. *B. brachypoda* Maxim.
4a. Mature stems yellowish brown; leaf blade to 14 cm, margin spinose with 15 to 40 teeth on each side . . . .
. . . . 227. *B. salicaria* Fedde
4b. Mature stems purple, purplish red, or purplish brown; leaf blade to 6(–8) cm, margin entire or spinose with 1 to 50 teeth on each side 5
5a. Leaf blade obovate, suborbicular, oblong-elliptic, or broadly elliptic, 3–6(–8) × 2.5–4 cm, margin with 15 to 50 teeth on each side or sometimes entire; petiole almost absent or to 50 mm . . . . 132. *B. dasystachya* Maxim.
5b. Leaf blade lanceolate or obovate-lanceolate, narrowly oblanceolate, spatulate-oblanceolate, narrowly elliptic, obovate, or obovate-elliptic, 1–4 × 0.4–2 cm, margin entire or spinose with 1 to 15 teeth on each side; petiole almost absent or to 6 mm 6
6a. Leaf blade narrowly oblanceolate, spatulate-oblanceolate, obovate-lanceolate, or narrowly elliptic, 1.5–3.5 × 0.4–0.8 cm, margin entire . . . . 122. *B. caroli* C. K. Schneid.
6b. Leaf blade lanceolate or obovate-lanceolate, obovate, or obovate-elliptic, 1–4 × 0.4–0.8 cm, margin spinose or spinulose or sometimes entire 7
7a. Leaf blade lanceolate or obovate-lanceolate, 1–4 × 0.4–0.8 cm, margin spinose with (2 to)9 to 15 teeth on each side or sometimes entire . . . . 219. *B. purdomii* C. K. Schneid.
7b. Leaf blade obovate or obovate-elliptic, 1–4 × 0.8–2 cm, margin spinulose with 1 to 6 teeth on each side or entire . .
. . . . 258. *B. wanhuashanensis* Y. J. Zhang
8a. [1b] Leaf blade more than 6 cm 9
8b. Leaf blade 6 cm or less 13
9a. Mature stems dark red, purple, or reddish brown; flowers per inflorescence (9 to)20 to 40 10
9b. Mature stems shades of yellow, yellowish gray, or pale yellowish brown; flowers per inflorescence 3 to 30 11
10a. Leaf blade elliptic or oblong-obovate, 2.5–8 × 1–4 cm; berry initially creamy white, ultimately pale red or pink, translucent . . . . 170. *B. jamesiana* Forrest & W. W. Sm.
10b. Leaf blade ovate to elliptic or sometimes oblong-lanceolate, 2–7 × 1–3 cm; berry scarlet . . . .
. . . . 152. *B. francisci-ferdinandi* C. K. Schneid.
11a. Leaf blade oblong-rhombic; flowers 3 to 15 per inflorescence; inflorescences rarely sub-umbels . . . .
. . . . 257. *B. virgetorum* C. K. Schneid.
11b. Leaf blade obovate-elliptic, elliptic, oblong-obovate, broadly lanceolate, or ovate; flowers 10 to 30 per inflorescence; inflorescences never sub-umbels 12
12a. Pedicel 3–4 mm; outer sepals oblong to lanceolate-oblong; petals oblong-elliptic to obovate . . . .
. . . . 236. *B. subsessiliflora* Pamp.
12b. Pedicel (2–)5–10 mm; outer sepals obovate or ovate; petals elliptic . . . . 97. *B. amurensis* Rupr.
13a. [8b] Mature stems shades of yellow, pale yellowish, pale reddish, or grayish brown 14
13b. Mature stems brown, reddish or purplish brown, red, or purple 24
14a. Shrubs to 1 tall; leaf blade obovate, 0.5–1.4 × 0.2–0.5 cm; racemes 1.5–2 cm . . . . 248. *B. tomentulosa* Ahrendt
14b. Shrubs to 3 m tall; leaf blade obovate to elliptic, suborbicular, oblanceolate, or linear-oblanceolate, 0.5–6 × 0.2–3 cm; racemes 2–9 cm 15
15a. Leaf blade narrowly oblanceolate, linear-oblanceolate, or narrowly obovate, 1.5–6 × 0.4–1 cm, base attenuate 16
15b. Leaf blade obovate to elliptic or suborbicular, 0.6–6 × 0.4–3.2 mm, base attenuate or cuneate 18
16a. Leaf blade narrowly oblanceolate or linear-oblanceolate, base often elongated up to 2.2 cm . . . .
. . . . 203. *B. ninglangensis* Harber
16b. Leaf blade very narrowly oblanceolate or very narrowly obovate, not elongated 17
17a. Inflorescences sometimes partly paniculate or subumbellate at apex; flowers per inflorescence 8 to 30; ovules 1 . . .
. . . . 123. *B. chinensis* Poir.

17b. Inflorescences sometimes with fascicles at base, sometimes with fascicles at apex of stems; flowers (2 to)5 to 10 per inflorescence; ovules 2 to 4. . . . .*109. B. biguensis* Harber
18a. Leaf blade oblong-obovate or oblong-elliptic, 0.6–2(–2.5) × 0.4–0.8 cm; racemes 2–3 cm. . . . .
. . . . 165. *B. humidoumbrosa* Ahrendt
18b. Leaf blade orbicular, obovate to elliptic, or oblanceolate, 1.5–6 × 0.5–3 cm; racemes 2.5–9 cm 19
19a. Leaf blade suborbicular, broadly elliptic, or elliptic-obovate, 2–6 × 1.5–3 cm; inflorescence 10- to 27-flowered; species of Anhui and Zhejiang . . . .*99. B. anhweiensis* Ahrendt
19b. Leaf blade narrowly to broadly obovate, elliptic, or oblanceolate, 1.5–6 × 0.5–2(–3.2) cm; inflorescence 4- to 20-flowered; species of Sichuan, Yunnan, and Xizang (Tibet). 20
20a. Leaf blade obovate to elliptic, 2–6 × 0.5–1.5 cm; species of S Xizang (Tibet). . . . . 100. *B. aristata* DC.
20b. Leaf blade obovate or narrowly obovate, 1.5–4.5 × 0.6–3.5 cm; species of Sichuan and Yunnan 21
21a. Raceme without fascicles at base; pedicel 3–6 mm . . . .*131. B. daochengensis* T. S. Ying
21b. Raceme with fascicles at base; pedicel 4–15(–20) mm. 22
22a. Mature stems yellow; leaf margin spinose with (3 to)10 to 15 teeth on each side, rarely entire; flowers per inflorescence 6 to 20. . . . .*190. B. mekongensis* W. W. Sm.
22b. Mature stems pale reddish or pale yellowish brown; leaf margin entire or rarely spinose with 1 to 6(to 8) teeth on each side; flowers per inflorescence (4 to)7 to 13 23
23a. Spines mostly absent, solitary when present, rarely 3-fid, (0.6–)1.2–2.5 cm; racemes 6–9 cm . . . .151. *B. forrestii* Ahrendt
23b. Spines 3-fid, 0.4–1(–1.6) cm; racemes 2.5–6 cm . . . . .212. *B. polybotrys* Harber
24a. [13b] Leaf blade obcordate or subcuneate. . . . . 226. *B. retusa* T. S. Ying
24b. Leaf blade elliptic, oblong, obovate, obovate-elliptic, lanceolate, obovate-oblanceolate, or oblanceolate. 25
25a. Leaf blade mainly elliptic. 26
25b. Leaf blade oblong, oblong-obovate, obovate, obovate-elliptic, lanceolate, obovate-oblanceolate, or oblanceolate 28
26a. Racemes 2–6 cm; flowers per inflorescence 10 to 20; spines 1–3 cm, mostly 3-fid. . . . . .162. *B. henryana* C. K. Schneid.
26b. Racemes 1–4.5 cm; flowers per inflorescence 4 to 12; spines absent or 0.3–1.6 cm, mostly solitary 27
27a. Leaf blade entire, rarely spinulose with 1 to 7 teeth on each side; species of WC Sichuan. . . . .
. . . . .145. *B. emeishanensis* Harber
27b. Leaf blade spinulose with 8 to 20 teeth on each side, sometimes entire; species of S Qinghai, NW Sichuan, and NE Xizang (Tibet). . . . . .144. *B. elliptifolia* Harber
28a. Species of Xizang (Tibet) 29
28b. Species not in Xizang (Tibet) 34
29a. Raceme 4–8 cm; spines 0.8–2.5 cm 30
29b. Raceme 1–4 cm; spines 0.3–1.2 cm 31
30a. Leaf blade obovate, 1.2–2.8 × 0.7–1.3 cm; pedicel 5–7 mm; ovules 1 or 2. . . . . 94. *B. agricola* Ahrendt
30b. Leaf blade oblanceolate, 1.5–4.75 × 0.5–1 cm; pedicel 10–20 mm; ovules 4. . . . . .177. *B. kongboensis* Ahrendt
31a. Mature stems reddish brown; abaxial surface of leaf blade slightly glaucous; pedicel 3–10 mm; species of S Xizang (Tibet). . . . . .256. *B. virescens* Hook. f.
31b. Mature stems reddish purple or dark red; abaxial surface of leaf blade green or pale grayish green; pedicel 6–14 mm; species of SE Xizang (Tibet). 32
32a. Mature stems partially pruinose; flowers per inflorescence 5 to 20; peduncle to 0.6 cm . . . . . . .107. *B. baxoiensis* Harber
32b. Mature stems not pruinose; flowers per inflorescence 12 or fewer; peduncle 0.8–2.5 cm 33
33a. Mature stems reddish purple; leaf blade obovate, obovate-elliptic, or obovate-oblanceolate, (1.5–)2–3 × 0.5–0.8(–1) cm; inflorescence rarely partially paniculate; whorls of sepals 2. . . . . 140. *B. dispersa* (Ahrendt) Harber
33b. Mature stems dark red; leaf blade obovate, rarely oblanceolate, 1.3–2.6 × 0.6–1.2 cm; inflorescence sometimes an umbel and solitary flowers at the tips of stems; whorls of sepals 3 . . . . .210. *B. pluvisylvatica* Harber
34a. [28b] Leaf blade obovate-lanceolate or narrowly obovate, 1.5–3.5 × 0.5–0.8 cm; inflorescence to 10 cm, sometimes 1 to 3 racemes fascicled at base; flowers per inflorescence 20 to 30; pedicel ca. 2 mm . . . .171. *B. jiulongensis* T. S. Ying
34b. Leaf blade obovate-elliptic, obovate-lanceolate, oblong-elliptic, or lanceolate, 0.3–4 × 0.3–1.8 cm; inflorescence to 5.5 cm, sometimes with fascicles at base; flowers per inflorescence 4 to 20(to 25); pedicel 3–22 mm 35
35a. Spines absent or solitary, 0.2–0.3 cm; inflorescences loose racemes to 5.5 cm; pedicel (5–)12–22 mm . . . .
. . . . .260. *B. wenchuanensis* Harber
35b. Spines 1- to 3(to 5)-fid, 0.2–5 cm; inflorescences racemes, rarely sub-racemes or sub-panicles, sometimes fascicled at base, 1–5 cm; pedicel 3–12 mm (pedicels at base up to 18 mm). 36
36a. Leaf blade thickly leathery, adaxially distinctly shiny, margin often dentate; spines (1.5–)2–6 cm, often longer than internodes. . . . . 213. *B. potaninii* Maxim.
36b. Leaf blade papery or thinly leathery, adaxially not or slightly shiny, margin entire or spinose; spines 0.2–4.75 cm, shorter than internodes. 37
37a. Mature stems, abaxial leaf surface, rachis, and pedicel partially pruinose; flowers per inflorescence 1 to 7. . . . .
. . . . .181. *B. leptoclada* Diels
37b. Mature stems, abaxial leaf surface, rachis, and pedicel all epruinose; flowers per inflorescence 3 to 25 38
38a. Spines 1–4.75 cm, mostly solitary; flowers per inflorescence (10 to)20 to 25; species of Xinjiang . . . .
. . . . 167. *B. integerrima* Bunge
38b. Spines 0.6–2.5 cm, mostly 3-fid (5-fid in *B. dubia*) or sometimes solitary or absent toward apex of stems; flowers per inflorescence 3 to 18; species not found in Xinjiang. 39
39a. Leaf blade obovate, obovate-elliptic, or oblong-elliptic, 0.5–2(–3.5) × 0.3–1.3 cm 40
39b. Leaf blade narrowly obovate, obovate-oblanceolate, or narrowly oblong-obovate, 1.5–5 × 0.4–1.8 cm. 42
40a. Leaf blade 0.3–0.6 cm wide, margin entire; flowers per inflorescence 4 to 8; seeds 2 . . . . 110. *B. boschanii* C. K. Schneid.

40b. Leaf blade 0.3–1.2 cm wide, margin spinose with 1 to 25 teeth on each side or sometimes entire; flowers per inflorescence 3 to 15; seeds/ovules 2 to 4 41

41a. Mature stems reddish brown; racemes sometimes partially paniculate, without fascicles at base, to 4.5 cm; flowers per inflorescence 6 to 15. . . . . 234. *B. sichuanica* T. S. Ying

41b. Mature stems dark purplish red; racemes not partially paniculate, often with fascicles at base, to 3.5 cm; flowers per inflorescence 3 to 10. . . . . 241. *B. tenuipedicellata* T. S. Ying

42a. Leaf blade narrowly obovate, 1.5–3 × 0.5–1.8 cm, margin spinose with 6 to 14 teeth on each side, but leaves at apex of stems often entire; species of Gansu, Nei Mongol, Ningxia, Qinghai, and N Sichuan . . . . . 142. *B. dubia* C. K. Schneid.

42b. Leaf blade narrowly oblong-obovate, narrowly obovate to obovate-oblanceolate, (1.5–)2.2–5 × (0.4–)0.8–1.5 cm, margin entire, spinose, or coarsely spinose-serrulate with 1 to 5 teeth on each side; species of NW Yunnan 43

43a. Leaf blade 2.5–5 × 0.6–1.5 cm; inflorescences racemes or subumbellate racemes, rarely partially paniculate, without fascicles at base; peduncle 0.5–2 cm, pedicel 5–8 mm; flowers per inflorescence 3 to 11 . . . . 218. *B. pseudotibetica* C. Y. Wu

43b. Leaf blade (1.5–)2.2–4 × (0.4–)0.8–1.3 cm; inflorescences loose racemes, without panicles, sometimes with fascicles at base; peduncle to 1.2 cm, pedicel 4–8 mm (pedicels at base to 16 mm); flowers per inflorescence 5 to 18. . . . . .276. *B. zhaoi* Harber

KEY 6. FLOWERS IN PANICLES

1a. Shrubs semi-evergreen or deciduous; leaf blade abaxially pruinose 2

1b. Shrubs deciduous; leaf blade not abaxially pruinose. 3

2a. Shrubs semi-evergreen, to 40 cm tall; mature stems pale brown; leaf blade obovate, narrowly ovate, or obovate-lanceolate, 0.8–3 × 0.4–1 cm; flowers per inflorescence 10 to 43; rachis to 14 cm, pedicel 5–8(–12) mm . . . . .208. *B. pingbaensis* M. T. An

2b. Shrubs deciduous, to 1.5 m tall; mature stems purplish or reddish brown; leaf blade oblanceolate 1–3 × 0.4–0.8 cm; flowers per inflorescence 14 to 24; rachis to 9 cm, pedicel 6–10 mm . . . . 116. *B. brevipaniculata* C. K. Schneid.

3a. Panicle dense and congested or narrow and rigid; flowers densely spaced 4

3b. Panicle open; flowers distinctly spaced. 5

4a. Rachis to 1.5 cm; pedicel 1.5–2 mm; flowers per inflorescence 10 to 14; mature stems pale orange-brown; leaf blade obovate-oblong or elliptic-obovate, 0.8–2 × 0.4–1 cm . . . . .93. *B. aggregata* C. K. Schneid.

4b. Rachis to 15 cm; pedicel 2.5–4 mm; flowers per inflorescence 15 to 80; mature stems pale reddish brown; leaf blade obovate-elliptic or obovate, 1–3(–4) × 0.5–1.5 cm . . . . 215. *B. prattii* C. K. Schneid.

5a. Mature stems purplish brown or dark reddish brown 6

5b. Mature stems pale yellow, pale yellowish brown, or pale reddish brown 8

6a. Rachis to 16 cm, peduncle 0.5–4 cm; pedicel 5–10 mm; flowers per inflorescence 15 to 70; ovules 2 or 3; species of S Xizang (Tibet) . . . . .176. *B. koehneana* C. K. Schneid.

6b. Rachis to 7.5(–16) cm, peduncle 0.5–1.2 cm; pedicel to 5 mm; flowers per inflorescence 10 to 30(to 50); ovules 1 to 5; species of SE Xizang (Tibet) 7

7a. Spines solitary, 0.4–1 cm, but 3-fid and to 1.4 cm at base of stems; flowers per inflorescence 10 to 20; rachis to 5 cm; sepals in 3 whorls, outer sepals oblong, median sepals oblong-elliptic, inner sepals obovate; ovules 1 or 2(or 3). . . . . 231. *B. sherriffii* Ahrendt

7b. Spines 1- to 3-fid, 0.4–1 cm; flowers per inflorescence 15 to 30(to 50); rachis to 7.5(–16) cm; sepals in 2 whorls, outer sepals oblong-obovate, inner sepals oblong-suborbicular; ovules 3 to 5 . . . . 159. *B. gyalaica* Ahrendt

8a. Leaf blade broadly obovate; adaxial veins markedly reticulate, dense, and conspicuous; flowers per inflorescence 30 to 100; rachis 5–15 cm . . . . .211. *B. polyantha* Hemsl.

8b. Leaf blade obovate, narrowly obovate, oblong-obovate, or obovate-elliptic; adaxial veins not markedly reticulate or dense; flowers per inflorescence 10 to 50; rachis 2.5–13 cm. 9

9a. Flowers per inflorescence (20 to)30 to 50; rachis 4–13 cm . . . . .187. *B. luhuoensis* T. S. Ying

9b. Flowers per inflorescence 10 to 40; rachis 2.5–6(–9) cm 10

10a. Ovules 3 or 4; flowers per inflorescence 10 to 20 . . . . 100. *B. aristata* DC.

10b. Ovules 1 to 3; flowers per inflorescence 10 to 40 11

11a. Leaf blade spinose with (2 to)8 to 15 teeth on each side, sometimes entire toward apex of stems and on young shoots . . . . 264. *B. xiangchengensis* Harber

11b. Leaf blade entire, rarely spinose or spinulose with 1 to 4 teeth on each side 12

12a. Reticulation on adaxial surface of leaves inconspicuous; ovules (1 or)2 or 3; flowers per inflorescence 15 to 30; apex of petals deeply incised; species of Lhünzê (Longzi) Xian, SE Xizang (Tibet). . . . . 263. *B. xanthophlaea* Ahrendt

12b. Reticulation on adaxial surface of leaves conspicuous; ovules 1 or 2; flowers per inflorescence 10 to 40; apex of petals slightly incised or emarginate; species of Dêqên (Deqin) Xian, NW Yunnan, and Markam (Mangkang) Xian, E Xizang (Tibet) . . . . 125. *B. concolor* W. W. Sm.

KEY 7. MIXED INFLORESCENCES (ONE OR MORE OF FASCICLES, SUB-FASCICLES, UMBELS, AND SUB-UMBELS PLUS ONE OR BOTH OF RACEMES AND SUB-RACEMES)

1a. Leaves abaxially pruinose. 2

1b. Leaves not abaxially pruinose. 5

2a. Shrubs evergreen 3

2b. Shrubs deciduous 4

3a. Leaf blade obovate, obovate-spatulate, or oblanceolate, 0.6–2.5 × 0.2–0.6 cm; inflorescences fascicles, sub-fascicles, stalked corymbs, sub-racemes, racemes, umbels, sub-umbels, or very rarely panicles; berry salmon red, translucent, globose or subglobose . . . . 261. *B. wilsoniae* Hemsl.

3b. Leaf blade narrowly obovate-elliptic or narrowly elliptic, 1–1.6 × 0.3–0.4 cm; inflorescences sub-racemes. umbels, or sub-umbels; berry red, opaque, oblong or ellipsoid . . . . 96. *B. amoena* Dunn

4a. Shrubs to 1.2 m tall; mature stems dark reddish brown; leaf blade obovate, 1.2–1.5 × 0.5–0.6 cm; inflorescences sub-racemes, rarely fascicles; flowers per inflorescence 2 to 4(to 6) . . . . 229. *B. saxatilis* Harber

4b. Shrubs to 2.5 m tall; mature stems dark purplish red; leaf blade narrowly or broadly obovate, 1–2.5 × 0.4–0.9 cm; inflorescences sub-umbels or sub-racemes; flowers per inflorescence 2 to 12. . . . . 118. *B. calcipratorum* Ahrendt

5a. Ovules/seeds more than 3 6

5b. Ovules/seeds 3 or fewer 25

6a. Ovules more than 6 7

6b. Ovules more than 3 but not more than 6 9

7a. Mature stems brownish red; inflorescences sub-umbels or sub-racemes; flowers per inflorescence 2 to 7; species of Zayü (Chayu) Xian, SE Xizang (Tibet) . . . . 274. *B. zayulana* Ahrendt

7b. Mature stems dark blackish purple or purplish red; inflorescences fascicles, sub-umbels, umbels, or sub-racemes; flowers per inflorescence 1 to 8; species of Sichuan or NW Yunnan 8

8a. Mature stems dark blackish purple; leaf blade oblong-obovate or elliptic, 2–4 × 1–2 cm; inflorescences fascicles, occasionally sub-umbels or sub-racemes; flowers per inflorescence 1 to 5; pedicel 20–30 mm; species of Sichuan . . . . 92. *B. aemulans* C. K. Schneid.

8b. Mature stems purplish red; leaf blade narrowly or broadly obovate, (2.5–)3–5.5 × 1.5–2.5(–3.3) cm; inflorescences fascicled at base; flowers per inflorescence 3 to 8; pedicel 8–18 mm; species of Gongshan Xian, NW Yunnan . . . . 188. *B. mabiluoensis* Harber

9a. Leaf blade 2 cm or less 10

9b. Leaf blade to 5 cm 13

10a. Leaf blade narrowly obovate, oblanceolate, linear-oblanceolate, or narrowly rhomboid-elliptic, 0.4–1.3 × 0.1–0.4 cm. 11

10b. Leaf blade obovate, 1–1.9 × 0.5–1.2 cm. 12

11a. Shrubs to 1 m tall; mature stems reddish purple; spines 0.6–1.5 cm; leaf blade linear-oblanceolate or narrowly rhomboid-elliptic; inflorescences fascicles, sometimes sub-racemes or sub-umbels; flowers per inflorescence 3 to 6; species of W Xinjiang and NW Xizang (Tibet) . . . . 254. *B. ulicina* Hook. f. & Thomson

11b. Shrubs to 3.5 m tall; mature stems pale reddish yellow; spines 0.8–2.5 cm; leaf blade narrowly obovate or oblanceolate; inflorescences fascicles, sub-fascicles, or sub-umbels; flowers per inflorescence 1 to 4(to 6); species of SE Xizang (Tibet) . . . . 156. *B. glabramea* (Ahrendt) Harber

12a. Leaf blade spinose with 1 to 4 teeth on each side or sometimes entire; inflorescences sub-fascicles or sub-racemes; flowers per inflorescence 2 to 5; whorls of sepals 3; ovules 2 to 5 . . . . . . 169. *B. jaeschkeana* var. *usteriana* C. K Schneid.

12b. Leaf blade entire; inflorescences sub-racemes or sub-umbels; flowers per inflorescence 5 to 7; whorls of sepals 2; ovules 4. . . . . 175. *B. kartanica* Ahrendt

13a. Leaf blade obovate-elliptic or elliptic-orbicular, 2–6 × 1–4 cm; species of Xinjiang . . . . 163. *B. heteropoda* Schrenk

13b. Leaf blade elliptic, oblong-obovate, obovate, obovate-elliptic, or obovate-orbicular, 0.9–3.6(–4.7) × 0.6–2.3 cm; species of Sichuan, Yunnan, Gansu, and Xizang (Tibet) 14

14a. Species of Gansu and Xizang (Tibet). 15

14b. Species of Sichuan and Yunnan 20

15a. Mature stems shades of dark purple 16

15b. Mature stems shades of yellow or brown 18

16a. Leaf blade margin dentate with 2 to 7 coarse teeth on each side, rarely entire . . . . 199. *B. multiserrata* T. S. Ying

16b. Leaf blade margin entire or minutely spinulose with 3 to 12(to 18) teeth on each side. 17

17a. Leaf blade obovate, (1.4–)2–3(–4) × 0.6–1.5 cm, margin entire; inflorescences fascicles, sub-fascicles, or short sub-racemes; pedicel 13–17 mm . . . . 249. *B. trichohaematoides* Ahrendt

17b. Leaf blade oblong-obovate, obovate-elliptic, or obovate-orbicular, 0.9–1.7(–3.8) × 0.6–1.4(–1.8) cm, margin entire or minutely spinulose with 3 to 12(to 18) teeth on each side; inflorescences sub-racemes, sometimes with fascicles at base; pedicel 6–9 mm. . . . . 114. *B. brachystachys* T. S. Ying

18a. Leaf blade elliptic to obovate-elliptic, occasionally obovate-orbicular, (0.8–)1.6–3 × (0.3–)0.6–1.2 cm, margin spinulose with 6 to 15 teeth on each side, or sometimes entire especially toward the tip of stems; ovules 3 or 4; species of Gansu . . . . 238. *B. taoensis* Harber

18b. Leaf blade obovate or elliptic-obovate, 1.5–4.5 × 0.8–2.5 cm, margin entire, rarely spinulose with 2 to 6 teeth on each side; ovules 2 to 5; species of Xizang (Tibet) 19

19a. Mature stems reddish or yellowish brown; leaf blade obovate or elliptic-obovate, 1.5–3 × 0.8–2.5 cm; inflorescences without fascicles at base; whorls of sepals 2 . . . . 172. *B. johannis* Ahrendt

19b. Mature stems pale grayish yellow or brown; leaf blade obovate, 2–4.5 × 1–2 cm; inflorescences sometimes with fascicles at base; whorls of sepals 4 . . . . 243. *B. thomsoniana* C. K. Schneid.

20a. [14b] Leaf blade leathery, elliptic or obovate-elliptic, 1.5–2.7 × 0.5–1 cm; inflorescences sub-umbels, sub-racemes, or sub-panicles; flowers per inflorescence 6 to 12. . . . . 232. *B. shunningensis* (Ahrendt) Harber

20b. Leaf blade papery, obovate, obovate-elliptic, or oblong-obovate, 1.6–4 × 0.7–2.3 cm; inflorescences fascicles, sub-fascicles, sub-umbels, or sub-racemes, sometimes with fascicles at base; flowers per inflorescence 2 to 10(to 15). 21

21a. Spines absent, rarely 0.2–0.5 cm; leaf blade narrowly obovate, 1.6–2.8 × 0.7–1.1 cm; pedicel 3–8 mm (to 20 mm when from base) . . . . 265. *B. xiaozhongdianensis* Harber & Xin Hui Li

21b. Spines 0.6–3.5 cm; leaf blade oblong-obovate, obovate, or obovate-elliptic, 1.5–4(–4.7) × 1–2.3 cm, pedicel 6–18 mm (to 36 mm when from base) 22
22a. Abaxial and adaxial leaf surfaces glaucous, margin entire; whorls of sepals 3 . . . . 141. *B. dokerlaica* Harber
22b. Abaxial leaf surface pale green, adaxial surface dark green or dull gray green; margin sometimes spinose with 3 to 10 teeth on each side; whorls of sepals 2 23
23a. Mature stems reddish brown; spines mostly absent; leaf margin entire or spinulose with 3 to 6 teeth on each side; pedicel 6–12 mm (to 20 mm when from base); ovules 2 to 4; berry ellipsoid or oblong, apex not attenuate . . . .
. . . . 174. *B. kangwuensis* Harber
23b. Mature stems pale brown or pale yellowish brown; spines (0.5–)1.2–3.5 cm; leaf margin spinulose with 3 to 12(to 15) teeth on each side; pedicel 8–18 mm (to 36 mm when from base); ovules 3 to 5, berry narrowly ovoid, apex attenuate
. . . . 24
24a. Leaf blade adaxially dull gray-green; outer sepals broadly elliptic or elliptic-ovate, 3–4 × 3.5–4 mm; inner sepals broadly elliptic or elliptic-ovate, 4.5–5 × 3.5 mm; apex of petals entire; anther connective extended, apiculate; ovules 3 or 4; species of W Sichuan . . . . 237. *B. tachiensis* (Ahrendt) Harber
24b. Leaf blade dark green; outer sepals narrowly obovate, 3.5–4 × 2–3 mm; inner sepals broadly obovate or elliptic-obovate, 4–6 × 3–4.5 mm; apex of petals distinctly incised; anther connective not extended, truncate; ovules 3 to 5; species of N Sichuan . . . . 247. *B. tischleri* C. K. Schneid.
25a. [5b] Mature stems purple, reddish or brownish purple, dark red, or dark reddish brown 26
25b. Mature stems grayish yellow, pale yellow, pale brown, pale yellowish brown, or pale reddish brown 46
26a. Peduncle 1–3 cm 27
26b. Peduncle absent or less than 1 cm 32
27a. Inflorescences solitary flowers or loose racemes or sub-umbels; pedicel 13–25 mm . . . . . 217. *B. pseudothunbergii* P. Y. Li
27b. Inflorescences umbels, sub-umbels, sub-racemes, or subumbellate racemes; pedicel 2–12 mm 28
28a. Leaf blade obovate or oblanceolate, 1–1.6 × 0.6–1.1 cm; inflorescences sub-umbels or sub-racemes; flowers per inflorescence 4 to 10; pedicel 2–3 mm . . . . 102. *B. atroviridiana* T. S. Ying
28b. Leaf blade elliptic, elliptic-obovate, obovate, or oblanceolate, 0.8–2.8 × 0.2–1 cm; inflorescences sub-umbels, sub-racemes, or semi-umbellate racemes; flowers per inflorescence 2 to 8; pedicel 4–12 mm. 29
29a. Mature stems partially pruinose; leaf blade obovate, 1.2–1.5 × 0.5–0.6 cm, margin spinose with 1 or 2(to 4) widely spaced coarse teeth on each side; peduncle to 1.8 cm; pedicel 2–6 mm . . . . 229. *B. saxatilis* Harber
29b. Mature stems epruinose; leaf blade narrowly obovate, elliptic, obovate-elliptic, or oblanceolate, 0.8–2.8 × 0.2–1 cm, margin entire or rarely spinose with 1 to 5 inconspicuous teeth on each side; peduncle to 3 cm; pedicel 6–10 mm. . . . 30
30a. Spines 3-fid, 0.5–1 cm; leaf blade narrowly oblanceolate or narrowly elliptic, 1–2 × 0.3–0.4 cm, margin rarely spinose . . . . 178. *B. leboensis* T. S. Ying
30b. Spines solitary, 0.5–3.2 cm; leaf blade elliptic, obovate-elliptic, or narrowly obovate, 0.8–2.8 × 0.4–1 cm, margin entire 31
31a. Mature stems purple; peduncle of inflorescence 1–1.6 cm; whorls of sepals 2; species of SE Gansu . . . .
. . . . 168. *B. integripetala* T. S. Ying
31b. Mature stems reddish brown; peduncle of inflorescence to 2.6 cm; whorls of sepals 3; species of Tengchong Xian, W Yunnan . . . . 240. *B. tengchongensis* Harber
32a. [26b] Leaf blade margin spinose or spinulose with 10 to 24 teeth on each side 33
32b. Leaf blade margin entire or spinulose with 1 to 8(to 10) teeth on each side 35
33a. Leaf blade with reticulation indistinct adaxially; ovules 3 . . . . 143. *B. dulongjiangensis* Harber
33b. Leaf blade with reticulation distinct adaxially; ovules 2 34
34a. Leaf blade obovate or spatulate-obovate, 2–4.5 × 0.8–2 cm; inflorescences dense subumbellate or sub-corymbose racemes; pedicel 4–7 mm . . . . 224. *B. reticulata* Bijh.
34b. Leaf blade elliptic, elliptic-obovate, or obovate, 0.5–4 × (0.2–)0.7–1.7 cm; inflorescences sub-fasciculate racemes; pedicel 8–14 mm . . . . 137. *B. dictyoneura* C. K. Schneid.
35a. Leaf blade 0.5–2 × 0.3–0.5 cm 36
35b. Leaf blade 1–3.5(–4) × 0.4–2.2 cm 37
36a. Leaf blade narrowly elliptic or obovate, 0.5–1.5 × 0.3–0.5 cm; flowers per inflorescence 4 to 7; pedicel 2–7 mm; berry black . . . . 135. *B. derongensis* T. S. Ying
36b. Leaf blade narrowly obovate, 1–2 × 0.4–0.5 cm; flowers per inflorescence (3 to)5 to 10; pedicel 6–18 mm; berry scarlet . . . . 194. *B. moloensis* (Ahrendt) Harber
37a. Species of W and NC Sichuan and S Gansu 38
37b. Species of NW Yunnan, SW Sichuan, Xizang (Tibet), and Xinjiang 41
38a. Leaf blade obovate, often narrowly so, 1.5–2.6 × 0.4–0.7 cm; inflorescences fascicles (often dense), sub-fascicles, or sub-racemes, to 1.7 cm; pedicel to 8 mm . . . . 269. *B. yarigongensis* Harber
38b. Leaf blade obovate to obovate-elliptic, 1–4.5 × 0.4–1.8(–2) cm; inflorescences fascicles, sub-fascicles, sub-umbels, or sub-racemes, sometimes with fascicles at base, to 4.5 cm; pedicel more than 8 mm 39
39a. Leaf blade margin mostly spinose with 4 to 8(to 10) teeth on each side; pedicel 8–16 mm. . . . . . .266. *B. yaanica* Harber
39b. Leaf blade entire, rarely with 1 to 8 teeth on each side; pedicel 5–18 mm 40
40a. Inflorescences fascicles, sub-fascicles, umbels, sub-umbels, or sub-racemes, to 2.5 cm; pedicel 10–18 mm; flowers per inflorescence 3 to 8. . . . . 235. *B. silva-taroucana* C. K. Schneid.
40b. Inflorescences sub-racemes, sometimes with fascicles at base or sub-fascicles toward apex of stems, to 4.5 cm; pedicel 5–12 mm; flowers per inflorescence 6 to 12 . . . . 197. *B. mouillacana* C. K. Schneid.
41a. Inflorescences 1.5–4.5 cm; flowers per inflorescence 5 to 20 . . . . 206. *B. pallens* Franch.
41b. Inflorescences 1–3 cm; flowers per inflorescence 3 to 10 42

42a. Mature stems partially pruinose; species of SW Xinjiang . . . . 119. *B. calliobotrys* Bien. ex Koehne
42b. Mature stems not pruinose; species of NW Yunnan, SW Sichuan, and Xizang (Tibet) 43
43a. Mature stems dark or blackish purple; leaf blade narrowly obovate or oblanceolate, 1.4–4 × (0.4–)0.6–1 cm; inflorescences sub-umbels or sub-racemes, sometimes with fascicles at base . . . . 220. *B. purpureocaulis* Harber
43b. Mature stems dark red, reddish purple, or reddish brown; leaf blade oblanceolate, narrowly elliptic, obovate, or oblong-elliptic, 1.1–2.5 × 0.5–0.8 cm; inflorescences fascicles, sub-fascicles, or sub-umbels or sub-racemes without fascicles at base 44
44a. Leaf blade oblanceolate or very narrowly elliptic, 1–2.5 × 0.5–0.7 cm; flowers per inflorescence 1 to 4(or 5) . . . . . .
. . . . 161. *B. hemsleyana* Ahrendt
44b. Leaf blade obovate or oblong-elliptic, 1–2 × 0.5–0.8 cm; flowers per inflorescence 4 to 8 45
45a. Mature stems dark reddish brown; spines 1–2 cm; leaf margin spinose with 1 to 4 teeth on each side or sometimes entire; pedicel 8–15(–20) mm; apex of petals emarginate . . . . 252. *B. tsarongensis* Stapf
45b. Mature stems purplish red; spines 0.6–1.2 cm; leaf margin entire or rarely spinulose with 1 to 5 teeth on each side; pedicel 4–10 mm; apex of petals entire . . . . 204. *B. nyingchiensis* Harber
46a. [25b] Species of Xizang (Tibet) 47
46b. Species not in Xizang (Tibet) 48
47a. Mature stems pale pink or yellowish pink; spines 0.8–2 cm; leaf blade obovate or obovate-elliptic, 0.8–2.3 × 0.3–1.4 cm; inflorescences fascicles, sub-fascicles, or sub-umbels; pedicel 4–8 mm . . . . 205. *B. obovatifolia* T. S. Ying
47b. Mature stems pale brownish yellow; spines 0.4–0.8 cm; leaf blade narrowly obovate to obovate-elliptic, (1.5–)2–3 × (0.6–)0.8–1 cm; inflorescences fascicles, sub-fascicles, or sub-racemes; pedicel 9–13 mm . . . .
. . . . 105. *B. basumchuensis* Harber
48a. Leaf blade obovate, 0.5–4 × (0.2–)0.7–1.7 cm, venation and reticulation dense, margin spinose with 5 to 12 teeth on each side . . . . 115. *B. bracteata* (Ahrendt) Ahrendt
48b. Leaf blade elliptic, obovate, oblong-obovate, oblong-spatulate, or oblanceolate, 0.4–6 × 0.2–2.7 cm, venation and reticulation not dense, margin entire or spinulose with 1 to 25 teeth on each side 49
49a. Species of Gansu, Guizhou, and Qinghai 50
49b. Species of Sichuan and Yunnan 52
50a. Leaf blade elliptic or obovate, occasionally oblong-elliptic, (0.7–)2–6 × (0.4–)1–2.7 cm, margin entire, sometimes spinulose with 12 to 25 teeth on each side; inflorescences sub-umbels, fascicles, sub-fascicles, or racemes, sometimes compound at base; species of Gansu . . . . 246. *B. tianshuiensis* T. S. Ying
50b. Leaf blade oblanceolate or obovate, narrowly obovate-elliptic or narrowly elliptic, 0.4–2 × 0.2–0.5 cm, margin entire or with 1 to 6 spinulose teeth on each side; inflorescences sub-umbels, fascicles, or sub-racemes, sometimes with fascicles at base; species of Guizhou and Qinghai 51
51a. Inflorescences sub-umbels or sub-racemes; pedicel 3–4 mm; ovules 3; species of W Guizhou . . . .
. . . . 259. *B. weiningensis* T. S. Ying
51b. Inflorescences fascicles, sub-umbels, or short sub-racemes, sometimes with fascicles at base; pedicel 5–8 mm; ovules 2; species of S Qinghai . . . . 274. *B. yushuensis* Harber
52a. Leaf blade obovate or narrowly obovate to oblanceolate, (0.6–)1.6–4.5 × (0.4–)0.5–1(–1.2) cm 53
52b. Leaf blade obovate, obovate-elliptic, oblong-obovate, or spatulate-oblong, 1–4.5 × 0.5–1.8 cm 58
53a. Leaf blade 1.7–4.5 × 0.2–0.5 cm; berry black-purple, ovoid-oblong, 8–11 × 5–7 mm . . . . 180. *B. lepidifolia* Ahrendt
53b. Leaf blade (0.6–)0.8–3 × 0.3–1(–1.2) cm; berry red, oblong, ellipsoid, or ovoid, 6–8 × 3–5 mm . . . .
. . . . 54 [see also multi-access key]
54a. Inflorescences fascicles, sub-fascicles, sub-umbels, and sub-racemes; whorls of sepals 3 55
54b. Inflorescences sub-fascicles, sub-umbels, and sub-racemes or fascicles, sub-fascicles, and sub-racemes; whorls of sepals 2 56
55a. Mature stems pale yellowish or orangish brown; outer sepals narrowly oblong-ovate, median sepals narrowly oblong ovate, inner sepals obovate; ovules 2 or 3 . . . . 277. *B. zhongdianensis* Harber
55b. Mature stems pale reddish brown; outer sepals narrowly elliptic, median sepals elliptic-obovate, inner sepals elliptic-orbicular; ovules 2 . . . . 272. *B. yulongshanensis* Harber
56a. Inflorescences sub-fascicles, sub-umbels, and sub-racemes; pedicel 4–8 mm . . . . 179. *B. lecomtei* C. K. Schneid.
56b. Inflorescences fascicles, sub-fascicles, and sub-racemes; pedicel 5–16 mm 57
57a. Mature stems pale yellow; spines 0.2–0.7 cm or mostly absent; pedicel 5–12 mm; outer sepals oblong-ovate, inner sepals obovate; anther connective conspicuously apiculate; style not persistent . . . . . 192. *B. microtricha* C. K. Schneid.
57b. Mature stems pale yellowish brown; spines 0.6–2 cm; pedicel 7–16 mm; outer sepals broadly ovate, inner sepals broadly obovate or elliptic-obovate; anther connective not or slightly extended, truncate; style persistent . . . .
. . . . 270. *B. yingii* Harber
58a. [52b] Petiole sometimes to 13 mm 59
58b. Petiole almost absent, or rarely to 4 mm 61
59a. Mature stems pale reddish brown; flowers per inflorescence 3 to 12; pedicel 4–8 mm; whorls of sepals 2; inner sepals elliptic . . . . 202. *B. ngawaica* Harber
59b. Mature stems very pale brownish yellow; flowers per inflorescence 3 to 8(or 9); pedicel 6–15 mm; whorls of sepals 3; inner sepals obovate 60
60a. Leaf blade broadly obovate, obovate-elliptic, or obovate-orbicular, (0.5–)0.8–1.4(–2) × (0.4–)0.6–1.2(–1.8) cm, margin entire, rarely spinose with 1 to 6 teeth on each side; petiole to 11 mm or sometimes almost absent; flowers per inflorescence 4 to 8; pedicel 8–15 mm; outer sepals narrowly ovate; petals narrowly obovate . . . .
. . . . 160. *B. heishuiensis* Harber

60b. Leaf blade obovate or obovate-elliptic, 1.5–2.8(–3.5) × (0.4–)0.8–1.1(–1.6) cm, margin entire; petiole almost absent or sometimes to 13 mm; flowers per inflorescence 3 to 7(to 9); pedicel 6–10 mm; outer sepals narrowly elliptic; petals obovate-elliptic . . . . .184. *B. lixianensis* Harber

61a. Leaf blade obovate or obovate-elliptic, rarely obovate-oblanceolate, 1–2.5 × 0.4–0.9 cm, margin spinulose with 9 to 16 often conspicuous teeth on each side; whorls of sepals 3 . . . . 271. *B. yui* T. S. Ying

61b. Leaf blade obovate, narrowly oblong-obovate, or obovate-spatulate, 1–4.5 × 0.7–1.7 cm, margin entire or with 3 to 15 mostly inconspicuous teeth on each side; whorls of sepals 2 62

62a. Leaf blade narrowly obovate to narrowly oblong-obovate or obovate-spatulate, 1–3.5 × 0.7–1.1 cm; spines 0.6–1.5 cm. . . . . .133. *B. dawoensis* K. Mey.

62b. Leaf blade obovate, 1.5–4.5 × 0.8–1.7 cm; spines (0.5–)0.8–3 cm 63

63a. Leaf blade mostly broadly obovate, 1.4–2.5 × 1–1.4 cm, reticulation conspicuous and dense, margin spinulose with 2 to 5 widely spaced teeth on each side or entire . . . . .134. *B. deqenensis* Harber

63b. Leaf blade obovate, 2–4.5 × 0.8–1.7 cm, reticulation inconspicuous and not dense, margin spinose with 3 to 15 teeth on each side or entire 64

64a. Petiole almost absent; leaf blade margin spinose with 3 to 15 teeth on each side or entire particularly toward apex of stems; inflorescences loose racemes or sub-racemes sometimes with a few fascicles at base, sometimes sub-fascicles or sub-umbels toward apex of stems, to 4 cm; pedicel 6–12 mm (to 25 mm when from base); berry orange-red . . . . .
. . . . .164. *B. hubianensis* Harber

64b. Petiole almost absent, rarely to 15 mm; leaf blade margin entire; inflorescences sub-racemes, sub-umbels, or umbels, sometimes with fascicles at base, 1.5–6 cm; pedicel 4–8 mm (to 18 mm when from base); berry red. . . . .
. . . . 158. *B. gyaitangensis* Harber

KEY 8. PLANTS EVERGREEN AND SEMI-EVERGREEN

1a. Shrubs semi-evergreen, to 40 cm tall 2

1b. Shrubs evergreen, more than 50 cm tall 3

2a. Shrubs prostrate, to 20 cm tall; inflorescences 1-flowered. . . . . 250. *B. tsangpoensis* Ahrendt

2b. Shrubs upright, to 40 cm tall; inflorescences panicles, 10- to 43-flowered . . . . .208. *B. pingbaensis* M. T. An

3a. Leaf blade obovate, obovate-elliptic, or oblanceolate, 2–6(–9) × 1–3(–5) cm, thickly leathery, rigid; flowers per inflorescence (8 to)15 to 25(to 35) . . . . .101. *B. asiatica* Roxb. ex DC.

3b. Leaf blade elliptic, obovate, obovate-elliptic, oblanceolate, or lanceolate, 0.4–3(–4) × 0.2–1.2 cm, leathery but not thick and rigid; flowers per inflorescence 4 to 12(to 15) 4

4a. Mature stems pale yellow or brownish yellow 5

4b. Mature stems reddish brown, dark red, or dark purplish red. 6

5a. Leaf blade narrowly obovate-elliptic, narrowly elliptic, or obovate, 0.4–2 × 0.2–0.5 cm; inflorescences sub-umbels; flowers per inflorescence 3 to 6; species of W Guizhou. . . . . .259. *B. weiningensis* T. S. Ying

5b. Leaf blade elliptic or obovate-elliptic, 1.5–2.7 × 0.5–1 cm; inflorescences sub-umbels, sub-racemes, or sub-panicles; flowers per inflorescence 6 to 12; species of C Yunnan. .232. *B. shunningensis* (Ahrendt) Harber

6a. Inflorescences fascicles, sub-fascicles, stalked corymbs, sub-racemes, racemes, umbels, sub-umbels, or very rarely panicles; berry salmon red, translucent, globose or subglobose . . . . 261. *B. wilsoniae* Hemsl.

6b. Inflorescences sub-racemes, umbels, or sub-umbels; berry red, opaque, oblong or ellipsoid 7

7a. Spines 0.4–1 cm; leaf blade narrowly obovate-elliptic or narrowly elliptic, 1–1.6 × 0.3–0.4 cm, abaxially dull, margin entire or occasionally spinulose with 1 or 2 teeth on each side. . . . . .96. *B. amoena* Dunn

7b. Spines (1.5–)2–6 cm; leaf blade lanceolate, narrowly obovate, or obovate, 1–3(–4) × 0.25–1.2 cm, adaxially shiny, margin spinose or dentate with 1 to 4(to 6) teeth on each side or rarely entire . . . . 213. *B. potaninii* Maxim.

Table 4.   Characters for Key 9.

Code	Characteristics	Code	Characteristics
Mature stems		Outer sepal shape	
A	yellow or yellowish gray	c	oblong-lanceolate
B	pale brownish yellow or yellowish brown	d	obovate-oblong
C	pale orangish brown	e	ovate/broadly ovate
D	pale reddish brown or reddish brown	f	narrowly elliptic
Spines		g	elliptic-ovate
E	3-fid, mostly present	h	oblong-ovate/narrowly oblong-ovate
F	mostly absent	Median sepal shape	
Leaf shape		i	oblong-ovate/narrowly oblong-ovate
G	obovate	j	elliptic-orbicular
H	narrowly obovate	Inner sepal shape	
I	obovate-oblanceolate	k	oblong-ovate
J	oblanceolate	l	oblong-elliptic
K	obovate-oblong or oblong-obovate	m	obovate-elliptic
L	spatulate	n	obovate/broadly obovate
Leaf margin		o	elliptic
M	always entire	p	elliptic-orbicular
N	entire and spinose up to 5 teeth	Petal shape	
O	entire and more than 10 teeth	q	obovate/broadly obovate
Inflorescence		r	orbicular-obovate
P	fascicles	Petal apex	
Q	sub-fascicles	s	entire
R	umbels	t	emarginate/slightly emarginate
S	sub-umbels	u	crenate or slightly notched
T	sub-racemes	Anther connective	
Maximum number of flowers		v	conspicuously apiculate
U	6	w	distinctly extended, obtuse
V	10	x	slightly extended, obtuse
W	18	y	not or slightly extended, truncate
Pedicels		Fruiting style	
X	to 10 mm	z	persistent
Y	to 16 mm	ω	absent
Sepal whorls			
a	2 whorls		
b	3 whorls		

Key 9 (Multi-Access Key). Deciduous Species from NW Yunnan and SW and W Sichuan with Mostly Narrowly Obovate Leaves; Inflorescences Mostly Fascicles, Sub-fascicles, and Sub-umbels; Flowers with 2 Ovules (see Table 4 for code to formulas)

A, E/F, H/J, M, P/Q/T, V, Y, a, h, n, q, s, y, ω. . . . . 192. *B. microtricha* C. K. Schneid.
B, E, K/L, M, P/R/S, U, Y, b, ?n, q, t, z . . . . .207. *B. papillifera* (Franch.) Koehne
B, E, H/J, N, P/Q/T, V, Y, a, e, m, q, s, y, z . . . . 270. *B. yingii* Harber
B/C, E, H/I, O, P/Q/S/T, W*, b, h, i, n, q, s, x, z. . . . . 277. *B. zhongdianensis* Harber
B/D, E, H, N, P/Q, V, Y, a, d, n, s, w, z. . . . . 112. *B. bowashanensis* Harber
D, E, H/J, N, P/Q, U, Y, a, g, l, q, s, y, z. . . . . .139. *B. difficilis* Harber
D, E, H, M, Q/S/T, V, X, a, f, o, q, t, y, z . . . . .179. *B. lecomtei* C. K. Schneid.
D, E, H/J, M, P/Q/S/T, U, X, b, f, j, p, q, u, y, z . . . . 272. *B. yulongshanensis* Harber
D, E, G/H/I, N, P/Q, V, Y, b, c, i, k/m, q, t, v, z . . . . .91. *B. abbreviata* (Ahrendt) Harber
* ovules 2 or 3

Key 10. Plants of Taiwan

1a.   Shrubs deciduous . . . . 196. *B. morrisonensis* Hayata
1b.   Shrubs evergreen 2
2a.   Leaf blade adaxially dull green. 3
2b.   Leaf blade adaxially ± shiny green 5

3a. Leaf blade densely white pruinose . . . . .54. *B. morii* Harber & C. C. Yu
3b. Leaf blade not or only very rarely pruinose. 4
4a. Leaf blade elliptic to elliptic-lanceolate, 2–4 × 0.6–1.2 cm, abaxial surface not pruinose; flowers per inflorescence 2 to 7. . . . . .34. *B. hayatana* Mizush.
4b. Leaf blade narrowly elliptic, elliptic-lanceolate, or lanceolate, 5.5–11 × 2.4–2.9 cm, abaxial surface rarely pruinose; flowers per inflorescence 5 to 15. . . . . .5. *B. aristatoserrulata* Hayata
5a. Mature stems purplish red or dark purplish . . . . .63. *B. ravenii* C. C. Yu & K. F. Chung
5b. Mature stems brown, brownish red, reddish brown, pale reddish brown, or pale brownish yellow 6
6a. Leaf blade 1–3.5 × 0.5–1 cm 7
6b. Leaf blade 0.4–10.5 × 0.8–3.3 cm 9
7a. Leaf blade often abaxially pruinose, margin conspicuously revolute; inflorescences densely congested fascicles . . . . .55. *B. nantoensis* C. K. Schneid.
7b. Leaf blade not abaxially pruinose, margin sometimes slightly revolute; inflorescences open fascicles 8
8a. Shrubs to 0.5 m tall; leaf blade narrowly elliptic, narrowly obovate, or oblanceolate, 2–3.5 × 0.8–1 cm; berry dark purple, partially or mostly pruinose; ovules 5 to 8 . . . . 74. *B. tarokoensis* S. Y. Lu & Y. P. Yang
8b. Shrubs to 1.2 m tall; leaf blade elliptic to oblong-elliptic, 1–2.5 × 0.7–1 cm; berry black, not pruinose; ovules 4 to 6 . . . . .2. *B. alpicola* C. K. Schneid.
9a. Inflorescences congested fascicles . . . . 45. *B. kawakamii* Hayata
9b. Inflorescences open fascicles 10
10a. Leaf margin densely spinose with 20 to 64 teeth on each side . . . . .67. *B. schaaliae* C. C. Yu & K. F. Chung
10b. Leaf margin spinose with 2 to 27 widely spaced teeth on each side 11
11a. Shrubs to 4 m tall; ovules 6 to 8; berry globose. . . . . 57. *B. pengii* C. C. Yu & K. F. Chung
11b. Shrubs to 1.5 m tall; ovules 1 to 5; berry ellipsoid 12
12a. Leaf blade elliptic, 1.4–5.7 × 0.8–1.7 cm; ovules 1 to 4. . . . . 17. *B. chingshuiensis* T. Shimizu
12b. Leaf blade narrowly elliptic, elliptic-lanceolate, lanceolate, oblong-elliptic, or oblanceolate-elliptic, (0.4–)2.3–9 × 0.8–2.5 cm; ovules 3 to 6 13
13a. Upright shrubs; leaf blade narrowly elliptic, oblong-elliptic, or oblanceolate-elliptic, (0.4–)2.3–4.2(–5.5) × 0.8–1.2(–1.7) cm, margin spinose with 1 to 5 teeth on each side. . . . . 10. *B. brevisepala* Hayata
13b. Decumbent shrubs; leaf blade narrowly elliptic, elliptic-lanceolate, or lanceolate, 5.5–9 × 1.3–2.5 cm, margin spinose with 5 to 16 teeth on each side . . . . 53. *B. mingetsensis* Hayata

KEY 11. PLANTS OF VIETNAM

1a. Leaves abaxially pruinose; species of S Vietnam . . . . 80. *B. vietnamensis* Harber
1b. Leaves abaxially green; species of N Vietnam. 2
2a. Spines largely absent, when present 1- to 3-fid, 0.1–0.5 cm . . . . 38. *B. hypoxantha* C. Y. Wu ex S. Y. Bao
2b. Spines always present, 1- to 3-fid, 0.4–1.5(–2.2) cm. 3
3a. Leaf blade narrowly elliptic, elliptic-oblanceolate, or elliptic-lanceolate, 4–10 × 1.5–2.5 cm, thickly leathery, margin spinulose with 35 to 60 teeth on each side; ovules 1. . . . . 28. *B. ferdinandi-coburgii* C. K. Schneid.
3b. Leaf blade elliptic-lanceolate, lanceolate, or narrowly elliptic, 3–9 × 1.2–2.2 cm, thinly leathery, margin spinulose with 8 to 15 teeth on each side; ovules 2 or 3 . . . . 71. *B. subacuminata* C. K. Schneid.

Berberis sect. **Wallichianae** C. K. Schneid., Bull. Herb. Boissier, sér. 2, 5: 400. 1905. TYPE: *Berberis wallichiana* DC., Prodr. 1: 107. 1824 (designated by Chamberlain & Hu [1985: 532]). TYPE: "Napalia 1819," [E. Gardiner via N. Wallich] *s.n.* (lectotype, designated by Adhikari et al. [2012: 510], G-DC G00201760!). Figure 2.

Leaves evergreen, thinly or thickly coriaceous. Flowers shades of yellow, rarely red and white, partially pink, or purple, solitary, in fascicles or very rarely partially racemose.

As Adhikari et al. (2012: 510) note, there has been some confusion about the type of *Berberis wallichiana*. The protologue simply recorded material from Nepal that had come from Wallich. Ahrendt (1961: 71) and Chamberlain and Hu (1985: 549) stated the type to be *Wallich 1478*, Nepal, Mt. Sheopor (K K001113189), which is dated 1821. However, the specimen in the Herbier Prodrome in Geneva is dated 1819, and since Wallich did not visit Nepal until 1820, it cannot have been collected by him. Prior to this, however, Wallich had received specimens from Edward Gardiner, then resident in Nepal.

1. **Berberis acuminata** Franch., Bull. Soc. Bot. France 33: 387. 1886. TYPE: China. NE Yunnan: [Shuifu Xian], Tchen-fong-chan, 4 May 1882, *J. M. Delavay 494* (lectotype, designated here, P P00716538!; isolectotypes, A fragm. 00038718!, LE fragm.!, P P00716539!, P002682365!; possible isolectotypes, *J. M. Delavay s.n.* [P P00716537!] Tchen-fong-chan, 4 May 1882, *J. M. Delavay s.n.* [MPU ex P MPU013533] s. loc., s.d., image!).

Berberis weixinensis S. Y. Bao, Bull. Bot. Res., Harbin 5(3): 13. 1985, syn. nov. TYPE: China. NE Yunnan: Weixin Xian, 1450 m, 17 June 1980, *S. Y. Bao 286* (holotype, KUN 1207800!; isotype, KUN 1204061!).

Berberis xingwenensis T. S. Ying, Acta Phytotax. Sin. 37(4): 311. 1999, syn. nov. TYPE: China. SE Sichuan: Xingwen Xian, Jianfengshan, 1800 m, 14 May 1959, *Yibin Division, Sichuan Economic Plant Exp. 437* (holotype, PE 00935212!; isotypes, KUN 0175593!, PE 00935213!, SITC 00001603 image!, SZ [3 sheets] 00289797, 00289801–02 images!).

Shrubs, evergreen, to 2.5 m tall; mature stems brownish yellow, terete; spines absent or 3-fid, concolorous, 0.3–0.6(–1.2) cm. Petiole almost absent; leaf blade abaxially mid-green, adaxially dark dull green, lanceolate, 7–20 × 1–3.5 cm, leathery, midvein raised abaxially, slightly impressed adaxially, lateral venation and reticulation conspicuous on both surfaces, base attenuate, margin spinose, with 10 to 30 teeth on each side, apex narrowly acuminate, sometimes slightly bent. Inflorescence a fascicle, 3- to 24-flowered; pedicel 15–25 mm. Sepals in 3 whorls; outer sepals ovate, ca. 2 × 1.5 mm, apex acuminate; median sepals oblong-elliptic, 3.5 × 2.5 mm; inner sepals broadly elliptic, 4.5 × 3.5 mm, apex rounded; petals obovate, 4–5 × 3–4 mm, base clawed, with separate glands, apex emarginate. Stamens ca. 2.5 mm; anther connective extended. Ovules 2. Berry black, ellipsoid or obovoid, ca. 6 × 4 mm; style not persistent or persistent and short.

Phenology. *Berberis acuminata* has been collected in flower from April to May and in fruit between June and September.

Distribution and habitat. *Berberis acuminata* is known from southeast Sichuan, northeast Yunnan, and northwest Guizhou. It has been collected from thickets, roadsides, forests, and rocky slopes at ca. 950–2500 m.

The protologue of *Berberis acuminata* cited only *Delavay 494* of 4 May 1882. There are three specimens with this number at P (one of which, 002682365, is dated simply May 1882). The 4 May specimen has the best floral material and has been chosen here as the lectotype. The collector's notes do not record where in northeast Yunnan the specimens were collected beyond "Tchen-fong-chan." From Bretschneider (1898b), this can be located in Shuifu Xian near the border with Sichuan.

The account of flowers given above is from the type collection. The differing account given by Ahrendt (1961: 80), the source of which is unclear, should be discounted. The description of flowers given in the protologue of *Berberis xingwenensis* differs little from those of *B. acuminata* (which was not noticed by Ying [2001, 2011]), except that the number of whorls of sepals is given as sometimes being four.

"*Berberis acuminata*" Stapf (1908) is *B. veitchii* (for further details, see under that species). *Cavalerie 1944* (E E00017972, K) Guizhou, "Lou-Mong-Touan," Nov. 1904, cited by Schneider (1918: 145) as *B. acuminata*, is *B. bicolor*. Specimens from Emei Shan in Sichuan identified as *B. acuminata* (e.g., by Ahrendt, 1961: 80) are referable to *B. simulans*.

The relationship between *Berberis acuminata* and *B. caudatifolia* requires further investigation.

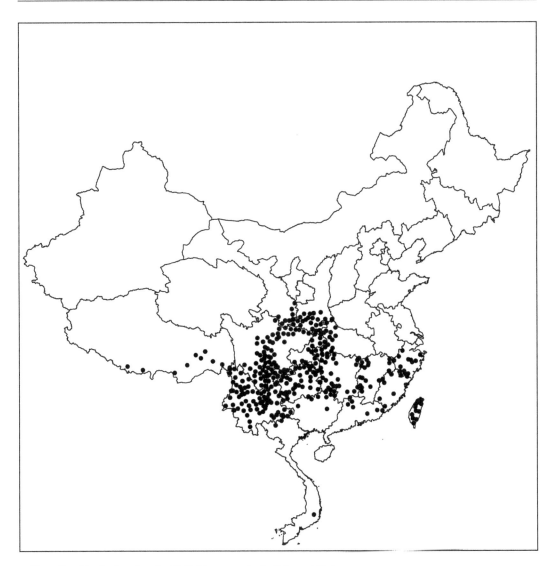

Figure 2.   Distribution of section *Wallichianae* species in China and Vietnam.

Selected specimens.

NW Guizhou. "Yang kia tchong," 9 Aug. 1907, *J. Cavalerie 3040* (P P02482748). **Anlung (Anlong):** 20 Nov. 1930, *Y. Tsiang 9363* (IBSC 0092016, 0092484, NY). **Tsunyi Hisian (Zunyi Xian):** Liang Feng Yah, 1100 m, 8 Aug. 1931, *A. N. Steward, C. Y. Chiao & H. C. Cheo 250* (A 00280167, BM, N 093057001, NY, P P02313364).

SE Sichuan. Gulin Xian: Deyue, Huangjin Gongshe, near Bajie Cave, 950 m, 24 July 1976, *General Survey Team Gu 838* (SM SM704800304). **Junlian Xian:** Daxue Shan, 1600 m, 26 Apr. 1959, *Sichuan Economic Plant Exp. 0173* (PE 00935214–15, SM SM704800508); Minzhu Gonghse, 30 June 1977, *General Survey Team 758* (SM SM704800511); Gonggu Gongshe, Laojun Shan, 1700 m, 15 July 1977, *General Survey Team 0567* (SM SM704800512). **Pingshan Xian:** "Ping-shan Hsien," 2500 m, Apr. 1931, *F. T. Wang 22785* (A 00279301, IBSC 0091478, KUN 0175599, LBG 00064139,

NAS NAS00314330, P P02313362, WUK 0045148); Baiyan, 124 Lin Chang, 1400 m, 21 June 1959, *Yibin Division Sichuan Economic Plant Exped (59) 1129* (CDBI CDBI0027976, KUN 0175732, PE 01030733–34, SITC 00016229, SM SM704800218). **Xingwen Xian:** Xianfeng Xiang, 1290 m, 12 May 1959, *Yibin Division Sichuan Economic Plant Exped 342* (CDBI CDBI0027975, KUN 0175602, PE 00935216–17, SM SM704800508, SZ 00289802). **Xuyong Xian:** Shuiwei Qu, Guangmu Gongshe, Xixi Dui, 26 June 1977, *Xuyong Team 0405* (SM SM704800507).

NE Yunnan. Suijiang Xian: near Luohan Ping, 1300 m, 10 May 1973, *B. X. Sun et al. 183* (KUN 0175595, YUKU 02065679–80). **Yiliang Xian:** Long-ki (Longjie), Sep. 1893, *J. M. Delavay 5024 bis* (P P02313363, P02313358, VNM); near Tianmalin Qu, 1973, *Z. Y. Wu 96* (KUN 0175594); near Xiaocaba, 1870 m, 4 Sep. 1991, *S. G. Wu et al. 162* (KUN 0178283).

2. Berberis alpicola C. K. Schneid., Repert. Spec. Nov. Regni Veg. 46: 253. 1939. TYPE: China. Taiwan: Prov. Kagi (Chiayi Hsien), Mt. Morrison (Yushan), 3666 m, 24 Oct. 1918, *E. H. Wilson 10952* (holotype, A 00038721!; isotypes, B B10 0365257!, BM BM000559458!, K K000644916!, US 00103858 image!).

Shrubs, evergreen, to 1.2 m tall; mature stems pale brownish yellow, sulcate, sparsely verruculose; internodes 7–10(–12) mm; spines 3-fid, concolorous, 0.6–1.2 cm, slender, terete. Petiole almost absent; leaf blade abaxially pale green, shiny, adaxially dark green, shiny, elliptic to oblong-elliptic, 1–2.5 × 0.7–1 cm, leathery, abaxially epruinose, midvein raised abaxially, impressed adaxially, lateral veins and reticulation indistinct or obscure on both surfaces, base cuneate, margin sometimes slightly revolute, spinulose with 2 to 4 widely spaced teeth on each side, apex acute, mucronate. Inflorescence a fascicle, (1- to)4- to 8-flowered; pedicel reddish, 8–16 mm; bracteoles reddish, triangular-ovate, ca. 2 × 1 mm, apex acuminate. Flowers mid-yellow. Sepals in 3 whorls; outer sepals with reddish vertical stripe, ovate, ca. 3.75–4 × 2 mm, apex acute; median sepals obovate-elliptic, 5.5–7 × 3.5–4 mm; inner sepals obovate, 8–9 × 6 mm; petals broadly obovate or obovate-elliptic, 5–7 × 4.5 mm, base shortly clawed, apex entire, glands separate, oblong, ca. 1 mm. Stamens ca. 3 mm; anther connective slightly extended, truncate or obtuse. Pistil 2.5 mm; ovules 4 to 6. Berry black, epruinose, ellipsoid, ca. 8 × 5 mm; style persistent, to 0.9 mm; seeds black, ca. 3 × 2 mm.

Phenology. *Berberis alpicola* is known to flower in May and has been collected in fruit from September to November.

Distribution and habitat. *Berberis alpicola* is known from Yushan in Chiayi Hsien. It has been collected by streams and in *Abies* forest at ca. 3300–3660 m.

The protologue of *Berberis alpicola* cited only the type collection, which has fruit but no flowers. In the protologue, Schneider described it as "apparently a distinct small species" similar to *B. kawakamii*, while cautioning it might be simply a high-mountain form of the latter. This caution was understandable in that it is often extremely difficult to distinguish herbarium specimens from those of *B. kawakamii* that have only upper branches with small leaves and are sterile or with fruit. This is so much the case that initially I would have been inclined to accept T. S. Liu's (1976: 16) treatment of *B. alpicola* as a synonym of *B. kawakamii* (a synonymy followed by Chamberlain & Hu [1985: 539]) had I not grown the plants from *B. & S. Wynn-Jones 6939*,

referred to below. These have uniformly small leaves and a very different flower structure from *B. kawakamii*. These plants are the source of the description of flowers used above. Information on the flowering season of *B. alpicola* in the wild is from C. C. Yu (pers. comm. 23 Mar. 2013).

An annotation to the sheet of the holotype in Wilson's hand describes it as being a "Shrub 1–4 ft" (30–120 cm) tall and common. The seeds of *B. & S. Wynn-Jones 6939* were from a plant 30 cm high. The heights of the wild-collected specimens cited below are not recorded on the specimen sheet, but whatever the maximum height of *Berberis alpicola* might be, from the evidence of cultivated plants, at least, it appears to grow very slowly; in 2018 the largest of my plants was 85 cm high and was growing at the rate of less than 10 cm a year.

S. Y. Lu and Yang (1996: 576) and Yu and Chung (2014) treated *Berberis alpicola* as a synonym of *B. brevisepala*. However, the leaves of this latter species both from the type area and from elsewhere are significantly longer and narrower than plants from Yu Shan and are frequently abaxially pruinose. The flower structure of *B. alpicola* is also different and the style of fruit of *B. brevisepala* appears to be either absent or markedly shorter than that in *B. alpicola*.

Selected specimens.
Taiwan. Chiayi Hsien, Yushan: 3350 m, 18 Oct. 1906, *T. Kawakami & U. Mori s.n.* (TAIF 9895); Nitakayama (Yushan), 3350 m, 27 Oct. 1918, *R. Kanehira & S. Sasaki s.n.* (TAIF 9896, 9897); Paiyun Lodge, ca. 3400 m, 17 Jan. 2009, *C. C. Yu 216* (TAI); 1 km to Paiyun Lodge, ca. 3300 m, 18 Jan. 2009, *C. C. Yu 217* (TAI); 1.1 km to Paiyun Lodge, ca. 3310 m, 18 Jan. 2009, *C. C. Yu 218* (TAI); 1.1 km to Paiyun Lodge, 3330 m, 20 Jan. 2009, *C. C. Yu 219* (TAI); 1 km after Paiyun Lodge, ca. 3500 m, 20 Jan. 2009, *C. C. Yu 220* (TAI); same details, *C. C. Yu 221* (TAI); 0.5 km to Paiyun Lodge, ca. 3500 m, 20 Jan. 2009, *C. C. Yu 222* (TAI).

Cultivated material:
Living cultivated plants.
Foster Clough, Mytholmroyd, West Yorkshire, U.K., and Royal Botanic Garden Edinburgh, from *B. & S. Wynn-Jones 6939*, Chiayi Hsien, Yushan, Paiyun Lodge area, 3410 m, 15 Nov. 1999.

Cultivated specimens.
Foster Clough, Mytholmroyd, West Yorkshire, U.K., 12 May 2015, from *B. & S. Wynn-Jones 6939* (details above), *J. F. Harber 2015-13* (A, E, HAST, PE, TAI).

3. Berberis amabilis C. K. Schneid., Repert. Spec. Nov. Regni Veg. 46: 257. 1939. TYPE: N Myanmar (Burma). Adung Valley, 2100–2400 m, 27 Mar. 1931, *F. Kingdon-Ward 9324* (lectotype, designated by Ahrendt [1961: 43], BM BM000794119!; isolectotype A 00038711!).

Berberis taronensis Ahrendt, J. Bot. 79(Suppl.): 23. 1941, syn. nov. TYPE: China. NW Yunnan: [Gongshan Xian], Upper

Kiukiang (Dulong) Valley, (Clulung) Chöherton, 2600 m, 5 Aug. 1938, *T. T. Yu 19658* (holotype, E E00117391!; isotypes, A 00038808!, KUN 1204870!, PE [2 sheets] 00935263–64!).

Shrubs, evergreen, to 2.5 m tall; mature stems pale yellow or brownish yellow, subterete; spines 3-fid, concolorous, 0.5–1.5 cm, abaxially slightly sulcate. Petiole almost absent or 1–2 mm; leaf blade abaxially pale yellow-green, adaxially dark green, shiny, narrowly oblong-elliptic, oblong-lanceolate, or lanceolate, 2.5–7 × 0.6–1.5 cm, leathery, abaxially mostly white or glaucous pruinose, sometimes densely so, midvein raised abaxially, impressed adaxially, lateral veins slightly raised abaxially, slightly visible adaxially, reticulate veins obscure abaxially, inconspicuous adaxially, base cuneate, margin often revolute particularly so in winter months, spinose or sometimes coarsely dentate, with 2 to 5(to 7) teeth on each side, sometimes entire toward the ends of branches, apex acute, mucronate. Inflorescence a fascicle, 2- to 20(to 25)-flowered; pedicel dark red, 6–20(–25) mm, slender; bracteoles red, ovate, ca. 1 mm, apex acute. Sepals in 2 whorls; outer sepals broadly ovate, 5 × 4 mm; inner sepals oblong-obovate, ca. 7 × 4.5 mm; petals obovate, 7 × 4 mm, base clawed, with oblong glands, apex emarginate. Anther connective distinctly extended. Ovules 2. Berry black, slightly pruinose at first, otherwise epruinose, obovoid or ellipsoid, ca. 7–8 × 4 mm; style persistent.

Phenology. *Berberis amabilis* in China has been collected in flower in April and in fruit between August and January the following year.

Distribution and habitat. In China *Berberis amabilis* is known from the Dulong Jiang in Gongshan Xian in northwest Yunnan. It has been collected from forest margins, thickets on river banks, among boulders on field edges, and from ruderal vegetation in felled areas at ca. 1400–2600 m. *Berberis amabilis* is also known from the Adung Valley in north Myanmar (Burma).

The protologue of *Berberis amabilis* cited three gatherings but did not designate a type. Later, Schneider (1942: 35) referred to *F. Kingdon-Ward 9324* as the type but did not cite a herbarium. Ahrendt (1961: 43) lectotypified the specimen at BM. The syntype *J. F. Rock 7388* cited in the protologue of *B. amabilis* is not this species and appears to be *B. holocraspedon*.

It has proved impossible to locate exactly where on the Dulong Jiang the type of *Berberis taronensis* was collected. The Dulong rises in Zayü (Chayu) Xian in southeast Xizang then flows into Gongshan Xian in Yunnan and thence into Myanmar. Though the protologue states *T. T. Yu 19690* was collected in Zayü, the collection

details on the sheet give Yunnan. The coordinates given for the place of collection in the protologue appear to be speculation by Ahrendt since the collector's notes do not record this.

Chamberlain and Hu (1985: 548) noted that *Berberis amabilis* and *B. taronensis* were very similar but accepted them as different species on the basis that the latter had leaves with more marked lateral veins and margins that were only slightly revolute, longer pedicels, and pruinose fruit. Their publication appeared before most of the specimens from the Dulong Jiang cited below were collected. These show considerable variation in leaf shape, venation, and margin and length of pedicel. Fruit of specimens collected from late August to January the following year are epruinose.

The collector's notes to *Kingdon-Ward 9324* state "the leaves curl up in winter," and this is confirmed by those specimens from the Dulong Jiang cited below, collected between November and January, which evidence particularly revolute leaf margins.

Interestingly, another section *Wallichianae* species, *Berberis incrassata*, is also found both in the Dulong Jiang and in the Adung valley in Myanmar, both rivers being tributaries of the Irrawaddy.

Ahrendt (1961: 43) cited *T. T. Yu 17208* (A 00279303, E E00395970, KUN 0175674–75, PE 01031584–85, west Yunnan, Chengkang, Hsiaoshuishan [Zhenkang Xian, Xiaoxueshan], 3200 m, 5 Aug. 1938) as *Berberis amabilis*, but this is *B. sublevis*. Ahrendt (1961: 79) also recognized a *B. taronensis* var. *trimensis* Ahrendt with a type from that part of Arunachal Pradesh, India, where there is a conflicting territorial claim by China. There would seem no reason to associate this taxon with *B. amabilis*.

Ying (2001: 129, 2011: 742) treated *Berberis amabilis* var. *holophylla* C. Y. Wu & S. Y. Bao from Dali, northwest Yunnan, as a synonym of *B. amabilis*. There would, however, also seem to be no evidence to associate this taxon with *B. amabilis* (for further details of variety *holophylla*, see Taxa Incompletely Known).

Berberis taronensis is on the current IUCN Red List of Threatened Species (China Plant Specialist Group, 2004f).

Selected specimens.
Syntypes. Berberis amabilis. Myanmar, Adung Valley, 2440 m, 14 Mar. 1931, *F. Kingdon-Ward 9299* (BM BM001015546); China, W Yunnan, betw. Tengyueh (Tengchong) & Burmese border, near Changlifang, en route to Sadon (Sadung), Nov. 1922, *J. F. Rock 7388* (A 00038688, US 0946050).

Other specimens.
NW Yunnan. Gongshan Xian, Dulong Jiang: Xianjiudang, 1700 m, 24 Aug. 1982, *Qinghai-Xizang Team 9602* (KUN 0176280, PE 01030836); Dulongjiang Xiang 1900 m, 2 Sep. 1982, *Qinghai-Xizang Team 9797* (KUN 0176922, PE

01840098); Xianjiudang, 1550 m, 22 Nov. 1990, *Dulong Jiang Investigation Team 2138* (KUN 0176765–66); Longyuan, 1690 m, 23 Nov. 1990, *Dulong Jiang Investigation Team 2174* (KUN 0176763–64); Dizhendang, 1950 m, 28 Nov. 1990, *Dulong Jiang Investigation Team 2204* (KUN 0176769–70, 0176793); Kongdang, 1420 m, 2 Jan. 1991, *Dulong Jiang Investigation Team 1573* (KUN 0176767–68); Longdongwan, 1850 m, 2 Apr. 1991, *Dulong Jiang Investigation Team 6222* (KUN 0176784–85); Longyuan, 1700 m, 11 Apr. 1991, *Dulong Jiang Investigation Team 5499* (KUN 0176776–77); Dizhenggang, 1880 m, 13 Apr. 1991, *Dulong Jiang Investigation Team 5702* (CAS, KUN 0176775), *5705* (KUN 0176780–81); Dizhenggang, 1760 m, 14 Apr. 1991, *Dulong Jiang Investigation Team 5802* (KUN 0176778–79); Dizhenggang, 1800 m, 16 Apr. 1991, *Dulong Jiang Investigation Team 5988* (KUN 0176782–83); Xianghong, 1650 m, 20 Apr. 1991, *Dulong Jiang Investigation Team 6435* (KUN 0176789–90); Dulongjiangshan, Lilaqia, 2400 m, 22 Apr. 1991, *Dulong Jiang Investigation Team 6360* (KUN 0176786, 0176788); E side of valley, vic. of Dizhendang (Lengdang), 28.079722°N, 98.329167°E, 1810 m, 28 Oct. 2004, *Gaoligong Shan Biodiversity Survey 21326* (CAS, E E00258552, GH 00300034, HAST 124211, TNM S105819); W side of valley, 0.5 km N of Dizhendang (Lengdang), 28.083333°N, 98.326944°E, 1850 m, 28 Oct. 2004, *Gaoligong Shan Biodiversity Survey 21390* (CAS 00120175, E E00261688); W side of valley, along N side of Shilawa river, 0.5 km SW of Dizhenggang, 28.075556°N, 98.3225°E, 1910 m, 29 Oct. 2004, *Gaoligong Shan Biodiversity Survey* 21432 (CAS 00120176, E E00320111).

N MYANMAR (BURMA). Nam Tamai Valley (Adung Wang), 28.166667°N, 97.5°E, 2130–2440 m, 14 Oct. 1937, *F. Kingdon-Ward 13421* (BM).

4. Berberis arguta (Franch.) C. K. Schneid., Bull. Herb. Boissier, sér. 2, 8: 197. 1908. *Berberis wallichiana* DC. f. *arguta* Franch., Bull. Soc. Bot. France 33: 388. 1886. TYPE: China. NE Yunnan, [Shuifu Xian], "Les bois à Tchen-fong-chan (Ta Kuan)," 4 May 1882, *J. M. Delavay 2354* "A" (lectotype, designated here, P P00716541!; isolectotypes, A fragm. 00038818!, B fragm. B10 0250733!, K missing, P P00716542!).

Shrubs, evergreen, to 1.5 m tall; mature stems probably grayish yellow, sulcate, verruculose; spines absent or 3-fid, pale brownish yellow, 0.4–0.8 cm, weak. Petiole almost absent or to 1.5 mm; leaf blade abaxially pale green, adaxially dark green, elliptic, elliptic-lanceolate, or broadly lanceolate, 3–6.5 × 1.2–2.2 cm, leathery, midvein raised abaxially, impressed adaxially, both surfaces with conspicuous lateral veins and dense reticulation, base cuneate or obtuse, margin spinulose with 12 to 50 teeth on each side, apex acute, mucronate. Inflorescence a fascicle, (1- to)5- to 40-flowered. Flowers unknown; immature fruiting pedicel pale brownish yellow, 8–12 mm. Immature berry 4 mm; seeds 3.

Phenology. *Berberis arguta* has been collected with immature fruit in May. Its flowering season is unknown.

Distribution and habitat. *Berberis arguta* is known from only the type collection from woods in Shuifu Xian in northeast Yunnan at an unrecorded elevation and from a collection from nearby in Pingshan Xian in southeast Sichuan made on a forest margin at 1150 m.

There has been some confusion in relation to *Berberis arguta*. The protologue of *B. wallichiana* f. *arguta* gave a brief description including stating it was from Yunnan "in monte Tsang-chan" [Cang Shan]. From the evidence of Franchet's entry for *B. wallichiana* f. *parvifolia* on the same page, this appears to be Cang Shan above Dali in northwest Yunnan. Subsequently, Franchet (1889: 38) cited "Yun-nan in monte Tsang-chan et in silvis montis Tchen-fong chan ad Ta-kouan; 4 maj. 1882 (Delavay, n. 2354)." From Bretschneider (1898b), Tchen-fong chan can be located in Shuifu Xian in northeast Yunnan near the border with Pingshan Xian in Sichuan. In publishing *B. arguta* as a "nov. spec.," Schneider repeated Franchet's collection details of 1889 and cited *Delavay 2354* as being at P. Later, Ahrendt (1961: 60) cited *Delavay 2354* at K of 4 May 1882 as being from "Mt Tsang-chan and Tchenfongchan" (this is missing), while Chamberlain and Hu (1985) cited *Delavay 2354* from Tchen-fong-chan at P as the holotype. In fact, there are three specimens of *Delavay 2354* at P. Two of 4 May 1882 are annotated as being from Tchen-fong-chan, while the third is undated and has no collection details beyond Yunnan. This third is not the same species as the other two and, as an annotation by C. M. Hu of 1982 suggests, is *B. deinacantha*. Whether this is the specimen referred to as being from Tsang-chan is impossible to say, but if it is, then either this may not be the Cang Shan near Dali since *B. deinacantha* is not otherwise recorded from there (though it is found elsewhere in northwest Yunnan) or Tsang-chan is a mistake. In any event, the specimens from Tchen-fong-chan are designated here as *2354* "A" and the third specimen as *2354* "B." Had the specimen of *2354* "B" been annotated as being from Cang Shan, then there would have been a case for this being the type of *B. arguta*, in which case *B. deinacantha* would be a synonym and *2354* "A" would have needed another name. But since it is not so annotated, one of the *2354* "A" specimens is lectotypified here. P P00716541 has been chosen here because it was annotated by Schneider on 18 June 1906 as "spec. nov."

The description of *Berberis arguta* given above is based on the type specimens and *Sichuan University Department of Biology Herbarium 110028* (collected some 35 km to the north and which is sterile); no other examples of the species having been found. As such, it differs from that given by Schneider in the protologue which at times conflates elements of *Delavay 2354* "B" with those of *2354* "A." It differs even more from that given by Ying (2001: 130, 2011: 742) who included a description of flowers. Ying also stated that the species is found in Guizhou. No evidence for this has been found (specimens at HGAS annotated as *B. arguta* are, in some cases, *B. deinacantha*, in others, *B. bicolor*).

Selected specimens.
SE Sichuan. Pingshan Xian: Jinping Shan, 1150 m, 19 May 1973, *Sichuan University Department of Biology Herbarium 110028* (SZ 00291212).

5. **Berberis aristatoserrulata** Hayata, Icon. Pl. Formosan. 3: 13. 1913, as "*aristato-serrulata.*" TYPE: China. Taiwan: "*Montibus Centralibus*" (Central Mtns.), Apr. 1910, *s. coll.* [*U. Mori*] *s.n.* (holotype, TI 02616 image!, TAIF fragments 9893!).

Shrubs, evergreen, to 2 m tall; mature stems pale brown or reddish brown, subterete, not verruculose; spines 2- or 3-fid, rarely 4-fid, concolorous, 0.4–1(–1.9) cm, weak, often absent. Petiole almost absent or to 5 mm; leaf blade abaxially pale green, adaxially dull green, narrowly elliptic, elliptic-lanceolate, or lanceolate, (3.8–)5.5–11 × (1.6–)2.4–2.9(–3.4) cm, thickly leathery, abaxially rarely lightly pruinose, midvein raised abaxially, impressed adaxially, lateral veins and reticulation slightly raised and inconspicuous abaxially, slightly impressed and inconspicuous adaxially, base attenuate, margin spinulose with 12 to 32 teeth on each side, apex acute, sometimes obtuse, rarely retuse, mucronate. Inflorescence a fascicle, 5- to 15-flowered; pedicel 7–16 mm; bracteoles triangular, red, 1.5–2 × 1 mm. Flowers yellow. Sepals in 3 whorls; outer sepals obovate, sometimes with reddish stripe, 2.5 × 2 mm; median sepals obovate to narrowly obovate, 5 × 2 mm; inner sepals elliptic or narrowly obovate, 6.5 × 4 mm; petals elliptic or obovate-spatulate, ca. 5–6 × 2.5–4 mm, base clawed, glands narrowly ovoid, close together, apex incised to emarginate. Stamens 3–4 mm; anther connective distinctly extended, shortly apiculate. Pistil 3.5 mm; ovules 2 to 4. Berry black, subglobose, 5 × 3 mm; style persistent.

Phenology. *Berberis aristatoserrulata* has been collected in flower in April and in fruit from June to the following February.

Distribution and habitat. *Berberis aristatoserrulata* is known from Hualien and Kaohsiung Hsien in Taiwan. It has been found in the understories of coniferous and broadleaf mixed forests at ca. 1500–2300 m.

Hayata's protologue of *Berberis aristatoserrulata* cited only the type, gave no collector, and had only the vaguest of collection details. Schneider (1918: 146) stated that the collector was Hayata himself, and this was re-

peated by Ahrendt (1961: 61), H. L. Li (1963: 169), Chamberlain and Hu (1985: 539), and S. Y. Lu and Yang (1996: 576). In fact, it appears that the collector was actually the Japanese ethnographer U. Mori, who was commissioned by the Regenerative and Products Bureau of the Japanese Governor General's Office to collect plant specimens on various of his expeditions. From the evidence from his travel reports (Mori, 2000: 407–425), in April 1910, Mori crossed part of the central mountain range via the late Qing dynasty Guanmen Trail. In April 2014, I accompanied Chih-Chieh Yu of the National Taiwan University (NTU) on an expedition to this remote area which confirmed this is where the type was likely to have been collected. Specimens we collected are listed below. For a report of our expedition, see Harber (2015). Interestingly, very near where our specimen no. 7 was collected (details below), we found a sterile plant which appeared to be *B. aristatoserrulata* × *B. kawakamii*.

The protologue gave the number of ovules as two. From the evidence of the 2014 Guanmen Expedition specimens, the number can be up to four. The description of berries above comes from an image taken in the field by C. C. Yu at Dagui Hu, Kaohsiung Hsien, on 12 February 2015.

Selected specimens.
Taiwan. Hualien Hsien: 77 Compt. of Mukwashan, 2100–2330 m, 15 Aug. 1956, *Liu et al. 250* (PH 00066430); Mt. Luanshan to Mt. Patolushan, 2000–2100 m, 3 Aug. 1963, *M. Tamura et al. 21556* (E E00395986, HAST 53698, KUN 0175730); Yenping Logging Trail, 25 July 1973, *C. H. Ou 1941* (TCF [3 sheets]); Juisui Forest Rd., 6 Apr. 2000, *K. C Yang et al. 05981* (TNM S91817); Hsiulin Hsiang, en route from first river bed of Pingfengshan to main river, 24.173333°N, 121.312222°E, 2410 m, 4 Sep. 2009, *C. I. Huang 4151* (HAST 125895); Qing dynasty Guanmen Trail, 1973 m, 10 Apr. 2014, *J. F. Harber & C. C. Yu Guanmen Exped. 1* (E); same details, but 2300 m, *J. F. Harber & C. C. Yu Guanmen Exped. 7* (E); 27 km marker, Guangfu Logging Trail, 1500 m, 12 Apr. 2014, *J. F. Harber & C. C. Yu Guanmen Exped. 12* (TAI).

6. **Berberis asmyana** C. K. Schneid., Pl. Wilson. (Sargent) 1(3): 357. 1913. TYPE: China. WC Sichuan: Mupin (Baoxing Xian), 1825 m, June 1908, *E. H. Wilson 2873* (lectotype, designated by Ahrendt [1961: 46], K K000077344!; isolectotypes, A 00038716!, B B10 0250728!, BM BM000559453!, E E00217969!, HBG HBG-506680 image!, LE!, US 00103860 image!, W 1914–0004806!).

Shrubs, evergreen, to 1.2 m tall; mature stems yellow, sulcate, sparsely black verruculose; spines 3-fid, concolorous, 0.2–0.6(–0.9) cm, terete, weak. Petiole almost absent; leaf blade abaxially pale green, adaxially dark green, shiny, elliptic or obovate-elliptic, 1–3 ×

0.6–1 cm, thinly leathery, midvein raised abaxially, impressed adaxially, lateral veins and reticulation obscure abaxially, inconspicuous adaxially, base cuneate, margin spinulose with 1 to 3 teeth on each side, sometimes entire, apex acute, mucronate; pedicel 20–40 mm, slender; bracteoles ovate, ca. 1.8 mm. Flowers 1(to 3). Sepals in 2 whorls; outer sepals ovate, ca. 3 × 2.5 mm; inner sepals obovate-orbicular, ca. 5 × 4.5 mm; petals obovate, ca. 4.5 × 3.2 mm, base clawed, glands very close together, apex slightly emarginate. Stamens ca. 3.2 mm; anther connective truncate. Ovules 4 or 5, sessile. Berry immature, green, slightly pruinose, ellipsoid, (6.5–)7–8 × 4–5 mm; style not persistent.

Phenology. *Berberis asmyana* has been collected in flower in June and with immature fruit in July or August.

Distribution and habitat. *Berberis asmyana* is known from Baoxing and Tianquan Xian in west-central Sichuan. It has been collected in mountain thickets and along a roadside from 600–1825 m.

The lectotypification of the specimen at K was by Ahrendt (1961: 46). No other specimens were located beyond the type and those listed below. The collection details for *Z. L. Wu 12185* were deduced from Z. L. Wu specimens of other genera at PE with close numbers in the same sequence.

Berberis verruculosa, which has somewhat similar leaves to *B. asmyana* (although these are usually abaxially pruinose), but much shorter pedicels and three whorls of sepals, is also found in the Baoxing-Tianquan area but (where this is recorded) at higher elevations than that of *B. asmyana* (for examples, see entry for *B. verruculosa*). The relationship between the two species in this locality needs further investigation. In this respect, the following collection is interesting: *Sichuan Economic Plants Expedition Liangshan Team 1777* (KUN 0177281, PE 01037893–94, SM SM704800413), Sichuan, Meigu Xian, Houbutuo Qu, E'kou, 19 Aug. 1959. These specimens have very similar leaves to *B. asmyana* but are very distinctly abaxially pruinose. Like *B. asmyana*, they are 1- to 3-fascicled, but the pedicels are shorter.

Selected specimens.
WC Sichuan. Tianquan Xian: Tienchuan-Lingkwan (Tianquan-Lingguan) hwy., 600 m, July–Aug. 1939, *S. Y. Hu 946* (A 00279358–59, SZ 00289841); s. loc., s.d. [but probably Tianquan Xian, Aug. 1940], *Z. L. Wu 12185* (PE 01030739).

7. **Berberis atrocarpa** C. K. Schneid., Pl. Wilson. (Sargent) 3: 437. 1917. TYPE China. WC Sichuan: Mupin (Baoxing), 1200–1800 m, Nov.

1908, *E. H. Wilson 1284* (lectotype, designated by Ahrendt [1961: 77], K missing, new lectotype, designated here, A 00038717!; isolectotypes, B B10 000365256!, BM BM001015569!, E E00386195!, HBG HBG-506742 image!, P [2 sheets] P00716535!, P02482733!, US 00103861 image!).

Berberis atrocarpa C. K. Schneid. var. *subintegra* Ahrendt, J. Linn. Soc., Bot. 57: 77. 1961. TYPE: WC Sichuan: Lushan Xian, 1100 m, 18 Oct. 1936, *K. L. Chu 4040* (holotype, E E00217970!; isotypes, BM BM000895063!, IBSC [2 sheets] 0091552–53 images!, K K000644910!, P P02465475!, PE [3 sheets] 01031477–79!).

?*Berberis silvicola* C. K. Schneid. var. *angustata* Ahrendt, J. Linn. Soc., Bot. 57: 62. 1961. TYPE: WC Sichuan: Tianquan Xian, 1900 m, 20 Apr. 1936, *K. L. Chu 2368* (holotype, E E00217971!; isotypes, BM BM001010628!, IBSC 0091547 image!, PE [3 sheets] 01031473–75!).

Shrubs, evergreen, to 3 m tall; mature stems pale yellow, angled to sulcate, finely black verruculose; spines 3-fid, concolorous, (2–)3–5 cm, stout, abaxially sulcate. Petiole almost absent; leaf blade abaxially pale green, adaxially dark green, slightly shiny, lanceolate or narrowly oblong-elliptic, 3–6(–8) × 0.7–2 cm, leathery, midvein raised abaxially, impressed adaxially, lateral veins and reticulation inconspicuous or indistinct on both surfaces, base cuneate, margin spinulose with 5 to 10 teeth on each side, rarely entire, apex acute or acuminate, mucronate. Inflorescence a fascicle, (2- to)8- to 20-flowered; pedicel reddish, 5–10 mm. Sepals in 2 whorls; outer sepals oblong-obovate, ca. 4 × 2 mm; inner sepals obovate, ca. 7 × 4 mm; petals obovate, 6 × 4.5 mm, base cuneate, glands separate, apex rounded, deeply narrowly incised. Stamens ca. 4 mm. Ovules 2 or 3(?or 4), sessile or very shortly funiculate. Berry at first with blue bloom, finally jet black, shiny, epruinose, ovoid, ca. 5 × 4 mm; style persistent.

Phenology. *Berberis atrocarpa* has been collected in flower in March and early April and in fruit between May and November.

Distribution and habitat. *Berberis atrocarpa* is known from west and west-central Sichuan. It has been collected from thickets, field sides, roadsides, and trailsides at ca. 600–2200 m.

The protologue of *Berberis atrocarpa* did not specify a herbarium for the type. Ahrendt (1961: 77) lectotypified a specimen of *Wilson 1284* at K, but this is missing (see introduction for further details). The specimen at A annotated as type by Schneider on 25 July 1916 has been chosen here as a new lectotype.

The protologue of *Berberis atrocarpa* cited only the type specimen and *Wilson 4287*, both of which have only fruit, noting that they contained two or three seeds. The description of flowers used above is from Ahrendt (1961: 77) who cited a cultivated plant grown at Kew from seeds of the type. This does not differ substantially from the incomplete description given by Schneider (1942: 15) based on cultivated material sent to him from a plant at the Golden Gate Park in San Francisco grown from *Wilson 4287*, both authors noting the apex of the petals to be deeply incised.

Chamberlain and Hu (1985: 555) synonymized *Berberis atrocarpa* var. *subintegra* and *B. silvicola* var. *angustata*. However, the incomplete description of flowers given by Ahrendt (1961: 62) for the latter differs substantially from his description of the flowers of *B. atrocarpa*. It has proved impossible to verify this description since there is now no floral material on the holotype. There is one complete flower on the isotype at BM and three between the various isotypes at PE, but I have not sought permission to dissect them. Complicating matters further is an annotation by C. M. Hu on the sheet of the holotype recording four ovules. Clearly more research is needed here.

The report by Bao (1997: 65) of *Berberis atrocarpa* in Zhengxiong Xian in northeast Yunnan was based on *G. M. Feng et al. 447* and *S. Y. Bao 106*. However, these are *B. zhenxiongensis*. The reports of the species in Guizhou by He and Chen (2004: 29) and in Hunan by L. H. Liu (2000: 710) are likely to be based on misidentifications.

Selected specimens.
NC & WC Sichuan. Baoxing Xian: "Thibet oriental Moupin," s.d., *A. David s.n.*, (P P02313624); Lingguan Qu, Xinxing Gongshe, Linguan Daqiao, 1320 m, 19 Apr. 1983, *D. Y. Peng 47377* (CDBI CDBI0027121–23). **Kuan Hsien:** (Guan Xian, now Dujiangyan Shi), Mt. Tsing-Cheng (Qingcheng Shan), 5 Apr. 1944, *W. P. Fang 19317* (PE 01031487–88). **Guan Xian:** Wuyan Shan, 880 m, 24 Oct. 1956, *X. Li 46991* (IBSC 0091549, PE 01031472, SZ 00289046). **Hongya Xian:** Wawu Shan, near Jionglingci, 2 Aug. 1939, *Z. W. Yao 2655* (PE 01031492). **Mabian Xian:** Meihua Shan, 1180 m, 26 July 1934, *S. L. Sun 5714* (IBSC 0091470, PE 01031213). **Pujiang Xian:** Ganxi, Jidong Cave, ca. 600 m, 24 Apr. 1986, *T. Naito et al. 18* (PE 01031329, TUS). **Qingshen Xian:** Tianche Gongshe, San Dadui, Chejing Shan, 650 m, 15 Aug. 1979, *Qingshen Team 459* (SM SM704800047). **Tianquan Xian:** Erlang Shan, Longda Xi, 2200 m, 29 Apr. 1953, *X. L. Jiang 34049* (IBK IBK00012845, IBSC 0092717, PE 01031786). **[Xiaojin Xian–Li Xian border]:** "W of Kuan Hsien," Pan-lan-shan (Bawangshan), 1830–2130 m, Oct. 1910, *E. H. Wilson 4287* (A 00279360). **Yingjing Xian:** Qingfeng Xiang, 28 Nov. 1959, *W. G. Hu & Z. He 12111* (PE 01031485, SZ 00294083).

8. Berberis bergmanniae C. K. Schneid., Pl. Wilson. (Sargent) 1: 362. 1913. TYPE: China. NC Sichuan: "Wa-sen country," Wen-chuan Hsien (Wenchuan Xian), 1600–2000 m, Aug. 1908, *E. H. Wilson 2877* (lectotype, designated by Ahrendt [1961: 67], K missing, new lectotype, designated here, A 00057548!).

Berberis bergmanniae var. *acanthophylla* C. K. Schneid., Pl. Wilson. (Sargent) 1: 362. 1913, syn. nov. TYPE: NC Sichuan: "Wa-sen country," Wen-chuan Hsien (Wenchuan Xian), 2000–2500 m, Nov. 1910, *E. H. Wilson 4149* (lectotype, designated by Ahrendt [1961: 68], K missing, new lectotype, designated here, A 00057549!).

Shrubs, evergreen, to 2 m tall; mature stems yellow, angled, sparsely black verruculose; spines 3-fid, pale yellow, 1.5–2.5 cm, stout. Petiole almost absent; leaf blade abaxially pale green, adaxially dullish green, oblong-elliptic or elliptic, (3–)5–8 × 1.6–2.5 cm, thickly leathery, midvein raised abaxially, impressed adaxially, lateral veins and reticulation obscure abaxially, inconspicuous adaxially, base cuneate, margin thickened, slightly revolute, sometimes undulate, spinose with 6 to 10 sometimes coarse teeth on each side, apex acute or acuminate, mucronate. Inflorescence a fascicle, 5- to 20-flowered; pedicel 7–15 mm. Sepals in 2 whorls; outer sepals ovate, ca. 5.5 × 3.5 mm; inner sepals obovate, ca. 7 × 5 mm; petals obovate, ca. 6 × 5 mm, base clawed, with separate glands, apex rounded, incised. Stamens ca. 4.5 mm; anther connective truncate. Ovules 1 or 2. Berry black, blue pruinose, ovoid-ellipsoid or ovoid-globose, 8–9 × ca. 6 mm; style persistent and markedly conspicuous.

Phenology. *Berberis bergmanniae* has been collected in flower from March to May and in fruit between the end of May and November.

Distribution and habitat. *Berberis bergmanniae* is known from west-central, north-central, and north Sichuan and Wen Xian in south Gansu. It has been collected from thickets, forests, and mountain slopes at ca. 1100–2600 m.

The protologue of *Berberis bergmanniae* cited three gatherings, *Wilson 2876* and *2877* from west Sichuan and *Wilson 2878* from west Hubei. Subsequently, Schneider (1917: 438) stated that *Wilson 2877* "should be regarded as the type" but did not indicate any herbarium. Ahrendt (1961: 67) cited a specimen at K as the type but that is missing (see introduction for further details). The specimen of *Wilson 2877* at A has therefore been chosen here as the new lectotype. Later, Schneider (1918: 138) mistakenly cited *Wilson 2876* as the type, a mistake reproduced both by Ahrendt (1961: 67) and Chamberlain and Hu (1985: 546). The syntypes *E. H. Wilson 2878* "A" and "B" listed below are not *B. bergmanniae* but are as Schneider later stated (1917: 438) *B. julianae* (subsequently *Wilson 2878* was cited by Ahrendt as the type of *B. julianae* var. *patungensis*; for further details, see under *B. julianae*).

The protologue of *Berberis bergmanniae* var. *acanthophylla* cited only *Wilson 4149* but without citing a herbarium. Ahrendt (1961: 68) cited a specimen at K as the type but that is also missing. The specimen at A has therefore been chosen here as a new lectotype.

The justification Schneider gave in the protologue for distinguishing variety *acanthophylla* was that it has leaves that are abaxially bright green with four or five coarse marginal spines versus the abaxially pale or pruinose leaves with six to 12 more delicate marginal spines of *Berberis bergmanniae*. However, not only do the leaves of *Wilson 2877* have the same coarse marginal spines of *Wilson 4149*, but both gatherings were made in the same area. Subsequently, Schneider (1918: 138) reduced variety *acanthophylla* to a synonym of *B. bergmanniae*, only to restore it again (1942: 21) as a separate variety, now differentiating it from *B. bergmanniae* on the basis that it had leaves that somewhat resembled those of *Ilex*. From the evidence of the various specimens cited below, it appears that the leaves of *B. bergmanniae* are very varied even on the same plant. This is confirmed by a living plant at the Royal Botanic Gardens, Kew, grown from *Wilson 4149*.

H. Smith 13494, cited below, was originally identified by Schneider (1939: 260) as *Berberis deinacantha* var. *valida* and subsequently (1942: 40) as *B. valida*.

The reports of *Berberis bergmanniae* in Hubei by Fu (2001: 392) and in Guizhou by He and Chen (2004: 26) are likely to be based on misidentifications.

Selected specimens.
Syntypes. Berberis bergmanniae. Sichuan: Ching-Chi-Hsien (Hanyuan Xian), 15 Nov. 1908, *E. H. Wilson 2876*, (A, fruit only, 00057547, E E00020982, K missing, US 00516862); W Hubei, Patung Hsien (Badong Xian), 1300 m, May 1907, *E. H. Wilson 2878* "A" (A 00038767, B B10 000250732, BM BM001010630, E E00217947, HBG HBG-506679, K K000077319, US 00103863); W Hubei, Patung Hsien (Badong Xian), 1300 m, July 1907, *E. H. Wilson 2878* "B" (A 00338445, B, BM BM001010629, HBG HBG-506679, K K000077318, US 00103863).

Other specimens.
S Gansu. Wen Xian: Danbao Gongshe, Liujiagou, 872 m, 16 July 1959, *Z. Y. Zhang 9119* (HNWP 16965, WUK 0146507, 0371330); same location, but 1460 m, 30 July 1959, *Z. Y. Zhang 9854* (HNWP 17212, WUK 0153578); Shifang Gongshe, Minbao Gou, 2000 m, 28 July 1959, *Z. Y. Zhang 7161* (HNWP 16124, WUK 0154915).
WC & NC Sichuan. Baoxing Xian: "Thibet Oriental, Moupine, Thibet Oriental," Apr. 1869, *A. David s.n.* (P P06868323); "Pao-hsin-hsien," 1954, *Z. P. Song 38146* (KUN 0176654, PE 01031383); Gongling Xiang, Zhuangzi, 1400 m, 13 May 1958, *Sichuan Agricultural College 4652* (CDBI CDBI0027698–99, PE 01031382); Yangbi Cun, Xiao Gou, 1500 m, 25 May 1958, *X. S. Zhang & Y. X. Ren 4969* (CDBI CDBI0027697, PE 01031386). **Ganluo Xian:** Dayao Shan, 2600 m, 15 Sep. 1979, *General Survey Team 908* (SM SM704800179). **Hanyuan Xian:** Sandao Qiao, 2300 m, 11 Oct. 1938, *K. Pei 8297* (FUS 00014168, PE 01031387); 9 Aug. 1939, *Y. C. Yang 3594* (FUS 00014236, PE 01031519);

Xinhua Gongshe, 1070 m, 15 Mar. 1983, *D. Y. Peng 47001* (CDBI CDBI0027124–125). **Mao Xian:** Ciliu Gou, 8 June 1959, *S. Jiang et al. 00611* (PE 01031391); Tumen Gongshe, 1100 m, 8 Aug. 1975, *Sichuan Botany Team 8601* (CDBI CDBI0027136–37, PE 01031903). **Shimian Xian:** 1955, *Z. J. Xie 39956* (IBSC 0092019, PE 01031435); 1955, *Z. J. Xie 39992* (IBSC 0092023, PE 01031437). **[Yingjing Xian]:** Tahsiangling (Daxiangling), ca. 2000 m, 19 Nov. 1934, *H. Smith 13494* (MO 4367277, PE 01293893, S 12-25257, UPS BOT:V-040863).

N Sichuan. Beichuan Xian: Xiaozhaizi Gou Nature Reserve, 1500 m, 25 Aug. 1984, *C. Tang et al. 42* (CDBI CDBI0027134–35). **Pingwu Xian:** Laohe Gou Nature Reserve, 19 Apr. 2013, *L. Zhang et al. 250* (PE 01994324).

Cultivated material:
Living cultivated plants.
Royal Botanic Gardens, Kew, U.K., from *E. H. Wilson 4149*.

9. Berberis bicolor H. Lév., Repert. Spec. Nov. Regni Veg. 9: 454. 1911. TYPE: China. SC Guizhou: N of Longli Xian, Marjo, Kouy-yong, May 1908, *J. Cavalerie 3043* (lectotype, designated here, K K000077325!; isolectotypes, A [2 sheets] 00038731–32!, E E00217997!, K K000077326, P P02790690!).

Shrubs, evergreen, to 2 m tall; mature stems pale yellowish brown, terete; spines 3-fid, concolorous, 1–1.5 cm. Petiole almost absent; leaf blade abaxially green, rarely pruinose, adaxially green, narrowly elliptic or elliptic-lanceolate, 7–10 × 2–2.5 cm, thinly leathery, midvein raised abaxially, impressed adaxially, lateral veins and reticulation inconspicuous on both surfaces, base cuneate, margin spinulose with 10 to 20 teeth on each side, apex acuminate, mucronate. Inflorescence a fascicle, 3- to 7-flowered; pedicel 10–12 mm, slender; bracteoles triangular-ovate, ca. 2 mm, apex acute. Flowers pale pinkish red outside, white within. Sepals in 2 whorls; outer sepals ovate, ca. 3 × 1.75 mm, apex acute; inner sepals obovate, ca. 4.5 × 3.5 mm; petals obovate, ca. 3.75 × 2 mm, base clawed, glands very close together, apex emarginate or subentire. Stamens ca. 3 mm; anther connective not extended, truncate. Ovules 2. Berry black, ellipsoid, 7 × 5 mm; style persistent.

Phenology. *Berberis bicolor* has been collected in flower in April and May and in fruit between July and September.

Distribution and habitat. *Berberis bicolor* is known from central, southwest, south, and northeast Guizhou. It has been collected from roadsides and forested areas at ca. 700–1800 m.

The protologue of *Berberis bicolor* cited only *Cavalerie 3043* but without citing a herbarium. Ahrendt

(1961: 62) cited the type as being at K. There are two specimens of the gathering there. His lectotypification has been completed here by choosing the specimen with the most floral material.

Schneider (1918: 144) treated *Berberis subacuminata* from south Yunnan as a synonym. Had the synonymy been justified, it would have been the other way around since *B. subacuminata* was published first. However, though both taxa have similar leaves, the flowers are very different and they are separate species as Schneider later (1942: 41–42) concluded while at the same time correcting to *B. bicolor* his earlier (1918: 145) identification of *Cavalerie 1944* as *B. acuminata*.

Berberis bicolor appears to be one of the only species in the whole genus with red and white flowers (though *B. triacanthophora* has flowers that are pinkish and whitish green). Interestingly, some forms of *Mahonia gracilipes* have similar red-and-white coloring. Léveillé noted that *B. bicolor* was very rare. It appears that it is restricted to highly localized habitats in widely separated areas of Guizhou.

Specimens from Taiwan treated as *Berberis bicolor* by T. S. Liu (1976: 15) are referable to *B. mingetsensis*. Specimens from south Yunnan identified as *B. bicolor* are likely to be referable to *B. subacuminata* as are reports of *B. bicolor* in Myanmar by Kress et al. (2003: 172). Conversely, specimens from Guizhou identified as *B. subacuminata* are *B. bicolor*.

Berberis bicolor is on the current IUCN Red List of Threatened Species (China Plant Specialist Group, 2004a).

Selected specimens.
Guizhou: Lou-mong-touan, 20 Sep. 1904, *J. Cavalerie 1944*, (E E00217972, E00395991, K). **Anlong Xian:** Longshan Gongshe, 1600 m, 16 June 1960, *Guizhou Team 4629* (HGAS 013688, IBSC 0092727, NAS NAS00314355, PE 01031242), **Guanling Xian:** Yongningcha, 12 Sep. 1985, *Guanling General Investigation 394* (GZTM 0011421–22). **Guiding Xian:** Chengguan, Qingshan, 1480 m, 14 June 1986, *D. X. Li & R. You 17-18* (GZTM 0011337). **Jiangkou Xian:** Fanjingshan, Huguo Si, 5 May 1959, *Guizhou Team 625* (GF 09001043, HGAS 013660, KUN 0178431, PE 01031244–45); Macaohe, betw. Dahe & Yu'ao, 700 m, 22 Apr. 1964, *Z. S. Zhang et al. 400460* (PE 01031248). **Leishan Xian:** Wudong Xiang, 3 km E of Nanxiang, 1300 m, 30 Apr. 1959, *Qiannan Team 995* (HGAS 013631, KUN 0175766, NAS NAS00314350–51, PE 01030721–22); Wudong, near Xumu Chang, 27 Apr. 1959, *Qiannan Team 1023* (HGAS 013606, KUN 0178377, NAS NAS00314354, PE 01037999–800); near Wudong Xiang, 3 km W of Xumu Chang, 1400 m, 6 May 1959, *Qiannan Team 1067* (HGAS 013628, 013631, KUN 0175765, NAS NAS00314352–53, PE 01030719–20); Wudong, Xumu Chang, 1400 m, 23 May 1965, *Z. P. Jian et al. 50427* (HGAS 013629, KUN 0178432, PE 01037995, 01037998); betw. Wudong Linchang & Daomao Ping, 26.375°N, 108.175°E, 1450 m, 3 June 1965, *Z. P. Jian et al. 50694* (HGAS 013630, KUN 0175767, PE 01037996–97). **Pu'an Xian:** Qingshan Gongshe, Pubei Linchang, 1800 m, 8 Sep. 1958, *Anshun Team 1380* (HGAS 013659, KUN 0175745,

PE 01031269–70). **Qinglong Xian:** Bifeng, 18 Apr. 1987, *C. J. Yang et al. 54* (GZTM 0011359). **Shiqian Xian:** Fuyan, Foding Shan, 1600 m, 2 Aug. 1988, *Wulingshan Botany Team 2925* (GFS 0010939, KUN 0179110–11, PE 01802417). **Xingren Xian:** Longchang Gongshe, Shanheguan Qu, 1400 m, 21 Aug. 1960, *Guizhou Team – Z. S. Zhang & Y. T. Zhang 8750* (IBSC 0092729, NAS NAS00314356, PE 01031264). **Yinjiang Xian:** Fanjingshan, 1350 m, 20 July 1959, *Guizhou Team 2116* (HGAS 013665, PE 01031251).

10. Berberis brevisepala Hayata, Icon. Pl. Formosan. 3: 14. 1913. TYPE: Chi\na. Taiwan: "Mt Morrison" (protologue) "Central Mountains" (type sheet), 15 Apr. 1910, *U. Mori s.n.* (holotype, TI 02620 image!).

Shrubs, evergreen, to 1 m tall; mature stems pale brown, subterete, sparsely verruculose; spines 3(or 4)-fid, concolorous, 0.4–1.3 cm, weak, abaxially slightly sulcate. Petiole almost absent, leaf blade abaxially pale green, shiny, adaxially green, shiny, narrowly elliptic, oblong-elliptic, or oblanceolate-elliptic, (0.4–)2.3–4.2(–5.5) × 0.8–1.2(–1.7) cm, leathery, abaxially sometimes pruinose, midvein raised abaxially, slightly impressed adaxially, lateral veins and reticulation obscure abaxially, inconspicuous adaxially, base attenuate, margin slightly revolute, spinulose with 2 to 5 widely spaced teeth on each side, apex acute, mucronate. Inflorescence a fascicle, 2- to 8-flowered; pedicel 3–8(–15) mm. Flowers bright yellow. Sepals in 4 whorls; outer sepals mostly reddish, triangular, 2 × 2–2.5 mm; outer median sepals yellow with reddish vertical stripe, broadly ovate, 3 × 3–3.5 mm; inner median sepals orbicular-elliptic, 4.5–5 × 4–4.5 mm; inner sepals broadly obovate, 5.5–6 × 4–5 mm; petals obovate, 4 × 3 mm, base clawed, glands contiguous, ovoid, ca. 1 mm, apex incised. Stamens ca. 3.5 mm; anther connective distinctly extended, truncate. Pistil 4 mm. Ovules 3 to 4. Berry black, epruinose, ellipsoid, ca. 5 × 3.5 mm; style absent or persistent and very short.

Phenology. Berberis brevisepala has been collected in flower in April and in fruit in December.

Distribution and habitat. Berberis brevisepala is known from Hualien, Nantou, Pingtung and Taitung Hsien, and Kaohsiung Shi in Taiwan. It has been found in subalpine meadows and the understories of coniferous forests at ca. 2200–3300 m.

Hayata cited only the type in the protologue of *Berberis brevisepala*. This has leaves (which are sometimes abaxially pruinose) and pedicels, none of which (despite the name given to the species) have flowers and only one of which has a very immature fruit (the Mori collection of the same date at TAIF cited below has three immature fruit but fewer leaves). Although the protologue stated "*Flores non visi*," Schneider (1918: 136–137) gave an incomplete description of flowers based on immature floral fragments apparently from the type sent to him by Hayata. Though indicating the number of ovules to be one, Schneider queried whether this was likely to always be the case. Schneider's description was repeated by Ahrendt (1961: 72) though the number of ovules was given as one without qualification. If Schneider retained the fragments, they would have been lost along with rest of his personal herbarium in the destruction the Berlin-Dahlem herbarium in 1943. Schneider's description was either not noticed or ignored by S. Y. Lu and Yang (1996: 576), who gave a very different description of the flowers of *B. brevisepala*, the source of which was not given.

The uncertainties concerning the flowers of *Berberis brevisepala* could be resolved only by obtaining further specimens from the area where the type was collected. This is given as "Mt Morrison" [Yu Shan] in the protologue, but on the type sheet itself in Japanese simply as "Central Mountains." In 2013 through a close reading of Mori (2000: 407–425), Chih-Chieh Yu of NTU established that, in April 2010, Mori was not on Yu Shan, but had crossed part of the central mountain range via the late Qing dynasty Guanmen Trail. In April 2014, I was privileged to join Chih-Chieh on an expedition to this remote area where apparently no botanists had collected specimens since Mori's visit. Here we found plants in flower with exactly the same leaves as those of the type. These are the source of the description of flowers given above (and as such differ somewhat from the description given by Schneider). For an account of our expedition, see Harber (2015).

Yu and Chung (2014: 76), who give a different account of flowers from that given above, treated *Berberis alpicola* as a synonym and stated that the species is very variable with a wide distributional and elevation range citing various specimens as evidence. Their synonymy of *B. alpicola* is not accepted here because its leaves and flower structure are different from *B. brevisepala*. However, some evidence of its variability comes from a plant I have grown from seed collected on Shinkangshan in Hualien Hsien by C. C. Yu (details below) which flowered for the first time in 2015 and which produced flowers more or less the same as the *B. brevisepala* we found on Guanmen Shan, but with pedicels only 2–6 mm long versus ones of up to 15 mm from Guanmen Shan. The fruit from which the Shinkangshan seed came is pictured in Yu and Chung (2014: 77, fig. 8J).

Yu and Chung (2014) cite as *Berberis brevisepala* specimens from Hsinchu, Taichung, Taitung, and Yilan Hsien. Those that I have seen images of appear to be of other species.

Selected specimens.
Taiwan. Hualien Hsien: Tonkuran River, 15 Apr. 1910, *U. Mori s.n.* (TAIF 9898); Mt. Yangtou, 2800 m, 13 Oct. 2008, *C. C. Yu 162* (TAI); Qing dynasty Guanmen Trail, Guanmen Shan, 2750 m, 12 Apr. 2014, *Guanmen Exped. J. F. Harber & C. C. Yu 18* (E); same details, but 2859 m, *Guanmen Exped. J. F. Harber & C. C. Yu 17* (E); same details, but 2900 m, *Guanmen Exped. J. F. Harber & C. C. Yu 15* (E). **Kaohsiung Hsien (now Shi):** Kuanshanlingshan, Taoyuan Hsiang, trail to Takuanshan, 2940 m, 20 May 1992, *C. C. Wang 1077* (E E00612558, HAST 20790); near Kuanshan, 3100–3300 m, 11 Oct. 1996, *C. K. Liou 407* (TAIF 84335); Taoyuan Hsiang, on the way from campsite to the top of Kuanshan, 23.216667°N, 120.85°E, 3026–3150 m, 16 May 1995, *K. Y. Wang 1096* (HAST 53045). **Nantou Hsien:** Hsinyi Hsiang, Shalihsienhsi forest rd., 23.480278°N, 120.894444°E, 2550–2680 m, 5 May 1998, *C. M. Wang & Y. H. Tsai 03169* (IBSC 0181573, TNM). **Pingtung Hsien:** the last water source along the trail to Peitawushan, 22.615°N, 120.734722°E, 2700 m, 8 May 2009, *C. C. Yu 304* (TAI).

Photographs. Fruit of *Berberis brevisepala* (photographs by C. C. Yu).

Cultivated material:
Living cultivated plants.
Foster Clough, Mytholmroyd, West Yorkshire, U.K., from seed collected by C. C. Yu, Hualien Hsien, E slope of Shinkanshan, ca. 2400 m, 25 Dec. 2011.

Cultivated specimens.
Foster Clough, Mytholmroyd, West Yorkshire, U.K., 28 Apr. 2015, *J. F. Harber 2015-02*, from above plant (E).

11. Berberis calliantha Mulligan, Gard. Chron., ser. 3, 97: 394. 1935. TYPE: China. SE Xizang (Tibet): [Bomi Xian], Tsangpo Gorge, Pemakochung, 2400–2700 m, 21 Nov. 1924, *F. Kingdon-Ward 6308* (holotype, K K00007369!).

Shrubs, evergreen, to 1 m tall; mature stems red-brown, slightly angled; spines 3-fid, brownish yellow, 1–2 cm, sulcate. Petiole 2–5 mm; leaf blade adaxially dark green, very shiny, oblong-elliptic, 2–6 × 1–2.2 cm, leathery, abaxially white pruinose, midvein raised abaxially, impressed adaxially; lateral veins raised abaxially, impressed adaxially, reticulation indistinct on both surfaces, base attenuate, margin spinose with 5 to 10 conspicuous teeth on each side, apex acute, subacute, or subobtuse, mucronate. Inflorescence a fascicle, (1- or)2- or 3-flowered; pedicel 20–40 mm; bracteoles triangular-ovate, 4 × 4 mm. Flowers bright yellow, ca. 2.5 cm diam. Sepals in 3 whorls; outer sepals orbicular-elliptic, 6–7 × 6 mm; median sepals obovate-elliptic, 9–10 × 7–8 mm; inner sepals broadly obovate, 10–12 × 8–10 mm; petals obovate, ca. 8 × 6.5 mm, base clawed, apex entire, glands separate, ovate, marginal, situated toward apex of claw. Stamens 5 mm; anther connective not extended, truncate. Ovules 10 to 15, shortly funiculate. Berry black, pruinose, ovoid, 11–14 × 6–9 mm; style not persistent or persistent and short.

Phenology. *Berberis calliantha* has been collected in flower in May and in fruit in November.

Distribution and habitat. *Berberis calliantha* is known only from Pemakochung at the head of the Tsangpo gorge in Bomi Xian in southeast Xizang at 2400–2700 m and from Tsari in Nang (Lang) Xian at ca. 2400 to 3050 m. The collector's notes to the type describe it as "growing in masses round the marshes in *Rhododendron* thickets and *Alder* copse." The collectors' notes to *Ludlow, Sherriff & Elliot 13605* describe it as "sprawling over the ground in the shade in forests."

Berberis calliantha is an extremely attractive species with unusually large flowers and apparently is known from only the type and the collections cited below. Its sprawling nature is confirmed by an ancient plant apparently grown from seed of the type I saw at Exbury Gardens, Hampshire, U.K., in August 2009 which was some 25 m long and 7 m wide. It was still continuing to spread through suckering, a rare characteristic in *Berberis*.

Though Ying (1985: 143) included *Berberis calliantha* as a species of Xizang, it was not included in Ying (2001, 2011).

Selected specimens.
SE Xizang (Tibet). [**Bomi Xian**]: Tsangpo Gorge, Pemakochung, 2590 m, 2 May 1947, *F. Ludlow, G. Sherriff & H. H. Elliot 13605* (BM, E E00395929). [**Nang (Lang) Xian**]: Migyetun, 2750–3050 m, 6 Aug. 1935, *F. Kingdon-Ward 11911* (BM); Tsari, Migyetun, 28.666667°N, 93.633333°E, 2895 m, 23 May 1936, *F. Ludlow & G. Sherriff 1671* (BM).

12. Berberis candidula (C. K. Schneid.) C. K. Schneid., Bull. Herb. Boissier, sér. 2, 5: 402. 1905. "*Berberis wallichiana* [var.] *pallida* ?," M. Vilm. & Bois, Frutic. Vilmor. 15. 1904. *Berberis hookeri* Lem. var. *candidula* C. K. Schneid., Ill. Handb. Laubholzk. 1: 303. 1904. TYPE: Cultivated, Royal Botanic Gardens, Kew, U.K., "516–1905 Vilmorin," 10 May 1911, from plant supplied by M. Vilmorin, Les Barres, France, from seed collected by P. G. Farges probably at Tchen-kéou tin (Chengkou), NE Sichuan (Chongqing) in 1896 (neotype, designated here, K K000395243!).

Shrubs, evergreen, to 45 cm tall; mature stems yellow, terete, verruculose; spines 3-fid, concolorous, 1–1.5 cm, subterete. Petiole almost absent; leaf blade abaxially pale green, white pruinose, adaxially dark green, very shiny, oblong-lanceolate to lanceolate, 1.2–2.6(–3) × 0.4–1 cm, thickly leathery, midvein raised abaxially, impressed adaxially, lateral veins and reticulation obscure abaxially, inconspicuous adaxially, base cuneate, margin conspicuously revolute, spinulose with 3 or 4 teeth on each side, sometimes entire, apex acuminate, mucronate. Inflorescence 1-flowered; pedicel 4–10 mm. Sepals in 3 whorls; outer sepals sometimes with red tinge to margins, oblong-ovate, 4.5–5 × 2.25–3 mm, apex acute; median sepals oblong-obovate, 5–6.5 × 4–5 mm; inner sepals obovate or obovate-elliptic, 7–8 × 6–7 mm; petals broadly obovate, 6 × 5–6 mm, base cuneate, glands close together, ovate, apex slightly emarginate. Stamens ca. 5 mm; anther connective slightly extended, truncate or obtuse. Ovules 4 or 5. Berry purplish black, thinly pruinose, ellipsoid, 8–9 × 4–5 mm; style not persistent.

Phenology. *Berberis candidula* has been collected in flower in May. Its fruiting season is unknown.

Distribution and habitat. *Berberis candidula* is known from Chengkou Xian in Chongqing (northeast Sichuan) and Fang Xian in northwest Hubei. *E. H. Wilson 2880* (see below) was found on rocks at 2300 m.

Berberis candidula has a somewhat convoluted history. As can been seen above, it first appeared in 1904 as "*Berberis wallichiana, pallida* ?" in a nursery catalogue of Vilmorin as a cultivated plant grown from seed collected by Farges. These plants were named successively by Schneider as *B. hookeri* var. *candidula* and *B. candidula*. Neither then, nor subsequently (1908: 196, 1918: 22, 1942: 55–56) did Schneider indicate any herbarium where a specimen from one of these plants could be found. Nor did Ahrendt (1961: 45) who reported that the seeds were from *Farges 637* (though no evidence to support this has been found). The issue was sidestepped by Chamberlain and Hu (1985: 536) who reported the type to be "*Farges s.n.* (n.v.)." Under these circumstances, a specimen at K, which from the evidence on the sheet appears to have been from a plant supplied by Vilmorin in 1905, has been chosen as a neotype. It is possible that the Vilmorin Herbarium may hold a specimen with better claim to be the type, but currently it is inaccessible. Images of its specimens are scheduled to be eventually available on the Paris Herbarium website.

Schneider (1918: 22) identified *Wilson 2880* as *Berberis candidula*. The collector's notes to this, which are the only source of the species' habitat, describe it as rare.

Wilson 2880 at A has only a few very immature flowers. Restrictions on the destructive dissection of type specimens precludes a description from the neotype at K. The description of flowers above, therefore, comes from a cultivated plant at the Dawyck Garden Royal Botanic Garden Edinburgh (details below). This is the only existing plant located in the U.K. directly supplied by Vilmorin. Why this plant is particularly important is plants described as *Berberis candidula* were once widely found in cultivation in the U.K. and were known to hybridize with *B. verruculosa* producing various cultivars. The description of flowers differs slightly from that given by Ahrendt (1961: 45) and reproduced by Ying (2001: 72, 2011: 723).

Ahrendt (1961: 45) described the height of the species as up to 0.9 m whereas the collector's notes to *Wilson 2880* describe it as "prostrate or nearly so" and only 0.15–0.45 m tall. In 2015, the cultivated plant at Dawyck was 0.6 m tall.

The relationship between *Berberis candidula* and *B. laojunshanensis*, which has very similar leaves but is fascicled rather than single-flowered, needs further investigation, but that requires finding living plants of a species apparently not seen in the wild since 1907.

Berberis candidula is on the current IUCN Red List of Threatened Species (China Plant Specialist Group, 2004b). It is to be hoped it is not extinct.

Selected specimens.
W Hubei. Fang Hsien (Xian): 2300 m, May 17 1907, *E. H. Wilson 2880* (A 00280186, US 00945873).

Cultivated material:
Living cultivated plants.
Royal Botanic Garden Edinburgh, Dawyck, Scotland, from Vilmorin plant (accession *19061029*).

13. Berberis caudatifolia S. Y. Bao, Bull. Bot. Res., Harbin 5(3): 5. 1985. TYPE: China. NE Yunnan: Daguan Xian, Tian Xing Dang, Mugan Linchang, 1950 m, 17 Aug. 1972, *N. E. Yunnan Exp. 0041* (holotype, KUN 1204016!; isotypes, KUN 1204015!, PE 01031273!, YUKU 02065678 image!).

Berberis atrocarpa C. K. Schneid. var. *suijiangensis* S. Y. Bao, Bull. Bot. Res., Harbin 5(3): 5. 1985, syn. nov. TYPE: NE Yunnan: Suijiang Xian, Suijiang, 1500 m, 5 Oct. 1973, *K. M. Feng 73-179* (holotype, KUN 1204013!).

Shrubs, evergreen, to 1 m tall; mature stems pale yellowish brown, terete; spines absent or 1- to 3-fid, pale brown, 0.4–0.6 cm, weak. Petiole almost absent; leaf blade abaxially yellowish green, adaxially dark dull green, narrowly lanceolate, 7–14 × 0.6–2.5 cm, thinly leathery, midvein raised abaxially, impressed adaxially, lateral venation conspicuous abaxially, inconspicuous adaxially, reticulation indistinct abaxially and adaxially, base attenuate, margin spinulose with 2 to 18(to 25) teeth on each side, sometimes entire, apex caudate-acuminate. Inflorescence a fascicle, 1- to 9-flowered; pedicel 14–23 mm. Flowers unknown. Berry black, epruinose, ellipsoid or obovoid, 7–8 × 4–5 mm; style persistent, short; seeds 1 or 2.

Phenology. *Berberis caudatifolia* has been collected in fruit in August and October. Its flowering period is unknown.

Distribution and habitat. *Berberis caudatifolia* is known only from one collection from Daguan Xian, and one from Suijiang Xian in northeast Yunnan. The first was from a mixed forest at 1950 m, the second from an unrecorded habitat at 1500 m.

Ying (2001: 114, 107, 2011: 735, 737) treated *Berberis caudatifolia* as a synonym of *B. gagnepainii*, and *B. atrocarpa* var. *suijiangensis* as a synonym of *B. insolita*. Both of these species, however, have very different shaped leaves.

14. Berberis cavaleriei H. Lév., Repert. Spec. Nov. Regni Veg. 9: 454. 1911. TYPE: China. Guizhou: Kouan Chan (near Guiyang), Apr. 1907, *J. Cavalerie 3209* (lectotype, designated by Ahrendt [1961: 67], K K000077300!; isolectotypes, A fragm. (2) 00038723–24!, E [2 sheets] E00217973!, E00373497!)

Berberis emilii C. K. Schneid., Repert. Spec. Nov. Regni Veg. 46: 255. 1939. TYPE: China. Guizhou: near Guiyang, 30 Mar. 1898, *E. M. Bodinier 2145* (holotype, E E00217980!; isotypes, P [2 sheets] P00716552–53!).
Berberis nemorosa C. K. Schneid., Repert. Spec. Nov. Regni Veg. 46: 246. 1939, syn. nov. TYPE: China. Guangxi: Dar-Siar Ping, Miu Shan, N Luchen, 1200 m, 20 June 1928, *R. C. Ching 6192* (holotype, NY 00000041!; isotypes, NAS [2 sheets] not seen, photographs of NAS specimens E!).
Berberis dolichostemon Ahrendt, J. Linn. Soc., Bot. 57: 59. 1961. TYPE: China. Guizhou: s. loc., s.d., *J. Cavalerie 7852* (holotype, K K000077331!; isotype, E E00217977!).

Shrubs, evergreen, to 2 m tall; mature stems brownish yellow, very sulcate, sparsely black verruculose; spines 3-fid, concolorous, 1–2.5 cm, abaxially slightly sulcate or flat. Petiole 1–2.5 mm; leaf blade abaxially pale whitish green with pinkish tinge on young leaves, drying to pale yellow-green, sometimes pruinose, adaxially yellowish green, slightly shiny, elliptic, oblanceolate, or obovate, 2.5–6 × 1–1.5 cm, thickly leathery, midvein raised abaxially, impressed adaxially, lateral veins slightly raised abaxially, impressed adaxially, reticulation obscure abaxially, inconspicuous adaxially, base cuneate, margin slightly revolute, spinulose with 6 to 15(to 18) teeth on each side, apex obtuse or subacute, mucronate. Inflorescence a fascicle, 5- to 20-flowered; pedicel (5–)8–20 mm, slender; bracteoles narrowly ovate or lanceolate-ovate, 2.5–3 × 1–1.5 mm, apex acuminate. Sepals in 2 whorls; outer sepals ovate or narrowly ovate, 2.75–3 × 1–1.25 mm, apex obtuse or acute; inner sepals obovate or narrowly obovate, 3–4.5

× 2–3 mm, apex subacute; petals obovate, ca. 5–6 × 3.5–4 mm, base cuneate or slightly clawed, glands separate, oblong, ca. 0.75 mm, apex entire or slightly emarginate. Stamens 3.5–4.5 mm; anther connective extended, truncate or obtuse. Ovules 1, shortly funiculate. Berry black, epruinose, oblong, 7–8 × ca. 4 mm; style persistent.

Phenology. *Berberis cavaleriei* has been collected in flower in April and May and in fruit between July and October.

Distribution and habitat. *Berberis cavaleriei* is known from north Guangxi and central, east, and northeast Guizhou. It has been collected in forested areas on limestone hills at 400–1860 m.

The protologue of *Berberis cavaleriei* cited only *Cavalerie 3209* but without indicating the herbarium. Ahrendt (1961: 67) lectotypified the specimen at K.

The protologue of *Berberis cavaleriei* gave only a sparse description of flowers. The most complete description of its flower structure was given by Ahrendt (1961: 67). I have augmented it based on flowers of a cultivated plant in my own collection grown from cuttings of the plant in John Simmons's garden grown from *GUIZ 286* (details below).

The protologue of *Berberis nemorosa* cited only the type gathering and, on the basis of its fruit, gave the number of ovules as one. From the evidence of the holotype, it would appear to differ from the type of *B. cavaleriei* only by having leaves that are abaxially lightly pruinose (only the adaxial surface of the leaves are visible on the photographs of the two isotypes). It has proved impossible to determine the location of Dar-Siar Ping in Guangxi, though it appears from collections of other genera made by R. C. Ching around that time that it is on the border with Guizhou.

The specimens from Suiyang and Zheng'an Xian in Guizhou (details below) have leaves that are abaxially densely pruinose and, from the evidence of a dissection of the flowers of *Y. Wang 503300* by its collector (pers. comm. 17 June 2015), appears to be *Berberis cavaleriei*.

Chamberlain and Hu (1985: 553) were the first to synonymize *Berberis emilii* and *B. dolichostemon* under *B. cavaleriei*. They (Chamberlain & Hu, 1985: 553) also treated two taxa from the Kunming area of Yunnan, *B. praecipua* var. *major* and *B. liophylla* var. *conglobata*, as synonyms of *B. cavaleriei*. These latter two are, in fact, likely to be *B. ferdinandi-coburgii* × *B. pruinosa* (see "*Berberis* sect. *Wallichianae* from the Kunming Area" below). The reports of *B. cavaleriei* in Yunnan by Bao (1997: 42, 44) and Ying (2001: 138, 2011: 745) are presumably based on Chamberlain and Hu synonymizing these two taxa.

Berberis cavaleriei H. Lév. (1915: 48) is an illegitimate name, later corrected to *Gymnosporia esquirolii* in H. Léveillé (1916: 18).

Selected specimens.
N Guangxi. Lingyun Xian: Luolou Gongshe, 1971, *Q. H. Lu 4* (IBK IBK00012847). **Longsheng Xian:** Jizhua Xiang, Xiejiawan, Qingshan, 29 July 1957, *H. F. Qin & Z. T. Li 70878* (IBK IBK00012751, IBSC 0091837).

Guizhou. Pin-fa, s.d., *J. Cavalerie & J. J. Fortunat 2666* (P P02682008); Pin-fa, s.d., *J. Cavalerie 7854* (K K000644862); "inter dganschim et dganping pluries," 1300–1400 m, 25 May 1917, *H. R. E. Handel-Mazzetti 138 (10446)* (A 00280241, W 1924–2192, WU 039235). **Anshun Shi:** near W side of Anshun, 1300 m, 19 Sep. 1959, *Anshun Team 1933* (HGAS 013622, PE 01031632–33). **Guiyang Shi:** Juchang, Haozhiliang, 1300 m, 17 Oct. 1956, *Sichuan-Guizhou Team 2280* (PE 01031621, SZ 00294083); Shuikou Si, 1400 m, 11 Mar. 1959, *Qiannan Team 47* (HGAS 013624, KUN 0178427–28, PE 01031664–65). **Jiangkou Xian:** Niuweihe, Huangqian Gou, 1780 m, 29 May 1964, *Z. S. Zhang et al. 402101* (HGAS 013636, IBSC 0092733, PE 01031628–29). **Kaiyang Xian:** 6 Oct. 1979, *Kaiyang Survey Team 286* (GZTM 0011334–5). **Pingba Xian:** Pingba Wu Qu, Qingchong Gongshe, 1500 m, 22 Sep. 1959, *Anshun Team 1492* (HGAS 013648, KUN 0175804, PE 01031618–19). **Shibing Xian:** Fodingshan, 7 July 1959, *Qiannan Team 2656* (GF 09001041, HGAS 013637, KUN 0178430, PE 01031622–23). **Suiyang Xian:** Kuankuoshui, 640–1300 m, 19 May 2009, *T. Chen 093074* (ZY 0002212); Rangshui Gou, 4 June 2013, *Q. S. Yang s.n.* (GUES 003280); Kuankuoshui, Guochangping, 12 Apr. 2015, *Y. B. Yang & Y. Y. Pu SY-20150263* (GZAC 0031394–96, 0031723). **Yinjiang Xian:** Fanjing Shan, Pingsuo Lin Chang, 2310 m, 19 May 1959, *Guizhou Team 930* (GF 09001044, HGAS 013635, KUN 0178429, 0178892, PE 01031620, 01293904); Suqia Yan, 22 Aug. 1963, *Z. P. Jian et al. 31048* (PE 01031624–25). **Zheng'an Xian:** Bifeng Xiang, Jan. 1987, *s. coll. 86-0495-96* (GZTM); Banzhu, Dingmu, 1715 m, 6 June 2014, *W. L. Shi 520324140606007LY* (GZTM 0043304-05). **Zunyi Shi:** Shanpen Qu, Xianren Shan, Yangshi Yan, 10 Apr. 1959, 1860 m, *Guizhou Team 0111* (HGAS 013650, PE 01031671–73).

Cultivated material:
Living cultivated plants.
Royal Botanic Gardens, Kew, U.K., garden of John Simmons, Gresham, Norfolk, U.K., and Foster Clough, Mytholmroyd, West Yorkshire, U.K., all from *GUIZ 286*, Guizhou, Huaxi, S of Guiyang, 400 m, 28 Oct. 1985.

15. Berberis centiflora Diels, Notes Roy. Bot. Gard. Edinburgh 5: 167. 1912. *Berberis pruinosa* Franch. var. *centiflora* (Diels), Hand.-Mazz., Symb. Sin. 7: 325. 1931. TYPE: China. NW Yunnan, E flank of Dali Range, 2700–3300 m, June–Aug. 1906, *G. Forrest 4689* (lectotype, designated here, E E00217983!; isolectotype E E00217984!).

Shrubs, evergreen, to 1.5 m tall; mature stems pale yellow, terete, not verruculose; spines 3-fid, concolorous, ca. 1 cm, slender. Petiole 1–4 mm; leaf blade abaxially pale green, adaxially yellow-green, shiny, oblong-obovate or oblanceolate, 3–7 × 1–3 cm, leathery, midvein raised abaxially, slightly impressed adaxially, lateral veins and reticulation obscure abaxially, inconspicuous adaxially, base cuneate, margin sometimes slightly revolute, spinose with 3 to 7 teeth on each side, apex obtuse, mucronate. Inflorescence a fascicle, (15- to)20- to 30-flowered, often mixed with 2- or 3-flowered racemes, 5–6 cm overall; pedicel 15–25 mm; bracteoles ca. 2 × 1.5 mm. Sepals in 2 whorls; outer and inner sepals obovate, equal in size, ca. 5.5 × 4 mm; petals obovate, ca. 6.5 × 5 mm, base clawed, glands separate and lanceolate, apex slightly retuse. Stamens ca. 5 mm; anther connective extended, short-apiculate. Ovules 4 or 5. Berry unknown.

Phenology. The type of *Berberis centiflora* was collected in flower sometime between June and August. Its fruiting season is unknown.

Distribution and habitat. Berberis centiflora is known only from the type. The collector's notes described it as being from "open situations in *Pinus* and *Rhododendron* forests" at 2700–3300 m.

Chamberlain and Hu (1985: 550) cited "*Forrest 4689* (holo. E)." However, there are two specimens of *Forrest 4689* at E. I have lectotypified the one with the most material.

Berberis centiflora is the only known *Berberis* sect. *Wallichianae* species whose inflorescence is not either single-flowered or fascicled. As Schneider (1939: 251) noted, the presence of flowering racemes may indicate that this is a natural hybrid (possibly between *B. pruinosa* and a non-section *Wallichianae* species). I have located no similar specimens beyond the type, even after examining all the *Berberis* sect. *Wallichianae* specimens from the Dali area at KUN. However, *S. E. Liu 20587* (KUN 0175810–11), Dali, 12 Mar. 1946, identified as *B. centiflora* but without any racemes and appearing, in fact, to be *B. pruinosa*, does have an unusually large number of fascicles for that species. *T. T. Yu 8113* (A 00279384, KUN 0176833, 0176955, PE 01031741–42), Lijiang Xian, Wenbi Shan, 2500 m, 12 Apr. 1957, also has an unusually large number of fascicles, though again without any racemes. However, interestingly an annotation to the A specimen records

four ovules (vs. the two or three of *B. pruinosa*). One specimen at PE identified as *B. centiflora*, *B. Y. Qiu 55164* (01033759), central Yunnan, Songming Xian, Shaodian Qu, 2400 m, 21 Sep. 1957, lacks evidence of racemes but does have an inflorescence which is partially very distinctly umbellate. This, too, may be a natural hybrid of some sort but may be just an unusual form. However, both it and *Forrest 4589* may simply be unusual forms of *B. pruinosa* which is a very varied species.

16. Berberis chingii W. C. Cheng, Contr. Biol. Lab. Sci. Soc. China, Bot. Ser. 9: 191. 1934; *Berberis cavaleriei* H. Lév. var. *pruinosa* Bijh., J. Arnold Arbor. 9: 132. 1928. TYPE: China. S Anhui: [? N Jiangxi], Wuyuan, Chang gon shan, 17 Aug. 1925, *R. C. Ching 3248* (holotype, A 00038722!; isotypes, E E00217985!, IBSC 0000584 image!, K K000567990!, NAU 0000584 image!, PE [2 sheets] 01031947!, 01863962!, SYS SYS00052171 image!).

Shrubs, evergreen, to 2.5 m tall; mature stems pale yellow, subterete or slightly angled, sparsely black verruculose; spines 3-fid, concolorous, 1–2.5 cm, stout. Petiole 2–4 mm; leaf blade abaxially yellowish green, often white pruinose, adaxially dark green, sometimes shiny, oblong-oblanceolate, elliptic-oblanceolate, or narrowly obovate-elliptic, 2–8 × 0.8–2.5 cm, thinly leathery, midvein raised abaxially, impressed adaxially, lateral veins and reticulation obscure abaxially, inconspicuous adaxially, base cuneate or attenuate, margin entire or spinulose with 2 to 10(to 20) teeth on each side, sometimes confined to upper half of leaf, apex acute, mucronate. Inflorescence a fascicle, 4- to 14-flowered; pedicel 7–12 mm; bracteoles triangular, 3 × 2.5–3 mm. Sepals in 2 whorls; outer sepal broadly elliptic, 5–6 × 3.5–4 mm; inner sepals obovate-oblong, 6.5–7 × 5–6 mm; petals obovate, 5–5.5 × 2.5–3 mm, base clawed, glands 0.5 mm, very close together, apex emarginate. Stamens 4.5–5 mm; anther connective extended to ca. 0.5 mm, obtuse. Ovules 2 or 3. Berry dark blue-black, pruinose at first. Mature fruits epruinose, ellipsoid-oblong, ca. 7 × 4 mm; style persistent.

KEY TO SUBSPECIES OF *BERBERIS CHINGII*

1a. Pedicel 7–12 mm; leaves oblong-oblanceolate, elliptic-oblanceolate, or narrowly oblong-elliptic, abaxially sometimes pruinose, margin entire or spinulose with 1 to 4 teeth on each side usually confined to upper half of leaf . . . . subsp. *chingii*
1b. Pedicel 12–18 mm; leaves consistently oblong-elliptic, abaxially epruinose, margin spinulose with up to 20 teeth on each side along full length . . . . subsp. *wulingensis*

16a. Berberis chingii subsp. **chingii**

Berberis chingii W. C. Cheng subsp. *subedentata* C. M. Hu, Bull. Bot. Res., Harbin 6(2): 9. 1986. TYPE: China. SE

Hunan: Yanling Xian, Qingshigang Linchang, Qingshihongchang, Tiedingba, 900 m, July 1977, *Hunan Forestry College 77-0055* (holotype, IBSC 0092068 image!; isotype, CSFI CSFI004684 image!).

Pedicel 7–12 mm; leaves oblong-oblanceolate, elliptic-oblanceolate, or narrowly oblong-elliptic, abaxially sometimes pruinose, margin entire or spinulose with 1 to 4 teeth on each side usually confined to upper half of leaf margin.

Phenology. *Berberis chingii* subsp. *chingii* has been collected in flower from March to April and in fruit between May and November.

Distribution and habitat. *Berberis chingii* subsp. *chingii* is known from Jiangxi, northwest Fujian, Chongyang and Tongcheng Xian in southeast Hubei, Pingjiang Xian in northeast Hunan, and Yanling Xian in southeast Hunan. It has been found in mixed forests, thickets, on runlet sides, and on limestone slopes at ca. 500–1700 m.

The protologue of *Berberis cavaleriei* var. *pruinosa* which designated *R. C. Ching 3248* as the type was clear that this was a specimen at A. Since W. C. Cheng published *B. chingii* as a nom. nov. for this taxon, it follows that the holotype is the specimen at A. Chamberlain and Hu's description of the specimen at PE as the holotype was therefore mistaken. The collection details of *R. C. Ching 3248* stated it was collected in Chang gon Shan, Wanyuan, south Anhui. I have not located Chang gon Shan but Wanyuan is in northeast Jiangxi and I have followed Hu (1986: 8) in assuming that south Anhui was a mistake. Ying (2011: 748) mistakenly gave the author of *B. chingii* as S. S. Cheng rather than W. C. Cheng.

The reason given by Hu for recognizing *Berberis chingii* subsp. *subedentata* was that the leaves are adaxially particularly shiny with margins that are either entire or with few teeth; however, many of the leaves on the type of *B. chingii* exhibit exactly these characteristics. *M. H. Nieh 8401* (collection details below), cited in the protologue as another example of subspecies *subdentata*, would also seem to be a typical gathering of *B. chingii*. Subspecies *subedentata* was synonymized by Ying (2001: 146).

The protologue of *Berberis chingii* cited only the type which has only fruit. Hu (1986: 9) gave a description of flowers for subspecies *wulingensis*. I have augmented this from flowers of the MO specimens of *C. M. Tan 9604118* (details below).

The leaves of *Berberis chingii* subsp. *chingii* are very variable. This is evidenced by the six specimens of the type *R. C. Ching 3248* cited above which not only display a variety of shapes—oblong-oblanceolate, elliptic-oblanceolate, and narrowly oblong-elliptic—but include leaves which abaxially are on a range from densely pruinose through epruinose.

Berberis chingii subsp. *subedentata* was listed in error as a variety of *B. lempergiana* rather than of *B. chingii* by Jin and Chen (1994: 162).

Selected specimens.
Fujian. Chong'an Xian (now Wuyishan Shi): Santa, Xinglu Shan, 1700 m, 20 Apr. 1955, *M. J. Wang 3444* (NAS NAS00313891–93, PE 01031948). **Dehua Xian:** Shiniushan, 1700 m, 2 Dec. 1999, *G. S. He 9537* (CAS 1020114). **Taining Xian:** Xinqiao Gongshe, 1400 m, 18 June 1978, *G. L. Cai 570* (FJSI 008973, IBSC 0092069, KUN 0178639).

SE Hubei. Chongyang Xian: 30 Mar. 1959, *S. Y. Liang 700* (HIB 00129555); 24 Nov. 1972. *s. coll. 22* (HIB 00129467, 0129529); Hongxia, Songbai Qidui, Chenjiawan, 510 m, 11 July 1976, *South Hubei Collection Group 3507* (WH 06008747). **Tongcheng Xian:** Shadui, Shouxiandong, 760 m, 4 May 1959, *S. Y. Liang 1018* (HIB 00129554); 12 Nov. 1972, *s. coll. 28* (HIB 00129397).

NE Hunan. Pingjiang Xian: Sanshi Gongshe, 13 Mar. 1976, *Drug Testing Team; Medicinal Plant Collecting Group 35* (HUFD 0002543).

SE Hunan. Yanling Xian: Zhidu Zhen, Mihua Cun, 700–800 m, 17 Oct. 1998, *B. Xiong 4911* (MO 04552983); Zhidu Zhen, 600–700 m, 15 May 1999, *B. Xiong 7043* (MO 04691693). **Zixing Xian:** Yanping Xiang, 900–1000 m, 24 July 1999, *B. Xiong 8056* (MO 04696173).

Jiangxi. Anfu Xian: Wugong Shan, Yangshiji, 1600 m, 16 Aug. 1963, *J. S. Yue et al. 3577* (IBSC 0091465–66, KUN 0175823, PE 01031941); Wugong Shan, 4 Aug. 1999, *Y. G. Xiong 8587* (MO 04692713). **Dexing Shi:** Sanqing Shan, 1600 m, 2 June 1991, *G. Yao & R. P. Jiang 11787* (NAS NAS00315816–17). **Fengxin Xian:** Gangfan, Tongji West Pagoda, 11 Oct. 1982, *S. S. Lai 001059* (LBG 00113739). **Ji'an Shi:** Donggu, Kenzhichang, 30 May 1959, *S. S. Lai 793* (PE 01031935). **Jing'an Xian:** Zhouken, Sanzhualun, 800 m, 21 Oct. 1982, *S. S. Lai 001274* (LBG 00113737); Sanzhualun, 800 m, 27 May 1995, *C. M. Tan 95320* (MO 04542798, PE 01815904). **Jingganshan Shi:** Bamian Shan, 1050 m, 27 Oct. 1963, *J. S. Yue et al. 5485* (NAS NAS00313889, PE 01031946). **Lichuan Xian:** 1200 m, Desheng Guan, Kenzhichang, 1200 m, 24 June 1958, *M. X. Nie & S. S. Lai 3041* (FUS 00014252, KUN 0175819, 0175821, LBG 00013842, PE 01031936, SHM 0014253); Jiufang, 1100 m, 24 Mar. 1963, *X. X. Yang et al. 650022* (IBSC 0092244, PE 01031934). **Shangyou Xian:** Guangushan, 760 m, 23 May 1965, *M. H. Nie 8401* (IBSC 0696705, KUN 0175822). **Suichuan Xian:** Dafen Qu, Linyang, Xiaoshui, 1200 m, 24 Sep. 1963, *J. S. Yue et al. 4291* (NAS NAS00313888, NAS00313904, PE 01031940). **Tonggu Xian:** Zhixi Xiang, 4 May 1998, *B. Xiong 1563* (MO 04552086). **Wuning Xian:** Jiulingshan, 500 m, 20 Apr. 1996, *C. M. Tan 9604118* (IBSC 9604118, MO 04754982, 0543035, NAS NAS00315803). **Xingguo Xian:** Xingfozi Shan, 760 m, 18 Aug. 1964, *W. H. Wan & Z. R. Yu 1837* (PE 01031942, 01031945); Junfu Shan, Kenzhichang, 840 m, 10 May 1960, *Institute of Botany demoted cadres 201* (PE 01031943–44); Junfu Gongshe, Laocaowo, 8 Sep. 1960, *Institute of Botany; demoted cadres 644* (PE 01031939). **Xiushui Xian:** Longgang, 900 m, 10 Nov. 1995, *C. M. Tan 951079* (IBSC 0091825, MO 04543552); Huangluo, 400 m, 15 Nov. 1996, *L. X. Li 9611001* (JJF 00002188, KUN 0175820, PE 01839982, TNM S52216). **Yanshan Xian:** Wuyi Shan, Yejiachang, 14 June 2009, *C. M. Tan 09430* (CCAU 0001992, JJF 00012958, SZG 00006584). **Yifeng Xian:** Gongshan Nature Reserve, 800 m, 24 Apr. 1997, *S. S. Lai & H. R. Shan 1735* (MO 04516543).

16b. Berberis chingii subsp. **wulingensis** C. M. Hu, Bull. Bot. Res., Harbin 6(2): 9. 1986. TYPE: China. N Guangdong: Ruyuan Xian, Dalingjiao, 23 Mar. 1934, *S. P. Ko 53901* (holotype, IBSC 0000585 image!; isotype, IBK IBK00012817 image!).

Pedicel 12–18 mm; leaves consistently oblong-elliptic, abaxially epruinose, margin spinulose with up to 20 teeth on each side along full length.

Phenology. *Berberis chingii* subsp. *wulingensis* has been collected in flower in March and April and in fruit between May and December.

Distribution and habitat. *Berberis chingii* subsp. *wulingensis* is known from north Guangdong, southeast Guizhou, and north, northwest, and west Hunan. It has been collected on open and rocky areas on valley sides, roadsides, sparse forests, and limestone and sandy areas at ca. 200–800 m.

Hu differentiated *Berberis chingii* subsp. *wulingensis* from *B. chingii* subsp. *chingii* largely on the basis that it had longer pedicels. However, the differences are more substantial than this. These include the leaves being consistently oblong-elliptic with teeth along the length of the margin versus the more variably shaped and often narrower leaves of *B. chingii* subsp. *chingii*, which are sometimes entire or mostly with spines on the upper half. There is also a very distinct geographical separation with subspecies *chingii* being found in Fujian and Jiangxi and in Hunan only in the southeast on the border with Jiangxi. Subspecies *wulingensis* is not recorded growing above 800 m and mostly at elevations lower than this, while *B. chingii* subsp. *chingii* is recorded largely above 800 m.

H. R. E. Handel-Mazzetti 11272/438 and *11825/552* listed below were cited as syntypes of *Berberis praecipua* by Schneider (1939: 248) along with two collections of *R. E. Cooper* from Bhutan and *J. F. Rock 3089* from west Yunnan. This was to conflate three different species. Ahrendt (1961: 42) lectotypified one of the Bhutan species thus making *B. praecipua* the name of a species which is endemic to that country. *J. F. Rock 3089* is *B. sublevis*. Schneider rarely made mistakes of this order.

Subspecies *wulingensis* was synonymized as *Berberis chingii* by Ying (2001: 146). It was listed in error as a variety of *B. lempergiana* rather than of *B. chingii* by Jin and Chen (1994: 162).

Selected specimens.
N Guangdong. Ruyuan Xian: Daiqiao Qu, Yuanjiu (or Zhijiu) Tang, 4 June 1933, *S. P. Ko 52783* (A 00280236, IBSC 0091614, 0091618); Dalingjiao, Dalikou, 20 Oct. 1933. *S. P.*

Ko 53500 (IBK IBK00012843, IBSC 0991617); Xishan Xiang, near Zhedong, 13 Dec. 1957, *Z. Huang 44436* (IBSC 0091835, KUN 0176415, MO 3748153).
NE Guangxi. Quanzhou Xian: Dongshan Xiang, Jinchashuang Shui, 617 m, 24 Apr. 2005, *Y. D. Peng & B. Y. Huang 16125* (GXMG 19583).
SE Guizhou. Liping Xian: Maogong Qu, Gongzhai Gongshe, 680 m, 24 June 1981, *J. M. Yuan 00345* (HGAS 013619). **Tianzhu Xian:** Shidong Xiang, Huaiyang, 7 Oct. 1986, *H. G. Yang 109* (GZTM 0011369).
N Hunan. Cili Xian: Qi Qu, Fuwu Xiang, 18 Apr. 1955, *G. X. Zhu 035* (IBK IBK00012773, IBSC 0091841); Nanshan Xiang, Sanhe Cun, 21 Apr. 1955, *G. X. Zhu 054* (IBK IBK00012842, IBSC 0091842). **Taoyuan Xian:** 6 Oct. 1974, *s. coll. 1211* (PE 01031321).
NW Hunan. Fenghuang Xian: Liangtouyang, Tianxing Shan, 700 m, 20 Sep. 1988, *Wuling Team 1441* (IBSC 0091852, PE 01376125, WUK 0091852). **Huayuan Xian:** near Huayuan, 10 Aug. 1953, *West Xiang Survey Team 0552* (HIB 00129561, IBSC 0091847, PE 01582518). **Longshan Xian:** Tingzibao, 650 m, 18 Apr. 1958, *L. H. Liu 1442* (IBSC 0092251, PE 01031511). **Yongshun Xian:** Zejia Xiang, Baiyang Cun, 28.815556°N, 109.741278°E, 277 m, 21 Mar. 2015, *K. D. Lei 4331271503211589* (JIU 12252).
W Hunan: betw. Linling & Sinning (Xinning) in vic. of Wangdjiapu (Huangtianpu) & Tjintiesse, 200–500 m, 15 Aug. 1915, *H. R. E. Handel-Mazzetti 11272/438* (A 00280249, W 1924–0002193, WU 0039265); Hsikwangschan (Xikuangshan) mines near Hsinhwa (Xinhua), 600–800 m, 14 May 1918, *H. R. E. Handel-Mazzetti 11825/552* (A 00280248, W 1924–0002194, WU 0039266). **Dong'an Xian:** Shunhuang Shan, Gaoyan, 500 m, 13 Sep. 2004, *J. K. Liu 691* (H 1761113, LD 1390683, P P03396411, PE 01839967, US 00945895). **Lingling Qu:** Yangming Shan, Ma'an Ling, 250 m, 4 Apr. 1942, *S. Q. Chen 399* (IBK IBK00012752, IBSC 0091620, MO 4098909, URV 004268). **Qianyang Xian (now Hongjiang Shi):** Er Qu Qigong Tan, 12 May 1954, *Z. T. Li 2051* (IBSC 0091846, PE 01031806). **Shaoyang Shi:** Beita Qu, 450 m, 12 Mar. 2004, *L. D. Duan 0640* (PE 01839954, TNM S117759). **Wugang Shi:** Jiupo Ling, 600 m, 2 Aug. 1987, *s. coll. 901* (PE 01605719). **Xinning Xian:** Ziyunshan, Tongmuchong, 26.4°N, 110.8°E, 400 m, 24 Apr. 1996, *Z. C. Luo 1317* (MO 5330488, P P02313669, PE 01840086, UPS BOT.V-152967, US 00946012). **Xupu Xian:** Dajiangkou, 300 m, Apr. 1980, *Q. Z. Lin 10754* (CSFI CSFI004648, CSFI004654–60, IBSC 0091622). **Zhijiang Xian:** Dashu'ao Xiang, Zhupo, 450 m, *Wuling Shan Survey Team 2459* (IBSC 0091851, PE 01376126).

17. Berberis chingshuiensis T. Shimizu, J. Fac. Text. Sci. Techn., Shinshu Univ. No. 36, Biol., No. 12 (Stud. Limest. Fl. Jap. & Taiwan, Pt. 2): 29. 1963. TYPE: China. Taiwan: Pref. Hualien, summit of Mt. Chingshui, ca. 2400 m, 1 May 1961, *T. Shimizu 12520* (holotype, KYO KYO00022300 image!; isotypes, SHIN not seen, TI 02621 image!).

Shrubs, evergreen, to 1 m tall; mature stems brown, subterete, not or scarcely verruculose; spines 3-fid, semi-concolorous, 0.5–1.2 cm, terete. Petiole almost absent; leaf blade abaxially pale green, adaxially green, elliptic, 1.4–5.7 × 0.8–1.7 cm, leathery, abaxially sometimes lightly pruinose, midvein raised abaxially, slightly impressed adaxially, lateral veins slightly raised and inconspicuous abaxially, slightly raised and inconspicuous adaxially, reticulation obscure abaxially, indistinct adaxially (inconspicuous when dry), margin sometimes slightly revolute, spinulose with 3 to 10 teeth on each side, base attenuate or cuneate, apex acute, mucronate. Inflorescence a fascicle, 2- to 6(to 10)-flowered; pedicel reddish, 8–14 mm; bracteoles sometimes reddish tinged, triangular, ca. 1.5 × 1 mm. Flowers yellow. Sepals in 3 whorls; outer sepals ovate, partially red, 2.5 × 2 mm; median sepals elliptic; inner sepals obovate, 7 × 6 mm; petals obovate, 5.4 × 4 mm, base clawed, glands ovoid, close together, apex emarginate. Stamens ca. 4 mm; anther connective extended, truncate. Pistil 4 mm; ovules 1 to 4, funiculate. Berry black, ellipsoid, ca. 5.5 × 4 mm, epruinose; style persistent.

Phenology. *Berberis chingshuiensis* has been collected in flower in April and May and with immature fruit in May.

Distribution and habitat. *Berberis chingshuiensis* is known from a small number of mountain peaks in the Taroko area of north Hualian Hsien in Taiwan. It has been found on exposed limestone outcrops and in crevices at ca. 1400–2550 m.

The description of the color, shape, and size of the fruit of *Berberis chingshuiensis* is taken from Yu and Chung (2014: 78), though they cite no relevant specimens.

Berberis chingshuiensis was not noticed by Chamberlain and Hu (1985). It was treated as a synonym of *B. kawakamii* by T. S. Liu (1976: 16) and Ying (2001: 133; 2011: 744).

Selected specimens.
Taiwan. Hualien Hsien: Chingshuishan, 2400 m, 28 Apr. 1989, *S. Y. Lu s.n.* (E E00085144, HAST 61119, TNM); Chingshuishan, 2300 m, 11 Apr. 2009, *S. W. Chung 9577* (TAIF 311213); Great Cliff Mtn. area, 24.184675°N, 121.538361°E, 1400 m, 28 ? 2009, *C. C. Yu 484* (TAI); Shioulin Village, Badagun Shan, 24.196161°N, 121.566536°E, 1910 m, 6 Apr. 2012, *C. C. Yu 846* (TAI); Shioulin Village, near Triangle Peak, 24.208319°N, 121.586683°E, 2550 m, 8 Apr. 2012, *C. C. Yu 852* (TAI).

18. Berberis chrysosphaera Mulligan, Bull. Misc. Inform. Kew 1940: 77. 1940. TYPE: China. SE Xizang (Tibet): Zayü (Chayu) Xian, Sri-La, Rong-to valley, 2700–3000 m, 10 Dec. 1933, *F. Kingdon-Ward 11036* (holotype, BM BM000559450!).

Shrubs, evergreen, to 60 cm tall; mature stems reddish brown, sulcate, very sparsely verruculose; spines 3-fid, concolorous, 1–2 cm, sulcate. Petiole almost absent; leaf blade abaxially white pruinose, waxy at first, adaxially dark green, very shiny, narrowly elliptic or narrowly obovate-elliptic, 1.5–4 × 0.4–1 cm, leathery, midvein raised abaxially, impressed adaxially, lateral veins and reticulation largely obscure on both surfaces, base attenuate, margin conspicuously revolute, spinulose with 5 to 12 teeth on each side, apex acuminate, mucronate. Pedicel 18–25 mm. Flowers 1. Sepals in 3 whorls; outer sepals ovate, ca. 4 × 3 mm, apex acute; median sepals obovate-elliptic, ca. 7 × 6 mm; inner sepals ca. 10 × 8 mm; petals obovate, ca. 8 × 6 mm, glands oblong, apex slightly emarginate. Stamens ca. 5 mm; anther connective slightly extended, truncate. Ovules 9 to 12. Berry black, slightly pruinose, ovoid, ca. 10 × 6 mm; style not persistent or persistent and short.

Phenology. The flowering time of *Berberis chrysosphaera* is unknown as is the extent of the fruiting season outside December.

Distribution and habitat. *Berberis chrysosphaera* is known from only the type collection in Zayü (Chayu) Xian in southeast Xizang and is described in the collector's notes as growing on granite cliffs in the open or among scrub and trees at 2700–3000 m.

This is a very distinctive and attractive species. An account of its collection is in Kingdon-Ward (1934: 277). The description of flowers is from cultivated plants in the U.K. grown from the seeds of the type. It appears there have been no *Berberis* collections made in this remote area since Kingdon-Ward.

19. Berberis coxii C. K. Schneid., New Fl. & Silva x. 257. 1938, in obs., anglice; et in Fedde, Repert. Spec. Nov. Regni Veg. 46. 261. 1939, latine. TYPE: Burma (Myanmar). Hpimaw (Pianma) Pass [now part of China], 3200 m, 13 June 1919, *R. Farrer 1030* (lectotype, designated by Ahrendt [1961: 50], E E00077797!).

Shrubs, evergreen, to 2 m tall; mature stems pale brownish yellow, terete, finely brown verruculose; spines 3-fid, concolorous, 0.6–2 cm. Petiole 3–10 mm; leaf blade abaxially very pale green, white pruinose, adaxially dark green, very shiny, elliptic to ovate-elliptic, (3–)4–6(–7) × (1.2–)1.6–2.6(–3) cm, leathery; midvein raised abaxially, impressed adaxially, lateral veins inconspicuous abaxially, conspicuous adaxially, reticulation largely obscure abaxially, inconspicuous adaxially, base attenuate, margin slightly revolute, spinulose with 8 to 11 teeth on each side, apex subacute or obtuse, sometimes minutely mucronate. Inflorescence a fascicle, 3- to 6-flowered; pedicel 10–15(–20) mm. Sepals in 3 whorls; outer sepals sometimes reddish on the outside, oblong-ovate, ca. 4 × 2.2 mm, apex acute to subacuminate; median sepals oblong-ovate or oblong-obovate, 6–7 × ca. 3.5 mm; inner sepals obovate, 8–9 mm; petals obovate, ca. 6.5 × 5.5 mm, base clawed, glands very close together, oblong. Stamens 3–3.5 mm; anther connective not extended, truncate. Ovules 2 to 4. Berry purplish black, slightly blue pruinose, oblong-ellipsoid, 11–12 × 5–6 mm.

Phenology. *Berberis coxii* has been collected in flower in May and June and in fruit in October.

Distribution and habitat. In China *Berberis coxii* is known from Lushui Xian in northwest Yunnan on the border with Myanmar. The extent of its distribution in Myanmar is unknown. In China it has been found on hemlock forest margins at 2400–3500 m.

The protologue of *Berberis coxii* cited *R. Farrer 1030* and *F. Kingdon Ward 3175*. Ahrendt (1961: 50) lectotypified the former at E.

Pianma had been part of China, but was incorporated into Burma after being occupied by British troops in 1913. It was not restored to China until 1960, see Woodman (1962: 494–517) and Maung (2011: 49). Four of the additional collections from Lushui Xian cited below are from near where *Farrer 1030* was collected. The collectors' notes for no. *311* describe it as locally common.

The description of the color of the outer sepals comes from photographs taken in the field at Pianma on 9 May 2015 by Yong Wang of GZTM.

Berberis coxii was not included by Ying (2001, 2011).

Syntype. Myanmar, valley of the Chawng Maw Hka, 2440 m, 4 June 1919, *F. Kingdon-Ward 3175* (E E00117987).

Selected specimens.

NW Yunnan. Bijiang Xian (now partially incorporated into Lushui Xian): Daimuchang, 3300 m, 14 June 1976, *Nujiang Botany Team 556* (KUN 0175671–72, YUKU 02065786). **Lushui Xian:** W Gaoligong Shan, Pianmagang, 2800 m, 27 May 1981, *Beijing Hengduan Mountains Botanic Team 144* (PE 01293929, 02046497); W Gaoligong Shan, 3500 m, 31 May 1981, *Beijing Hengduan Mountains Botanic Team 311* (PE 0129330, 02046548); Gaoligong Shan, 3 June 1981, *K. Y. Zhao 140* (IGA 0002433); Luzhang Zhen, Pianma Yakou, pass betw. Pianma & Lushui, crest of Gaoligong Shan, 3250 m, 15 Oct. 2002, *Gaoligong Shan Biodiversity Survey 15956* (CAS 00120138, E E00625280); Pianma, 3000 m, 9 May 2015, *Y. Wang 0502* (GZTM).

20. **Berberis davidii** Ahrendt, J. Linn. Soc., Bot. 57: 56. 1961, replacement name for *Berberis densa* C. K. Schneid., Repert. Spec. Nov. Regni Veg. 46: 254. 1939, not Planch. & Linden (1862). *Berberis wallichiana* DC. f. *parvifolia* Franch., Bull. Soc. Bot.

France 33: 388. 1886. TYPE: China. NW Yunnan, "Au pied du Mt Tsang-chan [Cangshan], S./ Tali [Dali]," 2000 m, 28 Mar. 1884, *J. M. Delavay 1124* (lectotype, designated here, P P00716575!; isolectoypes, A fragm. 00038819!, K missing, LE fragm.!, P [2 sheets] P00716576–77!).

Shrubs, evergreen, to 1.5 m tall; mature stems pale yellow, sulcate, not verruculose; spines 3-fid, concolorous, 1–2 cm, abaxially sulcate. Petiole ca. 1 mm; leaf blade abaxially pale green, slightly shiny, adaxially green, narrowly elliptic or elliptic-lanceolate, rarely lanceolate, 1–3.5 × 0.2–0.6 cm, leathery, midvein raised abaxially, impressed adaxially, lateral veins and reticulation inconspicuous on both surfaces, base cuneate, margin revolute, spinulose with 3 to 6(to 8) teeth on each side, apex obtuse, mucronate. Inflorescence a fascicle, 2- to 5(to 8)-flowered; pedicel 5–12 mm; bracteoles triangular-ovate, apex aristate, mucronate. Sepals in 2 whorls; outer sepals broadly ovate, ca. 2 × 2 mm; inner sepals broadly ovate, 4–4.5 × ca. 4 mm; petals obovate, ca. 3 × 2 mm, base attenuate, glands separate, apex rounded, entire. Stamens ca. 2 mm; anther connective extended, notched. Ovules 2. Berry purple-black, gray pruinose at first, ellipsoid, 8–9 × ca. 7 mm; style not persistent.

Phenology. *Berberis davidii* has been collected in flower in March and April and in fruit between May and March the following year.

Distribution and habitat. *Berberis davidii* has been collected from the foot of the eastern flank of the Cangshan range in Dali Shi, and Weixi, Yongping, and possibly Gongshan Xian (see discussion below). It has been found on streamside and roadside banks at 2000–ca. 3350 m.

The protologue of *Berberis wallichiana* f. *parvifolia* cited only *Delavay 1124* of 28 Mar. 1884 and this was repeated in Franchet (1889: 38). Subsequently, Schneider (1939: 254) published *B. densa* as a "spec. nov., - B. Wallichiana* f. *parvifolia* Franchet, Pl Delav. 38 (1889)" not designating a type but citing *Delavay 1124* and the four other collections listed under syntypes below. Later (1942: 32), Schneider stated *Delavay 1124* to be the type. However, *B. densa* was actually an illegitimate name hence its subsequent replacement by Ahrendt as *B. davidii*. There are three specimens of *J. M. Delavay 1124* at P. The lectotype has been chosen because it has both floral material as well as fruit from the previous year. Ahrendt (1961: 56) cited a specimen of *1124* at K, but this is missing. It is possible that *Delavay s.n.* (P P06868280), listed below and which is labelled *B. wallichiana* DC. var. *parvifolia*, is an isolectotype, but this has no collection details beyond "1883–1885."

There are identification issues in relation to *Berberis davidii*. The type was collected on the plain between the Erhai Lake and Cang Shan as was *Sino-British Exp. Cangshan 1105*. Whether the species is still there is uncertain given the massive urban development of the area since no. *1105* was collected. Various specimens at higher elevations on Cang Shan itself have been identified as *B. davidii*, but those investigated have proved to be *B. wui* whose leaves at the apex of stems of the form on Cang Shan can resemble those of *B. davidii*, but whose flowers are different including having three or four ovules. It is in this context that the syntype *Forrest 7276* listed below should be seen, which was collected on Cang Shan much higher than the type, has longer pedicels, and may, in fact, be *B. wui*.

Of the other syntypes, both *Forrest 23599* and *Forrest 28167* have two seeds per fruit and appear to be *Berberis davidii*, though interestingly the former was collected at a higher elevation than *Forrest 7276*. However, *C. W. Wang 66201* from southeast Tibet is not *B. davidii* and is treated here as the type of a new species, *B. nujiangensis*.

Delavay 1124 and *Sino-British Exp. Cangshan 1105* are recorded as being 60 cm and 20–30 cm respectively, while *Forrest 23599* is recorded as being up to 1.5 m high. Unfortunately, the collection details of *Z. D. Fang et al. W0251* (see below), which has flowers exactly the same as the type of *Berberis davidii*, does not record the height of the plant.

It is likely that the living plant of *M. Foster 96129* (details below) is *B. davidii*, though its flower structure is slightly different being as follows: "bracteoles in 1–2 whorls, triangular-ovate with vertical red stripe, 1.75–3 × 1.5–2.5 mm. Sepals in 2 whorls; outer sepals elliptic-ovate, 3.5–4.75 × 3–4 mm; inner sepals obovate-elliptic, 6–7 × 4.5–5 mm; petals obovate, 5–5.5 × 3.5–4 mm, base slightly clawed, glands separate, ovoid, ca. 0.4 mm, apex entire. Stamens 4.5 mm; anther connective extended, truncate or obtuse. Ovules 2 or 3" (my dissection). The collector's notes record the seed as being from a plant 1 m tall.

Chamberlain and Hu (1985: 543) stated that the leaves of *Berberis davidii* can be up to 5 × 1.5 cm. This was based on *C. K. Schneider 388* (A 00279904, K K000567935, WU 039244), Yunnan, north of Kunming, between Hsiaodsang and Loheitang, 800–2100 m, 12 Mar. 1914, which has two ovules and which Schneider (1939: 260–261) identified as *B. fallax* (for further details, see under that species). There is a similar collection from the area north of Kunming: *B. Y. Qiu 58776* (IBK IBK00357392, LBG 00064095, ZM ZMNH0006692), Fumin Xian, Longtannao, Dajian Shan, Madoulingqing, 2700 m, 7 Mar. 1964. Neither

collection is *B. fallax* but they appear not to be *B. davidii* either.

Selected specimens.
Syntypes. **NW Yunnan:** E flank of the Tali (Dali) Range, 25.666666°N, 2745–3050 m, 15 July 1910, *G. Forrest 7276* (E E00424136, K K000644861); [Yongping Xian], on the Chienchuan-Mekong divide, 25.50°N, 99.33333°E, 3050–3350 m, Aug. 1923, *G. Forrest 23599* (BM BM000895066, E E00424138, IBSC 0091645, K K000644860, PE 01031010); "West Yunnan," 1929, *G. Forrest 28167* (E E00424137); SE Xizang (Tibet), [Zayü (Chayu) Xian], "Me-kong, Tsa-wa-rung [Cawarong/Chawalong], 3000 m, Sep. 1935, *C. W. Wang 66201* A 00038691, KUN 0177061, NAS NAS00314333, PE [3 sheets] 01031007–09, WUK 0047683).

Other specimens.
NW Yunnan. 1883–1885, *J. M. Delavay s.n.* (P P06868280). **Dali Shi:** Cangshan, Qingbixu, 2200 m, 23 May 1981, *Sino-British Exp. Cangshan 1105* (A 00279496, E E00612451, K, KUN 0175866, 0176108). **Weixi Xian:** Badixiang, Luoma He, 2510 m, 8 Apr. 2003, *Z. D. Fang et al. W0251* (TNM S136740).

Cultivated material:
Living cultivated plants.
Howick Hall, Northumberland and Foster Clough, Mytholmroyd, West Yorkshire, U.K., from *M. Foster 96129*, NE Yunnan, Gongshan Xian, Yima Di, ca. 30 km SW of Gong Shan in the Salween drainage in the Mekong/Salween Divide, ca. 2550 m, autumn 1996.

21. Berberis deinacantha C. K. Schneid., Repert. Spec. Nov. Regni Veg. 46: 259. 1939. TYPE: China. NW Yunnan: [Lanping Xian/Yunlong Xian border], Chienchuan/Mekong divide, 26.42°N, 99.33°E, 2750–3050 m, Aug. 1923, *G. Forrest 23556* (lectotype, designated by Ahrendt [1961: 60], K K000077329!; isolectotypes, BM BM001015539!, E E00217988!, HBG HBG-506738 image!, PE 02193905!).

Berberis deinacantha var. *valida* C. K. Schneid., Repert. Spec. Nov. Regni Veg. 46: 260. 1939. *Berberis valida* (C. K Schneid.) C. K Schneid., Mitt. Deutsch. Dendrol. Ges. 55: 40. 1942, syn. nov. TYPE: China. S Sichuan: Ning Yuen fu, (Xichang), Lo tieh shan, ca. 2500 m, 16 Apr. 1914, *C. K. Schneider 918* (lectotype, designated here, A 00038806!; isolectotype, Herb. Dendrol. C. Schneider†).

Shrubs, evergreen, to 4 m tall; mature stems pale reddish or yellowish brown, sometimes purplish when dry, terete or slightly angled, black verruculose; spines 3-fid, concolorous, 2.5–5 cm, stout, sometimes retrorse, abaxially sulcate. Petiole 2–5 mm; leaf blade abaxially yellow-green, adaxially dark green, oblong-elliptic, elliptic, elliptic-lanceolate, or lanceolate, 3–13 × 1.5–3.5 cm, leathery, midvein raised abaxially, impressed adaxially, lateral veins conspicuous and slightly raised abaxially, conspicuous and slightly impressed adaxially, reticulation inconspicuous abaxially (but conspicuous when dry), conspicuous adaxially, base cuneate, margin sometimes slightly revolute, spinulose with (8 to)20 to 35 teeth on each side, apex obtuse, mucronate. Inflorescence a fascicle, (3- to)6- to 15(to 20)-flowered; pedicel (6–)12–15(–20) mm. Flowers deep yellow. Sepals in 2 whorls; outer sepals ovate or suborbicular, ca. 4 × 3.5 mm, apex rounded, obtuse; inner sepals obovate, obovate-orbicular, or elliptic, 5.5(–7) × ca. 5 mm; petals oblong-obovate or obovate, ca. 4.5 × 3 mm, base not clawed, with separate glands, apex slightly emarginate or entire. Stamens 3–4.5 mm; anther connective extended, slightly retuse. Ovules 1. Berry purplish black, not or slightly pruinose, ellipsoid, 6–8 × ca. 4 mm; style not persistent.

Phenology. Berberis deinacantha has been collected in flower between March and May and in fruit between June and November.

Distribution and habitat. In China *Berberis deinacantha* is known from west and southwest Guizhou, east, north, northeast, and northwest Yunnan, and south Sichuan. It has been found in thickets on rocky slopes, and in oak, pine, and mixed and deciduous forests at ca. 1800–3200 m. It is also known from Myanmar (Burma).

The protologue of *Berberis deinacantha* cited *G. Forrest 23556* at E and *C. W. Wang 66415*, but did not use the term type. Subsequently, Schneider (1942: 40) designated the Forrest collection as type but did not designate the herbarium. Ahrendt's designation (1961: 60) of *Forrest 66415* at K as type was an effective lectotypification. Chamberlain and Hu (1985: 458–459) mistakenly described the specimen at E as a holotype. The specimen of the syntype *C. W. Wang 66415* at A was annotated by Schneider as only probably *B. deinacantha*. From the evidence of the better specimens at NAS and PE, it appears more likely to be *B. dumicola*.

In the protologue of *Berberis deinacantha* var. *valida*, Schneider cited three collections but did not designate a type. They included *Schneider 918* in the "Herb. Dendrol. C. Schneider" which, as noted in the introduction, was his personal herbarium. Subsequently (1942: 40), when he elevated the taxon to a separate species, *B. valida*, he designated *Schneider 918* as the type, but without indicating a herbarium. His own specimen was destroyed in WWII. The specimen at A, the only other one located, has been designated the lectotype.

In the protologue of variety *valida*, Schneider noted that, if *H. Smith 13494* (UPS BOT:V-040863, but also MO 4367277, PE 01293893, S 12–25257), central Sichuan, Yingjing Xian, Tahsiangling (Daxiangling), ca. 2000 m, 19 Nov. 1934, with different fruit from *Berberis*

deinacantha, was the same taxon as the syntypes he cited for variety *valida*, then variety *valida* was perhaps a separate species. In designating *B. valida* as a separate species, he again cited *H. Smith 13494* as *B. valida*, this time without qualification. However, this gathering with two seeds per fruit (M. Hjertson of UPS, pers. comm. 7 Oct. 2009) is *B. bergmanniae*. Schneider's protologue of variety *valida* mistakenly reported the syntype *H. T. Tsai 50864* as *50684*.

Schneider (1939) differentiated variety *valida* from *Berberis deinacantha* essentially because it had yellow or yellowish brown mature stems versus the purplish brown mature stems of the latter, while Ahrendt (1961: 59), who followed Schneider in accepting *B. valida* as a separate species, described the stems of *B. deinacantha* as being purple and those of *B. valida* as yellow. Chamberlain and Hu (1985: 548–549), who relegated *B. valida* back to being a variety of *B. deinacantha*, described the difference as being only that the stems of the former were yellowish and the latter purplish. In fact, the color of mature stems of the various specimens of the type of *B. deinacantha* is varied, those of the specimens at E being purplish brown, those at HBG being reddish or yellowish brown, that at BM shades of brownish yellow, while those at K are purplish yellow. Moreover, none of the mature stems (where present) on any of the herbarium specimens cited below are purplish, all being various shades of yellowish brown. Observations made of the young and mature stems of my living plant of *A. Clark 1010*, cited below, might shed some light on this problem in that, whereas these stems are respectively green and pale reddish brown on the living plant, when dried they can turn distinctly purplish. Given all this, differentiation of variety *valida* on the basis of stem color would seem untenable, hence my synonymy.

It is worth noting that, while the leaves of the living plants cited below grown from *M. Hird 81* and *SICH 1449* are almost identical to those of the plants of *A. Clark 1010*, those of the latter are noticeably thicker. The flowers and fruit, however, appear not to differ. Further investigation is needed here.

If all of the identifications made below are correct, then it would seem that *Berberis deinacantha* has a very unusual distribution pattern, occurring in four separate areas—northwest Yunnan, south Sichuan and adjoining areas of Yunnan, east Yunnan and the adjoining area of Guizhou, with one collection recorded from Zhenxiong in the far northeast of Yunnan. No other *Berberis* species in this part of China appears to have such a disjunctive range.

Syntype. Berberis deinacantha. SE Xizang: [Zayü (Chayu) Xian], Tsa-wa-rung, Dzer-nar, 3200 m, Aug. 1935, *C. W. Wang 66415* (A 00279406, KUN 0176100, NAS NAS00314034, PE [2 sheets] 01031188, 01293901).

Syntypes. *Berberis deinacantha* var. *valida*. S Sichuan: Huili chou (Huili Xian), Lung tchu (Longzhu) Shan, ca. 3000 m, 25 Mar. 1914, *C. K. Schneider 571* (A 00038686, K K000567934). NE Yunnan: Chaotung Hsien, (Zhaotong Xian), 2800 m, 1 May 1932, *H. T. Tsai 50864* (A 00038687, IBK IBK00301326, IBSC [3 sheets] 0091477, 0091650, 0092293, KUN [2 sheets] 0160519, 0175777, PE 01293897, SZ 00291225, WUK 0037544).

Selected specimens.
W & SW Guizhou. Pan Xian: Bada Shan, 2300 m, 29 Aug. 1959, *Anshun Team 889* (HGAS 13604, KUN 0175926, PE 01293902–03). **Xingyi Xian:** Qishe, 2200 m, 17 Apr. 1987, *Z. Y. Deng 1629* (KUN 0179230); 4 Apr. 1989, *s. coll. D 2460* (KUN 0179232).
S Sichuan. Dechang Xian: Luoji Shan, Xumu Chang, Laochang Bu, 2700 m, 2 July 1976, *Southwest Normal University Department of Biology class of 1974 11832* (CDBI CDBI0028225, PE 01031517). **Puge Xian:** Shanggu, Jiaping, Liangzi, 2200 m, 23 Apr. 1959, *Z. T. Guan 8049* (IBSC 0091472, PE 01031522); Tuomu Gou Qu, Huangzhu Shan, 2500 m, 25 Aug. 1959, *Yibin Division Sichuan Economic Plant Exped (59)5513* (PE 01031523–24); Qingshui Gou, 2300 m, 5 Aug. 1976, *Sichuan Botany Team 14190* (CDBI CDBI0027988, CDBI0027991–2, CDBI0027996, PE 01031525). **Yanbian Xian:** Tianjia, Jiaodi, 3020 m, 23 May 1959, *S. G. Wu 578* (KUN 0177268, SZ 00294260, 00290485).
E Yunnan. Fuyuan Xian: Housuo Xiang, Laohei Shan, 2500 m, 17 June 1989, *Hongshui He Botany Team 2230* (KUN 0179203–5).
N & NE Yunnan. Dongchuan Shi: Yinmin, Wagangsai, Banbian Shan, 3000 m, 14 Apr. 1985, *S. B. Lan 356* (PE 01293898–900). **Luquan Xian:** San Qu, Shande Xiang, Sayingshan, Maoluqing, 3000 m, 27 Nov. 1952, *P. Y. Mao 1881* (HITBC 003644, KUN 0175775, PE 01293895–96). **Qiaojia Xian:** Jila, 2100 m, 12 Aug. 1974, *Kunming Botany Team s.n.* (KUN 0175929). **Zhenxiong Xian:** San Qu, Yaoduo Xiang, Laolin, 1800 m, 26 Sep. 1957, *P. H. Yu 924* (KUN 0175922–23, PE 01293908, WUK 0195332).
NW Yunnan. Lanping Xian: 104 Linchang, 3100 m, 28 June 1981, *Beijing Hengduan Team 0879* (PE 01293909). **Lushi Xian:** Luzhang Zheng, above Yiajiaping forest station, 25.983611°N, 98.706389°E, 2737 m, *Gaoligong Shan Biodiversity Survey 24446* (CAS 00120155, E E00269498 GII 00297773, HAST 124219, MO 6062086). **Weixi Xian:** Yi Qu, Yongsheng Xiang, Er Cun, Chengxin Chang, 2800 m, 1 Nov. 1956, *P. Y. Mao 00582* (KUN 0175933–4, PE 01293913); Zhichang, 2980 m, 3 May 1960, *Nanshuibei Investigation Team 8648* (KUN 0175932, PE 01293910, 01293912, YUKU 02004168). **Yunlong Xian:** Yizu Cun, Tianchi, 2740 m, 15 Oct. 1994, *Z. S Yue et al. 4* (KUN 0179263).
MYANMAR (BURMA). Hpare Pass, "along rocky crest of the ridge," 3050 m, 25 Mar. 1939, *F. Kingdon-Ward, Vernay-Cutting Exp. 449* (A 00342317).

Cultivated material:
Living cultivated plants.
Sir Harold Hillier Gardens, Hampshire, U.K., from *M. Hird 81*, W Yunnan: Baoshan Shi, Longyang Qu, near Weshang Lizu, 25.47°N, 99.06°E, 2400 m, 13 Nov. 1994; Royal Botanic Garden Edinburgh, and Royal Botanic Gardens, Kew, from *SICH 1449*, S Sichuan: Xichang Xian, N flank of Luoji Shan, ca. 2705 m, 7 Oct. 1994; Foster Clough, Mytholmroyd, West Yorkshire, U.K., from *A. Clark 1010*, NE Yunnan: near Zhaotong, 1995.

22. Berberis delavayi C. K. Schneid., Pl. Wilson. (Sargent) 1: 364–365. 1913. TYPE: China. NW Yunnan: [Heqing Xian], "Les bois de Houang le pin au dessus de Tapintze," 1800 m, 21 Mar. 1887, *J. M. Delavay s.n.* (lectotype, designated here, P P00716569!; isolectotypes, P P00716570!; labelled "Hoang li ping": KUN 1221631, P [2 sheets] P02682332!, P02682367!).

Berberis paraspecta Ahrendt, J. Linn. Soc., Bot. 57: 47. 1961, syn. nov. TYPE: China. NW Yunnan; [Lijiang Shi], "in reg. Lichiangfu prope pagum Ngu-leh-keh," (Xuesong Cun), 2900 m, 27 July 1914, *C. K. Schneider 2028* (holotype, K K000077343!; isotypes, A 00279424!, GH 00279425!).

Shrubs, evergreen, to 1.5 m tall; mature stems yellow, slightly angled, scarcely verruculose; spines 3-fid, concolorous, 1–3 cm, abaxially sulcate. Petiole 2–5 mm; leaf blade abaxially pale green, shiny, adaxially dark green, shiny, ovate, oblong-ovate, or oblong-elliptic, 2.5–5.5 × 0.8–1.3(–1.7) cm, leathery, midvein raised abaxially, impressed adaxially, lateral veins and reticulation inconspicuous on both surfaces, base cuneate, margin undulate, slightly revolute, spinulose with 7 to 12 teeth on each side, apex acute, mucronate. Inflorescence a fascicle, 3- to 15-flowered; pedicel 3–13 mm, slender; bracts ovate-triangular, ca. 2.5 × 2 mm. Sepals in 2 whorls; outer sepals ovate, ca. 4 × 3.5 mm, apex obtuse; inner sepals obovate, ca. 7 × 5.5 mm; petals obovate, 5.5–6 × ca. 4.5 mm, base clawed, glands separate, apex subround, obtusely emarginate. Stamens ca. 3.5 mm; anther connective not extended, obtuse. Ovules 2 or 3. Berry black-purple, blue pruinose, ovoid, ca. 9 × 6 mm; style not persistent.

Phenology. *Berberis delavayi* has been collected in flower in March and in fruit between April and November.

Distribution and habitat. *Berberis delavayi* is known from Heqing, Eryuan, and Jianchuan Xian and Lijiang Shi in northwest Yunnan. It has been found in forests at ca. 1800–3400 m.

Schneider did not indicate a herbarium for the type of *Berberis delavayi*. There are two specimens of *Delavay s.n.*, 21 Mar. 1887, at Paris. The one with an annotation by Schneider dated 18 June 1906 has been chosen as the lectotype. Ahrendt (1961: 61) mistakenly cited *J. D. Delavay 485* as the type. This is not *B. delavayi* but *B. wui*.

Besides the type, the protologue of *Berberis delavayi* also cited *J. M. Delavay s.n.*, Fang-yang-Tchong, 14 Oct. 1887 (P P02313660), and this was the basis for reporting that the pedicels of the species were up to

20 mm long. However, this specimen is also *B. wui* and the pedicels of *B. delavayi* are no longer than 13 mm and mostly shorter than this.

Schneider (1942: 42) cited *C. K. Schneider 2028* as an example of *Berberis delavayi*. On the basis of a specimen at K, Ahrendt (1961: 46–47) determined this to be a new 1-flowered species, *B. paraspecta*. However, the holotype at K actually has a set of paired fruit while the isotypes at A and GH are fascicled with up to four flowers. Chamberlain and Hu (1985: 543), while accepting *B. paraspecta* as a valid 1-flowered species, pointed out that Ahrendt was mistaken in describing the pedicels of the holotype as 2–3 cm long; they are actually 2–3 mm long. The pedicels of the isotypes are 3–10 mm. H. X. Li et al. (2015b), while noting the holotype has paired fruit, maintained *B. paraspecta* as a separate species. They appeared unaware of the isotypes at A and GH and that Schneider had identified *Schneider 2028* as *B. delavayi* (a species to which they made no reference). Their account of the flowers of *B. paraspecta* is more or less the same as that of *B. delavayi* given above, except it includes ovate-triangular bracteoles.

Berberis delavayi was treated as a synonym of *B. phanera* by Ying (2001: 113), but *B. delavayi* was published first (this was corrected in Ying [2011: 736] where the synonymy is reversed). They are, however, different species, the leaves of *B. delavayi* being adaxially shiny and mostly narrower versus the wider adaxially dull leaves of *B. phanera*. The flower structures are also different with *B. phanera* having three whorls of sepals. On current evidence, *B. delavayi* appears to flower significantly earlier than *B. phanera*. *Berberis phanera* also occurs farther north.

Selected specimens.
NW Yunnan. E slope of pass betw. Sung Kwen (Heqing Xian) & Teng chuen (Eryuan), 3300 m, 29 Sep. 1914, *C. K. Schneider 2891* (A 00279422). **Eryuan Xian:** Nov. 1959, *Institute for Drug Control 226* (KUN 0175944). **Heqing Xian:** Ma'er-shan, 2500 m, Oct. 22 1889, *J. M. Delavay s.n.* (P P02313620, 02465453); Xiangshui He, 3200 m, 2 Sep. 1929, *R. C. Qin 24277* (KUN 0175885–87); Shiang-shu-ho (Xiangshui He), by Ma-erh-shan (Ma'er Shan) near Sung-kwei (Songgui), 18 Apr. 1939, *G. M. Feng 725* (A 00279890, KUN 0175878–79, 0177153); Songgui Qu, Ma'er Shan, Chamujing, 2900 m, 21 July 1963, *Northwest Yunnan Jinsha River Team 4772* (KUN 0176714–15, PE 01030977–78). **Jianchuan Xian:** Laojunshan, 2800 m, 13 Oct. 1958, *W. C. Wang 460* (KUN 0175911). **Lijiang Xian:** "Lichiang-fu," 2745–3050 m, 22 Apr. 1906, *F. Kingdon-Ward 5023* (E E00258862); Yulong Shan, Yunshan Ping, 3280 m, 11 July 1962, *A. L. Zhang 100936* (KUN 0176719–21).

23. Berberis dongchuanensis T. S. Ying, Acta Phytotax. Sin. 37(4): 312–313. 1999. TYPE: China. N Yunnan: Dongchuan Qu, betw. Yinmin & Baoqing, 2600 m, 23 Apr. 1984, *S. B. Lan 665* (holotype, PE 00935138!; isotype, PE 00935139!).

Shrubs, evergreen, to 1 m tall; mature stems probably grayish yellow, terete; spines 3-fid, pale brownish yellow, 0.8–1.6 cm. Petiole almost absent; leaf blade abaxially pale green, densely white pruinose, adaxially yellow-green, oblanceolate or narrowly elliptic, 3.5–8 × 1–2 cm, thickly leathery, midvein raised abaxially, impressed adaxially, lateral veins and reticulation indistinct on both surfaces, base cuneate or attenuate, margin slightly revolute, aristate-dentate with 1 to 4 coarse, widely spaced teeth on each side, apex acute, mucronate. Inflorescence a fascicle, 2- to 8-flowered; pedicel 18–34 mm, slender. Sepals in 3 whorls; outer sepals triangular-ovate, ca. 2 × 1.5 mm, apex acuminate; median sepals ovate-elliptic, ca. 5.1 × 3 mm; inner sepals broadly elliptic, ca. 6 × 5 mm; petals obovate, ca. 5 × 4 mm, base clawed, glands separate, apex rounded, entire. Stamens ca. 3.5 mm; anther connective not extended, truncate. Ovules 4, shortly funiculate. Fruit unknown.

Phenology. Berberis dongchuanensis has been collected in flower in April. Its fruiting period is unknown.

Distribution and habitat. Berberis dongchuanensis is known only from the type collection made at the base of a cliff at 2600 m.

24. Berberis dumicola C. K. Schneid., Repert. Spec. Nov. Regni Veg. 46: 249–250. 1939. TYPE: China. NW Yunnan: [Gongshan Xian], Mekong-Salwin divide, 28.166667°N, Sep. 1914, *G. Forrest 13295* (lectotype, inadvertently designated by Chamberlain & Hu [1985: 555], E E00373491!; isolectotypes, K missing, PE 01031897!).

Shrubs, evergreen, to 2 m tall; mature stems pale yellow, terete, brownish black verruculose, young shoots bright red; spines 3-fid, brownish yellow, 1–2 cm, adaxially flat or slightly sulcate. Petiole 4–5 mm; leaf blade abaxially very pale green, adaxially slightly shiny, dark green, elliptic, sometimes narrowly so, 5–9.5 × 1–1.8 cm, leathery, midvein raised abaxially, impressed adaxially, lateral veins impressed and conspicuous abaxially, conspicuous adaxially, reticulation inconspicuous abaxially, conspicuous and dense adaxially, base cuneate, margin spinulose with 20 to 40 teeth on each side, apex acute, rarely obtuse, mucronate. Inflorescence a fascicle, 5- to 20-flowered; pedicel reddish, 8–15 mm; bracteoles ovate, apex obtuse or subobtuse. Sepals in 2 whorls; outer sepals slightly reddish, oblong-ovate, ca. 3 × 2.5 mm, apex obtuse; inner sepals oblong-obovate, ca. 6 × 5 mm; petals oblong-obovate, ca. 6.5 × 5 mm, base clawed, with glands separate and ovate, apex rounded, incised to emarginate. Stamens ca. 4.5 mm; anther connective extended, shortly apiculate. Ovules 1 or 2; funicles ca. as long as ovules. Berry purplish black, pruinose, ellipsoid or obovoid-ellipsoid, 8–9 × 4–5 mm; style persistent.

Phenology. Berberis dumicola has been collected in fruit between April and October. Its flowering season outside cultivation is unknown.

Distribution and habitat. Berberis dumicola is known from Dêqên (Deqin), Gongshan, Weixi, and Yulong Xian in northwest Yunnan and in the Chawalong area of Zayü (Chayu) Xian in southeast Xizang. It has been found among scrub by streams and in open, rocky hillsides, on the margins of pine forests, and (at lower elevations) in sub-tropical forest disturbed by agriculture and felling between 1500 and ca. 3350 m.

The protologue of Berberis dumicola cited eight gatherings but did not designate a type. Subsequently, Schneider (1942: 18) cited one of these, *Forrest 13295*, as the type but did not cite a herbarium. Chamberlain and Hu (1985: 555) incorrectly cited *Forrest 13295* at E as a holotype, but under Art. 9.10 of the Shenzhen Code (Turland et al., 2018), this can be treated as an inadvertent completion of lectotypification. Ahrendt, (1961: 69–70) did not notice Schneider's lectotypification of *Forrest 13295* and mistakenly cited *Forrest 19474* as the type.

One of the gatherings cited by Schneider was *Soulié 1597*, "Tsekou (?Tzuku) et Nekou," 6 Jan. 1898, but no herbarium was indicated. There are five specimens of *1597* at P. One (P02682352) from Nekou is dated 6 Jan. 1895, and the others from Tsekou simply 1895. I have assumed that all are from the same collection and that Schneider misreported 1898 for 1895. Tsekou can be identified as being very near present-day Cizhong in Gongshan Xian.

As noted under that species, Schneider (1939: 259) cited *C. W. Wang 66415* as a syntype of *Berberis deinacantha* based on the evidence from the specimen at A. From the evidence of the better specimens of *66415* at NAS and PE, it appears to be *B. dumicola*.

I have located no wild-collected specimens of Berberis dumicola with flowers. The description of flowers above is from cultivated material from plants grown in the U.K. from *Forrest 13925* and *Forrest 19474*.

Syntypes. **NW Yunnan:** [Dêqên (Deqin) Xian], on descent of the Kari pass, 28°N, 2740–3350 m, Sep. 1904, *G. Forrest 192* (E E00320572); [Gongshan Xian], Mekong-Salwin divide, 28°N, 3050 m, 17 Oct. 1914, *G. Forrest 13441* (E E00320575, PE 01031896); [Yulong Xian], E flank of Li-tiping, Mekong-Yangtze divide, 27.2°N, 99.666667°E, 3050 m, June 1921, *G. Forrest 19474* (A 00279405, E E00217992, K K000077317); [Gongshan Xian], Mekong-Salwin divide, 27.6°N, 98.716667°E, 2740–3050 m, Oct. 1921, *G. Forrest*

20931 (E E00320576); [Yulong Xian], Li-Ti-ping range, Mekong-Yangtze divide, E of Wei shi (Weixi), 1923, *J. F. Rock 8916* (A 00279843, US 00945910); "Tsekou et Nekou (Haut-Mekong)," 6 Jan. 1895, *J. A. Soulié 1597* (P [5 sheets] P02482746–47, P02682352, P03618386, P00580427); s. loc., 3–4000 m, 1 July 1907, *T. Monbeig 14* (E [2 sheets] E00327783–84). Probable syntype, *T. Monbeig s.n.* (W 1922–8670 ex-K).

Selected specimens.
NW Yunnan. Dêqên Xian: Cang Jiang, 2000 m, 5 Oct. 1959, *G. M. Feng 23941* (KUN 0176096–97, 0176099). **Gongshan Xian:** betw. Bingzhongluo & Nidadang, 1650 m, 23 June 1982, *Qinghai-Xizang Team 7358* (KUN 0178418–19, PE 01031148–49); Bingzhongluo, Bibili, 27.983889°N, 98.661667°E, 1560 m, 16 Apr. 2002, *H. Li et al. 14416* (CAS 00120156–57); Bingzhongluo, Sijitong Cun, near mouth of Niwaluo He, 28.045306°N, 98.5915°E, 1550 m, 29 Aug. 2006, *Gaoligong Shan Biodiversity Survey 34319* (CAS 00120153–54, GH 00292914). **Weixi Xian:** Badi Xiang, Luoma He, 2425 m, 8 Apr. 2003, *Z. D. Fang et al. W0277* (TNM S136746).

SE Xizang. [**Zayü (Chayu) Xian**]: Songta Xueshan, 3000 m, 17 June 1960, *Nanshuibei Investigation Team 8999* (KUN 0176098. PE 01031900).

Cultivated material:
Living cultivated plants.
Royal Botanic Garden Edinburgh, from *G. Forrest 19474* and *20931* (collection details above).

25. Berberis ebianensis Harber, sp. nov. TYPE: China. SC Sichuan: Ebian Xian, Yiping Xiang, 1600 m, 16 Sep. 2010, *X. J. Li 184* (holotype, TAIF 05632!; isotype, TAIF 405633!).

Diagnosis. Berberis ebianensis has leaves that are somewhat similar to *B. veitchii* but has dark reddish purple rather than yellow mature stems and leaves with fewer marginal spines.

Shrubs, evergreen, to 90 cm tall; mature stems dark reddish purple, subterete; spines 3-fid, pale yellowish brown, 0.4–1.6 cm, adaxially slightly sulcate, weak. Petiole almost absent; leaf blade abaxially pale green, adaxially green, lanceolate, rarely elliptic-lanceolate, 3–11 × 0.8–1.4(–2.4) cm, leathery, midvein raised abaxially, slightly impressed adaxially, lateral venation slightly raised and inconspicuous abaxially, inconspicuous adaxially, reticulation obscure abaxially, inconspicuous adaxially, base cuneate, margin spinose with 6 to 10(to 12) widely spaced coarse teeth on each side, apex acute, mucronate. Inflorescence a fascicle, 1- to 5-flowered. Flowers unknown; fruiting pedicel bright red, 5–10 mm. Immature berry turning dark purple or black, epruinose, ellipsoid, 8–9 × 4–5 mm; style persistent, short; seeds 1 to 3.

Phenology. Berberis ebianensis has been collected in fruit in September. Its flowering period is unknown.

Distribution and habitat. Berberis ebianensis is known from Ebian Xian in south-central Sichuan. The type collection was made on a mountainside at 1600 m.

IUCN Red List category. Berberis ebianensis is assessed as DD or Data Deficient, according to IUCN (2001) criteria.

Very few *Berberis* specimens are known from Ebian Xian. The only other evergreen species known are *B. gagnepainii* and *B. simulans*.

26. Berberis fallaciosa C. K. Schneid., Repert. Spec. Nov. Regni Veg. 46: 258. 1939. TYPE China. Chongqing (SE Sichuan): Nanchuan, Pen-cha'ai, autumn 1891, *C. Bock & A. von Rosthorn 681* (lectotype, designated by Chamberlain & Hu [1985: 541], B B10 000264917!; isolectotype, O image!).

Shrubs, evergreen, to 3 m tall; mature stems yellow, sulcate, scarcely verruculose; spines 3-fid, pale yellow, 1–4 cm. Petiole 2–4 mm; leaf blade abaxially yellow-green, shiny, adaxially dark green, shiny, lanceolate or narrowly elliptic-lanceolate, 3–7 × 1–1.8 cm, leathery, midvein raised abaxially, impressed adaxially, lateral veins slightly raised and conspicuous abaxially, conspicuous adaxially, reticulation inconspicuous on both surfaces, base attenuate, margin spinulose with 15 to 30(to 40) teeth on each side, apex subacuminate, mucronate. Inflorescence a fascicle, 2- to 7-flowered; pedicel reddish, 10–25 mm; bracteoles broadly ovate, apex obtuse. Flowers yellow. Sepals in 2 whorls; outer sepals oblong-elliptic, 3 × 2 mm; inner sepals suborbicular, 5 × 4 mm; petals obovate-oblong, 4.5 × 2.5 mm, base clawed, glands very close together, apex emarginate. Stamens ca. 3 mm; anther connective slightly extended, truncate or apiculate. Ovules 2, sessile. Berry black, obovoid, 6–9 × 5–6 mm; style persistent, short.

Phenology. Berberis fallaciosa has been collected in flower in early April and in fruit between May and October.

Distribution and habitat. Berberis fallaciosa is known from Nanchuan and Jiangjin Xian in Chongqing (southwest Sichuan), and possibly from southwest Hubei. It has been found in thickets, among bamboo on mountain slopes, forests, and trail and streamsides at ca. 1060–2750 m.

The protologue of *Berberis fallaciosa* cited "A. v. Rosthorn no. 68" at B and *W. P. Fang 842* at NY, but did not use the term type. Subsequently, Schneider (1942: 39) cited the former as type but did not desig-

nate a herbarium. However, it is reasonable to infer that he was referring to the specimen at B and this was assumed by Chamberlain and Hu (1985: 541) who, like Ahrendt (1961: 60), also gave the collection as being *Rosthorn 68*. However, the specimen at B, annotated by Schneider as type in 1938, is actually labelled *C. Bock & A. von Rosthorn 681*. This number was first listed by Diels (1901: 341) as an example of *B. wallichiana*.

The distribution of *Berberis fallaciosa* needs further investigation. Though there are numerous collections known from Nanchuan and two from Jiangjin Qu, whether it is found in Lichuan Shi is doubtful. The two collections cited below, if not *B. fallaciosa*, are likely to be of one or more undescribed taxa (the HIB specimens are identified on the sheets as unpublished *B. fallaciosa* var. *lichuanensis*). Specimens from northeast Chongqing identified as *B. fallaciosa* appear to be mostly *B. sargentiana*.

Syntype. Chongqing (Sichuan): Nanchuan, 2450–2750 m, 20 May 1928, *W. P. Fang 842* (A 00279487, E E00320619, IBSC 0091687, K K000077328, NY 1365275, P P00580432, PE (2) 01031536–37).

Selected specimens.
Chongqing (Sichuan). Jiangjin Xian (Qu): Meilin Qu, Simian Xiang, Honghai, 1320 m, 22 Apr. 1959, *s. coll. 426* (KUN 0176128); Simian, Honghai, Xiaowan, 1150 m, 25 Apr. 1959, *s. coll. 144* (KUN 0176129). **Nanchuan Xian:** SE of Daheba, 1060 m, 4 Apr. 1932, *D. H. Du 2778* (IBK IBK00013164, PE 01031540); Jinfoshan, Fenghuang Xi, 1950 m, 26 May 1957, *J. H. Xiong & Z. L. Zhou 90998* (HIB 00129334, IBSC 0091688, KUN 0176124, PE 01031201); Tougao Gou, 1200 m, 3 May 1957, *G. F. Li 60908* (IBSC 0092230, KUN 0176133, PE 01031542); betw. Liangyuanding & Qincaiba, 2150 m, 4 Oct. 1964, *Sichuan Team – K. J. Guan et al. 2238* (CDBI CDBI0027324, PE 01031552).
SW Hubei. Lichuan Shi: Moudao Qu, near Tiedou, 1300 m, 23 Sep. 1957, *G. X. Fu & Z. S. Zhang 1601* (FUS 00014200, HIB 00129633, IBSC 0091839, KUN 0178344, LBG 00064118, PE 01031581–83, SWCTU 00014200); Moudao Qu, near Maha Xiang, 25 Sep. 1957, *M. X. Nie 1731* (HIB 00129340, IBSC 0092224, 0092247, KUN 0176127, LBG 00064117, PE 01030728–30).

27. Berberis fallax C. K. Schneid., Repert. Spec. Nov. Regni Veg. 46: 260. 1939. TYPE: China. NW Yunnan: Che-tze-lo Xsien (Bijiang Xian, now divided betw. Fugong & Lushui Xian), 3200 m, 12 Sep. 1934, *H. T. Tsai 58521* (lectotype, designated by Ahrendt [1961: 49], A 00038743!; isolectotypes, IBSC 0091690 image!, KUN [2 sheets] 1204017!, 1204020!, LBG 0064109 image!, NAS NAS00070726 image!, PE 01031812!).

Shrubs, evergreen, to 2 m tall; mature stems very pale brownish yellow, sulcate, slightly verruculose; spines 3-fid, concolorous, 0.6–1.2 cm, slender, abaxially sulcate. Petiole almost absent; leaf blade abaxially pale green, adaxially dark green, very shiny, oblong-lanceolate or oblong-elliptic, 2–6 × 0.8–1.6 cm, thinly leathery, midvein raised abaxially, slightly impressed adaxially, lateral veins and reticulation inconspicuous on both surfaces, base cuneate, margin spinulose with 5 to 12 teeth on each side, apex acute, mucronate. Inflorescence a fascicle, (1- or)2- to 6-flowered, otherwise unknown. Fruiting pedicel 8–16 mm. Berry black, epruinose, ellipsoid, ca. 7 × 5 mm; style not persistent or persistent and short; seeds 4 or 5.

Phenology. *Berberis fallax* has been collected in fruit between June and October. Its flowering season is unknown.

Distribution and habitat. *Berberis fallax* is known from Lushui Xian and possibly Fugong Xian in northwest Yunnan. It has been found in broadleaf forest, on a streamside at the edge of degraded evergreen forest, and on a roadside at ca. 2650–3200 m.

The protologue of *Berberis fallax* cited *H. T. Tsai 58521*, *H. T. Tsai 54102*, and *C. K. Schneider 388*. Subsequently, Schneider (1942: 41) designated *H. T. Tsai 58521* as the type but did not designate a herbarium. The lectotypification of the specimen at A was by Ahrendt (1961: 49) though there is no evidence that he had seen the specimen.

The protologue included a description of the flowers, but it was based on the syntype *C. K. Schneider 388*, which Schneider conceded has only two ovules. It is not *Berberis fallax*. Chamberlain and Hu (1985: 543) identified this collection as *B. davidii*, but that also seems to be incorrect.

Specimens from elsewhere in northwest Yunnan at KUN and PE identified as *Berberis fallax* have a wide range of characteristics and all appear to be of other species.

There would seem to no evidence to associate *Berberis fallax* var. *latifolia* C. Y. Wu & S. Y. Bao from Jingdong Xian with *B. fallax*. For further details of this, see Taxa Incompletely Known).

Syntypes. NW Yunnan: Che-tze-lo, 3000 m, 3 Sep. 1933, *H. T. Tsai 54102* (A 00279491, IBSC 0000586, LBG [2 sheets] 00064110–11, KUN 0176104, NAS NAS00313977, PE 01031813, SZ 00291245, WUK 0041991); C Yunnan, N of Kunming betw. Hsiadsang & Loheitang, 1800–2100 m, 12 Mar. 1914, *C. K. Schneider 388* (A 00279904, K K000567935, WU 039244).

Selected specimens.
NW Yunnan. Bijiang Xian: Famu Chang, 2650 m, 23 June 1978, *Bijiang Investigation Team 702* (KUN 0176112–13, YUKU 02065796). **Lushui Xian:** near Jiaping Forest Station, 25.969722°N, 98.713611°E, 2650 m, 27 Oct. 1996, *Gaoligong Shan Expedition (1996) 8168* (E E00205895); near

Jiaping Forest Station, 25.969722°N, 98.713611°E, 2270 m, 28 Oct. 1996, *Gaoligong Shan Expedition (1996) 8232* (E E00205944).

28. Berberis ferdinandi-coburgii C. K. Schneid., Pl. Wilson. (Sargent) 1: 364. 1913. TYPE: China. S Yunnan: Mengzi [Xian], 1700 m, 21 Feb. s. anno, but pre 1898, *A. Henry 10257* "A," (lectotype, designated here, K K000077315!; isolectotypes, A 0038745!, E E00217993!, LE [2 sheets]!, MO 2270578!, US 00103878 image!).

Berberis ferdinandi-coburgii C. K. Schneid. var. *vernalis* C. K. Schneid., Repert. Spec. Nov. Regni Veg. 46: 249. 1939; *Berberis vernalis* (C. K. Schneid.) D. F. Chamb. & C. M. Hu, Notes Roy. Bot. Gard. Edinburgh 42: 554. 1985, syn. nov. TYPE: China. C Yunnan: betw. Kunming & Pushi (Puqi), 3 Mar. 1914, *C. K. Schneider 226* (lectotype, designated here, A 0038747!; isolectotype, Herb. Dendrol. C. Schneider†).

Berberis iteophylla C. Y. Wu ex S. Y. Bao, Bull. Bot. Res., Harbin 5(3): 7. 1985, syn. nov. TYPE: China. C Yunnan: Shuangbai Xian, Fa-biao, Bai-zhu shan, 2180 m, 3 Apr. 1952, *W. Q. Yin 524* (holotype, KUN 1204033!; isotypes, IBSC 0092120 image!, KUN 0176310!, LBG 00064077 image!, PE 01031392!, WUK 0205391 image!).

Berberis pectinocraspedon C. Y. Wu ex S. Y. Bao, Bull. Bot. Res., Harbin 5(3): 11. 1985, syn. nov. TYPE: China. S Yunnan: Xichou Xian, Fadou Cun, 29 Apr. 1959, *Q. A. Wu 7824* (holotype, KUN 1204046!; isotypes, IBSC 0092282 image!, KUN [2 sheets] 0176700–01!).

Berberis pingbienensis S. Y. Bao, Bull. Bot. Res., Harbin 5(3): 12. 1985, syn. nov. TYPE: China. S Yunnan: Pingbian Xian, Diwu Qu, Guilong Shan, 1900 m, 30 Apr. 1962, *K. M. Feng 22269* (holotype, KUN 0107659!; isotype, KUN 0176156!).

Shrubs, evergreen, to 2 m tall; mature stems grayish yellow, sulcate, scarcely black verruculose; spines 3-fid, pale brown, 0.7–1.5(–2.2) cm, slender, abaxially sulcate. Petiole 2–4 mm; leaf blade abaxially pale yellowish green, adaxially mid-green, shiny, often castaneous, narrowly elliptic, elliptic-oblanceolate, or elliptic-lanceolate, rarely linear-lanceolate, 4–10(–16) × 1.5–2.5 cm, thickly leathery, midvein raised abaxially, impressed adaxially, lateral veins raised and conspicuous abaxially, impressed and conspicuous adaxially, reticulation obscure abaxially, dense and conspicuous adaxially, base cuneate, margin sometimes slightly revolute, spinulose with 35 to 60 teeth on each side, apex acute, mucronate. Inflorescence a fascicle, 8- to 20(to 40)-flowered; pedicel 10–20 mm, slender; bracteoles reddish, ca. 1.5 mm. Flowers bright yellow. Sepals in 2 whorls; outer sepals lanceolate, ca. 2.25 × 0.75 mm; inner sepals ovate, ca. 5 × 3 mm; petals narrowly obovate, ca. 4.5 × 2.75 mm, base clawed, glands separate, oblong, apex emarginate. Stamens ca. 3 mm; anther connective truncate. Ovules 1, subsessile. Berry black, not or sometimes slightly pruinose, ellipsoid or ovoid,

7–8 × 5–6 mm; style persistent; seeds 1, black, ca. 5 × 3 mm.

Phenology. *Berberis ferdinandi-coburgii* has been collected in flower in February and March and in fruit between April and November.

Distribution and habitat. In China *Berberis ferdinandi-coburgii* is known across a wide area of central and south Yunnan. It has been found in woods, on rocky hills, in thickets, and by road and trailsides at ca. 1000–2700 m. In north Vietnam it is known from Sapa in Lào Cai Province and has been found on degraded cutover forest areas at 1585–1830 m.

The protologue of *Berberis ferdinandi-coburgii* cited *A. Henry 10267* as type, but without citing a herbarium. Although Schneider did not say so directly, the implication of his description was that this number has both fruit and flowers. In fact, there are two different gatherings with the same number (the time of year of these only being recorded on the specimens at US). At A, E, K, and US the gatherings are on different sheets. The gathering with flowers is designated here as *Henry 10257* "A" and that with fruit as *Henry 10257* "B." It is clear from annotations on the sheets that the only specimens for which it can be proven that Schneider had seen prior to the protologue were *10257* "A" and "B" at A and K. *Henry 10257* at K was cited as "type" by Ahrendt (1961: 70) and was lectotypified by Imchanitzkaja (2005: 271), but neither noted that this number covered two gatherings. The specimen of *Henry 10257* "A" at K has been chosen here as the lectotype because floral material is a better taxonomic guide than fruit and this is a better specimen than the one at A.

The protologue of *Berberis ferdinandi-coburgii* var. *vernalis* did not cite a type, but cited *Schneider 226* "im Herb. Dendrol. C. Schneider" and seven other gatherings (including *Schneider 234*), all without indication of any herbarium. Subsequently (1942: 17), Schneider implied that *Schneider 226* and *234* were syntypes as against the other five gatherings he had previously cited. Ahrendt (1961: 70) cited *Schneider 226* as type but without citing any herbarium, while Chamberlain and Hu (1985: 554) cited "*Schneider 226* (holo. Herb. Dendrol. Schneider, n.v.)." This latter specimen cannot be regarded as a holotype and, in any case, was lost with the destruction of the Berlin-Dahlem herbarium in 1942. In these circumstances, *Schneider 226* at A has been designated as the lectotype.

Schneider justified the recognition of a separate variety *vernalis* on the grounds that the leaf margin had fewer spines per centimeter than does the typical *Berberis ferdinandi-coburgii*, while Ahrendt (1961: 70) dif-

ferentiated it on the grounds that it had lanceolate leaves with 10 to 35 marginal spinules on each side versus the 35 to 60 spinules and oblanceolate leaves of *B. ferdinandi-coburgii*. Chamberlain and Hu (1985: 554) elevated variety *vernalis* to a separate species differentiating it from *B. ferdinandi-coburgii* solely on the basis that its leaves had four to five spines per centimeter versus six to eight per centimeter of the latter. In fact, from the numerous specimens available, it is clear that the leaves of *B. ferdinandi-coburgii* are unusually varied, particularly in shape but also in margin. A particularly good example of both variations are the two specimens of *C. W. Wang 82566* from Pingbian Xian cited below. Not only do they have a variety of leaf shapes, ranging from narrowly elliptic to elliptic-lanceolate, but they have some leaves with a high density of marginal spines per centimeter and some with a low density. There is also no significant difference in the flower structures of *B. ferdinandi-coburgii* and variety *vernalis*; hence my synonymy. It is for all these reasons that *B. iteophylla*, *B. pectinocraspedon*, and *B. pingbienensis* are also treated here as synonyms. *Berberis iteophylla* has some leaves that are linear-lanceolate (this is particularly the case with the isotype at LBG) whereas *B. pectinocraspedon* has leaves that are mostly narrowly elliptic. The type of *B. pingbienensis* is simply a typical example of *B. ferdinandi-coburgii*.

Berberis iteophylla is on the current IUCN Red List of Threatened Species (China Plant Specialist Group, 2004c).

It is perhaps worth noting here *A. Henry 10618* s.d. (K K000077323, MANCH) from I-men (Yimen Xian), Yunnan. This was successively identified as *Berberis julianae* (Schneider, 1913: 361), *B. zanlancianensis* (Ahrendt, 1961: 67), and *B. vernalis* (Chamberlain and Hu, 1985: 554). This is similar to *B. ferdinandi-coburgii* except in one particular in that it has two ovules rather than one.

Chamberlain and Hu (1985: 554) noted that Hubei specimens named *Berberis ferdinandi-coburgii* are referable to *B. julianae*.

As noted in the introduction to this study, *Berberis ferdinandi-coburgii* was first reported in Vietnam by Gagnepain (1938: 144) who, citing Petélot as the collector, reported *B. wallichiana* from Chapa (Sapa) in Lào Cai Province. This initial misidentification was almost certainly because both *B. ferdinandi-coburgii* and *B. wallichiana* have only one ovule, a relatively uncommon occurrence in *Berberis* sect. *Wallichianae*. However, given that *B. wallichiana* is otherwise found only in Nepal, this identification was always implausible, but it appears never to have been challenged. Indeed, it has been reproduced by successive authors, e.g., Petélot (1952: 45), Nguyên (1998: 45), Pham (1999: 326), Võ (2007: 120), and Hien et al. (2018).

Gagnepain reported the taxon as "assez rare," while Petélot, who collected the first specimens in 1929, reported it as "Assez commun à Chapa et aux environs." Almost certainly limited to a small geographical area, it subsequently seems to have gone into steep decline. The reason appears to have been its use in traditional medicine. Petélot (1952) noted that an infusion of the stems was used locally as a treatment for toothache, while the World Health Organization (1990: 65) reported "The roots, collected in autumn, are sun-dried or heat-dried" and are used for "treating diarrhoea, dysentery, ophthalmia and dyspepsia." Dinh (1999: 37) notes that between 1973 and 1975 there was mass exploitation of "*Berberis wallichiana*" in the Sapa area, so much so that it was approaching extinction. Osborn and Fanning (2003: 16) also reported the past "extreme over collection" of "*Berberis julianae*" in the Sapa area, but their accompanying photograph is clearly of *B. ferdinandi-coburgii*. It was undoubtedly this intensive and destructive collection that led "*Berberis wallichiana*" to be listed in the *Red Data Book of Vietnam* in 1996 and subsequently for the Government of Vietnam (Government of the Socialist Republic of Vietnam, 2006) to declare "*Berberis wallichiana*" a protected species.

Selected specimens.
Syntype. Berberis ferdinandi-coburgii. S Yunnan: Mengzi (Xian), 1700 m, 8 Nov. s. anno, but pre 1898, *A. Henry 10257* "B" (A 0038746, E E00217994, K K000077316, NY 01104640, US 00103877).

Syntypes. Berberis ferdinandi-coburgii var. *vernalis.* Yunnan fu (Kunming), above Pu chi (Puqi), 2000 m, 7 Mar. 1914, *C. K. Schneider 234* (A 00279497, WU 039246); [N of Kunming], betw. Ta sung chu & Lo che tan behind Lo ku (Luoqu), 12 Mar. 1914, *C. K. Schneider 387* (A 00279625!); near Kunming, Tschangtschung (Changchong) Shan, 1950–2400 m, 28 Feb. 1914, *H. R. E. Handel-Mazzetti 318* (W 1930–0004048, WU 039245); Kunming, Western Hills, ca. 2300 m, 9 Apr. 1922, *H. Smith 1604* (A 00279622, UPS BOT:V-040874); Wenshan Xian, 1800 m, 11 Jan. 1933, *H. T. Tsai 51500* (A 00279507, IBSC 0091717, KUN 0176153, NAS NAS00313968, PE 01031347, SZ 00291235); Wenshan Xian, 1600 m, 8 Feb. 1933, *H. T. Tsai 51704* (A 00279508, IBSC 0091716, KUN 0176147, 0176152, NAS NAS00313967, PE 01031348, SZ 00287826, WUK 0038141); Lu-feng Hsien (Lufeng Xian), 1700 m, 28 June 1933, *H. T. Tsai 53610* (A 00279509, IBSC 0091715, KUN 0177318, NAS NAS00314373, PE [2 sheets] 01031146–47, SZ 00287852, WUK 0035898).

Other specimens.
C & S Yunnan. Anning Xian: Caoxi Si, 20 Mar. 1976, *K. J. Guan 76044* (PE 01031145). **Chiu-Hsiung (Chuxiong Shi):** To-Tsu, 1950 m, 20 Sep. 1939, *M. K. Li 42* (KUN 0176154–55, PE 01031353, WUK 0279206). **Foo-ning (Funing Xian):** Lung-mai, 1000 m, 3 May 1940, *C. W. Wang 89111* (IBSC 0091915, KUN 0176143, 0176702, PE 01031150, WUK 0268063); same details, but *C. W. Wang 89129* (HITBC 003610, IBSC 0091916, KUN 160896, PE 01031151, WUK 0268078). **Jiangchuan Xian:** Xiongguan Xiang, 2050 m, 2 Nov. 1989, *X. D. Yu 1414* (KUN 0176137–38). **Jingdong Xian:** Longjie Xiang, 2400 m, 30 Apr. 1983, *s.*

coll. 46 (HITBC 074432). **Jinghong Shi:** Mengwang, 1520 m, 25 Mar. 1986, *S. R. Guo 0076* (IMDY 0001926). **Kunming Shi:** Xishan, E of Taihua Si, 22 Sep. 1955, *B. Y. Qiu 51059* (IBSC 0091710, KUN 0177303, 0177348, PE 01031421, SZ 00287776, WUK 0193781). **Kwang-Nan (Guangnan Xian):** Wang-ga-tang-tze, 1550 m, 13 Mar. 1940, *C. W. Wang 87789* (KUN 0176142, PE 01030725). **[Lincang Xian]:** Mienning, Poshang, 2700 m, 9 Oct. 1938, *T. T. Yu 17958* (A 00279505, E E00392877, PE 01031355–56). **Lufeng Xian:** Yipinglang Zhen, Zilin Shan, 1900 m, 4 May 1982, *H. Li et al. 532* (HITBC 074372, 074376). **Maguan Xian:** Bazhai, 1585 m, 7 Apr. 1940, *X. Wang & X. P. Gao 100499* (IBSC 0091709). **Malipo Xian:** Hwang-ging-ying, 1200 m, 18 Jan. 1940, *C. W. Wang 83961* (KUN 0176144, PE 01031393). **Mengzi Xian:** Mengjiao, Lao Huayuanqing, 1800 m, 4 July 1958, *Y. Y. Hu & S. K. Wen 580509* (KUN 0177331). **Nanjian Xian:** Leqiu Xiang, Dayao Reservoir, 25.00595°N, 100.345572°E, 2057 m, 2 July 2015, *E. D. Liu et al. 4263* (KUN 1278015–16). **Ping-bien Hsien (Pingbian Xian):** Ji-mu-te, ta-hei-shan, 1900 m, 16 Oct. 1939, *C. W. Wang 82566* (KUN 0176145, PE 01031350). **Shizong Xian:** near Longqin Gongshe, 1680 m, 2 May 1977, *Shizong Team 398* (KUN 0177312). **Shuangbai Xian:** Tuodian, Chaye Qing, 5 Oct. 1958, *S. Q. Huang 190* (KUN 0177580–81). **Simao:** 1525 m, s.d., *A. Henry 11617* (A 00279500, E E00392878, MANCH). **Simao Shi:** Dazhai reservoir, 1500 m, 16 Mar. 2002, *S. S. Zhou 4324* (HITBC 101595). **Tonghai Xian:** Lishan Xiang, Xiang Ping, 1900 m, 11 Aug. 1989, *X. D. Yu 709* (KUN 0176891–92). **Wenshan Xian:** Baozhu Shan, Pishuang Chang, 990 m, 9 Sep. 1992, *Y. M. Shui 546* (KUN 0177276). **Xichou Xian:** Xiaoqiao Gou, 1700 m, 5 Apr. 1959, *Q. A. Wu 7354* (KUN 0176157, WUK 0273364); Fadou, 23.383333°N, 104.816667°E, 28 Aug. 2002, *C. M. Wang & H. Y. Tzeng 06247* (TNM S083182). **Xinping Xian:** Diliu Qu, 1650 m, 28 Oct. 1958, *S. G. Wu 620* (KUN 0177574–75); Xinhua Xiang, 24.13675°N, 101.82036°E, 2 June 2012, *Xingping Xian Survey Team 5304270385* (IMDY IMDY0001919). **Yan-shan (Yanshan Xian):** Shui-tou-dzai, 12 Oct. 1939, *C. W. Wang 84509* (KUN 0176141, 0176149, PE 01031351, WUK 0270312). **Yimen Xian:** San Qu, Judini Xiang, 25 Mar. 1957, *W. Q. Yin 310* (IBSC 0092119, KUN 0177578–79, LBG 00064080, PE 0092119, WUK 0205609). **Yuanjiang Xian:** Yangchajie en route to Shijiaofu, 9 Apr. 1982, *Q. Lin 770494* (KUN 0177156–57). **Zhenyuan Xian:** Wenlong Xiang, Shanshen Si, 2740 m, 12 May 1983, *s. coll. 83* (HITBC 074436).

N VIETNAM. Lào Cai: Chapa (Sapa), ca. 1600 m, Feb. 1929, *A. Petélot 3.378* (A, P P06868292, UC UC378508, VNM); "entre Chapa et la garderie de Lo qui Ho," 1600 m, 1945, *s. coll. 0431* (HNU); Sapa, Mt. Fansipan, 1300 m, 4 June 1963, *N. T. Luong & N. A. Avrorin 12323* (LE); Shaba, Hanlong, 1830 m, 12 Feb. 1964, *Sino-Vietnam Exp. 390* (KUN 0176158); Sapa, s.d., *Thao s.n.* (HN). **s. loc.:** s.d., *s. coll. 1132* (HN 0000014906); s.d., *s. coll. 1608* (HN 0000014904–5); s.d., *s. coll. s.n.* (HN 0000014917).

29. **Berberis fujianensis** C. M. Hu, Bull. Bot. Res., Harbin 6(2): 5. 1986. TYPE: China. N Fujian: Chong'an Xian (now Wuyishan Shi), Huanggang Shan, 1900 m, 30 Apr. 1981, *Wuyi Exp. 2352* (holotype, FJSI 008982 image!; isotypes, FJSI 008981 image!, IBSC 0091751 image!, MO 04182378!).

Shrubs, evergreen, to 1.8 m tall; mature stems pale yellow, sulcate, sparsely verruculose; spines 3-fid, brownish yellows, 1–2 cm, subterete. Petiole 1–3 mm;

leaf blade abaxially pale green, adaxially dark green, slightly shiny, elliptic or elliptic-oblanceolate, 3–5(–7) × 1–2(–2.5) cm, leathery, midvein raised abaxially, impressed adaxially, lateral veins and reticulation inconspicuous on both surfaces, base cuneate, margin spinulose with 2 to 8(to 14) teeth on each side, occasionally entire, apex acute, sometimes minutely mucronate. Inflorescence a fascicle, (2- to)4- to 8-flowered; pedicel 4–7 mm; bracteoles triangular-ovate, ca. 1.5 mm. Sepals in 2 whorls; outer sepals lanceolate, elliptic, or elliptic-oblong, ca. 3 mm; inner sepals obovate or suborbicular, 3.5 mm, apex rounded; petals obovate, 3 mm, base slightly clawed, glands very close together, apex rounded, entire or slightly emarginate. Stamens ca. 2 mm; anther connective shortly apiculate. Ovules 2 or 3. Berry immature. epruinose, ellipsoid, ca. 6 mm; style not persistent.

Phenology. *Berberis fujianensis* has been collected in flower in April and with immature fruit between July and September.

Distribution and habitat. *Berberis fujianensis* is known only from Huangang Shan in Chong'an Xian in north Fujian and from one collection from Suichang Xian in south Zhejiang. It has been found by roadsides, in thickets, forested areas, and mountain summits at 1900–2146 m.

The report of *Berberis fujianensis* in Guizhou by He and Chen (2004: 324) is likely to be based on misidentification (three specimens from Guizhou at HGAS identified as *B. fujianensis* are *B. cavaleriei*).

Selected specimens.
  N Fujian. Chong'an Xian (now Wuyishan Shi): Huangang Shan, Yuhuanglian, 27.883333°N, 117.8°E, 1900 m, 11 Aug. 1964, *C. P. Jian et al. 400636* (PE 01037992–93); Huangang Shan, 27.883333°N, 117.8°E, 1900 m, 12 Aug. 1964, *C. P. Jian et al. 400662* (PE 01037990, 01037994); betw. Tongmuguan & Huanggang Shan, 2100 m, 28 Aug. 1979, *Wuyi Shan Exp. 1073* (IBSC 0091753); Huanggang Shan, near Shanding, 2100 m, 20 July 1980, *Wuyi Shan Exp. 80-0195* (IBSC 0091752, WUK 0427121); Huanggang Shan, 1900 m, 4 Sep. 1980, *Wuyi Shan Exp. 1786* (FJSI 008980, IBSC 0091754); Huanggang Shan, 27.862056°N, 117.784144°E, 2146 m, 26 May 2015, *E. D. Liu et al. WYS 0049* (KUN 1265364–65).
  S Zhejiang. Suichang Xian: Daixikeng, 20 Apr. 1979, *Z. C. Tang & Z. Y. Li 380* (PE 01030744).

30. Berberis gagnepainii C. K. Schneid., Bull. Herb. Boissier, sér. 2, 8: 196. 1908 [as "*gagnepaini*"]. TYPE: China. "Western China," [Sichuan]: 3050 m, July 1903, *E. H. Wilson (Veitch) 3148* (lectotype, designated here, P P00716536!; isolectotypes, A 00038755!, BM BM001015573!, HBG image!, K [2 sheets] K000077335–36!).

Berberis gagnepainii var. *filipes* Ahrendt, J. Bot. 79(Suppl.): 39. 1941. TYPE: China. W Sichuan: "SE of Tachien-lu" (Kangding), 2100–3050 m, July 1908, *E. H. Wilson 1137a* "A" (lectotype, designated here, A 00038756!; isolectotypes, K missing, US 00945897 image!).
Berberis gagnepainii var. *lanceifolia* Ahrendt f. *pluriflora*, J. Linn. Soc., Bot. 57: 53. 1961. TYPE: China. WC Sichuan: Mupin (Baoxing), Nov. 1908, *E. H. Wilson 1137* "B" (holotype, K missing; lectotype, designated here, E E00217936!; isolectotypes, A 00279538!, US 00945896 image!, W 1914–4807!).

Shrubs, evergreen, to 2 m tall; mature stems pale yellow, terete, scarcely verruculose; spines 3-fid, concolorous, 1–4 cm, stout, abaxially flat or sulcate. Petiole almost absent; leaf blade abaxially yellow-green, adaxially dull green, narrowly elliptic, elliptic-lanceolate, or lanceolate, 3.5–11 × 0.4–2.2 cm, leathery, midvein raised abaxially, slightly impressed adaxially, lateral veins inconspicuous abaxially, conspicuous or not adaxially, reticulation obscure or absent, inconspicuous on both surfaces, base cuneate, margin spinose with 6 to 20 teeth on each side, apex acuminate, mucronate. Inflorescence a fascicle, 3- to 8(to 15)-flowered; pedicel 10–35(–40) mm; bracteoles triangular, 2 × 1 mm. Flowers greenish yellow. Sepals in 3 whorls; outer sepals triangular-ovate, sometimes with pinkish vertical stripe, 3.5 × 3 mm; median sepals with greenish vertical stripe, broadly ovate, ca. 5–6 × 5.5 mm; inner sepals orbicular-obovate, ca. 7–8 × 6.5–8 mm; petals orbicular-obovate, ca. 7 × 6 mm, base cuneate or slightly clawed, glands separate, apex slightly emarginate. Stamens ca. 4 mm; anther connective not extended, truncate. Ovules (2 to)4 to 6. Berry dark blue, blue pruinose, oblong-ovoid, 8–10 × ca. 6 mm; style not persistent.

Phenology. *Berberis gagnepainii* has been collected in flower from April to July and in fruit between August and November.

Distribution and habitat. *Berberis gagnepainii* is known from west, west-central, and north-central Sichuan. It has been collected from thickets, forest margins, and understories at ca. 1100–3000 m.

The protologue of *Berberis gagnepainii* cited only *Wilson 3148* and stated "Typ Herb. Paris, Hamburg etc," this being the format Schneider used when he recognized that the specimens he had seen had duplicates elsewhere. The specimen at P has been chosen here as the lectotype rather than that at HBG because it has much better floral material. X. H. Li et al.'s (2015a: 32) description of P00716536 as a holotype was mistaken.

The situation in relation to *Berberis gagnepainii* var. *lanceifolia* f. *pluriflora* and *B. gagnepainii* var. *filipes* is somewhat complex in that there is a *Wilson 1137* and a

Wilson 1137a and both consist of two gatherings made at different times of the year. The gathering of *Wilson 1137* of June 1908 is designated here as *Wilson 1137* "A" and that of November 1908 as *Wilson 1137* "B," that of *Wilson 1137a* of July 1908 as *Wilson 1137a* "A" and that of October 1908 as *Wilson 1137a* "B." The protologue of *B. gagnepainii* var. *lanceifolia* f. *pluriflora* cited "*Wilson 1137* (in part)" as the type. Since Ahrendt described the type as having fruit, I have assumed that this refers to *Wilson 1137* "B." Ahrendt described both *Wilson 1137* "A" and *Wilson 1137a* being at K (presumably with *1137* "A" and "B" being on the same sheet as they are at A and W and *1137a* "A" and "B" are at A), but both are missing.

There have been confusions in relation to *Berberis gagnepainii*, mostly caused by Ahrendt. The collection details of the type give the place of collection simply as W China and the date as July 1903. From Briggs (1993: 33–37) it is clear that, during this month, Wilson was in west-central Sichuan, mostly in Kangding Xian. Ahrendt (1961: 52) and Ying (2001: 114), therefore, erred in stating the type was collected in west Hubei. Ahrendt (1941b: 39) further confused matters by publishing a *B. gagnepainii* var. *lanceifolia* whose type is a cultivated specimen from seed collected in west Hubei. This is not *B. gagnepainii* but *B. veitchii* (for further details, see under that species). In fact, contrary to Ahrendt, Chamberlain and Hu (1985: 535), Fu (2001: 391), and Ying (2001: 113, 2011: 737), *B. gagnepainii* has not been found in Hubei. It is, therefore, unfortunate that the Chinese name for the species is 湖北小檗 (hubei xiao bo).

Schneider's protologue, which cited only the type, noted its flowers were not well preserved and hence he was unable to give an account of the shape of the sepals and petals. However, C. M. Hu made a dissection of two of the flowers of K000077335 and a flower of K000077336 and these are on their sheets. They are of poor quality, but they appear to confirm the description given by Schneider (1942: 52–53) which, from the specimens cited, would seem to come from either or both *Wilson 1137* "A" or *Wilson 1137a* "A" except that, while Schneider gave the number of ovules as "3–5 (rarely 2)," Hu recorded four to six. Schneider's account is the same as mine from flowers of cultivated plants grown from *Howick 1586* collected in Luding Xian (details below) and, as such, in slightly expanded form is the source of the description used above modified by Hu's findings in relation to the number of ovules. This differs somewhat from the description given by Ying (2001: 113) which reproduces that given by Ahrendt (1961: 52) for variety *lanceifolia* where Ahrendt appears to have drawn on a cultivated plant of unspecified origin.

One issue remains and that is the fact that the leaves of the lectotype and the isolectotypes have a somewhat different shape from almost all other specimens identified as *Berberis gagnepainii* in that they are mostly narrowly elliptic or elliptic-lanceolate with a ratio of length to width of approximately 4:1, whereas the leaves of *Wilson 1137*, *Wilson 1137a*, *Howick 1586*, and all other specimens cited below with one exception are mostly lanceolate with a ratio of length to width of approximately 5 or 6:1. The exception is *Hsieh 41252* where the majority of the leaves are more like those of the type. But these also have some leaves which are lanceolate. I have therefore followed Chamberlain and Hu (1985: 535) in treating as synonyms Ahrendt's variety *filipes* and variety *lanceifolia*, forma *pluriflora*, the types of both of which were identified by Schneider (1942: 53) as simply *B. gagnepainii*.

The report of *Berberis gagnepainii* in Guizhou by He and Chen (2004: 37) and Ying (2001: 113, 2011: 737) appear to refer to a similar but different species, *B. tengii*. The report by Ying (2001: 113, 2011: 737) of *B. gagnepainii* in Yunnan is likely to be based on misidentification.

For *Berberis gagnepainii* var. *omeiensis* and variety *subovata*, see under *B. simulans*.

Selected specimens.
Syntype. *Berberis gagnepainii* var. *filipes*. W Sichuan: "SE of Tachien-lu" (Kangding), 2100–3050 m, Oct. 1908, *E. H. Wilson 1137a* "B" (A 00338444, K missing).

Other specimens.
Sichuan. Baoxing Xian: Mupin, June 1908, *E. H. Wilson 1137* "A" (A 00279534); "Pao-Hsing Hsien," 2700 m, 29 June 1936, *K. L. Chu 2998* (BM, IBSC 0091760, 0091762, K, P P02313111, PE 01031196–98); Lingguan Qu, Daxi Gongshe, Daping Shan, 1525 m, 21 Apr. 1983, *D. Y. Peng 47439* (CDBI CDBI0027751–52). **[Ebian Xian]:** Wa-shan, 2135–2745 m, Sep. & Nov. 1908, *E. H. Wilson 1344* (A 00279537, BM). **Ganluo Xian:** Dayao Shan, 2600 m, 15 Sep. 1979, *Survey Team 909* (SM SM704800412). **Hanyuan Xian:** Xiangling Shan, 2000 m, 9 Sep. 1938, *T. P. Wang 9632* (KUN 0175751–52). **Jiulong Xian:** Hongba, Yeren Gou, 2500 m, 26 June 1991, *Z. B. Liu & X. R. Yi 9 98* (Ganzi Institute of Forestry Herbarium). **Kangding Xian:** 11 Aug. 1930, *Z. P. Huang et al. 1751* (IBSC 0092459, KUN 0175753, PE 01031202). **Kuan Hsien:** (Guan Xian, now Dujiangyan Shi), 1070–1220 m, 14 July 1928, *W. P. Fang 2372* (A 00279540, IBSC 0091761, NAS NAS00314023, NAU 00024174, NY, P P02313108, PE 01031950). **Luding Xian:** Gongga Shan, E slope, Yanzi Gou, 2600 m, 1 July 1982, *Y. K. Lang et al. 374* (KUN 0178907, PE 01031951–52). **Muchuan Xian:** 1200 m, *Y. F. Xu 240042* (MO 04750890). **Shih-mien-hsien (Shimian Xian):** 1955, *C. C. Hsieh 41252* (HGAS 013641, IBSC 0091755, PE 01031193). **Tianquan Xian:** Erlang Shan, Tuanniuping, 2370 m, 24 Aug. 1953, *X. L. Jiang 35359* (IBSC 0092720, IBK IBK00012754, PE 01031204); Erlang Shan, 19 Oct. 1951, *W. G Hu & Z. He 11623* (PE 01031275). **[Yingjing Xian]:** Tahsiangling (Daxiangling), 2200 m, 28 June 1922, *H. Smith 2141* (PE 01031096, UPS BOT:V-040877);

Tahsiangling, 19 Nov. 1934, *H. Smith 13491* (MO 4367286, PE 01031098, UPS BOT:V-040875).

Cultivated material:
Living cultivated plants.
Howick Hall, Northumberland, U.K., and Royal Botanic Garden Edinburgh, U.K., from seeds of *C. Howick 1536*, Luding Xian, Hailuogou Glacier Park, Gongga Shan, 2160 m, 7 Oct. 1991.

31. Berberis griffithiana C. K. Schneid., Bull. Herb. Boissier, sér. 2, 5: 403. 1905. TYPE: Bhutan. [1838], *W. Griffith 125* (holotype, W 0015487!; possible isotypes, *W. Griffith 125*, GH 00038828!, P P00716586!, *W. Griffith 1741/125* K ex Herb. Hook. K000340243!, *W. Griffith 1741*, E E00170024!, G G00226026!, K ex Herb. Hook. K000340244!).

Berberis bhutanensis Ahrendt, J. Bot. 79(Suppl.): 17. 1941, syn. nov. TYPE: Bhutan. 1838, *W. Griffith 1741/125* (lectotype, designated here, K ex Herb. Hook. K000340243!).
Berberis wallichiana DC. var. *pallida* Hook. f. & Thomson, Fl. Ind. 1(2): 226. 1855. TYPE: Bhutan. 1838, *W. Griffith 1742* (lectotype, designated here, ex Herb. Hook. K K000340245!; isolectotypes, BM BM001015551!, CAL CAL17262 image!, E E00170030!, G G00303108!).
Berberis replicata W. W. Sm. var. *dispar* Ahrendt, J. Bot. 79(Suppl.): 20. 1941. TYPE: China/India. Border betw. Arunachal Pradesh, [Tawang] & Xizang (Tibet), Cona (Cuona) Xian, betw. Pangchen (27.683333°N, 91.8°E) & Le (27.783333°N, 91.833333°E), Nyam Jang Chu, 2300 m, 4 Apr. 1936, *F. Ludlow & G. Sherriff 1274* (holotype, BM BM000559452!).
Berberis griffithiana var. *pallida* D. F. Chamb. & C. M. Hu, Notes Roy. Bot. Gard. Edinburgh 42: 547. 1985. *Berberis wallichiana* DC. var. *pallida* Hook. f. & Thomson, Fl. Ind. 1(2): 226. 1855, p.p., syn. nov. TYPE: Bhutan. 1838, *W. Griffith 1741* (lectotype, designated here, K ex Herb. Hook. K000340244!).

Shrubs, evergreen, to 3 m tall; mature stems pale yellowish brown, terete; spines 3-fid, 1.2–2.5 cm, concolorous. Petiole almost absent; leaf blade abaxially pale green and glaucous or white pruinose at first, becoming mainly epruinose and shiny, adaxially dark green, narrowly elliptic-lanceolate, 1.2–5 × 0.4–0.9 cm, leathery, midvein raised abaxially, slightly raised adaxially, lateral veins inconspicuous on both surfaces, reticulation absent on both surfaces, base cuneate, margin revolute, slightly undulate, spinulose or spinose with 2 to 4(or 5) teeth on each side, apex acuminate, mucronate. Inflorescence a fascicle, 2- to 6-flowered; pedicel (7–)11–22 mm; bracteoles ovate, ca. 2 × 1 mm, apex acuminate. Sepals in 2 whorls; outer sepals ovate, 3.5–5 × 2–4 mm, apex acute; inner sepals oblong-obovate, ca. 6–8 × 4–6 mm, apex rounded, obtuse; petals obovate, ca. 5–7 × 3–4 mm, base clawed, glands separate, narrowly ovate, apex emarginate. Stamens 4–5 mm. Ovules 2 to 4. Berry purplish black, slightly pruinose, ellipsoid, 7–9 × 5–6 mm; style persistent.

Phenology. In China *Berberis griffithiana* has been collected in flower in April and in fruit between July and September.

Distribution and habitat. In China *Berberis griffithiana* is known from Cona (Cuona) Xian in southeast Xizang (Tibet). It is also known from Bhutan and from Tawang and West Kameng in that part of Arunachal Pradesh in northeast India where there is a conflicting territorial claim by China. In China it has been collected by roadsides, in open jungle scrub, and forest margins at 2300–3300 m.

There has been considerable but understandable confusion in relation to the type of *Berberis griffithiana*. This is rooted in two separate but related issues. The first concerns *B. wallichiana* var. *pallida* Hook. f. & Thomson (1855: 226), the second the complexities of the numbering of Griffith specimens. The protologue of *B. wallichiana* var. *pallida* reads as follows:

Berberis Wallichiana . . . δ pallida; foliis anguste lanceolatis 2–3 pollicaribus, spinuloso-dentatis, subtus pallidis, glaucisve, fasciculis paucilfloris . . . δ Bhotan, Griffith (!), (Flore vere). Of this variety we have two forms from Griffith, of which one differs conspicuously from the ordinary form of *B. Wallichiana* in the distinctly glaucous under-surface of the leaves, approaching *B. asiatica* in this respect, from which it differs in the long slender spines and lanceolate leaves, which are not lacunose. It is very probable that its glaucous hue is due to the bushes having grown in dry places. The other specimens have not the glaucous under-surface but agree in every other respect; and indeed considering how variable the glaucous character it is quite possible these two forms grew on the same bush.

It appears from this that the authors based their observations on at least three specimens and, from the reference to "bushes" rather bush in relation to the first of these forms, probably to four or more. Two of these specimens can be identified as *Griffith 1741* (K K000340244) and *Griffith 1742* (K K000340245). These are ex-Herb. Hooker (which was sold to Kew in 1867) and are both annotated in Hooker's hand as "B. Wallichiana DC var. δ pallida, Fl. Ind." Importantly, both of these specimens have some leaves that appear to be abaxially pruinose, glaucous in the case of *1741*, whitish in the case of *1742*. Equally importantly, there seems no reason to doubt they are of the same species.

It is useful to report at this stage that these leaf characteristics are exactly the same as those of the living cultivated plants cited below. These plants have leaves that start out as abaxially green and partially or wholly white or glaucous pruinose, a condition which remains on some lower leaves where the glaucous covering tends to become more whitish while those further up the branches become abaxially pale green and epruinose.

Berberis wallichiana "var. 4 *pallida*" reappeared in Hooker and Thomson (1875: 111) and was described as having "leaves 1–2 in, narrow lanceolate, spinulose pale and glaucous beneath, fascicles few flowered – Bhotan," with no commentary but a reference to *B. asiatica* Griff. and to plate DCXLVIII in Griffith (1849). Plate DCXLVIII is entitled *"Berberis asiatica"* but, in fact, clearly depicts a species of section *Wallichianae*.

Lamond (1970: 163–165) appears to have been the first to have noted that Griffith's collections have three different numbering systems. The first system was published in Griffith (1848) and largely follow the date order in which the specimens were collected and which Lamond usefully refers to as "itinerary numbers." The second system is Griffith's subsequent attempt to arrange his specimens (whether collected in Bhutan or elsewhere) systematically in families. *Griffith 1741* and *1742* (the numbers being in Griffith's hand) are from this second numbering. The third system is not Griffith's, but were the numbers given by Hooker to Griffith specimens in Griffith's personal herbarium (which Thomson may have seen in Griffith's lifetime) bequeathed on his death in 1845 to the East India Company and inaccessible between then and 1858 when it was transferred into Hooker's charge. A catalogue of these specimens is in Hooker (1865). Specimens with either or both the first two sets of numbers (and some without either) were variously distributed including to CAL, the Herb. Lemann, and the Herb. Hooker. Specimens with the third set of numbers were widely distributed by Hooker from Kew, though it is clear he retained some which he added to the Herb. Hooker.

Schneider (1905: 403) reported variety δ *pallida* to be a distinct variety from Bhutan of *Berberis wallichiana* (found in Nepal), but cited no specimens, while on the same page publishing a new species, *B. griffithiana*, citing only *Griffith 125*, east Himalayas, "Typus in Herb. Hofmuseum, Wien." The printed label on this exceptionally poor specimen reads "Herbarium of the late East India Company . . . Herb. Griffith. Distributed at the Royal Gardens, Kew, 1861–2." The number *125* is handwritten and is an example from the third numbering system. It is listed in Hooker (1865: 2) as *Berberis* "Wallichiana D. C. var."

Later, Schneider (1913: 364) cited as a synonym of *Berberis griffithiana*, "*B. wallichiana* var. *pallida*, Hooker f & Thomson Fl. Brit. Ind. 1 III (pro parte). (1872) secundum specimen originale," commenting that *Griffith 125* was the same as *Griffith 1741* with a reference to a specimen at Kew. This was to K K000340243, which is labelled both *1741* in Griffith's hand but also *125* on a separate label identical to that of the Griffith specimen in Vienna. An annotation to the sheet itself in Hooker's hand states the specimen is from "Hb. Griff."

Matters might have rested there had not Schneider (1918: 29–31) returned to the subject. Here in commentaries under *Berberis wallichiana* var. *pallida* and *B. griffithiana*, he tried to show why he had been mistaken in 1913 and that, while *Griffith 125* was *B. griffithiana*, *Griffith 1741* was the different *B. wallichiana* var. *pallida*. It is clear from these commentaries that what had mainly prompted this change of mind was that he had now read Griffith (1847, 1848, 1849). From this, he speculated that *Griffith 125* was the *B. asiatica* depicted in the plate referred to above and also the *B. asiatica* mentioned in Griffith (1847: 211) as being in Khegumpa on 25 January 1838 and which he stated was listed in Griffith (1848: 122) as "Nr. 38" (this is a misprint; Griffith's itinerary number for this is actually 388). Schneider, however, admitted to being perplexed as to why this list did not include any *Berberis* with the number *125* but (clearly unaware of Hooker's Catalogue of 1865) concluded this was because numbers listed in 1848 had not always been put on the right specimens. Interestingly, he failed to ask why *1741* and *1742* were not to be found in Griffith (1848) either, given that his itinerary list of Bhutanese specimens end at no. *1191*.

Schneider, however, then changed his opinion yet again and, in annotations to the specimens of *Griffith 1741* at Kew dated 14 March 1928, identified them both as *Berberis griffithiana*, only to retract this in his last treatment of *Berberis* sect. *Wallichianae* (1942: 14, 28–29), where he stated that *B. wallichiana* var. *pallida* was not a variety of *B. wallichiana*, but possibly a variety of another species from Bhutan, *B. praecipua* (Schneider, 1939: 248), while stating of *B. griffithiana* "Griffith sammelte [collected] den Typ am 25 Januar 1838 zu Khegumpa (Nr. *125* und *1742*)," which implied that *Griffith 1742* was a parallel number for *Griffith 125*.

From the above, it can be seen the importance of the identity of the specimen at Kew numbered both *1741* and *125*. For, if it is the same as the other *Griffith 1741* at Kew and the *Griffith 125* at Vienna, then specimens of *Griffith 1741* are also isotypes of *Griffith 125*. I can find no reason to doubt that the two Kew specimens of *Griffith 1741* are the same species, but whether they are of the same gathering as *Griffith 125* at Vienna, it is impossible to say, given it is unknown on what basis Hooker gave all three specimens of *Griffith 125* cited above the same number as the specimen he kept himself, which is the only one also labelled *1741*. Indeed, it is possible that the holotype of *Griffith 125* is actually *Griffith 1742* rather than *1741* as Schneider appeared to suggest in 1942. The same may also be true of either or both the specimens of *Griffith 125* at GH and P. For this reason, I have treated all specimens of *Griffith 125* other than the holotype and all specimens of *Griffith*

1741 as possible rather than definite isotypes of *B. griffithiana*.

Ahrendt (1941b: 17) published *Berberis bhutanensis* with *B. wallichiana* var. *pallida* as a synonym and the type *Griffith 1741* at Kew. No mention was made of Schneider's previous treatments of *Griffith 1741*, nor that there were two specimens with this number at Kew (both being annotated by Ahrendt on 9 April 1940 as types) and that one of them was also numbered *Griffith 125*. K000340243 has been chosen here as the lectotype since this has leaves that most closely resemble Ahrendt's description.

Chamberlain and Hu (1985: 547), who made no reference to Schneider's successive changes of mind in relation to *Griffith 125* and *1741* nor to the work of Lamond, differentiated between a *Berberis griffithiana* var. *griffithiana* and a *B. griffithiana* var. *pallida*, the latter as a comb. nov. for *B. wallichiana* var. *pallida*. They stated that variety *griffithiana* had two syntypes, *Griffith 125* (W) and *Griffith 1742* (E, K) while the holotype of variety *pallida* was *Griffith 1741* (K). The two varieties were differentiated on the basis that variety *griffithiana* had green leaves that were epruinose whereas variety *pallida* had leaves that were glaucous and abaxially more or less pruinose.

All of this needs to be unpicked. First, as noted above, Schneider cited only *Griffith 125* (W) as type so *Griffith 1742* is not a syntype. Secondly, as also noted above, the protologue of *Berberis wallichiana* var. *pallida* was clearly based on more than one specimen. It therefore follows there can be no holotype of *B. griffithiana* var. *pallida*. Nevertheless, their holotype designation of *Griffith 1741* at K can be regarded as a first-step lectotypification. It is completed here by selecting K K000340244 rather than K K000340243, because the former was certainly available to Hooker and Thomson before 1855 while the latter may not have been. As to their differentiation between their two varieties, again as noted above, leaves on the same plants of *B. griffithiana* can be pruinose or epruinose.

There remains the question of where the specimens of *Griffith 1741* and *1742* were collected. By a process of elimination of the variously numbered Griffith specimens of other *Berberis* species from Bhutan in CAL, E, G, K, and M, I deduce that one of them is, indeed, *Griffith 388* from Khegumpa (identified by Long [1979] as Keri Gompa in east Bhutan, 27.033333°E, 91.416667°N). As to the other, it is possible that it was collected at the same time but just looked different to Griffith when he came to catalogue it, or is of a gathering that did not have an itinerary number. Griffith (1847: 219, 221, 245, 247) recorded his further observations of "Berberis asiatica" on 29 and 31 January and on 15 February 1838.

Chamberlain and Hu (1985: 547) also treated *Berberis subpteroclada* Ahrendt and *B. subpteroclada* var. *impar* Ahrendt (Ahrendt, 1941b: 21–22), both from Chendebi in central Bhutan, as synonyms of *B. griffithiana* var. *griffithiana* and *B. replicata* var. *dispar* (Ahrendt, 1941b: 20), *B. leptopoda* (Ahrendt, 1941b: 33), and *B. taronensis* var. *trimensis* (Ahrendt, 1961: 79) from Arunachal Pradesh as synonyms of *B. griffithiana* var. *pallida* (*B. replicata* var. *dispar* had already been treated as a synonym of *B. griffithiana* by Ahrendt [1961: 43]). However, the types of *B. subpteroclada* and *B. subpteroclada* var. *impar* (both BM) do not appear to be *B. griffithiana*, while *B. leptopoda* is treated here as a separate species.

Husain et al. (1994, 1997) compared the type of *Berberis taronensis* var. *trimensis* with that of *B. replicata* var. *dispar* and concluded they were not the same, and that the former was a synonym of *B. griffithiana* var. *pallida* and the latter indeed a variety of *B. replicata*. However, although they are different, *B. taronensis* var. *trimensis* is not *B. griffithiana*, but *B. replicata* var. *dispar* is (there appears to be no reason to associate *B. taronensis* var. *trimensis* with *B. taronensis*, the type specimen of which is from northwest Yunnan and which is treated here as a synonym of *B. amabilis*).

Grierson (1984: 326) suggested that *Berberis praecipua* is "possibly only subspecifically distinct from *B. griffithiana*." However, living plants of *B. praecipua* are easily distinguishable from those of *B. griffithiana* (pers. obs. western Bhutan in 1997 and from a living plant in my own collection grown from seeds of *R. Liddington s.n.*, western Bhutan, Thimphu Chu, 1999). In particular, the branches of *B. griffithiana* are more delicate than those of *B. praecipua* and the leaves narrower, darker, and more revolute. As noted above, the leaves of *B. griffithiana* are often abaxially pruinose, especially when young, whereas those of *B. praecipua* never are. The young shoots and new leaves of *B. praecipua* are usually bright pink, whereas those of *B. griffithiana* are pale green. The flower structure and the number of flowers per fascicle of the two species are also different.

Ahrendt (1941b: 17) cites as an example of *Berberis bhutanensis Kingdon-Ward 13421* (BM) from north Myanmar. This is *B. amabilis*. Ying (2001: 119–120, 2011: 738) accepted Chamberlain and Hu's varietal division of *B. griffithiana*.

Syntypes. Berberis wallichiana var. *pallida*. Bhutan. [1838] *Griffith 1742* (BM BM001015551, E E00170030, G G00303108, K K000340245); *Griffith s.n.* (CAL CAL17263).

Selected specimens.
SE Xizang (Tibet). Cona (Cuona) Xiang: Mazhi Xiang, 3030 m, 6 Aug. 1974, *Qinghai-Xizang Team, Botany Group 2261* (PE 00049419); betw. Mama Xiang & Jiba Xiang, 3200 m,

19 July 1975, *Qinghai-Xizang Supplementary Collections 75-1086.* (HNWP 54089, KUN 0178798–99, PE 01031863–64); Mama, 3000 m, 26 Aug. 1975, *Qinghai-Xizang Supplementary Collections 75-1599* (HNWP 52218, KUN 0178800–01, PE 01840016, 01031862); Ma, 2900 m, 2 Sep. 1975, *Qinghai-Xizang Supplementary Collections 75-1814* (HNWP 52428, KUN 0178271–72, PE 01031860, 01031867); Sesen Gou, Jibian Gou, 3300 m, 11 Sep. 1975, *s. coll. 75-1754* (PE 01031861); Bule Gou, Mama Xiang, 27.908472°N, 91.8085°E, 3082 m, 17 Sep. 2010, *Y. D. Tang & Q. W. Lin 2010-099* (PE 02036564).

INDIA. **Arunachal Pradesh:** W Kameng, Bomdila, on way to Dirang, 2530 m, 4 May 1993, *T. S. Rama & Party 210335* (LWG). E BHUTAN. Takhtoo, Gamri Chu, 27.383333°N, 91.783333°E, 2285 m, 15 Mar. 1936, *F. Ludlow & G. Sherriff 1187* (BM); Bhutan-Monyül [West Kameng] frontier, Changpu, 27.5°N, 91.666667°E, 21 Mar. 1936, *F. Ludlow & G. Sherriff 1221* (BM); Tobrang, Yangsi Chu, 27.75°N, 91.466667°E, 2590 m, 2 May 1949, *F. Ludlow, G. Sherriff & J. H. Hicks 20585* (BM).

Cultivated material:
Living cultivated plants.
Sherwood Gardens, Devon, and Foster Clough, Mytholmroyd, West Yorkshire, U.K., from *A. Clark 5260,* Arunchal Pradesh, W Kameng, near Chander, Oct. 2004; Foster Clough from plant supplied by Seaforde Gardens, County Down, Northern Ireland, from *SF 06008,* Arunchal Pradesh, Tawang, near Urgelling monastery, 24 Nov. 2006; Exbury Gardens, Hampshire, U.K., from *K. Rushforth 9438,* Arunachal Pradesh, W Kameng, betw. Tongri (27.442444°N, 92.381306°E, 3145 m) & Chori campsite (27.434111°N, 92.351056°E, 3004 m), 28 Oct. 2008.

Cultivated specimens.
Foster Clough, Mytholmroyd, West Yorkshire, U.K., 1 May 2015, from *SF 06008* (details above), *J. F. Harber 2015-19* (A, E, MO, PE, TI); same details, but 7 Feb. 2016, *J. F. Harber 2016-01* (A, E, TI).

32. Berberis grodtmanniana C. K. Schneid., Oesterr. Bot. Z. 67: 32. 1918, as "*grodtmannia.*" TYPE: China. S Sichuan: Yen Yüan Hsien (Yanyuan Xian), betw. Kalapa & Liuku, ca. 3500 m, 7 May 1914, *C. K. Schneider 1268* (lectotype, designated by Ahrendt [1961: 44], K K000077349!; isolectotypes, A 0038760!, E E00217940!).

Berberis grodtmanniana var. *flavoramea* C. K. Schneid., Repert. Spec. Nov. Regni Veg. 46: 256. 1939, syn. nov. TYPE: China. NW Yunnan: Yungning (Ninglang Xian), 3000–3300 m, 22 Apr. 1922, *F. Kingdon-Ward 5078* (lectotype, designated by Schneider [1942: 33], E E00217939!).
Berberis jinshajiangensis X. H. Li, J. Trop. & Subtrop. Bot. 15(6): 553. 2007, replacement name for *Berberis micropetala* T. S. Ying, Acta Phytotax. Sin. 37: 313. 1999, not C. K. Schneid. (1939), syn. nov. TYPE: China. N Yunnan: Dongchuan Shi, Yinmin, Yubaoqing, 2800 m, 1 Apr. 1986, *S. B. Lan 636* (holotype, PE 00935167!; isotype, PE 00935168!).

Shrubs, evergreen, to 3 m tall, rigidly upright; mature stems reddish brown turning yellow (sometimes purplish when dry), stout, very sulcate, not verruculose; spines 3-fid, reddish brown, 1–3 cm, stout, abaxially terete or flat. Petiole almost absent; leaf blade abaxially pale green, slightly shiny, adaxially dull green, lanceolate or narrowly elliptic, 3–7(–14) × 0.5–1.2(–2.4) cm, leathery, midvein raised abaxially, impressed adaxially, lateral veins inconspicuous on both surfaces, reticulation obscure on both surfaces, base cuneate, margin conspicuously revolute, especially on upper leaves, spinulose with 7 to 12(to 15) teeth on each side, apex acute or acuminate, mucronate. Inflorescence a fascicle, 5- to 10-flowered; pedicel (1–)3–4 mm; bracteoles reddish in upper part, lanceolate, ca. 2 × 1 mm, apex acute. Flowers bright yellow. Sepals in 2 whorls; outer sepals ovate or ovate-elliptic, sometimes with faint reddish stripe, ca. 4 × 2 mm; inner sepals obovate or obovate-elliptic, ca. 5 × 2–4 mm; petals obovate, 4.5–5 × 2.5–3 mm, apex emarginate or entire. Stamens ca. 3 mm; anther connective slightly extended, truncate. Ovules 1(to 3), sessile. Berry blue, epruinose, ellipsoid, 8–9 × ca. 4 mm; style persistent, short.

Phenology. *Berberis grodtmanniana* has been collected in flower in April and in fruit between May and October.

Distribution and habitat. *Berberis grodtmanniana* is known from south Sichuan with a few collections from adjoining areas of Yunnan. It has been collected from thickets, open grassy slopes, forests, and trail and runlet sides at ca. 2100–3500 m.

The protologue of *Berberis grodtmanniana* indicated *Schneider 1268* to be the type but did not indicate the herbarium. Ahrendt (1961: 44) lectotypified the specimen at K. In the protologue of *B. grodtmanniana* var. *flavoramea,* Schneider cited four gatherings, including *Kingdon Ward 5078* at E, which he subsequently (1942: 33) designated as the type.

In the protologue of *Berberis grodtmanniana,* Schneider describes the mature stems of the type as being reddish purple, while these were reported as "dark purple" by Ahrendt (1961: 44), "purplish" by Chamberlain and Hu (1985: 552), and "dark grey" by Ying (2001: 124, 2011: 740). It is certainly true that the isolectotype at A has one stem that is very distinctly purple though another is more brownish purple. Both the lectotype at K and the isolectotype at E, however, have stems which are more a reddish brown. In his protologue to variety *flavoramea,* Schneider's sole reason for distinguishing this as a variety is that it has young branches (ramulus juvenilibus) that are yellow or distinctly yellowish. However, while the mature stems of the holotype at E are yellow, the mature stems of the syntypes at W (*Handel-Mazzetti 2138, 3275,* and *7184,* as well as other Handel-Mazzetti specimens at WU, including *Handel-Mazzetti 1439* and *2738*) have mature stems ranging from yellow to reddish brown. The cultivated plants grown from *E. D. Hammond (China) 97063* and *B. & S. Wynne-Jones 7889* cited below have mature stems that are very distinctly reddish brown, but which turn yellow on aging. My deduction from all of this is that the color variations in the various herbarium specimens cited here are likely to be a result of either their age or, in the case of purple, from the drying process (in relation to the drying process, see discussion under *B. deinacantha* for something similar for that species).

The only significant difference between *Berberis jinshajiangensis* and *B. grodtmanniana* would seem to be that the former is reported to have three whorls of sepals rather than two. *S. B. Lan 348* (details below), cited in the protologue of *B. micropetala*, is also *B. grodtmanniana*.

The flowers of *E. D. Hammond (China) 97063* and *B. & S. Wynne-Jones 7889* have up to three ovules versus the one of dissected flowers of both *Schneider 1268* and *Kingdon-Ward 5978*.

The description of fruit used above comes from nos. *97063* and *7889*. It is also worth noting that those plants have lower leaves that, when in shade, are slightly shiny and non-revolute. The flat nature of some leaves was noted by Schneider (1939: 256) in relation to *Handel-Mazzetti 3275*.

Though *Berberis grodtmanniana* can grow up to 3 m in height, *Boufford, Harber & Li 43507* cited below and others I saw growing in the same exposed area in Qiaojia Xian were no more than 50 cm tall.

Syntypes. Berberis grodtmanniana var. *flavoramea.* **S Sichuan**: [Yanyuan Xian], Yalung river toward Yenyüen [Yanyuan] above Daliaopingdse village, Dadjin Mtns., 27.516667°N, 2550 m, 11 May 1914, *H. R. E. Handel-Mazzetti 2138* (W 1925–0003931); [Muli Xian], Muli monastery area toward Yungning, 3200–3330 m, 24 July 1915, *H. R. E. Handel-Mazzetti 7184* (W 1925–0003933, WU 0039248). **Yunnan**: [Ninglang Xian], betw. town of Yungbei & village of Yungning, in Hungguwo Mtns., near Hsinyingpan, 3110–3400 m, 28 June 1914, *H. R. E. Handel-Mazzetti 3275* (WU 0039252).

Selected specimens. **S & SW Sichuan. Huili Xian:** Lungdschu-schan (Longzhou Shan), near town of Huili, 2700–3600 m, 26 Mar. 1914, *H. R. E. Handel-Mazzetti 887* (WU 0039249); southern mtns. near Hui li chou, 3500 m, 26 Mar. 1914, *C. K. Schneider 588* (A 00279532); Longzhou Shan 3300 m, 24 May 2007, *Y. Y. Geng et al. 20070160* (WCSBG 015099). **Jinyang Xian:** 3100 m, 6 May 1959, *Sichuan Economic Plants Exp. Liangshan Group (1959) 303* (PE 01031342–43). **Leibo Xian:** near Gudui, 2100 m, 1 July 1959, *Sichuan Economic Plants Exp. Liangshan Group (1959) 1038* (PE 01031325–26). **Meigu Xian:** Chenguan, 13 July 1959, *Sichuan Economic Plants Exp. Liangshan Group (1959) 1088* (PE 01031322). **Muli Xian:**

913 Linchang, Zhani Gou, 2700 m, 23 Aug. 1983, *Qinghai Xizang Team 13311* (KUN 0178879–81, PE 01031323, 01031332). **[Xichang Shi]:** Lose-schan S of Ningyüen (Xichang), 16 Apr. 1914, 2900–3500 m, *H. R. E. Handel-Mazzetti 1439* (WU 0039250); [Yichang-Shi border with Zhaojue Xian], "Daliang-schan, Lolo, east of Ningyüen" 3000–3300 m, 21 Apr. 1914, *H. R. E. Handel-Mazzetti 1488* (E E00612494, US 01121968, WU 0039251); S of Ningyüan fu, Lolololand, near Lolo ku, 3400 m, 21 Apr. 1914, *C. K. Schneider 951* (A 00279533); Luoji Shan, 2700 m, 2 Apr. 1985, *Y. J. Li 599*, (CDBI CDBI0027710–11). **Yanbian Xian:** Yankou Xiang, 2600 m, 28 July 2001, *Z. M. Tan 326* (SZ 00294154). **Yanyuan Xian:** Yuanbao Qu, Dalin Xiang, 3200 m, 20 July 1983, *Qinghai-Xizang Team 12103* (KUN 0178995–96, PE 01031327–28). **Zhaojue Xian:** Jiefanggou Qu, 3000 m, 17 May 1983, *M. Y. He & Q. S. Zhao 116810* (CDBI CDBI0028023, SZ 00294165).

N, NE & NW Yunnan. Dongchuan Shi: Chonghengqing, 3200 m, 7 Apr. 1985, *S. B. Lan 348* (PE 00935169–70); Yinmin Shui, Maching, Xiaozhuqing, 2200 m, 4 May 1985, *S. B. Lan 685* (PE 01031872–73). **Huize Xian:** Dahai, 2880 m, 1 June 1995, *H. Peng & L. G. Lei 2246* (KUN 0178745). **Luquan Xian:** Wumeng Shan, betw. Daheiching & Da Cun, 3300 m, 7 July 1990, *R. Z. Rui & Z. W. Lu 54* (KUN 0178959). **Ninglang Xian:** Xichuan Xiang, 2900 m, 26 Apr. 2007, *Y. Y. Geng et al. 20070235* (WCSBG 015093–95). **Qiaojia Xian:** Wuming Shan on way to Yao Shan along Xiaoshui section of hwy. X250, 27.091389°N, 102.99°E, 2950 m, 9 Sep. 2013, *D. E. Boufford, J. F. Harber & X. H. Li 43507* (A 00914423, CAS, E E00770756, K, KUN 1278348, PE).

Cultivated material:
Living cultivated plants.
Foster Clough, Mytholmroyd, West Yorkshire, U.K., from *E. D. Hammond (China) 97063*, Yunnan: Longzhou Shan, 1997, and *B. & S. Wynne-Jones 7889*, Sichuan: Huili Xian, Longzhou Shan, 3260 m, 9 Oct. 2000.

33. Berberis guizhouensis T. S. Ying, Acta Phytotax. Sin. 37: 320. 1999. TYPE: China. NW Guizhou: Bijie Xian, Shi Qu, Shengji Xiang, 1350 m, 2 Sep. 1957, *P. H. Yu 214* (holotype, PE 00935226!; isotypes, IBSC 0091598 image!, KUN [2 sheets] 0175736–37!, LBG 00064097 image!, WUK 0196312 image!).

Shrubs, evergreen, to 1.2 m tall; mature stems pale yellowish brown, subterete; spines 1- to 3-fid, concolorous, 0.5–1.6 cm. Petiole 2–3 mm; leaf blade abaxially pale green, adaxially dark green, elliptic, narrowly elliptic, or oblong, 2–6.5 × 0.8–1.6 cm, leathery, midvein raised abaxially, impressed adaxially, lateral veins raised and conspicuous abaxially, conspicuous adaxially, reticulation inconspicuous abaxially, conspicuous adaxially, base cuneate, margin conspicuously revolute, undulate, spinose with 13 to 20 prominent teeth on each side, apex acute or subacuminate, mucronate. Inflorescence a fascicle, 2- to 6-flowered, otherwise unknown; fruiting pedicel pale brown, 10–15 mm. Immature berry epruinose, ellipsoid, 7–9 × 4–4.5 mm; style not persistent; seeds 2 or 3.

Phenology. Berberis guizhouensis has been collected in fruit in September. Its flowering season is unknown.

Distribution and habitat. Berberis guizhouensis is known from Bijie and Weining Xian in northwest Guizhou. It has been found in thickets in dry places on slopes at 1350 m.

Although *K. M. Lan 00447* is sterile, it appears to be *Berberis guizhouensis*.

Selected specimens.
Guizhou. Bijie Xian: Shi Qu, Shengji Xiang, 1350 m, 2 Sep. 1957, *P. H. Yu 200* (KUN 0175734–35, PE 01840071). **Weining Xian:** Xi Liangzi, 21 Nov. 1973, *K. M. Lan 00447* (GZAC GZAC14011888).

34. **Berberis hayatana** Mizush., Misc. Rep. Res. Inst. Nat. Resources 35: 31. 1954. TYPE: China. Taiwan: Taihoku Province [Yilan Hsien], Mt. Taiheizan (Taipingshan), Ratô-gun, 17 May 1917, *B. Hayata s.n.* (holotype, TI image!).

Berberis formosana H. L. Li, J. Wash. Acad. Sci. 42: 41. 1952 (non Ahrendt 1941b). TYPE: China. Taiwan: Yilan Hsien, Nanshantsun Mtns. near Muroroahu, 17 July 1938, *T. Suzuki 7258* (holotype, TAI 156164!; isotype, PH 0066417 image!).

Shrubs, evergreen, to 1.6 m tall; mature stems brownish red, sulcate, not verruculose; spines 3-fid, concolorous, 1–2.3 cm, abaxially sulcate. Petiole almost absent; leaf blade abaxially pale green, slightly shiny, adaxially dull mid-green, elliptic to elliptic-lanceolate, 2–4 × 0.6–1.2 cm, leathery, midvein raised abaxially, slightly impressed adaxially, lateral veins indistinct or obscure abaxially, indistinct adaxially, reticulation obscure on both surfaces, base attenuate or cuneate, margin sometimes slightly revolute, spinulose with 3 to 11 teeth on each side, apex acute or acuminate, mucronate. Inflorescence a fascicle, 2- to 7-flowered; pedicel red, 8–10 mm, slender; bracteoles brownish red, triangular-ovate, ca. 1.5 mm, apex acute. Flowers bright yellow. Sepals in 2 whorls; outer sepals broadly ovate or obovate-elliptic, 3.5 × 2 mm, apex acute; inner sepals obovate-elliptic or obovate, 4.5 × 3 mm, apex rounded; petals obovate, 4.5 × 3 mm, base attenuate, glands ovoid, separate, apex emarginate. Stamens ca. 2.5 mm; anther connective slightly extended, truncate. Pistil 4 mm; ovules 1(or 2 to 4), funiculate. Berry black, initially pruinose, ellipsoid or ovoid, 4–6 × ca. 2 mm; style not persistent.

Phenology. Berberis hayatana has been collected in flower between February and May and in fruit between July and November.

Distribution and habitat. Berberis hayatana is known from Hsinchu, Taoyuan, and Yilan Hsien and New Taipei City (formerly Taipei Hsien) in northern Taiwan. It has been collected in forest understories and stunted woodlands on moist mountain ridges and slopes at ca. 700–2200 m.

The protologue of *Berberis hayatana* synonymized *B. formosana* H. L. Li (1952), non Ahrendt (1941b). H. L. Li (1963: 170) cited an isotype of this taxon as being at "M." Chung et al. (2009: 161, 166) reported this as being a reference to MOAR. The specimen has since been transferred to PH.

There can be significant variations in the leaf shape of *Berberis hayatana*, variations which appear to have no geographic basis. The protologue stated the number of ovules as one and this was repeated by Ahrendt (1961: 358), Chamberlain and Hu (1985: 538), and Ying (2001: 138, 2011: 745). In fact, these can be up to four.

S. Y. Lu and Yang (1996: 576) treated *Berberis hayatana* as a synonym of *B. mingetsensis*. However, the latter species has shiny leaves and a different flower structure and does not occur in northern Taiwan.

Selected specimens.
Taiwan. Hsinchu Hsien: Yuanyanghu, 1600–2400 m, 1 Dec. 1972, *C. C. Hsu & C. S. Kuoh 14214* (TAI 200665); Litungshan, 24.695667°N, 121.293833°E, 1600–1850 m, 12 July 1999, *S. H. Su 416* (TAI 280201). **New Taipei City (formerly Taipai Hsien):** Kabosan (Chiamushan), ca. 1300 m, 20 Feb. 1935, *T. Susuki 12799* (TAI 069649–50); Peichatenshan, 1400–1727 m, 28 Sep. 1984, *R. T. Lee 3580* (TAI 196185–86); Lupeishan, 1400 m, 1 Mar. 1989, *C. F. Hsieh et al. 687* (TAI 214344); Rarashan, Fushan, 7 Apr. 1983, *K. C. Yang s.n.* (TAI 185811); Mt. Chulu, 1200–1350 m, 29 Mar. 2014, *P. F. Lu 26587* (TAIF 44495). **Taoyuan Hsien:** Lalashan, 20 Mar. 1977, *C. T. Wang 974* (TAI 181281); Palin, Tamanshan, 6 Apr. 1983, *K. C. Yang 1257* (TAI 188589); Fuhsing Hsiang, en route from Meikueihsimoshan to Papokulushan, 24.685556°N, 121.466944°E, 1840 m, 11 Apr. 2002, *W. C. Leong 2880* (HAST 96029); border betw. Taoyuan & Yilan Hsien, Lalashan, 24.72575°N, 121.442°E, 1728 m, 21 Apr. 2014, *J. F. Harber, C. J. Huang & H. H. Huang s.n.* (E). **Yilan Hsien:** betw. Mururoahu (Chililo) & Kyanrawa (Hsulawa), 16 July 1932, *T. Susuki 7164* (TAI 047155); Sanhsing Hsiang, Songluhu, 700–1300 m, 28 Mar. 2009, *P. F. Lu 18081* (HAST 124162).

35. **Berberis holocraspedon** Ahrendt, J. Bot. 79(Suppl.): 22. 1941. TYPE: China. W Yunnan: Shunning [Fengqing Xian], Snow Range, 3000 m, 22 Nov. 1938, *T. T. Yu 18228* (holotype, E E00217944!; isotypes, A 00038764!, KUN [2 sheets] 1204027!, 1204030!, PE [2 sheets] 00935259–60!).

Berberis subholophylla C. Y. Wu ex S. Y. Bao, Bull. Bot. Res., Harbin 5(3): 13. 1985, syn. nov. TYPE: China. W Yunnan: Shunning [Fengqing Xian], Tehloching, 2800 m,

6 June 1938, *T. T. Yu. 16168* (holotype, KUN 0161652!; isotypes, A 00038799!, E E00373490!, KUN 0161653!, PE [2 sheets] 01030829–30!).

Shrubs, evergreen, to 2 m tall; mature stems pale yellow, terete, sparsely black verruculose; spines 3-fid, pale yellow, 0.7–2 cm. Petiole 2–3 mm; leaf blade abaxially pale green, often pruinose, adaxially dark green, shiny, oblong-lanceolate or oblong-elliptic, 4–10 × 1.5–2.5 cm, leathery, midvein raised abaxially, slightly impressed adaxially, lateral veins inconspicuous or obscure abaxially, slightly conspicuous adaxially, reticulation obscure abaxially, inconspicuous adaxially, base cuneate, margin very slightly revolute, entire, or rarely with 1 to 5(to 8) spinulose teeth on each side, apex obtuse, mucronate. Inflorescence a fascicle, 3- to 12-flowered, otherwise unknown. Pedicel reddish, 10–15 mm; bracteoles partially with reddish tinge, triangular-ovate, 4.5 × 2.5 mm, apex acute. Sepals in 2 whorls; outer sepals narrowly elliptic, 6 × 3 mm, inner sepals broadly obovate, 7 × 5.5 mm, apex rounded; petals obovate, 5.5–6 mm, base cuneate, glands separate, ca. 0.5 mm, apex entire. Stamens 4.5 mm; anther connective distinctly extended, truncate with a toothlike point on either side. Pistil 5 mm; ovules 2(or 3). Berry black, pruinose at first, ellipsoid, 7–10 × ca. 6 mm; style not persistent.

Phenology. Berberis holocraspedon has been collected in flower in May and in fruit between May and November.

Distribution and habitat. Berberis holocraspedon is known from Fenqing, Jingdong, Yangbi, and Yongde Xian and probably Tengchong Xian in west Yunnan. It has been collected from thickets on open slopes, forests, and forest margins at ca. 2600–3100 m.

In the protologue of *Berberis holocraspedon*, Ahrendt cited *T. T. Yu 16168* as well as the type, but Bao (1985: 13) decided it was a separate species, *B. subholophylla*. However, Bao's description of the latter differs little from Ahrendt's description of *B. holocraspedon*, except that the fruit are described as pruinose, instead of epruinose as in Ahrendt's description. This is likely to be because they were immature, unlike the mature fruit of *T. T. Yu 18228*. The number of seeds per fruit is described as being three or four, whereas it is recorded as two in the protologue of *B. holocraspedon*. However, though an annotation to the sheet of the holotype of *B. holocraspedon* records the seed number as two, dissection of two fruit of the isotype at A produced two and three seeds, while dissection of two fruit from the isotype of *B. subholophylla* at A both produced two seeds.

The only wild-collected specimens of *Berberis holocraspedon* I have located that have floral material are those of *R. C. Chin 22509* cited below, but I have not had the opportunity to examine these. The description of flowers used above, therefore, is from material grown from *K, Rushforth 7458*, Yunnan, Yongde Xian, Wumulong Xiang, 24.125°N, 99.6725°E, 3000 m, 24 Oct. 2012, kindly sent to me by Keith Rushforth in April 2019. The flowers I was sent had two ovules. The anther connective of these flowers was very unusual. It is somewhat similar to that recorded for *B. pendryi* Bh. Adhikari, endemic to Nepal (see Adhikari et al., 2012: 279).

The leaves of *Berberis holocraspedon* appear to be variable. While the leaves of the holotype of *B. holocraspedon* and *B. subholophylla* are rarely abaxially pruinose, those of many of the leaves of the isotypes at PE are markedly so. While the type gatherings have leaves that are largely entire, there are other gatherings that have leaves that are markedly spinose. This particularly seems to be the case with some specimens from Yangbi Xian. *J. F. Rock 7388* was cited by Schneider (1939: 258) as a syntype of *Berberis amabilis*, but it appears to be *B. holocraspedon* even though it was collected a considerable distance from where the species is otherwise recorded.

Selected specimens.
W Yunnan. Jingdong Xian: "Ching-tung," Ta-Tun-Tzu-Shan, Yen-Tung-Tou, 2700 m, 2 Nov. 1939, *M. G. Li 1055* (KUN 0177560–61); Wuliangshan, Huangcaoling, Bai'ao Shan, 3100 m, 18 Nov. 1956, *B. Y. Qiu 53779* (KUN 160904–05, 0176240, 0176272). **[Tengshong Xian]:** betw. Tengyueh (Tengshong) & Burmese border, near Changlifang, en route to Sadon, Nov. 1922, *J. F. Rock 7388* (A 00038688, US 0946050). **Yangbi Xian:** Malutang, 2 May 1929, *R. C. Chin 22509* (KUN 0176273–75); above Yangbi, Shangchang, 2700 m, 8 May 1981, *Sino-British Cangshan Plant Exp. 342* (A 00279556, KUN 0176276–77); W side of Diancang Shan range, en route from Xueshanhe to Dapingzi, 25.71666667°N, 100.03333333°E, 2600–3000 m, 17 June 1984, *Sino-Amer. Bot. Exp. 226* (A 00279558, KUN 0175827, US 00946026). **Yongde Xian:** Baohu Qu, Muguadi, 3000 m, 24 June 2002, *W41657* (KUN 0175794).

36. Berberis hookeri Lem., Ill. Hort. 6: t. 207. 1859. TYPE: Illustration of cultivated plant of unknown origin bought by A. Verschaffelt of Belgium from an English nurseryman (lectotype, designated by Adhikari et al. [2012: 513]).

Shrubs, evergreen, to 1.5 m tall; mature stems yellow, angled or sulcate, verruculose; spines 3-fid, concolorous, 1–2.5 cm, terete. Petiole almost absent; leaf blade abaxially pale green, sometimes white or glaucous pruinose, adaxially green, shiny, lanceolate-elliptic or oblong-elliptic, sometimes obovate, 3–7 × 0.8–3 cm, midvein raised abaxially, impressed adaxially, lateral veins and reticulation inconspicuous abaxially, conspicuous adaxially, base cuneate or attenuate, margin spinose with 2 to 7(to 15) teeth on each side, apex acute, mucronate. Inflorescence a fascicle, 1- to 7(to 11)-flowered; pedicel 15–30 mm, stout; bracteoles red, ovate, ca. 2.5 × 1.5 mm, apex acute. Flowers yellowish green, ca. or less than 1.5 cm diam. Sepals in 3 whorls; outer sepals ovate or oblong-ovate, ca. 4 × 2 mm; me-dian sepals oblong-elliptic, ca. 8 × 5 mm; inner sepals obovate, ca. 9 × 7 mm; petals obovate, ca. 4.5 × 3.5–6 mm, base cuneate, ca. 1 × 0.6 mm, apex obtuse or emarginate. Stamens 3–4.5 mm; anther connective extended, truncate or slightly retuse. Ovules 3 to 6. Berry black, glaucous pruinose, oblong-ovoid, 10–15 × 6–8 mm; style not persistent.

KEY TO SUBSPECIES OF *BERBERIS HOOKERI*

1a.   Flowers 3 to 8(to 14) per fascicle; pedicel 15–25 mm; flowers 1.5 cm or more diam. . . . . .subsp. *hookeri*
1b.   Flowers 1 to 3 per fascicle; pedicel 25–30 mm; flowers less than 1.5 cm diam. subsp. *longipes*

36a. Berberis hookeri subsp. hookeri.

Berberis wallichiana sensu Hooker in Bot Mag. 78: t. 4656. 1852, non DC. 1824. TYPE: Illustration of plant based on cultivated material from seed collected in Nepal by T. Lobb and in eastern Himalayas by J. D. Hooker (holotype).
Berberis hookeri var. *viridis* C. K. Schneid., Bull. Herb. Boissier, sér. 2, 8: 197. 1908. TYPE: India. Sikkim: Lachan Valley, 6 July 1849, *J. D. Hooker s.n.* (holotype, K K000567817!).
Berberis hookeri var. *platyphylla* (non *B. hookeri* var. *latifolia* Bean, Trees and Shrubs Brit. Isles 1. 243, 1914) Ahrendt, J. Linn. Soc., Bot. 57: 40. 1961. *Berberis wallichiana* var. [γ] *latifolia* Hook. f. & Thomson, Fl. Ind. 1: 226. 1855. TYPE: India. Sikkim: "10,000 ped," s.d., *J. D. Hooker s.n.* (lectotype, designated here, K twigs other than *Hooker 40* on sheet K000567815!; isolectotype, GH 00872736!, P00580433!).
Berberis buchananii C. K. Schneid. var. *tawangensis* Ahrendt, J. Bot. 79(Suppl.): 37. 1941. TYPE: India. NW Arunachal Pradesh: Monyul, Tawang, 2900 m, 22 Oct. 1934, *F. Ludlow & G. Sherriff 1089* (holotype, BM BM000559454!).
Berberis hookeri var. *microcarpa* Ahrendt, J. Linn. Soc., Bot. 57: 40. 1961. TYPE: India. Sikkim, 24 Oct. 1874, *C. B. Clarke 25561* (holotype, K K000567818!).
Berberis parapruinosa T. S. Ying, Fl. Xizang. 2: 145. 1985, syn. nov. TYPE: China. S Xizang (Tibet): Nyalam (Nielamu) Xian, Zhangmu, Kangbajiba, 2750 m, 6 May 1966, *Y. T. Zhang & K. Y. Lang 3415* (holotype, PE 00935232!; isotype, PE 00935233!).

Flowers 3 to 8(to 14) per fascicle; pedicel 15–25 mm; flowers 1.5 cm or more diam.

Phenology. In China *Berberis hookeri* subsp. *hookeri* is known to flower from May to June and to fruit in August.

Distribution and habitat. In China *Berberis hookeri* subsp. *hookeri* is known from areas very near the border with Nepal in Dinggyê (Dingjie) and Nyalam (Nielamu) Xian in Xizang (Tibet). It is also known from that part of northwest Arunachal Pradesh in India where there is a conflicting territorial claim by China. It has been collected in *Tsuga* forests and open hillsides at ca. 2600–3400 m. Besides Sikkim and Bhutan, *B. hookeri* subsp. *hookeri* is also known from Nepal.

Ahrendt (1961: 39) stated that *J. D. Hooker 255* (K K000644893), India, Sikkim, 1848, was the type of *Berberis hookeri*, but the illustration in Lemaire's *L'Illustration Horticole* (1859) fulfills the requirement of the Shenzhen Code, Art. 9.3 (Turland et al., 2018), and was lectotypified by Adhikari et al. (2012: 513).

In publishing *Berberis hookeri* var. *platyphylla* as a nom. nov. for *B. wallichiana* var. *latifolia*, Ahrendt (1961: 40) stated the type to be *Hooker 40* at K. However, *Hooker 40* (K000567815) was collected at "Tonglo, Sinchal, 8–10,000[ft]" whereas the protologue of *B. wallichiana* var. *latifolia* cited only var. γ "Sikkim, alt. 10,000 ped." The label with these details is under a different twig on the same sheet as *Hooker 40* and it is clear that this should be regarded as the type. The specimen at K has been chosen as the lectotype.

All the various varieties of *Berberis hookeri* listed above were differentiated on the basis of variation in leaf morphology except for variety *microcarpa* which was recognized by Ahrendt on the basis of its slightly smaller fruit. As Chamberlain and Hu (1985: 533), who synonymized varieties *microcarpa*, *platyphylla*, and *viridis* and *B. buchanii* var. *tawangensis*, noted, *B. hookeri* subsp. *hookeri* is a species variable in leaf shape, size, color, and density of marginal spines.

In his protologue of *Berberis buchanii* var. *tawangensis*, Ahrendt noted that *F. Ludlow & G. Sherriff 1089* resembled *B. hookeri* var. *viridis* but had three or four seeds. In fact, the usual number of ovules of *B. hookeri* is three to six; Ahrendt (1961: 39) was mistaken in reporting the number to be six to nine. There is an ancient plant at St. Andrews Botanic Garden, Scotland, grown from seeds of *F. Ludlow & G. Sherriff 1089*. Material from this plant kindly supplied to me by L. Cunningham of the Gardens in March 2009 confirmed that it is *B. hookeri*. That *B. hookeri* var. *hookeri* grows in West Kameng is confirmed by the living plant grown from *A. Clark 5405*, cited below.

The description in the protologue of *Berberis parapruinosa* is almost exactly the same as that of *B. hookeri* subsp. *hookeri* and this is confirmed by the specimen itself.

Selected specimens.
S Xizang (Tibet). Dinggyê (Dingjie) Xian: en route from Gengxin to Chentang, 3400 m, 5 June 1975, *Qinghai-Xizang Team 5520* (HNWP 48800, KUN 0178674, PE 01840011). **Nyalam (Nielamu) Xian:** Zhangmu, 2900 m, 3 May 1966, *Y. T. Zhang & K. Y. Lang 3113* (PE 01037747–49); near Zhangmu, 2600 m, 18 Aug. 1972, *Xizang Herbal Medicine Survey Team 1290* (HNWP 30952, PE 01037750).
BHUTAN. 1838, *W. Griffith 1740* (G, K, P); near Laya, Mo Chu Valley, 3505 m, 1 July 1949, *F. Ludlow, G. Sherriff & J. H. Hicks 17374* (BM); Tongsa, Yuto, 2895 m, 3 June 1960, *S. A. Bowes-Lyon 3305* (BM).

Cultivated material:
Living cultivated plants.
Sherwood Gardens, Devon, U.K., from *A. Clark 5405*, India, Arunachal Pradesh, West Kameng, Nov. 2004.

36b. Berberis hookeri subsp. **longipes** D. F. Chamb. & C. M. Hu, Notes Roy. Bot. Gard. Edinburgh 42: 533. 1985. TYPE: China. SE Xizang (Tibet): [Nyingchi (Linzhi) Xian, now Bayi Qu], Pomé, Tongyuk Dzong, 29.966667°N, 94.833333°E, 2750 m, 20 May 1947, *F. Ludlow, G. Sherriff & H. H. Elliot 13713* (holotype, BM BM001010956!; isotype, E E00259018!).

Flowers 1 to 3 per fascicle; pedicel 25–30 mm; flowers less than 1.5 cm diam.

Phenology. *Berberis hookeri* subsp. *longipes* has been collected in flower from May to June and in fruit between June and December.

Distribution and habitat. *Berberis hookeri* subsp. *longipes* is known from Bomê (Bomi) Xian and Bayi Qu (formerly Nyingchi (Linzhi) Xian in southeast Xizang. It has been collected in *Pinus*, *Picea*, and *Juniperus* forests at ca. 2060–3050 m.

Berberis hookeri subsp. *longipes* occurs some 400 kilometers northeast of the nearest location of *B. hookeri* subsp. *hookeri* in Arunachal Pradesh and, as such, represents an interesting disjunct. Its relationship with *B. leptopoda* from farther south in Mêdog (Motuo) Xian needs further investigation.

Though the type specimens of subspecies *longipes* and the other *Ludlow, Sherriff & Elliot* specimens cited below have 1 to 3 long pedicels, a number of the other cited specimens have more pedicels (in the case of *s. coll. 1521*, up to eight) that are sometimes shorter. This suggests that either the inflorescence is more varied than suggested above or that the specimens represent two taxa rather than one.

Chamberlain and Hu (1985: 533) noted that *F. Ludlow, G. Sherriff & H. H. Elliot 12026, 13713*, and *13980* were all cited by Ahrendt (1961: 48) as examples of *Berberis buchananii* C. K. Schneid. var. *tawangensis*.

Selected specimens.
SE Xizang (Tibet). Bomê (Bomi) Xian: Dem, Po Tsangpo, 30.033333°N, 95.25°E, 2285 m, 6 June 1947, *F. Ludlow, G. Sherriff & H. H. Elliot 13080* (A 00279559, BM BM001010954, E E00395939, P P02682342); Tongmai, near army depot, 20 July 1965, *Xizang Team – Y. T. Zhang & K. Y. Lang 725* (PE 01031876–78); betw. Tongmai & Daqiao, 1960 m, 20 July 1965, *s. coll. 2246* (KUN 0178793); Tongmai, 2059 m, 28 June 1974, *s. coll. 1521* (HNWP 88895, KUN 0178273–4, PE 01031874–5, 02068292); Tongmai, 2000 m, 14 May 1986, *T. Naito et al. 1001* (PE 020623420). **Nyingchi (Linzhi) Xian (now Bayi Qu):** Pomé, Tongyuk Dzong, 29.966667°N, 94.833333°E, 3050 m, 22 Dec. 1946, *F. Ludlow, G. Sherriff & H. H. Elliot 12026* (BM BM001010955, E E00395940); near tea plantation, 2600 m, 19 June 1972, *Chinese Herbal Medicine, Xizang Survey Team 73491* (PE 00049411, 01031865).

37. Berberis hsuyunensis P. K. Hsiao & W. C. Sung, Acta Phytotax. Sin. 12: 388. 1974. TYPE: China. S Sichuan: Xuyong Xian, 2 June 1964, *s. coll. s.n.* (holotype, IMM image!).

Shrubs, evergreen, to 2 m tall; mature stems purplish brown, slightly angled; spines 3-fid, orange-brown, to 1 cm, slender. Petiole 2–3 mm; leaf blade abaxially yellow-green, pruinose, adaxially green, oblong-elliptic, elliptic, or broadly elliptic, 6–9 × 3–5 cm, thickly leathery, midvein raised abaxially, slightly impressed adaxially, lateral veins raised and conspicuous abaxially, conspicuous adaxially, reticulation largely obscure abaxially, conspicuous adaxially, base cuneate, margin spinose with 5 to 9 teeth on each side, apex acute, mucronate. Inflorescence a fascicle, 30- to 50-flowered; pedicel 15–20 mm; bracteoles triangular, ca. 1 × 1 mm. Sepals in 2 whorls; outer sepals triangular-ovate, ca. 3 × 2.5 mm; inner sepals broadly ovate, ca. 4 × 4 mm, apex obtuse; petals obovate, ca. 4.5 × 3.5 mm, base clawed, with separate glands, apex slightly emarginate to subrounded. Stamens ca. 2 mm; anther connective extended, truncate. Ovules 5 or 6, shortly funiculate or subsessile. Berry black, pruinose, subglobose, ca. 7 × 6 mm; style not persistent.

Phenology. *Berberis hsuyunensis* has been collected in flower in May and June and in fruit from July to September.

Distribution and habitat. *Berberis hsuyunensis* is known from Junlian, Xuyong, and Yanbian Xian in south Sichuan and Weixin Xian in northeast Yunnan. It has been collected in forests and streamsides at 1200–2800 m.

Berberis hsuyunensis is a very distinctive species with an apparently very limited distribution.

Selected specimens.
S Sichuan. Gulin Xian: Longmei Gongshe, Heping Dadui, 9 July 1976, *Gulin Team 324* (SM SM704800282). **Junlian**

Xian: Xiaoxueshan, 1600 m, 26 Apr. 1959, *Yibin Division, Sichuan Economic Plant Exp. 173* (PE 00935215). **Xuyong Xian:** Luohan Linchang, 1200 m, 9 May 1959, *Z. C. Zhong 363* (HGAS 013924, KUN 0176289), 3 Oct. 1964, *Xuyong Team 1444* (SM SM704800279); Fenshui Qu, 1400 m, 14 Aug. 1977, *Xuyong Team 963* (SM SM704800281). **Yanbian Xian:** Dayuzi Qu, Baoshi Shan, 2800 m, 29 June 1983, *Qinghai-Xizang Team 11684* (KUN 0178863, 0179010, PE 01293935–36).

Sichuan. S. loc., 19 Sep. 1978, *L. S. Chen 259* (IMM).

NE Yunnan. Weixin Xian: Gaotian, 1450 m, 9 June 1960, *P. Zhai 1101* (KUN, 0176282, 0176287–88).

38. **Berberis hypoxantha** C. Y. Wu ex S. Y. Bao, Bull. Bot. Res., Harbin 5(3): 6. 1985. TYPE: China. SE Yunnan: Xichou Xian, Nanchang Gongshe, 19 May 1959, *Q. A. Wu 8025* (holotype, KUN 0160907!; isotype, KUN 0160906!, WUK 0270468 image!).

Shrubs, evergreen, to 3 m tall; mature stems purple, shiny, terete, densely verruculose; spines largely absent, sometimes 1- to 3-fid, concolorous, 0.1–0.5 cm, weak. Petiole almost absent or to 4 mm; leaf blade abaxially pale green, adaxially dark green, elliptic, oblong-elliptic, or elliptic-obovate, 3–7 × 2–2.5 cm, leathery, abaxially sometimes pruinose, midvein raised abaxially, impressed adaxially, lateral veins inconspicuous or obscure on both surfaces, margin entire, sometimes undulate, rarely spinulose with 2 to 4 teeth on each side, base cuneate, apex rounded or acute, sometimes mucronate. Inflorescence a fascicle, 4- to 20-flowered; pedicel 18–20 mm; bracteoles triangular, 1.3–1.7 × 1–1.5 mm. Flowers yellow. Sepals in 2 whorls; outer sepals ovate or elliptic-ovate, 2.5–3 × 2–2.5 mm; inner sepals obovate or oblong-ovate, 5–6.5 × 4–4.5 mm; petals obovate to orbicular, 4.5–5.5 × 2–3.3 mm, base very slightly clawed, glands close together, ca. 1 mm, apex distinctly notched. Stamens 3.5–4 mm; anther connective extended to ca. 0.5 mm, truncate. Pistil ca. 4 mm; ovules 1 to 3. Berry black or dark purple, narrowly obovoid or ellipsoid, 9–12 × 5–7 mm; style not persistent or persistent and short.

Phenology. In China *Berberis hypoxantha* has been collected with immature fruit in May and mature fruit in October, its flowering season being unknown. In Vietnam the species has been collected in flower from March to May and in fruit in October.

Distribution and habitat. In China *Berberis hypoxantha* is known from only the type collection from a mountain slope at an unrecorded elevation in Xichou Xian and a specimen from Pingbian Xian collected on a rocky slope at 1700 m, both in southeast Yunnan. The only four collections known from Vietnam were collected in Ha Giang Province in wet evergreen mixed forest on karst limestone at 1500–1600 m and in Lai Cai Province in dense forest at 2300 m.

The protologue of *Berberis hypoxantha* cited only the type. The specimen *C. W. Wang 82523* (details below) was annotated as an unpublished *B. sulcata* by C. Y. Wu.

The description of flowers above is from the MO specimen of *CBL 1779* and *HNU 022576*, and the description of fruit is from *NTH 3483* (details of all three below).

Berberis hypoxantha is a quite unusual species. Although it appears to belong to section *Wallichianae*, its lack of significant spines and the obscure venation of the leaves are a particular combination not recorded elsewhere in the section. The collectors' notes to *CBL 1888* describe it as being a "climbing shrub" and those of *NTH 3483* as a "climbing or pendant shrub," a feature not recorded for any other *Berberis* from southeast Asia.

All five collections recorded here were found from a relatively small area straddling the China–Vietnam border.

The unidentified *Berberis* species from Vietnam referred to in Averyanov et al. (2002: 35) is *B. hypoxantha*.

Selected specimens.

SE Yunnan. Pingbian Xian: Je-mu-te, 15 Oct. 1939, *C. W. Wang 82523* (KUN 161655).

N VIETNAM. Hà Giang Province: Dong Van Distr., Ho Quang Mun., vic. of Ta Xa village, 23.266667°N, 105.366667°E, 1550–1600 m, 28 Apr. 1999, *P. K. Loc, P. H. Hoang & L. Averyanov CBL 1779* (HN, LE [2 sheets], MO MO-2331499, 5157592 [specimen on left of sheet], P P02313634); Yen Minh Distr., La Va Chai mun., vic. of Ngan Chai village, 6 km E of Yen Minh town, 23.1167°N, 105.1334°E, 1500–1600 m, 1 May 1999, *P. K. Loc, P. H. Hoang & L. Averyanov CBL 1888* (HN, LE, MO MO-2331500, 5181983, P P02313635); Yen Minh Distr., La Va Chai mun., vic. of La Va Chai village, 23.116667°N, 105.133333°E, 1500–1550 m, 9 Oct. 1999, *N. T. Hiep, B. Q. Binh, L. Averyanov & P. Cribb NTH 3483* (LE). **Lao Cai Province:** Tam Durong Distr., Hoang Lien Son Mtn. range, 2300 m, 15 Mar. 2016, *HNU 022576 & 022577* (HNU).

39. **Berberis impedita** C. K. Schneid., Repert. Spec. Nov. Regni Veg. 46: 263. 1939. TYPE: China. N Guangxi: "Bin Long, Miu Shan, N. Luchen border of Kweichow," 1500 m, 17 June 1928, *R. C. Ching 6053* (lectotype, designated by Ahrendt [1961: 60], W 0014943!; isolectotypes, CQNM 0005207 image!, IBSC 0000587 image!, IBSC 0000588 fragm., image!, NAS [2 sheets] NAS 00314067 image!, NAS 00314095 image!, NY 1365271!, PE [2 sheets] 01293914–15!).

Shrubs, evergreen, dense, to 2 m tall; mature stems pale yellow, angled or sulcate, not verruculose; spines

1- to 3-fid, pale yellow, 0.3–1.7 cm, very weak, absent toward apex of stems. Petiole 5–8 mm; leaf blade abaxially glaucous pruinose at first, finally greenish, adaxially dark green, elliptic or oblong, 4–6.5 × 1–2.5 cm, leathery, midvein raised abaxially, impressed adaxially, lateral veins raised and conspicuous abaxially, conspicuous adaxially, reticulation inconspicuous abaxially, conspicuous adaxially, base attenuate, margin spinulose with 8 to 15 teeth on each side, apex obtuse or acute, mucronate. Inflorescence a fascicle, (1- or)2- to 4-flowered; pedicel 8–15 mm; bracteoles ovate, ca. 2.5 mm, apex acute. Sepals in 2 whorls; outer sepals elliptic-oblong, 3.5–4.5 × 1.8–2.5 mm; inner sepals elliptic, 5–5.5 × 3–3.5 mm, apex rounded; petals obovate, ca. 4 × 2.5 mm, apex emarginate. Stamens ca. 3 mm; anthers oblong, connective slightly thickened with 2 small teeth. Ovules 4 to 6. Berry black, epruinose, oblong, 8–9 × 5–6 mm; style not persistent.

Phenology. *Berberis impedita* has been collected in flower in April and May and in fruit between June and October.

Distribution and habitat. *Berberis impedita* is known from northeast Guangxi, north Guangdong, and south Hunan with one collection from west Jiangxi. It has been collected from sunny places on mountain summits and woody moorlands at ca. 1400–1960 m.

The protologue of *Berberis impedita* cited both *R. C. Ching 6053* at W and *C. Wang 40094*, but did not designate a type. Schneider (1942: 50) designated the former as the type but without indicating the herbarium. Ahrendt (1961: 50) completed the lectotypification by designating the specimen at W, though he clearly had not seen the specimen.

Distinguishing between herbarium specimens of *Berberis impedita* and *B. xanthoclada* can be very difficult since, as Chamberlain and Hu (1985: 549) note, the main difference lies in the structure of the flowers. However, in addition, the leaves of *B. impedita* are often abaxially pruinose whereas those of *B. xanthoclada* are not. There is also a very distinct geographical separation between the species.

The description of flowers used above comes from Chamberlain and Hu (1985: 549) who based it on *Z. Z. Chen 50818*, cited below.

The report by Ying (2001: 143, 2011: 747) of *Berberis impedita* in Sichuan is likely to be based on misidentification possibly of the very similar *B. jinfoshanensis.*

Syntype. Guangxi. Dayaoshan: Yong-Gu, Chen Gang, 19 Oct. 1936, *C. Wang (Z. Huang) 40094* (A 00280049, IBK IBK0001275, IBSC 0091812).

Selected specimens.

N Guangdong. Ruyuan Xian: Niujiaolong, 27 Oct. 1934, *X. P. Gao 54639* (IBK IBK00013089, IBSC 009792, 0091795); Ruyuan Xian–Yangshan Xian border, 1400 m 30 Nov. 1957, *X. G. Li 201315* (CSFI CSFI004673, HHBG HZ009046, HIB 00129469, IBK IBK00012775, IBSC 0091791, PE 01293937). **Yangshan Xian:** 1750 m, 1 June 1956, *L. Deng 1157* (HITBC 003652, IBSC 0091790, PE 01293938). **Yingde Xian:** Shigutang Zhen, 18 Sep. 1998, *F. W. Xing 1518* (IBSC 0091798).

NE Guangxi. Dayaoshan (now Jinxiu) Xian: Wuzhi Shan, 29 June 1934, *Zhongshan University Dept. of Biology 23377* (IBK IBK00013101, IBSC 0091811); Yaoshan, Tseunyuen, 27 June 1936, *C. Wang (Z. Huang) 39564* (A 00280231, CAS 691594, IBK IBK00012770, IBSC 0091813). **Guanyang Xian:** Dupang Ling, Guguai Chong, 3 Oct. 1958, *Z. Z. Chen 52401* (IBK IBK00012761, IBSC 0091821). **Lingui Xian:** Huaping, 1420 m, 1 June 1958, *Z. Z. Chen 50818* (IBK IBK00012766, IBSC 0091803, KUN 0176291, PE 01293924, WUK 0259216). **Longsheng Xian:** Liu Qu, Dadi Xiang, Huoya Tang, 14 May 1955, *Guilin Qu Collection Team 176* (HITBC 003625, IBK IBK00012771, IBSC 0091807, SZ 00290822). **Quanzhou Xian:** Baoding, Qingshi Ling, 26 June 1936, *J. X. Zhong 83348* (A 00279990, IBK IBK00012774, IBSC 0091820, PE 01293923). **Rongshui Xian:** Antai Qu, Xiaoxiang, Yuanbao Shan, 27 Oct. 1958, *S. Q. Chen 16963* (IBK IBK00012781, IBSC 0091804, KUN 0177559, NAS NAS00314442, PE 01293918). **Xing'an Xian:** Tangdong, Leigong Tian, Mao'er Shan, 20 July 1958, *Z. Z. Chen 51168* (IBK IBK00012769, IBSC 0091814, KUN 0176290, PE 01293921). **Ziyuan Xian:** Guali Xiang, Yinhanping, 30 Aug. 1958, *Z. Z. Chen 52006* (IBK IBK00012764, IBSC 0091815).

S Hunan. Chengbu Xian: Erbaoding, 2000 m, 19 Aug. 1981, *Q. Z. Lin 12* (CSFI CSFI004670–71); Nanshan Muchang, 1550 m, 29 Apr. 1985, *Q. Z. Lin 054032* (CSFI CSFI004675–76). **Yizhang Xian:** Jinquan Xiang, Mangshan, Lingzi Ping, 30 Sep. 1942, *B. H. Liang 83757* (IBK IBK00012816, IBSC 0091802, MO 04104504); Mangshan, Lingzi Ping, 20 Oct. 1942, *S. Q. Chen. 2752* (IBK IBK00012815, IBSC 0091801, MO 4099014); Mangshan, Yanzi Shi, 3 Nov. 1944, *B. H. Liang 85186* (IBK IBK00012813, IBSC 0091800); Mangshan, Yanzi Shi, 1680 m, 24 Aug. 1958, *G. Xiao 4157* (PE 01293926–27); Mangshan Linchang, 1960 m, 1 Apr. 1977, *H. L. Liao 011* (PE 01293925).

W Jiangxi. Anfu Xian: Wugong Shan, 1600 m, 22 Aug. 1963, *J. S. Yue et al. 3704* (PE 01293928).

40. Berberis incrassata Ahrendt, Gard. Chron., ser. 3, 105: 371. 1939. *Berberis insignis* Hook. f. & Thomson subsp. *incrassata* (Ahrendt) D. F. Chamb. & C. M. Hu, Notes Roy. Bot. Gard. Edinburgh 42: 537. 1985. TYPE: Burma (Myanmar). Adung Valley, 2100 m, 6 May 1931, *F. Kingdon-Ward 9358* (holotype, BM [2 sheets] BM00571173–4!; isotype, A 00872739!).

Berberis incrassata var. *bucahwangensis* Ahrendt, J. Bot. 79(Suppl.): 11. 1941. TYPE: China. NW Yunnan: [Gongshan Xian], Taron-Taru divide, Bucahwang valley, 1500 m, 6 Sep. 1938, *T. T. Yu 20138.* (holotype, E E00117364!; isotypes, A 00279574!, KUN 1204866!, PE [2 sheets] 01031101!, 01037183!).

Berberis incrassata var. *fugongensis* S. Y. Bao, Bull. Bot. Res. Harbin 5(3): 7. 1985. TYPE: China. NW Yunnan: Fugong

Xian, 2350 m, 1 Aug. 1979, *Q. Lin 791935* (holotype, KUN 1204034!; isotype, KUN 1204031!).

Shrubs, evergreen, to 2.5 m tall; mature stems pale yellowish brown, terete; spines 3-fid, concolorous, 0.4–1.2 cm, weak, but mostly absent particularly toward apex of stems. Petiole almost absent; leaf blade abaxially pale green, slightly shiny, adaxially green, elliptic-lanceolate, sometimes oblong-elliptic, 9–15 × 2–6 cm, thinly leathery, midvein raised abaxially, impressed adaxially, lateral veins and reticulation inconspicuous abaxially, conspicuous adaxially, base cuneate, margin spinose with (6 to)10 to 17 coarse teeth on each side, apex acuminate, sometimes acute. Inflorescence a fascicle, 5- to 20-flowered; pedicel 10–22 mm, slender, thickening to 1.5 mm toward apex; bracteoles triangular or triangular-ovate, 4 × 3–4 mm. Flowers 12–13 mm diam. Sepals in 3 whorls; outer sepals ovate, greenish yellow on inside, mixture of pinkish red and green on outside, 4 × 3.5–4.5 mm; median sepals broadly elliptic, obovate-elliptic, or orbicular-elliptic, very pale greenish yellow with pink patches on inside, pinkish red surrounded by very pale whitish margin on outside, 5–6 × 5.5–6.5 mm, inner sepals obovate, similar coloring both inside and outside to median sepals, 6.5–7 × 6.5 mm; petals obovate, very pale whitish yellow, 5.5 × 5.5 mm, base cuneate, glands separate, yellow, oblong, ca. 1.5 mm, apex entire. Anther connective slightly extended, truncate. Ovules 4 to 6. Berry purplish black, epruinose, globose, sometimes ellipsoid, 6–7 × 5–6 mm; style persistent.

Phenology. In China *Berberis incrassata* has been collected in fruit from April to January the following year. In Myanmar it has been collected in flower in early May; in China its flowering period is unknown.

Distribution and habitat. Berberis incrassata is known from Gongshan, Fugon, and Yulong Xian in northwest Yunnan. It has been collected in evergreen broadleaf, bamboo, and mixed forests and on riversides at ca. 1300–2500 m. It is also known from north Myanmar (Burma).

Berberis incrassata was reduced to a subspecies of *B. insignis* (a species from Nepal, Bhutan, and Sikkim and Arunachal Pradesh in India) by Chamberlain and Hu (1985: 437) and this was followed by Ying (2001: 109–110, 2011: 735). Chamberlain and Hu based their decision on "differences too slight to maintain them as a separate species," the only differences noted being that *B. incrassata* had pedicels that thickened markedly toward the apex and globose fruit versus the usually oblong fruit of *B. insignis*. However, although the leaves of the two taxa are certainly very similar (though those of *B. incrassata* are somewhat thinner), there are other differences beyond those noted by Chamberlain and Hu. These include the fruit of *B. incrassata* having a distinct style and, most importantly, the flower structure and color being very different. An account of flowers was given in the protologue, but this failed to note the quite unusual flower color detailed above. This color is apparent from the flowers of the lectotype (though confusingly, the collector's notes on the sheet describe them quite inaccurately as "bright yellow") and is confirmed both by a plant in my own living collection grown from *A. Clark & S. Newman 5742* from northwest Yunnan and a living plant at the Royal Botanic Garden Edinburgh (details of both below). The flower structure and color are so radically different from the *B. insignis* as described by Adhikari et al. (2012: 516–517) that I recognize *B. incrassata* a separate species herein.

Chamberlain and Hu (1985: 437) and Ying (2001: 109–110) synonymized *Berberis incrassata* var. *bucahwangensis* and *B. incrassata* var. *fugongensis*, respectively, as *B. insignis* var. *incrassata*. I have treated this synonymizing as also applying to *B. incrassata*.

It should be noted that, though none of specimens cited below have spines, both the living plant at the Royal Botanic Garden Edinburgh and my own plant do. These are the source of their length.

The report of *Berberis incrassata* in India (Sikkim and West Kameng, Arunachal Pradesh) by Rao at al. (1998a: 38) is likely to be based on misidentifications of *B. insignis*.

Selected specimens.
NW Yunnan. Fugong Xian: 25.916667°N, 98.883333°E, 2500 m, s.d., *s. coll. s.n.* (HITBC 081858). **Gongshan Xian:** Kiukiang (Dulong) Valley, Bucahwang, 1600 m, 20 Nov. 1938, *T. T. Yu 21045* (A 00279573, E E00117363, KUN 0176305–06, PE 01031099, 01037184); Dulong Jiang, E of Ansan Cun, 1300 m, 16 Nov. 1959, *G, M. Feng 24738* (KUN 0176302–04, PE 01031102); Dulong Jiang, Gongshe, 1500 m, 22 Aug. 1982, *Qinghai-Xizang Team 9396* (HITBC 074302, KUN 0176301, PE 01031100); Dulongjiang Xiang, along Dandangwang He, NW of Bapo, 1400 m, 14 Jan. 1991, *Dulong Jiang Investigation Team 3140* (CAS 00120159, KUN 0176299–300); same details, but 19 Jan. 1991, *Dulong Jiang Investigation Team 3295* (CAS 00120160, KUN 0176297–98); Dulongjiang, Kongdang, 1550 m, 28 Apr. 1991, *Dulongjiang Investigation Team 6670* (KUN 0176295–96); Dulongjiang Xiang, Kongdang, on trail from Bapo to Dizhengdang, 27.877222°N, 98. 335556°E, 1550 m, 21 July 2002, *H. Li et al. 15149* (CAS 00120161).
N MYANMAR (BURMA). Adung Valley, vic. of Tahawndam, 2135 m, 7 Nov. 1931, *F. Kingdon-Ward 10148* (A 00279572, BM); Black Rock, Ngawchang Valley, 1370 m, 27 Feb. 1939, *F. Kingdon-Ward, Vernay-Cutting Exp. 358* (A 00342312).

Cultivated material:
Living cultivated plants.
Royal Botanic Garden Edinburgh, accession number 19687164, provenance unrecorded but accessioned as *Berberis*

incrassata var. *bucahwangensis*, possibly grown from *T. T. Yu 21045*; in my own collection from *A. Clark & S. Newman 5742*, NW Yunnan, Yulong Xian, between Lijiang and Laojun Shan, 2050 m, autumn 2007.

Cultivated specimens.
Royal Botanic Garden Edinburgh, from accession number 9687164 (details below), 30 Oct. 1986, *s. coll. s.n.* (E E00112283); Foster Clough, Mytholmroyd, West Yorkshire, U.K., from *A. Clark & S. Newman 5742* (details above), 28 Apr. 2015, *J. F. Harber 2015-11* (A, E)

41. Berberis insolita C. K. Schneid., Repert. Spec. Nov. Regni Veg. 46: 257. 1939. TYPE: China. S Sichuan: E of Ning Yuan fu (Xichang), Tailang Shan (Daliang Shan), betw. Sikwai (Xikuai) & Lamba, ca. 2500 m, 25 Apr. 1914, *C. K. Schneider 1029* (lectotype, designated by Ahrendt [1961: 56], K K000077334!; isolectotypes, A 00263206!, G G00237530!, GH 00038765!, Herb. Dendrol. C. Schneider†).

Shrubs, evergreen, to 3 m tall; mature stems pale yellow, slender, angled, sparsely verruculose; spines 3-fid, concolorous, 0.5–2 cm. Petiole almost absent; leaf blade abaxially dull pale green, adaxially dull dark green, linear-oblong, 3–8 × 0.3–0.4 cm, thinly leathery, midvein raised abaxially, impressed adaxially, venation obscure on both surfaces, base attenuate, margin spinulose with 5 to 10 widely spaced teeth on each side, apex acuminate, mucronate. Inflorescence a fascicle, 3- to 6-flowered; pedicel dark purplish red, 2–5 mm; bracteoles with reddish margin, ovate, 1.8–2 × 1.5–1.8 mm. Sepals in 2 whorls; outer sepals ovate, 3–4 × 2–3 mm; inner sepals broadly ovate to elliptic, 4–4.5 × 3–3.5 mm; petals obovate, ca. 3.5 × 3–3.5 mm, apex emarginate, base with separate glands. Stamens ca. 2.5 mm; anther connective not extended. Ovules 1 or 2. Berry black, epruinose, ellipsoid or oval, ca. 7 × 4 mm; style persistent, short.

Phenology. *Berberis insolita* is known to fruit from May to October. Its flowering time outside cultivation is unknown.

Distribution and habitat. *Berberis insolita* is known from south and southwest Sichuan. It has been collected in forests and in broadleaf woods on steep mountain slopes at ca. 1800–2800 m.

The lectotypification of *Schneider 1029* at K was by Ahrendt (1961: 56). It is almost certain that *Handel-Mazzetti 1745* (details below) was collected from the same plant as the type (for the joint expedition of Handel-Mazzetti and Schneider to China, see H. Handel-Mazzetti [1925, English translation 1996]).

None of the specimens of the type gathering have flowering or fruiting material. The description of fruit and flowers comes, respectively, from the living plants at Kew and Howick Hall, cited below.

A second specimen of *Q. S. Zhao, Y. B. Yang & K. H. Mou 7514*, cited below (CDBI CDBI0027411), with different collection details is *Berberis grodtmanniana*.

Ying (2001: 107, 2011: 735) treated *Berberis atrocarpa* var. *suijiangensis* from northeast Yunnan as a synonym of *B. insolita*. I treat it here as a synonym of *B. caudatifolia*.

Berberis insolita was reported from Guizhou by Ying (2001: 107, 2011: 735) and He and Chen (2004: 29). This was based on specimens identified by Harber and He in Harber (2012: 117–120) as belonging to a new species, *B. wuchuanensis*. S. Y. Bao (1997: 47) reported *B. insolita* from the Suijiang area of northeast Yunnan. This appears to be based on the specimen he had previously (1985: 5) identified as the type of *B. atrocarpa* var. *suijiangensis*.

Selected specimens.
S & SW Sichuan. E of Ning Yuan fu (Xichang), Tailang Shan (Daliang Shan), betw. Sikwai (Xikuai) & Lamba, ca. 2500 m, 25 Apr. 1914, *H. R. E. Handel-Mazzetti 1745* (W 1925–0003930, WU 0039254). **Huidong Xian:** Xinjie, Taojia Gou, 1800 m, 27 May 1978, *Huidong Team 0002* (CQNM 0005174, SM SM704800286). **Leibo Xian:** Huanglang Qu, Mahu, Tangjia Shan, 24 May 1959, *Sichuan Economic Plants Exp. Liangshan Group (1959) 0333* (PE 01031094, 01037179). **Muli Xian:** Dingdong Shan Gou, 2800 m, 13 Aug. 1978, *Q. S. Zhao, Y. B. Yang & K. H. Mou 7514* (CDBI CDBI0027420, SZ 00291196–97); Daliang Shan betw. Xichang & Zhaojue, ca. 2460 m, 15 Oct. 1992, *SICH 1200* (K). **Puge Xian:** Lian Xiang, 2550 m, 7 Aug. 1979, *Puge Team 503* (SM SM704800287). **Yanyuan Xian:** Deshi Gongshe, Yunchang Gou, 2300 m, 15 June 1978, *Yanyuan Team 290* (SM SM704800285). **Zhaojue Xian:** Sikai, 2500 m, 30 June 1976, *Sichuan Botany Team 12795* (CDBI CDBI002743–45, PE 01031093).

Cultivated material:
Living cultivated plants.
Quarry Hill Botanical Garden, California, U.S., Royal Botanic Gardens, Kew, from *SICH 1200* (see above); Howick Hall, Northumberland, U.K., from *C. Howick 1636*, Muli Xian, Gangou, N of Baiyangping pass, Nongsha Shan, 3230 m, 3 Oct. 1992.

42. Berberis jiangxiensis C. M. Hu, Bull. Bot. Res., Harbin 6(2): 9. 1986. TYPE: China. W Jiangxi: Anfu Xian, Wu Gong Shan, under Baihefeng, 1700 m, 30 Apr. 1954, *Y. G. Xiong 7885* (holotype, HLG [LBG], but currently on loan to IBSC 0696703 image!).

Berberis jiangxiensis var. *pulchella* C. M. Hu, Bull. Bot. Res., Harbin 6(2): 10. 1986, syn. nov. TYPE: China. E Jiangxi: Nanfeng Xian, Shan Qu (Yi Qu), Junfeng Xiang, 1600 m, 6 May 1958, *M. H. Nie 2489* (holotype, (HLG

[LBG], but currently on loan to IBSC 0696702 image!; isotype, KUN 0179078!).

Shrubs, evergreen, to 2 m tall; mature stems yellowish brown, very sulcate, not verruculose; spines 3-fid, concolorous, 1–2.5 cm, subterete. Petiole 1–3 mm; leaf blade abaxially pale green, pruinose, adaxially green, shiny, elliptic, elliptic-oblanceolate, or oblong, 1.5–4 × 0.5–1.2 cm, thinly leathery, midvein raised abaxially, slightly impressed adaxially, lateral veins in 2 to 4 pairs, inconspicuous or obscure abaxially, inconspicuous adaxially, base attenuate, margin spinulose with 4 to 7 teeth on each side, apex acuminate, mucronate. Inflorescence a fascicle, 2- to 4-flowered; pedicel 6–10 mm, slender or stout; bracteoles ovate, ca. 3 × 2 mm, apex acuminate. Sepals in 2 whorls; outer sepals ovate-elliptic, ca. 4 mm; inner sepals broadly elliptic, ca. 4.5 mm; petals obovate, ca. 3.5 mm, base shortly clawed, glands very close together, apex emarginate, lobes rounded, obtuse. Stamens ca. 3 mm; anther connective truncate. Ovules 2 to 4. Berry blue-black, slightly pruinose, ellipsoid, ca. 9 × 5.5 mm; style persistent, short.

Phenology. *Berberis jiangxiensis* has been collected in flower in April and May and in fruit between July and September.

Distribution and habitat. *Berberis jiangxiensis* is known from west Jiangxi and from one collection each from east Jiangxi and Hunan. It has been collected from rocky areas, trailsides, streamsides, and forests at 1550–2060 m. The Hunan collection was made from the foot of a mountain at 300 m.

As can be seen below, the *Jiangxi Investigation Team 00136* specimen at PE has the same collection details as the type of *Berberis jiangxiensis*. An annotation by T. S. Ying to PE 01031611 suggests that it is from the same gathering, but under a different collection name and number.

The reasons for recognizing a separate variety *pulchella* were the oblong rather than lanceolate or elliptic-oblanceolate leaves and that it has one fewer flower per fascicle and a more robust pedicel. In fact, the isotype of variety *pulchella* at KUN (not cited in the protologue), *M. H. Nie 2489*, also has some elliptic-oblanceolate leaves. The description of the leaves of *Berberis jiangxiensis* as lanceolate is misleading since neither the type specimens nor those cited below (including *Y. G. Xiong 8587*, cited in the protologue) exhibit these characteristics. The other differences seem trivial.

The protologue of *Berberis jiangxiensis* also cited *Y. G. Xiong 8569*, again from Anfu Xian, Wu Gong Shan, *S. S. Lai et al. 487* from Wanzai Xian, and *J. Xiong 3022* from Suichuan Xian, Jing Gang Shan, 1500 m, all also cited as being at LBG, but I have been unable to locate these.

The collection from Hunan is interesting being from a much lower elevation than those collected in Jiangxi.

Selected specimens.
Hunan. Wugang Shi: Yunshan, 300 m, 24 Oct. 1963, *H. L. Liu & Q. Z. Zhou 16455* (IBSC 0091844, KUN 0178381, MO 04770704, WUK 0257817).
Jiangxi. Anfu Xian: Wugong Shan, 7 Apr. 1954, *Y. G. Xiong 8587* (IBSC 0696704); Wugong Shan, under Baihefeng, 1700 m, 30 Apr. 1954, *Jiangxi Investigation Team 00136* (PE 01031611–12). **Ji'an Xian:** Jingganshan, Pingshui Shan, 1780 m, 12 Nov. 1982, *J. L. Wang & S. C. Zhang 8220651* (LBG 00121701). **Luxi Xian:** Wanlongshan Xiang, Wugong Shan, 5 Nov. 2015, *W. Y. Zhao et al. LXP-13-09870* (SYS 00181815). **Shangyou Xian:** Wuzhifeng Gongshe, Yinpan Shan, Liaowang Tai, 1550 m, 18 July 1971, *Jiangxi Team 0307* (PE 01031606–07); Huangshakeng, Yingpan Shan, 1700 m, 19 May 1965, *M. X. Nie et al. 8302* (IBSC 0092248, KUN 0178797); Guangu Shan, Qiyunfeng, 2060 m, 22 May 1965, *M. X. Nie et al. 8367* (KUN 0178796). **Suichuan Xian:** Dafen, Linyang, Daba Ling 1500 m, 27 Sep. 1963, *J. S. Yue et al. 4346* (IBSC 0092240, KUN 0179166, NAS NAS00313884, PE 01031617); Qianmo Cun, Nanfengmian, 30 Mar. 2011, *Q. Fan et al. JGS4148* (SYS 00173050).

43. Berberis jinfoshanensis T. S. Ying, Acta Phytotax. Sin. 37: 316. 1999. TYPE: China. Chongqing (SE Sichuan): Nanchuan Xian, Jinfoshan, 1650 m, 18 Apr. 1988, *Z. Y. Liu 10913* (holotype, PE 00935144!; isotype, IMC 00022538 image!).

Shrubs, evergreen, to 2 m tall; mature stems pale yellow, terete; spines 3-fid, concolorous, 1–3.5 cm, abaxially slightly sulcate. Petiole 2–4 mm; leaf blade abaxially yellow-green, slightly pruinose, adaxially dark green, elliptic, 3–7 × 1.3–2.5 cm, leathery, midvein raised abaxially, impressed adaxially, lateral veins in 5 to 7 pairs, raised and conspicuous abaxially, conspicuous adaxially, reticulation inconspicuous abaxially, conspicuous adaxially, base cuneate, margin spinulose with 3 to 10 teeth on each side, apex acute, mucronate. Inflorescence a fascicle, 5- to 8-flowered; pedicel 15–22 mm; bracts ovate-triangular, ca. 2 × 1 mm; bracteoles ovate, ca. 2 × 2 mm. Sepals in 2 whorls; outer sepals obovate, ca. 5 × 3.5 mm; inner sepals oblong, ca. 5.5 × 4 mm; petals obovate, 5–6 × 4–5.5 mm, base clawed, glands close together, ovate, apex entire. Stamens 4.2–5 mm; anther connective slightly extended, obtuse. Ovules 2 or 3. Berry black, ellipsoid, 7–8 × 5–6 mm, pruinose at first; style persistent.

Phenology. *Berberis jinfoshanensis* has been collected in flower in April and in fruit between May and September.

Distribution and habitat. *Berberis jinfoshanensis* appears to be endemic to Jinfoshan in Nanchuan Xian in Chongqing (southeast Sichuan). It has been collected from roadsides, mountain summits, and forests at ca. 1500–2380 m.

The protologue stated that *Berberis jinfoshanensis* was similar to *B. lijiangensis* but, in fact, the greatest similarities are with *B. xanthoclada* and *B. impedita*, both of which are also found on mountain summits. It differs from the former species in having leaves that are sometimes abaxially pruinose, and from both by its longer pedicels and fewer ovules. The relationship between *B. jinfoshanensis* and these two species needs further investigation.

Selected specimens.
Chongqing (SE Sichuan). Nanchuan Xian, Jinfoshan: Fenhuang Si, 2380 m, 19 May 1932, *D. H. Du 3686* (CQNM 0005081, IBSC 0092221, PE 01293932); betw. Fenhuang Si & Gufudong, 2150 m, 19 May 1957, *J. H. Xiong & Z. L. Zhou 90880* (IBSC 0092032, PE 01031433); Wang Xiang, Tai Jingfoshan, Yangyu Ping, 1850 m, Aug. 1972, *Sichuan Institute of Traditional Chinese Medicine 721247* (PE 01293931); Guwudong, 2100 m, 28 July 1978, *Plant Geological Survey Team 619* (CDBI CDBI0027700, PE 01293933–34); Yangyu Ping, 1650 m 13 Aug. 1978, *S. Y. Liu 784193* (IMC 00012093, PE 00935145); Shangzi Ping, 1500 m, 18 Apr. 1996, *Z. Y. Liu 16705* (IMC 00022524–27, MO 4471348, TAIF 183986).

44. Berberis julianae C. K. Schneid., Pl. Wilson. (Sargent) 1: 360–361. 1913. TYPE: China. W Hubei: "North & South of Ichang," 900–1200 m, May 1907, *E. H. Wilson 417* "A" (lectotype, designated here, A 00038769!; isolectotypes, BM BM000559459!, K K000077322!).

Berberis zanlancianensis Pamp., Nuov. Giorn. Bot. Ital. n. s. 22: 293–294. 1915, syn. nov. TYPE: China. N Hubei: [Zhushan Xian], Zan-lan-scian (Canglang Shan), Apr. 1912, *C. Silvestri 4110a* (lectotype, designated here, FI image!).
Berberis julianae C. K. Schneid. var. *oblongifolia* Ahrendt, J. Linn. Soc., Bot. 57: 68. 1961. TYPE: China. W Hubei: Changyang, May 1900, *E. H. Wilson (Veitch) 535* (holotype, K K000077321!; isotypes, A 00038766!, E E00217948!, K K000077322!, P P00716527!, W 0015479!).
Berberis julianae C. K. Schneid. var. *patungensis* Ahrendt, J. Linn. Soc., Bot. 57: 69. 1961. TYPE: China. W Hubei: Patung Hsien (Badong Xian), 1300 m, May 1907, *E. H. Wilson 2878* "A" (lectotype, designated here, K K000077319!; isolectotypes, A 00038767!, B B10 000250732, BM BM001010630!, E E00217947!, HBG HBG-506679 image!, US 00103863 image!).

Shrubs, evergreen, to 3 m tall; mature stems yellow, sulcate, scarcely black verruculose; spines 3-fid, concolorous, 1–4 cm, stout, abaxially sulcate. Petiole 1–4 mm; leaf blade abaxially pale green, shiny, adaxially dark green, slightly shiny, elliptic, lanceolate, oblanceolate, obovate, or obovate-elliptic, 4–10 × 1–3 cm, thickly leathery, midvein raised abaxially, impressed adaxially, lateral veins and reticulation inconspicuous on both surfaces, base cuneate, margin spinose with (5 to)10 to 20 teeth on each side, apex acuminate, mucronate. Inflorescence a fascicle, 10- to 25(to 30)-flowered; pedicel 8–15 mm; bracteoles ovate, ca. 2.5 × 1.5 mm, apex acute. Sepals in 2 whorls; outer sepals ovate, ca. 4.5 × 2.5 mm, apex acute; inner sepals oblong-elliptic, ca. 6.5 × 3.5 mm, apex rounded; petals oblong-elliptic, ca. 6 × 3 mm, base clawed, glands oblong, apex emarginate. Anther connective not extended. Ovules 1. Berry blue-black, white pruinose, oblong, 7–8 × 3.5–4 mm; style persistent.

Phenology. *Berberis julianae* has been collected in flower in May and in fruit between May and November.

Distribution and habitat. *Berberis julianae* is known from north, southwest, and west Hubei, Chongqing (east Sichuan), northwest Guizhou, north Hunan, and southeast Shaanxi. It has been collected in mixed forests, thickets, bamboo groves, and streamsides at ca. 900–2200 m.

The protologue of *Berberis julianae* cited *Wilson 417*, "May and October 1907," as the type. The former has flowers, the latter fruit. That of May is designated here as *417* "A" and that of October as *417* "B." The specimen of *417* "A" at A has been chosen as the lectotype because floral material is a better guide to identification than fruit, and Schneider's protologue was based on Wilson material sent to him by Harvard to identify.

The protologue of *Berberis julianae* var. *oblongifolia* states "Chang-yong [sic] . . . *Wilson (Veitch) 535* (Type, K)." There are two specimens of this at Kew, neither of which was annotated by Ahrendt. Only one has collection details stating it is from Changyang and so can be treated as the holotype.

The protologue of *Berberis julianae* var. *patungensis* stated "1907, *Wilson 2878* (Type, K)." However, there are two specimens with this number on the same sheet at Kew, one of May 1907 with flowers and one of July 1907 with fruit. The former is designated here as *Wilson 2878* "A" and the latter *Wilson 2878* "B." The former has been chosen as the lectotype because floral material is a better guide to identification than fruit.

Ahrendt published both *Berberis julianae* var. *oblongifolia* and *B. julianae* var. *patungensis* solely on the basis of differences in the shape and breadth of their leaves, but as Chamberlain and Hu (1985: 553–554), who synonymized both varieties, noted, the shape and breadth of the leaves in *B. julianae* are variable.

As noted under that species, *Wilson 2878* was originally identified by Schneider (1913: 362) as a syntype of *Berberis bergmanniae*.

Berberis zanlanscianensis has a curious history. It was published with a cursory protologue by Pampanini on the basis of *Silvestri 4110* and *Silvestri 4110a*, but despite its relatively early publication, it was not noticed at all by Schneider in any of his various publications. It was noticed by Ahrendt (1961: 67) who designated *Silvestri 4110* as the type, although it is clear that, not only had he never seen this, but did not know where it was held. His choice of type was thus arbitrary and based on *4110* being cited first in the protologue rather than on a comparison of the specimens with the original description. As such, Ahrendt's typification should be regarded as a largely mechanical method of selection as defined by Art. 10.6 of the Shenzhen Code (Turland et al., 2018) and discounted. *Silvestri 4110a* has been chosen as the lectotype since it has more fruits than *4110*. Ahrendt also complicated matters by modifying Silvestri's description on the basis of *A. Henry 10618* (K, also at MANCH) from the Imen (Yimen) area of Yunnan, which he identified as *B. zanlancianensis*. *Henry 10618* was identified subsequently by Chamberlain and Hu (1985: 554) as *B. vernalis* (treated here as a synonym of *B. ferdinandi-coburgii*; under which see for further details). Neither having seen the type nor knowing where it was, Chamberlain and Hu (1985: 556) included *B. zanlancianensis* in their Taxa Not Seen.

The protologue of *Berberis zanlancianensis* noted that it was similar to *B. julianae* (which Pampanini also reported from Zan-lan-scian on the basis of *C. Silvestri 4100* and *4100a*, cited below), but differed from it mainly by having narrowly lanceolate leaves. The leaves of *Silvestri 4110* and *4110a*, however, are broadly rather than narrowly lanceolate and those of *4110a* include obovate-elliptic leaves (as is predominantly the case with *Silvestri 4100* and *4100a*). In any case, the leaves of *B. julianae* are variable. Fruit of *Silvestri 4110* kindly provided by FI proved to be 1-seeded, as in *B. julianae*; hence my synonymy.

The occurrence of *Berberis julianae* in west Guizhou at a very considerable distance from anywhere else reported here is particularly interesting. I am grateful to Shenzhi He of GZTM for drawing my attention to his collection from Weining Xian (details below).

As noted in the introduction, *Berberis julianae* appears to be a frequent "dustbin species" for plants of section *Wallichianae* that cannot otherwise be identified. No evidence has been found to support the reports of *B. julianae* in Anhui (S. C. Li, 1987: 347), Fujian (Jiang, 1985: 45), Guangxi (Wang, 1991: 313), Dengzhou Xian in Henan (B. Z. Ding et al., 1981: 489), and Jiangxi (G. F. Zhu, 2004: 195), all of which are likely based on misidentifications. Reports of the species in Vietnam (e.g., Pham, 1999: 326; Võ, 2007: 120; Hien et al., 2018) are mistaken and are referable to *B. subacuminata* (for further information, see under that species).

Plants under the name *Berberis julianae* are widely cultivated in Europe and North America. Although their descent is most likely from original Wilson-collected seeds, many are, in fact, hybrids.

Selected specimens.
Syntype. Berberis julianae. W Hubei: "North & South of Ichang," 900–1200 m, Oct. 1907, *E. H. Wilson 417* "B" (A 00038769, BM BM000559459, K K000644917, US 0013885).

Syntype. Berberis julianae var. *patungensis.* W Hubei: Patung Hsien (Badong Xian), 1300 m, July 1907, *E. H. Wilson 2878* "B" (A 00338445, B B10 000250732, BM BM001010629, HBG HBG-506679, K K000077318, US 00103863, W 1914–0004808).

Syntype. Berberis zanlancianensis. NW Hubei: [Zhushan Xian], Zan-lan-scian (Canglang Shan), Apr. 1912, *C. Silvestri 4110* (FI).

Other specimens.
Chongqing (E Sichuan). Fengjie Xian: Xinglong, 2100 m, 18 June 1958, *Z. R. Zhang 25294* (IBSC 0091863, KUN 0175761, NAS NAS00314349, PE 01031454). **Wushan Xian:** Dangyang Xian, 1900 m, 3 July 1958, *G. H. Yang 58719* (IBSC 0092599, PE 01031455, 01031591). **Wuxi Xian:** Tongcheng Qu, Shuangyang Gongshe, 2100 m, 28 June 1962, *N. P. Chi 00436* (CDBI CDBI0027459).

N & W Guizhou. Daozhen Xian: Guanyinyan, 19 Aug. 1994, *L. L. Deng 84-92* (GZAC GZAC0030006). **Weining Xian:** vic. of Xiaohai Zhen, 2200 m, 14 Apr. 2014, *S. Z. He 404196* (GZTM).

Hubei. Badong Xian: "Patung Hsien," 10 May 1934, *H. C. Chow 137* (A 00280244, PE 01031634); Nanping, 1600 m, 29 June 1939, *T. P. Wang 11235* (PE 01031635–37, WUK 0057153). **Baokang Xian:** Fenlubei, 1100 m, 2 Aug. 1975, *Y. J. Ma 2146* (HIB 00129407–08). **Danjiang Qu:** Wudang, Shibapan, 1400 m, 30 May 1986, *Z. E. Zhao 2072* (HIB 00129270). **Enshi Shi:** Banqiao Xiang, Qingyanwan, 1400 m, 18 June 1958, *M. Y. Fang 24311* (FUS 00014394, HIB 00129447, IBSC 0091858, KUN 0176439, NAS NAS00314478, PE 01031640). **Hefeng Xian:** Huping Ling, Baowan, 1350 m, 13 Aug. 1958, *H. J. Li 5512* (HIB 00129452, IBSC 0091857, KUN 01/6418, PE 01031642, SZ 00294183, WH 06008782). **Jianshi Xian:** Dangyang He, Jiaping Gou, 21 Sep. 1951, *L. Y. Dai & Z. H. Qian 1315* (HIB 0129445, PE 01031638, WH 06008668). **Lishuan Shi:** Li-chuan, Tuan-Pao-Hsiang (Tuanbao Xiang), 1125 m, 16 Oct. 1947, *C. T. Hwa 137* (PE 01031639). **Shennongjia Lin Qu:** along Miaogou canyon, 32.5°E, 110.5°N, 1800 m, 28 Aug. 1980, *Sino-American Bot. Exp. 232* (A 00280233, HIB 00129413–14, KUN 0176416, NAS NAS00314473, PE 00996101, UC UC1559323). **Wufeng Xian:** Yuanhe Linchang, Yangjiahe, 1060 m, 8 May 1980, *T. Q. Xu 0238* (HIB 00129267). **Xianfeng Xian:** Xiao Shaping, Xiang Shui Dong, 24 Sep. 1956, *H. J. Li 9225* (HIB 00129220–21, IBSC 0091854, KUN 0176419, PE 01031644). **Xingshan Xian:** Wanchaoshan, 1800 m, 28 Sep. 1956, *Q. M. Hu 217* (IBSC 0092254, LBG 00064122). **Xuan'en Xian:** Maobatang, Laowuji, 8 June 1958, *H. J. Li 2676* (HIB 00129463). **[Zhushan Xian]:** Zan-lan-scian (Canglang Shan), Apr. 1913, *C. Silvestri 4100 & 4100a* (FI).

N Hunan. Shimen Xian: Xiao Xizhi, 1800 m, 12 July 1987, *Huping Shan Research Team 1369* (PE 01376129–31).

SE Shaanxi. Pingli Xian: betw. Zhongbao & Maozi Miao, 29 May 1962, *B. C. Ni 00165* (CDBI CDBI0027461).

45. Berberis kawakamii Hayata, J. Coll. Sci. Imp. Univ. Tokyo 30(1): 24–25. 1911. TYPE: China. Taiwan: Mt. Morrison (Yushan), ca. 2750 m, Oct. 1906, *T. Kawakami 1941* (holotype, TI 02622 image!).

Berberis formosana Ahrendt, J. Bot. 79(Suppl.): 24. 1941. *Berberis kawakamii* Hayata var. *formosana* (Ahrendt) Ahrendt, J. Linn. Soc., Bot. 57: 65. 1961. TYPE: China. Taiwan: betw. Mt. Arisan (Alishan) & Mt. Morrison (Yushan), 2600–3600 m, 25 Oct. 1918, *E. H. Wilson 10910* (holotype, BM BM001015554!; isotype, A 00038750!, K [2 sheets] K000644914–15!, US 00956032 image!).

Shrubs, evergreen, to 3 m tall; mature stems pale brownish yellow, sulcate, verruculose; spines 3-fid, concolorous, (0.6–)1.2–2.7 cm, abaxially flat or slightly sulcate. Petiole almost absent, sometimes to 3 mm; leaf blade abaxially pale green, shiny, adaxially mid to dark green, elliptic, oblong, or oblong-lanceolate to obovate-oblong or oblanceolate, 2.5–5.5(–7) × (0.8–)1.2–2 cm, leathery, abaxially epruinose, midvein raised abaxially, impressed adaxially, lateral veins indistinct abaxially, inconspicuous adaxially (conspicuous when dry), reticulation obscure abaxially, indistinct adaxially (inconspicuous when dry), base attenuate or cuneate, margin spinulose with 5 to 12 teeth on each side, apex acute or acuminate, mucronate. Inflorescence a congested fascicle, (2- to)4- to 15-flowered; pedicel 6–12 mm. Flowers yellow. Sepals in 3 whorls; outer sepals with reddish stripe, narrowly triangular-ovate, rarely linear, 3–6 × 1–2 mm; median sepals ovate or elliptic-obovate, 4–5 × 1.5–3 mm; inner sepals narrowly obovate, 4–5.5 × 2.5–3 mm; petals oblong, oblong-obovate or oblong-orbicular, (2–)4–6 × 1.5–3 mm, base clawed, glands ovoid, separate, apex entire or slightly incised. Stamens ca. 3 mm; anther connective slightly extended, truncate. Pistil 4 mm; ovules 2 or 3. Berry dark blue, slightly pruinose, ovoid, oblong, or ellipsoid, 5–6 × ca. 2.5–3 mm; style persistent to 1 mm.

Phenology. *Berberis kawakamii* has been collected in flower in March and April and in fruit from May to November.

Distribution and habitat. *Berberis kawakamii* has been found in all the major mountain areas of Taiwan except those in Pingtung Hsien. It has been found in subalpine and alpine meadows and forests at ca. 2000–3200 m.

Berberis kawakamii var. *formosana* was first placed in synonymy under typical *B. kawakamii* by T. S. Liu (1976: 16).

Berberis kawakamii has the widest distribution of any the section *Wallichianae Berberis* of Taiwan. The shape of its leaves varies considerably both between plants and often on the same plant. Its unusually long outer sepals make its flowers very distinctive.

Selected specimens.

Taiwan. Chaiyi Hsien: en route from Tatachia Saddle to Paiyunshanchuang, 2700–3000 m, 9 Nov. 1985, *C. I. Peng 8972* (HAST 2828); Pa-yuan Lodge to Ta-ta-ka, 2900 m, 12 Aug. 1999, *W. H. Hu 1324* (HAST 16332); Alishan Hsiang, Mt. Tata, 2400–2663 m, 23 Sep. 2000, *S. M. Kuo et al. 98* (HAST 93655). **Hsinchu Hsien:** Wufeng Hsiang, Sheipa Natl. Park, Chiuchiushanchuang ("99 Lodge"), 24.468889°N, 121.208333°E, 2694 m, 6 Sep. 1993, *C. L. Huang et al. 45* (HAST 43243). **Hualien Hsien:** Chingshuishan, 23.271111°N, 121.266667°E, 2400 m, 28 Apr. 1989, *S. Y. Lu s.n.* (HAST 6119, TAIF 76250); Taroko Natl. Park, Hsiulin Hsiang, hiking trail beside the Eagle hostel, 24.158490°N, 121.326098°E, 2867 m, 20 Oct. 2004, *M. L. Weng et al. 2439* (HAST 129200). **Kaohsiung Hsien:** Taoyuan Hsiang, from Chinching Bridge to a campsite by trail to Kuanshan 23.216667°N, 120.85°E, 2500–2700 m, 17 May 1992, *C. C. Wang 1122* (HAST 53200); Taoyuan Hsiang, Kuanshan, 23.220833°N, 120.901944°E, 3100 m, 11 Oct. 1996, *C. K. Liou et al. 408*, (HAST 70347, TAIF 84413); Kuanshanling, 23.275°N, 120.95°E, 2900–3175 m, 12 Aug. 2012, *Z. W. Lee 244* (HAST 93211, TAIF 179056). **Miaoli Hsien:** Lo-shan (Lu-chang-ta-shan) to Kwan-wu, 2000–2300 m, 24 July 1987, *J. C. Wang & C. Yang 4685* (TAI 217089); Sheipa Natl. Park, Chiuchiushanchuang ("99 Lodge"), 24.468889°N, 121.208333°E, 2694 m, Sep. 6 1993, *C. L. Huang 45* (HAST 43242); Taian Hsiang, Chiuchishanchuang, 2600–2800 m, 13 Nov. 1995, *C. C. Chen 1458* (HAST 71122). **Nantou Hsien:** Hohuanshan, 3050 m, 26 Apr. 1985, *S. Y. Lu 15953* (TAIF 161660); Jenai Hsiang, Yuanfeng, hwy. 14, 24.117778°N, 121.245278°E, 2800 m, 8 Apr. 1993, *C. I. Peng 15341* (HAST 72689, PE 01839980, TAIF 100452); Jenai Hsiang, Yuanfeng, adjacent to Prov. Rd. 14, 24.121111°N, 121.241667°E, 2700–2800 m, 28 Mar. 1994, *C. H. Chen et al. 469* (HAST 43752, US 01121981); Hsinyi Hsiang, en route from Huankao to Chungyanchinkuang, 23.497222°N, 120.997222°E, 2600–2900 m, 22 Dec. 1995, *C. H. Chen et al. 99* (HAST 26501); along trail from Tien-Chih to Nanhua Shan, 24.046944°N, 121.272778°E, 2860 m, 13 Sep. 1998, *S. L. Kelley et al. 207-98* (HAST 75560); along trail from Tienluanchih to Hohuanpeifeng, 24.200833°N, 121.276667°E, 2950 m, 16 May 2002, *Y. Y. Huang 1089* (HAST 90035). **Taichung Hsien:** Wuling, en route from Wuling Lodge to Taoshan, 2600 m, 24 Aug. 1988, *C. I. Peng's collectors (Y. K. Chen & Y. J. Chen) 12018* (HAST 2830); Hsuehshan, 2500–3200 m, 10 July 2001, *S. M. Kuo 358* (HAST 98470, TAIF 178054). **Taitung Hsien:** Southern Cross hwy. to Yilan Hsien, Szuyuanyakou, 2600 m, 30 Mar, 1996, *S. Y. Lu 25003* (TAIF 245122); Haituan Hsiang, en route to Kuanshanling, 3000 m, June 20 2002, *C. H. Liu 102* (HAST 9015). **Yilan Hsien:** Mt. Nanhuta-shan, betw. Nanshan (Piyanan) & Kirettoi, ca. 2250 m, 21 July 1963, *M. Tamura & T. Shimizu 20598* (E E00612557); Nanhutashan, 2800–3000 m, 13 May 1989, *S. Y. Lu 24993* (TAIF 76289–90).

46. Berberis laojunshanensis T. S. Ying, Acta Phytotax. Sin. 37(4): 318. 1999. TYPE: China. NW Hubei: Xingshan Xian [that part now in Shennongjia Lin Qu], Laojunshan, 28 May 1957, *Y. Liu 577* (holotype, PE 00935227!).

Shrubs, evergreen, to 1 m tall; mature stems pale yellow, not verruculose; spines 3-fid, concolorous, 0.8–2.4 cm, weak. Petiole almost absent; leaf blade abaxially pale green, mostly pruinose, adaxially dark green, elliptic or narrowly elliptic, 1–2.3 × 0.4–07 cm, leathery, midvein raised abaxially, indistinct adaxially, veins indistinct on both surfaces, base attenuate, margin slightly revolute, spinulose with 3 to 7 teeth on each side, apex acuminate, mucronate. Inflorescence a fascicle, 3- or 4-flowered; pedicel 10–15 mm. Sepals in 3 whorls; outer sepals triangular-ovate, ca. 2.1 × 1.2 mm; median sepals ovate, ca. 4 × 2 mm, apex acuminate; inner sepals oblong, 4.6–5 × 2.8–3.5 mm; petals obovate, ca. 4 × 2 mm, base attenuate, with approximate glands, apex slightly retuse. Stamens ca. 3 mm; anther connective truncate. Ovules 4. Berry black, epruinose, ellipsoid or subglobose, 4–6 × 4–5 mm; style persistent.

Phenology. *Berberis laojunshanensis* has been collected in flower in May and in fruit in September and October.

Distribution and habitat. *Berberis laojunshanensis* is known from Shennongjia Lin Qu in west Hubei and Wufeng Xian in southwest Hubei. Where recorded, it has been found in thickets on mountainsides at 1650–1760 m.

For the possible relationship between *Berberis laojunshanensis* and the single-flowered *B. candidula*, see under the latter species.

Selected specimens.
SW Hubei. Wufeng Xian: Hongyu Ping, 1760 m, 12 Oct. 1990, *F. S. Peng 5032* (HIB 001295530); Wanzi He, 1650 m, 29 Sep. 1991, *F. S. Peng 5729* (HIB 00129280–81).

47. Berberis lempergiana Ahrendt, Gard. Chron., ser. 3, 109: 101. 1941. TYPE: Cultivated. From plant supplied to Ahrendt by Hillier's Nursery, Hampshire, U.K. (raised from seed, sent to Dr. Fritz Lemperg of Hatzendorf, Austria, from Nanjing Botanic Garden at an unrecorded date), 22 Oct. 1942, *L. W. A. Ahrendt 150* (neotype, designated here, OXF!).

Berberis chunanensis T. S. Ying, Acta Phytotax. Sin. 37: 315. 1999; T. S. Ying in S. Y. Jin & Y. L. Chen, Cat. Type Spec. Herb. China (Suppl. 2), 52. 2007, syn. nov. TYPE: China. W Zhejiang: Chun'an Xian, Wangfu, 29 Aug. 1958, *X. Y. He 30227* (holotype, PE 00935135!; isotypes, HHBG HBG-05659 image!, IBSC 0092262 image!).

Shrubs, evergreen, to 2 m tall; mature stems pale yellow or grayish yellow, terete, sparsely black verruculose; spines 3-fid, concolorous, 1–4 cm, stout, subterete. Petiole 1–5 mm; leaf blade abaxially pale green, slightly shiny, adaxially dull gray-green or blue-green, oblong-elliptic or broadly lanceolate, 3.5–6(–8) × 1–3 cm, leathery, midvein raised abaxially, impressed adaxially, lateral veins indistinct abaxially, inconspicuous adaxially, reticulation indistinct or obscure on both surfaces, base cuneate, margin spinose with 5 to 14 teeth on each side, apex acuminate. Inflorescence a fascicle, 3- to 15-flowered; pedicel reddish, 6–10(–15) mm; bracteoles red, ovate, ca. 1.3 mm. Sepals in 3 whorls; outer sepals ovate-elliptic, ca. 2.5 × 2 mm; median sepals ovate-elliptic, ca. 5.5 × 4 mm; inner sepals obovate, ca. 7 × 6 mm; petals oblong-obovate, ca. 6 × 4 mm, base cuneate, glands contiguous, apex emarginate with rounded lobes. Stamens ca. 5 mm; anther connective distinctly extended, truncate. Ovules 2 or 3, subsessile. Berry deep purple, pruinose, oblong-ellipsoid or ellipsoid, 7–10 × 5–5.5 mm; style persistent; seeds 2 or 3, obovoid-globose or ellipsoid.

Phenology. *Berberis lempergiana* has been collected in flower in February and in fruit between June and December.

Distribution and habitat. *Berberis lempergiana* appears to be endemic to Zhejiang. It has been collected on sandy and forested slopes, forest margins, and streamsides at ca. 580–1800 m.

The protologue of *Berberis lempergiana* cited no type, simply referring to a plant in Ahrendt's living collection received from "Messrs. Hillier of Winchester." Subsequently, Ahrendt (1961: 78) cited "Cultivated: fl. 9 May 1939, 27 Apr. 1941, 13 May 1942; fr. 26 Oct. 1940 (Type O)." By "O," Ahrendt was referring to what is now OXF, but no specimens of any of these dates are at OXF, only one of 22 Oct. 1942. Although it is annotated as the type by Ahrendt, it is dated subsequent to the protologue; therefore, it is designated here as a neotype.

In describing the origin of his plant, Ahrendt noted that it was not recorded whether the original seed was of cultivated or wild-collected origin, and "so it cannot yet be decided whether the plant is a hybrid or a new species." Later, he (Ahrendt, 1961: 78) decided it was the latter, and this is confirmed by the various specimens cited below. The description of the flowers above is from Ahrendt and is based on cultivated material.

Ying (in Jin & Chen, 2007: 52) validated *Berberis chunanensis*, whose name was not validly published by Ying (1999: 315) because no type was indicated (Shenzhen Code, Art. 40.1 [Turland et al., 2018]). The validation by Jing et al. (2008) is a later isonym. The name of the collector of the type was not included in the protologue and comes from the isotypes at HHBG and IBSC.

Ying's justification for a separate species, *Berberis chunanensis*, is largely centered on floral characteristics (including its two whorls of sepals). However, the type (and only gathering cited) collected in late August has only fruit. It is possible that the source of the description of the flowers are *s. coll. 18* and *132* (both PE and annotated by Ying as *B. chunanensis*), but these are from cultivated plants in the Hangzhou Botanic Garden whose provenance appears to be unknown (X. L. Chen of HHBG, pers. comm. 2 Feb. 2010) and which differ from each other both in leaf shape and length of pedicels. The type of *B. chunanensis* appears little different from that of *B. lempergiana*; hence my synonymy.

Selected specimens.
Zhejiang. s. loc., 18 June 1933, *S. Chen 1603* (A 00280246, FUS 00014247, N 093057033, NAS NAS00314092, SZ 00294225); S Zhejiang, "Region of King Yuan," 800–1200 m, 7 Aug. 1924, *R. C. Ching 2335* (IBSC 0091891, S 12-25471, SYS SYS00052172, UC UC281281, US 00945867). **Dongyang Xian:** "Tung-yang Hsien," 1 Aug. 1927, *Y. L. Keng 927* (IBSC 0091892, NAS NAS00314070, NAU 00024173, PE 01840057, UC UC361951). **Jingning Xian:** Kengdi, 880 m, 3 Dec. 1958, *Z. Y Zhang 24474* (HHBG HZ009040, NAS NAS00313901). **Kaihua Xian:** Suzhuang, Lingyun Si, 28 May 1959, *S. Y. Zhang et al. 26153* (HHBG HZ008985, HZ009044, NAS NAS00313902). **Lin'an Shi:** Yong'an, Linjiakeng, 26 Sep. 1964, *H. X. Zhou 325* (HHBG HZ009005); Changhua, Daoshi Gongshe, Qingshuikeng, 30 July 1987, *L. Hong 1355* (HHBG HZ009002). **Longquan Xian:** Fengyanshan, 1800 m, 1 Aug. 1958, *S. Y. Zhang 3317* (HHBG HZ009037, PE 01031726–28); Fengyanshang, Huangmaojian, 1800 m, 31 May 1981, *Z. B. Hang 1767* (HHBG HZ009004, HTC HTC0006085). **Pan'an Xian:** Shuangfeng Xiang, Gaopeng Ping, 18 Aug. 1988, *L. Hong 2196* (HHBG HZ009000); Dapa Shan Nature Reserve, Wangyin, 580 m, 3 Aug. 2005, *F. G. Zhang DPS 543* (Dapan Shan Nature Reserve Herbarium). **Qingyuan Xian:** Jiulong Shan, s. anno, *s. coll. 849* (HHBG HZ008984). **Suichang Xian:** Zhedaikou Xiang, Daxikeng, 17 Aug. 1985, *F. G. Zhang 4477* (ZM ZMNH0006660). **Taishun Xia:** Liguan, 1200 m, 30 Nov. 1958, *D. X. Zuo 23980* (HHBG HZ009036, NAS NAS00313897–98, 00313903). **Tientai Xian:** Tientai, Feb. 1890, *E. Faber 179* (K, P P06868366); Tientai. 1891, *E. Faber s.n.* (B); Tientai Shan, 1065 m, May 5 1924, *R. C. Ching 1426* (A 00280250, NA 0082381, UC UC281360, US 00945866); Tientai Shan, Huating, 920 m, 21 July 1927, *C. Y. Chiao 14406* (A 00280245, N 057075, UC UC281360, US 00945865); Tantou, 25 July 1959, *s. coll. 28505* (HHBG HZ009042, NAS NAS00314094, PE 01031729); Huading, Ximaopeng, 16 Sep. 1964, *H. X. Zhou 101* (HHBG HZ008986). **Wencheng Xian:** Xicheng Qu, Shiyang Gongshe, Xinyan, 16 Sep. 1972, *Zhejiang Medicinal Plant Recording Group 3513* (ZM ZMNH0006637). **Yinzhou Qu:** Zhangshui Gongshe, Zhangxi Dadui, Yanzike, 20 Sep. 1960, *Ningbo Shi Medicinal Resources Survey Team 242* (NAS NAS00313899, ZM ZMNH0006627). **Yiwu Shi:** Huaxi, Beishan, 26 Nov. 1971, *H. J. Wang & J. C. Yin s.n.* (FUS 00014242). **Yunhe Xian:** Shenkeng Kou, 27 Oct. 1934, *X. Y. He 3550* (NAS NAS00314091).

48. Berberis leptopoda Ahrendt, J. Bot. 79(Suppl.): 33–34. 1941. TYPE: India. [Arunachal Pradesh],

Pachakshiri Distr., Lhalung, 28.7°N, 94.2°E, 2133 m, 29 Apr. 1938, *F. Ludlow, G. Sherriff & G. Taylor 3697* (holotype, BM BM000559449!).

Shrubs, evergreen, to 2 m tall; mature stems pale yellow, terete; spines 3-fid, concolorous, 1–2.3 cm, abaxially sulcate. Petiole almost absent; leaf blade abaxially pale green, slightly pruinose, adaxially dark green, narrowly elliptic, oblong-elliptic, ovate-elliptic, or lanceolate-elliptic, 3.3–6.5(–9) × 0.8–1.8 cm, leathery, midvein raised abaxially, impressed adaxially, lateral veins largely obscure abaxially, inconspicuous adaxially, reticulation obscure abaxially, indistinct adaxially, base cuneate, margin spinose with 2 to 6 widely spaced teeth on each side, sometimes entire, apex acute, mucronate. Inflorescence a fascicle, (2- to)6- to 10(to 18)-flowered; pedicel 15–25 mm; bracteoles reddish, ovate-triangular, 2–2.5 × 1.5–2 mm, apex subacute. Sepals in 2 whorls; outer sepals broadly oblong-elliptic, 4 × 3.25 mm, apex rounded; inner sepals broadly obovate, 5–6 × 3 mm; petals obovate, 4.5 × 3 mm, apex entire or slightly retuse, base cuneate, glands approximate or contiguous. Anther connective distinctly extended, apiculate. Ovules 2 or 3. Immature berry green, obovoid-ellipsoid, 8–10 × 3–4 mm; style persistent.

Phenology. *Berberis leptopoda* has been collected in China in flower in February and March and with immature fruit between April and August.

Distribution and habitat. In China *Berberis leptopoda* is known from Mêdog (Motuo) Xian in southeast Xizang (Tibet). It has been collected from thickets and forests on mountain slopes at 1800–2200 m. It is also known from that part of Arunachal Pradesh in India where there is a conflicting territorial claim by China.

Chamberlain and Hu (1985: 458) treated *Berberis leptopoda* as a synonym of a *B. griffithiana* var. *pallida* (itself treated here as a synonym of *B. griffithiana*). However, *B. leptopoda* differs from *B. griffithiana* both in leaf shape and flower structure. On current evidence, there is also a distinct geographical separation between the two species with *B. griffithiana* occurring farther west in Cona (Cuona) Xian, Xizang, and in eastern Bhutan.

The type specimen of *Berberis leptopoda* is from that part of Arunachal Pradesh claimed by both China and India. I have located no other specimens of *B. leptopoda* from Arunachal Pradesh beyond the type, although I have not investigated Indian herbaria.

Berberis leptopoda was included in Ying (1985) but omitted without explanation in Ying (2001, 2011).

Selected specimens.
SE Xizang (Tibet). Mêdog (Motuo) Xian: betw. Gedang & Xingkai, 2200 m, 31 Aug. 1974, *Qinghai-Xizang Team 74-4965* (KUN 0178308, PE 01031866); Gangde, Jiudamu Shan, 2200 m, 21 Feb. 1993, *H. Sun, Z. K. Zhou & H. Y. Yu 3573* (KUN 0177209–210); Gangde, 2200 m, 21 Feb. 1993, *H. Sun, Z. K. Zhou & H. Y. Yu 3646* (KUN 0177211); betw. Damu & Gedang, 1800 m, 14 Mar. 1993, *H. Sun, Z. K. Zhou & H. Y. Yu 4550* (KUN 0176257); Gedang, 2100 m, 16 Mar. 1993, *H. Sun, Z. K. Zhou & H. Y. Yu 4580* (KUN 0176247–48); Gedang, Xingkai, 2200 m, 17 Mar. 1993, *H. Sun, Z. K. Zhou & H. Y. Yu 4758* (KUN 0176249–50); Renchinpeng, 2000 m, 21 Apr. 1993, *H. Sun, Z. K. Zhou & H. Y. Yu 5752* (KUN 0176256); Renchinpeng, 2000 m, 22 Apr. 1993, *H. Sun, Z. K. Zhou & H. Y. Yu 5662* (KUN 0176254) & *5676* (KUN 0176255); Renchinpeng, 2000 m, 23 Apr. 1993, *H. Sun, Z. K. Zhou & H. Y. Yu 5299* (KUN 0176251–52).

49. Berberis levis Franch., Bull. Soc. Bot. France 33: 386. 1886. TYPE: China. NW Yunnan: [Binchuan Xian border with Heqing Xian], "ad Mao-Kou-Chan supra Tapin-tze," 23 Apr. 1883, *J. M. Delavay 495* "A" (lectotype, designated here, P P00716565!; isolectotypes [2 sheets] P P02682362–63!, probable isolectotype K K000077310!)

Berberis willeana C. K. Schneid., Oesterr. Bot. Z. 67: 141–142. 1918. TYPE: China. NW Yunnan: [Yongsheng Xian], "auf dem Passe zwischen Tai-nao ko und Lichiang," 3000 m, 4 July 1914, *C. K. Schneider 1763* (lectotype, designated by Ahrendt [1961: 69], K K000077308!; isolectotypes, A 00038821!, E E00217952!).
Berberis subcoriacea Ahrendt, Gard. Chron., ser. 3, 105: 371. 1939, syn. nov. TYPE: Cultivated. Royal Horticultural Society Garden, Wisley, Surrey, U.K., 2 Oct. 1939 (lectotype, designated here, K K00395242!).
Berberis willeana var. *serrulata* C. K. Schneid., Repert. Spec. Nov. Regni Veg. 46: 245. 1939. TYPE: China. NW Yunnan: Pin-chuan Hsien (Binchuan Xian), 2800 m, 19 July 1933, *H. T. Tsai 53000* (lectotype, inadvertently designated by Chamberlain & Hu [1985: 552], A 00038822!; isolectotypes, IBSC 0092190 image!, KUN 1204869!, LBG 00064119 image!, PE 01037166!, SZ 00287843 image!).
Berberis levis var. *brachyphylla* Ahrendt, J. Linn. Soc., Bot. 57: 75. 1961. TYPE: China. NW Yunnan: [Heqing Xian], on ascent of Sungkwei Pass from Langkong Valley, 26.5°N, 2740–3050 m, Apr. 1906, *G. Forrest 2012* (holotype, K K000077308!; isotypes, A 00038774!, E E00217953!, LE!).

Shrubs, evergreen, to 2.5 m tall; mature stems very pale yellow or yellowish gray, angled or subterete, black verruculose; spines 3-fid, pale brownish yellow, 1–4 cm, stout, terete or adaxially flat. Petiole almost absent; leaf blade abaxially pale yellow-green, slightly shiny, adaxially dull green, narrowly elliptic or elliptic-lanceolate, 3–8(–12.5) × 0.7–2(–3.5) cm, leathery, midvein raised abaxially, impressed adaxially, lateral veins indistinct or obscure abaxially, inconspicuous adaxially, reticulation obscure abaxially, indistinct adaxially, base cuneate, margin spinulose with 5 to 15(to 20)

teeth on each side, apex shortly acuminate, mucronate. Inflorescence a fascicle, 3- to 25-flowered; pedicel 10–20 mm; bracteoles lanceolate. Sepals in 2 whorls; outer sepals lanceolate, 3–5 × 1–1.5 mm; inner sepals obovate or lanceolate, 4–5 × 1–2 mm; petals obovate or broadly obovate, 5–6 × ca. 2.5 mm, base cuneate, glands separate, apex entire or slightly emarginate, rounded or mucronate. Stamens 3–4 mm; anther connective extended, obtuse. Ovules 1, shortly funiculate. Berry black, pruinose or not, ellipsoid, 7–8 × 5–6 mm; style persistent.

Phenology. *Berberis levis* has been collected in flower in April and May and in fruit between July and December.

Distribution and habitat. *Berberis levis* is known from northwest Yunnan. It has been collected in forests, from banksides, and among scrub at ca. 2220–3350 m.

The protologue of *Berberis levis* cited three Delavay collections, *495* of 23 April 1883, *893* of 14 April 1884, and *993* of 12 October 1885, but did not designate a type. Schneider (1918: 139) designated *Delavay 495* of 23 April 1883 as type, but did not indicate the herbarium. Imchanitzkaja (2005: 274), apparently unaware of Schneider's typification, lectotypified *Delavay 495* at P without giving any date of collection. In fact, there are seven specimens of *495* at P, three of 23 April 1883 which are designated here as *Delavay 495* "A," two of 12 March 1883 which are designated here as *Delavay 495* "B," and two undated ones which might be duplicates of either "A" or "B," but might be of neither and which are designated here as *Delavay 495* "C." The best of the specimens of *Delavay 495* "A" has been chosen as the lectotype. *Delavay 495* at K is undated, but has the same collection details as that of 23 April 1883.

A second specimen of the syntype *Delavay 993* at P (02465452), although from the same area as the collection of 12 October 1885, must have been collected at a different time since it has flowers. Therefore, that of 12 October 1885 is designated here as *Delavay 993* "A" and the undated specimen *Delavay 993* "B."

The protologue of *Berberis willeana* cited *Schneider 1763* but without indicating the herbarium. The lectotype specimen at K was designated by Ahrendt (1961: 69).

The protologue of *Berberis willeana* var. *serrulata* cited *H. T. Tsai 53000* at A and *J. F. Rock 3214* but did not designate a type. Subsequently, Schneider (1942: 12) designated *H. T. Tsai 53000* as the type, but without indicating the herbarium. Chamberlain and Hu (1985:

552) mistakenly cited *H. T. Tsai 53000* as the holo-type; under the Shenzhen Code, Art. 9.10 (Turland et al., 2018), this can be treated as inadvertent lecto-typification.

The synonymy of *Berberis willeana, B. willeana* var. *serrulata*, and *B. levis* var. *brachyphylla* was made by Chamberlain and Hu (1985: 552).

In the protologue of *Berberis subcoriacea*, Ahrendt described a living plant at the Royal Horticultural Society Garden at Wisley that "Mr Mulligan [of the Garden] tells me that he is almost certain that it was raised from Forrest seed, although there is no proof of this." Subsequently, Ahrendt (1961: 75) cited "Cultivated (from reputed Forrest seed): fl., fr. 1938, Wisley (Type)." It is difficult to interpret this. There are two specimens at WSY, WSY 0057684 and WSY 0057683 (collection details below). They were originally annotated as "*Berberis levis*," but then been crossed out and replaced by "*Berberis subcoriacea*." An additional specimen at K, which is ex-Wisley and which is dated 10 October 1939, was annotated as type by Ahrendt on 16 September 1941. In the absence of any other specimen so annotated, this has been chosen as the lectotype. Although it is sterile, from its leaves it appears to be *B. levis* (as do the two WSY specimens with flowers and fruit). Not having seen any of these specimens, Chamberlain and Hu (1985: 534) described *B. subcoriacea* as being possibly a synonym of *B. phanera*. Their basis for this was *Delavay s.n.* (K), Yunnan Fang-yang-tchang, by Mo-so-yn, 21 Oct. 1887, also cited by Ahrendt (1961: 75) as an example of *B. subcoriacea*. The K specimen has not been located, but *Delavay s.n.* (P P02313659) with the same collection details and annotated by Chamberlain and Hu as *B. phanera* is *B. wui*.

There is a significant difference in the size of the upper and lower leaves of *Berberis levis*, the upper leaves sometimes being more than twice the size of the lower ones. This is particularly usefully illustrated by *Schneider 1763* at A and is confirmed by the living plant in my own collection cited below.

No evidence has been found to support the report by Ying (2001: 122, 2011: 739) of *Berberis levis* in Sichuan.

Syntypes. Berberis levis. NW Yunnan, "In calcareis ad collum Pi-iou-se supra Tapin-tze," 2200 m, 14 Apr. 1884, *J. M. Delavay 893* P P007165623–63; Hee-chan-men (Heqing Xian), 12 Oct. 1885, *J. M. Delavay 993* "A" (P P00716566, possibly K K000077309).

Syntype. Berberis willeana var. *serrulata*. NW Yunnan: "high plateau between Talifu [Dali] to the foot of the Likiang [Lijiang] Snow Range," 6–11 May 1922, *J. F. Rock 3214* (US 00946039).

Selected specimens.
NW Yunnan. "Au Mt Che-te-ho-tse," 12 Mar. 1883, *J. M. Delavay 495* "B" (P P02313664, 02682364). **Binchuan**

Xian: "Pin-Chuan Xsien," 19 July 1933, *H. T. Tsai 53000* (IBSC 0092190, LBG 00064119, PE 01037166, SZ 00287843); Jizu Shan, betw. Zhusheng Si & Jinding Si, 21 Dec. 1946, *S. E. Liu 21999* (IBSC 0092324, PE 01031373–74); Jizu Shan, near Zhusheng Si, 2200 m, 4 July 1987, *S. Y. Bao 142* (KUN 0178710–11); Jizushan Zhen, NNW of Jizushan Zhen & Jizu Shan along path to Zhusheng Si, 25.960278°N, 100.390278°E, 2270 m, 15 Sep. 2013, *D. E. Boufford, Y. S. Chen & J. F. Harber 43557* (A, CAS, E, K, KUN, PE, TI). **Dali Xian:** Diancang Shan range, en route from Huadianba to Xizhou, 25.86666667°N, 100.03333333°E, 2800 m, 20 July 1984, *Sino-Amer. Bot. Exp. 1220* (A 00279676, KUN 0177378, US 00946038). [**?Dayao Xian**]: "Sou pin chao, via Pe yen Tsin [Shiyang] ad Pien kio," 19 Apr. 1917, *P. S. Ten 336* (A 00279679, E E00392909, US 0946005). **Dayao Xian:** Yunfeng Shan, 2900 m, 1 July 1963, *North West Yunnan Jinsha River Team 6727* (KUN 0177362–63, PE 01031368). **Eryuan Xian:** "Mo-so-yin, (Lankong)," 10 Apr. 1886, *J. M. Delavay s.n.* (KUN 1216482, P P02682357); Fengxiang, Shuangjia Ping, 11 July 1929, *R. C. Qin 23235* (KUN 0177368, PE 01031369). **Heqing Xian:** "Bois de Hee-chan-men," s.d., *J. M. Delavay 993* "B" (P P02465452); "in silvis mont. versus augustias prope Sung-queh," 3400 m, 29 Sep. 1914, *C. K. Schneider 2920* (A 00279982, GH 00279983); western flank of Sung-Kwei Pass, 26.2°N, 100.2°E, 2745 m, 15 May 1921, *G. Forrest 19393* (A 00279677, E E00392917, UC UC253195, US 00946006, W 1925–6268). **Jianchuan Xian:** "Chien-Chuan–Mekong Divide," 26.5°N, 99.666667°E, 3350 m, 15 Aug. 1922, *G. Forrest 23068* (A 00279678, E E00392911, K K000644919); Diannan Gongshe, Yuhua Dadui, 2550 m, 27 May 1981, *Qinghai-Xizang Team 137* (CDBI CDBI002741–2, HITBC 003586, KUN 0177375–76, PE 01031371–72). **Li-kiang Hsien (Lijiang Xian):** 2800 m, July 1935, *C. W. Wang 71430* (A 00279680, KUN 0177387, PE 01037170, WUK 0037488); Yongsheng Xian, 5 Aug. 1976, *Z. Y. Wu 4125* (KUN 0178259).

Cultivated material:
Living cultivated plants.
Foster Clough, Mytholmroyd, West Yorkshire, U.K., from seeds of *D. J. Hinkley (China) 140*, NW Yunnan: Lijiang Shi, Yulong Xian, Yulongxue Shan, 3230 m, 1996.

Cultivated specimens.
"*Berberis subcoriacea*," "Near Mr Blakey's house," Royal Horticultural Garden, Wisley, Surrey, U.K., Feb. 1934, *s. coll. s.n.* (WSY 0057684); Apr. 1936, *s. coll. s.n.* (WSY 0057683).

50. Berberis lijiangensis C. Y. Wu ex S. Y. Bao, Bull. Bot. Res., Harbin 5(3): 9. 1985. TYPE: China. NW Yunnan: Lijiang Xian, 3070 m, 25 Sep. 1958, *N. Li 101052* (holotype, KUN 1204038!).

Shrubs, evergreen, to 2 m tall; mature stems yellow, subterete, verruculose; young shoots densely white pruinose; spines 3-fid, concolorous, white pruinose on young shoots, 1.5–3 cm, stout. Petiole almost absent; leaf blade abaxially pale green, mostly densely white pruinose, adaxially dark green, oblong-elliptic or narrowly elliptic, (1.5–)3–5 × (0.8–)1.4–1.8 cm, leathery, midvein raised abaxially, impressed adaxially, lateral veins in 3 to 6 pairs, raised and conspicuous abaxially, slightly impressed and inconspicuous adaxially, base cuneate, margin thickened, sometimes slightly revolute, spinose with 3 to 5 teeth on each side, apex acute, mucronate. Inflorescence a fascicle, 3- to 6-flowered; pedicel 12–15(–35) mm, white pruinose; bracteoles ovate, ca. 2.5 mm. Sepals in 2 whorls; outer sepals broadly elliptic, ca. 8 × 7 mm, apex rounded; inner sepals oblong-elliptic, ca. 9 × 7 mm; petals oblong-obovate, ca. 6 × 4 mm, base attenuate, not clawed, glands separate and linear, apex rounded, entire or slightly retuse. Stamens ca. 5 mm; anther connective extended, truncate. Ovules 4 or 5, shortly funiculate. Berry shiny purplish black, lightly pruinose when immature, otherwise not, oblong, ca. 12 × 5–7 mm; style not persistent.

Phenology. Berberis lijiangensis has been collected in flower between May and July and in fruit in September.

Distribution and habitat. Berberis lijiangensis has been found on the eastern flank of the Yulongshan in Lijiang Shi with one collection from Zhongdian (Xianggelila) Xian. It has been found in thickets, forest margins, and shady rocky situations in pine forests from 2400–3400 m.

Berberis lijiangensis was independently determined to be undescribed and given the same name by Chamberlain and Hu (1985: 534) with *G. Forrest 2203* at E as the holotype. The name, however, was first published by Bao.

Despite *G. Forrest 2203* being collected as early as 1906, it took until 1985 before the species was recognized as distinct. The reason appears to be due to the similarities between herbarium specimens of *Berberis lijiangensis* and the very variable-leaved *B. pruinosa*. However, living plants of the two species are easily distinguished with the leaves of the former being adaxially significantly darker and with densely white pruinose young shoots and spines. The mature fruit of *B. lijiangensis* is also epruinose in contrast to the densely white fruit of *B. pruinosa*.

It has proved impossible to pinpoint where the only specimen located from Zhongdian Xian was collected.

Selected specimens.
NW Yunnan. Lijiang Shi: E flank of the Lijiang range, 27.166667°N, 2700–3100 m, May 1906, *G. Forrest 2203* (E E00217951, K K000644866, PE 01030833); Yangtze watershed, Prefectural Distr. of Likiang, eastern slopes of Likiang Snow Range, May–Oct. 1922, *J. F. Rock 3551* (A 00279807, US 0945954); Yulongshan, A'meiluo Guo, 2820 m, 14 May 1962, *Lijiang Botanic Garden 100069* (LBG 00064144, KUN 0176670–71, 0176978–79); Yulongshan, NE flank, 2800 m, 2 June 1981, *Qinghai-Xizang Team 244* (CDBI CDBI0027172–73, HITBC 003615, KUN 0176920–21, PE 01031870–71); Yulongshan, betw. Heishui & Sandaowan, 3300 m, 7 June

1985, *Kunming/Edinburgh Yulong Shan Expedition 630* (E E00392930, KUN 0176827); Yulong Shan, Gangheba, 3250 m, 22 May 1987, *Sino-British Expedition to Lijiang 020* (E E00392931, K). **Zhongdian (Xianggelila) Xian:** Shangjiuluo, 2400 m, 13 July 1963, *Zongdian Team 63-3479* (KUN 1204035–36).

Cultivated material:
Living cultivated plants.
Sir Harold Hillier Gardens, Hampshire, and Royal Horticultural Society, Wisley, Surrey, U.K., from *B. & S. Wynn-Jones 7771*, Lijiang, near Yuhu Cun, 3040 m, 29 Sep. 2000.

51. Berberis liophylla C. K. Schneid., Repert. Spec. Nov. Regni Veg. 46: 247. 1939. TYPE: China. S Sichuan: [Zhaojue Xian], betw. Sik wai (Sikai) & Tjiadjio (Zhaojue), E of Nin yuan fu (Xichang), ca. 2200 m, 22 Apr. 1914, *C. K. Schneider 969* (lectotype, designated here, A 00279668!; isolectotypes, G G00237535!, GH 00038773!, Herb. Dendrol. C. Schneider†).

Shrubs, evergreen, to 2 m tall; mature stems pale brownish yellow, sulcate; spines 3-fid, concolorous, 0.6–2.8 cm, terete, slender. Petiole almost absent; leaf blade abaxially pale yellowish green, adaxially bright green, shiny, oblanceolate, elliptic-lanceolate, or elliptic, 2–6 × 0.7–2 cm, leathery, midvein raised abaxially, flat or impressed adaxially, lateral veins slightly raised and inconspicuous abaxially, inconspicuous adaxially, reticulation indistinct or obscure on both surfaces, base attenuate, margin spinose with (3 to)5 to 10 teeth on each side, apex acute, mucronate. Inflorescence a fascicle, (3- to)5- to 14-flowered; pedicel 8–15 mm; bracteoles ovate, apex acute. Sepals in 2 whorls; outer sepals elliptic-ovate, ca. 3 mm, apex subacute; inner sepals obovate, ca. 4.5 mm; petals obovate, ca. 4 mm, base clawed, glands oblong, apex emarginate. Stamens ca. 3 mm; anther connective extended, obtuse. Ovules 1, subsessile. Berry black, sometimes pruinose, ellipsoid, 7–8 × 3–4 mm; style persistent.

Phenology. Berberis liophylla has been collected in flower in April and May and in fruit between May and October.

Distribution and habitat. Berberis liophylla is known from Xichang Shi, and Butuo, Jinyang, Leibo, Xide, Yuexi, and Zhaojue Xian in south Sichuan, and Qiaojia Xian in northeast Yunnan. It has been collected on the margins of forests and *Populus* woods at 1900–3400 m.

Schneider (1942: 13) cited four collections in the protologue of *Berberis liophylla* but did not designate a type. Ahrendt (1961: 74) cited the first of these listed, *Schneider 969*, as the type, although he had not seen the specimens and provided no evidence where they could be found (the protologue listed *Schneider 969* as being in "Herb. Dendrol. C. Schneider," which, as noted in the introduction, was destroyed in WWII). As such, Ahrendt's typification should be regarded as a largely mechanical method of selection as defined by Art. 10.6 of the Shenzhen Code (Turland et al., 2018) and discounted. *Schneider 969* at A has been chosen as the lectotype because it is the only surviving specimen listed in the protologue with floral material. Of the other three collections listed, one, *H. T. Tsai 50979* (details below) is not *B. liophylla* but *B. zhaotongensis*.

The description of flowers given above is taken from Schneider's protologue.

Selected specimens.
Syntypes. **S Sichuan:** Lose (Luosi) shan, near Nin-yu-an-fu, 2900–3300 m, 17 Apr. 1914, *H. R. E. Handel-Mazzetti 1438* (W 1930–0004053, WU 0039260); same details as *H. R. E. Handel-Mazzetti 1438*, but *C. K. Schneider 917* (A 00279669). **NE Yunnan:** Yung chan Hsien (Yongshan Xian), 2100 m, 2 June 1932, *H. T. Tsai 50979* (A 00279667, KUN 0177389, PE 01031466–67, SZ 00294121).

Other specimens.
S Sichuan. Butuo Xian: Lada Gongshe, 2259 m, 11 July 1976, *Sichuan Botany Team 137* (CDBI CDBI0027478–79, CDBI0027480). **Jinyang Xian:** Jinyangdi, Boluoliangzi, 3000 m, 16 May 1959, *J. L. Chuan 3043* (KUN 0177283, PE 01030986–87). **Leibo Xian:** Huangmaogeng, Ahe Gou, 3400 m, 21 June 1959, *Z. T. Guan 8667* (IBSC 0092128, PE 01031829); Ahe Gou Linchang, 3000 m, 14 Aug. 1972, *236 Task Group 0789* (PE 01037890–91). **Ningnan Xian:** Paoma, Yi Dadui, 17 July 1978, *Ningnan Team 0365* (CQNM 0005062, SM SM704800325). **Puge Xian:** Li'an Gongshe, 7 Aug. 1976, *West Sichuan Botany Investigation Team 13908* (PE, SWCTU 00042839). **Xichang Shi:** Luoji Shan, 2350 m, 28 May 1978, *Q. S. Zhao & G. Hu 4917* (CDBI CDBI0027475–77); betw. Xichang & Zhaojue, 2500 m, 17 May 1984, *W. L. Chen et al. 05247* (PE 01031463, 02044420, 02044439). **Xide Xian:** Liang He Kou, Tuanjie Dui, 16 May 1979, *Xide Team 275* (SM SM704800293). **Yuexi Xian:** 1900 m, 5 June 1932, *T. T. Yu 995* (A 00279374, IBSC 0092533, PE 01031831, 01293940); Dahua, 2000 m, 28 June 1959, *s. coll. 3527* (PE 01031832). **Zhaojue Xian:** near Zhaojue, 6 Aug. 1964, *s. coll. 129* (CDBI CDBI0028212, PE 01031461); S of Zhaojue, Rikeluo Shan, 2100 m, 27 June 1960, *W. H. Sun 0049* (SZ 00294129, 00294177).
NE Yunnan. Qiaojia Xian: Xiao He, 2800 m, 15 July 1973, *B. X. Sun et al. 937* (KUN 0178558).

52. Berberis lubrica C. K. Schneid., Repert. Spec. Nov. Regni Veg. 46: 265. 1939. TYPE: China. S Sichuan: [Yanyuan Xian-Muli Xian border], betw. Kua-pieh (Guabie) & Mo-lien (Molian), ca. 2800 m, 25 May 1914, *C. K. Schneider 1384* (holotype, Herb. Dendrol. C. Schneider†; lectotype, designated by Ahrendt [1961: 55], K K000567951!; isolectotypes, A 00267887!, E E00217956!).

Shrubs, evergreen, to 1.25 m tall; mature stems pale yellow, angled or subterete, verruculose; spines 3-fid,

concolorous, 1.5–2 cm, flat. Petiole 1–3 mm; leaf blade abaxially pale yellow-green, shiny, adaxially green, shiny, narrowly lanceolate, 6–9 × 1–1.5 cm, leathery, midvein raised abaxially, impressed adaxially, lateral veins and reticulation indistinct or obscure or both surfaces, base cuneate, margin undulate, revolute, spinulose with 12 to 24 teeth on each side, apex acuminate, mucronate. Inflorescence a fascicle, 6- to 10-flowered; pedicel 5–10 mm; bracteoles ovate, ca. 2.5 × 1.5 mm, apex acute. Sepals in 2 whorls; outer sepals ovate, ca. 3.5 × 2 mm, apex acute; inner sepals obovate, 6–7 × ca. 4.5 mm; petals oblong-obovate, ca. 3.5 × 2 mm, base cuneate, glands separate, apex subentire. Stamens ca. 3.5 mm; anther connective conspicuously extended, truncate. Ovules 4. Immature berry dark blue, ovoid; style persistent.

Phenology. *Berberis lubrica* has been collected in flower and with immature fruit in May.

Distribution and habitat. *Berberis lubrica* appears to be known from only the type collection from the Yanyuan Xian-Muli Xian border. This was collected in a thicket at 2800 m.

Schneider 1384 (K) was described as the type of *Berberis lubrica* by Ahrendt (1961: 55). Given the destruction of the holotype, this can be regarded as an obligate lectotypification.

53. Berberis mingetsensis Hayata, Icon. Pl. Formosan. 5: 4, pl. 2. 1915. TYPE: China. Taiwan: Chiayi Hsien, Mt. Arisan (Alishan), near Minget-sukei, 8 Apr. 1914, *B. Hayata s.n.* (holotype, TI 02625 image!).

Shrubs, evergreen, to 1.5 m tall, decumbent; branches slender, mature stems pale brownish yellow, subterete, not verruculose; spines 3-fid, concolorous, 0.5–2.2 cm, terete. Petiole almost absent, rarely to 3 mm; leaf blade abaxially pale green, adaxially dark green, shiny, narrowly elliptic, elliptic-lanceolate, or lanceolate, 5.5–9 × 1.3–2.5 cm, leathery, abaxially usually glaucous pruinose, midvein raised abaxially, impressed adaxially, lateral veins obscure abaxially, indistinct or inconspicuous adaxially, reticulation obscure abaxially, indistinct adaxially, base cuneate, margin spinulose with 5 to 16 teeth on each side, base cuneate, apex acuminate, mucronate. Inflorescence a fascicle, 2- to 8-flowered; pedicel 6–20 mm; bracteoles triangular, ca. 1.3 × 1 mm, apex acute. Flowers bright yellow. Sepals in 3 whorls; outer sepals oblong-ovate or ovate, usually tinged red at apex, 3 × 2 mm; median sepals ovate, 5 × 3 mm; inner sepals elliptic or broadly obovate, 6.5 × 4.5 mm; petals obovate, ca. 3 × 2.5 mm, base cuneate, glands oblong-ovate, separate, ca. 0.6 × 0.25 mm, apex incised. Stamens ca. 4 mm; anther connective slightly extended, truncate. Pistil 5 mm; ovules 3 to 6. Berry black or bluish black, mostly pruinose, ellipsoid, ca. 7 × 4 mm; style persistent, short.

Phenology. *Berberis mingetsensis* has been collected in flower in March and April and with immature fruit from March to June and has been recorded with mature fruit from September to October.

Distribution and habitat. It appears that the population of *Berberis mingetsensis* is extremely small, being confined to Mt. Alishan in Chiayu Hsien with a small disjunct population in Nantou Hsien. It has been found in the understory of coniferous and mixed forests and by the now disused Mienyueh Spur of the Alishan Forest Line at 2200–2800 m.

With its abaxially glaucous pruinose, dark green, shiny leaves, *Berberis mingetsensis* is a very distinctive species. The main area where it is found was subject to extensive damage by Typhoon Morakot in August 2009. Though the nearby railway was destroyed, the population of the species survived. Information on the fruit of *B. mingetsensis* has been kindly supplied by C. C. Yu of NTU.

Mizushima (1954: 28) mistakenly treated *Berberis mingetsensis* as a synonym of *B. bicolor*, a species endemic to Guizhou. In correcting Mizushima, Chamberlain and Hu (1985: 539) treated *B. mingetsensis* as a synonym of *B. aristatoserrulata*, but that species has a different flower structure and leaves that are adaxially dull green. Chamberlain and Hu's synonymy was followed Ying (2001: 139, 2011: 745). S. Y. Lu and Yang (1996: 576) mistakenly treated *B. hayatana* from northern Taiwan as a synonym of *B. mingetsensis*.

Selected specimens.
Taiwan. Chiayi Hsien: Mt. Arisan, 4 Apr. 1914, *s. coll. (?B. Hayata) s.n.* (TI); Alishan Museum, 1 Aug. 1957, *S. T. Lu s.n.* (HAST 101460); Mt. Alishan, by railway betw. Alishan & Mienyueh stations, ca. 2300 m, 20 Mar. 1983, *C. I. Peng 4566* (A 00415553, HAST 2827); Alishan Xiang, Mingets Railway 7.5 km, 23.542222°N, 120.804167°E, 2200 m, 10 Sep. 2008, *C. C. Yu 145* (TAI); same details, but 23.535833°N, 120.796389°E, 2200 m, 18 Dec. 2008, *C. C. Yu 251* (TAI). **Nantou Hsien:** Shinyi village, Siluantashan, 23.696389°N, 120.9425°E, 2800 m, 11 Apr. 2009, *C. C. Yu 398, 399* (TAI).

54. Berberis morii Harber & C. C. Yu, Taiwania 63(3): 236. 2018. TYPE: Cultivated. 20 Mar. 2016, *J. F. Harber & C. C. Yu 2016-01* from a cutting of *J. F. Harber & C. C. Yu Guanmen Exped. 19*, Taiwan. Hualian Hsien, Guangfu Logging Trail, 1576 m, 9 Apr. 2014 (holotype, HAST 141801!).

Shrubs, evergreen, to 1.4 m tall; mature stems purplish red turning pale brownish yellow, terete; spines 3-fid, concolorous, 0.6–2.4 cm. Petiole almost absent; leaf blade abaxially densely white pruinose, adaxially dull yellowish green, lanceolate, lanceolate-ovate, or lanceolate-elliptic, 5.5–9 × 2–3 cm, thickly leathery, midvein raised abaxially, slightly impressed or flat adaxially, lateral venation and reticulation largely obscure abaxially, inconspicuous adaxially, base attenuate, margin spinose with 5 to 12 widely spaced, often coarse teeth on each side, apex acuminate, mucronate. Inflorescence a fascicle, 8- to 15-flowered; pedicel 9–15 mm; bracteoles absent. Sepals in 3 whorls; outer sepals broadly ovate, sometimes with reddish stripe, 2 × 2 mm; median sepals broadly obovate or obovate-orbicular, 4–4.25 × 3.5–4 mm; inner sepals obovate or obovate-elliptic, 6 × 3.75–4 mm; petals obovate, 5.75 × 3.5–4 mm, base cuneate, glands contiguous, ovoid, ca. 1 mm, apex entire or slightly retuse. Stamens 3.5 mm; anther connective extended, truncate. Pistil 3.5 mm, ovules 3. Berry black, densely pruinose, subglobose or broadly ellipsoid, 7–8 × 6–7 mm; style persistent, short.

Phenology. *Berberis morii* has been found in flower in March. In cultivation it has fruited from June to November.

Distribution and habitat. *Berberis morii* is known in the wild from only two small colonies in Hualien Hsien in east Taiwan, one found among *Miscanthus sinensis* at the side of a disused and degraded logging trail in a cloud forest at 1576 m, the second at the summit of Mt. Wanwuta.

As the protologue noted, *Berberis morii* was first found on an expedition to find *B. aristatoserrulata* and *B. brevisepala* in the area where the type specimens of both were collected by Mori (for further details of the expedition, see Harber, 2015 and Yu and Chung, 2016). Only one small colony of four plants of *B. morii* was found, all of which were, unfortunately, sterile. A second colony, found subsequently by C. C. Yu on the summit of Mt. Wanwuta, was also of only sterile plants. There was a proposal for students from TAI to revisit the first site in 2016 to see if the plants found in 2014 had flowered, but the remote area concerned is highly susceptible to landslides and, following the earthquake in south Taiwan in February 2016, entry to it was refused by the Taiwanese government on the grounds of safety. The account of flowers used here is, therefore, taken from a plant grown from a cutting made on the 2014 expedition. This is why the species was published on the basis of a cultivated plant grown from a cutting of one of the plants we discovered in 2014. The holotype consists of all the flowering material produced by this plant in 2016. The account of fruit is from specimens from the same plant made in 2018.

In April 2019, Chih-Chieh Yu sent me photographs of the species in flower taken in March on the Guangfu Logging Trail by a TAIF expedition. From the photographs, this appeared to be of the same colony we found in 2014.

The large, thickly leathery, abaxially densely pruinose leaves make this a very distinctive species quite unlike any other known from Taiwan.

Selected specimens.
Taiwan. Hualian Hsien: Guangfu Logging Trail, 1576 m, 9 Apr. 2014, *J. F. Harber & C. C. Yu Guanmen Exped. 19* (E E00783683, TAI).

Cultivated material:
Cultivated specimens.
From same plant from which the holotype was made, 16 Nov. 2018, *J. F. Harber & C. C. Yu 2018-01* (A, A00934973, HAST143786).

55. Berberis nantoensis C. K. Schneid., Repert. Spec. Nov. Regni Veg. 46: 252. 1939. *Berberis densifolia* Bijh., J. Arnold Arbor. 9: 133. 1928, non Rusby (1920). TYPE: China. Taiwan: Nanto (Nantou Hsien), Mt. Kiraishui (Chilaishushan), 3500–3600 m, 6 Mar. 1918, *E. H. Wilson 10074* (holotype, A 00056594!; isotypes, B B00038729/00056594!, BM BM000810310!, K K000077348!, PNH image!, US 00103871 image!).

Shrubs, evergreen, compact, 1.5–2 m tall; mature stems brown, subterete, sparsely verruculose; spines 3-fid, concolorous, 0.8–1.2 cm, weak, abaxially sulcate. Petiole almost absent; leaf blade abaxially pale green, adaxially green, narrowly obovate or elliptic, 1–3 × 0.5–1 cm, leathery, abaxially often pruinose at first, midvein raised abaxially, impressed adaxially, lateral veins obscure on both surfaces, base attenuate or cuneate, margin conspicuously revolute especially when dry, spinulose with 1 to 4 teeth on each side, usually confined to upper part of leaf, apex acute or subacute, mucronate. Inflorescence a dense congested fascicle, 6- to 10(to 18)-flowered; pedicels 1–3 mm; bracteoles yellow with red markings, oblong-ovate, ca. 2 × 1.5 mm, apex acute. Flowers bright yellow. Sepals in 2 whorls; outer sepals elliptic-obovate, ca. 6 × 3.5 mm, apex rounded; inner sepals narrowly elliptic, ca. 5 × 2.5 mm, apex rounded; petals broadly obovate, ca. 4.5 × 3 mm, base cuneate, glands separate, elliptic, apex rounded, entire or slightly incised. Stamens ca. 3 mm; anther connective extended, truncate. Pistil 4 mm; ovules 3 to 5. Berry dark purplish, pruinose, ellipsoid, 5–7 × 4–5 mm; style persistent, short.

Phenology. *Berberis nantoensis* has been collected in flower from March to May and in fruit from July to January the following year.

Distribution and habitat. *Berberis nantoensis* is known from Hsinchu, Miaoli, Nantou, Taichung, and Yilan Hsien in Taiwan. It has been found in the understories of coniferous forests, open alpine meadows, and exposed rocky areas at ca. 2400–3400 m.

The type and only collection cited in the protologue has fruit. The description of flowers given above is from my living plant grown from *B. & S. Wynn-Jones 6936* (details below).

The combination of revolute leaves and densely congested fascicles make this a very distinctive and easily recognized species.

Selected specimens.
Taiwan. Hsinchu Hsien: Ta-pa-chian-shan (Dabajian Shan), 3 May 1972, *T. C. Hung 5925* (TAI 156167, PH 66424). **Miaoli Hsien:** Mt. Chung-shu shan, Tashushan Logging Station, 3000 m, 26 May 1958, *C. C. Kuo & M. T. Kao s.n.* (HAST 2831, PH 66426, TAI 047136); Chung Hsueh Shan, 2600–2800 m, 27 Sep. 1984, *T. Y. Yang 394* (TAI 196511), en route from "99 villa" to Mt. Tapachienshan (Dabajian Shan), 2700–2900 m, 11 Aug; 1985, *C. I. Peng 8482* (HAST 6210); Taian Hsiang, Shei-Pa Natl. Park, ca. 10 km from entrance on Hsuehshan no. 230 forest rd., 24.308056°N, 121.030556°E, 2400 m, 4 May 1999, *S. H. Wu 1201* (HAST 79356, KUN 0175779). **Nantou Hsien:** Mt. Noko (Nengkao Shan), 6 Mar. 1918, *S. Sasaki s.n.* (TAI 047122); Mt. Noko, 16 July 1930, *Y. Kudo & K. Mori 146* (TAI 047121); Hohuanshan, Luoyingshan Zhuang, 24.15°N, 121.266667°E, 2900 m, 10 Nov. 1989, *S. Y. Lu 24977* (HAST 61307, TAIF 76586–87, 111789, 111794); Jenai Hsiang, Tayuling 820 Forest Rd., ca. 1.4 km from entrance, 24.193056°N, 121.308333°E, 2530 m, 13 May 2004, *C. C. Wu et al. 588* (HAST 103201). **Taichung Hsien:** "Mt Siaoshushan," 2800 m, 12 Oct. 1957, *T. S. Liu et al. 279* (HAST2832); Hsiaohsuehshan, 24.297778°N, 121.028611°E, 19 Sep. 1995, *C. M. Wang 01785* (HAST 67995); Hoping Hsiang, hiking trail from Shenmachenshan to Yunling Lodge, 24.339444°N, 121.393778°E, 3400 m, 18 July 1996, *T. Y. Liu 1021* (HAST 64264); Hsiaohsuehshan, 24.3°N, 121.033333°E, 2900 m, 14 Jan. 1997, *S. Y. Lu 25178* (HAST 70018). **Yilan Hsien:** Mt. Taihei (Taipingshan), 24–29 Apr. 1930, *S. Sasaki s.n.* (TAI 190522, 190534); Nahutanshan, 26 May 1974, *C. M. Kuo 5175* (TAI 173447); Nanhutanshan, 24.411944°N, 121.443611°E, 2800–3100 m, 4 May 1995, *C. K. Liou 560* (HAST 69806, TAIF 84676, 149302–3); Tatung Hsiang, Chialohu, 24.477222°N, 121.471667°E, ca. 2660 m, 11 May 2002, *C. I. Huang 839* (HAST 90312).

Cultivated material:
Living cultivated plants.
Foster Clough, Mytholmroyd, West Yorkshire, U.K., and at Royal Botanic Garden Edinburgh, from *B. & S. Wynn-Jones 6936*, Chiayi, Nantou Hsien, Kaohsiung Shi border, Yushan, 2815 m, 15 Nov. 1999.

56. Berberis nujiangensis Harber, sp. nov. TYPE: China. SE Xizang (Tibet): [Zayü (Chayu) Xian],

"Si-Kang, Me-kong, Tsa-wa-rung" [Meikong, Cawarong/Chawalong], 3000 m, Sep. 1935, *C. W. Wang 66201* (holotype, A 00038691!; isotypes, KUN 0177061!, NAS NAS00314333 image!, PE [3 sheets] 01031007–09 images!, WUK image 0047683!).

Diagnosis. *Berberis nujiangensis* is somewhat similar to *B. davidii*, but has longer, narrower leaves and fewer-flowered fascicles.

Shrubs, evergreen, to 1 m tall; mature stems pale yellowish brown, subterete, sparsely verruculose; spines 3-fid, concolorous, 1.2–2.5 cm, sparsely verruculose. Petiole almost absent; leaf blade abaxially pale green, adaxially mid-green, shiny, lanceolate, 2.5–4 × 0.6–0.8 cm, leathery, midvein raised abaxially, impressed adaxially, lateral veins and reticulation inconspicuous on both surfaces, base cuneate, margin undulate, sometimes slightly revolute, spinulose with 3 to 9 teeth on each side, apex narrowly acuminate, mucronate. Inflorescence a fascicle, 1- to 3-flowered; pedicel 10–20 mm. Sepals greenish yellow, otherwise unknown. Immature berry ellipsoid, 8 × 6 mm; style not persistent; seeds ?1 or 2.

Phenology. *Berberis nujiangensis* has been collected in flower in June and with immature fruit in September.

Distribution and habitat. *Berberis nujiangensis* is known from the Nujiang valley in Zayü Xian in southeast Xizang (Tibet). It has been collected from slopes, under pine forest, and in a damp, degraded forest at ca. 3000–3200 m.

IUCN Red List category. *Berberis nujiangensis* is assessed as DD or Data Deficient, according to IUCN (2001) criteria.

The type of *Berberis nujiangensis*, *C. W. Wang 66201*, was first treated by Schneider (1939: 254) as one of the syntypes of *B. densa*. But *Wang 66201* is of a different species from the Delavay collection that Schneider subsequently (1942: 32) designated as the type of *B. densa* (itself an illegitimate name that was replaced by *B. davidii* by Ahrendt [1961: 56–57]). It is not completely clear exactly where *Wang 66201* was collected. Tsa-wa-rung was a Salween (Nujiang) valley sub-prefecture in Tibet (see Kingdon-Ward, 1923: 272). It might be thought that the "Me-kong" refers to the Mekong River, but the Chinese characters on the collector's notes on the type collection are 梅空 (Meikong) not 湄公 (Meigong), one of the Chinese names for the Mekong or Lancang Jiang, which suggests that Meikong applies to another geographical name.

Though *Qinghai-Xizang Team 7770* has flowers, I have not had an opportunity to examine them. The collectors' notes record the sepals as being greenish yellow.

Selected specimens.
SE Xizang (Tibet). Zayü (Chayu) Xian: Chawalong, Songta Xue Shan, 3200 m, 29 June 1982, *Qinghai-Xizang Team 7770* (HITBC 074266, KUN 0178157–58, PE 01030981–2).

57. Berberis pengii C. C. Yu & K. F. Chung, Phytotaxa 184(2): 85. 2014. TYPE: China. Taiwan: Pingtung Hsien, Taiwu, Kuaiku Lodge, 22.613333°N, 120.744167°E, 2150 m, 18 Apr. 2011, *C. C. Yu 683* (holotype, TAI 284283 image!).

Shrubs, evergreen, to 4 m tall; mature stems yellowish brown, sulcate, not verruculose; spines 3-fid, concolorous, 0.8–1.8 cm, terete. Petiole almost absent, sometimes to 5 mm; leaf blade abaxially pale green, sometimes pruinose, adaxially green, shiny, elliptic, narrowly elliptic, or elliptic-lanceolate, 4.4–8.9 × 1.4–2.6 cm, leathery; midvein raised abaxially, impressed adaxially, lateral veins indistinct on both surfaces, reticulation obscure abaxially, indistinct adaxially, base attenuate, margin spinulose with 13 to 27 widely spaced teeth on each side, apex acuminate, mucronate. Inflorescence a fascicle, 4- to 7-flowered; pedicel 4–16 mm. Flowers yellow. Sepals in 3 whorls; outer sepals sometimes with reddish tinge, narrowly triangular or triangular-oblong, 5 × 1 mm; median sepals sometimes with reddish tinge, ovate, 7.5 × 2.5 mm; inner sepals obovate, 8 × 4 mm; petals elliptic, 7 × 4 mm, base clawed, glands narrowly ovoid, very close to each other, apex acutely emarginate. Stamens ca. 5 mm; anther connective extended, truncate. Pistil ca. 5.5 mm; ovules 6 to 8. Berry black, globose or subglobose, ca. 10 × 10 mm, partially pruinose; style not persistent.

Phenology. *Berberis pengii* has been collected in flower in March and April and in fruit from June to November.

Distribution and habitat. *Berberis pengii* is known from Kaohsiung, Pingtung, and Taitung Hsien in south Taiwan. It has been found in coniferous and broadleaved mixed forests at 2100–2500 m.

Selected specimens.
Taiwan. Kaohsiung Hsien: Dagueii Lake, 2150 m, 11 Feb. 2009, *Yu 338* (TAI); near Camp Yukuting, 2400 m, 12 Feb. 2009, *C. C. Yu 355, 358* (TAI). **Pingtung Hsien:** en route to Peitawushan, 3000 m, 6 June 1988, *T. C. Huang et al. 13670* (TAI 213336); Wutai Hsiang, on the way from Kuaikushanchuang to peak of Peitawushan, 22.611389°N, 120.744722°E, 2250–2900 m, 2 Apr. 1994, *C. H. Chen et al.*

595 (HAST *45173*, KUN 0176447, TNM); Wutai Hsiang, en route from the first Lodge to Chih-pen-chu-shan, 1900 m, 10 Mar. 1990, *C. H. Lin 408* (HAST 17082); Pa-yu-p'ao (Xiao Gui Hu) to Lakalakashan, 14 Feb. 1993, *S. Z. Yang 30239* (HAST 25285, PE 01793790). **Taitung Hsien:** near Mt. Shishuitou, 2300 m, 11 Feb. 2009, *C. C. Yu 341* (TAI).

58. Berberis petrogena C. K. Schneid., Repert. Spec. Nov. Regni Veg. 46: 253. 1939. TYPE: China. NW Yunnan: [Fugong Xian], N'Maikha-Salwin divide, 26.5°N, 2750 m, July 1919, *G. Forrest 18195* (holotype, E E00117384!; isotypes, A 00279845!, K K000077333!, WSY 0057633!).

Shrubs, evergreen, to 1.5 m tall; mature stems brownish yellow, sulcate; spines 3-fid, concolorous, 0.7–1.3 cm. Petiole almost absent; leaf blade abaxially pale green, adaxially mid-green, shiny, ovate to elliptic, 1.1–2.2 × 0.6–1.2(–1.5) cm, leathery, abaxially sometimes slightly pruinose, midvein raised abaxially, impressed adaxially, both surfaces with indistinct 1 to 4 lateral veins and rectangular reticulation, base cuneate, margin spinose with 4 to 8(to 12) teeth on each side, apex obtuse. Inflorescence a fascicle, 2- to 6-flowered; pedicel reddish, 7–10 mm. Sepals in 2 whorls; outer sepals elliptic-ovate with reddish vertical stripe on outside, 4 × 3 mm, apex rounded; inner sepals obovate-elliptic, 7–8 × 4.5–5.5 mm, apex rounded; petals obovate-elliptic, 6 × 5 mm, base slightly clawed, glands separate, oblong, apex rounded, emarginate. Stamens ca. 4 mm; anther connective extended, truncate. Ovules 2 or 3. Berry purplish black, slightly pruinose at first, obovoid or oblong, ca. 7 × 4 mm; style persistent; seeds 2.

Phenology. Berberis petrogena has been collected with immature fruit between May and July. Its flowering season outside cultivation is unknown.

Distribution and habitat. Berberis petrogena has been collected in Fugong Xian in northwest Yunnan and Tengchong Xian in west Yunnan. The type was collected in open rocky situations on the margins of scrub at 2750 m. The only other specimen located was found in subtropical broadleaf forest dominated by *Lithocarpus* and *Tsuga* at 2750–2850 m.

The type collection and *Gaoligong Shan Biodiversity Team 29193* have only fruit. The description of flowers used here is from the cultivated plants cited below. They were grown from wild-collected seed given to representatives of the Royal Horticultural Society, U.K., on a visit to Kunming Botanic Garden. Unfortunately, not only are there no details for the seed collection, but various other *Berberis* seed collections (including at least one of a deciduous species) from the same source were all given the same number and widely distributed

in the U.K. Hence, on its own, *C. Brickell & A. Lesley 12060* is no guide to species identification. Despite the lack of collection details, the resemblance of the Wisley plants to the type collection is so striking that I have used them here.

Ahrendt (1961: 57) cited as *Berberis petrogena*, *J. Cavalerie 7854* (K K000644862) which he stated was collected from Yunnan-fu (Kunming). In fact, it was collected in Guizhou and is *B. cavaleriei*.

Berberis petrogena was not included by Ying (2001, 2011).

Selected specimens.
W Yunnan. Tengchong Xian: border with Lushui Xian, Minguan, Zizhi Cun, NE of Zizhi, Jiangao Shan, 25.798083°N, 98.624056°E, 2750–2850 m, 19 May 2006. *Gaoligong Shan Biodiversity Team 29193* (CAS 00120152, GH 00294455, HAST 124194).

Cultivated material:
Living cultivated plants.
Royal Horticultural Society Garden, Wisley Surrey, U.K., from *C. Brickell & A. Lesley 12060* (W925385*A and W964885*A); plants grown from cuttings of the Royal Horticultural Society plants at Royal Botanic Garden Edinburgh and Howick Hall, Northumberland, U.K.

59. Berberis phanera C. K. Schneid., Oesterr. Bot. Z. 67: 22. 1918. TYPE: China. SW Sichuan: N of Yen-yuan Hsien (Yanyuan Xian), betw. Ouentin & Kalapa, ca. 2800 m, 4 June 1914, *C. K. Schneider 1460* (lectotype, designated by Ahrendt [1961: 54], K missing; new lectotype, designated here, A 00038779!; isolectotype, E E00217962!, W fragm. W 1936–5741a!).

Berberis delavayi C. K. Schneid. var. *wachinensis* Ahrendt, J. Bot. 79(Suppl.): 33. 1941. TYPE: China. SW Sichuan: Muli [Xian], Wachin, near Lamasery, 3100 m, 1 Oct. 1937, *T. T. Yu 14401* (holotype, E E00217963!; isotypes, A 00279747!, BM BM01015542!, KUN 1204864!, PE 01030993!).

Shrubs, evergreen, to 2 m tall; mature stems pale brownish yellow, subterete to terete, black verruculose; spines 3-fid, concolorous, 1–2.5 cm, stout, abaxially sulcate. Petiole almost absent, rarely to 4 mm; leaf blade abaxially pale green, slightly shiny, adaxially mid-green, slightly shiny, narrowly to broadly elliptic, 2–5 × 0.8–1.5 cm, leathery, midvein raised abaxially, slightly impressed adaxially, lateral veins and reticulation inconspicuous on both surfaces but conspicuous when dry, base attenuate, margin undulate, spinose with 3 to 5 widely separated teeth on each side, apex acute, mucronate. Inflorescence a fascicle, 2- to 5(to 7)-flowered; pedicel 6–15 mm; bracteoles broadly ovate, 4.5 × 4 mm, apex acute. Sepals in 2 whorls; outer sepals broadly obovate, 6–7 × 5.5–6.5 mm; inner sepals obovate-orbicular, 6.5–8 × 6.5–7.5 mm; petals obovate, 5.5 ×

4 mm, base clawed, glands separate, apex slightly emarginate. Stamens 3–4 mm; anther connective slightly or not extended, truncate or slightly obtuse. Ovules 3 or 4. Berry black, lightly pruinose, oblong, ellipsoid, or obovoid, 10–11 × 4–5 mm; style persistent.

Phenology. Berberis phanera has been collected in flower in May and June and in fruit between July and November.

Distribution and habitat. Berberis phanera is known from west and southwest Sichuan. It has been collected from thickets, forests, forest margins, and streamsides at ca. 2500–4000 m.

The protologue of *Berberis phanera* designated *Schneider 1460* as type but did not cite any herbarium. Ahrendt's citing of a specimen at K was an effective lectotypification, but this is missing. The specimen at A has been chosen as a new lectotype because it is annotated as type in Schneider's hand.

In the protologue, Schneider described the number of ovules of *Berberis phanera* as two, but later (1942: 54) amended this to four. The description of flowers used here is based on the protologue augmented by flowers of the living plant at Kew grown from seed of *SICH 1379* (collection details below).

Chamberlain and Hu (1985: 534) were the first to place *Berberis delavayi* var. *wachinensis* in synonymy under *B. phanera*.

Specimens from northwest Yunnan identified as *Berberis phanera* are mostly referable to *B. wui*.

Ying (2001: 113) treated *Berberis delavayi* as a synonym of *B. phanera*, but given that *B. delavayi* was published first, any synonymy should have been the other way around and this was corrected by Ying (2011: 736). However, these are two separate species.

Selected specimens.
W & SW Sichuan. Daocheng Xian: Riwa Xiang, Gangushe–Dingyashe, 3200 m, 14 July 1971, *Sichuan Economic Plants Exp. 0239* (CDBI CDBI0027174–76); Riwa Xiang, Gangushe, 3000 m, 9 Sep. 1971, *Sichuan Economic Plants Exp. 608* (CDBI CDBI0027177); Mengzi Xiang, 3500 m, 20 Aug. 1973, *Sichuan Province Botany Investigation Team 2593* (KUN 0178385, PE 01030988, SWCTU 00006313); following Yading River downstream to Chituy River, ca. 3435 m, 24 Sep. 1994, *SICH 1379* (K). **Jiulong Xian:** betw. Chahualingzi & Fangma Ping, 4000 m, 28 May 1960, *Nanshuibeidiao Exp., T. S. Ying 3634* (KUN 0179352, PE 01030989–90). **Kangding Xian:** Shade Qu, Shahalong Gou, 3060 m, 23 May 1961, *Nanshuibeidiao Exp. 02953* (KUN 0178673, PE 01030991). **Muli Xian:** Gu-mu-tian, 3100 m, 24 May 1937, *T. T. Yu 5703* (A 00279746, KUN 0176448); Shuiluo Xiang, 2400 m, 24 Oct. 1959, *J. X. Chuan 29* (KUN 0175951); Shanwang Miao, Liangzi, 3500 m, 10 May 1960, *Nanshuibeidiao Exp. 5656* (KUN 0178388, PE 01031000); Chabulang, 2900 m, 31 May 1984, *W. L. Chen et al. 5712* (PE 01030996); N of city of Muli, hwy. Z020, 28.115°N, 101.177778°E, 3650 m, *D. E. Bouf-*

ford, J. F. Harber & X. H. Li 43465 (A 00619273, E E00770755, KUN 1278427, PE). **Xide Xian:** Lake Gongshe, Lianhe Dui, 2500 m, 5 May 1979, *Xide Team 229* (SM SM704800519). [**Yanyuan Xian**]: betw. Hunka & Wo Lo ho, 3200–3400 m, 13 June 1914, *C. K. Schneider 1535* (A 00279753, E E00623059, K).

Cultivated material:
Living cultivated plants.
Royal Botanic Gardens, Kew, from *SICH 1379* (collection details above).

60. Berberis photiniifolia C. M. Hu, Bull. Bot. Res., Harbin 6(2): 4. 1986, as "*photiniaefolia.*" TYPE: China. E Guangdong: Zijin Xian, border with Huidong Shi, Niaoqinzhang, 940 m, 21 July 1958, *C. F. Wei 120949* (holotype, IBSC 0000590 image!; isotypes, IBK IBK00012844 image!, KUN 0175591!, MO 04138729!, PE 01432166!).

Shrubs, evergreen, to 2 m tall; mature stems reddish or purplish brown, terete; spines 3-fid, semi-concolorous, 1–2 cm. Petiole almost absent; leaf blade abaxially green, adaxially dark green, shiny, broadly lanceolate, 4–8.5 × 1.5–2.8 cm, leathery, midvein raised abaxially, slightly impressed adaxially, lateral veins and reticulation inconspicuous on both surfaces, base cuneate, margin spinulose with 8 to 15 teeth on each side, apex acuminate. Inflorescence a fascicle, (2- to)4- to 8-flowered; pedicel 10–15 mm; bracteoles triangular-ovate. Sepals in 2 whorls; outer sepals ovate-oblong, 3–3.5 mm; inner sepals obovate or suborbicular, ca. 4 × 3.2 mm; petals obovate, ca. 3 mm, base clawed, glands separate, apex subentire. Stamens ca. 2.5 mm; anther connective truncate. Ovules 3 or 4. Berry bluish black, slightly pruinose, ovoid-ellipsoid, 7–8 × 4–5 mm; style not persistent.

Phenology. Berberis photiniifolia has been collected in flower in April and in fruit between the end of May and August.

Distribution and habitat. Berberis photiniifolia is known from Fengshun, Raoping, and Zijin Xian in east Guangdong, and Pinghe and Shanhang Xian in southwest Fujian. It has been found on mountain summits and slopes at ca. 940–1400 m.

Berberis wallichiana, reported by Dunn and Tutcher (1912: 32) from Mt. Phoenix in Guangdong but without description and noted by Schneider (1916: 317), is likely to refer to *S. T. Dunn 6187* (details below) which is *B. photiniifolia*.

All the specimens from Fujian listed below are sterile but, from the evidence of their leaves, they appear to be *Berberis photiniifolia*.

Selected specimens.
SW Fujian. Pinghe Xian: Baishui Xiang, Daqin Shan, 19 Aug. 1958, *G. Wei 10242* (AU 007434). **Shanghang Xian:** Baijie Ling, 25 Oct. 1932, *R. Lin 4123* (PE 01037989); Buyun, Guizhuping, 1400 m, 6 Sep. 1987, *L. G. Lin 7248* (PE 01037991); Gutian Zhen, 9 Sep. 1987, *Xiamen Meihua Shan Collection Team 173* (AU 007427–28).
E Guangdong. Fenshun Xian: Tonggu Zhang, 1240 m, 7 July 2009, *X. F. Zheng ZXF6972* (CZH CZH0003706). **Raoping Xian:** Mt. Phoenix (Fenhuang Shan), Apr. 1909, *S. T. Dunn 6187* (HK HK1260); Mt. Phoenix, 30 May 1957, *X. G. Li 200837* (HHBG HZ009046–47, IBK IBK00012776, IBSC 0091922, MO 04192754); Mt. Phoenix, s.d., *s. coll. 30361* (IBSC 0091921).

61. Berberis pingshanensis W. C. Sung & P. K. Hsiao, Acta Phytotax. Sin. 12: 387. 1974. TYPE: China. S Sichuan: Pingshan Xian, Longhua Xiang, Bajian Shan, 800 m, 1 June 1959, *Sichuan Econ. Pl. Exped 737* (holotype, IMM not seen; isotypes, KUN 0176281!, PE [2 sheets] 00935234–35!, SZ 00290945 image!).

Shrubs, evergreen, to 2 m tall; mature stems purplish brown, sulcate; spines 3-fid, concolorous, ca. 0.5 cm. Petiole 2–3 mm; leaf blade abaxially green, adaxially dark green, elliptic, 8–17 × 3.5–7 cm, thinly leathery, midvein raised abaxially, impressed adaxially, lateral veins and reticulation slightly raised and conspicuous abaxially, inconspicuous adaxially, base broadly cuneate or subobtuse, margin entire, apex acute. Inflorescence a fascicle, 6- to 14-flowered; pedicel purplish brown, 15–30 mm; bracteoles ovate-triangular, ca. 1 mm. Sepals in 2 whorls; outer sepals broadly elliptic, ca. 4 × 3 mm; inner sepals suborbicular or reniform-orbicular, ca. 4 × 4–5 mm; petals suborbicular, base with glands separate, apex entire or obtusely emarginate. Stamens ca. 2 mm; anther connective truncate. Ovules 2; funicles ca. as long as ovules. Berry unknown.

Phenology. Berberis pingshanensis has been collected in flower and with very immature fruit in June. Its fruiting season is otherwise unknown.

Distribution and habitat. Berberis pingshanensis is apparently known from only the type from Pingshan Xian in south Sichuan collected by a trailside at 800 m.

The specimen at KUN was annotated by C. Y. Wu as an unpublished *Berberis holophylla*.

62. Berberis pruinosa Franch., Bull. Soc. Bot. France 33: 387. 1886. TYPE: China. NW Yunnan: [Eryuan Xian], "Les haies a Mo-so-yn, près Lankong, fl. Février, fruit Novembre. 1883" (lectotype, designated here, *J. M. Delavay 493* "A," P P00716556!; isotype, K K001092868!; possible

isolectotype, P specimen on sheet with fruit, P02682373!, same location, 8 Nov. 1883).

Berberis pruinosa Franch. var. *viridifolia* C. K. Schneid., Repert. Spec. Nov. Regni Veg. 46: 250. 1939. TYPE: China. C Yunnan: "Ad Yünnanfu-Dali (Talifu), ad rivulos prope vicum Tsaopu trans oppid. Nganning [An'ning']," 1850 m, 28 Apr. 1916, *H. R. E. Handel-Mazzetti 8642* (lectotype, designated here, WU 0039268!; isolectotypes, A 00279383!, E E00217999!, W 1930–4110!).
Berberis pruinosa Franch. var. *brevipes* Ahrendt, J. Bot. 79(Suppl.): 15. 1941. TYPE: China. NW Yunnan: Atunze (Dêqên [Deqin] Xian), Hungpoh (Hongpo), 2600–2700 m, 25 Nov. 1937, *T. T. Yu 15662* (holotype, BM BM001015553!; isotypes, A 00279818!, E E00218000!, KUN 1204865!).
Berberis hibberdiana Ahrendt, J. Linn. Soc., Bot. 57: 79. 1961. TYPE: China. NW Yunnan: [Heqing Xian], Sung Kuei (Songgui), no further details, but collected before Mar. 1933, *H. D. McClaren's Collectors C103* (holotype, K K000077304!; isotypes, A 00279798!, BM BM001010633!, E E00217998!).
Berberis pruinosa Franch. var. *punctata* Ahrendt, J. Linn. Soc., Bot. 57: 82. 1961, syn. nov. TYPE: Cultivated. Stonefield, Watlington, Oxfordshire, U.K., 4 May 1944, from *T. T. Yu 14938*, China. NW Yunnan: Chungtien (Zhongdian, now Xianggelila Xian), Chi-chih, 2600 m, 21 Nov. 1937 (lectotype, designated here, BM BM000544630!).
Berberis pruinosa Franch. var. *tenuipes* Ahrendt, J. Linn. Soc., Bot. 57: 91. 1961. TYPE: China. NC Yunnan: N of Kunming, betw. "Shinling" & "Da-sung-shu,"10 Mar. 1914, *C. K. Schneider 342* (holotype, K K000077301!; isotype, A 00038787!).

Shrubs, evergreen, to 2 m tall; mature stems pale yellow, shiny, terete, black verruculose; spines 3-fid, concolorous, 1–3.5 cm, stout, abaxially sulcate or flat. Petiole almost absent; leaf blade abaxially pale yellow-green, sometimes pruinose, adaxially dark or yellow-green, sometimes shiny or dull grayish green, elliptic to obovate, 2–6 × 1–2.5 cm, rigidly leathery, midvein raised abaxially, flat or slightly impressed adaxially, lateral veins and reticulation slightly raised and inconspicuous abaxially, slightly impressed and inconspicuous or indistinct adaxially, base cuneate, margin slightly revolute or flat, spinose with usually 1 to 6 coarse teeth on each side, occasionally entire or spinulose with 8 or 9 teeth on each side, apex obtuse or shortly acuminate, occasionally retuse, mucronate. Inflorescence a fascicle, (8- to)10- to 20-flowered; pedicel 10–30 mm, slender; bracteoles lanceolate, ca. 2 mm, apex acuminate. Sepals in 2 whorls; outer sepals oblong-elliptic, ca. 4 × 2 mm, apex obtuse, rounded; inner sepals obovate, ca. 6.5 × 5 mm, apex rounded; petals obovate, ca. 7 × 4–5 mm, base clawed, glands completely basal, apex deeply incised. Stamens ca. 6 mm; anther connective not extended, rounded to truncate. Ovules 2 or 3. Berry purplish black, densely white pruinose, ellipsoid or subglobose, 6–7 × 4–5 mm; style not persistent or persistent and short.

Phenology. Berberis pruinosa has been collected in flower in March and April and in fruit between May and November.

Distribution and habitat. Berberis pruinosa is known from central, east, and northeast Yunnan, south and southwest Sichuan, southeast Xizang (Tibet), and west Guizhou. It has been collected in thickets, forests, forest margins, by road and streamsides, and on limestone pavements at 1700–2700 m.

In the protologue of *Berberis pruinosa*, Franchet's reference to gatherings stated "Yun-nan, in sepibus ad Mo-so-yn prope Lankong; fl febr.; fr. Nov. 1883 (Delav. n. 493 et 1861)." As is often the case with Franchet's protologues of *Berberis*, correlating his statements with specimens at P is not straightforward.

There are six specimen sheets of *Delavay 493* and three of *1861* at P. None of these have a February date and none of *1861* are from 1883. Of the specimens of *Delavay 493*, one with mature fruit (P00716556) is labelled "Les haies a Mo-so-yn, près Lankong, fl. Février, fruit Novembre. 1883." One sheet (P02682373) has four specimens, two with flowers and two with fruit. There are three labels on the sheet. All indicate the specimens as being from Mo-so-yn; one label indicates the fruit as being from 8 November 1883, a second that the flowers are from March 1883, and the third has the same information as given on P00716556. A third sheet with fruit (P00580423) is dated November 1883 and is again from Mo-so-yn, but from "terrain inculti." A fourth specimen (P02465454), also from Mos-so-yn, has flowers but is undated, while two more (P00716557–58) also have flowers but have no collection details. I have assumed that both P00716556 and those of 8 November 1883 on P02682373 are from the same collection and is designated here as *Delavay 493* "A," while all those with flowers are designated here as *Delavay 493* "B." That of "terrain inculti" was only located on a visit to the Paris Herbarium in March 2018 and is designated here as *Delavay 493* "C." To complicate matters further, apparent duplicates of P02682373 were sent to K, KUN, and PE after the reorganization of the Herbarium in 2011. All three of these specimens are labelled *Delavay 493* and have the same collection data as to be found on P00716556. The K sheet has both floral and fruiting material and would, therefore, seem to consist of both *493* "A" and *493* "B," whereas the KUN specimen has only floral material and so is *493* "B." The PE specimen is sterile and, therefore, could be either "A" or "B."

There are three specimens of *Delavay 1861* at P (P00716559–61) all dated 1 April 1885. These specimens are all from Mo-so-yn but each has slightly different collection details, including one (P00716559) from

"Les halliers et les haies." Since sepes and haie are respectively the Latin and French names for hedge (halliers is thickets), it appears that, in terms of description of location, this latter specimen and those of *493* "A" are the ones that correlate most closely with Franchet's protologue. However, the dates of *493* "A" match the protologue, whereas that of *1861* does not. Since the protologue does not give a specific date for November 1883, *Delavay 493* "A" (P00716556) is designated here as the lectotype and that of 8 November is treated as a probable isolectotype. All three *Delavay 1861* specimens are treated as syntypes and a specimen of *Delavay 1861* without collection details at UC as a possible syntype.

Ahrendt (1961: 80) cited specimens of both *Delavay 493* and *1861* as being at K. However, the only Delavay specimen of *Berberis pruinosa* located at K (K000077303) is an unnumbered one dated 28 March 1887 and hence collected after the first publication of the name. An annotation on the sheet by Schneider identifying it as a type is therefore mistaken.

Chamberlain and Hu (1985: 551) stated that Schneider did not designate a type for *Berberis pruinosa* var. *viridifolia*. However, in publishing the taxon as a nom. nov. for "pro parte maxima" of *B. pruinosa* var. *centiflora* (Handel-Mazzetti, 1931), Schneider stated it covered all of Handel-Mazzetti's citations except for the type of *B. centiflora* Diels (*G. Forrest 4689*). There are three such gatherings cited by Handel-Mazzetti which should, therefore, be regarded as syntypes. All specimens of these gatherings at W and WU are annotated by Schneider as variety *viridifolia*. I have chosen the specimen with the most material as the lectotype.

The protologue of *Berberis pruinosa* var. *punctata* stated "Cultivated (*Yu 14938*), fl. 7 Apr. 1943, 4 May 1944; fr. 20 Nov. 1944 (Type O)." As with many similar treatments by Ahrendt of taxa whose types are cultivated rather than wild-collected, it is impossible to interpret this further for the simple reason that, having cited a particular collector's number as a type, he treated all specimens from cultivated plants with the same collector's number as also being types, irrespective of their origin or date of collection. O was the acronym Ahrendt used for OXF, but I was unable to locate specimens there. I have, therefore, chosen as the lectotype a specimen at BM whose date corresponds to one of those given by Ahrendt.

As Chamberlain and Hu (1985: 551) noted, *Berberis pruinosa* is a very variable species. The leaves on a single plant or plants in close proximity can exhibit very substantial variation. Attempts to distinguish different varieties on the basis of leaf shape, size, number of marginal spines, and whether or not the leaves are abaxially pruinose are, therefore, misplaced. It was for this reason Chamberlain and Hu (1985: 551–552) synonymized *B. hibberdiana* and *B. pruinosa* vars. *brevipes*, *tenuipes*, and *viridifolia*. They treated as excluded taxa all the other varieties of *B. pruinosa* that Ahrendt (1939: 266–267, 1941b: 15, and 1961: 81–82) had published solely on the basis of cultivated material. I have followed this with the exception of *B. pruinosa* var. *punctata* where the type grown from seed of *T. T. Yu 14938* corresponds to wild-collected specimens of the same number at A (00279781), BM (BM001017985), and E (E00623102). Ahrendt's justification for this variety was largely that it had stems that were markedly verruculose. This is not evident on the type specimen (although it is on wild-collected ones) but is, in any case, a trivial difference.

Though *Berberis pruinosa* is a very variable species, its rigid, leathery leaves are a consistent feature and are an important aid to identification (this characteristic would seem to be the reason why it is sometimes confused with the similarly rigid-leaved *B. bergmanniae* from central and north Sichuan and Wen Xian in Gansu).

Berberis pruinosa var. *viridifolia* was treated as a synonym of both *B. pruinosa* and *B. wangii* by Ying (2001: 110, 140, 2011: 735, 746).

Syntypes. *Berberis pruinosa.* NW Yunnan: Mar. 1883. *J. M. Delavay 493* "B" (P P02682373); "Les haies a Mo-so-yn, près Lankong, fl. Février, fr. Novembre. 1883" (K K001092868!, KUN 1206552; possible syntype, PE 01901813); "in sepibus ad Mo-so-yn, prope Lankong" s.d., *J. M. Delavay 493* "B" (P P02465454); s.d., *J. M. Delavay 493* "B" (B B100250730, P P00716557–58); Mo-so-yin, près Lankong, "terrain inculted," Nov. 1883, *J. M. Delavay 493* "C" (P P00580423); Mo-so-yin, near Lankong, 1 Apr. 1885, *Delavay 1861* (P P00716559–6); s. loc., s.d., *J. M. Delavay 1861* (UC UC1038331).

Syntypes. *Berberis pruinosa* var. *viridifolia.* C Yunnan: near Yunnanfu (Kunming), 1850–2100 m, 16 Feb. 1914, *H. R. E. Handel-Mazzetti 38* (W 1930–4054, WU 0039267); Yunnan: N of Kunming, near Hsinlung Village, beyond Pudu-ho river, 25.57°N, ca. 2000 m, 10 Mar. 1914, *H. R. E. Handel-Mazzetti 531* (E E00320618, W 1930–5052, WU 0039234 specimen marked by Schneider as "b").

Selected specimens.
W Guizhou. Weining Xian: Heishitou Qu, Gaole Xiang, 2100 m, 12 July 1959, *Bijie Team 272* (GF 09001037, HGAS 013652, KUN 0177001, PE 01031687–88).
S & SW Sichuan. Huidong Xian: Wu Qu, Daqiao, 1900 m, 14 June 1959, *S. K. Wu 259* (CDBI CDBI0027628, KUN 0177003, SZ 00289766). **Huei-li Xian (Huili Xian):** 1950 m, 31 Aug. 1932, *T. T. Yu 1399* (A 00279821, CQNM 005206, IBSC 0092531, PE 01031679–80, WUK 0092531). [**Muli Xian**]: S Mu-li, Ko-pa-tian, 2550 m, 4 July 1937, *T. T. Yu 7237* (A 00279825, KUN 0177004, PE 01031682–85). **Yanyuan Xian:** Yongza, Maji, 25 li, 3000 m, 29 Apr. 1960, *Nanshuibeidiao Exp. S. D. Jiang 5502* (PE 01031700, SZ 00294131); NW of Ganhaizi, 2400 m, 8 June 1960, *Nanshuibeidiao Exp. S. D. Jiang 5865* (KUN 0176996, PE 01031685).

SE Xizang (Tibet). Zayü (Chayu) Xian: near Zhuwagen, 2600 m, 2 July 1973, *Qinghai-Xizang Team 545* (KUN 0176997–98, PE 01031677–78).
C & N Yunnan. Anning Xian: N of An'ning City, S side of Bijia Shan, 2070–2150 m, 26 Aug. 2005, *D. E. Boufford et al. 34871* (A 00279808, HAST 115068). **Chengjiang Xian:** Zhongguan, 2100 m, 27 Mar. 1991, *s. coll. 4418* (KUN 0176771–72). [**Dongchuan Xian**]: plaine et coteaux à Tongtchouan, 2500 m, Apr., s. anno, *E. E. Maire 398* (A 00279795–96, K K000077302). **Fumin Xian:** betw. Tuodan & Jiunianping, 2200 m, 22 May 1964, *B. Y. Qiu 59087* (IBK IBK00357940, KUN 0176843–44, LBG 00064083). **Kunming Shi:** Yanjia Cun, 1946, *S. E. Liu 16267* (KUN 0176904, PE 01031782). **Luquan Xian:** Er Qu, E'mao Xiang, 2600 m, 30 Oct. 1952, *P. Y. Mao 1520* (HITBC 003657, IBSC 0092404, KUN 0176817, PE 01031842, WUK 0244819). **Shilin Xian:** vic. of Shilin, 24.883333°N, 103.4°E, 1800 m, 1 Aug. 1984, *Sino-American Bot. Exp. 1565* (A 00279828, KUN 0176969). **Shizong Xian:** SE of Luxin Cun, 2100 m, 22 Apr. 1977, *Shizong Team 49* (KUN 0176812). **Songming Xian:** Shaodian, Dazhuyuan Xiang, 22 Apr. 1956, *B. Y Qiu 51615* (IBSC 0091962, KUN 0176860–61). **Wuding Xian:** Shizi Shan, 15 Sep. 1964, *J. M. Zeng & Q. Yang s.n.* (YUKU 02004150). **Xundian Xian:** 15 Apr. 1981, *H. Y. Yu & H. B. Shi 93* (KUN 0176868–69). **Yiliang Xian:** en route to Yangzonghai, 1800 m, 7 Sep. 1977, *B. Y. Qiu 771083* (CDBI CDBI0027606, KUN 0176814, 0176893). **Yimen Xian:** Jingle'an, 1770 m, 28 Sep. 1990, *s. coll. 3463* (KUN 0176916–17). **Yuanmou Xian:** side of Yaunmou to Wuding rd., km marker 96, 2100 m, 25 July 1958, *B. Y. Qiu 57202* (FUS 00014202, IBSC 0091952, KUN 0176846, 0176851). **Yuxi Shi:** Cidongguan, 1700 m, 12 Mar. 1991, *s. coll. 4004* (KUN 0176914–15).
E Yunnan. Fuyuan Xian: Dongshan Xiang, Luzu, 2200 m, 16 June 1989, *Hongshui He Botany Team 2176* (KUN 0179178–79). **Zhanyi Qu:** Zhujiangyuan Fengjingqu, 2003 m, 15 Aug. 2015, *E. D. Liu et al. 4528* (KUN 1263355).
NW Yunnan. Dali Shi: 1946, *S. E. Liu 15598* (KUN 0176900). **Dêqên (Deqin) Xian:** Cang Jiang, Huanfuping, 1900 m, 24 Sep. 1959, *G. M. Feng 23668* (KUN 0176811, 0176879–80, PE 01031856). **Eryuan Xian:** Mo-so-yin, near Lankong, 28 Mar. 1887, *J. M. Delavay s.n.* (K K000077303, LE, P P02313244–45, P P02313249); Menduan Ying, 3100 m, 30 July 1963, *Jinsha Jiang Team 6254* (KUN 0176896–97, PE 01031843). **Heqing Xian:** Lianping, Fengchui Ling, s.d., *R. C. Qin 23824* (KUN 0176972–74, PE 01031846). **Jianchuan Xian:** 3 km SW of suburbs, 2250 m, 9 June 1981, *Z. X. Tang 310* (PE 01605708). **Lijiang Xian:** Yongsheng, 2200 m, 11 June 1981, *W. H. Li & Y. Hu 159* (PE 01840106–07). **Weixi Xian:** en route to Jiajiatang, 2 May 1940, *Z. Y. Wu 3584* (KUN 0176948, 0177065–66). **Zhongdian (Xianggelila) Xian:** Yongsheng Xiang, 2350 m, 3 July 1963, *Zhongdian Team 3083* (KUN 0176796, 0176809).

63. Berberis ravenii C. C. Yu & K. F. Chung, Phytotaxa 184(2): 88. 2014. TYPE: China. Taiwan: Kaohsiung Hsien, Maolin, Shuangguie Lake Major Wild Life Habitat, Lake Upunuhu (Wan-shan-shen Lake), 22.914722°N, 120.828056°E, 2150 m, 7 Feb. 2009, *C. C. Yu 267* (holotype, TAI 284282 image!).

Shrubs, evergreen, ± decumbent, to 1 m tall; mature stems purplish red or dark purplish, terete, not verruculose; spines 3-fid, concolorous, 0.8–2.3 cm, terete.

Petiole almost absent; leaf blade abaxially green or dark green, adaxially green or dark green, shiny, elliptic to lanceolate, 5.5–9.5 × 1.2–2 cm, thinly leathery, midvein raised abaxially, impressed adaxially, lateral veins and reticulation obscure or indistinct abaxially, inconspicuous adaxially, base attenuate, margin sometimes slightly revolute, spinulose with 16 to 28 widely spaced teeth on each side, apex acuminate, mucronate. Inflorescence a fascicle, 4- to 7-flowered; pedicel 13–15 mm; bracteoles pale yellow or greenish yellow, triangular, 1 × 1 mm, sometimes absent. Flowers pale yellow or greenish yellow. Sepals in 3 whorls; outer sepals sometimes with reddish tinge, ovate, 2.5 × 2 mm; middle sepals yellow, ovate, 4 × 2.5 mm; inner sepals yellow, obovate, 5 × 4.5 mm; petals obovate, 4.5 × 3 mm, base clawed, glands ovoid, close to each other, apex incised. Stamens ca. 3 mm; anther connective extended, truncate. Pistil 4 mm; ovules 2 or 3. Berry black, ellipsoid, ca. 7 × 4 mm, epruinose; style persistent, short.

Phenology. *Berberis ravenii* has been collected in flower in April, with very immature fruit in July and mature fruit in February (presumably from the previous year). Its phenology is otherwise unknown.

Distribution and habitat. *Berberis ravenii* is known from Kaohsiung and Pingtung Hsien in south Taiwan. It has been found in coniferous and broadleaf mixed forests at 700 to 2300 m.

Selected specimens.
Taiwan. Kaohsiung Hsien: Chunyunshan, 700 m, 7 Mar. 1996, *C. K Liou et al. 5* (TAIF 076188). **Pingtung Hsien:** Chutunshan, Japanese Subpolice Office, 12 Aug. 1937, *Y. Yamamoto & K. Mori. 516* (TAI 47157); en route to Tawaushan, 2100–3090 m, 16–17 July 1988, *T. C. Huang et al. 13678* (TAI 206713, 213332).

64. Berberis replicata W. W. Sm., Notes Roy. Bot. Gard. Edinburgh 11: 200. 1920. TYPE: China. W Yunnan: [Tengchong Xian], Ma-Chang-kai valley, N of Tengyueh (Tengchong), 25.333334°N, Feb. 1913, *G. Forrest 9545* (lectotype, designated here, E E00117377!; isolectotypes, E E00117985!, K K000077356!).

Shrubs, evergreen, to 1.5 m tall; branches weak; mature stems pale yellow, terete, sparsely verruculose; spines 3-fid, concolorous, 1–2 cm, weak, abaxially sulcate. Petiole 1–2 mm; leaf blade abaxially pale green, pruinose, adaxially dull pale green, oblong-elliptic or oblong-lanceolate, 1.5–3.5(–4.5) × 0.3–0.5(–0.8) cm, leathery, midvein raised abaxially, impressed adaxially, lateral veins and reticulation indistinct or obscure on both surfaces, base attenuate, margin conspicuously revolute, entire or spinulose with 1 to 3 teeth on each side, apex narrowly acuminate, mucronate. Inflorescence a fascicle, 3- to 8-flowered; pedicel reddish, 5–13 mm; bracteoles ca. 2 mm. Sepals in 2 whorls; outer sepals reddish, ovate or suborbicular, 3.5–4 × ca. 3 mm; inner sepals obovate or suborbicular, 6–7 × 5–6 mm; petals obovate, 5–5.2 × 3.5–4 mm, base clawed, glands separate, apex entire. Stamens ca. 3 mm; anther connective truncate or obtuse. Ovules 2(to 4), sessile. Berry purplish black, epruinose, oblong, 6–8 × 3–5 mm; style persistent, short.

Phenology. *Berberis replicata* has been collected in flower in February and in fruit between April and December.

Distribution and habitat. Apart from *G. Forrest 18508* and *F. Du & P. Li 589* (see below), *Berberis replicata* appears to be known from only a small area in Tengshong Xian in west Yunnan. It has been found in thickets, on trailsides and slopes, and in openings in pine forests among scrub on lava beds in the volcanic area to the northeast of Tengshong at ca. 1800–2100 m. Whether the species is found in the neighboring area of Myanmar is unknown.

The protologue of *Berberis replicata* cited five gatherings, but did not designate a type specimen. Schneider (1942: 35) stated "G. Forrest sammelte [collected] den Typ im Februar 1913 'in the Ma-chang valley, north of Tenggyueh.'" The only one of the five specimens cited by Smith that fits this description is *G. Forrest 9545*, so this can be regarded as a first-step lectotypification. Chamberlain and Hu (1985: 547) mistakenly reported *Forrest 9545* at E as the holotype, but under Art. 9.10 of the Shenzhen Code (Turland et al., 2018), this can be treated as an inadvertent partial completion of lectotypification. It is only partial because two specimens of this number are at E. The lectotypification has been completed by selecting the specimen that was annotated by Schneider with ! in 1937. Ahrendt (1961: 43) did not notice Schneider's lectotypification and mistakenly cited *G. Forrest 7785* as the type.

Herbarium specimens of *Berberis replicata* without mature stems are sometimes confused with *B. taliensis* of northwest Yunnan. Schneider (1942: 35) made such a mistake by identifying as *B. replicata*, *J. F. Rock 9588* from Lijiang Shi in northwest Yunnan (further details under *B. taliensis*). Living plants of the two species, however, cannot be so confused, the weak, arched branches and dull light green leaves of *B. replicata* being in distinct contrast to the stiffly upright branches and dark, shiny green leaves of *B. taliensis*. There is also a very distinct geographical separation between the species.

Though *Berberis replicata* mostly has two ovules per fruit, a sample of 240 fruit analyzed by X. H. Li and Zhang (2014) found a small percentage with three seeds and one fruit with four.

Syntypes. W Yunnan: Tengchong Xian, lava bed W of Tengyueh, 25°N, 1800–2100 m, July 1912, *G. Forrest 7785* (E E00117378, K K000077353, N 093057086, S 08–778); Ma-Chang-kai valley, N of Tengyueh, 25.5°N, Dec. 1912, *G. Forrest 9457* (E E00117382, IBSC 0091986, K K000077355); hills to NE of Tengyueh, 25.166667°N, July 1912, *G. Forrest 8782* (E E00117383, IBSC 0091985, K K000077354); Yunnan, Shweli-Salween divide, 25.5°N, 19 Oct. 1917, *G. Forrest 16030* (E E00117379, K K000077352, P P02313242).

Selected specimens.
W Yunnan. Longling Xian: Daxue Shan, 2800 m, 28 May 2002, *F. Du & P. Li 589* (SWFC 00016312). **[Lushui Xian–Yunlong Xian border]:** N'Maikha Salwin divide, 26.333333°N, 15 Sep. 1919, 2130–2440 m, *G. Forrest 18508* (E E00117380, K, P P02313241, WSY 0057654). **Tengchong Xian:** Shweli River drainage basin and environs of Tengyueh, 2740 m, Feb. 1923, *J. F. Rock 7934* (A 00279833, US 00945929); Mashang valley, 1830 m, July 1924, *G. Forrest 24682* (E E00117381, US 00945946); Guyong Qu, betw. Houqiao Xiang & Heinitang, 19 May 1964, *S. G. Wu 6691* (HITBC 003651, KUN 0175669–70, LBG 00064143); Mazhan Qu, Baojia Xiang, 1850 m, 15 Apr. 1985, *Spice Investigation Team 231* (KUN 0177023–27); Mazhan Xiang, Xiaokong Shan, 25.216111°N, 98.496111°E, 1930 m, 25 Oct. 1998, *H. Li et al. 10925* (A 00279834, CAS 00120136, HAST 80992); Mazhan, Dakong Shan volcano, 25.221111°N, 98.500194°E, 2000 m, 2 June 2006, *Gaoligong Shan Biodiversity Survey 29875* (CAS 00120143); Mazhan, Dakong Shan volcano, 25.208056°N, 98.432139°E, 1940 m, 2 June 2006, *Gaoligong Shan Biodiversity Survey 29894* (CAS 00120144).

Cultivated material:
Living cultivated plants.
Sir Harold Hillier Gardens, Hampshire, U.K., from *G. Forrest 16030*.

65. Berberis sanguinea Franch., Nouv. Arch. Mus. Hist. Nat., sér. 2, 8: 194. 1886. TYPE: China. WC Sichuan: Mupin (Baoxing), Apr. 1869, *A. David s.n.* (lectotype, designated here, P P00716571!; isolectotype, P P00716572!).

Berberis panlanensis Ahrendt, Kew Bull. 1939: 265. 1939. TYPE: China. NC Sichuan: [Xiaojin Xian-Li Xian border], Pan-lan-Shan (Bawangshan), "west of Kuan Hsien" (Guan Xian, now Dujiangyan Shi), 2300 m, 21 June 1908, *E. H. Wilson 2875* (holotype, K missing; lectotype, designated here, A 00230188!).
Berberis multiovula T. S. Ying, Acta Phytotax. Sin. 37(4): 309. 1999. TYPE: China. WC Sichuan: Baoxing Xian, Puxi Gou, Xiaoxi, 2900 m, 3 May 1959, *Sichuan Economic Plant Exp. Yajiang Division, 293* (holotype, PE 00935173!; isotype, KUN 0177041!).
Berberis viridiflora X. H. Li, Pl. Diversity 39(2): 96. 2017, syn. nov. TYPE: China. WC Sichuan: Baoxing Xian, Longdong Zhen, Donglashan Gou, 30.425833°N, 102.560833°E, 2085 m, 6 May 2015, *X. H. Li, L. C. Zhang & W. H. Li 140511* (holotype, NAU image!).

Berberis sanguinea var. *viridisepala* X. H. Li, L. C. Zhang & W. H. Li, Pl. Diversity 39(2): 97. 2017, syn. nov. TYPE: China. WC Sichuan: Baoxing Xian, Longdong Zhen, Donglashan Gou, 30.4245°N, 102.559667°E, 2086 m, 6 May 2015, *X. H. Li, L. C. Zhang & W. H. Li 140513* (holotype, NAU image!).

Shrubs, evergreen, to 3 m tall; mature stems pale yellow, very sulcate, sparsely black verruculose; spines 3-fid, pale yellow, 1–3 cm. Petiole almost absent; leaf blade abaxially pale yellow-green, slightly shiny, adaxially dark green, shiny, linear-lanceolate, 1.5–6 × 0.3–0.6 cm, slightly leathery, midvein raised abaxially, impressed adaxially, lateral veins and reticulation indistinct or obscure on both surfaces, base cuneate, margin spinulose with 7 to 14 teeth on each side, apex acute or acuminate, mucronate. Inflorescence a fascicle, 2- to 7-flowered; pedicel reddish, 4–12(–15) mm; bracteoles purple, ovate, ca. 2.8 × 1.2 mm, apex acute. Sepals in 3 whorls; outer and sometimes median and inner sepals purple (dark reddish when dry), otherwise median and inner sepals very pale greenish yellow; outer sepals ovate, ca. 3 × 2 mm, apex acute; median sepals ovate or triangular-ovate, 3.5 × 3.5 mm; inner sepals broadly obovate to orbicular-obovate, 5–5.5 × 4–4.5 mm; petals pale greenish yellow, sometimes with purple markings, obovate, ca. 3.5–4 × 2.5–3 mm, base clawed, glands separate, apex emarginate. Stamens ca. 2 mm; anther connective not extended, truncate. Ovules 3 to 5. Berry dark purple, epruinose, ellipsoid, 7–12 × 4–5 mm; style not persistent.

Phenology. *Berberis sanguinea* has been collected in flower in April and May and in fruit between May and October.

Distribution and habitat. *Berberis sanguinea* is known from west-central, north-central, and west Sichuan. It has been found in thickets, by trail and streamsides, on dry wasteland, and on forest margins at ca. 1100–3800 m.

Two specimens of the type gathering of *Berberis sanguinea* are at P. P00716571 has been chosen as the lectotype because it has the most material. The statement by X. H. Li et al. (2017) that this is a holotype is inaccurate.

There has been some confusion about the color of the flowers of *Berberis sanguinea*. This should have been unnecessary since the collector's notes on the sheets of the type gathering state "fl. pourpre" (flowers purple). However, Franchet either did not notice these notes or chose not to regard them, and his protologue described the outer sepals as blood-red. Schneider (1942: 25–26) described the flowers as golden yellow, the outermost parts with a somewhat reddish tinge.

However, this description seems to have been based not on the type gathering, but on a cultivated plant at the Les Barres Arboretum, France, grown from seed received by Vilmorin from an unspecified part of China in 1898; and the source of the description appears not to be personal observation, but an article on *B. sanguinea* by Creasey (1935: 11) which was almost certainly describing a plant of some other species.

Ahrendt (1939: 266) first described the outer sepals of *Berberis sanguinea* as having a "distinctly red colouring" and later (1961: 56), when giving a complete description of flowers, described their color as "deep yellow and conspicuously red without." But his description clearly did not come from the type whose whereabouts he appeared not to know, but from "Sa-washan, *Henry 5763* (K)." This was undoubtedly a misprint for *Henry 5753* (K K000077340) which is *B. triacanthophora* and which also has bicolored flowers (for further details, see under that species). Ahrendt's description, which was largely repeated by Ying (2001: 107, 2011: 735), should therefore be discounted.

Confirmation that the collector's notes regarding the type gathering were essentially correct and that the outer sepals and sometimes the inner sepals of the species are indeed purple rather than red comes from plants grown in the U.K. from two recent seed collections (details below). The color of dry flowers from these plants correspond exactly to those of the type gathering, whose flowers I have also examined. These are the source of the description of the flowers given above. *Berberis sanguinea* appears to be the only species in the genus with partially purple flowers.

Schneider (1913: 359) cited *Wilson 2875* as an example of *Berberis sanguinea* noting that the flowers were "yellow and bronze" and the unripe fruit had three ovules. From the specimen at A, bronze refers to the outer sepals and appears to be a reflection of the fact that the flowers were nearly past anthesis, as demonstrated by the presence of immature fruit. Ahrendt (1939b: 266) recognized *Wilson 2875* as a separate species, *B. panlanensis*, citing as the type a specimen at K. Ahrendt's description was, however, internally incoherent in that he both quoted Schneider's description of the color of the flowers of *Wilson 2875* while also maintaining that the outer sepals have only the slightest reddish tinge and that the inner sepals and petals are greenish yellow. He also described the flowers in one place as being solitary and further on as sometimes in pairs while also claiming that *B. panlanensis* has three ovules versus two ovules in *B. sanguinea* (in fact, *B. sanguinea* normally also has three). In all this, it is clear he was drawing to a large extent not on *Wilson 2875* but on "*B. sanguinea* var. *microphylla*"—a cultivated plant of unknown origin which was also briefly referred to by Creasey (1935). That this was the case is

shown by Ahrendt's subsequent description of *B. panlanensis* (1961: 56) where the flowers are now described as "pale greenish . . . devoid of any red markings" and *B. sanguinea* var. *microphylla* is listed as a synonym. *Berberis panlanensis* was accepted by Schneider (1942: 26), though he noted Ahrendt's description also drew on cultivated material. *Berberis panlanensis* was reduced to a synonym of *B. sanguinea* by Chamberlain and Hu (1985: 543); however, with the holotype at K missing and without a knowledge of the specimen at A, they were unable to present the full case for the synonymy.

The type of *Berberis multiovula* is yet another gathering from Baoxing. The description of the flower structure is more or less the same as for *B. sanguinea*, except it gives the number of ovules as five versus the much more common three. Unreported in the protologue and in Ying (2001: 114) are the collectors' notes which record the flowers as being reddish purple on the outside and green on the inside. The synonymy was made by X. H. Li et al. (2017).

Berberis viridiflora (published on the basis of only one specimen with flowers and one with fruit, both from the same bush) was differentiated from *B. sanguinea* solely on the basis of it having greenish flowers with minor differences in the shape of its petals. This would seem simply to represent an unusual form of *B. sanguinea* and this is confirmed by the type of *B. sanguinea* var. *viridisepala* (collected some 200 m from the type of *B. viridifolia*) which, as the authors' figure 5a shows, has some flowers on the same twig that are purely greenish and some which are partially purplish.

The report by Ying (2001: 109; 2011: 735) of *Berberis sanguinea* being found in Hubei is mistaken and is likely to be referable to *B. triacanthophora*.

Selected specimens.
Sichuan. Baoxing Xian: Meiwangchuan, Xinjia Gou, 2400 m, 12 June 1936, *K. L. Chu 2900* (BM, E E00623105, IBSC 0091997, PE 01031114, 01031120–21); Yulong Xian, Haizi Gou, 1100 m, 14 May 1958, *Sichuan Agricultural College 4672* (CDBI CDBI0027671–72, PE 01031103). **Danba Xian:** Xiaopian Qiao, 2500 m, 20 July 1959, *Nanshuibeidiao Exp. 02206* (KUN 0177044, PE 01031084, SZ 00291038). **Jinchuan Xian:** Dusong, Bayi Gou, 2770 m, 9 July 1958, *Sichuan 8th Forest Brigade 4055* (IBSC 0092001). **Kangding Xian:** Tachien-lu (Kangding Xian), ca. 2130–2440 m, Sep. 1910, *E. H. Wilson 4637* (A 00279835, US 00945945); NE of Tachienlu, Xiaojin He valley, 2200 m, 17 Oct. 1931, *W. C. Cheng 3518* (CQNM 0005192, IBSC 0092004, 0092464, LBG 00064145, NAS NAS00314182, PE 01031082); "Kangting (Tachienlu); meridionem versus," 2700 m, 22 Oct. 1934, *H. Smith 12924* (MO 4367272, PE 01031090, S 12–25256, UPS BOT:V-040888); Chengjiao Gongshe, 3800 m, 24 May 1974, *Z. S. Qin 06043* (CDBI CDBI0027674, CDBI0027677, CDBI0027680, PE 01031078, SITC 00016216). **Luding Xian:** Lenqi Qu, Lukuang hwy., 2700 m, 28 May 1974, *Sichuan Botany Team 6851* (CDBI CDBI0027658–59, PE 01031126); Xinxing Gongshe, 1850 m, 9 July 1982, *K. Y. Lang et al. 564*

(KUN 0178908, PE 01031122–23). **Mianzhu Shi:** Suidao Gongshe, Shiwu Dadui, 18 June 1978, *Mianzhu Team 24* (SM SM704800075). **Tianquan Xian:** Erlang Shan, Ganhaizi, 2610 m, 20 Aug. 1982, *D. Y. Peng 46345* (CDBI CDBI0027663–64, IBSC 0092005). **Wenchuan Xian:** near Wolong, 1920 m, 4 Oct. 1986, *Z. Y. Wu et al. 1172* (KUN 0178476). **Xiaojin Xian:** near Chongde Gongshe, 2500 m, 31 Aug. 1975, *Sichuan Botany Team 9843* (CDBI CDBI0027535–36, CDBI0027660, IBSC 0092002, PE 01031083).

Cultivated material:
Living cultivated plants.
Sherwood Gardens, Exeter, Devon, U.K.; Foster Clough, Mytholmroyd, U.K., from *B. & S. Wynn-Jones 8172*, Baoxing Xian, 44 km N of Baoxing, 1490 mm, 2000; Howick Hall, Northumberland and Royal Botanic Garden Edinburgh, from *M. Foster 97115*, Baoxing Xian, Yaoji Village, ca. 2000 m, 1997.

66. Berberis sargentiana C. K. Schneid., Pl. Wilson. (Sargent) 1: 359. 1913. TYPE: China. W Hubei: Hsing-shan Hsien (Xingshan Xian), 1200–1500 m, 5 May 1907, *E. H. Wilson 564* "A" (lectotype, designated here, A 00036791!; isolectotypes, BM [2 sheets] BM001010627!, E E00259003!, HBG HBG-506688 image!, K K000644913!, LE!, US 00103905 image!, W 1914–0004809!).

Berberis recurvata Ahrendt, Gard. Chron., ser. 3, 124: 175. 1948. TYPE: China. W Hubei: Fang Hsien (Xian), 1500 m, 15 Nov. 1907, *E. H. Wilson 555* (holotype, K K000077313!; isotypes, A 00058268!, BM [2 sheets] BM001010637!, E E00259004!, HBG HBG-506689 image!, US 00945944 image!).

Shrubs, evergreen, to 3 m tall; mature stems pale reddish brown, terete, sometimes black verruculose; young shoots often bright red; spines 3-fid, light brown, (2–)3–6 cm, stout, abaxially sulcate. Petiole almost absent; leaf blade abaxially pale green, adaxially midgreen, slightly shiny, oblong-elliptic, 4–15 × 1.5–6.5 cm, thickly leathery, midvein raised abaxially, impressed adaxially, lateral veins and reticulation slightly raised and inconspicuous abaxially (but conspicuous when dry), slightly impressed and conspicuous adaxially, base cuneate, margin spinose with 10 to 25 teeth on each side, apex acute, mucronate. Inflorescence a fascicle, 4- to 10-flowered; pedicel 10–20 mm; bracteoles red, ca. 2 × 2 mm. Flowers pale greenish yellow. Sepals in 3 whorls; outer sepals with a red band along middle, ovate, ca. 3.5 × 3 mm, apex subacute; median sepals rhombic-elliptic, ca. 5 × 4.5 mm; inner sepals obovate, ca. 6.5 × 5 mm; petals obovate, ca. 6 × 4.5 mm, base cuneate, with contiguous, orange glands, apex emarginate with rounded lobes. Stamens ca. 4.5 mm; anther connective truncate. Ovules 2 to 4. Berry black, epruinose, oblong or oblong-ellipsoid, 6–8 × 4–6 mm; style not persistent.

Phenology. *Berberis sargentiana* has been collected in flower in April and May and in fruit between June and November.

Distribution and habitat. *Berberis sargentiana* is known from Chongqing (east and southeast Sichuan), southwest and west Hubei, and northwest Hunan. It has been collected in woods, *Pinus* forests, forest understories, and ravine sides at 800–1840 m.

The protologue cited *E. H. Wilson 564*, June and November 1907, as the type. The former has flowers, the latter fruit. The first (actually 5 May) is designated here as *564* "A" and that of 7 November as *564* "B." Wilson *564* "A" at A has been designated here as the lectotype because floral material is a better guide to identification than fruit, and Schneider's protologue was based on Wilson material sent to him by Harvard to identify.

In the protologue, Schneider cited *E. H. Wilson 555* as another gathering of *Berberis sargentiana* besides the type. Subsequently, Ahrendt recognized *Wilson 555* as a separate species, *B. recurvata*, largely on the basis that it had mature stems that were yellow rather than pale reddish brown. However, none of the specimens of *Wilson 555* referred to above evidence this and the isotype of *B. recurvata* at BM has mature stems that are clearly pale reddish brown. The synonymy with *B. sargentiana* was made by Chamberlain and Hu (1985: 541).

Chamberlain and Hu (1985: 541) mistakenly treated *Berberis simulans* found in west-central Sichuan as a synonym of *B. sargentiana* and this was followed by Ying (2001: 141, 2011: 746). For further details of why the synonymy was mistaken, see under *B. simulans*.

Syntype. *Berberis sargentiana*. W Hubei: *Wilson 564* "B," same details as *Wilson 564* "A," but 7 Nov. 1907 (A 00036791, E00259004, HBG HBG-506688, K K000077314, LE, US 00103905, W 1914–0004809).

Selected specimens.
Chongqing (E & SE Sichuan). Fengjie Xian: Xinhe Xiang, Hegenlin, 1000 m, 15 June 1958, *H. F. Zhou 23600* (KUN 0177058, PE 01031549). **Qianjiang Xian:** Huiqian television relay station, 1800 m, 18 Aug. 1988, *Z. C. Zhao 88-1720* (PE 01376133–34). **Shizhu Xian:** Shazi Qu, Shazi Gongshe, 1310 m, 21 July 1978, *W. H. Wang 2263* (CDBI CDBI0027682–84). **Wushan Xian:** Zhuxian Xiang, Xujia Gou, 1500 m, 18 May 1958, *G. H. Yang 58139* (IBSC 0092601, PE 01031602–04). **Youyang Xian:** Xinglong Xiaoxian Xiang, 10 May 1959, *s. coll. 02524* (KUN 0176243, PE 01031592, 01031702).
SW & W Hubei. Badong Xian: Luoping, Dongyu Kou, Xiaoxi Houwan, 1840 m, 6 Oct. 1952, *Q. L. Chen et al. 1918* (HIB 00129519, PE 01031714, 01031719); Xiagu Gongshe, Banqiao, 1200 m, 27 July 1977, *Hubei Plant Investigation Team 24165* (HIB 00129520–21, PE 01031715). **Fang Xian:** Shi Qu, Xiao Luoxi, 800 m, 23 Sep. 1958, *K. R. Liu 0232* (HIB 0129528, IBSC 0092253, LBG 00064116, PE 01031713). **Hefeng Xian:** 17 Apr. 1959, *F. S. Peng 149* (HIB 00129525).

Lichuan Shi: Hongsha Xisha, 1150 m, 21 Sep. 1951, *L. Y. Dai & Z. H. Qian 1177* (PE 01031721). **Shennongjia Lin Qu:** Panlong Gongshe, 900 m, 11 Aug. 1976, *Shennongjia Team 21556* (HIB 0129502–03, PE 01031717–18); Laojunshan Yaowan canyon on the W side of the Jiuchong River ca. 1 km S of Mucheng, 1250 m, 31 Aug. 1980, *Sino American Bot Exp. 472* (A 00279844, HIB 0129509–12, KUN 0177047, PE 00996102, UC UC1491159). **Xingshan Xian:** near Xiangping, 1240 m, 3 Nov. 1952, *Q. L. Chen 2207* (HIB 00129518, PE 01031723–24); Muyuping, 25 Oct. 1958, *K. R. Liu 0669* (HIB 00129514, LBG 00064115). **Xuan'en Xian:** Bada Gongshan, 1600 m, July 1958, *H. J. Li 5453* (HIB 0129530–31, IBSC 0091894, PE01031836).

NW Hunan. Sangzhi Xian: Baomaoxi Gongshe, Tianping Shan, 1800 m, 6 July 1975, *B. G. Li & S. B. Wan 750159* (IBSC 0091850, PE 01031381).

67. Berberis schaaliae C. C. Yu & K. F. Chung, Phytotaxa 184(2): 89. 2014. TYPE: China. Taiwan: Hualien Hsien, Sioulin, Mt. Tatuanyai (The Great Cliff), 24.204572°N, 121.579867°E, 1500 m, 8 Aug. 2008, *C. C. Yu 147* (holotype, TAI 284281 image!).

Shrubs, evergreen, to 1.5 m tall; mature stems brown, subterete, not verruculose; spines 3-fid, rarely 7-fid, concolorous, 0.2–1.2 cm, solitary, 2-fid or absent toward the apex of stems. Petiole almost absent; leaf blade abaxially pale green, adaxially green, slightly shiny, elliptic, ovate, or lanceolate, 5.4–10.5 × 1.4–3.3 cm, thickly leathery, midvein raised abaxially, impressed adaxially, lateral veins slightly raised and inconspicuous abaxially, slightly impressed and inconspicuous adaxially, dense reticulation obscure abaxially, inconspicuous adaxially, base attenuate, margins sometimes slightly revolute, spinulose with 20 to 64 teeth on each side, apex acute, mucronate. Inflorescence a fascicle, 3- to 12-flowered; pedicel 7–23 mm; bracteoles red, triangular, 0.5 × 0.5 mm. Flowers yellow. Sepals in 3 whorls; outer sepals partially reddish tinged, triangular, 2 × 1.5 mm; median sepals elliptic to ovate, 3 × 2 mm; inner sepals obovate, 6.5 × 4 mm; petals elliptic, 4.5 × 3 mm, base clawed, glands ovoid, close to each other, apex usually dentate. Stamens 3 mm; anther connective extended, truncate. Pistil 4 mm; ovules 3 to 4. Berry dark purple to black, ellipsoid, ca. 7 × 4 mm, epruinose; style persistent, short.

Phenology. *Berberis schaaliae* has been collected in flower in April and May and in fruit from March to January the following year.

Distribution and habitat. *Berberis schaaliae* appears to be endemic to limestone areas in Hualien Hsien in east Taiwan. It has been found in open areas and the understories of warm-temperate forests at 1100–2400 m.

The leaves of *Berberis schaaliae* are often remarkably similar to those of *B. arguta* found in northeast Yunnan and southeast Sichuan.

Selected specimens.
Taiwan. Hualien Hsien: around Mt. Chingshui, 600–1400 m, 29 Mar. 1961, *T. Shimizu & K. T. Kao 11747* (TAI 047117); Mt. Chingshui, 1400–2100, 31 Mar. 1961, *T. Shimizu & K. T. Kao 11822* (TAI 246075); Sheauchingshoei, 900 m, 24.583333°N, 121.583333°E, 17 May 1986, *S. Y. Lu 19302* (HAST 79482, TAIF 93884); Chingshuishan, 200–2400 m, 25 July 1986, *T. C. Huang et al. 12838* (TAI 215208, 215226); Sanchiaochuishan, 1800 m, 14 Jan. 1990, *S. Y. Lu 24956* (TAIF 94806); Hsiulin Hsiang, from peak of Chingshuishan to Shakatang Forest Rd., 24.236944°N, 121.625°E, 1500–2000 m, 6 June 1993, *W. P. Leu et al. 1806* (HAST 25499); Pilu, 2300 m, 9 Aug. 2000, *T. T. Chen et al. 10817* (TAIF 125240); Hsiulin Hsiang, Chingshuishan, 24.2425°N, 121.635556°E, 1800 m, 7 Dec. 2000, *J. J. Chen et al. 581* (HAST 89426); Hsiulin Hsiang, betw. the lower & upper branch of Yenhai Forest Rd., 24.1425°N, 121.505556°E, 1450 m, 14 Aug. 2002, *C. I. Huang 1180* (HAST 112644); en route from the last camp to Chingshuishan, 24.241389°N, 121.633889°E, 1910 m, 27 June 2005, *C. I. Huang 2102* (HAST); Mt. Chuilu, 1650 m, 30 Mar. 2007, *T. T. He 118* (TAIF 27454); Hsiulin Hsiang, Tatung to Chienliyenshan, 800–1500 m, 22 Mar. 2009, *P. F. Lu 18014* (HAST 124134); Shioulin Village, Mt. Great Chingshui area, 24.237153°N, 121.647725°E, 2100 m, 4 Apr. 2011, *C. C. Yu 674* (TAI).

68. Berberis silvicola C. K. Schneid., Pl. Wilson. (Sargent) 3: 438–439. 1917. TYPE: China. W Hubei: Hsieng shan Hsien (Xingshan Xian), 1800–2400 m, 31 May 1907, *E. H. Wilson 2879* (lectotype, designated by Ahrendt [1961: 52], K K000077327!; isolectotypes, A 00038795!, BM BM000559456!, E E00259006!, HBG HBG-506758 image!, US 00103907 image!, W 1914–0004812!).

Shrubs, evergreen, to 1 m tall; mature stems yellow, terete; spines 3-fid, concolorous, 0.3–0.6(–1) cm, slender, sometimes absent. Petiole 1–3 mm; leaf blade abaxially pale green, adaxially dark green, shiny, elliptic or oblong, 2–5 × 1–2 cm, thinly leathery, midvein raised abaxially, impressed adaxially, lateral veins and reticulation slightly raised and conspicuous abaxially, indistinct adaxially, base cuneate or broadly attenuate, margin spinose with 12 to 16 teeth on each side, apex acute, minutely mucronate. Inflorescence a fascicle, 2- to 5-flowered; pedicel 5–8 mm, slender; bracteoles ovate, ca. 2.5 × 1 mm, apex acute. Sepals in 2 whorls; outer sepals ovate, ca. 4 × 1.8 mm, apex obtuse; inner sepals obovate, ca. 6 × 3.5 mm, apex rounded; petals obovate, ca. 5 × 3 mm, base cuneate, glands separate, apex incised. Stamens 5–8 mm; anther connective slightly extended, truncate. Ovules 2. Immature berry reddish purple, ellipsoid, ca. 6 × 3 mm; style persistent.

Phenology.   *Berberis silvicola* has been collected in flower in May and in fruit in August.

Distribution and habitat.   *Berberis silvicola* is known from west Hubei and Chongqing (northeast Sichuan). It has been collected in woods and forests at ca. 1800–2400 m.

The protologue of *Berberis silvicola* cited only *Wilson 2879* but did not indicate the herbarium. Ahrendt's (1961: 52) citation of the specimen at K as type was an effective lectotypification.

Berberis silvicola was treated as a synonym of *B. petrogena* by Bao (1997: 63), but the structure of the flowers of these two species differs.

Berberis silvicola is on the current IUCN Red List of Threatened Species (China Plant Specialist Group, 2004e).

Selected specimens.
Chongqing (NE Sichuan). Chengkou Xian: Distr. de Tschen-kéou-tin, s.d., *P. G. Farges 594* (P P06868556). **Wushan Xian:** Yangzhao Ping, 2050 m, 3 May 1958, *G. H. Yang 57919* (FUS 00014384, MO 04721698, PE 01031827–28); Yangzhao Ping, Xiao Yanzhao Ping, 2050 m, 5 May 1958, *G. H. Yang 57960* (FUS 00014324, IBK IBK00282333, MO 04721932, PE 01031457, 01031460).
    W Hubei. Shennongjia Lin Qu: Guanmen Shan, 1800 m, 6 Aug. 1976, *Shennongjia Botany Team 10857* (HIB 00129542–43, PE 01031837–8); Qianjiaping, 28 Sep. 1987, *X. X. Liu IV3050558* (CCAU 0002118–19); Yantian scenic area, 2400 m, 23 Aug. 1996, *S. G. Shi S-0889* (TAIF 316262). **[Xingshan Xian]:** Wan Tsao (Wanchao) Shan, 17 Aug. 1922, *W. Y. Chun 3941/4286* (N 093057064, NAU 00024184, US 0945906).

69. Berberis simulans C. K. Schneid., Repert. Spec. Nov. Regni Veg. 46: 258. 1939. TYPE: China. WC Sichuan: Emei Shan, 1850 m, 19 Apr. 1932, *T. T. Yu 414* (holotype, A 00038796!; isotypes, HIB 00096933 image!, PE [2 sheets] 01031709–10!).

Berberis gagnepainii C. K. Schneid. var. *omeiensis* C. K. Schneid., Repert. Spec. Nov. Regni Veg. 46: 264. 1939, syn. nov. TYPE: China. WC Sichuan: Emei Shan, 2600–2750 m, 13 Aug. 1928, *W. P. Fang 2916* (lectotype, designated here, E E00217966!; isolectotypes, A 00038757!, E E00217937!, P P00716540!).
Berberis gagnepainii C. K. Schneid. var. *subovata* C. K. Schneid., Repert. Spec. Nov. Regni Veg. 46: 264. 1939, syn. nov. TYPE: China. WC Sichuan: Ching-chi Hsien (Hanyuan Xian), "nr Mt. Wa-wu," 2400 m, 16 Oct. 1908, *E. H. Wilson 2874* (lectotype, designated by Ahrendt [1961: 52–53], K missing, new lectotype, designated here, A 00038758!).

Shrubs, evergreen, to 3 m tall; mature stems pale yellowish brown, terete, sparsely verruculose; spines 3-fid, semi-concolorous, shiny, 0.5–1 cm, weak, abaxially sulcate. Petiole almost absent; leaf blade abaxially

pale green, rarely pruinose, adaxially dark green, very shiny, broadly or narrowly ovate-elliptic, 6–10 × 2–3(–4) cm, thinly leathery, sometimes slightly crinkled, midvein raised abaxially, impressed adaxially, lateral veins slightly raised and inconspicuous abaxially, slightly impressed and inconspicuous adaxially, reticulation indistinct abaxially, inconspicuous adaxially, base attenuate, margin slightly revolute, spinulose with 5 to 30 teeth on each side, apex subacuminate, mucronate. Inflorescence a fascicle, 2- to 8-flowered; pedicel slightly reddish, 12–20 mm, slender; bracteoles ovate, apex acute. Sepals in 2 whorls; outer sepals elliptic, ca. 3 × 1.5 mm; inner sepals similar, elliptic-obovate, ca. 6 × 4 mm; petals oblong-obovate, ca. 5 × 2.75 mm, base clawed, slightly emarginate. Ovules 2 or 3. Berry blue-black and slightly pruinose, ellipsoid or semiglobose, 1–1.2 × 0.6–0.8 cm; style persistent or not.

Phenology.   *Berberis simulans* has been collected in flower between April and June and in fruit between August and November.

Distribution and habitat.   *Berberis simulans* is known from Emeishan Shi and Ebian, Hanyuan, Hongya, Mabian, and Muchuan Xian in west-central Sichuan. It has been found in woodland and forest understories and margins at ca. 1600–2800 m.

Berberis simulans was reduced to synonymy under *B. sargentiana* by Chamberlain and Hu (1985: 541) and by Ying (2001: 141). This was understandable given that herbarium specimens of the two species sometimes look similar; however, living plants cannot be so confused. The thin, leathery, lustrous dark green leaves of *B. simulans* with indistinct reticulation adaxially contrast with the thick, leathery, yellowish green leaves, with distinct impressed reticulations adaxially of *B. sargentiana*.

A living plant I have grown from seed of *D. J. Hinkley (China) 811* collected at 2600 m on Emei Shan in 1996 has slightly crinkled leaves and ellipsoid fruit, whereas living plants at the Royal Botanic Gardens, Kew, and Howick Hall, Northumberland, U.K., grown from seeds of *SICH 1646* collected at 2660 m on 5 October 1995 near the top of the east side of Wa Shan in Wawu Shan National Forest Park, Hongya Xian, have flat leaves and more globose fruit. This latter was collected in an open *Abies fabri* forest among bamboo and *Rubus*.

The protologue of *Berberis gagnepainii* var. *omeiensis* cited *W. P. Fang 2916* at E and "Ferner wohl auch [also probably]" *Y. S. Liu 2232* at A, but did not use the word type. Subsequently, Schneider (1942: 53) designated *Fang 2916* as the type but without citing a herbarium. There are two specimens of *Fang 2916* at E.

E00217966, which was annotated as type by Schneider on 8 August 1937, has been chosen here as the lectotype. From the research of Callaghan (2011), it appears that this gathering was not collected by Fang on Emei Shan, but by one of his collectors on Wa Shan.

The protologue of *Berberis gagnepainii* var. *omeiensis* also refers to *W. P. Fang 2372* from Guan Xian (Dujiangyan Shi) as being between *B. gagnepainii* and variety *omeiensis*. In fact, this would seem to be simply *B. gagnepainii* (for further details of this gathering, see under *B. gagnepainii*).

The protologue of *Berberis gagnepainii* var. *subovata* cites three specimens. Subsequently, Schneider (1942: 53) designated *T. T. Yu 414* as the type, but did not indicate the herbarium. The description by Ahrendt (1961: 52–53) of the specimen at K as type was an effective lectotypification, but this is missing. The specimen at A has therefore been designated as a new lectotype. *T. T. Yu 414* appears to differ from *B. simulans* only in having some leaves that are abaxially pruinose. The significance of this would seem impossible to judge given the apparent absence of any other similar specimens from the same or any other area.

The two gatherings in addition to *T. T. Yu 414* cited in the protologue of variety *subovata* were *C. W. Wang 63774* and *63889* from northwest Yunnan (details under syntypes below). Chamberlain and Hu (1985: 556) stated that the former is referable to *Berberis phanera*. However, Schneider's description of the flowers of *B. gagnepainii* var. *subovata* which was based on *C. W. Yang 63774*, though brief, differs from that of *B. phanera* and is treated here together with *C. W. Wang 63889* as an example of *B. wui*. *Berberis gagnepainii* var. *subovata* was not noticed by Ying (2001, 2011).

Berberis gagnepainii var. *subovata* was reported from Wen Xian, Gansu, by Instituto Botanico Boreali-Occidentali Academiae Sinica (1974: 309). Specimens from Wen Xian at WUK identified as this taxon represent two different species: *B. bergmaniae* and *B. soulieana*.

The vast majority of herbarium specimens of *Berberis simulans* located are from Emei Shan, but this may simply reflect its popularity as a collecting area. Reports of *B. acuminata* and *B. sargentiana* from Emei Shan are referable to *B. simulans*.

Syntypes. *Berberis gagnepainii* var. *omeiensis*. Sichuan: O-pen Hsien (Ebian Xian), 1300–1800 m, Sep. 1937, *Y. S. Liu 2232* (A 00279878, LBG 00064136–37).

Syntypes. *Berberis gagnepainii* var. *subovata*. NW Yunnan: Wesi Xsien (Weixi Xian), 3000 m, June 1935, *C. W. Wang 63774* (A 00279748, IBSC 0091919, KUN 0177060, LBG 00064134, NAS NAS00314205, PE [2 sheets] 01030970–71, WUK 0037190); same details, but 3500 m, *C. W. Wang 63889* (A 00279749, LBG 00064135, PE [2 sheets] 01030972–73, WUK 0040453).

Selected specimens.

Sichuan. Ebian Xian: 1300–1800 m, Sep. 1937, *Y. S. Liu 2232* (A 00279878, LBG 00064136–37, NA 0082465); Shaping, s.d., *Z. S. Zheng 64* (KUN 0175606); "O-pien Hsien," 28 July 1939, *S. L. Sun 0837* (KUN 0175605; SWCTU 00006255); Jianzhi Gou, 29 July 1939, *Z. W. Yao 4304* (PE 01031281–82). **Emeishan Shi:** Leidongping, 11 June 1974, *Sichuan Botany Team 269* (CDBI CDBI0027378, CDBI0027380). **Hanyuan Xian:** 1934, *Z. S. Liu 532* (PE 01031299). **Hongya Xian:** Wawu Shan, Taizidian, 21 July 1938, *Z. W. Zhao 2407* (KUN 0177051–52, PE 01031284); Wawu Shan, Luohan Ge, 24 June 1939, *Z. W. Zhao 3807* (FUS 00014238, PE 01031283, SZ 00328072); 26 May 1955, *Sichuan Forest Research Institute 2169* (CDBI CDBI0027070). **Mabian Xian:** Huanlian Shan, 1600 m, 19 Aug. 1978, *Mabian Team 1242* (SM SM704800257). **Muchuan Xian:** Jianquan Gongshe, 9 Aug. 1979, *Muchuan Team 0725* (SM SM704800148). **Omei Hsien (Emei Xian):** Mt. Omei, Dacheng Temple, 1800 m, 4 Aug. 1938, *W. P. Fang 12925* (A 00279873, 00279877, KUN 0175604, N 093057074, PE 01031308, US 00946007); Mt. Omei, 2110 m, Oct. 1941, *C. L. Chow 4854* (A 00279874, US 00945968); Emei Shan, Xixiang Chi, 2250 m, 11 June 1957, *G. H. Yang 55308* (IBSC 0091769, KUN 0175607, NAS NAS00314335, PE 01031301); Mt. Omei, 12 Nov. 1946, *W. K. Hu 8429* (A 00279837, E E00623111).

70. Berberis soulieana C. K. Schneid., Bull. Herb. Boissier, sér. 2, 5: 449. 1905, replacement name for *Berberis stenophylla* Hance, J. Bot. 20: 257. 1882, not Lindl. (1864). TYPE: China. Chongqing (Sichuan): "ad Chung King" (Chongqing), 1881, *E. H. Parker s.n.* (holotype, BM ex Herb. H. F. Hance no. *21774*, 000559460!).

Berberis soulieana var. *paucinervata* Ahrendt, J. Linn. Soc., Bot. 57: 78. 1961. TYPE: China. "C Shensi" (Shaanxi): 17 Aug. 1916, *E. Licent 2588* (lectotype, inadvertently designated by Chamberlain & Hu [1985: 546], K K000077306!; isolectotypes, BM BM000895061!, TIE 00009534 image!, W 1931–0008293!).

Shrubs, evergreen, to 3 m tall; mature stems very pale yellowish brown, terete, scarcely verruculose; spines 3-fid, concolorous, 1–4 cm, terete, stout. Petiole 1–2 mm; leaf blade abaxially yellow-green, adaxially dark green, shiny, narrowly oblong, oblong-ovate, narrowly oblong-elliptic, or narrowly oblong-obovate, 3.5–10 × 1–1.5 cm, thickly leathery, midvein raised abaxially, impressed adaxially, lateral veins and reticulation indistinct on both surfaces, base cuneate, margin spinose with (4 to)6 to 10(to 18) teeth on each side, apex acute, mucronate. Inflorescence a fascicle, 5- to 20-flowered; pedicel 5–9 mm; bracteoles 2, reddish, ovate-triangular, ca. 2.2 × 1.5 mm, apex acute. Sepals in 2 whorls; outer sepals oblong-elliptic, ca. 5 × 3.5 mm; inner sepals obovate-oblong, ca. 7 × 5 mm; petals obovate, ca. 6 × 5 mm, base shortly clawed, glands separate, apex emarginate. Stamens ca. 3 mm; anther connective slightly extended, rounded. Ovules 2 or 3. Berry bluish black, pruinose at first, obovoid or subglobose, 7–8 × ca. 5 mm; style persistent.

Phenology. Berberis soulieana has been collected in flower between March and May and in fruit between May and November.

Distribution and habitat. *Berberis soulieana* is the dominant section *Wallichianae Berberis* in south Shaanxi. It is also known from south Gansu, northwest Hubei, north Sichuan, and Chongqing (east, southeast, and northeast Sichuan). It has been found on mountain slopes, cliffs, forest margins and understories, and river, road, and field sides at ca. 210–1800 m.

Schneider's approach to *Berberis soulieana* was somewhat confusing. In proposing it as a nom. nov. for "? *B. stenophylla* Hance," he indicated in a footnote (Schneider, 1905: 449) that he had been prompted to do so by a specimen sent to him by Vilmorin as *B. sanguinea* that matched Hance's description of *B. stenophylla*. Subsequently, Schneider (1917: 437), while stating "*Berberis soulieana* is as I now believe a distinct species and *B. stenophylla* Hance is a synonym of it," also stated that the cultivated plant he had referred to in the original protologue as "raised in Hort. Vilmorin from seeds collected by R. Farges in the region of 'Tchen keou tin' [Chengkou] in Eastern Sichuan" was the type, compounding the confusion by giving no details as to where this specimen might be (however, a specimen at P of *B. soulieana*, *Farges 754* from Chengkou, is annotated by Schneider as "B. sanguinea? 16. VI. 6"). At the same time, Schneider asserted that the Parker gathering could not have come from near Chongqing because "there is no evergreen Berberis at all in this locality according to Wilson's observations." Subsequently, Schneider (1918: 137, 1942: 15) retreated from this position implying that the Parker gathering actually *was* the type, although in both articles also asserting (without presenting evidence) that this gathering must have been made north of Chongqing. In fact, *B. soulieana* occurs south of Chongqing in Nanchuan Xian, an area not visited by Wilson.

The protologue of *Berberis soulieana* var. *paucinervata* cited the specimens of *Licent 2588* at BM and K as "Types." Chamberlain and Hu (1985: 546), who synonymized the taxon, were therefore mistaken in describing the specimen at K as a holotype. Nevertheless, this can be treated as an inadvertent lectotypification.

The report by C. S. Ding (1986: 313) of *Berberis soulieana* being found in Zhejiang is likely to be based on misidentification. I have found no evidence to support the reports of the species in Guizhou by He and Chen (2004: 34) or in Hunan by L. H. Liu (2000: 710).

Selected specimens.
Chongqing (E, NE & SE Sichuan). Chengkou Shi: Tchen-kéou-tin, s.d., *P. G. Farges 754* (P); Tchen-kéou-tin, s.d., *P. G. Farges s.n.* (A 00280190); Gigan Shan, 950 m,

31 Aug. 1958, *T. L. Dai 102284* (CDBI CDBI0027716, CDBI0027720, PE 01031996). **Fengjie Xian:** betw. Zhuyuan Tao Gonshe & Jiannong, 1200 m, 29 Sep. 1964, *H. F. Zhou & H. Y. Li 110641* (PE 01293953, WUK 0241069). **Kai Hsien (Kai Xian):** N of Wen-tong-ching, 8 June 1932, *W. P. Fang 10209* (LBG 00064127, NAS NAS00314344, PE 01031908–9). **Nanchuan Xian:** Xiao He, Tongjia Gou, 850 m, 19 Oct. 1957, *Z. H. Xiong 94060* (IBSC 0091874, KUN 0176441, PE 01031792); [near Xuetangwan], 29.016667°N, 107°E, 870 m, 5 Mar. 1996, *Z. Y. Liu 15231* (A 00280051, BM, CAS 943697, E E00087364, GZU 000216570, IMC 000017475, KUN 0177573, MEXU 853990, P P00248098, TAIF 317078, TNM S48519, UBC V213157, UPS BOT.V-152968). **Wan Xian (now Wanzhou Qu):** Dayakou, 1040 m, 21 May 1985, *Q. Q. Wang 0325* (CDBI CDBI0028218–19). **Wushan Xian:** Zhuxian Xiang, Yue'er Xi, 1400 m, 20 Oct. 1958, *G. H. Yang 59913* (CDBI CDBI0027714, CDBI0027714, FUS 00014321, PE 01031914, WUK 0302013).

S Gansu. Near Kwa-ka, 1370 m, 9 Nov. 1914, *F. N. Mayer 1823* (A 00279846). **Hui Xian:** Yushu Gongshe, Shibei, 1700 m, 8 May 1984, *Q. R. Wang et al. 11479* (PE 01031888). **Kang Xian:** near Qing He Lin Chang, 1420 m, 5 May 1963, *Y. Q. He & C. L. Tang 318* (KUN 0177106, WUK 0215064). **Tianshui Shi:** Dangchuan, Baojia Gou, Guanyinya, 1400 m, 2 Aug. 1963, *Q. X. Li 707* (PE 01031881). **Wen Xian:** Bikou Gongshe, Zhenggou, 950 m, 14 Sep. 1959, *Z. Y. Zhang 13745* (MO 04468689, WUK 0155072). **Wudu Xian:** Luotang Gongshe, 1300 m, 2 May 1959, *Z. Y. Zhang 1142* (LBG 00064132, WUK 0143706).

NW Hubei. Yunxi Xian: Sanguan Gongshe, 750 m, 22 May 1973, *Z. Zheng 1189* (HIB 00129399–400). **Zhuxi Xian:** Fengxi Gongshe, Xiaochao Qu, 10 June 1959, *P. Y. Li 2858* (KUN 0176126, PE 01031932).

S Shaanxi. Tai-pei Shan, 1910, *W. Purdom 7* (A 00279847). **Ankang Xian:** Zuolong Gongshe, Huabaguan Qu, Lazhu Shan, 1400 m, 6 Aug. 1959, *P. Y. Li 10818* (KUN 0177105, WUK 0139402). **Chenggu Xian:** Panlong Gongshe, 4 Apr. 1973, *S. M. Su 424* (PE 01031921, WUK 0294347). **Foping Xian:** Hetaoping, Shaoren Gou, 1550 m, 21 June 1952, *K. J. Fu 4665* (KUN 0177111, PE 01031923, WUK 0059773). **Hu Xian:** Tang Yu, Zoulupo, 29 Mar. 1962, *P. Y. Li 10103* (WUK 0019810). **Lan'gao Xian:** Jishengliang, 1 July 1959, *D. Xi 312* (WUK 0304176). **Lantian Xian:** Lanqiao Xiang, Wang Cun, 1400 m, 20 Oct. 1958, *G. X. Su 601* (KUN 0178346). **Liuba Xian:** Caizi Ling, 1550 m, 17 Oct. 1952, *K. J. Fu 6280* (IBK IBK00013116, PE 01031929, WUK 0060289). **Lueyang Xian:** betw. Luotuo Xiang & Lianghekou, 870 m, 2 Nov. 1958, *C. L. Tang 967* (KUN 0177108). **Mei Xian:** Tangyu Gou, 1200 m, 22 Apr. 1957, *Z. B. Wang 17730* (KUN 0177103, WUK 0091010, 0170536). **Mian Xian:** Daochaling, 1200 m, 14 May 1959, *Nanshuibao Team 00233* (KUN 0177104, PE 01031924). **Nanzheng Xian:** Shihe Gongshe, Tianchizi, 1100 m, 9 Apr. 1973, *X. X. Hou 71* (FJSI, FJSI021940, IBSC 0092045, WUK 0293988). **Ningqiang Xian:** Liejinba, 750 m, 26 Sep. 1958, *P. Y. Li 53* (WUK 0105726). **Ningshan Xian:** Xihe Gongshe, Gangtie Xiang, 1350 m, *J. Q. Xing 7258* (IBK IBK00013109). **Pingli Xian:** betw. Miaozi & Songmiaozi, 1300 m, 26 Nov. 1958, *C. L. Tang 1327* (KUN 0177109, WUK 0111268). **Shangnan Xian:** Kaihe to Shanyang, 1300 m, 18 Oct. 1958, *B. Z. Guo 4441* (IBSC 0092038, WUK 0112306). **Shanyang Xian:** Yuwang Gongshe, Yurou He, 980 m, 7 June 1964, *J. X. Yang & Y. M. Liang 2867* (PE 01031926, PEM 00042337, WUK 0228485). **Shiquan Xian:** Gangtie Gongshe, Lianghe, 1700 m, 19 June 1960, *s. coll. 334* (KUN 0177110). **Taibai Xian:** Huangbaiyuan, Dongchang Gou, *C. G. Ma et al. 00318* (PE 01533604–06). **Xunyang Xian:** Xuezi Gou, 1820 m, 6 June 1959, *J. Q.*

Xing 5759 (SZ 00287802, WUK 0126393). **Yang Xian:** Huayang, 33.066667°N, 107.533333°E, 1200 m, 3 June 1999, *Zhu, Chen, Xu & Wang 1085* (KUN 0179280). **Zhen'an Xian:** Heiyaogou Lin Chang, 1250 m, 4 June 1973, *X. X. Hou & Y. H Guo 765* (FJSI FJSI021938, IBSC 0092046, WUK 0295190). **Zhenping Xian:** near Zhongguan, 1800 m, 11 June 1959, *P. Y. Li 3233* (PE 01031931, WUK 0143118). **Ziyang Xian:** Shuangqiao Zhen, Liuhe Xiang, 1100 m, 12 Oct. 2006, *Y. S. Chen et al. 4771* (WUK 0502850). **Zhouzhi Xian:** Yinjiapo, 4 July 1951, *B. Z. Gou 10* (WUK 0024645); Tabeishan, Houzhingzhi, 28 July 1999, *Zhu, Chen, Xu & Wang 1758* (KUN 0179277).

E & N Sichuan. An Hsien (Xian) (now Anzhou Qu): 15 May 1938, *S. C. Huang s.n.* (N 093057114). **Anyue Xian:** Shiyang Qu, Shiyang Gongshe, Jiu Dadui, 5 May 1978, *Anyue Team 238* (SM SM704800041); same details, but 210 m, 12 May 1978, *Anyue Team 1047* (SM SM704800040). **Cangxi Xian:** Jiulongguan Shan, 1000 m, 2 May 1959, *M. X. Wang 07624* (PE 01031905). **Guangyuan Shi:** Chaotian Qu, Pingxi Xiang, 850 m, 9 July 1994, *Z. J. Ren s.n.* (TAIF 350251). **Jiange Xian:** near Jianmenguan, 11 Aug. 1939, *T. N. Liou & C. Wang 368* (PE 01031675, 01031907, WUK 0228485). **Jiangyou Shi:** Dakang Gongshe, Shi er Dadui, 9 Aug 1978, *Jiangyou Team 57* (SM SM704800072). **Nanjiang Xian:** Zhaipo Xiang, 1290 m, 8 June 1959, *X. Y. Zhou, Sichuan Economic Plants Exp. Daxian Group 2932* (CDBI CDBI0027133). **Pingwu Xian:** Zhenliu Zheng, 1030 m, 13 Apr. 1958, *s. coll. 10018* (PE 01031480–84). **Qingchuan Xian:** Qingxi, Gongnang Gongshe, Tangjia He, 17 Aug. 1978, *Qingchuan Team 0215* (SM SM704800378). **Santai Xian:** Qiulin, San Dadui, 23 Sep. 1977, *Santai Team 370* (SM SM704800071). **Tongjiang Xian:** Jincheng Shan, 1050 m, 27 Apr. 1959, *Sichuan Economic Plants Exp. 0073* (CDBI CDBI0027721). **Wangcan Xian:** Micangshan Nature Reserve, 1786 m, 30 June 2011, *Bashan Collection Team 5068* (PE 01873009). **Wanyuan Xian:** Hua'e Shan, 1730 m, 2 June 1959, *Sichuan Economic Plants Exp. Daxian Group 2273* (CDBI CDBI0027707, FUS 00014269, KUN 0176436). **Zhaohua Qu:** Rentou Shan, 1000 m, 6 Nov. 1958, *Y. Q. He 1728* (WUK 0166628).

71. Berberis subacuminata C. K. Schneid., Pl. Wilson. (Sargent) 1: 363. 1913. TYPE: China. S Yunnan: Yuanchiang (Yuanjiang Xian), 1525 m, s.d., but before 1901, *A. Henry 13267* (lectotype, designated by Ahrendt [1961: 62], K K000077324!; isolectotypes, A 00038798!, BM [2 sheets] 001015574–75!, E E00259007!, MO 2176016!, NY 00000042!).

Berberis malipoensis C. Y. Wu & S. Y. Bao, Bull. Bot. Res., Harbin 5(3): 10. 1985, syn. nov. TYPE: China. SE Yunnan: Malipo Xian, Babu Qu, 1000 m, 28 Nov. 1964, *Q. A. Wu 9921* (holotype, KUN 1204037!; isotype, KUN 1204040!).

Shrubs, evergreen, to 2.5 m tall; mature stems pale yellow, sulcate; spines 1- to 3-fid, pale yellow, 0.4–1.5 cm, slender, sometimes absent. Petiole 2–5 mm; leaf blade abaxially yellow-green, adaxially dark green, ovate-lanceolate, ovate-elliptic, or narrowly elliptic, 3–9 × 1.2–2.5 cm, thinly leathery, midvein raised abaxially, impressed adaxially, lateral veins slightly raised and inconspicuous abaxially, slightly impressed and conspicuous adaxially, reticulation indistinct abaxially,

inconspicuous adaxially, base cuneate or attenuate, margin spinulose with 8 to 15 teeth on each side, apex acuminate, sometimes minutely mucronate. Inflorescence a fascicle, 2- to 6(to 17)-flowered; pedicel 18–22 mm; bracteoles triangular-ovate, ca. 2 × 1.5 mm. Flowers yellow. Sepals in 2 whorls; outer sepals elliptic, 4 × 3.5 mm; inner sepals broadly obovate, ca. 7 × 5 mm; petals obovate, ca. 5 × 2.5 mm, base clawed, glands very close together, apex emarginate. Stamens ca. 4.5 mm; anther connective not or slightly extended, truncate. Ovules 2 or 3. Berry bluish black, pruinose, subglobose, 6–7 × ca. 5 mm; style not persistent.

Phenology. *Berberis subacuminata* has been collected in flower in March and in fruit between August and January the following year.

Distribution and habitat. In China *Berberis subacuminata* is known from south, southeast, and southwest Yunnan. It is also known from north Vietnam and is possibly found in Myanmar (Burma). In China it has been collected in scrub, dry slopes, streamsides, and in mixed forests at ca. 1000–2800 m. In Vietnam it has been collected in Hà Giang and Lào Cai provinces at 2200–3100 m.

The protologue did not cite a herbarium for the type of *Berberis subacuminata*. The lectotype specimen at K chosen by Ahrendt (1961: 62) was annotated as "spec. nov.?" by Schneider on 1 May 1906.

The protologue of *Berberis malipoensis* ignored *B. subacuminata*, choosing rather to differentiate it from *B. acuminata* whose type is from northeast Yunnan. In fact, the holotype of *malipoensis* (though not the isotype) appears to differ from most specimens of *B. subacuminata* in having some leaves that are narrowly elliptic rather than the more generally found lanceolate-elliptic or lanceolate ones. But the isolectotype of *B. subacuminata* at NY has some leaves that are elliptic. Besides the type, the protologue of *B. malipoensis* also cited two other collections from Malipo Xian, *C. Y. Wang 83927* and *K. M. Feng 12823*. Both of these exhibit the more characteristic leaves of *B. subacuminata*. Moreover, fruit of *K. M. Feng 12823* at A produced three seeds versus the two reported for *B. malipoensis* in the protologue.

Subsequent to his publication of *Berberis subacuminata*, Schneider (1918: 144) decided it was, in fact, a synonym of *B. bicolor*, later (1942: 42) re-asserting it as a separate species. It is certainly the case that *B. bicolor*, which is endemic to Guizhou, has very similar leaves, but it has very different red-and-white flowers with shorter pedicels. Despite these differences, the confusion between the two species has persisted (see, e.g., Mizushima, 1954: 28–30) and is likely to be the reason why *B. subacuminata* was reported in Guizhou by Ying

(2001: 132, 2011: 743) and He and Chen (2004: 34). Conversely, reports of *B. bicolor* being found in Myanmar (e.g., Kress at al., 2003: 172) may also be referable to *B. subacuminata*.

The report by Ying (2001: 132, 2011: 743) of *Berberis subacuminata* in Hunan is likely to be based on misidentification.

As noted in the introduction to this study, *Berberis subacuminata* from Vietnam was first reported as *B. julianae* in the *Red Data Book of Vietnam* (Government of the Socialist Republic of Vietnam, 1996) and subsequently under the same name in Nguyên (1998: 45) and Võ (2007: 120). As *B. julianae*, it was awarded protected status by the Government of Vietnam (Government of the Socialist Republic of Vietnam, 2006). This latter reported it as growing at 1600 m, but this is mistaken as it is found at much higher elevations. This is the *Berberis* growing in what Kouznetsov and Phan Loung (1999: 16–17) describe as the "top adjacent band (2800–3143 m)" of the Fansipan Mountain, and Tordoff et al. (1999: 23) describe as dwarf bamboo forest, though the former identify it as "*Berberis wallichiana*" (for which in Vietnam, see under *B. ferdinandi-coburgii*) and the latter simply as *Berberis*.

I am grateful to Nguyen Quynh Nga of NIMM for information about specimens from Hà Giang Province.

Selected specimens.
S, SE & SW Yunnan. Jingdong Xian: Wuliangshan, Huang Caoling, Zhangyao Shan, 2650 m, 19 Nov. 1956, *B. Y Qiu 53738* (KUN 0177137–38, PE 01031794). **Jinping Xian:** Wutaishan, 2500 m, 2 Oct. 1996; *S. K. Wu et al. 3728* (KUN 0177552, PE 01868973). **Lincang Xian:** Mengwang Qu, Zhangtuo Xiang, 1950 m, 3 Sep. 1957, *J. S. Xin 665* (IBSC 0092061, KUN 0177147–48, PE 01031701). **Longchuan Xian:** Husa Xiang, 27 Aug. 1976, *C. J. Pei 14128* (HITBC 003650). **Malipo Xian:** Hwang-ging-ying (Huangjinyin), 1300 m, Jan. 1940, *C. Y. Wang 83927* (HITBC 003597, IBK IBK00357884, IBSC 0091899, KUN 0176681, LBG 00064113, PE 01031880, WUK 0270002); Chung-dzai, 1600–1800 m, 3 Nov. 1947, *K. M. Feng 12823* (A 00279698, KUN 0176679–80, PE 01031879, WUK 0207974). **Shiping Xian:** Niujie Qu, Daleng Shan, 2400 m, 10 Oct. 1958, *S. G. Wu 948* (KUN 0177159–60). **Wenshan Xian:** Maa-lu-tarng, 2000 m, 16 Aug. 1947, *K. M. Feng 11260* (A 00279699, KUN 0177163–64, PE 01031317, WUK 0203350). **Yuanjiang Xian:** Er Qu, Yangchajie, 1930 m, 24 Oct. 1964, *Y. H. Li 5716* (HITBC 003653, KUN 0177158). **Chen Kang Hsien (Zhenkang Xian):** 2800 m, 19 Mar. 1936, *C. W. Wang 72435* (A 00279891, IBSC 0091691, KUN 0176105, PE 01031802, 01031826, WUK 0042847).

N VIETNAM. Hà Giang: Xín Mần Distr., Thu Tà, Tây Côn Lĩnh Range, ca. 2200 m, Apr. 2015, *Phan Van Truong et al. HG-262* (NIMM [3 sheets]); Hoàng Su Phì Distr., Hồ Thầu, Tây Côn Lĩnh Range, 2248 m, Mar. 2018, *Phan Van Truong B2* (NIMM [3 sheets]). **Lào Cai:** Chapa (Sapa), "Massif du Fan Si Pan," 2900 m, Aug. 1942, *A. Petélot 7.998* (P PO580441–42); "Phan Xi Pang," 3100 m, 12 Dec. 1964, *Nguyen Van Bien 168* (HNU); Fansipan, 2800 m, 26 Sep. 2004, *A. de Rouw 451* (E 00533949); Shaba, Huanglian Shan, s.d., *Sino-Vietnamese Exped. s.n.* (KUN 0176731); Shaba,

Huanglian Shan, s.d., *Sino-Vietnamese Exped. s.n.* (KUN 0176731); Sapa, s.d., *Sino-Vietnamese Expedition s.n.* (KUN 0177325–26); s. loc., s.d., *L. P. Pham 0008* (KUN 0178426).

72. Berberis sublevis W. W. Sm., Notes Roy. Bot. Gard. Edinburgh 9: 83. 1916. TYPE: China. W Yunnan: [Tengchong Xian], hills W of Tengyueh (Tengchong), 25°N, 1825 m, Feb. 1913, *G. Forrest 9559* (lectotype, inadvertently designated by Chamberlain & Hu [1985: 555], E E00117375!; isolectotype, IBSC 0092073 image!).

Berberis wallichiana DC. var. *microcarpa* Hook. f. & Thomson, Fl. Ind. 1: 226. 1855. *Berberis sublevis* var. *microcarpa* (Hook. f. & Thomson) Ahrendt, J. Linn. Soc., Bot. 57: 58. 1961. TYPE: India. Assam: "Khasia alt. 5–6000 ped," s.d., *T. Thomson & J. D. Hooker s.n.* (lectotype, inadvertently designated by Chamberlain & Hu [1985: 555], E E00438534!; isolectotypes, BM BM001015552!, G [2 sheets] G00343562!, G00343596!, CGE [3 sheets] CGE32535–7!, K, M M01264279 image!, NY 1365273!, P P02797837!, P00580429!, TCD TCD0017858 image!, U 1156222 image!, W 0032264!).
Berberis sublevis var. *grandifolia* C. K. Schneid., Repert. Spec. Nov. Regni Veg. 46: 253. 1939. TYPE: China. W Yunnan: near Tengyueh, Shweli river basin, Feb. 1923, *J. F. Rock 7930* (lectotype, designated by Ahrendt [1961: 58], A 00038800!; isolectotype, US 00103908 image!).
Berberis sublevis var. *exquisita* Ahrendt, J. Linn. Soc., Bot. 57: 58. 1961. TYPE: China. W Yunnan: hills NW of Tengyueh, 1800–2100 m, Feb. 1925, *G. Forrest 26196* (holotype, K K000077332!; isotypes, AU 038833 image!, BM [2 sheets] BM001015540–41!, E [2 sheets] E00117368–69!, IBSC 0092071041 image!, NY 01104637!, P P02313554!, S 08–754 image!, US 00945930 image!, W 1933–3748!).

Shrubs, evergreen, to 3 m tall; mature stems yellow, terete, scarcely verruculose; spines 3-fid, concolorous, 1–2 cm, slender, abaxially flat or sulcate. Petiole almost absent; leaf blade abaxially pale green, adaxially dark green, narrowly lanceolate, 4–12 × 1–1.5 cm, thinly leathery, midvein raised abaxially, impressed adaxially, lateral veins slightly raised and inconspicuous abaxially, inconspicuous adaxially, reticulation indistinct on both surfaces, base cuneate, margin slightly revolute, spinulose with 10 to 20(to 30) teeth on each side, sometimes entire, apex acuminate. Inflorescence a fascicle, 5- to 15-flowered; pedicel 7–15 mm, slender. Sepals in 2 whorls; outer sepals reddish, ovate, ca. 2.5 × 2 mm, apex subacute; inner sepals obovate to oblong-elliptic, ca. 5 × 3 mm; petals obovate, 5–5.5 × ca. 3 mm, base clawed, glands separate, apex emarginate. Stamens ca. 3.5 mm; anther connective obtuse. Ovules 1; funicles 3–6× longer than ovules. Berry purplish black, epruinose, ovoid, 6–7 × 3–3.5 mm; style persistent, short.

Phenology.   In China *Berberis sublevis* has been collected in flower in February and in fruit between April and December.

Distribution and habitat. In China *Berberis sublevis* appears to be restricted to Longling, Lushui, Tengchong, and Zhenkang Xian in west Yunnan. In India it has been collected in Meghalaya, and in Burma (Myanmar) in the Ngawchang Valley. In China *B. sublevis* has been collected in and on the margins and clearings of subtropical evergreen broadleaf forests, in ravines, and by streams and riversides at 1500–2600 m.

The protologue of *Berberis sublevis* cited six Forrest specimens, but did not designate a type. Schneider (1918: 142–143) cited *Forrest 9559* as the type, but did not indicate the herbarium. Chamberlain and Hu (1985: 555) reported the specimen at E to be the holotype, but in accordance with Art. 9.10 of the Shenzhen Code (Turland et al., 2018), this can be treated as an inadvertent completion of lectotypification.

The protologue of *Berberis wallichiana* var. *microcarpa* cited only "Khasia alt. 5–6000 ped" without giving a collector, but contextually this clearly refers to a collection of Hooker and Thomson. Schneider (1918) cited a Hooker and Thomson collection with this description as the type, but did not indicate the herbarium. Chamberlain and Hu (1985: 555) cited the specimen at E as an isotype. This can be treated as an inadvertent lectotypification.

In the protologue of *Berberis sublevis* var. *microcarpa*, Ahrendt (1961: 58) cited *B. wallichiana* var. *microcarpa* as a synonym and cited *Simons 114*, listed below, as the type. The type, however, can only be the Hooker and Thomson gathering from Khasia.

The protologue of *Berberis sublevis* var. *grandifolia* cited *J. F. Rock 7930* and *H. T. Tsai 53605* at A, but did not designate a type. Later, Schneider (1942: 30) designated *J. F. Rock 7930* as the type, but without indicating the herbarium. Ahrendt's citing of the specimen at A as type (Ahrendt, 1961: 58) can be treated as a lectotypification, although there is no evidence that he had seen it.

Chamberlain and Hu (1985: 555) synonymized *Berberis wallichiana* var. *microcarpa*, *B. sublevis* var. *grandifolia*, and *B. sublevis* var. *exquisite* under typical *B. sublevis*. They also synonymized the name *B. sublevis* var. *gracilipes* with a reference to Ahrendt (1941b: 23) with *Simms* [sic] *114* as the type. But the Ahrendt reference is, in fact, to *B. wallichiana* var. *gracilipes* Ahrendt, whose type is *G. Watt 6449* from Manipur, India, and which Chamberlain and Hu (1985: 554) themselves treated as a synonym of *B. victoriana* D. F. Chamb. & C. M. Hu.

In September 1916, Schneider annotated the sheet at K with *Simons 114, 119* and *T. Lobb s.n.* as *Berberis prainiana*, but never published this name, noting (Schneider, 1918: 143) that the three collections were the same as *B. sublevis*. Stapf (1928) used the name for a species from Khasia, but without a description. As a nomen nudum, the name is invalid.

Ahrendt (1961: 58) reported the fascicles of *Berberis sublevis* to have up to 50 flowers. No evidence has been found to support this. Chamberlain and Hu (1985: 555) note that *B. sublevis* "is unique in Sect. *Wallichianae* in its long funicle."

Syntypes. *Berberis sublevis.* China. W Yunnan: Tengchong Xian, Tengyueh, 1825 m, Aug. 1912, *G. Forrest 8635* (E E00117371, K K00724000); S of Tengyueh, 25°N, Feb. 1913, *G. Forrest 9560* (E E00117367, K K000644863, S 08–753); NW of Tengyueh, 25.166667°N, 2125 m, Feb. 1913, *G. Forrest 9693* (A 00279902, E E00117376, IBSC 0092070, K K000644865); Tengyueh, 25°N, 1825 m, May 1913, *G. Forrest 7621* (A 00279903, E E00117370, IBSC 0092072, K K000644864); Tengyueh, 25°N, 1825 m, Mar. 1914, *G. Forrest 12198* (E E00117372).

Syntypes. *Berberis sublevis* var. *grandifolia.* China. W Yunnan: Lung-ling Hsien (Longling Xian), 1800 m, 1 Apr. 1934, *H. T. Tsai 55605* (A 00279893, IBSC 0092074, KUN 0177191, 1204867, PE 01031184–85, SZ 00287854).

Selected specimens.
W Yunnan. Longling Xian: Huoshan, 18 Aug. 1941, *Q. W. Wang 90152* (KUN 0177194); Longling, 1510 m, 27 Nov. 1958, *J. Chen 605* (KUN 0177195–96); Lameng, 25 July 1987, *S. Y. Bao et al. 929* (KUN 0178725–26). **Lushui Xian:** Pianma He, 1900 m, 28 July 1978, *Bijiang Investigation Team 1439* (HITBC 074263, KUN 0177192–93, YUKU 02065791); Pianma Xiang, along river to E of Pianma, W side of Gaoligong Shan, 29.009722°N, 98.632222°E, 1990 m, 16 May 2005, *Gaoligong Shan Biodiversity Survey 23004* (CAS, GH 00352710, HAST 124221, MO 5798166). **Tengshong Xian:** Shweli River drainage basin and environs of Tengyueh, on summit of Ta yui shan, Feb. 1923, *J. F. Rock 7916* (A 00279895, US 0946965); Houqiao Xiang, 18 May 1964, *S. G. Wu 6602* (HITBC 003648, KUN 0177197–98, LBG 00064108); Yunhua Gongshe, Daying Shan, 2500 m, 20 Nov. 1978, *780 Team 801* (HITBC 074271, PE 01479902); Mingguan Gongshe, 2000 m, 2 Aug. 1980, *S. T. Li 443* (KUN 0178269–70); Jietu Xiang, Zhoujiapo, rd. betw. Jietu & Datang, 25.553056°N, 98.668333°E, 1680 m, 28 Oct. 1998, *H. Li et al. 11156* (CAS 00120150, E E00228960, GH 00279488, HAST 80812); Km 24.2 on hwy. S 317, rd. to Tenglang village, Xiaodifang river drainage, W side of Gaoligong Shan, 24.905°N, 98.761944°E, 2146 m, 27 May 2005, *Gaoligong Shan Biodiversity Survey 25039* (CAS 00120140, E E00263627, GH 00297448, HAST 124195, MO 6076081); Houqiao, Danzha Cun, vic. of Zhaobitan Linchang, 25.545111°N, 98.219278°E, 2600 m, 29 May 2006, *Gaoligong Shan Biodiversity Survey 30769* (CAS 00120147, GH 00275929, HAST 124213). **Zhenkang Xian:** Chengkang, Hsiaoshuishan (Zhenkang Xian, Xiaoxueshan), 3200 m, 5 Aug. 1938, *T. T. Yu 17208* (A 00279303, KUN 0175674–75, PE 01031584–85).

INDIA. **Assam:** Khasia Hills, Motling, Feb. 1850, *C. J. Simons 114* (K K000644887); same details, *C. J. Simons 119* (K K000644888), Khasya, s.d., *T. Lobb s.n.* (K K000644886).

MYANMAR (BURMA). Bhamo Division, 1675 m, Mar. 1909, *G. E. S. Cubitt s.n.* (E E00623254); Kang-fang, ca. 1830 m, 3 Dec. 1938, *F. Kingdon-Ward, Vernay-Cutting Exp. 72* (NY); Black Rock, Ngawchang Valley, 27 Feb. 1939, 1220 m, *F. Kingdon-Ward, Vernay-Cutting Exp. 348* (A 00342314–15);

same details but 1200–1700 m, *F. Kingdon-Ward, Vernay-Cutting Exp. 354* (BM).

73. Berberis taliensis C. K. Schneid., Repert. Spec. Nov. Regni Veg. 46: 252. 1939. TYPE: China. NW Yunnan: [Dali Shi], E flank of Dali range, l25.666667°N, 100.2°E, 3000 m, May 1921, *G. Forrest 19417* (lectotype, inadvertently designated by Chamberlain & Hu [1985: 547], E E00320570!; isolectotypes, A 00058269!, CAL CAL0000027195 image!, K K000077351!, US 00103911 image!).

Shrubs, evergreen, to 1.2 m tall; branches stiffly upright, mature stems pale reddish brown turning pale yellow, stout, sulcate, scarcely verruculose; spines 3-fid, concolorous, 0.8–2 cm, adaxially conspicuously sulcate. Petiole almost absent; leaf blade abaxially pale green, pruinose, adaxially dark green, shiny, narrowly elliptic or oblanceolate, 1.2–3 × 0.3–0.5 cm, leathery, midvein raised abaxially, impressed adaxially, lateral veins and reticulation indistinct on both surfaces, base cuneate, margin conspicuously revolute, spinose with 1 to 3(or 5 or 6) teeth on each side, occasionally entire, apex acuminate, mucronate. Inflorescence a fascicle, 2- to 5-flowered; pedicel brownish gray, (12–)15–20 mm; bracteoles ovate, ca. 3 mm. Sepals in 2 whorls; outer sepals oblong-elliptic, ca. 4.5 × 3 mm; inner sepals broadly elliptic, ca. 6 × 3.5–4 mm; petals obovate, ca. 5.5 × 3.5–4 mm, base cuneate, slightly clawed, glands separate, apex entire, slightly emarginate. Stamens ca. 3.5 mm; anther connective not extended, truncate. Ovules 4, sessile. Berry black-purple, white pruinose, oblong, 8–12 × ca. 5 mm; style not persistent.

Phenology. *Berberis taliensis* has been collected in flower in May and in fruit in July and August.

Distribution and habitat. *Berberis taliensis* is known from Dali and Lijiang Shi, and Dayao, Jianchuan, and Lanping Xian in northwest Yunnan. It has been collected from the ledges of cliffs and rocky slopes in ravines at 3000–3600 m.

Schneider (1942: 25) cited *G. Forrest 19417* as the type but did not indicate a herbarium. Chamberlain and Hu (1985: 547) incorrectly cited *G. Forrest 19417* at Edinburgh as a holotype. Under Art. 9.10 of the Shenzhen Code (Turland et al., 2018), this can be treated as an inadvertent completion of lectotypification. Ahrendt (1961: 44) did not notice Schneider's lectotypification and mistakenly cited *G. Forrest 21968* as the type. *C. W. Wang 71708*, cited below under Syntypes, is not *Berberis taliensis* but *B. wui*.

Herbarium specimens of *Berberis taliensis* without mature stems are sometimes confused with *B. replicata*

from western Yunnan. Living plants of the two species, however, cannot be confused; the stiffly upright branches and dark, shiny green leaves of *B. taliensis* being in distinct contrast to the weak, arched branches and dull pale green leaves of *B. replicata*. There is also a distinct geographical separation between the species.

Syntypes. **NW Yunnan. [Jianchuan Xian]:** Chinchuan–Mekong divide, 26.5°N, 99.666667°E, 3050 m, Aug. 1922, *G. Forrest 21968* (A 00058270, CAL CAL0000027193, E E00320571, K K000077350, US 00946062); Chinchuan–Mekong divide, 26.5°N, 99.666667°E, 3050–3350 m, July 1923, *G. Forrest 23499* (BM BM001015545, E E00612601, K K000567919, PE 01030838); Ta-li Hsien (Dali Xian), 2800 m, Dec. 1935, *C. W. Wang 71708* (A 00279489, KUN 1204881, PE [2 sheets] 01031808–09).

Selected specimens.
NW Yunnan. Dali Shi: Cangshan, Xiao Huadianba, 3200 m, 21 May 1981, *Sino-British Cangshan Expedition 1033* (A 00279906, E E0129599, KUN 0177202, 0177207); Xiao Huadianba, 3300 m, 18 May 1984, *Sino-German Exp. 0343* (KUN 0179300). **Dayao Xian:** Santai Qu, Duodiohe Gongshe, 13 July 1965, *Woody Oil Plant Resource Exp. 491* (KUN 0177203, 0177206). **Lanping Xian:** Laojunshan, 3600 m, 7 July 1960, *Northwest Yunnan Team 9816* (KUN 177205, PE 01030837, YUKU 02065677); Mt. Lauchünshan (Laojunshan), SW of the Yangtze bend at Shiku (Shigu), 1923, *J. F. Rock 9588* (A 00279905, N 093057087, UC UC328260, US 0945947).

Cultivated material:
Living cultivated plants.
Royal Botanic Gardens, Kew, U.K., from *G. Forrest 21968*.

74. Berberis tarokoensis S. Y. Lu & Y. P. Yang, Fl. Taiwan, ed. 2, 2: 581. 1996. TYPE: China. Taiwan: Hualien Hsien, Taroko Natl. Park, Yenhai Logging Trail, 24.166667°N, 121.5°E, 20 July 1988, *S. Y. Lu 23713* (holotype, TAIF 092818!).

Shrubs, evergreen, to 50 cm tall; mature stems reddish brown, sulcate; spines 3-fid, concolorous, 0.3–1.4 cm, weak, terete. Petiole almost absent; leaf blade abaxially pale green, epruinose, adaxially mid-green, shiny, narrowly elliptic, narrowly obovate, or oblanceolate, 2–3.5 × 0.8–1 cm, leathery, midvein raised abaxially, impressed adaxially, lateral veins and reticulation obscure abaxially, indistinct or obscure adaxially, base attenuate or cuneate, margins sometimes slightly revolute, spinulose with (1 to)3 to 5(to 8) teeth on each side, apex acute, mucronate. Inflorescence a fascicle, 3- to 7-flowered; pedicel 12–14 mm; bracteoles triangular-ovate, partially red, 1.5–2 × 1–1.5 mm. Flowers bright yellow. Sepals in 3 whorls; outer sepals elliptic to broadly ovate, usually partially reddish, 3 × 2.5 mm; median sepals elliptic, 4.5 × 3 mm; inner sepals obovate, 5 × 4.5 mm; petals obovate, 3.5–4 × 5 mm, apex acutely emarginate, base clawed, glands obovoid, very close together. Stamens ca. 4 mm, apex incised or acutely emarginate; anther connective extended, truncate or slightly apiculate. Pistil 5 mm; ovules 5 to 8. Berry dark purple, partially or mostly pruinose, ellipsoid, 7–8 × 5.5 mm; style persistent, 1–1.3 mm.

Phenology. *Berberis tarokoensis* has been collected in flower in March and April and in fruit from May to November.

Distribution and habitat. *Berberis tarokoensis* is mostly found in a restricted limestone area of north Hualien Hsien, but is also recorded from Chiayi, Hsinchu, and Nantou Hsien. It has been collected on exposed rocks and scree slopes at ca. 1000–2600 m with one collection at 500–800 m.

K. C. Yang 02843 from Hsinchu (details below) is distinctly disjunct.

Berberis tarokoensis was not noticed by T. S. Ying (2001, 2011).

Selected specimens.
Taiwan. Chiayi Hsien: Nanhsi Logging Trail, 1700–2600 m, 23 July 2003, *Y. L. Yung et al. s.n.* (TAIF 215831). **Hsinchu Hsien:** Chienshih Hsiang, Chenhsipao, 2200 m, 20 Apr. 2003, *K. C. Yang 02843* (TNM). **Hualien Hsien:** Batagan, Uchi Taroko, 17 Apr. 1917, *s. coll. s.n.* (TI); Chuilu, 24.166667°N, 121.6°E, 800 m, 18 Nov. 1989, *S. Y. Lu 24978* (TAIF 076154–55); Hsiolin, Yenhai-Lindao, 1150 m, 3 Apr. 1991, *T. Y. Yang et al. 5471* (TAI 223380); Hsiulien Hsiang, terminus of Yenhai Forest Rd., 24.166111°N, 121.511389°E, 1090 m, 29 July 1999, *C. I. Huang 537* (HAST 11461); Yenhai Logging Trail, 1000–1300 m, 9 May 2005, *S. W. Chung 8012* (TAIF 231627); Yenhai Logging Trail, 300–900 m, 25 May 2005, *P. F. Lu 9686* (TAIF 225082); Shioulin Hsiang, Daduanya Shan, 24.185581°N, 121.539375°E, 1450 m, 8 Aug. 2008, *C. C. Yu 149* (TAI); terminus of cable way of the Yenhai Forest Rd., 24.165833°N, 121.510278°E, 1200 m, 15 Mar. 2009, *C. I. Huang 3717* (HAST 122903); Hsiulin Hsiang, Zhuilu old rd., 500–800 m, 22 May 2011, *P. Lu 22029* (HAST 130284); Yenhai Logging Trail, aerial cableway Sec. 1, 11 Apr. 2012, *C. F. Chen 3323* (TAIF 408397). **Nantou Hsien:** Tatachia Saddle to Shalihsien Stream, 2500–2600 m, 13 Aug. 2011, *P. F. Lu 22642* (TAIF 374368).

75. Berberis tengii Harber, sp. nov. TYPE: China. Guizhou: Tsingchen (Qingzhen), Mou-po, 16 Apr. 1936, *S. W. Teng 90155* (holotype, A 00280084!; isotypes, IBK IBK00012848 image!, IBSC 0092096 image!, L L0831786 image!, MO 2205171 image!).

Diagnosis. *Berberis tengii* has similarities with *B. gagnepainii* from Sichuan, but differs from it, inter alia, in the color of its mature stems, its shorter leaves, and a different flower structure with a smaller number of ovules. The leaves of *B. tengii* can be somewhat similar to those of *B. bicolor* also found in Guizhou, but the teeth of the margins are more conspicuous.

Shrubs, evergreen, to 1.5 m tall; mature stems pale brownish yellow, terete or subterete; spines 3-fid, semi-

concolorous, 0.8–3 cm, stout, abaxially flat or sulcate. Petiole almost absent or to 3 mm; leaf blade abaxially green, adaxially dark green, lanceolate or narrowly lanceolate, 2.5–6(–8.5) × 0.5–2 cm, leathery, midvein raised abaxially, slightly impressed adaxially, lateral veins and reticulation slightly raised and conspicuous abaxially, inconspicuous adaxially, base narrowly attenuate, margin spinose with 9 to 16(to 24) conspicuous teeth on each side to 2 mm, apex narrowly acuminate, mucronate. Inflorescence a fascicle, (1- or)2- to 6(to 8)-flowered; pedicel 20–25(–32) mm, yellowish brown. Flowers yellow. Sepals in 3 whorls; outer sepals triangular-ovate, 2.5–3.5 × 2–3 mm; median sepals ovate, 3.5–4.5 × 3–4.5 mm; inner sepals obovate-elliptic, 4.5–6.5 × 3.5–4.5 mm, apex rounded; petals obovate, 3.5–5 × 3.5–4.5 mm, base not clawed, glands close together, apex slightly emarginate. Stamens ca. 3.5 mm; anther connective slightly extended, truncate. Ovules 2 or 3. Berry dark blue, epruinose, oblong-ovoid, ca. 8 × ca. 5 mm; style persistent; seeds (1 or)2, very dark brown.

Phenology. *Berberis tengii* has been collected in flower in April and in fruit in July and August.

Distribution and habitat. *Berberis tengii* is known from north Guizhou and northwest Hunan. It has been collected in sparse woods, dry areas, and mountain sides at 450–1950 m.

Etymology. *Berberis tengii* is named after the collector of the type, Shi Wei Teng (Deng in pinyin), who worked for the Institute of Agricultural and Forestry Botany of Sun Yat-Sen University and who died during this particular collecting expedition (see Z. G. Hu et al., 2015: 244).

IUCN Red List category. *Berberis tengii* is assessed as DD or Data Deficient, according to IUCN (2001) criteria.

As noted above, *Berberis tengii* has similarities with *B. gagnepainii* endemic to central and west Sichuan, and most of the specimens cited here were initially identified as such. These identifications may have been encouraged by the mistaken belief that *B. gagnepainii* is found in Hubei (for details as to why this is not so, see under that species).

There appears to be some uncertainty as to where *Y. Xiao & B. Zhao LS-2231* from northwest Hunan, cited below, was collected in that the only Bamian Shan located is not in Longshan Xian but in the neighboring Fenghuang Xian.

Selected specimens.
Guizhou. Bijie Xian: Baohe Xiang, 1300 m, 20 Aug. 1957, *P. H. Yu 439* (IBSC 0091560–61, KUN 0175749–50,

PE 01031262, WUK 0204240); Baohe Xiang, Hangxiaohe, 1350 m, 23 Aug. 1957, *P. H. Yu 540* (IBSC 0092722, 0092294, KUN 0175768–69, WUK 0195605). **Dafang Xian:** Baina Qu, Jiulongshan, 1950 m, 16 Aug. 1959, *Bijie Team 875* (GF 09001097, GZTM 0011930, HGAS 013661, KUN 0175746, PE 01031271–72); Bali Dujuan, 16 Apr. 2005, *J. X. Deng 4150* (GF 09003496). **Daozhen Xian:** Dashahe, 4 Aug 2003, *Q. W. Sun & S. H. Wei 030804039* (GZTM 0011347). **Fenggang Xian:** Mahuang Gou, 14 Apr. 1991, *T. L. Zhang 03* (GZTM 0011556). **Hezhang Xian:** Shuitang Lincheng, 11 May 1984, *R. B. Jiang 548* (GF 0091563, 09001100); Qingshan Qu, Goutou, 1300 m, 10 June 1986, *P. B. He 40* (GZTM 0011339, 0011345). **Nayong Xian:** Juren Qu, Luzui, 1750 m, 29 July 1959, *Bijie Team 582* (GF 09016507, GZAC 09016507, HGAS 013663, KUN 0175747, PE 01031267–68); Juren Qu, 1850 m, 16 Aug. 1959, *Bijie Team 765* (HGAS 013662, KUN 0175748, PE 01031265–66). **Shiqian Xian:** Zhongba, Fuyan, 9 Apr. 1991, *T. L. Zhang 01* (GZTM 0011538–42). **Songtao Xian:** Tianma Si, 1300 m, 4 June 1988, *Wuling Shan, Botany Team 47* (KUN 0179091, PE 01376119). **Xishui Xian:** Wangxiantai, 21 May 2014, *O. Li XS14052376* (ZY 0002206). **Yinjiang Xian:** 600 m, 26 Dec. 1930, *Y. Tsiang 7886* (IBSC 0092097, NY); Zhangjiaba, 950 m, 7 July 1988, *Wuling Shan Botany Team 2062* (KUN 0179088, PE 01376120). **Zunyi Shi:** Jinding Shan, 1400 m, 8 Aug. 1956, *Sichuan-Guizhou Team 1145* (PE 01031243, 01031261, SZ 00294206).

NW Hunan. Longshan Xian: Bamian Shan, 27 July 2013, *Y. Xiao & B. Zhao LS-2231* (CSH CSH0003875). **Yongshun Xian:** Songbai Xiang, Sanping Cun, 28.939167°N, 110.114167°E, 956 m, 13 Sep. 2015, *K. D. Lei 331271509131641* (JIU 12253).

76. Berberis triacanthophora Fedde, Bot. Jahrb. Syst. 36 (Beibl. 82): 43. 1905. TYPE: China. Chongqing (E Sichuan): Wushan, [1885–1888], *A. Henry 5681* (lectotype, designated by Ahrendt [1961: 51], K K000077341!; isolectotypes, B† [photo A!], BM BM000895069!, DBN image!, E E00259010!, G [2 sheets] G00343571! LE!, P P00716554!, TI 02631 image!, US 00103912 image!).

Shrubs, evergreen, to 2 m tall; mature stems reddish brown, terete, scarcely verruculose; spines 3-fid, pale yellowish brown, 1–2.5 cm. Petiole almost absent; leaf blade abaxially gray-green, densely papillose, adaxially dark green, shiny, linear-lanceolate, oblong-lanceolate, or narrowly elliptic, 2–4.5(–6) × 0.25–0.5 cm, leathery, midvein raised abaxially, impressed adaxially, lateral veins and reticulation obscure abaxially, indistinct adaxially, base cuneate, margin spinulose with 2 to 6 teeth on each side, occasionally entire, apex acuminate or acute, mucronate. Inflorescence a fascicle, 2- to 6-flowered; pedicel 15–30 mm; bracteoles ovate, 1 × 0.8 mm. Sepals in 3 whorls, salmon pink; outer sepals ovate-orbicular, ca. 2 × 1.8 mm; median sepals ovate, ca. 3.5 × 2.5 mm, apex acute; inner sepals obovate, ca. 5 × 4 mm, apex obtuse; petals very pale whitish yellow, obovate, ca. 3.75 × 2.5 mm, base cuneate, glands sep-

arate, oblong, apex emarginate. Stamens ca. 2 mm; anther connective extended, truncate. Ovules 2 or 3. Berry blue-black, slightly pruinose, ellipsoid, 6–8 × 4–5 mm; style not persistent.

Phenology. *Berberis triacanthophora* has been collected in flower in May and June and in fruit between July and October.

Distribution and habitat. *Berberis triacanthophora* is known from north Guizhou, west and southwest Hubei, north Hunan, south Shaanxi, north Sichuan, and Chongqing (east, northeast, and southeast Sichuan). It has been found in forest areas at ca. 400–2300 m.

Though no herbarium for the type was cited in the protologue, Fedde was based at Berlin when he published *Berberis triacanthophora*. Proof that a specimen of *Henry 5681* was there comes from a photograph at A taken by Rehder sometime in the interwar period. The specimen would have been destroyed in WWII. The sheet is annotated "Typus" by Schneider with the date "3 X 07." Ahrendt (1961: 51) citing the specimen at K was an effective lectotypification.

Berberis triacanthophora is one of four species in section *Wallichianae* that are recorded as having flowers that are not wholly or almost wholly yellow. The others are *B. bicolor*, *B. incrassata*, and *B. sanguinea*. The salmon-pink sepals were not reported in the protologue, but were noted by Marchant (1937: 27) and by Schneider (1942: 34), although not by Ahrendt (1961: 51) or by Ying (2001: 115).

Selected specimens.
Chongqing (E, NE & SE Sichuan). Chengkou Xian: Caimeng Qu, Yanmai Xiang, 2000 m, 9 Oct. 1958, *T. L. Dai 106440* (KUN 0177230, NAS NAS00070727, PE 01031162, 01031177). **Fengjie Xian:** Herui Gongshe, Huangjin Dadui, 1400 m, 6 May 1964, *H. F. Zhou & H. Y. Su 107974* (IBSC 0092631, 0092667, NAS NAS00314264, PE 01031164). **Nanchuan Xian:** Jinfo Shan, above Xiao He, 1650 m, 13 May 1957, *J. H. Xiong & Z. L. Zhou 90798* (HIB 0129600, IBSC 0092088, KUN 0177228, PE 01037174). **Shizhu Xian:** Sanhui Gongshe, Lengshui Xi, 1550 m, 20 July 1978, *Y. Chen 2848* (CDBI CDBI0027732, CDBI0027736–37). **Wushan Xian:** Pingzhuxian Gongshe, Hong Shui Gou, 1250 m, 25 Aug. 1964, *H. F. Zhou & H. Y. Su 110289* (IBSC 0092604, PE 01031181). **Wuxi Xian:** Bailu Qu, Longquan Gongshe, Xiaojian Shan 1850 m, 12 June 1960, *Sichuan Economic Plants Exp. 00553* (CDBI CDBI0027458). **Zhong Xian:** Wuyang Qu, Shizi Gongshe, Niutou Shan, 1570 m, 15 May 1959, *G. L. Wang 1368* (KUN 0177240, SZ 00289844).

N Guizhou. Tongzi Xian: Heiwan, 1450 m, 23 Apr. 1989, *Y. K. Li 11190* (HGAS 13760).

W & SW Hubei. [1885–1888], *A. Henry 5861A* (GH 00280070); Changyang, [1885–1888], *A. Henry 5681B* (K K000077342, P P00716555); Patung (Badong), June 1900, *E. H. Wilson 952* (A 00280077, DBN, E E00612610, K [4 sheets], NY, P P06868538, W 1907–8601). **Badong Xian:** (area now part of Shennongjia Lin Qu), Xiagu Gongshe, Banqiao, 1100 m, 27 July 1977, *Hubei Shennongjia 24157* (HIB 00129583–84, PE 01031234). **Baokang Xian:** Nan Xiyan, Xiongjia Da Gou, 19 Apr. 1988, *Z. Y. Chen 8820* (CCAU 0002137). **Enshi Shi:** 400–425 m, Nov. 1958, *H. J. Li 8768* (HIB 0129596–97, IBSC 0092085, NAS NAS00314270, PE 01031233, WUK 0199685). **Hefeng Xian:** Huping Linchang, 1250–1350 m, 25 Aug. 1958, *H. J. Li 5818* (HIB 0129596–97, IBSC 0092086, KUN 0177237, PE 01031235, SZ 00289808, WUK 0199463). **Jianshi Xian:** Dangyang, 24 Sep. 1951, *L. Y. Dai & Z. H. Qian 1392* (PE 01031236, WH 06008559). **Lichuan Shi:** Moudao Qu, near Mahe Xiang, 25 Sep. 1957, *G. D. Fu & Z. S. Zhang 1735* (FUS 0014201, HIB 0129590, KUN 0178383, LBG 00064148, NAS NAS00314267, NAS00314269, PE 01031238–39, 01037172, SWCTU 00014201). **Shennongjia Lin Qu:** Laojunshan, Chongli Gou, 1800 m, 3 July 1976, *Hubei Shennongjia Botany Investigation Team 30953* (HIB 0129568–69, PE 01031237). **Wufeng Xian:** Liziping, 1550 m, 14 May 2008, *Hou He Research Institute 080514068* (HHE). **Xingshan Xian:** (area now part of Shennongjia Lin Qu), Laojunshan, 1450 m, 1 June 1957, *Y. Liu 659* (HIB 0129571, KUN 0177229, LBG 00064150–51, PE 01031230–32, WUK 0234162); along the trail from Qiujiaping toward Lao Jun Mtn., 31.5°N, 110.5°E, 2100–2300 m, 3 Sep. 1980, *Sino-American Bot. Exp. 643* (A 00280072, HIB 0129570, KUN 0177236, NA 0016557, NAS NAS00314268, PE 00996103). **Xuan'en Xian:** Maoshan Xiang, Yangquxi, 1000 m, 21 July 1988, *Y. M. Wang 5519* (HIB 0129599, PE 01376137).

N Hunan. Sangzhi Xian: Xiaoyundongfu Ping, Guangdongkui, 26 May 1984, *Sangzhi Xian Forestry Institute 0694* (KUN 0178220, PE 01376139); Bamaoxi Gongshe, Tianping Shan, 1650 m, 10 July 1975, *B. G. Li 720216* (PE 01031229).

S Shaanxi. Lan'gao Xian: Zhanghe Xiang-Kuihuashe Xiang, 1350 m, 29 Oct. 1957, *Z. P. Wei & S. H. Luo 582* (WUK 0097666); Taohe Gongshe, Qingping, 1700 m, 20 July 1959, *P. Y. Li 7644* (KUN 0177235, 0177238, WUK 0136931).

N Sichuan. Wanyuan Xian: Hua'e Shan, 1850 m, 3 June 1959, *X. Z. Wang 2271* (FUS 00014267, KUN 0177241, SZ 00289843).

77. Berberis uniflora F. N. Wei & Y. G. Wei, Guihaia 15: 218. 1995. TYPE: China. N Guangxi: Huanjiang Xian, Mulun Xiang, Xiayi Cun, 850 m, 27 Oct. 1991, *YGG Expedition 70273* (holotype, IBK IBK00190345 image!).

Shrubs, evergreen, to 1.7 m tall; mature stems purple turning reddish brown, subterete; spines 1- or 2-fid, pale brownish yellow, 0.3–0.7 cm, largely absent. Petiole almost absent; leaf blade abaxially pale green, adaxially green, narrowly elliptic, oblanceolate, obovate, or narrowly obovate, 3–7 × 1–2 cm, leathery, midvein raised abaxially, impressed adaxially, lateral veins and reticulation indistinct abaxially, conspicuous adaxially, base narrowly attenuate, margin sometimes revolute, entire or spinulose with 1 to 3 inconspicuous teeth on each side, apex acute, subacute, rarely obtuse, mucronate. Inflorescence a fascicle, 4- to 30-flowered; pedicels ca. 25 mm; bracteoles triangular-ovate, ca. 2 × 1 mm. Sepals in 3 whorls; outer sepals ovate, ca. 3 × ca. 2 mm; median sepals ovate, 4.5–5 × ca. 3 mm, apex sometimes concave; inner sepals oblong-ovate or elliptic, 6–6.5 ×

ca. 4 mm, apex sometimes concave. Petals oblong-ovate or narrowly oblong-ovate, ca. 5.5 × 2–2.5 mm, base subobtuse, glands separate, ovate, ca. 0.75 mm, apex narrowly incised. Stamens ca. 3 mm; anther connective slightly extended, truncate. Ovules 2. Immature berry dark red, oblong, 10 × 6 mm; style persistent.

Phenology. *Berberis uniflora* has been collected in flower in April and May and in fruit between April and November.

Distribution and habitat. *Berberis uniflora* is known from Huanjiang Xian in north Guangxi and Libo Xian in south Guizhou. It has been found in forest understories on rocky limestone slopes at 640–1000 m.

Berberis uniflora has had an unusual history. The protologue, which cited only the type and *YGG Expedition 70271* (details below), stated the species could be easily distinguished by its single-flowered inflorescence, the evidence for this being the single pedicel with fruit on the type (no *70271* is sterile). F. L. Chen et al. (2012: 605–607) re-examined the type and found residual evidence of fascicles. They also cited a more recent collection from the type area, *L. Wu et al. ML 0101*, a specimen with fascicled berries. Noting the similarity between the leaves of the type and this new specimen with those of *B. ziyunensis*, which had been published subsequently, they concluded that *B. ziyunensis* should be treated as a synonym of *B. uniflora*. However, this synonymy was based solely on the apparent similarity of the leaves of the two taxa, the authors giving no information about the color of the mature stems of *L. Wu et al. ML 0101* (the type of *B. ziyunensis* is a distinctive blackish purple) nor the number of seeds per fruit. On inquiry to one of the authors of the synonymy, Y. F. Deng of IBSC, I was kindly provided with an image of *ML 0101* as well as images of *ML 2013* and *y0169* cited below. They are all the same species. In addition, Deng (pers. comm. 23 Apr. 2013) informed me that fruit from *0101* contained two seeds. However, this was insufficient to prove the synonymy since *B. ziyunensis* has three ovules and the mature stems of *ML 2013* are purple turning reddish brown. To prove the synonymy, an account of the flowers of *B. uniflora* was kindly provided to me by S. Z. He of GZTM on the basis of *Y. Wang 4041001* (details below) collected some 40 km north from where the type of *B. uniflora* was collected. This showed that the flower structure of *B. uniflora* was significantly different from that of *B. ziyunensis*. The account of the flowers given above is thus that of S. Z. He. It is regrettable that *B. uniflora* has such an inappropriate name.

As F. L. Chen et al. noted, *Berberis uniflora* was not noticed by Ying (2001, 2011).

Selected specimens.

S Guizhou. Libo Xian: Shui Yao, 850–880 m, 14 Nov. 1981, *Y. K. Li 9370* (HGAS 013684, IBSC 0091972); Shui Yao, 970 m, 5 Sep. 1982, *Y. K. Li 9951* (HGAS 013683); Yongkang, near school, 640 m, 16 May 1983, *Y. K. Li 11449* (HGAS 013685); Maolan, 24 June 1983, *C. H. Yang 4259* (GF 09001035–36); Weng'ang Gongshe, Mogan Zhai, Yipian, 800–1000 m, 13 Apr. 1984, *Y. K. Li L1350* (IBSC 0091973); 1000 m, 10 May 1991, *S. Z. He & T. L. Zhang 92510* (GZTM 0011465–68); Jialang Xiang, 1000 m, 10 Apr. 2014, *Y. Wang 4041001* (GZTM); 24 July 2008, *D. G. Zhang 080724028* (JIU 30268).

N Guangxi. Huanjiang Xian: Mulun Nature Reserve, Mulun Xiang, Xiayi Cun, 850 m, 27 Oct. 1991, *YGG Expedition 70271*(IBK); Donglai lookout point, 700 m, 15 Jan. 2011, *W. B Xu & L. Wu 11009* (IBK IBK00309547); Hongdong, 880 m, 1 May 2011, *Y. S. Huang et al. y0169* (IBK IBK00309548); Mingli Kairong, 700 m, 9 Aug. 2011, *L. Wu et al. ML 0101* (IBK IBK00309554); on rd. to Gaochao, 900 m, 18 Apr. 2012, *Y. S. Huang et al. Y1245* (IBK IBK00309551); Minglitun, Lanhua Shan, 700 m, 19 Apr. 2012, *R. H. Hong et al. 11339* (IBK IBK00309550); Tonglai, lookout point, 910 m, 25 July 2012, *R. Peng et al. ML 2013* (IBK IBK00309260).

78. Berberis veitchii C. K. Schneid., Pl. Wilson. (Sargent) 1: 363. 1913. TYPE: China. W Hubei: s. loc., 1525 m, June 1900, *E. H. Wilson (Veitch) 1138* "A" (lectotype, designated here, K K000077337!; isolectotypes, A [2 sheets] 00038809!, 00038810 specimen with flowers!, K [2 sheets] 000077338–39!, NY 01365274!, P P00716544!, W 1907-0008583!).

Berberis gagnepainii C. K. Schneid. var. *lanceifolia* Ahrendt, J. Bot. 79(Suppl.): 39. 1941. TYPE: Cultivated. Coombe Wood, U.K., 16 May 1907, *s. coll. s.n.*, from seeds of [*E. H. Wilson*] *W1503* collected in W. Hubei, "Grassy mountains," 1525–1830 m, 1900 (holotype, K K001044703!).

Shrubs, evergreen, to 1.5 m tall; mature stems yellow, terete, verruculose; spines 3-fid, pale yellow, 1.5–3 cm, abaxially sulcate. Petiole almost absent; leaf blade abaxially pale yellow-green, shiny, adaxially dull gray-green, lanceolate, sometimes narrowly elliptic, 5–15 × 1–2 cm, thinly leathery, midvein raised abaxially, impressed adaxially, lateral veins slightly raised and inconspicuous abaxially, inconspicuous adaxially, reticulation indistinct on both surfaces, base cuneate, margin slightly undulate, slightly revolute, spinose with 10 to 24 widely separated, coarse teeth on each side, apex acuminate, mucronate. Inflorescence a fascicle, (2- to)4- to 8(to 12)-flowered; bracts ovate-triangular, red; pedicel 14–35 mm; bracteoles triangular-ovate, ca. 2 × 2 mm. Sepals in 2 whorls, pale pinkish with whitish margins; outer sepals obovate-rhomboid, ca. 5 × 4 mm; inner sepals widely ovate-orbicular, ca. 4–5 × 4 mm; petals whitish, obovate, 1.5 × 0.8 mm, base clawed, glands very close together, apex rounded and narrowly incised. Stamens ca. 3 mm; anther connective slightly extended, obtuse. Ovules 4 or 5. Berry blue to dark blue, blue pruinose, ovoid to ellipsoid, ca. 9 × 6 mm; style not persistent.

Phenology. *Berberis veitchii* has been collected in flower from April to June and in fruit between July and October.

Distribution and habitat. *Berberis veitchii* has been collected in northeast Chongqing (Sichuan) and western Hubei. It has been collected in forests and forest margins at 1360–1700 m.

Schneider's protologue gave the type of *Berberis veitchii* as *Wilson (Veitch) 1138* and the date of the gathering as June 1900. However, there are two different gatherings of this number, one with flowers and one with fruit (clearly collected later in the year). The gathering with flowers is designated here as *Wilson (Veitch) 1138* "A" and that with fruit as *Wilson (Veitch) 1138* "B." There is a fragment of *Wilson 1138* at LE, but it is unclear whether this is from *1138* "A" or *1138* "B." As Schneider noted in the protologue, he had first (1908: 197) identified "Wilson Nr. 1138, W. Hupeh" as an example of *B. acuminata*. The specimen of *Wilson 1138* "A," K000077337 at Kew, has been chosen as the lectotype because it has annotation by Schneider dated 26 April 1906 identifying it as *B. acuminata*.

There has been considerable confusion in relation to the provenance of the type of *Berberis gagnepainii* var. *lanceifolia*. The protologue stated "syn *B. acuminata* Stapf (Bot Mag, t. 8185) non Franch." with the collection details "W. Hupeh . . . on grassy mountains, 5000–6000 ft, May 16th 1907, *Wilson 1503* (Typus in Herb Kew)." These details were repeated by Ahrendt (1961: 53) and by Chamberlain and Hu (1985: 535), although the latter reported that they were unable to locate the specimen at K. In his article about *B. acuminata*, Stapf (1908) made it clear that he was describing not a wild-collected specimen, but a cultivated one from a plant grown from seed collected by Wilson in western Hubei for Veitch's Nursery (although he gave no collector's number for the seed). That it was a cultivated specimen is confirmed by the collection details on the Kew sheet and annotated as "Wilson *1503* (cult.) *B. gagnepainii* Schneid. var. *lanceifolia*" by Ahrendt on 12 April 1941. These details do not indicate the date the seeds were collected, but this can be established as 1900 from Veitch (1906: 391), which reports as *B. acuminata* the plants that were subsequently described as such by Stapf. The collector's number on the Kew sheet is given not as *Wilson 1503*, as Ahrendt reported, but as *W1503* (*Wilson 1503* collected in Jiangsu in 1907 is *Acer buergerianum* [see A 00232025]). Unfortunately, number *1503* in the "Numerical list of seeds collected

1899–1901" (Arnold Arboretum Wilson Papers, Series WII, Box 4, Folder 3) is blank, but it is likely the seeds were from *Wilson (Veitch) 1138* "B" (there is no *1138* in Wilson's seed list). It should be noted that, in misidentifying as *B. gagnepainii* the species reported by Stapf as *B. acuminata*, Ahrendt was following Schneider (1918: 23) who made the same mistake and who repeated it (Schneider, 1942: 52) even though five pages previously (1942: 47) he determined the *B. acuminata* reported by Veitch in 1906 to be *B. veitchii*. Fu (2001: 391) synonymized *B. gagnepainii* var. *lanceifolia* under *B. veitchii*, though he presented no evidence to justify this synonymy.

It should also noted that, confusingly, the *Berberis gagnepainii* var. *lanceifolia* f. *pluriflora* (Ahrendt, 1961: 53) from Baoxing Xian in west-central Sichuan is *B. gagnepainii* and not *B. veitchii* (for further details, see under that species).

Berberis veitchii has upper leaves that are somewhat similar not just to those of *B. gagnepainii* but also *B. tengii* from Guizhou, but has coarser marginal spines. This is even more evident in larger leaves lower down the branches which, as Schneider (1918: 137) noted, are somewhat similar to the leaves of *B. insignis*, though they are much narrower. Schneider's protologue states the number of ovules to be two, rarely three, subsequently (1942: 47) revising this to three or four. However, flowers of the type specimens at Kew have four or five ovules. The color of the flowers used above is from a photograph of *Hou He Research Institute 080428070* on the Hou He National Nature Reserve website accessed on 1 November 2016 via http://www.nsii.org.cn/. For a further species from Sichuan with somewhat similar leaves to *B. veitchii*, see *B. ebianensis*.

Wilson 1138 is simply described as being from west Hubei. However, there appears to be no information as to exactly where Wilson was in June 1900. There is a similar difficulty with identifying where *Henry 703*, cited below, was collected because, as noted under *Berberis brevipaniculata*, Patung was a general term used by Henry for Changyang and Badong districts and the mountainous area southwest of Yichang.

Berberis veitchii was reported from Guizhou by He and Chen (2004: 37), but I have found no evidence to support this. Schneider (1942: 47) cited *Y. Tsiang 5101*, north Guizhou: Tungtze (Tongzi Xian), 450 m, 26 Apr. 1930, at B, while noting that an unripe fruit of this had only two ovules. This specimen was destroyed in WWII, but may be the same as, or similar to *Kweichow Exp. 5101B* (IBSC 0092098), Guizhou, 1930. This is not *B. veitchii* and is possibly *B. bicolor*.

The report of *Berberis veitchii* in Baoxing Xian in west Sichuan by X. H. Li et al. (2017) is likely to be based on misidentifications of *B. gagnepainii*.

Seeds from *Wilson W1530* are likely to be the source of cultivated plants in the U.K. referred to by Ahrendt (1961: 51), though at least some of these were likely to have been hybrids.

Selected specimens.
NE Chongqing (Sichuan). Wuxi Xian: Banxi Xiang, Changcao, 1500 m, 8 Sep. 1958, *G. H. Yang 65328* (CDBI CDBI0028221, FUS 00014313, IBSC 0091475, KUN 0178389, PE 01030725); Shuangyang Xiang, Baiguo Cun, 1700 m, 13 July 2004, *Y. S. Chen et al. 1591* (WUK 0489042–45); Baiguo Linchang, Qingshui Qiao, Longdong Gou, 1700 m, 15 Sep. 2007, *Three Gorges Plant Investigation Team 0258* (PE 01788887–8).
W Hubei. s. loc., specimen with fruit, s.d. [1900], *E. H. Wilson (Veitch) 1138* "B" (A 00038810, K K000077338); "Ichang: Patung [Badong]" Distr., before Mar. 1886, *A. Henry 703* (K). **Hefeng Xian:** Sep. 1958, *H. J. Li 6729* (HIB 0129206–8, IBSC 0092095, KUN 0177049, PE 01031725, SZ 00289800, WUK 0198331). **Wufeng Xian:** Gaojia'ao, 1100 m, 9 Sep. 1990, *F. S. Peng 0250* (HIB 00129222); Beifengwu Linchang, 1300 m, 5 Oct. 1990, *F. S. Peng 4986* (HIB 00129205); Hupingshan, 1100 m, 24 Apr. 2008, *Hou He Research Institute 080428070* (HHE); Jiepai, Wantan, 1360 m, 30 Mar. 2014, *M. Ogisu 656* (E). **Zhuxi Xian:** Yunwuxi, 1680 m, 29 Sep. 1991, *Z. Zheng 323* (KUN 0178284) and *333* (KUN 0178289).

79. Berberis verruculosa Hemsl. & E. H. Wilson, Bull. Misc. Inform. Kew 1906: 151. 1906. TYPE: China. W Sichuan: Tatsien lu (Kangding), July 1903, *E. H. Wilson (Veitch) 3150* (lectotype, inadvertently designated by Chamberlain & Hu [1985: 536], K K000077346!; isolectotypes, A 00038813!, BM BM001015571!, HBG HBG-506683 image!, P P00716528!).

Shrubs, evergreen, to 1 m tall; mature stems brownish yellow, terete, densely verruculose; spines 3-fid, pale yellow, 1–2 cm, abaxially subsulcate. Petiole almost absent; leaf blade abaxially grayish green, pruinose, sometimes becoming epruinose on old leaves, adaxially dark green, shiny, elliptic, ovate-elliptic, or rhomboid-elliptic, 1–2 × 0.6–1.1 cm, leathery, midvein raised abaxially, impressed adaxially, lateral veins in 3 or 4 pairs, indistinct abaxially, inconspicuous adaxially, reticulation indistinct abaxially, inconspicuous adaxially, base cuneate, margin slightly revolute, spinose with 2 to 4 widely spaced teeth on each side, apex acute, mucronate. Inflorescence 1-flowered; pedicel 4–10 mm. Sepals in 3 whorls; outer sepals ovate, ca. 4 × 3 mm; median sepals ovate, ca. 6 × 5 mm; inner sepals obovate, ca. 10 × 8 mm; petals elliptic or obovate, 5.5–6 × ca. 3 mm, base cuneate, glands separate, apex emarginate or retuse with rounded lobes. Stamens ca. 3.5 mm; anther connective slightly extended, rounded. Ovules 4 to 6. Berry purplish blue to black, pruinose, oblong-ovoid, 10–12 × 6–7 mm; style not persistent.

Phenology. *Berberis verruculosa* has been collected in flower in May and June and in fruit between July and November.

Distribution and habitat. *Berberis verruculosa* is known from west, central, and north Sichuan and south Gansu. It has been collected in semi-open, thicketed and forested areas on mountain slopes, particularly in rocky places at ca. 1900–3400 m.

The protologue cited *Wilson (Veitch) 3150* and *3150a*. Subsequently, Schneider (1918: 22) declared no. *3150* to be the type, but did not indicate the herbarium. Chamberlain and Hu (1985: 536) mistakenly reported the specimen at Kew to be the holotype. Under Art. 9.10 of the Shenzhen Code (Turland et al., 2018), this can be treated as inadvertently completing the lectotypification.

I have found no evidence to support the report by Ying (2001: 88, 2011: 729) of *Berberis verruculosa* being found in Yunnan.

Selected specimens.
Syntype. *Wilson (Veitch) 3150a*. W Sichuan: same details as lectotype, but with fruit, s.d. (A [2 sheets] 00280118–19, BM BM001015572, HBG HBG-506684, K [2 sheets] K000077345, K000077347, P P00716529).

Other specimens.
S Gansu. Hui Xian: Yanpingtou Tan, Dayan Gou, 1910 m, 12 Aug. 1965, *X. Z. Peng 5607* (LZU 00076357, 00076638). **Wen Xian:** Baishuijiang Qiujiaba, 2400 m, 25 June 2006, *Bailong Jiang Exp. 0236* (PE 01359676); Motianling Shan, Baishui Jiang Nature Reserve, 2000–2230 m, 16 May 2007, *D. E. Boufford & Y. Jia 37670* (GH 00288634, PE 01814637). **Wudu Xian:** Luotang, Maya, 2750 m, 11 May 1959, *Z. Y. Zhang 1872* (LBG 00064153–54, WUK 0160910). **Zhouqu Xian:** Tieba, Wen Xian Gou, 2300 m, 24 May 1990, *Bailong Jiang Exp. 1284* (PE 01556126).
WC & W Sichuan. An Xian: Hung-si, 6 Sep. 1952, *W. G. Hu 13386* (SZ 00289784); Chaping Gongshe, Qianfo Shan, 2500 m, 16 July 1973, *An Xian Team 236* (SM SM704800422). **Baoxing Xian:** Jiandao Shan Gou, 2400 m, 10 June 1958, *X. S. Zhang 5273* (CDBI CDBI0027781, GF 09001033, PE 01037870). **Emei Xian:** Mt. Omei, 3000–3200 m, Sep. 1937, *Y. S. Liu 1517* (A 00280114, LBG 00064155, NA 0082468). **Jiulong Xian:** Hongba, 19 June 1939, *K. Y. Yao 3827* (NAS NAS00314295–96, PE 01037887). **Kangding Xian:** Ta-tsien-lou, 1893, *J. A. Soulié 176* (G, P P02682324, 02682333–34); Tachienlu, 12 Nov. 1922, *H. Smith 4938* (A 00280116, PE 01037879, UPS BOT:V-041026); S outskirts of Kangding, 2700 m, 28 Apr. 1990, *J. S. Yang 91-046* (PE 01037899). **Li Xian:** Suoluo Gou, Guaipeng Gou, 2700 m, 13 Oct. 1956, *X. Li 46848* (IBSC 0092126, PE 01037895, SZ 00287757). **Luding Xian:** Reshui Gou, 2300 m, 2 June 1993, *Botany Group 30878* (CDBI CDBI0027805–06). **Mao Xian:** Xijiao Xishan, 2200 m, 13 June 1958, *Mao Xian Team: S. Y. Chen et al. 5214* (NAS NAS00314300, SM SM704800416, SM704800520. **Mianning Xian:** Yele Gongshe, Shihuiyao, 2900 m, 3 July 1978, *Mianning Team 0275* (SM SM704800292). **Tianquan Xian:** Erlang Shan, 3400 m, 13 June 1953, *L. X. Jiang & H. J. Xiong 34413* (IBK IBK00012749, IBSC 0092123, PE 01037872).

Wenchuan Xian: "Wen chuan hsien," Schu ling kou, 2400 m, 12 May 1914, *H. W. Limpricht 1440* (WU 0039272); "W. Wenchuan Hsien," May 1930, *F. T. Wang 20992* (KUN 0177277, NAS NAS00314298, PE 01037898, WUK 0045138); 2100 m, 2 May 1959, *Maowen Group 2012* (CDBI CDBI0027816).
N Sichuan. Nanping (now Jiuzhaigou) Xian: Guoyuan Gongshe, Xinchiping 2300 m, 6 June 1979, *Nanping Team 0387* (SM SM704800418). **Pingwu Xian:** Wangba Chu, 2100 m, 23 May 1961, *X. N. Tang et al. 00237* (CDBI CDBI0027793). **Qingchuan Xian:** Qingxi Xiang, 14 May 1959, *s. coll. 3134* (KUN 0177278).

Cultivated material:
Living cultivated plants.
Howick Hall, Northumberland, U.K., from seeds of *SICH 62*, Baoxing Xian, Ming River Gorge (N of Yinxiu), 2 km N of Mianci, 1360 m, 12 Sep. 1988.

80. Berberis vietnamensis Harber, sp. nov. TYPE: Vietnam. [Lâm Đồng Province]: "Annam: massif du Bi-Doup, prov. du Haut Donaï. Près du signal," 2287 m, 12 Oct. 1940, *M. Poilane 30762* (holotype, P P02482743!; isotypes, L 0790124 image !, VNM not seen).

Diagnosis. *Berberis vietnamensis* has leaves that are somewhat similar in shape and size to *B. photiniifolia*, but are abaxially pruinose, and its mature stems are pale brownish yellow versus the reddish or purplish brown of *B. photiniifolia*.

Shrubs, evergreen, to 4 m tall; mature stems pale brownish yellow, angled or sulcate, sparsely or not verruculose; spines 3-fid, concolorous, 1–2.6 cm, abaxially flat or slightly sulcate. Petiole almost absent; leaf blade abaxially green, adaxially bronze-green, shiny, lanceolate, rarely narrowly elliptic, (2.3–)4.4–6.6 × 1–1.5 cm, leathery, abaxially frequently glaucous or grayish white pruinose, midvein distinctly raised abaxially, slightly impressed adaxially, lateral veins and reticulation slightly raised on both surfaces, inconspicuous abaxially, conspicuous adaxially, base attenuate, margin slightly revolute, entire or spinulose with 3 to 12 teeth on each side, apex acuminate, mucronate. Inflorescence a fascicle, 1- to 3-flowered, otherwise unknown; fruiting pedicel 10–15 mm. Berry black, ellipsoid or ovoid-ellipsoid, 6–7 × 4 mm; style persistent.

Phenology. *Berberis vietnamensis* has been collected in fruit in October. Its flowering season is unknown.

Distribution and habitat. *Berberis vietnamensis* is known from Bidoup in Lâm Đồng Province of south Vietnam. It has been collected in an unrecorded habitat at 2287 m.

IUCN Red List category. *Berberis vietnamensis* is assessed as DD or Data Deficient, according to IUCN (2001) criteria.

Following the reorganization of the Paris herbarium in 2011, duplicates of *Poilane 30862* were sent from P to L and VNM, and duplicates of *30861* (see below) to HN, L, and VNM. I have been unable to locate the ones at VNM.

Berberis vietnamensis is found some 1350-km distance from the nearest other area in Vietnam where *Berberis* is found and is the second most southerly section *Wallichianae* species, the most southerly being *B. xanthoxylon* Hassk. ex C. K. Schneid. from Indonesia. Given that Bidoup Nui Ba is now a national park, it is to be hoped the species is still extant.

Selected specimens.
S VIETNAM. [**Lâm Đồng Province**]: "Annam: massif du Bi-Doup, prov. du Haut Donaï. Près du signal géodésique," 2287 m, 13 Oct. 1940, *M. Poilane 30861* (HN, L 0790123, P P02482742, VNM not seen).

81. Berberis wuchuanensis Harber & S. Z. He, Bot. Mag. 29(2): 120. 2012. TYPE: China. N Guizhou: Wuchuan Xian, Jushishan, 800–1100 m, 3 May 2004, *S. Z. He 0405035* (holotype, GZTM!; isotype, GZTM!).

Shrubs, evergreen, to 2 m tall; mature stems dark reddish brown, terete; spines 3-fid, pale yellow, 0.3–1.5 cm, weak. Petiole almost absent or to 5 mm; leaf blade abaxially pale yellowish green, adaxially mid-green, slightly shiny, linear-lanceolate, (3.5–)6–10.5 × (0.2–)0.4–0.8 cm, thinly leathery, midvein raised abaxially, impressed adaxially, lateral veins obscure abaxially, inconspicuous adaxially, base attenuate, margin spinulose with 2 to 11 teeth on each side or entire, apex narrowly acuminate, mucronate. Inflorescence a fascicle, 2- to 18-flowered; pedicel yellow or pale pink, 10–14 mm; bracteoles triangular-ovate, ca. 1.5 × 1.2 mm, apex acute. Flowers pale yellow. Sepals in 2 whorls; outer sepals sometimes with partial pinkish blush, elliptic, ca. 3.5 × 2.5 mm, apex rounded; inner sepals broadly elliptic, ca. 5 × 3.5 mm, apex rounded; petals obovate, ca. 3.5 × 3 mm, base slightly clawed with approximate elliptic glands, apex convex. Stamens ca. 3 mm; anther connective truncate. Ovules (2 or)3. Immature berry dark blue, slightly pruinose, ellipsoid or oval, 4–6 × 3–4 mm; style not persistent.

Phenology. *Berberis wuchuanensis* has been collected in flower in May and with immature fruit in August.

Distribution and habitat. *Berberis wuchuanensis* is known from Dejiang, Wuchuan, and Zheng'an Xian in north Guizhou. It has been collected from thickets in scrub on mountain slopes at ca. 800–1100 m.

Information on the color of flowers comes from the holotype and from images taken by S. Z. He in the field.

The specimens at PE of *Guizhou Team 1723* and *s. coll. 71324* (the latter being sterile) were annotated by T. S. Ying as *Berberis insolita* and are presumably the source of that species being reported from Guizhou in Ying (2001: 107, 2011: 725) and probably also the source of the report by He and Chen (2004: 29).

Selected specimens.
Guizhou. Dejiang Xian: Yanmenkou, 1000 m, 15 Aug. 1959, *Guizhou Team 1723* (PE 01037178). **Yanhe Xian:** Mayang He Protected Area, 875 m, 16 Sep. 2007, *H. H. Zhang 173* (Mayang He Herbarium 1723–24). **Zheng'an Xian:** Luoyuan, Nov. 1971, *s. coll. 71324* (PE 01031092).

82. Berberis wui Harber, sp. nov. TYPE: Cultivated. Royal Botanic Garden Edinburgh, 6 June 2012, *P. Brownless 183*, from seed of *Alpine Garden Society China Expedition (ACE) 2521*, China. NW Yunnan: Dali Shi, Cangshan, above Dali television mast, 25.691833°N, 100.104167°E, 3300–3400 m, 25 Oct. 1994 (holotype, E E00615601!).

Diagnosis. *Berberis wui* is similar to *B. phanera* but with smaller, narrower, slightly revolute leaves, longer pedicels, and a different flower structure. It is somewhat similar to *B. davidii*, but is a taller plant with larger leaves, longer pedicels, and a different flower structure.

Shrubs, evergreen, to 2 m tall; mature stems very pale yellowish brown, subterete, not verruculose; spines 3-fid, concolorous, (1–)2–4.5 cm, stout, slightly retrorse on lower part of stems, adaxially sulcate. Petiole almost absent; leaf blade abaxially pale green, adaxially dark green, slightly shiny, narrowly elliptic to lanceolate-elliptic, 3–4.5 × 1–1.5 cm, leathery, midvein raised abaxially, impressed adaxially, lateral veins and reticulation obscure abaxially, slightly impressed and inconspicuous adaxially, conspicuous when dry, base cuneate or attenuate, margin slightly revolute, spinulose with 3 to 6 widely separated teeth on each side, apex acute, mucronate. Inflorescence a fascicle, 2- to 4(to 7)-flowered; pedicel 18–32 mm; bracteoles triangular, 2–3 × 1.5–2 mm, sometimes with reddish apex. Sepals in 2 whorls; outer sepals oblong-ovate, 4.5–5 × 3–4 mm; inner sepals broadly obovate, 7–8 × 5.5–6 mm; petals obovate, 5.5–6.5 × 4.5–5 mm, base slightly clawed, glands close together, oblong-ovate, ca. 1.5 mm, apex obtuse, entire. Stamens ca. 3.5 mm; anther connective not extended, truncate. Pistil 4.5 mm; ovules 2 to 4. Berry black, ellipsoid, 8 × 5 mm; style persistent.

Phenology. *Berberis wui* has been collected in flower in May and June and in fruit from July to October.

Distribution and habitat. *Berberis wui* is known from northwest Yunnan. It has been found on forest margins, in open *Abies delavayi* forest, on trailsides, and in thickets on ravine sides at ca. 2800–3490 m.

Etymology. *Berberis wui* is named after Cheng Yih Wu (pinyin Zheng Yi Wu) in recognition of his contribution to the study of *Berberis* in Yunnan.

IUCN Red List category. *Berberis wui* is assessed as DD or Data Deficient, according to IUCN (2001) criteria.

I first became aware of *Berberis wui* through a plant in my own collection grown from seeds gathered as *Alpine Garden Society China Expedition (ACE) 2521* which I have studied over many years. Although there are wild-collected specimens of *B. wui* (see below), a specimen from a cultivated plant of *ACE 2521* at E has been chosen as the holotype because it has flowers, which are more informative than fruit.

The specimens cited below have been mostly identified as *Berberis davidii* or *B. phanera*. The confusion with *B. phanera* probably had its origin with Ahrendt (1961: 54), who misreported the pedicels of that species as being up to 30 mm long, whereas their maximum length is ca. 15 mm long.

C. W. Wang 71708 (details below) was cited as a syntype of *Berberis taliensis* by Schneider (1939: 252). The specimen at KUN is annotated by C. Y. Wu as an unpublished *B. tsangshanica*.

Selected specimens.
NW Yunnan. Dali Shi: Cangshan, E flank, Zhonghe Si, 3 May 1929, *R. C. Qin 22825* (KUN 0175884, 0175893–94, PE 01031004); Ta-li Hsien (Dali Xian), 2800 m, Dec. 1935, *C. W. Wang 71708* (A 00279489, KUN 1204881, PE 01031808–09); Cangshan, by footpath from Longquan to Dali, 3100 m, 15 May 1981, *Sino-British Exp. Cangshan (1981) 0633* (A 00279810, E E00623062, KUN 00279810, 0176711); Wutaishan, above Huadianba, N end of Cangshan, 21 May 1981, 3100 m, *Sino-British Exp. Cangshan (1981) 950* (A 00279811, E E00612441, KUN 0176716–17); Diancang Shan, vic. of Huadianba herbal medicine farm, 25.88333333°N, 100.01666667°E, 2900–3300 m, 18 July 1984, *Sino-Amer., Bot. Exp. 1136* (A 0027942, B, E E0061244, KUN 0175902, US 00956862); Cangshan, above Dali television mast, 25.691833°N, 100.104167°E, 3300–3400 m, 25 Oct. 1994, *Alpine Garden Society China Expedition (ACE) 2521* (E E00039585!, LIV 2005.15.1077, WSY 0057764). **Dêqên (Deqin) Xian:** Cang Jiang, Yongzi Huoshan, 2900–3300 m, 8 Aug. 1940, *G. M. Feng 6412* (KUN 0177062–64, PE 01030979); side valley betw. Deqin & Mekong River, 28.466667°N, 98.8°E, 3100 m, 2 June 1993, *Kunming, Edinburgh, Gothenburg Exp. 685* (E E00623053). **Eryuan Xian:** Fang-yang-Tchong, Oct. 14 1887, *J. M. Delavay s.n.* (P P02313660); bois de Fang-yang tschang, 21 Oct. 1887, *J. M. Delavay s.n.* (P P02313659). **Lijiang Shi (now Yulong Xian):** Yulongshan, 3100 m, 4 Aug. 1981, *Beijing Hengduan Mountains Botany Team 02614* (PE 01030968–69); Daba en

route to Wenkou, 2950 m, 4 June 1985, *Q. Lin et al. 771737* (KUN 0178482–83); Yulongshan, above Wenghai, 3200 m, 5 June 1985, *Kunming Edinburgh Yulongshan Exped 561* (E E00623054, KUN 0176730). **Ninglang Xian:** Yangjiancao, Cuiyi Linchang, 3480 m, *W. F. Han & K. M. Deng 81-1122* (PE 01840113–14). **Wesi Xsien (Weixi Xian):** 3000 m, June 1935, *C. W. Wang 63774* (A 00279748, IBSC 0091919, KUN 0177060, LBG 00064134, NAS NAS00314205, PE 01030970–71, WUK 0037190); same details, but 3500 m, *C. W. Wang 63889* (A 00279749, LBG 00064135, PE 01030972–73, WUK 0040453); Badi Xiang, Najiluo He, 2600 m, 12 May 2004, *D. Z. Fu & H. Li 62* (TNM S137955). **Yangbi Xian:** Meiweng Gongshe, Xiao Malutang, 3100 m, 25 July 1963, *Northwest Yunnan, Jinsha River Team 4082* (KUN 0175949–50, PE 01030974); W side of Diancang Shan, vic. of Dapingzi. 25.71666667°N, 100.03333333°E, 3000 m, 19 June 1984, *Sino-Amer., Bot. Exp. 290* (A 00279417, E E00612446, KUN 0175898, L L.0831880, US 00945861); W side of Diancang Shan, vic. of Baiyunfeng Peak above Malutang, 25.76666667°N, 100.01666667°E, 3450 m, 26 June 1984, *Sino-Amer., Bot. Exp. 564* (A 00279418, E E00612447, KUN 0175899, US 00945860). **Zhongdian (now Xianggelela Xian):** Chungtien, Haba, 2800 m, 23 Nov. 1937, *T. T. Yu 14944* (A 00280061, BM, E E00316013, KUN 0176724, PE 01030983); Haba Shan, 27.374444°N, 100.097222°E, 3400 m, 16 June 1994, *Alpine Garden Society China Expedition (ACE) 314* (E E00198412).

Cultivated material:
Living cultivated plants.
Foster Clough, Mytholmroyd, West Yorkshire, U.K.; Royal Botanic Garden Edinburgh, 19943817A, both from seeds of *Alpine Garden Society China Expedition (ACE) 2521*. Also Royal Botanic Garden Edinburgh from *D. S. Paterson & J. D. Main 80*, Dali Shi, Cangshan ca. 3490 m, 11 Oct. 1994.

Cultivated specimens.
Royal Botanic Garden Edinburgh, 19943817A, from seeds of type, 6 June 2012, *P. Brownless 183* (E E00615601); 19 June 2012, *H. S. Cubey 350* (E E00615602).

83. Berberis wuliangshanensis C. Y. Wu ex S. Y. Bao, Bull. Bot. Res., Harbin 5(3): 14. 1985. TYPE: China. SW Yunnan: Jingdong Xian, Wuliangshan, Modao He, Jiaoxi Shan, 2400 m, 23 Oct. 1956, *B. Y. Qiu 53014* (holotype, KUN 0203031!; isotypes, KUN [2 sheets] 0177550!, 0203046!, IBSC 0091647 image!, PE 01031890!, WUK 0275519 image!).

Berberis jingguensis G. S. Fan & X. W. Li, J. Trop. Subtrop. Bot., 5(3): 1. 1997. TYPE: China. SW Yunnan: Jinggu Xian, 2500–2900 m, 17 Jan. 1995, *G. S. Fan & X. W. Li 9510018* (lectotype, designated here, SWFC 0036138!; isolectotype, SWFC 0001031!).

Shrubs, evergreen, to 3 m tall; mature stems very dark purplish brown, terete, sulcate; spines 3-fid, concolorous, 1–3.5 cm, stout, weak or absent at the end of branches, abaxially sulcate. Petiole 3–5 mm, sometimes almost absent; leaf blade abaxially pale green, adaxially green, both sides usually reddish brown when dry, elliptic, mostly narrowly so, or elliptic-lanceolate,

6–13 × 2–3.5 cm, leathery, midvein raised abaxially, impressed adaxially, lateral veins and reticulation inconspicuous on both surfaces, base cuneate, margin spinulose with 15 to 35 inconspicuous teeth on each side, apex acuminate, sometimes subobtuse. Inflorescence a fascicle, 6- to 12-flowered; pedicel 15–20 mm. Sepals in 2 whorls; outer sepals suborbicular, ca. 4 × 3.5 mm, apex rounded; inner sepals elliptic, ca. 7 × 5 mm; petals obovate, ca. 6 × 3.5 mm, base cuneate, with separate glands, apex entire, rounded. Stamens ca. 4.5 mm; anther connective extended, truncate. Ovules 1. Berry black, oblong-ellipsoid, 7–8 × ca. 3 mm, not or slightly pruinose; style not persistent.

Phenology. *Berberis wuliangshanensis* has been collected in flower in April and May and in fruit between July and November.

Distribution and habitat. *Berberis wuliangshanensis* is known from Jinggu, Jingdong, Nanjian, and possibly Zhenyuan Xian in southwest Yunnan. It has been collected from the margins of evergreen broadleaf forests and on mountain slopes at ca. 1800–2600 m.

The protologue of *Berberis jingguensis* cited *S. Fan & X. W. Li 9510018* at SWFC as the holotype. However, there are two specimens of this both marked "typus." The better specimen has been chosen as the lectotype. The protologue also cited *G. S. Fan et al. 951091*, Zhenyuan Xian, 2 Oct. 1995, also at SWFC, but this has not been found. X. H. Li (2008) synonymized *B. jingguensis* while noting that the protologue's description of the taxon as a vine up to 7 m tall was likely to be the result of mistaken recording in the field.

Selected specimens.
SW Yunnan. Jingdong Xian: Wuliangshan, Huangcaoling, Changyao Shan, 2550 m, 16 Nov. 1956, *B. Y. Qiu 53740* (IBSC 0091648, KUN 203034, PE 01031895); Wuliangshan, Luoshui Dong, 1800 m, 26 Apr. 1959, *S. G. Xu 4743* (KUN 203031, 203033, 0177545, PE 01031893); Beiwashan, Guanyin Si, 2300 m, 25 May 1963, *Q. A. Wu 9332* (KUN 0177541–4); Modao He, 2400 m, 16 Apr. 1982, *Q. Lin 770527* (KUN 0177555–56); Dazhong Gongshe, Xujiaba, 2500 m, 28 Apr. 1991, *S. Y. Bao 379* (KUN 0177557–8); Ai'laoshan, 2490 m, 18 July 2003, *G. P. Yang 42* (HITBC 106722). **Nanjian Xian:** Baohua Zhen, Yongzhen Cun, Salaqing, 24.842933°N, 100.383500°E, 2616 m, 27 June 2015, *E. D. Liu et al. 4149* (KUN 1278013–14).

84. Berberis wuyiensis C. M. Hu, Bull. Bot. Res., Harbin 6(2): 7. 1986. TYPE: China. NE Jiangxi: Yanshan Xian, Huanggang Shan, s.d., *Q. H. Li, S. A. Luo et al. 231* (lectotype specimen with three twigs designated here; isolectotype specimen with two twigs, both HLG (LBG) currently on loan to IBSC images!).

Shrubs, evergreen, to 1.5 m tall; mature shoots pale brownish yellow, sulcate, sparsely verruculose; spines 3-fid, concolorous, 1–2 cm, subterete. Petiole 1–3 mm; leaf blade abaxially pale green, adaxially dark green, shiny, oblanceolate or elliptic-obovate, 3.5–7 × 0.8–1.6 cm, leathery, midvein raised abaxially, impressed adaxially, lateral veins and reticulation inconspicuous on both surfaces, base cuneate, margin spinulose with 2 to 4(to 6) teeth on each side, occasionally entire, apex acute, mucronate. Inflorescence a fascicle, 6- to 12-flowered; pedicel reddish, 8–10 mm; bracteoles lanceolate, 1.8–2.5 mm. Sepals in 2 whorls; outer sepals lanceolate or ovate-lanceolate, 3–3.5 mm; inner sepals oblong, 3.5–4.5 mm; petals obovate, 3–4.5 × 1.5–1.75 mm, base slightly clawed, apex emarginate. Stamens 2.5–3.5 mm; anther connective obtuse or apiculate. Ovules (1 or)2. Immature berry elliptic-oblong, ca. 7.5 mm, epruinose; style persistent.

Phenology. *Berberis wuyiensis* has been collected in flower in April and possibly in fruit in July and August.

Distribution and habitat. *Berberis wuyiensis* appears to be endemic to Huanggang Shan which straddles the border between Fuijian and Jiangxi. It has been found in thickets and sparse forests on mountain summits at 1900–2150 m.

The protologue of *Berberis wuyiensis* cited the type and three other collections, all from Mt. Huanggang: *Wuyi Expedition 1073A*, *Wuyi Expedition 2339*, and *Q. H. Li, S. A. Luo et al. 52*. The "A" of no. *1073* appears to be a later addition to the specimen sheet at FJSI to differentiate it from no. *1073* at IBSC (0091753), which is labelled (correctly) in C. M. Hu's hand as *B. fujianensis*.

Though the protologue omitted to indicate this there are, in fact, two specimens of *Li, Luo et al. 231*, the one with the most twigs has been chosen as the lectotype.

The protologue differentiated *Berberis wuyiensis* from *B. chingii* mainly on the basis that its leaves are oblanceolate and epruinose and the fruit is epruinose. Given the variability of the leaves of *B. chingii* and that its fruit is only initially pruinose, this would seem to be insufficient justification for a separate species. However, given the account of the flowers in the protologue (and confirmed by an image of dissection kindly sent to me by C. M. Hu), it is sufficiently different from *B. chingii* to suggest that it is distinct.

I have located further specimens from Huanggang Shan and listed them below. Some of them may, in fact, be *Berberis chingii*, as may be *Li, Luo et al. 52*, which is sterile. This suggests the need for further investigation of *Berberis* on Huanggang Shan. C. C. Yu of NTU visited the area in January 2016 and found only three plants, all of which were sterile (pers. comm. 8 Jan. 2016).

Selected specimens.
NW Fujian. Chong'an Xian (now Wuyishan Shi): Wuyi Shan, Huanggang Shan, Ximian Gou, 1900 m, 11 Aug. 1964, *C. P. Jian 400653* (MO 004468029, PE 01031949, 01031965); Tongmu Guan en route to Huanggang Shan, 2100 m, 28 Aug. 1979, *Wuyi Exped. 01073A* (FJSI FJSI008991); Huanggang Shan, 2000 m, 30 Apr. 1981, *Wuyi Exp. 2339* (FJSI FJSI008990, FNU 0003657, IBSC 0092220, MO 4183637); Huanggang Shan, 27.86125°N, 117.783753°E, 2150 m, 26 May 2015, *E. D. Liu et al. WYS 0033* (KUN 1265358–9).
NE Jiangxi. Yanshan Xian: Huanggang Shan, 1900 m, s.d., *Q. H. Li, S. A. Luo et al. 52* (currently at IBSC); Wuyi Shan, Caoping, Jiamukeng, 2000 m, 30 Aug. 1964, *C. P. Jian et al. 400986* (LBG 00013844, PE 01031937–38).

85. Berberis xanthoclada C. K. Schneid., Repert. Spec. Nov. Regni Veg. 46: 261. 1939. TYPE: China. NE Guizhou: Fang Ching Shan (Fanjing Shan), Laoshan, ca. 2100 m, 30 Sep. 1931, *A. N. Steward, C. Y. Chiao & H. C. Cheo 482* (holotype, A 00038825!; isotypes, BM BM000559457!, E [2 sheets] E00259014–15!, FUS 00014170 image!, K K000723999!, L L.1742591 image!, N [7 sheets] 093057020–21, 093057023–26, 093057105 images!, NAS NAS00070729 image!, NY 01085873!, P P00716543!, PE 01031964!, PEY PEY0002947 image!, S 08–755 image!, US 00945872 image!, W 1933–5775!).

Shrubs, evergreen, to 2.5 m tall; mature stems reddish brown, sulcate; spines 3-fid, retrorse, orange-brown, 1–3 cm, abaxially sulcate. Petiole 2–4 mm; leaf blade abaxially pale yellow-green, adaxially dark green, shiny, elliptic or broadly elliptic, occasionally ovate, 4–8 × 1.5–3 cm, thinly leathery, midvein raised abaxially, impressed adaxially, lateral veins slightly raised and inconspicuous abaxially, slightly impressed and conspicuous adaxially, reticulation indistinct abaxially, conspicuous adaxially, reticulate veins indistinct abaxially, distinct adaxially, base attenuate, margin spinulose with 7 to 18(to 25) teeth on each side, apex subacute or acute, rarely obtuse, mucronate. Inflorescence a fascicle, 2- to 6-flowered; pedicel reddish, 7–13 mm; bracteoles lanceolate, 4–5 × 1.5–2 mm. Sepals in 2 whorls; outer and inner sepals obovate-oblong, 6–10 × 2.5–5 mm, apex obtuse; petals obovate, 4–5 × 2.2–3 mm, base clawed, with separate glands, apex emarginate. Stamens 2–5 mm; anther connective extended, truncate. Ovules 3 to 5, shortly funiculate. Berry black, epruinose, ellipsoid, 8–9 × 5–6 mm; style persistent. short.

Phenology. *Berberis xanthoclada* has been collected in flower in April and May and in fruit between June and October.

Distribution and habitat. *Berberis xanthoclada* is known in Guizhou from Fanjinshan in Jiankou and Yingjiang Xian and Leigongshan in Leishan Xian with one collection from Xishui Xian. It has been collected from thickets, scrub, bamboo groves, and among *Rhododendron* on and near mountain summits at ca. 1000–2700 m.

The type specimen of *Berberis xanthoclada* has only fruit. The description of flowers by Chamberlain and Hu (1985: 546) was based on *Qiannan Team 911* (IBSC), cited below, although they gave the collector as Z. Y. Gao. Their description has been further augmented on the basis of flowers from the Kew plant of *GUIZ 150*, also cited below. *Berberis xanthoclada* appears to be limited to very few areas.

Selected specimens.
Guizhou. Jiankou Xian: Fanjinshan betw. Jinding & Juchi Shan, 2200 m, 23 June 1988, *Wulingshan Botany Team 733* (KUN 0179117, PE 01802418); Fanjin Shan Nature Reserve, 1600 m, 23 Apr. 1983, *Z. D. Yang 104* (FAN); vic. of Jinding along the crest of the Fanjing Shan mtn. range, 2700 m, 27 Aug. 1986, *Sino-American Bot. Exp. 477* (A 00280050, BM, PE 01031974). **Leishan Xian:** Xijiang Gongshe, Leigong Ping, 1300 m, 20 May 1959, *Qiannan Team 2054* (HGAS 013692, 013697, KUN 0178889, PE 01031976–77); Wudongmu Chang, Leigongshan, 1850 m, 29 Apr. 1959, *Qiannan Team 911* (HGAS 013690, IBSC 0092222, KUN 0178890, PE 01031979–80); SE of Leigongshan, 1350 m, 13 May 1959, *Qiannan Team 1403* (HGAS 013699, KUN 0178891, NAS NAS00314443–44, PE 01031982–83). **Xishui Xian:** Guanping, 1000 m, 6 July 1963, *Southwest Guizhou Comprehensive Research Team s.n. (170)* (PE 01805213–15). **Yingjiang Xian:** Fanjinshan, Jiulongchi, 2185 m, 8 Oct. 1963, *Z. P. Jian et al. 31849* (PE 01031968, 01031975); Juchi, 2300 m, 4 May 1964, *Beijing Botany Exp. 400618* (IBSC 0092731, PE 01031966–67).

Cultivated material:
Living cultivated plants.
Royal Botanic Gardens, Kew, U.K., from *Fanjingshan Guizhou Exp. 150*, Guizhou, Fanjingshan, betw. Boo Yen & Lanchatin, ca. 2100 m, 19 Oct. 1985; Royal Botanic Garden Edinburgh, U.K., from *A. Clark 4280*, Guizhou, Leishan Xian, summit of Leigongshan, ca. 2180 m, 1999.

86. Berberis yiliangensis Harber, Bot. Mag. 33(1): 35. 2016. TYPE: Cultivated. Royal Horticultural Society Wisley Gardens, Surrey, U.K., from W20020037-A, 18 June 2013, *s. coll. s.n.*, from seeds of *A. Clark 4332*, NE Yunnan: Zhaotong Shi, Yiliang Xian, Xiaocaoba, 1700 m, autumn 1999 (holotype, PE [2 sheets] 02014579–80!; isotypes, A [2 sheets] 00589098–99!; WSY [2 sheets] WSY01083530–54!).

Shrubs, evergreen, height in cultivation to 2.5 m tall; mature stems pale brownish yellow, subterete, verruculose; spines 3-fid, semi-concolorous, 1.2–3 cm, abaxially sulcate. Petiole almost absent or to 4 mm; leaf blade abaxially pale gray-green, yellow-green when dry, shiny, adaxially mid-green, shiny, narrowly elliptic or elliptic, rarely obovate-elliptic, (3–)4.5–8.5 × (0.8–)1.5–2(–2.5) cm, leathery, midvein raised abaxially, slightly impressed adaxially, lateral veins and reticulation indistinct abaxially, inconspicuous adaxially (more conspicuous when dry), base narrowly attenuate, margin spinulose with 5 to 15 teeth on each side, apex subacuminate, mucronate. Inflorescence a fascicle, 2- to 9-flowered; pedicel pale brown, 12–26 mm; bracteoles triangular, ca. 1.25 × 0.25 mm. Sepals in 2 whorls; outer sepals oblong-ovate, 2 × 1 mm; inner sepals oblong-elliptic, 3 × 1.75 mm; petals obovate, 3–4 × 2.5–3 mm, apex rounded, incised, base clawed, glands ca. 0.75 mm, separate. Stamens ca. 3.5 mm; anther connective not extended, truncate. Ovules 1 or 2. Berry purplish black, partially pruinose, oblong, 7 × 4 mm; style persistent, short.

Phenology. The flowering and fruiting period of *Berberis yiliangensis* is unknown outside cultivation. In cultivation it flowers between April to June and fruits from July to December.

Distribution and habitat. *Berberis yiliangensis* is known from only cultivated plants grown from seeds collected in Yiliang Xian in northeast Yunnan from a plant growing in heavy soil on the edge of a cultivated area. The collector's notes described it as uncommon.

The only other species of *Berberis* recorded for Yiliang Xian is *B. acuminata*. Six additional species from section *Wallichianae* are apparently endemic to northeast Yunnan and the immediately adjacent areas of Sichuan.

As noted in the protologue of *Berberis yiliangensis*, Yiliang Xian (彝良县) in Zhaotong Shi should not be confused with Yiliang Xian (宜良县) in Kunming Shi.

Cultivated material:
Living cultivated plants.
W20020037-A, Royal Horticultural Society Wisley Gardens, Surrey, U.K., and in my own collection, both from *A. Clark 4332.*

Cultivated specimens.
Foster Clough, Mytholmroyd, West Yorkshire, U.K., from *A. Clark 4332* (details above), 12 May 2015, *J. F. Harber 2015-17* (A, E, K, KUN); same details, but 7 Feb. 2016, *J. F. Harber 2016-02* (A, E, I, K, KUN, PE).

87. Berberis yingjingensis D. F. Chamb. & Harber, Bot. Mag. 29(2): 115. 2012. TYPE: China. W Sichuan: Yingjing Xian, Niba Shan, N side, 1800 m, 28 Sep. 1988, *C. Erskine, H. Fliegner, C. Howick & W. McNamara SICH 255* (holotype, K K000395201!).

Shrubs, evergreen, to 2 m tall, narrowly erect; branches arching at apex; mature stems very pale yellowish brown, sulcate, sparsely black verruculose; young shoots green; spines 3-fid, pale yellow-brown, 0.5–2 cm, terete, weak. Petiole almost absent; leaf blade abaxially pale green, shiny, adaxially mid-green, shiny, very narrowly lanceolate, 4.5–12(–18) × 0.8–1.5(–2) cm, thinly leathery, midvein raised abaxially, impressed adaxially, lateral veins in 10 to 18 pairs, indistinct on both surfaces, base attenuate, margin spinose with 15 to 30 teeth on each side, apex acute, mucronate. Inflorescence a fascicle, 2- to 5-flowered; bracts yellow with reddish tinge, triangular, ca. 2 × 1 mm; pedicel reddish, 7–13 mm; bracteoles mostly reddish, ovate, ca. 3.5 × 2 mm, apex acute. Flowers pale greenish yellow. Sepals in 2 whorls; outer sepals with vertical reddish stripe, oblong-elliptic, ca. 5 × 3.5 mm, apex rounded; inner sepals broadly elliptic, ca. 7 × 5 mm, apex obtuse; petals elliptic, ca. 4 × 3 mm, base slightly clawed with approximate elliptic glands, apex entire. Stamens ca. 3 mm; anther connective truncate. Ovules (1 to)3 or 4. Berry dark blue-black, slightly pruinose, obovoid-ellipsoid, 8–10 × 5–7 mm; style persistent, short; seeds purplish black, ca. 6 × 2 mm.

Phenology. *Berberis yingjingensis* has been collected in fruit from August to October. Its flowering season outside cultivation is unknown.

Distribution and habitat. *Berberis yingjingensis* is known from Yingjing Xian, the border between Hanyuan and Shimian Xian, and Tianquan Xian in Sichuan at ca. 1800 m. The type was collected on sandy soil and reaching out from under shade in an open valley of dense secondary shrubs on a limestone mountainside. The seed of *Compton, D'Arcy & Rix 2497*, referred to below, was collected from a plant growing among *Euonymus quinquecornutus*, *Daphne acutiloba*, and *Niellia thibetica*. *J. I. Leon et al. S11578* was collected in a temperate forest at 2243 m.

Berberis yingjingensis is known only from the type, one other collection, and a cultivated plant. The description of the flowers is from the living plants at Kew grown from seeds of the type collections. Collection details of *Compton, D'Arcy & Rix 2497* is from personal communication with J. Compton, Wiltshire, U.K., 5 Feb. 2010.

Selected specimens.
W Sichuan. Tianquan Xian: Lianglu Xiang, Erlang Shan, 29.874694°N, 102.306528°E, 2243 m, 3 Aug. 2010, *J. I. Leon et al. SI1578* (KH KHB1296875–76, PE 01899361, SNUA SNUA00016609).

Cultivated material:
Living cultivated plants.
Royal Botanic Gardens, Kew, Howick Hall, Northumberland, U.K.; Foster Clough Mytholmroyd, West Yorkshire, U.K.; Quarryhill Botanical Garden, California, U.S., from *SICH 255*; Sherwood Gardens, Devon, U.K., from *J. Compton, J. D'Arcy & E. M. Rix 2497*, Hanyuan Pass, above Shimian, 1830 m, 5 Oct. 1995.

Cultivated specimens.
Foster Clough Mytholmroyd, West Yorkshire, U.K., from *SICH 255* (details above), 29 Apr. 2015, *J. F. Harber 2015-06* (A, E, KUN, PE); same details, but 7 Feb. 2016, *J. F. Harber 2016-03* (A, E, KUN, PE).

88. Berberis zhaotongensis Harber, Bot. Mag. 34(2): 103. 2017. TYPE: Cultivated. Royal Botanic Garden Edinburgh, 17 Apr. 2012, *H. S. Cubey 346* from seeds of *P. Cox & P. Hutchison 7140*, China. NE Yunnan: Zhaotong Prefecture, Daguan Xian, side rd. E of rd. from Zhaotong to Daguan (S of Daguan), 3200 m, 9 Oct. 1995 (holotype, E E00615081!).

Shrubs, evergreen, to 1.8 m tall; mature stems reddish brown, very sulcate, not or sparsely verruculose; spines 3-fid, concolorous, 0.3–1.6 cm, weak. Petiole almost absent; leaf blade abaxially pale green, adaxially dark green, narrowly elliptic or obovate-elliptic 1–4.5(–5.5) × 0.4–0.8(–1.4) cm, thinly leathery, midvein raised abaxially, impressed adaxially, lateral veins obscure on both surfaces, base attenuate, margin slightly revolute on leaves at apex of stems, more pronounced when dry, spinulose with 6 to 12 teeth each side, apex acute, minutely mucronate. Inflorescence a congested fascicle, (2- to)10- to 15(to 22)-flowered; pedicel 4–10 mm; bracteoles triangular, 2 × 0.5 mm, apex with reddish or brownish tinge. Flowers ca. 5 mm diam. Sepals in 2 whorls; outer sepals triangular-ovate, 3 × 1.5 mm, apex narrowly acute with reddish or brownish tinge; inner sepals oblong-obovate, 4–5 × 3 mm, apex acute; petals obovate, ca. 4 × 3 mm, apex obtuse, entire; base cuneate or slightly clawed, glands separate, ellipsoid, ca. 1 mm. Stamens ca. 2 mm; anther connective extended, obtuse or apiculate. Ovules 2 or 3. Berry black, shiny, elliptic-oblong, 7–8 × 4–5 mm; style persistent; seeds purplish black.

Phenology. *Berberis zhaotongensis* is known to fruit from July to October. Its flowering season outside cultivation is unknown, but appears to be before May.

Distribution and habitat. *Berberis zhaotongensis* is known from Daguan, Yongshan, and Zhaotong Xian in northeast Yunnan, and Weining Xian in northwest Guizhou. It has been found in open shrubbery on a steep slope above mine workings, among bamboo thickets, and on the margin of *Populus* woods at 2000–3200 m.

As I noted in the protologue I first came across this species growing in the Royal Botanic Garden Edinburgh and subsequently at Howick Hall, Northumberland, both grown from seeds of *Cox and Hutchison 7140*. The wild-collected herbarium specimen is exceptionally poor, so I chose a specimen obtained from the plant at E with flowers as the type.

The type of *Berberis zhaotongensis* is one of only three species of section *Wallichianae* from Daguan Xian, the others being *B. deinacantha* and the type of the very different *B. caudatifolia*.

H. T. Tsai 50979 (details below) was listed as a syntype of *Berberis liophylla* by Schneider (1939: 247).

Y. Q. Huang & L. T. Zhang 91031 from Guizhou (see below) was published as "*Berberis weiningensis*, nom. ined.," in He et al. (1995: 647), the name *B. weiningensis* being subsequently used by Ying (1999: 326) for a validly published deciduous species.

Selected specimens.
NW Guizhou. Weining Xian: Heixhitou Qu, 2500 m, 9 July 1959, *Bijie Team 113* (GF 09001098, HGAS 013607, KUN 0175754, PE 01031257–58); Yangjie Qu, Songling Xian, Jiamashi, 2000–2200 m, 30 May 1991, *Y. Q. Huang & L. T. Zhang 91031* (GZTM 0004287, IBSC 0092189).
NE Yunnan. Daguan Xian: side rd. E of rd. from Zhaotong to Daguan (S of Daguan), 3200 m, 9 Oct. 1995, *P. Cox & P. Hutchison 7140* (E E00073198). **Yongshan Xian:** "Yung chan Hsien," 2100 m, 2 June 1932, *H. T. Tsai 50979* (A 00279667, KUN 0177389, PE 01031466–67, SZ 00294121). **Zhaotong Xian:** "Chao Tung Hsien," 2800 m, 1 May 1932, *H. T. Tsai 50860* (A 00279531, KUN 0175788, PE 01031464–65, SZ 00291225, 00294063).

Cultivated material:
Living cultivated plants.
Howick Hall, Northumberland, U.K.; Royal Botanic Garden Edinburgh, U.K., both grown from seeds of the type. Foster Clough, Mytholmroyd, West Yorkshire, U.K., from a cutting from the Howick Hall plant.

Cultivated specimens.
Royal Botanic Garden Edinburgh, from plant grown from seed of *P. Cox & P. Hutchison 7140*, 7 Nov. 2011, *H. S. Cubey 328* (E E00421943–44).

89. Berberis zhenxiongensis Harber, sp. nov. TYPE: China. NE Yunnan: Zhenxiong Xian, Machang Linchang, 8 June 1980, *S. Y. Bao 106* (holotype, KUN 0175742!; isotype, KUN 0175743!).

Diagnosis. *Berberis zhenxiongensis* has leaves somewhat similar to those of *B. atrocarpa* (a species of west and west-central Sichuan) although they are not as wide and have significantly more marginal spines. The mature stems are yellowish brown and subterete, whereas those of *B. atrocarpa* are pale yellow and angled to sulcate. Inflorescences of *B. atrocarpa* can be up to 20-flowered, whereas those of *B. zhenxiongensis* are 1- to 5-flowered. On current evidence, *B. zhenxiongensis* is a much smaller shrub than *B. atrocarpa*. The leaves of *B. zhenxiongensis* are somewhat similar to *B. zhaotongensis*, but the pedicels and spines are longer.

Shrubs, evergreen, to 1.5 m tall; mature stems pale yellowish brown, subterete, not or sparsely verruculose; spines 3-fid, concolorous, (0.8–)2–3.2 cm, narrow. Petiole almost absent; leaf blade abaxially pale green, adaxially dark green, narrowly lanceolate or very narrowly elliptic, (2.5–)4–9 × 0.8–1.4 cm, thinly leathery, midvein raised abaxially, impressed adaxially, lateral veins and reticulation conspicuous or indistinct on both surfaces, base attenuate, margin sometimes slightly undulate, spinulose with (2 to)8 to 20(to 30) inconspicuous teeth on each side, apex acute, mucronate. Inflorescence 1- to 5-flowered. Flowers unknown; fruiting pedicel 10–15 mm. Berry black, elliptic-oblong, 5–7 × 3–4 mm; style persistent, short; seeds 2.

Phenology. *Berberis zhenxiongensis* has been collected with immature fruit from June to October. Its flowering period is unknown.

Distribution and habitat. *Berberis zhenxiongensis* is known from Zhenxiong Xian in northeast Yunnan. It has been collected in *Pinus* forest and scrub at 1620–2400 m.

IUCN Red List category. *Berberis zhenxiongensis* is assessed as DD or Data Deficient, according to IUCN (2001) criteria.

S. Y. Bao 106 and *G. M. Feng et al. 447* were first identified by S. Y. Bao as *Berberis atrocarpa* and subsequently reported by him (1997: 65) as such.

I am grateful to Tu Feng of BJ for images of his collections listed below prior to mounting.

Selected specimens.
NE Yunnan. Zhenxiong Xian: Wufeng Shan, 2300–2400 m, 29 Oct. 1973, *G. M. Feng et al. 447* (KUN 0175744); Shibanqiao, 1620 m, 23 July 2011, *T. Feng 2011072307* (BJ), same details, but 24 July 2011, *T. Feng 2011072308* (BJ).

90. Berberis ziyunensis P. K. Hsiao & Z. Yu Li, Acta Bot. Yunnan. 21(1): 30. 1999. TYPE: China. SW Guizhou: Ziyun Xian, Baiyun Xiang, 1300 m, 10 Apr. 1991, *S. Z. He et al. 9109* (holotype, GZTM 0013782!; isotypes, GZTM 0013780!, IMD not seen, PE [2 sheets] 00935261–62!).

Shrubs, evergreen, to 1.6 m tall; mature stems purplish black, shiny, terete; spines 1- to 3-fid, pale yellow, 0.3–0.8 cm, very weak, mostly absent. Petiole almost absent or 2–3 mm; leaf blade abaxially pale green, adaxially dark green, narrowly elliptic, oblanceolate, or narrowly obovate, 4–10 × 1–3 cm, thickly leathery, abaxially slightly pruinose, midvein raised abaxially, impressed adaxially, lateral veins and reticu-lation obscure abaxially, indistinct or inconspicuous abaxially, base narrowly attenuate, margin entire, rarely spinulose with 1 or 2 inconspicuous teeth on each side, apex acute, mucronate. Inflorescence a fascicle, 4- to 10-flowered; pedicels ca. 15 mm; bracteoles triangular-ovate, ca. 1.5 × 0.75 mm. Sepals in 2 whorls; outer sepals ovate, ca. 3 × ca. 2.6 mm; inner sepals oblong-ovate or ovate, ca. 4 × ca. 3 mm; petals obovate, ca. 3.5 × 3 mm, base slightly clawed, glands widely separate, ca. 0.7 mm, apex entire. Stamens ca. 2.5 mm; anther connective extended-truncate. Ovules 3. Immature berry ellipsoid, ca. 4 × 2 mm; style persistent.

Phenology. *Berberis ziyunensis* is known to flower and produce immature fruit in April. Its phenology is otherwise unknown.

Distribution and habitat. *Berberis ziyunensis* is known from only the type and one other collection in Ziyun Xian in south Guizhou. The type was collected from mountainside thickets at 1300 m. *Ziyun Survey Team 394* was collected from a mountain forest understory at an unrecorded elevation.

The protologue of *Berberis ziyunensis* drew both on the type collection and *S. Z. He & T. L. Zhang 92510* (GZTM) from Libo Xian in south Guizhou. However, this latter is *B. uniflora* whose type is from Huanjiang Xian in north Guangxi some 40 km to the south of where no. *92510* was collected. As noted under that species, F. L Chen et al. (2012: 605–607) reduced *B. ziyunensis* to a synonym of *B. uniflora*. It is certainly the case that, even when *B. ziyunensis* is described solely on the basis of the type collection, the descriptions of the color of mature stems and leaf shape differ hardly at all from that of *B. uniflora*. However, as reported under this latter species, S. Z. He undertook a detailed investigation of the flower structure of *B. uniflora* and discovered that it differed from *B. ziyunensis* in having three rather than two whorls of sepals, petals whose apex is incised rather than entire, and two rather than three ovules. This, when combined with the fact that the type of *B. ziyunensis* was found some 180 km distant from where the nearest example of *B. uniflora* has been found, would seem sufficient evidence to continue to maintain two separate species. However, more research is needed here, particularly since currently, *B. ziyunensis* is known from only the type and two other collections (both of which are sterile).

Selected specimens.
SW Guizhou. Ziyun Xian: Maizi Qing, 15 June 1986, *Ziyun Survey Team 394* (GZTM 0011487–88); Maoying Zhen, Cedong Cun, 26.269750°N, 105.355222°E, 1011 m, 3 June 2017, *X. Q. Hou 520425170603236LY* (GZTM 0064589-90).

BERBERIS SECT. WALLICHIANAE FROM THE KUNMING AREA

One of the marked features of *Berberis* sect. *Walli-chianae* of Yunnan is the number of different taxa from the area around Kunming. In various herbaria there are substantial numbers of two of these species, *B. pru-inosa* and *B. ferdinandi-coburgii* (often identified as *B. vernalis*, which is treated here as a synonym), but other taxa are largely restricted to the type collection. The exception is *B. wangii*, but in this case, none of the specimens I have seen at KUN or PE from the Kunming area that are identified as *B. wangii* evidence one of the key characteristics of the type, namely abaxially pru-inose leaves. The paucity of specimens identified to be other than *B. pruinosa* and *B. ferdinandi-coburgii* is in itself odd, given the number of botanists who have made collections in the area over the years and the presence nearby of the Kunming Institute of Botany.

In May 2007, I spent one day on Xi Shan (Western Hills) near Kunming and report the following: on karst rocks on the higher slopes, the only species of section *Wallichianae* in evidence was *Berberis pruinosa*. Some 150 meters lower down among woods, I observed only *B. ferdinandi-coburgii*. Between them, however, was an area where not only both species, but a variety of plants that appeared to be neither, occurred. They included a plant with leaves identical to those of *B. ferdinandi-coburgii*, but were abaxially pruinose. It is likely that these were natural hybrids between the two species. Only a detailed molecular study can clarify the situa-tion, but it is likely that the following taxa are all forms of *B. ferdinandi-coburgii* × *B. pruinosa*.

Berberis kunmingensis C. Y. Wu ex S. Y. Bao, Bull. Bot. Res., Harbin 5(3): 8. 1985. TYPE: China. Yunnan: Kunming, no further details, *F. T. Wang 1284* (holotype, KUN missing; isotype, IBSC 0093717 image!).

The description in the protologue has elements of both *Berberis pruinosa* (three whorls of sepals) and of *B. ferdinandi-coburgii* (one ovule). The paratype *D. Lian. R. N. 19* (KUN 0176960), Kunming, no further details, cited by Wu does not resemble the description in the protologue and is *B. pruinosa*. The specimen at IBSC which gives the collectors as F. T. Wang & Y. Liu and the place of collection as "Kunming, Dapuji" has leaves resembling *B. ferdinandi-coburgii* but is sterile.

Berberis liophylla C. K. Schneid. var. **conglobata** Ahrendt, J. Linn. Soc., Bot. 57: 74. 1961. TYPE: China. Yunnan: Kunming, no further details but col-lected before Nov. 1906, *E. E. Maire 1738* (holotype, K K000077312!; isotypes, BM BM001015576!, E [2 sheets] E00217975–6!).

The protologue cited only the type. The taxon was mistakenly treated as a synonym of *Berberis cavaleriei* by Chamberlain and Hu (1985: 553). There are two twigs (E00217974 and E00217976) on the sheet of one of the isotypes at E. A dissection from both by C. M. Hu shows a flower of the former to have two ovules and one of the latter to have one. Hu's annota-tion on the sheet concludes that the former is *B. pru-inosa* and the latter *B. cavaleriei*, but it may be that the twigs are actually from the same plant. If so, they may provide further evidence of natural hybrids in the Kunming area.

Berberis mairei Ahrendt, J. Linn. Soc., Bot. 57: 76. 1961. TYPE: China. Yunnan: Kunming, no details except collected before Nov. 1906, *E. E. Maire 1999* (holotype, K K000077307!; isotypes, E [2 sheets] E00279477–78!).

The protologue cited only the type. The specimens have no flowers and only very immature fruit with up to four seeds. E E00279477 is annotated by Schneider as *Berberis bergmanniae* var. *acanthophylla* forma, but this is unsustainable. The gathering was considered by Chamberlain and Hu (1985: 551) to be close to a form of *B. pruinosa* with abaxially epruinose leaves. *Berberis mairei* was not treated by Ying (2001).

X. H. Li (2010: 441–442) treated *Berberis mairei* as a synonym of *B. wangii* despite the fact that, unlike the latter, its leaves are not abaxially pruinose.

Berberis praecipua C. K. Schneid. var. **major** Ahrendt, J. Linn. Soc., Bot. 57: 43. 1961. TYPE: China. Yunnan: Kunming, no details except col-lected before Nov. 1906, *E. E. Maire 1998* (holo-type, K K000077330!; isotypes, E [2 sheets] E00217978–9!).

The protologue cited only the type. There is no evi-dence to associate this gathering with *Berberis prae-cipua*, which is endemic to Bhutan. The taxon was also mistakenly treated as a synonym of *B. cavaleriei* by Chamberlain and Hu (1985: 553).

Berberis schneideriana Ahrendt, J. Linn. Soc., Bot. 57: 76. 1961. TYPE: China. Yunnan: Kunming, near "Shi-lung-pa," open hillside, 21 Feb. 1914, *C. K. Schneider 164* (holotype, K K000077307!; isotypes, A 00038792!, G G00343598!).

The protologue cited only the type. *Berberis schnei-deriana* was treated as a synonym of *B. wangii* by Chamberlain and Hu (1985: 553), despite not having abaxially pruinose leaves.

Berberis wangii C. K. Schneid., Repert. Spec. Nov. Regni Veg. 46: 246. 1939. TYPE: China. Yunnan: Kunming, Apr. 1935, *C. W. Wang 62639* (holotype, A 00038820!; isotypes, KUN 0754349!, NAS NAS00070728 image!, PE [2 sheets] 01031135–36!, WUK 0037320 image!).

As noted above, apart from the type, no specimens identified as *Berberis wangii* have been found with abaxially pruinose leaves. Interestingly, in a number of cases, some specimens from a gathering have been identified as *B. wangii* while others from the same gathering as *B. vernalis*, e.g., *B. Y. Qiu 50024* (KUN 0177302, 0177349, 0177351–53, PE 01031419), Kunming, Xishan, 2300 m, 20 Aug. 1953.

Taxa Incompletely Known

Berberis amabilis C. K. Schneid. var. **holophylla** C. Y. Wu & S. Y. Bao, Bull. Bot. Res., Harbin 5(3): 5. 1985. TYPE: China. NW Yunnan: Dali, 1934, *M. Chen 1891*, no further details (holotype, KUN 1204014!).

The protologue cited only the type. The specimen has no floral material and only one complete fruit. There is insufficient evidence to associate this with *Berberis amabilis*. It is possibly a form or hybrid of *B. pruinosa*.

Berberis fallax C. K. Schneid. var. **latifolia** C. Y. Wu & S. Y. Bao, Bull. Bot. Res., Harbin 5(3): 6. 1985. TYPE: China. W Yunnan: Jingdong Xian, Lo Shui-tung, 2100 m, 7 Dec. 1939, *M. K. Li 2284* (holotype, KUN 1204022!; isotypes, IBSC 0093700 image!, KUN 1204019!, WUK 0244152 image!).

The protologue cited the holotype as being *M. K. Li 1968* (KUN 0176103), but that appears to be an error; the collection details given are those from the label on the sheet of *M. K. Li 2284* (KUN 1204022), which also has a label identifying it as the type. There would seem to be no reason to associate this gathering with *Berberis fallax*, which has much narrower leaves than *Li 2284* and four or five seeds per fruit versus the two or three seeds per fruit given in the protologue of variety *latifolia*. The leaves of *B. fallax* var. *latifolia* are most like those of *B. petrogena*, which has two or three ovules, though the collector's notes for *Li 2284* record the specimens as being from a plant 10 ft. (3 m) tall, whereas *B. petrogena* is recorded as being up to 1.5 m tall. *Li 2284* appears to be the same species as a collection from the neighboring Shuangbai Xian: *W. G. Yin 515* (IBSC 0091896, KUN 0175787, 0176094, LBG 00064098), Si Qi, Fabiao, Baizhu Shan, 2400 m, 2 Apr.

1957. This is recorded as being 2.5 m high. Clearly more research is needed here.

The KUN specimen of *M. K. Li 1968*, Jingdong Xian, Hwang-Cha Lia, 1900 m, 29 Mar. 1940, is annotated as *Berberis fallax* var. *latifolia*, but from the duplicates at IBSC (0091705), PE (01031805), and WUK (0271919), it appears to be *B. wuliangshanensis*.

Excluded Taxa

Berberis phanera C. K. Schneid. var. **glaucosubtusa** Ahrendt, Gard. Chron., ser. 3, 105: 371–372. 1939. TYPE: Cultivated. Royal Horticultural Garden, Wisley, Surrey, U.K., 27 Oct. 1939, *s. coll. s.n.* (holotype, WSY WSY0029354!).

The protologue gave no provenance for the specimen cited as the type (the label on a specimen from the plant at OXF states "origin uncertain"). *Berberis phanera* var. *glaucosubtusa* was not treated in Ahrendt (1961).

Berberis praecipua C. K. Schneid., Repert. Nov. Regni Veg. 46: 248. 1939.

In the protologue, Schneider cited as syntypes *Cooper 2524* and *Cooper 1914*, Chapcha, both from western Bhutan, *Handel-Mazzetti 11272* and *11825* from Hunan, and *J. F. Rock 3089* from western Yunnan. Subsequently, Ahrendt (1961: 42) chose *Cooper 2424* as the lectotype. The Handel-Mazzetti collections at W and A are *Berberis chingii* var. *wulingensis*; the Rock collections at A and US are *B. sublevis*. For further information about *B. praecipua*, see the discussion under *B. griffithiana*.

Berberis pruinosa Franch. var. **barresiana** Ahrendt, Kew Bull. 1939: 266. 1939. TYPE: Cultivated. Royal Botanic Gardens, Kew, "*Berberis* Dell, Flagstall Mound, under *Tilia Vulgaris*," 7 May 1959, *s. coll. s.n.* from *No. 4775 MV, Barres 156–32* (neotype, designated here, K K001092827!; isoneotype K K001092826!).

The protologue cited only the type at K. The two specimens listed above collected 20 years after the protologue was published were the only ones found and the best of these has been designated as a neotype. Beyond the fact that the plant concerned was supplied by Vilmorin of Les Barres in France, its provenance is unknown. The taxon was included by Ying (2001: 140, 2011: 746), although with a description of the inflorescence at variance with that of the type. I have followed Chamberlain and Hu (1985: 556) in excluding it.

91. Berberis abbreviata (Ahrendt) Harber, comb. et stat. nov. Basionym: *Berberis tischleri* C. K. Schneid. var. *abbreviata* Ahrendt, J. Linn. Soc., Bot. 57: 125. 1961. TYPE: China. WC Sichuan: around Tachien-lu (Kangding), 2440–3050 m, June 1908, *E. H. Wilson 2854* (holotype, K K000077298!; isotypes, A 00038804!, BM BM001015568!, E E00259008!, HBG HBG-506745 image!, LE!, US 00946057 image!, W 1914–4786!). Figure 3.

Shrubs, deciduous, to 3 m tall; mature stems very pale reddish brown, sulcate; spines 3-fid, solitary or absent toward apex of stems, yellowish brown, (0.8–)1.1–2.2 cm, terete or abaxially slightly sulcate. Petiole almost absent; leaf blade abaxially pale green, papillose, adaxially green, obovate, often narrowly so or obovate-oblanceolate, 1.4–2.4 × 0.5–0.8 cm, papery, midvein raised abaxially, lateral veins and reticulation conspicuous abaxially, inconspicuous adaxially, base attenuate, margin entire, rarely spinulose with 1 to 8 teeth on each side, apex subacute or obtuse, mucronate. Inflorescence a fascicle or sub-fascicle, 2- to 5(to 9)-flowered; pedicel 8–16 mm; bracteoles oblong-lanceolate, 3 × 0.8 mm. Sepals in 3 whorls; outer sepals oblong-lanceolate, 3–4 × 2 mm; median sepals oblong-ovate, 4.5–5 × 3.25–3.5 mm; inner sepals oblong-ovate or obovate-elliptic, 5 × 3.5 mm; petals obovate, 4–5 × 3.5 mm, base clawed, glands narrowly separate, ca. 0.8 mm, apex obtuse, slightly emarginate. Stamens ca. 3 mm; anther connective conspicuously apiculate. Ovules 2. Berry red, oblong, 8 × 4 mm; style persistent.

Phenology. *Berberis abbreviata* has been collected in flower from April to July and in fruit from July to September.

Distribution and habitat. *Berberis abbreviata* is known from Kangding and Yajiang Xian in west Sichuan. In the former, it has been found in thickets and forests and on roadsides at ca. 2440–3380 m; in the latter, it has been found in a dry meadow at 3985–4270 m.

Wilson 2854 was originally identified as *Berberis tischleri* by Schneider (1913: 355–356) together with *Wilson 2853* and *4134*, both also from Kangding. Ahrendt (1961: 125) who, noting that *2854* has shorter pedicels than the latter two collections, published it as the type of *B. tischleri* var. *abbreviata*. None of these Wilson collections cited by Schneider are *B. tischleri* and no. *2854* differs from nos. *2853* and *4134* (treated here as examples of a new species, *B. tachiensis*) in having different leaves and inflorescence and flowers with only two ovules versus the four of *B. tachiensis*.

All of the type specimens and almost all of those cited below have uniformly gray twigs. Evidence for the color of the mature stems is from the specimen of *J. F. Rock 17544* at N and the IBK specimens of *Jiang 35627* and *35759*.

Selected specimens.
W Sichuan. Kangding Xian: "Ta-tsien-lou (Principauté de Kiala), 23 June 1893, *J. A. Soulie 165* (P P02682321, P02682336, P02682368); Djesi-La and Djesi-Longba, S of Tatsienlu, 3380 m, July 1929, *J. F. Rock 17544* (A 00279518, E E00612604, IBSC 0092083, N 093057051, P P02313535, US 00946060); Yulin Xiang, Simaqiao, behind Catholic Church, 3100 m, 20 May 1953, *X. L. Jiang 35627* (IBK IBK00012855, IBSC 0091630, PE 01033823, SWCTU 00006319); Yulin Xiang, above Simaqiao, 3000 m, 28 May 1953, *X. L. Jiang 35759* (IBK IBK00012833, IBSC 0092414, PE 01033927); Simaqiao Linchang, 2850 m, 22 May 1974, *Q. S. Zhao & Y. T. Wu 111039* (CDBI CDBI0028139, PE 01033926); Hongwei Gongshe, Wangmu, 3100 m, 26 Apr. 1975, *Z. S. Yu 06186* (CDBI CDBI0027467–69); Kangding Town, southern suburbs, 2800 m, 20 May 1991, *J. S. Yang 76* (KUN 0178573); Zheduo Pass, km post 2855, 2990 m, 25 Sep. 1991, *Chengdu Edinburgh Exped. 330* (E E00612606). **Yajiang Xian:** on unmarked rd. originating betw. km markers 2913 & 2914 of hwy. 318, just W of pass (border of Kangding and Yajiang Xian) at Gao'er Shan, 29.998056°N, 101.381111°E, 3985–4270 m, 30 July 2010, *D. E. Boufford et al. 42424* (A, E).

92. Berberis aemulans C. K. Schneid., Pl. Wilson. (Sargent) 3: 434. 1917. TYPE: China. C Sichuan: [Ebian Xian], "Near Wa Shan," 2750–3050 m, June 1908, *E. H. Wilson 930* "A" (lectotype, designated here, A 00038719!; isolectotypes BM BM000559588!, E E00217967!, K missing, P P02482727!, US 00945927 image!).

Berberis faberi C. K. Schneid., Oesterr. Bot. Z. 67: 215. 1918, syn. nov. TYPE: China. C Sichuan: [Ebian Xian], Mt. Omei (Emei Shan), 3050 m, s.d. but before Dec. 1887, *E. Faber 229* (lectotype, designated by Ahrendt [1961: 313], K K000077297!; isolectotype, NY 00000038!, probable isolectotype, NY 01365276!).

Shrubs, deciduous, to 2 m tall; mature stems very dark blackish purple, sub-sulcate, black verruculose; spines 3-fid, orange-red, 0.6–1 cm, weak, abaxially slightly sulcate. Petiole 2–5 mm; leaf blade abaxially pale green, papillose, adaxially green, oblong-obovate or elliptic, 2–4 × 1–2 cm, papery, midvein and lateral veins raised abaxially, impressed adaxially, reticulation indistinct or obscure on both surfaces, base cuneate or attenuate, margin spinose with 5 to 12 teeth on each

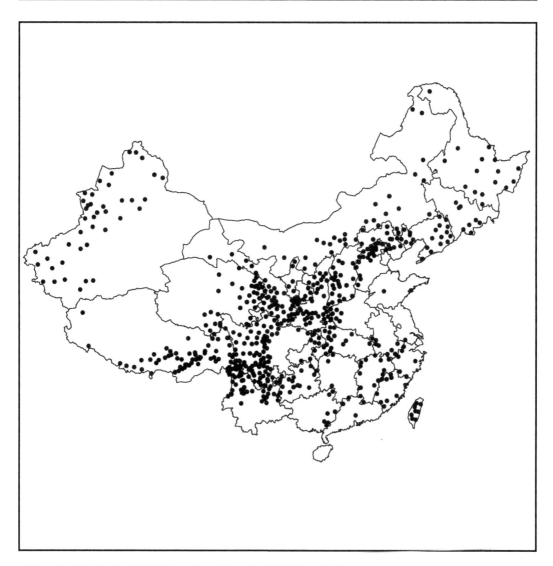

Figure 3. Distribution of *Berberis* species not in section *Wallichianae* by Xian.

side, sometimes entire, apex rounded. Inflorescence a fascicle, occasionally a sub-umbel or sub-raceme, 1- to 5-flowered; pedicel 20–30 mm. Sepals in 2 whorls; outer sepals oblong-elliptic, 7–7.5 × 4–4.5 mm; inner sepals oblong-elliptic, 7.5–8 × 4.5–5 mm; petals oblong, ca. 5 × 3.7 mm, glands separate, apex entire. Stamens ca. 4 mm; anther connective extended, apiculate. Ovules 7 to 11. Berry orange-red, sometimes slightly pruinose at first, ovate-elliptic, sometimes slightly bent at tip, 15–16 × 7–8 mm; style not persistent.

Phenology. *Berberis aemulans* has been collected in flower from May to June and in fruit from July to October.

Distribution and habitat. *Berberis aemulans* is known from Emei Shi, Ebian, Hanyuan, and Mianning Xian in central Sichuan. It has been found in thickets, grassy areas, forests, and trailsides on mountain slopes at ca. 2000–4000 m.

The protologue of *Berberis aemulans* simply stated the type to be *Wilson 930.* However, Wilson gave the same number to two separate gatherings of 1908, one of June and one of September. The former is designated here as *E. H. Wilson 930* "A" and the latter as *Wilson 930* "B." At BM and US, specimens of both are on the same sheet. The sheet at E has only *930* "A," while that at HBG is sterile. Ahrendt (1961: 124) stated the type to be "*Wilson 930*" at K, but this cannot be treated as a lectotypification; the sheet at K is, in any case, missing. *Wilson 930* "A" at A has been chosen as the lectotype because the sheet has an annotation by Schneider dated 5 August 1916 identifying it as "Type" and because flowers are more useful for diagnosis than fruit.

The protologue of *Berberis faberi* cited only the type gathering, but without indicating a herbarium. Ahrendt's (1961: 213) citation of the specimen at K as the type was an effective lectotypification. 01365276 at NY is ex K and has exactly the same collection details as *Faber 229* but is unnumbered. The taxon was not noticed by Ying (2001).

Berberis aemulans is somewhat similar to *B. diaphana*, but can be distinguished from that species by its very dark blackish purple mature stems and its larger number of ovules. There is also a geographical separation with *B. diaphana* being found farther north.

Almost all herbarium specimens of *Berberis aemulans* located are from Emei Shan, but this may simply be a reflection of the numerous visits by collectors to this area. The location in Hanyuan Xian of *H. Smith 2100*, cited below, is based on Herner (1988: 307) who located "Ta-hsiang-ling" at 29.55°N, 102.833333°E.

Selected specimens.
Syntype. Berberis aemulans. **Sichuan. [Ebian Xian]:** "Near Wa Shan," 2750–3050 m, Sep. 1908, *E. H. Wilson 930* "B" (A 0038726, BM BM1050022, US 00945927!); same details, but June and Sep. 1908, *E. H. Wilson 930* (HBG HBG-506735) sterile.

Other specimens.
Sichuan. Baoxing Xian: Xuecao Shan, 3100 m, 15 June 1958, *X. Y. Shang & Y. X. Ren 5397* (CDBI CDBI0027513, PE 01037485, SZ 00291540). **Ebian Xian:** Emei Shan, Jin Ding, 3060 m, Nov. 1928, *Y. Chen 7143* (IBSC 0091490, NAS NAS00314077); Jianzhi Gou, 27 June 1939, *C. W. Yao 4205* (FUS 00014304, PE 01032802); Emei Shan, betw. Linzi Ping & Puxian Pagoda, 13 July 1952, *X. L. Jiang & X. S. Zhang 31458* (IBK IBK00012827, IBSC 0091490, PE 01037114); Emei Shan, Jin Ding, 3100 m, 1 Oct. 1957, *G. H. Yang 57448* (KUN 0175612, NAS NAS00314329, PE 01030944, WUK 00286414); Emei Shan, betw. Jing Ding & Wanfo Ding, 3100 m, 3 Aug. 1971, *Sichuan 236 Task Group 0681* (PE 01030956). **[Hanyuan Xian]:** Ta-hsiang-ling (Daxiangling), 2000 m, 28 May 1922, *H. Smith 2100* (A 00279972, UPS BOT:V-040915). **Mianning Xian:** Tuowu Qu, Yele Xiang, 3600 m, 25 July 1959, *S. G. Wu 2121* (KUN 0175613, PE 01037118, 01037120); Lamagetou Nature Reserve, Niuchang, upstream from Yele Dam (Yele Xiang, above Liziping Xiang), 28.96°N, 102.103611°E, 3800–4000 m, 9 July 2005, *D. E. Boufford et al. 32730* (A 00279953) and *32731* (A 00279954, CAS 1078423).

93. Berberis aggregata C. K. Schneid., Bull. Herb. Boissier, sér. 2, 8: 203. 1908. TYPE: China. N Sichuan: [Jiuzhaigou (formerly Nanping) Xian], "Li cha p'u fl, Valle fl. Peishiu," 19 June 1885, *G. N. Potanin s.n.* (lectotype, designated here, LE!).

Berberis aridocalida Ahrendt, Bull. Misc. Inform. Kew 1939: 273. 1939, as "*arido-calida*," syn. nov. TYPE: Cultivated. Royal Horticultural Society Garden, Wisley, Surrey, U.K., from *R. Farrer 355*, *s. coll. s.n.*, 2 Oct. 1936 (holotype, WSY WSY0012759!).

Shrubs, deciduous, to 1.5 m tall; mature stems pale orange-brown, sulcate; spines 3-fid, concolorous, 0.8–2 cm, terete. Petiole almost absent; leaf blade abaxially pale gray-green, papillose, adaxially dull green, obovate-oblong or elliptic-obovate, 0.8–2 × 0.4–1 cm, papery, midvein raised abaxially, lateral veins and dense reticulation inconspicuous on both surfaces, conspicuous adaxially when dry, base cuneate, margin spinulose with 2 to 6 teeth on each side or entire especially toward the tips of branches, apex rounded or subacute, sometimes mucronate. Inflorescence a dense congested panicle, sometimes subcymose or fascicled at base, 10- to 14-flowered, 1–1.5 cm; peduncle almost absent; pedicel 1.5–2 mm; bracteoles ovate, 1 × 0.6 mm, apex acute. Flowers ca. 6 mm diam. Sepals in 2 whorls; outer sepals elliptic, 2.5 × 1.75 mm; inner sepals elliptic, 3.5 × 2.5 mm; petals obovate, ca. 3.5 × 2 mm, base clawed, glands oblong, apex scarcely emarginate. Stamens 2–2.5 mm. Ovules 2. Berry translucent white then pink turning red, subglobose, ca. 7 × 6 mm; style persistent, 0.5 mm.

Phenology. *Berberis aggregata* has been collected in flower in June and July and in fruit from August to November.

Distribution and habitat. *Berberis aggregata* is known from three separate areas: south Gansu, east Qinghai, and north Sichuan; Chongqing and northeast Sichuan; and Shanyang Xian in east Shaanxi. It has been found on dry roadside banks and shrubby hillsides, hillside meadows, and lower areas of mountain valleys at ca. 1200–3300 m.

The protologue of *Berberis aggregata* cited two Potanin specimens. These are on the same sheet at LE. The specimen with floral material has been chosen as the lectotype. These specimens were first identified by Maximowicz (1891: 40) as examples of *B. sinensis* var. *crataegina* (for which, see excluded taxa below). The protologue of *B. aggregata* describes the specimens as being from east Gansu, but it is clear from Bretschneider (1898a: 1013–1014) that they were collected in north Sichuan on the border with south Gansu, an area which is now in Jiuzhaigou (formerly Nanping) Xian.

The protologue of *Berberis aridocalida* cited as type a specimen at WSY from a plant at Wisley grown from seeds of *Farrer 355*. This is at WSY and is clearly identified as the holotype. Later, Ahrendt (1961: 213) mis-leadingly cited a number of cultivated specimens of *Farrer 355* with various dates without citing any herbarium as type. This was in line with his policy of regarding all specimens with the same collector's number equally as types irrespective of date or place of collection. Seeds of *Farrer 355* are listed in Farrer (1916: 61, 1917, vol. 2: 319). In the former, they are described as being from "the open upper alpine turf of the Siku-Satanee ranges." From the map in Farrer (1917, vol. 1), it appears the seeds were collected near Zhugqu in south Gansu. The protologue described *B. aridocalida* as having a fascicled or racemose–sub-fascicled inflorescence, but that of the type (and of the earlier-dated cultivated specimen at E, cited below, also grown from *Farrer 355*) is also partially paniculate. The taxon was not noticed by Ying (2001).

Both the size and shape of the leaves and the length of the inflorescence of *Berberis aggregata* vary considerably. The species is sometimes confused with the evergreen species *B. wilsoniae* with which, as Ahrendt (1961: 212) noted, it frequently hybridizes in cultivation. However, it differs from that species in being deciduous and frequently having leaves that are spinose. The only areas where the two species coincide appears to be Barkam (Ma'erkang), Heishui, Li, and Mao Xian in Sichuan. (On 5 August 2018, I found *B. aggregata* and *B. wilsoniae* growing within a few meters of each other in Heishui Xian; see *D. E. Boufford, B. Bartholomew, J. F. Harber, Q. Li & J. P. Yue 44436*, detailed below, and *44437*, detailed under *B. wilsoniae*).

J. F. Rock 12797, 12956, 13530, 13564, 13744, 13745, and *14911* (all at A and with duplicates elsewhere) from Gansu listed by Byhouwer (1928: 44) as *Berberis parvifolia* (treated here as a synonym of *B. wilsoniae*) are all *B. aggregata*.

There has been confusion in relation to the distribution of *Berberis aggregata*. Schneider (1913: 375), as well as citing specimens from north Sichuan, also cited *E. H. Wilson 4286* (A 00279324), west Sichuan, Mupin (Baoxing Xian), October 1910. Subsequently, Ahrendt (1961: 203) published a *B. aggregata* var. *integrifolia* with the type *K. L. Chu 2939*, also from Baoxing Xian. Both, however, appear to be *B. prattii* (for further details, see under that species). Ahrendt also cited *E. E. Maire 1997* (K), Yunnan-sen (Kunming) as an example of variety *integrifolia*. This is *B. wilsoniae* as are other specimens from the Kunming area of Yunnan identified as either *B. aggregata* or *B. aggregata* var. *integrifolia*. Ying (2001: 204) mistakenly treated variety *integrifolia* as a synonym of *B. aggregata*. The report of *B. aggregata* in Baoxing Xian by X. H. Li et al. (2017) is likely to be based on specimens identified as *B. aggregata* var. *integrifolia*.

The occurrence of *Berberis aggregata* in Shanyang Xian in east Shaanxi is interesting. It should be noted that *B. potaninii*, found in some of the same areas in Sichuan where *B. aggregata* is found, is also recorded for Shanyang Xian.

The report of *Berberis aggregata* in Guizhou by He and Chen (2004: 15) was based on misidentification of *B. wilsoniae*. The report of *B. aggregata* in Hubei by Fu (2001: 388) appears to be based on treating as a synonym *B. brevipaniculata*, whose type specimen was published as being from Badong in west Hubei. This synonymy, which is also made by Ying (2001), is mistaken. For further information about *B. brevipaniculata*, see discussion under that species.

Selected specimens.
Syntype. *Berberis aggregata*. **N Sichuan. [Jiuzhaigou (formerly Nanping) Xian]:** "Valle fl. Heiho," 22 July 1885, *G. N. Potanin s.n.* (LE).

Other specimens.
Chongqing (NE Sichuan). Chengkou Xian: 1500 m, 14 June 1958, *T. L. Dai 105201* (HNWP 5023, IBK IBK00382108, PE 01037379, SZ 00289641–42); Shizipu, 1600 m, 13 Aug. 1958, *T. L. Dai 104068* (IBSC 0093766, PE 01037377, 01037382, SZ 00289603–04); Caimeng Qu, Yanmai Xiang, 2000 m, 19 Oct. 1958, *T. L. Dai 106764* (PE 01037380, SZ 00289717). **Wuxi Xian:** Xining Xiang, Mao'erbei, Liangfeng, 1900 m, 19 Sep. 1958, *G. H. Yang 65443* (CDBI CDBI0027078, PE 01037375).

S Gansu. Hsiaho Hsien (Xiahe Xian): Chingshui, 2500 m, 7 July 1937, *K. T. Fu 1092* (PE 01037363). **Hui Xian:** Jialing Xiang, Taihuoshe, Boji Gou, 1550 m, 7 Sep. 1958, *Z. Y. Zhang 631* (IBK IBK00322524, KUN 0175668). **Jonê (Zhuoni) Xian:** Qihuang Gou, 2820 m, 30 June 1957, *Pan, Shen & Li B0156* (IBSC 0092028). **Kang Xian:** Changba Gongshe, Lijia Gou, 1550 m, 26 Sep. 1963, *Z. Y. Zhang 17097* (PE 01037370, WUK 0377010). **Kangle Xian:** Lianhua Shan, Mo Gou, 2450 m, 6 July 1992, *Tao He Team 95076* (KUN 0178278). **Li Xian:** 1800 m, 20 June 1936, *T. P. Wang 4409* (KUN 0178635, PE 01037342, WUK 0066935). **Linhsia Hsien (Linxia Xian):** near Kwanmen, 2200 m, 30 June 1937, *T. P. Wang 6891* (PE 01037357). **Lintan Xian:** Zhuoluoba, Dajia Shan, 3300 m, 11 June 1957, *Tao He Team 3330* (KUN 0176688, NAS NAS00314124, WUK 0091307). **Lintao Xian:** 2060 m, 19 June 1951, *Lanzhou University 14* (WUK 00166875). **Min Xian:** 2160 m, 3 July 1957, *Tao He Team 3649* (KUN 0175 666, PE 01037356). **Tianshui Xian:** foot of Maiji Shan, 19 Sep. 1950, *Z. L. Wu 20266* (PE 01037364). **Wen Xian:** Dongkou, 33.034167°N, 103.6575°E, 1805 m, 26 June 2000, *Sino-British Qinghai Alpine Garden Society Exped. 457* (E E00274456). **Wushan Xian:** Tange Zhen, 1 June 1956, *Huang He Team 4380* (PE 01037359). **Wutuhsian (Wudu Xian):** 3000 m, 19 June 1930, *K. S. Hao 460* (PE 01037354). **Zhouqu Xian:** Jiao'erqiao, 1800 m, 15 Sep. 1958, *Z. P. Wei 2457* (HNWP 90983, SZ 00289631, WUK 00107768); Duo'er valley, 33.856389°N, 103.816111°E, 2110 m, 19 June 2000, *Sino-British Qinghai Alpine Garden Society Exped. 294* (E E00274438).

E Qinghai. Menyuan Xian: Migu Gou, 2440 m, 19 June 1960, *Guizhou Team 2675* (PE 01037151, WUK 0167401). **Xunhua Xian:** Mengda, 2200 m, 25 June 1981, *B. Z. Guo & W. Y. Wang 25373* (HNWP 98845); Mengda, Cheng valley, 1950 m, 6 Nov. 1984, *Z. D. Wei s.n.* (HNWP 109074).

E Shaanxi. Shanyang Xian: Tianzhu Shan, Tiezhongping, 1810 m, 8 July 1964, *Y. M. Liang & J. X. Xiang 3230* (PE 01037055); Tianzhu Shan, near Tiezhongping, 2000 m, 31 Aug. 1982, *Z. B. Wang 16371* (HNWP 0178350, KUN 0178350. PE 01037056).

N Sichuan. Aug. 1903, 1980 m, *E. H. Wilson (Veitch) 3155* (A 00038689–90, BM, K K00007379); Sep. 1904, *E. H. Wilson (Veitch) 3155a* (K K00007380); Min valley, 1220–2130 m, Oct. 1908, *E. H. Wilson 1050* (A 00279322, BM, US 00945885). **Barkam (Ma'erkang) Xian:** A'mujiao Gou, 2600 m, 18 July 1957, *X. Li 23064* (IBSC 0092518, KUN 0177117, NAS NAS00314239, PE 01036919, WUK 00242190); near Lielie Tu, 14 Oct. 1957, *X. Li 72506* (IBSC 0092489, 0092653, KUN 0177118, NAS NAS00314238, NAS00314240, PE 01037536, WUK 00252582). **Heishui Xian:** 32.019225°N, 103.024236°E, 10 Aug. 2009, *L. Gu & Z. R. Li Gao XF1753* (A 00353755, CDBI); WNW of the city of Heishui and NW of Shashiduo-Xiang on side rd. off hwy. S302 near Sibugong, 32.10472222°N, 102.8025°E, 2800 m, 5 Aug. 2018, *D. E. Boufford, B. Bartholomew, J. F. Harber, Q. Li & J. P Yue 44436* (A, CAS, E, K, KUN). **Li Xian:** descending to Li fan Hsien, 3800 m, July 1930, *F. T. Wang 21563* (PE 01037350–51). **Mao Xian:** Tudiliangzi, ca. 2000 m, 30 June 1990, *M. Tamura et al. KS0553* (CDBI CDBI0027093). **Nanping (Jiuzhaigou) Xian:** near top of 124 Lin Chang, 2350 m, 9 Sep. 1975, *Sichuan Botany Team 8838* (CDBI CDBI0027088–90, PE 01037401); N of Songpan, 2400 m, 26 Sep. 1986, *P. Cox et al. 2571* (E E00125292). **Pingwu Xian:** Si'er Xiang, 2510 m, 29 Mar. 1958, *X. L. Jiang 11060* (IBSC 0092456, 0092536, PE 01037133, 01037135, 01037143, 01037146–47). **Songpan Xian:** Huanglong Gongshe, 2750 m, 11 Oct. 1983, *Y. L. Cao et al. 0110* (CDBI CDBI0027091, 0027100). **Wanyuan Xian:** Hua'e Shan, 1500 m, 30 July 1959, *P. Y. Li 4427* (PE 0103737–38, WUK 00140950). **Zoigê (Ruo'ergai) Xian:** Tiebu Qu, near border with Gansu, 2600 m, 3 July 1977, *Botany Team 10141* (CDBI CDBI0027912, CDBI0027943, IBSC 0092216–17).

Cultivated material:
Living cultivated plants.
Howick Hall, Northumberland, U.K.; from *SICH 110*, N Sichuan, rd. from Hongyuan to Songpan, 15 Sep. 1988; from *Shanghai Botanic Garden 05/21*, Gansu, Maiji Shan, ca. 1200 m, 2005.

Cultivated specimens.
Royal Botanic Garden Edinburgh, from *Farrer 355*, 19 Sep. 1924 (E E00112509).

94. Berberis agricola Ahrendt, J. Linn. Soc., Bot. 57: 192. 1961. TYPE: China. SE Xizang (Tibet): [Bomê (Bomi) Xian], Province Kongbo, Dzala, Pasum (Basum) Chu, 30.25°N, 94.033333°E, 3810 m, 3 July 1947, *F. Ludlow, G. Sherriff & H. H. Elliot 14071* (holotype, BM BM000565452!; isotype, A 00038720!, E E00373487!).

Shrubs, deciduous, to 1.5 m tall; mature stems purplish or brownish red, sulcate; spines 1- to 3-fid, pale yellow, 1–2 cm, abaxially sulcate. Petiole almost absent; leaf blade abaxially gray-green, papillose, adaxially deep green, obovate, 1.2–2.8 × 0.7–1.3 cm, papery, midvein raised abaxially, lateral veins and reticulation

raised and conspicuous abaxially, flat and conspicuous adaxially, base cuneate or attenuate, margin entire, apex rounded, minutely mucronate. Inflorescence a raceme, often verticillate, rarely compound at base, 16- to 24-flowered, 4–7.5 cm; bracts ca. 1.5 mm; pedicel 5–7 mm, weak. Flowers ca. 1 cm diam. Sepals in 2 whorls; outer sepals ovate, 3–3.5 × 1.5–2 mm; inner sepals obovate, ca. 5.5 × 4 mm; petals obovate, ca. 4.5 × 3 mm, base clawed, glands separate, apex entire. Stamens ca. 3 mm; anther connective slightly extended, rounded to truncate. Ovules 1 or 2. Berry unknown.

Phenology. *Berberis agricola* has been collected in flower in June and July. Its fruiting period is unknown.

Distribution and habitat. *Berberis agricola* is known from Pasum Chu, Bomê (Bomi) Xian in southeast Xizang (Tibet). It has been found on field edges and hedges bordering cultivation at 3600 and 3810 m.

Berberis agricola is known from only the type and *F. Ludlow, G. Sherriff & H. H. Elliot 13937*. According to Fletcher (1975: 291), Ludlow described this as one of the most handsome *Berberis* he had ever seen and, although it was common in the hedges of Pasum Chu, he found it nowhere else.

Selected specimens.
SE Xizang (Tibet). [**Bomê Xian**]: Kongbo, Kyonka, Pasum Chu, 3660 m, 19 June 1947, *F. Ludlow, G. Sherriff & H. H. Elliot 13937* (A 00279302, BM BM00939709, E E00395924).

95. Berberis ambrozyana C. K. Schneid., Pl. Wilson. (Sargent) 1: 356. 1913. TYPE: China. [W Sichuan, Kangding Xian]: 2440–2745 m, 25 May 1904, *E. H. Wilson (Veitch) 3146a* (lectotype, designated here, K K000077295!; isolectotypes, A 00038712!, BM BM000559590!, K K000567986!).

Berberis dictyophylla Franch. var. *epruinosa* C. K. Schneid., Pl. Wilson. (Sargent) 1: 353. 1913, syn. nov. TYPE: China. W Sichuan: NE of Kangding, Ta-p'ao-shan (Dapaoshan), 3350–3960 m, 7 July 1908, *E. H. Wilson 2866* (lectotype, designated by Ahrendt [1961: 128], K missing, new lectotype, designated here, A 00038739!; isolectotype, US 00103873 image!).

Shrubs, deciduous, to 2 m tall; mature stems purplish red, slightly angled; spines 3-fid, sometimes solitary toward apex of stems, pale yellow-brown, 0.5–1.2 cm, terete. Petiole almost absent; leaf blade abaxially pale gray-green, papillose, sometimes slightly gray pruinose at first, adaxially mid-green, narrowly obovate or oblong-obovate, 0.8–1.2 × 0.15–0.4 cm, papery, lateral veins and reticulation indistinct or obscure on both surfaces, base attenuate, margin entire, sometimes spinulose with 1 to 5 teeth on each side, apex obtuse or sub-acute, minutely mucronate. Inflorescence 1-flowered; pedicel 2–3 mm. Sepals in 2 whorls; outer sepals ovate, ca. 6 × 3 mm; inner sepals lanceolate, 7–9 × 3.75 mm, apex acute; petals narrowly oblanceolate, 7 × 3.75 mm, base slightly clawed, glands separate, oblong, apex slightly emarginate. Stamens 4 mm; anther connective distinctly apiculate. Ovules 5. Immature berry ellipsoid, ca. 5 × 4 mm; style persistent.

Phenology. *Berberis ambrozyana* has been collected in flower in April and June and in fruit in July and August.

Distribution and habitat. *Berberis ambrozyana* is known from Kangding, Litang, and Yajiang Xian in west Sichuan. It has been found among scrub on rocky mountain slopes at ca. (?2440–)3350–4250 m.

The protologue of *Berberis ambrozyana* cited only *Wilson 3146a* without indicating a herbarium. It also gave inaccurate information about the height of the type plant and the date and elevation of its collection (these being later corrected by Schneider [1916: 320]). Ahrendt (1961: 129), who reproduced Schneider's inaccurate information about the date and elevation, cited the type as being at K. There are two specimens at K. I have completed Ahrendt's lectotypification by selecting the specimen that he annotated as type on 21 March 1941.

All of the specimens of the type gathering give the date of collection simply as May 1904 except K K000567986, which is ex HK and has an additional label in Wilson's hand giving the date as "25/5/04." Wilson's diary (Wilson Papers, Arnold Arboretum) records that, on this date, he crossed the Ya-chia (Yajia) Pass which is in Kangding Xian, southwest of Kangding town and, indeed, the diary includes "*Berberis sp.*" in the list of specimens collected on that day. Given the Yajia pass is at an elevation of some 3800 m, this would suggest that Wilson's estimate of the elevation of where the type was collected was inaccurate.

The protologue of *Berberis dictyophylla* var. *epruinosa* designated *Wilson 2866* as the type but without citing a herbarium. Ahrendt's citing of a specimen at K was an effective lectotypification; this, however, is missing. I have, therefore, lectotypified the specimen at A, the protologue being based on material from there sent to Schneider for identification.

Berberis ambrozyana has had a checkered history. Having simultaneously published the taxon and *B. dictyophylla* var. *epruinosa*, Schneider later (1916: 320) decided that the former was a synonym of the latter. This was not followed by Ahrendt (1961: 128) who maintained them as separate taxa, but was followed by

Ying (2001: 75) who mistakenly stated he was making a new synonymy. There are two inter-related but separable issues here. The first is whether *Wilson 2866* is a variety of *B. dictyophylla*, the second whether the synonymy of Schneider and Ying is correct. The first would seem easier to answer than the second in that, as noted in the introduction and below, *B. dictyophylla* is endemic to northwest and northeast Yunnan and immediately adjoining areas of Sichuan and Xizang, but (as noted under that species below) has frequently been used as a taxonomic "dustbin" for a large variety of single-flowered or single-fruited specimens from a wide area of Sichuan stretching into southern Qinghai and northeast Xizang that could not otherwise be identified. *Wilson 2866* seems to be an example of this. The second issue is more difficult to answer since there is no account of the flowers of *Wilson 2866* (the account in Ahrendt [1961] should be discounted since it appears to be based on two specimens from Tibet, neither of which resemble *Wilson 2866*); and, given the small number of flowers on the type specimens, I have not sought permission to dissect any. Given this, it is not completely conclusive that they are the same species. Nevertheless, even without such dissection, it appears that the outer sepals of *Wilson 2866* have the same unusual lanceolate shape that is to be found in *Wilson 3146a*. Why whether the synonymy is correct is particularly important in that *Wilson 3146a* is the source of the color of the mature stems as these are absent from the type collection of *B. ambrozyana*. In this context, it should be noted that, where present, the mature stems of the specimens from Zheduo Shan, cited below, are of a similar color to those of *Wilson 2866*.

Beside the type, the protologue of *Berberis dictyophylla* var. *epruinosa* cited *Wilson (Veitch) 3146* (for which, see below) and *Wilson (Veitch) 3145* (A 00279715, BM, K [2 sheets]). This latter, however, appears to be *B. aemulans*.

Ahrendt (1961: 128) gave a description of the fruit of *Berberis ambrozyana* based on *W. P. Fang 3645*. This is treated here as an example of *B. epedicellata* and its collection details can be found under that species.

Selected specimens.

W Sichuan. Kangding Xian: W China, [probably Zhedou Shan], 4115 m, June 1904, *E. H. Wilson (Veitch) 3146* (A 00279433, BM, K[2 sheets], P P06868554); Zhedou Shan, 3850 m, 31 July 1963, *Sichuan Team – K. G. Guan & W. C. Wang 1196* (PE 01037903, 01037907–8) and *1199* (PE 01037910–11, 01037916); Zhedou Shan, rd. maintenance station 26, 4200 m, 3 June 1974, *Z. S. Qing 06234* (CBDI 0027266, 0027279, 0027303, PE 01840603); Zhedou Tang, 3350 m, 30 Apr. 1986, *T. Naito et al. 534* (PE 01037906, 01037913, TUS); E side of Zeduo pass, 3960 m, 21 Sep. 1991, *SICH 555* (K). **Litang Xian:** hwy. 217 from Litang to Xinlong, 30.311667°N, 100.295556°E, 3630–3675 m, 26 Aug. 2006, *D. E. Boufford et al. 37332* (A, E). **Yajiang Xian:** pass on the rd. from Yajiang to Xinduqiao, 4100 m, 1 June 1991, *J. S. Yang 232* (KUN 0178566).

96. Berberis amoena Dunn, J. Linn. Soc., Bot. 39: 422. 1911. *Berberis sinenis* Desf. var. *elegans* Franch., Pl. Delavay. 35. 1889. *Berberis elegans* (Franch.) C. K. Schneid., Bull. Herb. Boissier, sér. 2, 5: 463. 1905, not H. Lév. (1904), nor K. Koch (1869). *Berberis schneideri* Rehder, J. Arnold Arbor. 17(4): 323. 1936, nom. superfl. TYPE: China. NW Yunnan: [Eryuan Xian], "in collibus calcareis ad pedem montis Yang-in-chan, supra Mo-so-yn," 2500 m, 18 June 1887, *J. M. Delavay s.n.* (lectotype, designated here, P P00716547!; isolectotypes, K K000644927!, P [2 sheets] P00716548–49!).

Berberis amoena Dunn var. *umbelliflora* Ahrendt, J. Linn. Soc., Bot. 57: 154. 1961. TYPE: China. SW Sichuan: Muli Xian, "mountains around Muli," 28.2°N, 3350 m, June 1918, *G. Forrest 16323* (holotype, K K000077415!; isotypes, A 00038713!, E E00373494!, WSY 0057422!).

Shrubs, evergreen, to 1 m tall; mature stems dark red, angled; spines solitary or 3-fid, semi-concolorous, 0.4–1 cm, abaxially sulcate. Petiole almost absent; leaf blade abaxially pale green, pruinose, adaxially dark green, narrowly obovate-elliptic or narrowly elliptic, 1–1.6 × 0.3–0.4 cm, papery, midvein raised on both surfaces, lateral veins and reticulation inconspicuous abaxially, indistinct or obscure adaxially, base cuneate, margin slightly thickened, entire, occasionally spinulose with 1 or 2 teeth on each side, apex obtuse, mucronate. Inflorescence a sub-raceme, sometimes an umbel or sub-umbel, 4- to 9-flowered, 2–5 cm overall including peduncle 0.5–2 cm; bracts ca. 1 mm; pedicel 4–7 mm; bracteoles lanceolate, ca. 1 mm. Flowers orange-yellow. Sepals in 3 whorls; outer sepals ovate, 2 × 1.25 mm; median sepals obovate to oblong-elliptic, 2–3 × 1–2 mm; inner sepals obovate-elliptic, 4–4.5 × 3–3.5 mm; petals obovate, 3.5–4 × ca. 2.5 mm, base cuneate or slightly clawed, glands separate, ca. 0.5 mm, apex emarginate. Stamens ca. 2.5 mm; anther connective slightly extended, truncate. Ovules 1 or 2. Berry red, not or slightly pruinose, oblong, ca. 6 × 3 mm; style persistent.

Phenology. *Berberis amoena* has been collected in flower between May and July and in fruit from July to January the following year.

Distribution and habitat. *Berberis amoena* is known from southwest Sichuan and northwest Yunnan. It has been found on ledges, cliffs, stony places, calcareous rocks, and in open areas at ca. 2700–3300 m.

Though *Berberis sinensis* is an illegitimate name (see discussion under *B. chinensis*), variety *elegans* was validly published. Schneider (1905: 463) renamed the taxon *B. elegans*, but this invalid name was replaced by *B. amoena* by Dunn (1911: 422), a change that was acknowledged by Schneider (1918: 221–222), but not noticed by Rehder (1936: 323) who gave it the superfluous name of *B. schneideri*.

Franchet's original protologue cited three Delavay gatherings: *s.n.*, 18 June 1887; no. *827*, 23 January 1885; and no. *1087*, 8 October 1884. In renaming the taxon *Berberis elegans*, Schneider listed these gatherings and stated "Typus in Herb Berlin," without indicating whether he was referring to one or more of them. His subsequent (1918: 222) citing of *Delavay s.n.* 18 June 1887 as the type did not indicate any herbarium. In any event, any such specimen or specimens at B were destroyed in WWII. Imchanitzkaja (2005: 279) did not refer to Schneider's first-step typification and, in lectotypifying *Delavay s.n.* 18 June 1887 at P, appeared to believe she was completing a lectotypification begun by Ahrendt (1961: 154) who cited this collection as the type but cited no herbarium. Imchanitzkaja, however, appears to have been unaware that there are three specimens of *Delavay s.n.* 18 June 1887 at P. The specimen designated here as the lectotype has been chosen because it includes a dissected flower.

In relation to the other Delavay specimens cited by Franchet in his protologue, the specimen of 23 January 1885 is designated here as *Delavay 827* "A." This is because there is another specimen of *Delavay 827* at P; "Hee-chan-men Lankong," 2400 m, 21 April 1884, which is designated here as *Delavay 827* "B." The specimen of 8 October 1884 is designated here as *Delavay 1087* "A" because there are two specimens at P of *Delavay 1087* of 15 May 1884 which are designated here as *Delavay 1087* "B." Neither *Delavay 827* "B" nor *Delavay 1087* "B" are *Berberis amoena* but are *B. papillifera* although, confusingly, one of the sheets of *Delavay 1087* "B" (P P06868328) also includes a sterile specimen of *B. amoena*. This latter is possibly *Delavay 1087* "A," in which case it is a syntype. There is a sheet of specimens at BM and a specimen at K of *Delavay 1087* "in monte, Hee chan-men, ad Lankong 1883–1885" whose relationship to *1087* "A" and *1087* "B" is impossible to determine.

The holotype of variety *umbelliflora* and the isotype at A have no collection details beyond "1917–1919" and "Yunnan." The details recorded above come from the isotypes at E and WSY. Bao (1997: 75) synonymized the taxon. There appear to have been no collections of *Berberis amoena* from Muli Xian since Forrest's.

No description of the flowers of *Berberis amoena* appears to have been published before the partial one given by Ahrendt (1961: 154), a description reproduced in a slightly expanded form by Bao (1997) and Ying (2001: 168). All three authors inaccurately describe the number of whorls of sepals as two rather than three.

Berberis amoena is sometimes confused with some smaller forms of *B. wilsoniae* but differs from that species in flowering and fruiting earlier and having one or two rather than three to five ovules. From the evidence of *J. M. Delavay 827* "A," Ying (2001) was mistaken in describing the species as deciduous.

Selected specimens.
Syntypes.   *Berberis amoena.* NW Yunnan. Eryuan Xian: "in faucibus ad Kien-min-keou, supra Mo-so-yn," 23 Jan. 1885, *J. M. Delavay 827* "A" (P P00716551); "in monte Yang-in chan," 8 Oct. 1884, *J. M. Delavay 1087* "A" (P P00716550).

Other specimens.
NW Yunnan. Dali Shi: Zhoucheng, E flank Cangshan, Baijian Gou, 22 June 1929, *R. C. Qin 22990* (KUN 0175689, 0175700, PE 01032575); Zhoucheng, E flank Cangshan, Huajian Shan, 24 June 1929, *R. C. Qin 23013* (KUN 0175702–03, PE 01032576, 0175686). **Eryuan Xian:** "Au pied du Monte Yang-in-chan, près de Lang-kong," 30 July 1883, *J. M. Delavay 142* (P P02682379, P06868303); "Yan-in-chan supra Mo-so-yn, June 1887, *J. M. Delavay s.n.* (P P02465482, PE 01901816, SING); "Coteaux calcaires de Yang-in Chan au dessus de Mo-so-yn," 2800 m, 17 June 1887, *J. M. Delavay s.n.* (K, KUN 1206551, NAS NAS00314198, P P02682378); "Coteaux rocailleux de Yang-in Chan au dessus de Mo-so-yn," 19 Oct. 1887, *J. M. Delavay s.n.* (P P06868308); Sanying Qu, Yangyu Shan, 2900 m, 29 July 1963, *Zhongdian Team 63–6246* (KUN 0175681, 0178404, PE 01032577). **Lijiang Shi:** "In callibus calcar. prope Lichiang fu," 2800 m, 8 July 1914, *C. K. Schneider 1783* (A 00279353, GH 00279354, K); "Lidjiang," 2600–2800 m, 13 July 1914, *H. R. E. Handel-Mazzetti 3481* (IBSC 0091682, US 00946067, W 1930/4057, WU 039243); "Hills S.W. of Lichiang," 26.8°N, 100.2°E, 3050–3350 m, May 1922, *G. Forrest 21166* (A 00279357, CAL, E E00395976, N 093057002, P P00580444, UC UC253197, US 00945907, W 1925/6270); E slopes of Likiang Snow Range, May–Oct. 1922, *J. F. Rock 3784* (A 00279352, F E00395979, N 093057125, NY, SYS SYS00052212, US 00945908); Yulong Shan, Baiyang Yan, 3100 m, 17 July 1959, *G. M. Feng 22424* (KUN 0175690–91, 0177245–46, PE 01032581); Yulong Shan, E slope, Muzhu Gou, 2700–2800 m, 5 June 1981, *Qinghai-Xizang Team 359* (CDBI CDBI0028064–65, HITBC 003617, KUN 0175684–85, PE 01032583–84); Yulong Shan, Heishui He Gu, 7 June 1985, *Kunming/Edinburgh Yulong Shan Expedition 614* (E E00623007, KUN 0175631). **Zhongdian (Xianggelila) Xian:** betw. Haba & Xiao Zhongdian, 24 June 1939, *K. M. Feng 1390* (A 00279348, KUN 0175692–93, 0175697); "Alpine meadows north of Chung-tien in Tonwa Territory," 3050 m, Apr.–May 1932, *J. F. Rock 24688* (A 00279350, E E00395973, NY); Donghuan lu, betw. Haba-Cun & Jiangbian Xincun, Yongke at km marker 142 at Mianshaba village, 27.353611°N, 100.161111°E, 2725–2750 m, 19 July 2010, *D. E. Boufford et al. 42056* (A, E).

97. Berberis amurensis Rupr., Bull. Cl. Phys.-Math. Acad. Imp. Sci. Saint-Petersbourg Sér. 2, 15: 260. 1857. *Berberis vulgaris* L. var. *amurensis* (Rupr.)

Regel, Tent. Fl. Uss. 46. 1861. TYPE: Russia (Far East). Amur River, 27 June 1856, *A. G. Schrenk s.n.* (lectotype, designated here, LE!).

Berberis hersii Ahrendt, Gard. Ill. 64, 426. 1944, syn. nov. TYPE: Cultivated. Royal Botanic Gardens, Kew, U.K., 21 May 1939, reputedly from *J. Hers 2783*, China. W Shanxi: [Jiaocheng Xian], Nan-yang Shan, 22 Sep. 1922 (lectotype, designated here, BM BM000794123!).
Berberis amurensis var. *licentii* Ahrendt, J. Linn. Soc., Bot. 57: 194. 1961, syn. nov. TYPE: China. Shanxi: "King hien" [?Xing Xian], 5 May 1916, *E. Licent 1770* (holotype, K K000567989!; isotypes, BM BM000810311!, P P02482735!, PE 01036947!, TIE 00009528 image!).

Shrubs, deciduous, to 3.5 m tall; mature stems pale yellowish brown, pale yellow, or yellowish gray, terete or sub-sulcate; spines 3-fid, sometimes solitary or absent, red-brown, 1–2 cm, abaxially sulcate. Petiole 5–10 mm; leaf blade abaxially very pale green, adaxially yellow-green, obovate-elliptic, elliptic, or ovate, 5–10 × 2.5–5 cm, papery, midvein and lateral veins slightly raised abaxially, impressed adaxially, reticulate veins slightly conspicuous abaxially, inconspicuous adaxially, base cuneate, margin spinulose with 15 to 60 teeth on each side, apex acute or rounded. Inflorescence a raceme, 10- to 30-flowered, 4–10 cm overall; peduncle 1–3 cm; pedicel reddish, (2–)5–10 mm. Sepals in 2 whorls; outer sepals obovate or ovate, ca. 3 × 2 mm; inner sepals obovate, 5.5–6 × 3–3.4 mm; petals elliptic, 4.5–5 × 2.5–3 mm, base slightly clawed, glands separate, apex emarginate. Stamens ca. 2.5 mm; anther connective not extended, truncate or retuse. Ovules 2(to 5). Berry red, not or slightly blue pruinose at base, oblong or ellipsoid, ca. 10 × 6 mm; style not persistent.

Phenology.   *Berberis amurensis* has been collected in flower between April and June and in fruit between May and October.

Distribution and habitat.   In China *Berberis amurensis* is known from Hebei, Heilongjiang, Jilin, Liaoning, Nei Mongol, Shandong, and Shanxi. It is also known from Mongolia, Russia (Far East), and mainland Korea with varieties from the Korean islands of Jeju and Ulleungdo and possibly from Japan. In China it has been found in forest areas and margins, thickets on mountain slopes, streamsides, and trailsides at 700–2100 m.

The protologue of *Berberis amurensis* did not use the term type, but cited specimens from the Amur Valley area collected at the end of June 1856 by Schrenk and of 13 July (year not given) by an unnamed collector. There are specimens corresponding to both at LE, the latter identified as being collected by R. K. Maak. Schneider (1918: 291) cited the Schrenk specimen as type, but did not specify a herbarium. Fedtschenko (1937: 557) cited the type as being in LE but did not cite any specimen, hence my lectotypification. Shi et al. (2016) mistakenly stated that that a *Maximowicz s.n.* at LE is the holotype of *B. amurensis*.

The protologue of *Berberis hersii* cited the type as being *Hers 2783* "Shansi – Nan Yang Shan, 6,500 feet fr. 1923," but it is clear from the discussion that this was based on cultivated material from a plant or plants grown at Kew. This was confirmed by Ahrendt (1961: 194) who cited as the type specimens at OXF of cultivated material from Kew of *Hers 2783* of 12 May 1939 and 21 October 1939. These could not be found, so a cultivated specimen from Kew dated 12 May 1939 at BM has been chosen as the lectotype. It should be noted that *2783* is almost certainly an error in that the Kew Plant Database records only a *B. hersii* grown from *Hers 2730* and donated by the Arnold Arboretum. As noted below, there is a wild-collected herbarium specimen of *Hers 2730* at A from Nan Yang Shan.

Ahrendt (1961: 186) differentiated *Berberis hersii* from *B. amurensis* on the basis that the former has shorter spines and leaves that are more reticulate. This seems a trivial difference. His rationale (1961: 193–194) for a separate *B. amurensis* var. *licentii* was that it had smaller leaves and longer racemes and more flowers than *B. amurensis*, but this variability can often be found on specimens from the same plant.

Though it is easy to distinguish living plants of *Berberis amurensis* from *B. dasystachya*, there can be difficulties in relation to herbarium specimens, particularly those that are sterile or have only fruiting material. This is a particular problem in relation to specimens from Shanxi, for it is here that the most westerly extent of *B. amurensis* meets the most easterly extent of *B. dasystachya*. On current evidence, *B. dasystachya* seems limited in the south of Shanxi while *B. amurensis* is found further north (but see *Huang He Team 434* from Yongji Shi, listed below, which appears to be *B. amurensis*).

No evidence has been found to support the report of *Berberis amurensis* in Henan by B. Z. Ding et al. (1981: 491). The report of the species in Jiangxi by G. F. Zhu (2004: 198) is likely to be referable to *B. virgetorum*; the report of it in Xinjiang by Z. X. An (1995: 4) is likely to be referable to *B. heteropoda*. Specimens from Shaanxi and Gansu identified as *B. amurensis* are mainly referable to *B. dasystachya*.

Schneider (1905: 665) described the number of ovules of *Berberis amurensis* as two and this was repeated by Ahrendt (1961: 193) and Ying (2001: 190). However, a study by Li and Lu (2013) of 200 infructescences of the species from Taishan in Shandong found that while some 90% had one or two seeds per fruit, some 10% had more than this with 1.52% having four

or five seeds. In this context, it worth noting the specimen of *H. Smith 7195* from Jiaocheng Xian in Shanxi at UPS listed below. This has an accompanying undated line-drawing by Schneider recording that one of the berries had three seeds.

For the Korean *Berberis amurensis* var. *quelpaertensis* from Jeju, see Nakai (1936: 71); for variety or forma *latifolia* from Ulleungdo, see Nakai (1909: 41) and W. T. Lee (1996: 350); for both, see Hyun and Kim (2008). For Japan, see Terabayashi (2006: 342) who treated *B. vulgaris* L. var. *japonica* Regel, *B. amurensis* Rupr. var. *japonica* (Regel) Rehder, *B. japonica* (Regel) C. K. Schneid., *B. regeliana* (Regel) Koehne ex C. K. Schneid., *B. amurensis* var. *brevifolia* (Nakai) Ohwe, and *B. bretschneideri* Rehder as synonyms of *B. amurensis*. I have not investigated this matter further.

Selected specimens.

Syntype. *Berberis amurensis* var. *amurensis*. Russia (Far East). Amur Valley, Chutschir Churin Mtns., near mouth of Ussuri River, 13 July 1855, *R. K. Maack 544* (LE); possible syntype "Ad fluvium Ussuri," *R. K. Maak s.n.*, s.d. (P).

Other specimens.
Beijing. Changping Qu: Shisan Ling Gongshe, Xiaozhu Yu, 25 July 1972, *s. coll. Changping 336* (PE 01197744). **Fangshan Qu:** Huangshandian Gongshe, 20 June 1971, *Beijing Herbal Medicine Survey Team – Fangshan 299* (PE 01197746). **Huairou Qu:** Liulimiao Gongshe, 1485 m, 31 May 1964, *Beijing Comprehensive Mountain Survey Team 0106* (HNWP 114032, PE 01219994). **Mentougou Qu:** Qijiazhuang Gongshe, Jiangshui He, 20 July 1971, *Beijing Herbal Medicine Survey Team – Mentougou 294* (PE 01197745). **Miyun Xian:** Potou Linchang, 800 m, 22 May 1965, *s. coll. 819* (PE 01219999). **Yangching Xian:** Sihai Gongshe, 900 m, 25 June 1965, *s. coll. 877* (PE 01236001).

Hebei. Chengde Shi: Wenchang Ge, 800 m, 1 Sep. 1953, *Chengde Team 1522* (PE 01219995). **Chicheng Xian:** Dahaituo, Liufeng Gou, 2100 m, 11–12 Sep. 1959, *s. coll. 7394* (PE 01033596–97, 01033613). **Fuping Xian:** Longquanguan, 1500 m, 25 Sep. 1971, *s. coll. Fuping 062* (PE 01220000). **Huailai Hsien (Xian):** Sep. 1931, *H. F. Chow 42078* (PE 01033614). **Kangbao Xian:** Huangcheng, Zijun Machang, 26 Aug. 1973, *S. R. Liu s.n.* (HIMC 0039327). **Laishui Hsien (Xian):** Jiulongtan, Si'er Gou, 1200 m, 3 June 1934, *K. M. Liou L2296* (PE 01033463). **Laiyuan Xian:** Dianziliang, Hong Shanling, 1100 m, 7 Aug. 1959, *s. coll. 2566* (PE 01033620–21). **Longhua Xian:** Dayakou Shan, 1349 m, 30 May 1984, *L. Bei 0257* (CSFI CSFI004642). **Neiqiu Xian:** Xiaolingdi, Madao Guan, 1550 m, 17 Aug. 1950, *Y. Liu 13349* (PE 01033634). **Pingquan Xian:** Liaoheyuan Natl. Forest Park, 141.320981°N, 118.452917°E, 1330 m, 18 June 2014, *Z. T. Wang et al. 140471* (PE 02039502). **Pingshan Xian:** Muchang, 30 July 1971, *s. coll. 491* (PE 01033635, 01033471, 01236003). **Qinglong Xian:** Niuxin Shan, 340 m, 3 Sep. 1971, *s. coll. Qinglong 876* (PE 01236007). **Weichang Xian:** Huangtu Kan, 1460 m, 2 July 1971, *s. coll. 530* (PE 01219998). **Wu'an Xian:** Liejiang, Shenjiao, 950 m, 12 July 1972, *s. coll. Wu'an 83* (PE 01236145). **Xinglong Xian:** Wuling Shan, Qingliangjie, 1080 m, 23 July 1984, *W. X. Zhang et al. 0226* (PE 01539746–7). **Yu Xian:** Xiao Wutai Shan, Nantai, 1400 m, 14 July 1972, *s. coll. 370* (PE 01219992). **Zanhuang Xian:** Loudizhang Shiyan, 4 Sep. 1971, *Shijiazhuang*

Team 529 (PE 01236004–5, 01033644). **Zhuolu Xian:** Xiao Wutai Shan, 39.989333°N, 115.074722°E, 1600 m, 3 Sep. 2006, *B. Liu 416* (PE 01037588). **Zunhua Xian:** 1200 m, June 1931, *W. Y. Hsia 01887* (PE 01037022).

Heilongjiang. Acheng Qu: Yuquan, 11 May 1986, *G. L. Guo et al. s.n.* (NEAU 052001005000022). **Boli Xian:** Hongqi Linchang, 14 Sep. 1959, *Y. L. Zhang 2125* (FUS 00014261, IFP 05201005x0102–03, PE 01033544, WUK 00189617). **Dailing Qu:** Aug. 1952, *NE Teacher Training College s.n.* (NENU 0001254). **Hailin Xian:** S of Gaoling Zi railway station, 10 Sep. 1983, *G. L. Guo & L. J. Xu 83–580* (NEAU 052001005000010). **Hulin Xian:** Qiyuan, 12 Aug. 1980, *G. L. Guo 1980-177* (NEAU 052001005000008). **Jiage Daqi Shi:** 4 June 1983, *G. L. Guo s.n.* (NEAU 52001005000006). **Jinshantun Qu:** Xishan, 5 Aug. 2014, *Y. Wang 637* (JMSMC 000008712). **Jixan Xian:** Qixing Danlizi, 29 June 1984, *G. L. Guo 84-701* (NEAU 052001005000011). **Luobei Xian:** Baoquanling Farm Management Area, 15 June 1984, *G. L. Guo & L. J. Xu s.n.* (NEAU 052001005000025). **Mishan Xian:** Xiao Guangkai Hu, 24 Sep. 1952, *G. Z. Wang 761* (IFP 05201005x0096–98, PE 01033545, 01033665). **Mohe Xian:** 2 May 1982, *G. L. Guo & Q. Y. Xing s.n.* (NEAU 052001005000005). **Ning'an Xian:** Shihuanzhen, 13 Sep. 1950, *Y. L. Zhang et al. 1223* (IBK IBK00013153, IFP 05201005x0119–20, PE 01033556). **Raohe Xian:** en route to Dadai He, 26 Sep. 1952, *G. Z. Wang et al. 505* (IFP 05201005x0115–16, PE 01033558). **Shangzi Shi:** Mao'ershan, 16 Sep. 1950, *G. Z. Wang & Q. T. Li 108* (FUS 00014306, PE 01033555). **Tangyuan Xian:** Lianzihe Linchang, 314 m, 23 Sep. 2013, *Z. B. Wang 2013762* (JMSMC 00000401). **Tieli Xian:** 24 June 1982, *G. L. Guo & Q. Y. Xing s.n.* (NEAU 052001005000023).**Wudalianchi Shi:** Xiaoxing Anling, 11 July 1956, *S. E. Liu 7118* (KUN 0178697). **Yichun Xian:** N side of Nanshan, 26 July 1956, *S. E. Liu et al. 7754* (IBK IBK00013139, KUN 0178699, PE 01033548, 01033551). **Yilan Xian:** Heibei Shan, 310 m, 29 May 1991, *Harbin Teacher Training College 7297* (IBSC 0091524).

Jilin. Antu Xian: betw. Baishan Linchang & Erdaobai He, 1960, *C. S. Wang et al. 4569* (IBK IBK00013135, 00013143–44, 00013149, IFP 05201005x0053). **Changbai Xian:** Erbianbai He, 700 m, 16 July 1963, *W. Wang et al. 2026* (IFP 05201005x0058–59, PE 01033572). **Fusong Xian:** Lushui He, 9 Sep. 1959, *Tonghua Team 249* (PE 01033569). **Helong Xian:** Baili, 14 Sep. 1959, *Yanbian 2nd Group 838* (PE 01033561). **Yongji Xian:** Qidaji Shan, 7 July 1997, *Botany Group 9539* (PE 01840069)

Liaoning. Benxi Xian: Dongyingfang Gongshe, Xiaodong Gou, 24 Aug. 1965, *Z. Q. Lin 1260* (IFP 05201005x0023, PE 01033582). **Fengcheng Shi:** Aiyang Renmin Gongshe, Dadong Gou, 23 Sep. 1959, *W. Wang et al. 1672* (PE 01033584). **Gaizhou Shi:** 20 May 1950, *S. E. Liu 373* (IFP 05201005x0020–22). **Huanren Xian:** Mayuzi Gongshe, 23 Aug. 1964, *S. C. Cui & Y. C. Zhu 215* (IFP 05201005x0016–17, PE 01033577). **Kuandian Xian:** Baishizhezi Linchang, 30 Aug. 1964, *S. C. Cui & Y. C. Zhu 369* (IFP 05201005x0007–08, PE 01033587). **Lingyuan Xian:** Foyedong Renmin Gongshe, Dongzhuang Dadui, Matou Shan, 7 July 1959, *S. X. Li et al. 332* (IBK IBK00013134, IFP 05201005x0002, PE 01033579, 01033586, WUK 00186983).

Nei Mongol. Bairin (Bailin) You Qi: Hanshan Linchang, 10 July 1991, *W. Wang 517* (HIMC 0010809–11). **Chakar (Chahe'er) Qi:** "Hwa-Teh, Hwang-Hwa-No-Pau," 7 Sep. 1949, *Y. W. Tsui 860* (KUN 0175716–18, PE 01037018–19). **Chakar (Chaha'er) Youyi Zhong Qi:** Beida Shan, Shencong Gou, 17 Aug. 1990, *Y. Z. Zhao 3200* (HIMC 0010807). **Heixiten (Keshiketeng) Qi:** Dajuzi Linchang, 9 Aug. 1983, *W. S. Yang 829* (HIMC 0010739, 0010741).

Harchin (Kalaqin) Qi: Wangyedian Linchang, 18 July 1981, *W. S. Yang 203* (HIMC 0010740). **Hohhot (Huhehaote) Shi:** Picai Gou, Daqing Shan, 19 Aug. 1982, *D. Tao 28* (PE 01033588). **Horqin (Ke'erqin) Zouyi Hou Qi:** Daqingou, 12 July 1964, *C. P. Wang s.n.* (NMAC 000052210). **Liangcheng Xian:** Manhan Shan, 19 Aug. 1965, *Y. Q. Ma & Q. R. Wu 92* (HIMC 0010742). **Morin (Moli) Dawa Qi:** 15 Aug. 2016, *Z. L. Zhang 148* (YAK 002976). **Ningcheng Xian:** Heilihe Gongshi, Daoxu Gou, 6 June 1962, *Mengning Team 78* (PE 01536045). **Oroqen (Erlunchun) Qi:** 16 Aug. 2016, *Z. L. Zhang 150* (YAK 0003110). **Wuchuan Xian:** Daqingshan, 31 May 1974, *S. Q. Zhou s.n.* (NMAC 00040877). **Xilinhot (Xilinhaote) Shi:** July 2005, *S. Q. Zhou 05-05* (NMAC 00018385). **Xingcheng Qu:** Halageng, 18 May 1961, *C. P. Wang s.n.* (NMAC 00005218). **Xiwuzhumuqin Qi:** Beidashan, 25 Aug. 1992, *s. coll. 92-1217* (HIMC 0010806). **Zhenglan Qi:** Caozi Chang, 8 July 1974, *C. P. Wang 197* (NMAC 00032687). **Zhengxiangbai Qi:** Hailihao, Sep. 1974, *S. R. Liu s.n.* (HIMC 0039325). **Zhuozi Xian:** Liangshan, 24 July 1974, *Y. Q. Ma & Q. R. Ru 268* (HIMC 0010732).

Shandong. Laoshan [Qu]: Yuchu-an to Hwaliukow (Hualiukou), 700 m, 6 Oct. 1935, *Y. C. Wang W. 783* (PE 01033750, WUK 004072); Hualiukou, 800 m, 8 June 1959, *T. Y. Zhou et al. 1162* (NAS NAS00314052). **Tai'an:** Taishan Qu, Tai Shan, near Chaoyungdong, 1200 m, 31 May 1956, *Sino-German Team 654* (PE 01033645).

Shanxi. Fangshan Xian: Guandi Shan, Yangetai Linchang, Yangheiqu Gou, 1880 m, 15 July 1965, *P. Y. Li 10209* (HNWP 114041). **Heshun Xian:** 37.3°N, 113.5°E, 1300 m, 27 Aug. 1985, *T. W. Liu & Z. B. Zeng 2664* (TNM S36183). **Huozhou Xian:** Qiliyu, 2000 m, 7 June 1957, *Huang He Team 741* (KUN 0178687, 0179261, PE 01037260). **Jiaocheng Xian:** Nan yang Shan, 2000 m, 22–30 Sep. 1923, *J. Hers 2730* (A 00280180); "Chiao-ch-eng Distr. Yünting Shan," 1900 m, 19 Aug. 1924, *H. Smith 7195* (A 00280173, MO 4577179, PE 01033734, UPS BOT:V-040861);Yaojia Gou, 8 Sep. 1959, *S. Ma 15477* (HNWP 114036). **Lingchuan Xian:** Liuquan Gongeshe en route to Zhitou Cun, 3 July 1959, *S. Y. Bao 1352* (HNWP 114021, PE 01033719, 01033726). **Lingqiu Xian:** Shanmiao, Gouqulin, 25 May 1959, *K. J. Guan & Y. L. Chen 360* (HNWP 131302, PE 01033708, 01033713). **Lingshi Xian:** Shigao Shan, 6 July 1954, *Shanxi Team 366* (PE 01553246). **Ningwu Xian:** Sifanglin, Shuikou Ya, 2000 m, 22 Sep. 1953, *Shanxi Team 359* (HNWP 114037, KUN 0178980, PE 01033473–74, 01387568). **Qin Xian:** Qinyuan, Wayao Giu, 9 May 1959, *K. J. Guan & Y. L. Chen 55* (HNWP 114030). **Wutai Xian:** Yaozigou Cun, 11 Aug. 1959, *K. J. Guan & Y. L. Chen 02623* (PE 01033728, 01033732). **Wuzhai Xian:** Dianping, 1900 m, 17 Aug. 1955, *Huang He Investigation Team 2900* (HNWP 114077, PE 01601588). **Xi Xian:** Sijiao, Taikuan He, Shenlie Gou, 1140 m, 1 July 1962, *T. W. Liu 409* (HNWP 114040, PE 01033723). **Siao-yi Hsien (Xiaoyi Xian):** Shangtingshan, 1800 m, 26 Aug. 1935, *T. P. Wang 3312* (PE 01839951, WUK 004138). **Xing Xian:** near Angou, Xianlong Shan, 1800 m, 13 Sep. 1955, *Huang He Investigation Team 2752* (PE 01033712, WUK 0077346). **Yongji Shi:** Xuehua Shan, 1800 m, 25 May 1957, *Huang He Team 434* (KUN 0178230, 0178689, PE 01601597, 0178230). **Zhongyang Xian:** Yaoyao Cun, Magutou Gou, 1650 m, 1 July 1955, *Huang He Investigation Team 1981* (PE 01033718, 01601589, WUK 0077310).

MONGOLIA. Dornod Province, Khalkhgol (Sümber) Distr., SW slope of Mt. Bayan-Kher-Ul, 46.65°N, 119.783333°E, ca. 1200 m, 26 July 1985, *I. A. Gubanov 9939* (MW MW0179271).

NORTH KOREA. N Kankyo, Yuhyo-Sohyo, 9 Aug. 1917, *E. H. Wilson 8972* (A); S Kankyo, near Shinkou, 18 Sep. 1917, *E. H. Wilson 9187* (A); Pyeonyang, Man'gyŏngdae-guyŏk, Ryongak Mt., 38.633333°N, 128.1°E, 3 Sep. 1996, *s. coll. s.n.* (B B10 0356325).

SOUTH KOREA. Gangwon-do Inje-gun, 3 May 1997, *s. coll. s.n.* (SNU 00016708); Gyeonggi-do Gapyeong-gun, 27 June 1991, *Y. Lee 4243* (SKK); Gyeongsangbuk-do, Seongju-gun, Suryun-myeon, Baekun-gyo, 500–1200 m, 27 Apr. 2006, *H. T. Kim & S. Y. Park 065039* (CNU).

RUSSIA (FAR EAST). Primorje Province, Distr. Schkotovo, st. Kanguaz, 29 June 1972, *5775 T. Neczaeva s.n.* (B, E E00395899, LE, P P002559344, US 01121899).

Specimens not seen.
CHINA. **Heilongjiang:** Jiayin Xian, 4 June 1983, *T. Y. Ding 830445* (IFP 05201005x0134). **Jilin:** Hunjiang Shi, 8 Sep. 1963, *S. X. Li 1196* (IFP 05201005x0069–71). **Liaoning:** Zhuanghe Xian, 9 May 1961, *S. X. Li 172* (IFP 05201005x0035).

98. Berberis angulosa Wall. ex Hook. f. & Thomson, Fl. India 1: 227. 1855. TYPE: Nepal. Kumaon: Gosain Than, *Wallich Cat. 1475*, with later annotation *1475.1* (lectotype, designated by Harber in Adhikari et al. [2012: 461], K-W K000568013!; isolectotypes, BM BM000521634 [specimen on left of sheet!], CAL CAL0000006777 image!, K K000077366!).

Shrubs, deciduous, to 1.2 m tall; mature stems dark reddish brown, terete to sulcate; spines solitary or 3(or 5)-fid, concolorous, 0.4–3 cm, terete or slightly abaxially flat, stout. Petiole 2–4 mm, sometimes almost absent; leaf blade abaxially pale yellow-green, adaxially deep green, obovate to oblanceolate, 1–2.5 × 0.3–1 cm, thinly leathery, both surfaces shiny, midvein raised abaxially, impressed adaxially, lateral veins slightly raised abaxially, impressed adaxially, inconspicuous, base cuneate, margin entire, sometimes spinulose with 1 to 4 teeth on each side, apex rounded or acute, often minutely mucronate. Inflorescence 1-flowered; pedicel 5–20 mm. Flowers 15–20 mm diam. Sepals in 2 whorls; outer sepals ovate or spatulate, 6–10 × 3.5–4.5 mm; inner sepals obovate, 7–10 × 5–7 mm; petals obovate, 5.5–8.5 × 3.5–4.5 mm, base cuneate, glands separate, obovoid, apex rounded, entire. Stamens 4–5 mm; anther connective slightly extended, conical. Ovules 4 to 6. Berry bright red, subglobose, 8–10 × 6–8 mm; style not persistent.

Phenology. In China *Berberis angulosa* has been collected in flower in June and in fruit in August and September.

Distribution and habitat. Berberis angulosa is known in China from Gyirong (Jilong) and Nyalam (Neilamu) Xian in south Xizang (Tibet). It has been found in thickets, open places, and forest margins on mountain sides at 3700–4050 m. It is also found in Nepal, Bhutan, and India (Sikkim).

In the protologue of *Berberis angulosa*, Hooker and Thomson cited "*Wall. Cat. 1475* . . . in parte" and this was repeated by them subsequently (1875: 111) and by Ahrendt (1961: 113) and relates to the fact there are two gatherings with the same number. An annotation by Schneider of May 1906 to K K000077366 (whereas at BM specimens of both gatherings are on the same sheet) identified the other as being *B. umbellata.* I lectotypified K-W K000568013 in Adhikari et al. (2012: 461).

Adhikari et al. (2012) treated *Berberis parisepala* (Ahrendt 1941a: 100) as a synonym of *B. angulosa.* However, though the specimens from Nepal and at least some of the specimens from Bhutan cited by Ahrendt (1961: 114) are *B. angulosa*, the reputed provenance of the type of *B. parisepala* (the Arunachal Pradesh–Tibet border area) suggests this synonymy is mistaken. For details of the complications in relation to the provenance of *B. parisepala*, see excluded taxa below.

Ahrendt (1961: 113) mistakenly stated that *Berberis angulosa* has six to 10 ovules.

The report by Ying (2001: 77) of *Berberis angulosa* being found in Qinghai is mistaken.

Selected specimens.
S Xizang (Tibet). Gyirong (Jilong) Xian: Gyirongzhen, Lamamiao, 3700 m, 13 June 1972, *Xizang Chinese Herbal Medicine Team 226* (HNWP 29949, 79944, PE 00049434, 01037448); betw. Salexiang & Zaxiang, 4000 m, 4 Sep. 1990, *B. S. Li et al. 13502* (PE 01487667–69); 5 km S of Woma Basin, 4000 m, 17 Sep. 1990, *B. S. Li et al. 13829* (PE 01487661–63). **Nyalam (Neilamu) Xian:** near Quxiang, 3800 m, 27 Aug. 1972, *Xizang Herbal Medicine Survey Team 1552* (HNWP 30708, PE 00049435, 01037445); Xiao Gulinyin, 4050 m, 20 June 1975, *Qinghai-Xizang Botany Team 4218* (PE 01037438, 01037446, 01840013).
BHUTAN. Chesha La, Upper Pho Chu, 3930 m, 27 June 1949, *F. Ludlow, G. Sherriff & J. H. Hicks 16650* (BM).
INDIA. Sikkim, 3660–4270 m, 31 May 1936, *L. J. Townsend 85* (OXF).

99. Berberis anhweiensis Ahrendt, J. Linn. Soc., Bot. 57: 185. 1961. TYPE: China. SE Anhui: Wang Shan (Huang Shan), "Back of lion ridge," 1310 m, 12 July 1925, *R. C. Ching 2981* (lectotype, designated here, K K000077392!; isolectotypes, A [2 sheets] 00254039–40!, IBSC 0000583 image!, K K000077391!, NAU 00024182 image!, SYS SYS00052218 image!).

Shrubs, deciduous, to 3 m tall; mature stems yellow or brownish yellow, angled; spines 3-fid, solitary or absent toward apex of stems, concolorous, 1–1.5 cm, terete or abaxially slightly sulcate, weak. Petiole 5–25(–35) mm; leaf blade abaxially pale green, slightly papillose, adaxially bright green, suborbicular, broadly elliptic, or elliptic-obovate, 2–6 × 1.5–3 cm, papery, midvein and lateral veins raised abaxially, impressed adaxially, lateral veins and reticulation inconspicuous on both surfaces, conspicuous abaxially when dry, base narrowly attenuate, margin spinulose with 15 to 40 teeth on each side, apex rounded, obtuse. Inflorescence a raceme, 10- to 27-flowered, 3–7.5 cm overall including peduncle 1–1.5 cm; bracts ca. 1 mm; pedicel 4–7 mm; bracteoles ovate, ca. 1 mm. Sepals in 2 whorls; outer sepals oblong, 2.5–3 × 1.3–1.5 mm; inner sepals obovate, ca. 4.5 × 3 mm; petals elliptic, 4.8–5 × ca. 2.5 mm, base cuneate, glands separate, apex entire. Stamens ca. 3 mm; anther connective not extended, truncate. Ovules 2. Berry red, ellipsoid or obovoid, ca. 9 × 6 mm; style not persistent.

Phenology. *Berberis anhweiensis* has been collected in flower in May and in fruit from July to October.

Distribution and habitat. *Berberis anhweiensis* is known from south Anhui and northwest Zhejiang. It has been found in thickets and forests and by roadsides at 240–1780 m.

The protologue of *Berberis anhweiensis* cited "Type K," but in fact, there are two specimens of *Ching 2981* there. The one with the most material has been chosen as the lectotype.

Berberis anhweiensis is sometimes confused with the very similar *B. henryana* (to which it is probably related), but can be distinguished by the color of its mature stems, those of *B. henryana* being reddish brown, its weak or absent spines, and a somewhat different flower structure. It is also found at lower altitudes than *B. henryana* and appears to be limited to a small area on either side of the Anhui-Zhejiang border.

The report of *Berberis anhweiensis* in Hubei by Fu (2001: 387) is likely to be based on a misidentification of *B. henryana*. No evidence has been found to support the report of the species in Jiangxi by G. F. Zhu (2004: 198).

Selected specimens.
SE & SW Anhui. Huangshan Shi: Wangshan, Sze Tze Ling, 12 Aug. 1924, *A. N. Steward 7184* (A 00280184, 00280225, UC UC248739, US 00945881); Hwangshan (Huangshan), near Shihszefeng, 1700 m, 13 Aug. 1935, *T. N. Liou & P. C. Tsoong 2668* (PE 01033343, WUK 0004130); 31 Oct. 1951, *Chinese Academy of Sciences; East China Station 6444* (KUN 0175721, PE 01033335); 23 Sep. 1965, *H. X. Zhou 781* (HHBG HZ009049, PE 01033348). **Qimen Xian:** Lixi, Guniujiang, 1650 m, 25 May 1981, *Y. F. Xiao & W. Z. Xie 342* (IBSC 0092264). **She Xian:** Huangshan, 17 July 1975, *s. coll. 0610* (PE 01840073). **Yuexi Xian:** Baojia He, 1750 m, 24 Sep. 1953, *Chinese Academy of Sciences; East China Station 6940* (IBSC 0091542, NAS NAS00313861, PE 01033337); Anhui Yaoluoping Natl. Nature Reserve, 900 m, 7 May 2006, *R Zhang & Q. F. Gao 06489* (Yaoluoping Natl. Nature Reserve Herbarium NAU050888).

NW Zhejiang. Anji Shi: Xiaopingxi, 23 Oct. 1955, *X. Y. He 1912* (HHBG HZ009051, IBSC 0091535, WUK 00179375). **Lin'an Shi:** Changhua, 240 m, 1958, *S. Y. He 26623* (IBSC 0091535, NAS NAS00313873, NF 14500041, PE 01033358); W Tianmu Shan, 30 Aug. 1959, *Zhejiang Plant Resources Survey Team 29387* (HHBG HZ009052, PE 01033363); Tianmu Shan, 23 May 1964. *B. L. Qiu 384* (HHBG HZ008977).

100. Berberis aristata DC., Syst. Nat. 2: 8. 1821. TYPE: Nepal. "Berberis Chitria, Hamilt. Don," s.d., *F. Buchanan s.n.* (lectotype, designated by Adhikari et al. [2012: 479], ex Herb. Marti, ex Herb. Lambert, BR BR000000571934!).

?Berberis ceratophylla G. Don, Gen. Hist 1. 115. 1831. TYPE: Nepal. S.d., *N. Wallich s.n.* (no type identified).
Berberis chitria Buch.-Ham. ex Ker Gawl. var. *sikkimensis* C. K. Schneid., Bull. Herb. Boissier, sér. 2, 5: 453. 1905. *Berberis sikkimensis* (C. K. Schneid.) Ahrendt, J. Bot. 80(Suppl.): 85–86. 1942. TYPE: India. Sikkim: Lachoery, 2745 m, 27 Aug. 1849, *J. D. Hooker s.n.* (holotype, B†; lectotype, K K000077367!; isolectotype, CGE 32522!).
Berberis sikkimensis var. *baileyi* Ahrendt, J. Linn. Soc., Bot. 57: 99. 1961. TYPE: Nepal. Gyang, 3 June 1935, *F. M. Bailey's Collectors s.n.* (holotype, BM BM00897072!).

Shrubs, deciduous, to 2 m tall; mature stems pale reddish brown, sulcate or angled; spines 3-fid, solitary or absent toward apex of stems, semi-concolorous, 1–2 cm, terete, stout. Petiole almost absent; leaf blade abaxially pale green, adaxially dark green, obovate to elliptic, 2–6 × 0.5–1.5 cm, thickly papery, midvein raised abaxially, venation conspicuous and slightly raised on both surfaces, base cuneate or attenuate, apex acute or obtuse, mucronate, margin entire or spinulose with 2 to 10 teeth on each side. Inflorescence a raceme or panicle, 10- to 20-flowered, 4–6 cm overall including peduncle 0.5–2 cm; bracts ovate, reddish brown, 2–2.5 mm; pedicel 5–10 mm, slightly glaucous. Flowers ca. 1 cm diam. Sepals in 3 whorls; outer sepals ovate, 2–3 × 1–2 mm; median sepals elliptic or elliptic-obovate, 3–5 × 1.5–3 mm; inner sepals obovate, 6–8.5 × 3–5 mm; petals obovate, 5–8 × 3–5 mm, base cuneate, glands elliptic or obovoid, 0.7–1 mm, apex obtuse, rarely notched 0.2–0.3 mm deep, margin entire. Stamens 4–5.5 mm; anther connective scarcely extended, conical. Pistil 5–6 mm; ovules 3 or 4. Berry dark purple to black, oblong-ovoid, 8–10 × 5–6 mm, slightly glaucous pruinose; style persistent, 1–2.5 mm.

Phenology. In China *Berberis aristata* has been collected in flower between April and July and in fruit between May and October.

Distribution and habitat. *Berberis aristata* is known in China from south Xizang (Tibet). It is also known from India (Sikkim), Nepal, Bhutan, and Myanmar

(Burma). It has been found in Xizang in forest clearings and disturbed vegetation along forest edges and roadsides at 2000–3400 m.

For the somewhat complex issues in relation to the type and name of *Berberis aristata*, see Adhikari et al. (2012: 479–484) who lectotypified the specimen at BR and identified the specimen at LINN-SM as a likely syntype.

Don's protologue of *Berberis ceratophylla* contained a short description and stated "Native of Niapaul? B. floribunda? Wall. MSS," *B. floribunda* being a new taxon from Nepal described by Don on the same page, again with a reference to the Wallich Catalogue. However, since there was no reference to any number in the Catalogue for either taxon, it is impossible to identify which specimens Don was referring to. Despite this, Ahrendt (1945b: 1–2) identified an unnumbered Wallich specimen from Nepal as *B. ceratophylla* stating "Typus Dupl. in Herb. Oxon," which presumably meant he took this to be a duplicate of a specimen in the Wallich Catalogue at K. Ahrendt later (1961: 100) simply cited the specimen at OXF as the type. This OXF specimen is *B. aristata*. Noting that there is no evidence that Don saw this and being unable to discover any specimen that could be original material, Adhikari et al. (2012: 483) declined to designate a type.

Given the destruction of the holotype at Berlin, the citing by Ahrendt (1942: 85–86) of the specimen of *Berberis chitria* var. *sikkimensis* at Kew as the type of *B. sikkimensis* can be treated as an obligate lectotypification. *Berberis ceratophylla* and *B. chitria* var. *sikkimensis* were synonymized by Grierson (1984: 326) and this was followed by Adhikari et al. (2012: 479–482), although in the absence of any type specimen the synonymy of *B. ceratophylla* cannot be proved. Adhikari et al. also synonymized *B. sikkimensis* var. *baileyi*.

Ying (1985: 130) included *Berberis sikkimensis* as a species found in China, but did not include *B. aristata*. This was repeated in Ying (2001: 210) which included treating *B. sikkimensis* var. *glabramea* from Yunnan as a synonym. This latter is treated here as a separate species, *B. shunningensis*.

A particular source of confusion in relation to *Berberis aristata* has arisen in the U.K. where many plants in cultivation identified as *B. aristata* are, in fact, *B. glaucocarpa* Stapf, a semi-evergreen *Berberis* from the west Himalayas easily differentiated from *B. aristata* by its rigid racemose fruit stalk and its densely pruinose berries.

Selected specimens.
Syntype. Berberis aristata. Nepal. Chillong, 10 Apr. 1802, *F. Buchanan s.n.* (LINN-SM 622.7).

Other specimens.
S Xizang (Tibet). Dinggyê (Dingjie) Xian: Chentang Qu, N bank of Kama He, 2400 m, 7 June 1975, *Qinghai-Xizang Team, Botany Section 3808* (PE 01037579–81). **Gyirong (Jilong) Xian:** Mangmu, 2300 m, 4 Nov. 1967, *S. Jiang & C. F. Zhao 359* (PE 00049437, 01037569); Jilong Zhen, 2900 m, 1 June 1972, *Xizang Herbal Medicine Survey Team 59* (HNWP 29176, PE 00049440, 01037567); Jilong Zhen, 2970 m, 14 June 1972, *Xizang Herbal Medicine Survey Team 274* (HNWP 29983, PE 00049439, 01037568). **Nyalam (Nielamu) Xian:** Zhangmu, 2200 m, 2 May 1966, *Y. T. Zhang & K. Y. Lang 3060* (KUN 0179185–86, PE 01037572, 01037586–87, 01840045); Zhangmu hydro-electric power station, 2000 m, 13 May 1966, *Y. T. Zhang & K. Y. Lang 3373* (KUN 0178148, PE 01037584–85). **Tingri (Dingri) Xian:** 3400 m, July 1959; *s. coll. 840* (PE 01037570).
BHUTAN. Chendebi, 2285 m, 12 May 1937, *F. Ludlow & G. Sherriff 3031* (BM).
MYANMAR (BURMA). Chin State, trail to top of Mt. Victoria, Natma Taung Natl., 2750–3050 m, 24 Apr. 2003, *J. Murata et al. 029128* (MBK); Natma Natl. Park, betw. 10 mi. base camp & summit of Natma Taung (Mt. Victoria), 21.51911°N, 93.940333°E–21.243056°N, 93.90108°E, 2680–3053 m, 27 Aug. 2013, *K. Fujikawa et al. 094551* (E, MBK).

101. Berberis asiatica Roxb. ex DC., Syst. Nat. 2: 13. 1821. TYPE: Nepal. *N. Wallich s.n.* (lectotype, designated by Adhikari et al. [2012: 505], ex Herb. Lambert G-DC G00201781!).

Berberis dealbata Lindl., Edwards's Bot. Reg. 21: t. 1750. 1835, syn. nov. TYPE: Cultivated. S.d., *J. Lindley s.n.* (lectotype, designated here, CGE ex Herb. Lindl. 05328!).
Berberis hypoleuca Lindl., J. Hort. Soc. London 2: 246. 1847, syn. nov. TYPE: Cultivated. "raised from seeds received from Dr Royle and said to be from the North of India," "Garden of the Horticultural Society, Nepal, Royle 1844" (lectotype, designated here, GCE ex Herb. Lindl.! 32534).
Berberis asiatica var. *clarkeana* C. K. Schneid., Bull. Herb. Boissier, sér. 2, 5: 457. 1905, syn. nov. TYPE: India. Bihar: Paraṣnath, 1220 m, Hazarbagh, 11 Apr. 1884, *C. B. Clarke 34714B*, (lectotype, designated by Ahrendt [1961: 87], K missing, new lectotype, designated here, G G00343548!).
Berberis vinifera T. S. Ying, Fl. Xizang. 2: 142. 1985, syn. nov. TYPE: China. S Xizang (Tibet): Nyalam (Nielamu) Xian, Zhangmu, E side of hydroelectric power station, 2500 m, 12 May 1966, *Y. T. Zhang & K. Y. Lang 3344* (holotype, PE 00935253!; isotypes, KUN 0178151!, PE [2 sheets] 00935254!, 01840003!).

Shrubs, evergreen, to 3 m tall; mature stems pale grayish yellow, terete or angled, sparsely verruculose; spines 3-fid, usually solitary toward apex of stems, concolorous, 1–2 cm, stout. Petiole almost absent or to 8 mm; leaf blade abaxially pale gray-green or glaucous, papillose, adaxially pale green, obovate, obovate-elliptic, or oblanceolate, 2–6(–9) × 1–3(–5) cm, thickly and rigidly leathery (especially when dry), midvein raised abaxially, slightly impressed adaxially, lateral veins and dense reticulation inconspicuous on both surfaces, conspicuous when dry, base broadly cuneate or attenuate, margin spinose or dentate with 1 to 4 teeth on each side, rarely entire, apex obtuse, usually mucronate. Inflorescence a fascicle or a semi-umbellate raceme, (8- to)15- to 25(to 35)-flowered; bracts ovate-triangular, 1–2 mm; pedicel red, 15–20 mm, slender; bracteoles ovate-triangular, 1.5–2.5 × 1–1.5 mm. Sepals in 2 whorls; outer sepals ovate, ovate-triangular, or elliptic, 2.5–3.5 × 1.5–2.5 mm; inner sepals obovate, oblong-obovate, or elliptic, 4.5–6 × 2–4 mm; petals obovate, 4.5–6.5 × 3.5–4.5 mm, base cuneate or slightly clawed, glands separate, oblong-obovate, apex slightly emarginate, sometimes obtuse. Stamens 3.5–5 mm; anther connective slightly extended or not, obtuse. Ovules 3 to 6. Berry dark purple, glaucous pruinose, ovoid-globose or oblong-ovoid, 7–10 × 5–7 mm; style persistent, 1–1.5 mm.

Phenology. In China *Berberis asiatica* has been collected in fruit between May and July. Its flowering season in China is unknown, but outside China it has been collected in flower between February and April.

Distribution and habitat. *Berberis asiatica* is known in China from Gyirong (Jilong) and Nyalam (Nielamu) Xian in south Xizang (Tibet). It has been found in thickets, on roadsides, and in mixed forests at ca. 2200–2500 m. Besides India and Nepal, it is also found in Bhutan and Myanmar (Burma).

In the protologue of *Berberis asiatica*, de Candolle stated "Hab. in India orientali (Roxb.), in Napaulia (Wallich) . . . (v.s. sp. in h. Lamb.)." This reference was to the Herb. Lambert, though it is unclear whether this was referring to Roxburgh as well as Wallich specimens in this herbarium or whether the reference to Roxburgh was to any specific specimen at all. The lectotypification of the Wallich specimen at G-DC was by Adhikari et al. (2012: 505). Since Wallich did not go to Nepal until 1820, it is likely that the specimen was actually collected there by Edward Gardener and his team of collectors and sent by Wallich to Lambert. The Roxburgh specimen at BR cited below as a possible syntype is ex Herb. Marti. Since Martius was one of those who acquired some of the Herb. Lambert (see Miller 1970 passim), it is possible that de Candolle's reference to Roxburgh was to this specimen.

The protologue of *Berberis dealbata* stated "A native of Mexico whence it was obtained by the Horticultural Society." However, though *Mahonia* is found in Mexico, *Berberis* is not and the leaves and flower structure of the type are more or less as *B. asiatica*.

Two specimens of *Berberis hypoleuca* have been found. The one at CGE has been chosen as the lectotype because it was originally in Lindley's herbarium.

The protologue of *Berberis asiatica* var. *clarkeana* cited "Bengalen: Parasnath 1000 m, [Hooker f.; Clarke n. 34714 B]," but cited no herbarium. However, as noted in the introduction, all the new taxa published by Schneider in 1905 were from one or more of B, G, M, W, or WU. There are two Hooker specimens at W annotated by Schneider as *B. asiatica* var. *clarkeana* and one of *Clarke 34714 B* similarly annotated at G. However, Ahrendt (1961: 87) cited a specimen of *Clarke 34714 B* at K as type. This is missing. The specimen at G has therefore been chosen as a new lectotype. The Hooker specimens at E, GCE, and TCD have exactly the same collection details as the ones at W. There is a Hooker specimen at K (K000644878) annotated as variety *clarkeana* by Schneider in 1906, but this has somewhat different collection details.

Schneider (1905: 456–457) treated *Berberis hypoleuca* as a synonym of *B. asiatica* while, at the same time, publishing *B. asiatica* var. *clarkeana* on the basis that it was fascicled and had leaves that that were more round-elliptic with entire margins versus those of *B. asiatica* which were shortly racemose with broadly ovate leaves with margins that were two- or three-dentate. In fact, *B. asiatica* is a species very variable in leaf shape and margin and in inflorescence. This is illustrated by the type specimens of *B. vinifera* which are fascicled but have leaves with dentate margins. Such characteristics are also to be found on all other specimens from Xizang (Tibet) cited below, with the exception of one specimen of *Qinghai-Xizang Team 6920* (PE 01293941), which is both fascicled and racemose. Though the leaf margins and inflorescence are variable, the thick rigidity of the leaves, the unusually long style, and very early fruiting season are, however, consistent features. The protologue of *B. vinifera* noted all three of these features.

Selected specimens.
Possible syntype. *Berberis asiatica.* S.d., *Roxburgh s.n.* (BR BR0000006956875 ex Herb. Roxburgii, ex Herb. Marti).

Syntypes. *Berberis asiatica* var. *clarkeana.* India. Bihar, "Paras Nath 3–4000 ped," s.d., *J. D. Hooker s.n.* (C C10022123, CAL CAL0000004794, CGE [2 sheets] 32528–29), E E00612755, GH 00872735, P P02313306, TCD TCD0017856, VT UVMVT139806, W 002974–75).

Other specimens.
S Xizang (Tibet). Gyirong (Jilong) Xian: Jiangcun, 2300 m, 6 June 1972, *Xizang Herbal Medicine Survey Team 404* (HNWP 30154, 81320, PE 01293946); Jiangcun, 2400 m, 20 June 1975, *s. coll. 75-235* (PE 01293945); Jilong Qu, Yumu, 2450 m, 12 July 1975, *Qinghai-Xizang Team 6317* (HNWP 49236, KUN, 0175741, 0177583, PE 01293943–44); Jilong Qu, near Mangmu, 2230 m, 16 July 1975, *Qinghai-Xizang Team 6920* (HNWP 50417, KUN 0178309, 0178756, PE 01293941–42).

BHUTAN. Kinga Rapden, Mangde Chu, 1675 m, 2 Apr. 1949, *F. Ludlow, G. Sherriff & J. H. Hicks 18605* (A, BM, E E00168949).

MYANMAR (BURMA). Mandalay Division, Kyauk Padaung Township, Popa Mtn. Park, 1100–1300 m, 25 Feb. 2004, *T. R. Akiyama et al. 030208* (E, MBK); Popa Mtn. Park, 6 Mar. 2006, *S. Gale & K. M. Howe 045045* (E, MBK).

Cultivated material:
Cultivated specimens.
Berberis hypoleuca. "Garden (Lindley) July 1847" (K).

102. Berberis atroviridiana T. S. Ying, Acta Phytotax. Sin. 37: 336. 1999. TYPE: China. SE Xizang: [Zayü (Chayu) Xian], Tsa-wa-rung (Chawalong/ Tsawarong), Nar-jou, 3200 m, Sep. 1935, *C. W. Wang 66536* (holotype, PE 00935121!; isotypes, A 00279988!, IBSC 0091555 image!, LBG 00064315 image!, NAS NAS00314337 image!, PE 00935122!, WUK 0035567 image!).

Shrubs, deciduous, to 1 m tall; mature stems dark purplish red, terete, black verruculose; spines 3-fid, pale brown, 0.5–1 cm, terete, weak. Petiole 1–2 mm; leaf blade abaxially pale green, adaxially deep green, obovate or oblanceolate, 1–1.6 × 0.6–1.1 cm, thickly papery, midvein raised abaxially, impressed adaxially, lateral veins and reticulation inconspicuous on both surfaces, base cuneate, margin slightly revolute when dry, entire or spinulose with 1 to 5 teeth on each side, apex rounded, sometimes minutely mucronate. Inflorescence a sub-umbel or sub-raceme, 4- to 10-flowered, 3–5 cm overall; peduncle 1.7–3 cm; bracts ovate, apex acuminate. Flowers otherwise unknown; fruiting pedicel 2–3 mm, stout. Immature berry oblong-ellipsoid, 7–8 × 3–4 mm; style not persistent; seeds 2.

Phenology. *Berberis atroviridiana* has been collected with immature fruit in September. Its flowering period is unknown.

Distribution and habitat. *Berberis atroviridiana* is known from only the type from the Chawalong/Tsawarong area of Zayü (Chayu) Xian in southeast Xizang. This was collected in *Quercus* forest at ca. 3200 m.

The collector's notes to the type record it as "frequent." The very long peduncle in relation to the overall length of the inflorescence is very unusual.

103. Berberis baiyuensis Harber, sp. nov. TYPE: China. NW Sichuan: Baiyu Xian, side valley of Ou Qu, near Zhandu Xiang and A'se Cun, 31.118611°N, 98.974167°E, 3240 m, 21 Aug. 2006, *D. E. Boufford, B. Bartholomew, S. L. Kelley, R. H. Ree, H. Sun, L. L. Yue, D. C. Zhang,*

Y. H. Zhang & W. D. Zhu 36945 (holotype, A!; isotypes, CAS!, E!, KUN not seen, TI!).

Diagnosis. *Berberis baiyuensis* is somewhat similar to *B. epedicellata*, but its stems are epruinose and its leaves are smaller and abaxially densely reticulate.

Shrubs, deciduous, to 1.5 m tall; mature stems reddish brown, subterete; spines 3-fid, pale yellow or brownish yellow, 0.4–1.8 cm, abaxially sulcate, weak, sometimes solitary toward apex of stems. Petiole almost absent; leaf blade abaxially pale green, adaxially green, obovate, (0.7–)1–1.7 × 0.3–0.7 cm, papery, midvein slightly raised abaxially, largely obscure adaxially, lateral veins inconspicuous abaxially, largely obscure adaxially, dense reticulation inconspicuous abaxially, conspicuous adaxially, base attenuate, margin entire, sometimes spinose with 1 to 5 teeth on each side, apex obtuse, rounded, or acute, sometimes mucronate. Inflorescence 1-flowered, otherwise unknown; fruiting pedicel absent or to 2 mm. Immature berry turning red, partially pruinose, globose or ellipsoid, 5–6 × 5–6 mm; style persistent; seeds 3 or 4.

Phenology. *Berberis baiyuensis* has been collected in fruit in August. Its flowering period is unknown.

Distribution and habitat. *Berberis baiyuensis* is known from only the type and one other collection, both from Baiyu Xian in northwest Sichuan. The former was found on a disturbed river bank at 3240 m, the latter in dense thickets along a river at 3510 m.

IUCN Red List category. *Berberis baiyuensis* is assessed as DD or Data Deficient, according to IUCN (2001) criteria.

The obscure midvein and adaxially dense reticulation of the leaves combined with minimal or absent pedicels makes *Berberis baiyuensis* a very distinct species. The only other published Chinese species with the fruit epedicellate on the branches is *B. epedicellata*. The number of *Berberis* specimens of any description located from Baiyu Xian is currently tiny.

Selected specimens.
NW Sichuan. Baiyu Xian: along tributary of Ou Qu (Ou River) on S side of Ou Qu betw. Zhangdu-Xiang & Marong-Xiang off province hwy. S455, 3150 m, 31.05777778°N. 99.10333333°E, 23 Aug. 2018, *D. E. Boufford, B. Bartholomew, J. L. Guo & J. F. Harber 44868* (A, CAS, E, KUN, PE).

104. Berberis barkamensis Harber, sp. nov. TYPE: China. N Sichuan: Barkam (Ma'erkang) Xian, betw. Ma'erkang & Wolong, 31.815611°N, 102.275917°E, 2855 m, 23 Sep. 2005, *C. S. Chang, H. Kim, H. I.*

Lim, S. Park, C. Ho, W. B. Feng, S. X. Yu & D. M. Zheng SI0488 (holotype, SNUA SNUA00015986!; isotypes, KH [2 sheets] KHB1148121–22 images!, SNUA SNUA00015987 image!).

Diagnosis. *Berberis barkamensis* is somewhat similar to *B. daiana* but with entire leaves, much shorter pedicels, and far fewer seeds per fruit.

Shrubs, deciduous, to 2.5 m tall; mature stems dark reddish purple, sulcate; spines 1- to 3-fid, sometimes absent toward apex of stems, yellow, (0.2–)0.6–1.8 cm, terete. Petiole almost absent, sometimes to 3 mm; leaf blade abaxially pale gray-green, papillose, adaxially green, obovate or narrowly elliptic, 2–2.5 × 1–2 cm, papery, midvein and venation conspicuous and raised abaxially, inconspicuous adaxially, margin entire, apex obtuse or acuminate, mostly mucronate, base attenuate. Inflorescence 1-flowered. Flowers unknown; fruiting pedicel 2 mm. Berry red, ovoid, 10–15 × 7–10 mm; style persistent; seeds 2 or 3.

Phenology. *Berberis barkamensis* has been collected in fruit from June to September. Its flowering period is unknown.

Distribution and habitat. *Berberis barkamensis* is known from Barkam (Ma'erkang), Hongyuan, Xiaojin, and Zamtang (Rangtang) Xian in north Sichuan. It has been found on roadsides and forests on mountain slopes at 2650–3200 m.

IUCN Red List category. *Berberis barkamensis* is assessed as DD or Data Deficient, according to IUCN (2001) criteria.

The large single fruit on very short pedicels make *Berberis barkamensis* a very distinctive species.

Selected specimens.
N Sichuan. Barkam (Ma'erkang) Xian: near Zhuokeji, 2650 m, 6 June 1957, *X. Li 71304* (IBSC 0092662, KUN 0176086), *X. Li 71324* (IBSC 0092679, IFP 05299999w0167, PE 01037773); Emujiao Gou, 3000 m, 18 July 1957, *Z. Y. Zhang & H. F. Zhou 23066* (IBSC 0092517, SZ 00290520); Zhuokeji, Xisuo Gou, 2800 m, 31 July 1957, *Z. Y. Zhang & H. F. Zhou 23427* (IBSC 0092706, KUN 0176078, PE 01037774). **Hongyuan Xian:** near Shuajing Si, 3250 m, 11 Sep. 1957, *X. Li 74555* (IBSC 0092502, PE 01037799, SZ 00291077–78). **Xiaojin Xian:** Meiwo Gou, Zhi Gou, Huaniu Gou, 3200 m, 29 Aug. 1958, *X. S. Zhang & Y. X. Ren 6922* (CDBI CDBI0027295–96, PE 01037917). **Zamtang (Rangtang) Xian:** Hong Xing Gongshe (now Zongke Xiang), Yi Da Dui, E'ri Gou, 18 July 1975, *Sichuan Botany Team 9328* (CDBI CDBI0027291–93, IBSC 0091579, PE 01037802, SWCTU 00006193).

105. Berberis basumchuensis Harber, sp. nov. TYPE: China. SE Xizang (Tibet): Gongbo'gyamda

(Gongbujiangda) Xian, 3 km from Bahe (Beba) Zhen on rd. to Basong Hu (Basum Tso), 29.890556°N, 93.656389°E, 3275 m, 14 Aug. 2000, *D. E. Boufford, S. L. Kelley, R. H. Ree & S. K. Wu 30026* (holotype, A 00279936!; isotypes, CAS 1078425!, HAST 115114 image!, KUN not seen, TI image!).

Diagnosis. *Berberis basumchuensis* has leaves similar to (though narrower than) *B. koehneana* and the same mature stem color, but has a very different inflorescence.

Shrubs, deciduous, to 1.5 m tall; mature stems pale brownish yellow, sulcate, verruculose; spines 3-fid, solitary toward apex of stems, semi-concolorous, 0.4–0.8 cm, terete or abaxially slightly sulcate, weak. Petiole almost absent; leaf blade abaxially pale green, adaxially mid-green, narrowly obovate to obovate-elliptic, (1.5–)2–3 × (0.6–)0.8–1 cm, papery, midvein, lateral veins, and reticulation raised and conspicuous on both surfaces, base narrowly cuneate, margin spinose with 1 to 6 teeth on each side, sometimes entire apex acute or subacute, often mucronate. Inflorescence a fascicle, sub-fascicle, or sub-raceme, 2- to 4(to 7)-flowered, 1.8 cm overall; pedicel 9–13 mm. Flowers unknown. Immature berry oblong-ovoid or ellipsoid, 8 × 4–6 mm; style persistent; seeds 2.

Phenology. The flowering period of *Berberis basumchuensis* is unknown. It has been collected in fruit in August.

Distribution and habitat. *Berberis basumchuensis* is known from only the type collection from Gongbo'gyamda (Gongbujiangda) Xian, southeast Xizang. This was found in woods at 3275 m.

Etymology. *Berberis basumchuensis* is named after the Basum river (chu is Tibetan for river) where the type was collected.

IUCN Red List category. *Berberis basumchuensis* is assessed as DD or Data Deficient, according to IUCN (2001) criteria.

106. Berberis bawangshanensis Harber, nom. et stat. nov. Replaced synonym: *Berberis diaphana* Maxim. var. *uniflora* Ahrendt, J. Linn. Soc., Bot. 57: 124. 1961. TYPE: China. C Sichuan: [Xiaojin Xian-Li Xian border], Pan-lan shan (Bawang-shan), W of Kuan Hsien (Guan Xian, now Dujiangyan Shi), 3650–3950 m, 24 June 1908, *E. H. Wilson 2865* (holotype, K K000644859!; isotypes, A 0038737!, BM BM000559587!, E E00217989!, US 00945925 image!).

Shrubs, deciduous, to 1.5 m tall; mature stems pale yellowish gray, angled; spines 3-fid, pale brownish yellow, 0.5–1(–2.5) cm, abaxially sulcate, solitary or absent toward apex of stems. Petiole almost absent or to 4 mm; leaf blade abaxially pale green, papillose, adaxially mid-green, narrowly elliptic to obovate, 1.2–4 × 0.4–1.5 cm, papery, midvein raised abaxially, lateral veins and reticulation conspicuous abaxially, inconspicuous adaxially, base attenuate, margin entire or spinulose with 1 to 5(to 9) teeth on each side, apex obtuse or acute, sometimes mucronate. Inflorescence a fascicle, 1(to 3)-flowered; pedicel 10–16 mm. Sepals in 3 whorls; outer sepals oblong-lanceolate, 6 × 2.5 mm, apex narrowly acute; median sepals ovate, 6 × 3 mm; inner sepals elliptic-ovate, 5.5 × 4 mm; petals obovate, 4.5 × 2.5 mm, base distinctly clawed, glands separate, apex obtuse, entire. Stamens ca. 3 mm; anther connective extended, shortly apiculate. Pistil 4 mm; ovules 5 or 6. Berry red, ellipsoid or ovoid, 9–11 × 6–7 mm; style persistent.

Phenology. *Berberis bawangshanensis* has been collected in flower in June and in fruit in September.

Distribution and habitat. *Berberis bawangshanensis* is known from the type and two other collections from the Xiaojin-Li Xian border area of central Sichuan. The type was collected at 3650–3950 m, its habitat being unrecorded. *C. S. Chang et al. SI0339* was collected in a mixed forest at 3522 m and *F. T. Wang 21191* on a grassy slope at 3300 m.

Schneider (1913: 354) tentatively identified *Wilson 2865* as *Berberis diaphana*. On 21 March 1941, Ahrendt annotated the specimen at K as "Not B. diaphana . . . ?sp. nov." only later to publish it as the type of *B. diaphana* var. *uniflora*. Given that it differs from *B. diaphana* both in its inflorescence and flower structure, there appears to be no reason to associate it with that species. The taxon has been renamed *B. bawangshanensis* because the epithet *uniflora* is unavailable.

The description of flowers used here is from the isotype at A. As such, it differs somewhat from the description given by Ahrendt for *Berberis diaphana* var. *uniflora*, which may have drawn partially or wholly on a cultivated plant he cited whose provenance was not given and which is quite unlikely to have been *B. bawangshanensis*.

Ying (2001: 97) treated *Berberis diaphana* var. *uniflora* as a synonym of *B. diaphana*.

Selected specimens.

C Sichuan. Li Xian: E part of Zhegushan, 31.241917°N, 102.886194°E, 3522 m, 12 Sep. 2004, *C. S. Chang et al. SI0339* (PE 01839956, SNUA). **[Xiaojin Xian]:** Pa-Lang (Balang) Shan, 3300 m, 5 June 1930, *F. T. Wang 21191*

(A 00280062, KUN 0175619, NAS NAS00314046, PE 01037759–60).

107. Berberis baxoiensis Harber, sp. nov. TYPE: China. SE Xizang (Tibet): Baxoi (Basu) Xian, WNW of Rawu (Raog) on rd. to Bomê (Bomi) along Palongzang Bu (River), 29.481944°N, 96.616389°E, 3840 m, 28 July 2000, *D. E. Boufford, S. L. Kelley, R. H. Ree & S. K. Wu 29770* (holotype, A 00279931!; isotypes, CAS 1078914!, HAST 115004 image!, KUN not seen, TI image!).

Diagnosis. *Berberis baxoiensis* is somewhat similar to *B. dispersa*, but with a much darker mature stem color and four seeds per fruit rather than three. On current evidence, it would also appear to be a smaller shrub.

Shrubs, deciduous, to 1 m tall; mature stems dark reddish purple, sulcate, black verruculose, partially pruinose; spines 3-fid, pale brownish yellow, 0.6–1.2 cm, abaxially sulcate, weak, solitary or absent toward apex of stems. Petiole almost absent; leaf blade concolorous bright green on both surfaces, obovate to oblong-obovate, 1–2.3 × 0.4–0.7 cm, papery, midvein, lateral veins, and reticulation conspicuous and raised abaxially, conspicuous adaxially, base sub-attenuate, margin entire, rarely spinulose with 1 to 3(to 6) teeth on each side, apex acute or obtuse, mucronate. Inflorescence a raceme, sometimes fascicled at base, 5- to 20-flowered, 2.5–3 cm overall including peduncle to 0.6 cm; pedicel 7–13 mm. Flowers unknown. Immature berry turning red, oblong or oblong-ovoid, 10 × 3–4 mm, sometimes slightly bent at tip; style persistent, short; seeds 4.

Phenology. The flowering period of *Berberis baxoiensis* is unknown. It has been collected in fruit in July.

Distribution and habitat. *Berberis baxoiensis* is known from only the type collection in south Baxoi (Basu) Xian in southeast Xizang (Tibet). This was found on a boulder-strewn landscape along a river at the base of a mountain at 3740 m.

IUCN Red List category. *Berberis baxoiensis* is assessed as DD or Data Deficient, according to IUCN (2001) criteria.

108. Berberis beimanica (Ahrendt) Harber, comb. et stat. nov. Basionym: *Berberis muliensis* Ahrendt var. *beimanica* Ahrendt, Bull. Misc. Inform. Kew 1939: 261. 1939. TYPE: China. NW Yunnan: Dêqên (Deqin) Xian, Bei-ma Shan, 28.3°N, 99.166667°E, 3960 m, Oct. 1921, *G. Forrest 21002* (holotype, K K000077419!; isotype, E E00217961!).

Shrubs, deciduous, to 1.5 m tall; mature stems reddish brown, sulcate, verruculose; spines 3-fid, sometimes absent toward apex of stems, semi-concolorous, 0.8–2.2 cm, terete or abaxially slightly sulcate, weak. Petiole almost absent; leaf blade abaxially pale green, papillose, adaxially mid-green, oblong-oblanceolate or narrowly obovate, 2–2.5 × 0.4–1(–1.2) cm, papery, midvein raised abaxially, slightly impressed adaxially, lateral veins and reticulation inconspicuous on both surfaces, base cuneate or attenuate, margin spinose with (1 to)4 to 6(to 8) teeth on each side, occasionally entire, apex rounded or subacute. Inflorescence 1-flowered; pedicel 12–20 mm. Flowers 1–1.2 cm diam. Sepals in 2 whorls; outer sepals broadly ovate, 4–6 × 2–4 mm, apex subacute; inner sepals broadly elliptic, 7 × 4–5 mm, apex rounded; petals broadly elliptic, 4 × 3 mm, base slightly clawed, glands orange, separate, ovoid, ca. 1 mm, apex rounded, emarginate. Stamens ca. 3 mm; anther connective extended, truncate. Ovules 5 to 7. Berry scarlet red, ovoid or ellipsoid, 1.1–1.5 × 0.6–0.8 cm; style persistent.

Phenology. *Berberis beimanica* has been collected in fruit between July and October. Its flowering season outside cultivation is unknown.

Distribution and habitat. *Berberis beimanica* is known from only the type and three other collections, all from Beima Shan, Dêqên (Deqin) Xian in northwest Yunnan. The collection details of the type describe it being found on ledges of cliffs and along slopes at 3960 m; those of the latter three record it being found on open slopes at 4100–4400 m.

In the protologue, Ahrendt stated that *Berberis muliensis* var. *beimanica* was distinct from *B. muliensis* only in having oblong-oblanceolate leaves with spinose margins versus the narrowly obovate leaves with entire margins of *B. muliensis*. However, the differences are more substantial than this in that, while *B. muliensis* has three or four ovules and is 1- to 3-fascicled, *Forrest 21002* has up to seven ovules and appears to be exclusively single-flowered. *Forrest 21002* was also recorded as being only 0.6–0.9 m tall, whereas *B. muliensis* is recorded as being up to 1.5 m tall. These particular characteristics are confirmed by two more recent collections from Beima Shan cited below: *Alpine Garden Society China Exped. (ACE) 1371* (E) recorded as being 1 m tall, and *D. E. Boufford, J. F. Harber & Q. Wang 43132*. This latter was from a small colony, none of which were taller than 1 m.

I have a living plant from seeds of *ACE 1371*. This is the source of the description of flowers used above.

The protologue of variety *beimanica* also cited *G. Forrest 20264* (A, E, K). Subsequently, Ahrendt (1961: 130)

identified this gathering as *Berberis muliensis* var. *atuntzeana*. It is treated here as *B. capillaris*.

Selected specimens.
NW Yunnan. Dêqên (Deqin) Xian: Beima Shan, Shelong, 4100 m, 25 July 1937, *T. T. Yu 9145* (KUN 0176648, 0176650, PE 01037745–46); Beima Shan, 26 June 1965, *s. coll. 4573* (KUN 0176528); Beima Shan, N side of rd., 28.537778°N, 99.002222°E, 4400 m, 26 Sep. 1994, *Alpine Garden Society China Exped. (ACE) 1371* (E E00039598, LIV 2005.15.1083, WSY 0057770); Beima Shan, E slope, along rd. (hwy. G214) at km marker 1918, 28.327222°N, 99.115556°E, 4100 m, 19 Aug. 2013, *D. E. Boufford, J. F. Harber & Q. Wang 43132* (A, E, KUN).

Cultivated material:
Living cultivated plants.
Foster Clough, Mytholmroyd, West Yorkshire, U.K., from seeds of *ACE 1371*, cited above.

109. Berberis biguensis Harber, sp. nov. TYPE: China. NW Yunnan: Zhongdian (Xianggelila) Xian, Bigu, Tianchi lake, 27.525556°N, 99.6425°E, 3890 m, 31 Aug. 2013, *D. E. Boufford, J. F. Harber & X. H. Li 43333* (holotype, PE!; isotypes, A!, CAS!, E!, K!, KUN!, TI!).

Diagnosis. *Berberis biguensis* has leaves which are very similar in shape to *B. lepidifolia*, but a slightly different infructescence with red fruit with two to four seeds rather than black fruit with two seeds.

Shrubs, deciduous, to 1.3 m tall; mature stems pale grayish brown, sulcate, young shoots bright red; spines 1- to 3-fid, absent toward apex of stems, pale brownish yellow, 1–2.5 cm, terete. Petiole mostly most absent, rarely to 2 mm; leaf blade abaxially pale green, papillose, adaxially dark green, oblanceolate or narrowly obovate, (1.5–)2–4 × 0.4–0.7 cm, papery, midvein raised to mid-leaf abaxially, largely indistinct adaxially, lateral veins and reticulation largely indistinct on both surfaces, base attenuate, margin entire, apex acute, mucronate. Inflorescence a loose raceme, sometimes fascicled at base, sometimes a fascicle at apex of stem, (2- to)5- to 10-flowered, to 3.5 cm overall; pedicel 6–15 mm, to 20 mm from base. Sepals in 2 whorls; outer sepals lanceolate, 4 × 1 mm; inner sepals elliptic, 5 × 3.5 mm; petals obovate, 5 × 2.5 mm, glands 1 × 0.8 mm, apex incised. Stamens 2.8 mm; anther connective distinctly extended, apiculate. Pistil 3 mm; ovules 2 to 4. Immature berry yellow, turning red, oblong, 8–11 × 5–7 mm; style persistent; seeds 2 to 4.

Phenology. *Berberis biguensis* has been collected in flower in June and with immature fruit in August.

Distribution and habitat. *Berberis biguensis* is known from only the Bigu Tianchi lake area in Zhongdian (Xianggelila) Xian in northwest Yunnan. It has been found on trailsides at 3719–3890 m.

IUCN Red List category. *Berberis biguensis* is assessed as DD or Data Deficient, according to IUCN (2001) criteria.

The account of the flowers given above is from *Tibet-MacArthur 2317* (details below) at US undertaken on my behalf by Chih-Chieh Yu of NTU. His dissections produced only two ovules per flower, but investigations of many berries of the type collection showed the range to be two to four seeds per fruit.

Selected specimens.
NW Yunnan. Zhongdian (Xianggelila Xian): Xiaozhongdian, on rd. to Tianchi Lake, 27.592889°N, 99.686889°E, 3719 m, 16 June 2009, *Tibet-MacArthur 2317* (F, KUN, US 00863287).

110. Berberis boschanii C. K. Schneid., Pl. Wilson. (Sargent) 1: 369–370. 1913. TYPE: China. NC Sichuan: N Maochou (Mao Xian), Min valley, 1500–1800 m, Oct. 1908, *E. H. Wilson 1166* (lectotype, designated by Ahrendt [1961: 157] K missing, new lectotype, designated here, A 00038733!; isolectotype, US 00103864 image!).

Shrubs, deciduous, to 2.5 m tall; mature stems purplish, angled, slightly shiny; spines rarely 2- or 3-fid, 1–2 cm, terete. Petiole almost absent; leaf blade abaxially pale green, slightly shiny, adaxially dark green, obovate or oblong-elliptic, 0.6–2 × 0.3–0.6 cm, papery, venation indistinct on both surfaces particularly adaxially, base attenuate, margin entire, apex subacute or rounded. Inflorescence a raceme or sub-raceme, 4- to 8-flowered, 1.5–3 cm; pedicel 3–7 mm; bracts triangular, 1 mm. Flowers unknown. Berry red, broadly obovoid or ellipsoid, 6 × 4–5 mm; style absent or persistent and short; seeds 2.

Phenology. *Berberis boschanii* has been collected in fruit in October. Its flowering season is unknown.

Distribution and habitat. The type of *Berberis boschanii* was found in an unrecorded habitat at 1500–1800 m in Mao Xian in north-central Sichuan.

The protologue cited *Wilson 1166* as the type but without citing a herbarium. Ahrendt's citing (1961: 157) of a specimen at K was an effective lectotypification, but this is missing. The specimen at A has been chosen as a new lectotype because it is much better than that at US.

Berberis boschanii is an obscure species. The label on the specimen at US has no collection details beyond the date; the label on the Harvard specimen has "Min Valley Maochou." Wilson's fieldbook (Wilson Papers, Arnold Arboretum) is the source for the collection being

made in north Maochou. Beyond the type, only two possible collections of *B. boschanii* from Mao Xian have been found: *Wilson 3146*, cited in the protologue and whose collection details are even more imprecise than those of the type, and *Sichuan Economic Plant Investigation Team 01* (details below), the specimens of both collections being of poor quality.

Ying (2001: 194) treated *Berberis boschanii* as a synonym of *B. mouillacana*, whose type is from Kangding Xian. This would seem unlikely. It is, however, possible that *B. boschanii* is synonymous with *B. silva-taroucana*, though without information about the flower structure of *B. boschanii*, this cannot be proved. More research is certainly needed here.

Selected specimens.
Sichuan. Min Valley, 1830 m, Sep. 1904, *E. H. Wilson (Veitch) 3156* (A 00279340–41, BM BM000949959). **Mao Xian:** Qianfeng Gongshe, Sep. 1972, *Sichuan Economic Plant Investigation Team 01* (CDBI CDBI0027769–70).

111. Berberis bouffordii Harber, sp. nov. TYPE: China. S Qinghai: Nangqen (Nangqian) Xian, 31.893611°N, 96.596111°E, 4100–4200 m, 22 June 1995, *D. E. Boufford, M. J. Donoghue, X. F. Lu & T. S. Ying 26650* (holotype, A 00279964!; isotypes, CAS 26650!, HAST 73898 image!, TI 73898 image!).

Diagnosis. *Berberis bouffordii* is somewhat similar to *B. qinghaiensis* but with narrower, often oblanceolate leaves and a different flower structure.

Shrubs, deciduous, to 1.2 m tall; mature stems pale yellowish brown, sulcate; spines 3-fid, pale yellow, (0.4–)0.8–1.6 cm, terete or abaxially slightly sulcate. Petiole almost absent; leaf blade abaxially green, adaxially pale gray-green, oblanceolate or very narrowly elliptic, (0.7–)1.4–1.5 × 0.2–0.3 cm, papery, midvein raised adaxially, venation conspicuous abaxially, inconspicuous adaxially, base attenuate, margin spinulose with (1 or)2 to 4 widely separate teeth on each side, sometimes entire at the ends of branches, apex acute, mucronate. Inflorescence 1-flowered; pedicel 12 mm. Flowers yellow. Sepals in 3 whorls; outer sepals oblong to oblong-ovate, 4.5–4.7 × 2.5–3 mm; median sepals elliptic, 4.5–5 × 3–3.5 mm; inner sepals broadly obovate, 5.5–6 × 3.5–3.7 mm; petals obovate, 4.5–5 × 2.7–3 mm, base clawed, glands separate, oblong-ovoid, ca. 1 mm, apex incised. Stamens 3–3.5 mm; anther connective distinctly extended, truncate. Pistil 3–3.5 mm; ovules 4. Immature berry turning red, ellipsoid, 6 × 2 mm; style persistent.

Phenology. *Berberis bouffordii* has been collected in flower and with immature fruit in June.

Distribution and habitat. *Berberis bouffordii* is known from only the type from Nangqen (Nangqian) Xian in south Qinghai. This was found on the gravelly bottom of a west-facing ravine over limestone among sparse *Juniperus* (*Sabina*) and shrub vegetation at 4100–4200 m.

Etymology. *Berberis bouffordii* is named in honor of David Boufford of Harvard University in recognition of his contribution to the flora of China.

IUCN Red List category. *Berberis bouffordii* is assessed as DD or Data Deficient, according to IUCN (2001) criteria.

112. Berberis bowashanensis Harber, Bot. Mag. (Kew Mag.) 34(2): 109. 2017. TYPE: Cultivated. Royal Botanic Garden Edinburgh, Experimental Garden, E/23N, 30 May 2013, *P. Brownless 430*, from seeds of *C. Erskine, H. Fliegner, C. Howick & W. McNamara SICH 1324*, China. W Sichuan: Daocheng Xian, ca. 3 km S of pass on Bowashan, ca. 3780 m, 7 Oct. 1994 (holotype, E E00668747!; isotype, A!).

Shrubs, deciduous, to 2.5 m tall; mature stems pale reddish or yellowish brown, sulcate; spines 3-fid, solitary, paired or absent toward apex of stems, semi-concolorous, 1.5–2.5 cm, abaxially flat or sulcate. Petiole almost absent; leaf blade abaxially pale green, papillose, adaxially dark dull green, narrowly obovate, 1.2–2.4(–3) × 0.5–1 cm, midvein raised abaxially, flat adaxially, veins semi-camptodromous, conspicuous abaxially, inconspicuous adaxially, base attenuate, margin entire, rarely spinulose with 1 to 4 teeth on each side, apex subacute or obtuse, mucronate. Inflorescence a fascicle or sub-fascicle, 3- to 9-flowered; pedicel 9–14 mm; bracteoles absent. Sepals in 2 whorls; outer sepals obovate-oblong, 3 × 1.5 mm; inner sepals broadly obovate, 4 × 3 mm; petals broadly obovate or orbicular-obovate, 4 × 4 mm, base cuneate, glands obovoid, ellipsoid, or orbicular-ellipsoid, ca. 0.5 mm, widely separate, apex entire. Stamens 3.5 mm; anther connective distinctly extended, obtuse. Ovules 2, stipitate. Berry red, oblong or ovoid, 10–12 × 4–5 mm, apex slightly bent; style persistent.

Phenology. *Berberis bowashanensis* has been collected in fruit in October and possibly in flower in June.

Distribution and habitat. *Berberis bowashanensis* is known from only the type collection and possibly one other collection (see below) from Bowashan in Daocheng Xian in west Sichuan. The seeds of the type were collected in a regenerating area on a mountain slope at 3780 m; *Sichuan Vegetation Survey Team 1868* was collected from a roadside thicket at 3900 m.

As noted in the protologue, it is possible that *Sichuan Vegetation Survey Team 1868* (CDBI CDBI0027726–29, SWTC 00006216), Daocheng Xian, Mula Xiang, Kalong, 3900 m, 21 June 1973, from very near where *SICH 1324* was collected, is *Berberis bowashanensis*. However, these specimens are poor ones and I have seen only images of them, which are also poor.

Selected specimens.
W Sichuan. Daocheng Xian: ca. 3 km S of pass on Bowashan, ca. 3780 m, 7 Oct. 1994, *C. Erskine, H. Fliegner, C. Howick & W. McNamara SICH 1324* (K).

Cultivated material:
Living cultivated plants.
Royal Botanic Garden Edinburgh and Royal Botanical Gardens, Kew, from *SICH 1324*.

113. Berberis brachypoda Maxim., Bull. Acad. Imp. Saint-Pétersbourg, sér. 3, 23: 308. 1877. TYPE: China. W Gansu: near Lanzhou, 25 June 1875, *P. J. Piasetzki s.n.* (lectotype, sheet annotated "Typus" inadvertently designated by Imchanitzkaja [2005: 270], LE!; isolectotype, sheet annotated "Isotypus" LE!).

Berberis stenostachya Ahrendt, J. Linn. Soc., Bot. 57: 197. 1961, syn. nov. TYPE: Cultivated. Royal Botanical Gardens, Kew, U.K., 13 Oct. 1939, reputedly from seed of *J. F. Rock 15829*, Gansu (lectotype, designated here, BM BM000794123!).

Shrubs, deciduous, to 4 m tall; mature stems reddish brown, angled, sparsely black verruculose; spines 3-fid, rarely solitary, concolorous, 1–3 cm, abaxially sulcate. Petiole 4–10(–20) mm, sometimes almost absent, pubescent; leaf blade abaxially yellow-green, adaxially dark green, elliptic, obovate, oblong-elliptic, or lanceolate, 3–8(–14) × 1.5–3.5(–5) cm, thickly papery, adaxially rugose, midvein, lateral veins, and reticulation raised, conspicuous, and villous abaxially, impressed and conspicuous adaxially, base cuneate, margin spinose with 20 to 40 coarse teeth on each side, apex acute or subacute, rarely rounded or retuse. Inflorescence a spikelike raceme, 20- to 50-flowered, 5–12 cm overall; peduncle 1.5–4 cm; pedicel 2(–4) mm, pubescent; bracteoles in two series, red, ovate, apex acute. Flowers yellow. Sepals in 2 whorls; outer sepals elliptic, ca. 2 × 1 mm, apex acute; inner sepals obovate-elliptic, ca. 3.5 × 1.5 mm, apex obtuse; petals narrowly elliptic, ca. 5 × 3 mm, base clawed, glands separate, elliptic, apex emarginate, lobes acute. Stamens ca. 2 mm; anther connective not extended, truncate. Ovules 1 or 2. Berry shiny, scarlet, oblong, 6–9 × ca. 5 mm; style persistent.

Phenology. *Berberis brachypoda* has been collected in flower in May and June and in fruit between July and October.

Distribution and habitat. *Berberis brachypoda* is known from central and south Gansu, Ningxia, east Qinghai, northwest and southwest Shanxi, west Shaanxi, and central and north Sichuan. It has been found in thickets and on mountainsides, forest margins, and stream and trailsides at 900–2740 m.

The protologue of *Berberis brachypoda* cited only "prov. Kansu (Piasetzki, 1875)." Schneider (1918: 285) cited a Piasetzki gathering of 13 June 1875 as type, but gave no herbarium for this. I assume this is the same gathering as the two sheets in LE which are annotated as "13/25 VI 1875" (i.e., with the date according to both the Gregorian calendar, still in use in Russia in the 19th century, and the modern one) since one of these has an annotation by Schneider dated 1905. Imchanitzkaja's statement (2005: 270), repeated by Borodina-Grabovskaya (2010), that the specimen on the sheet annotated "Typus" by V. I. Grubov in 1957 is a holotype is inaccurate, but this is treated here as an inadvertent lectotypification.

Ahrendt (1961: 197) cited the type as being at W. However, no specimen could be found there nor any evidence that there was ever such a specimen.

Berberis stenostachya is known only from specimens from a plant grown at Kew. From the evidence of the Kew plant database, this was grown from seed sent by the Arnold Arboretum. The protologue cited "fl. 28 May; fr. 13 Oct. 1939 (both at K) (Type O)," "O" being Ahrendt's acronym for OXF. No specimen could be found there, so a specimen at BM corresponding to one of these dates has been chosen as the lectotype. No wild-collected specimen of *J. F. Rock 15829* has been found and it is likely that this was a mis-transcription of the *J. F. Rock 15029* cited below.

Ahrendt's reasons for differentiating *Berberis stenostachya* from *B. brachypoda* centered largely around minor differences in the length of bracteoles and pedicels and, as such, would seem trivial. At the time of writing, there is still a living plant at Kew grown from "*Rock 15829*" which is clearly *B. brachypoda*.

Even though the shape of the leaves of *Berberis brachypoda* are variable, their texture is very distinct and, as such, make specimens easily recognizable. Despite this, it has been confused with *B. subsessiliflora* from Hubei and its synonym *B. mitifolia*. This confusion seems to have originated with Schneider (1913: 375), who identified three Wilson specimens from that province as *B. brachypoda*. However, as Stapf (1931) pointed out, these specimens have a very different leaf structure.

J. Ma (1986: 490) reported *Berberis brachypoda* being found in Henan, and there are various specimens at PE from there identified as this species. The collection most similar to *B. brachypoda* is *Henan Team 2404* (PE 01032436–37), Song Xian, Longmao, 30 August 1960. Others are *B. subsessiliflora*. Further investigation is needed here.

Selected specimens.
C & S Gansu. Chongxin Xian: Chicheng Si, near Douwan, 1400 m, 4 July 1953, *S. Q. Zhong & P. L. Yang 69* (PE 01032492). **Heshui Xian:** Taibai Zhen, near Biangan Cun, 1320 m, 23 July 1954, *Huang He Team 740* (PE 01032465, 02081916, WUK 0069610). **Hsiaho Hsien (Xiahe Xian):** Tsingsui (Qingshui), 2400 m, 9 July 1937, *K. T. Fu 1120* (KUN 0178196, PE 01032478, SZ 00290946, WUK 0020761). **Huachi Xian:** 27 Sep. 1953, *C. Hou & Y. P. Yang 189* (PE 01032485, WUK 0069630). **Huating Xian:** Huating, Luzhu Gou, 1650 m, 13 July 1953, *Z. B. Wang 16894* (PE 01032486, WUK 0069631). **Jingchuan Xian:** near Er'shilipu, 1300 m, 14 June 1953, *Z. B. Wang 16732* (PE 01032466, WUK 0069632). **Jonê (Zhuoni) Xian:** Tao River basin, Poyuku (Boyu Gou), 2740 m, Sep.–Oct. 1925, *J. F. Rock 13545* (A 00279376, E E00395994, W 1933: 1897). **Kangle Xian:** Lianhua Shan, Mo Gou, 2200 m, 7 July 1992, *Tao He Team 100* (KUN 0178279). **Kaolan Hsien (Gaolan Xian):** 2100 m, 25 June 1937, *T. P. Wang 808* (PE 01032484, WUK 0020760). **Lanzhou Shi:** Xinglong Shan, 2205 m, 4 Aug. 1959, *Y. Q. He 5607* (PE 01032477, WUK 00159769, 0391217). **Lintao Xian:** Mianshan, 19 July 1956, *Academy of Sciences, Huang He Team, Gansu detachment 1649* (NAS NAS00313938–39, PE 01032487, WUK 0083361). **Pingliang Shi:** Kongdong Shan, 1600 m, 11 June 1953, *Shaanxi-Gansu Team 10261* (PE 01032464). **[Têwo (Diebu) Xian]:** "Lower Tebbu country," Wantsang ku (Wangzang Gou), 2590 m, Sep.–Oct. 1926, *J. F. Rock 15029* (A 00279381, AU 038767, E E00395992, NAS NAS00313944). **Tianshui Shi:** betw. Baiyanglin & Liqiao, 1680 m, 25 July 1951, *J. M. Liu 10258* (PE 01037252). **Tsin-shui Hsien (Qingshui Xian):** "inter Kwanschan et Tienshui add austro-or, opp Dungwei," 1500–2000 m, 31 May 1935, *G. Fenzel 1936* (W, WUK 0072640, 0171026). **Weiyuan Xian:** Laojun Shan, 15 July 1956, *Academy of Sciences, Huang He Team, Gansu detachment 1620* (KUN 0178637, NAS NAS00313936–37, PE 01032468). **Wushan Xian:** Tange Zhen, Chenjia Gou, 2 June 1956, *Huang He Team 4426* (PE 01032790, WUK 0085361). **[Xifeng Qu]:** Zhuan Qu, Xifeng Zhen, Xiaohe Gou, 1300 m, 16 June 1954, *Huang He Team 60* (WUK 0069609). **Yuzhong Xian:** Xinglongce, 2300 m, 16 Sep. 1980, *H. B. Cui et al. 80-357* (IBSC 0091592, PE 01032493). **Zhang Xian:** Hedi Cun, 2200 m, 10 June 1956, *Huang He Team 4627* (PE 01032491). **Zhengning Xian:** 1350 m, 22 July 1962, *S. S. Jin et al. 6022* (KUN 0178625). **Zhugqu (Zhouqu Xian):** Kaba, 2100 m, 1 June 1999, *Bailongjiang Exped. 1647* (PE 01556135).

Ningxia. Guyuan Xian (now Yuanzhou Qu): Dongjia Shan, 2000 m, Aug. 1956, *Chinese Academy of Science Yellow River Expedition, Gansu 1st Team 2307* (PE 01032467, WUK 0083128); 29 Sep. 1984, *J. X. Yang 5553* (MO 4764662, WUK 00451587). **Jingyuan Xian:** Longtan Linchang, 1950 m, 6 July 1984, *X. Liu s.n.* (LZU 0008846); 2200 m, 15 Aug. 1985, *Loess Team 4552* (WUK 00461068–69).

Qinghai. Haixi Zhou: Aug. 1985, *s. coll. 4526* (HNWP 110410–11). **Menyuan Xian:** Zhugu Gou, 2500 m, 9 Sep. 1980, *R. F. Huang & Y. C. Yang 1703* (HNWP 95173–74).

Minhe Xian: Xin'er, Beishan, 2500 m, 11 June 1970, *B. Z. Guo 7011* (HNWP 21514).

NW Shaanxi. Ganquan Xian: Er Qu, Laoshan Xiang, Liujiatun Cun, 22 Sep. 1954, *K. J. Fu 8166* (PE 01032452, WUK 0069628). **Wuqi Xian:** Yangguan Yaoxian, 22 Aug. 1956, *Huang He Team 8209* (PE 01032458, WUK 0087194). **Zhidan Xian:** Xinyaowan, 8 Oct. 1953, *Shaanxi-Gansu Team 10831* (PE 01032453–54). **C Shaanxi. Chencang Qu:** Xiangongzhen, Xuanwa Shan, 900 m, 11 May 2014, *Y. He & J. C. Hao GSL2014050086* (BNU 002224). **Feng Xian:** Linggongdian, 1230 m, 17 May 1959, *Nanshuibeidiao Exp. 00275* (KUN 0178808, PE 01032447, 01032462). **Huangling Xian:** 1200 m, 1 Aug. 2013, *Beijing Teachers' University SB-012* (BNU 009247). **Huanglong Xian:** Shibao Gongshe, Qianhu Dadui, 1510 m, 25 July 1983, *Loess Team 339* (WUK 00442469). **Mei Xian:** betw. Tangyu Wenquan & Sha Po, 760 m, 9 May 1978, *Z. Y. Zhang 17761* (IBSC 0091589). **Ningshan Xian:** Huoditang, Luanshi Gou, 1600 m, 28 May 1959, *J. Q. Xing 2574* (IBSC 0091746, NAS NAS00313999. WUK 0121711). **Taibai Xian:** near Renmin Gongshe, 1700 m, 14 July 1959, *J. X. Yang 1123* (PE 01032461, WUK 00133251). **Xunyi Xian:** Zhuanjiao, Dasxi Gou, 1710 m, 2 Oct. 1958, *C. L. Xiao 904* (WUK 00113376). **Zhen'an Xian:** Suzha Shan, 1900 m, 29 May 1983, *X. X. Hou 549* (FJSI 008969, IBSC 0091517).

W Shanxi. Pu Xian: Wulu Shan, Duanjie Liang, Dianjia He, 1375 m, 3 Aug. 1960, *X. Y. Liu 20930* (PE 01032430). **Si Hsien (Xi Xian):** Shik'owtze, 1500 m, 5 Sep. 1935, *T. P. Wang 3489* (IBSC 0091586, PE 01032433, WUK 0004083). **Xiangning Xian:** Guangwangmiao, 17 July 1960, *X. Y. Liu 20886* (HNWP 114049, PE 01032457). **Zhongyang Xian:** Wannianbao, Muhu Kou, 1600 m, 25 July 1955, *Huang He Survey Team 2077* (PE 01032434, 01601602, WUK 0077308).

N & NC Sichuan. Nanping (now Jiuzhaigou) Xian: Raona Gou, Shangpenzi, 2000–2100 m, 19 July 1959, *Nanping Group 4027* (CDBI CDBI0027139–40, 0027146). **Wenchuan Xian:** Longxi Xiang, A'er Gou, 7 June 1952, *Z. He 12612* (IBK IBK00012851, IBSC 0091587, NAS NAS00313947, PE 01032515).

114. Berberis brachystachys T. S. Ying, Fl. Xizang. 2: 137. 1985. TYPE: China. NW Xizang: SW Lhorong (Luolong) Xian, 4000 m, 8 July 1976, *Qinghai-Xizang Exped. 8985* (holotype, PE 01037106!; isotype, PE 02068554 image!).

Shrubs, deciduous, to 2.5 m; mature stems purple, angled; spines 1- to 3-fid, pale brown or brownish yellow, 0.6–2.6 cm, terete, weak. Petiole almost absent or to 5 mm; leaf blade abaxially pale green, papillose, adaxially green, oblong-obovate, obovate-elliptic, or obovate-orbicular, 0.9–1.7(–3.8) × 0.6–1.4(–1.8) cm, papery, midvein raised abaxially, lateral veins and dense reticulation conspicuous on both surfaces, base sub-attenuate or cuneate, margin entire or minutely spinulose with 3 to 12(to 18) teeth on each side, apex subacute, obtuse, or rounded, mucronate. Inflorescence a sub-raceme, sometimes fascicled at base, 4- to 10-flowered, 1.6–1.9 cm; pedicel 6–9 mm. Flowers unknown. Berry red, oblong-ovoid or oblong-obovoid, 7–9 × 5–6 mm; style persistent; seeds 2 to 4.

Phenology. *Berberis brachystachys* has been collected in fruit from July to September. It may flower in May (see discussion below).

Distribution and habitat. *Berberis brachystachys* is known from east and northeast Xizang. It has been found in upland meadows and on road, ravine, and mountain sides at ca. 3400–4300 m.

Berberis brachystachys is a little-known species. In the protologue, a detailed description including information about its distribution was given only in Chinese. Later, its author Ying (2001: 199) treated the taxon as a synonym of *B. dictyoneura* (the nearest collection of which is from Ngawa [Aba] Xian some 600 km from where the type of *B. brachystachys* was collected). A comparison of the two type specimens shows them to be different species, their only similarity being the very prominent reticulation of the leaves.

The type of *Berberis brachystachys* is, however, a poor one and it is unclear why it was selected, since some of the other specimens cited in the protologue are of better quality (as is the isotype, which was not cited in the protologue and at the time of writing is identified on the specimen simply as "sp."). Some of the description of the species used above, including the color of the mature stems and the range in size of the leaves, is from various of the other specimens cited.

It should be noted that there is uncertainty about the number of seeds of *Berberis brachystachys*. The protologue gave the number as being two to four. A note on the sheet of the holotype (which has only immature fruit) records "Seeds 2, ovules 2." A note on the sheet of *Zhang & Zheng 1712* records two seeds per fruit and this is the number found in *Boufford et al. 29650* (which are the best of all the collections cited below). However, notes on the sheet of *Qinghai-Xizang Team, Botany Group 9641* and *9642* record three seeds, while a note on *Qinghai-Xizang Exped. 12420* (PE 01037107), which also has only immature fruit, records "Ovules 4, seeds 2." Given this, it remains a possibility that two different species are being conflated here. Without doubt, more specimens are needed from this remote part of Xizang before any definitive account of *B. brachystachys* can be given.

Selected specimens.
E & NE Xizang. Baxoi (Basu) Xian: betw. Qamdo & Ranwu Gongshe, W side of Rd. Maintenance Station 78, 4190 m, 27 Aug. 1973, *Z. W. Zhang & D. Zheng 1712* (PE 01037105); SE of Bangda, on hwy. 214 to Zogang, 30.171111°N, 97.333611°E, 4100–4300 m, 20 July 2000, *D. E. Boufford et al. 29650* (A 00279702, CAS 1078424, HAST 89736, KUN, TI). **Gongjo (Gongjue) Xian:** Waba Gongshe, 3800 m, 20 Aug. 1976, *Qinghai-Xizang Team, Botany Group 9641* (PE 01037101) and *9642* (PE 01037102). **Jomda (Jiangda) Xian:** 3600 m, 22 July 1976, *Qinghai-*

Xizang Exped. 12420 (HNWP 61325, KUN 0178844, 0179147, PE 01037107, 01840007); Gangtuo Qu, 3200, 27 Aug. 1976, *Qinghai-Xizang Team, Botany Group* 9849 (PE 01037108); Gangtuo military depot, 3200 m, 28 Aug. 1976, *Qinghai-Xizang Team, Botany Group 9922* (PE 01033046). **Lhorong (Luolong) Xian:** betw. the cities of Luolong (Lhorong) & Bianba (Banbar) on hwy. 303, 30.7825°N, 95.595833°E, 3620 m, 20 July 2009, *D. E. Boufford et al. 40785* (A, E E00638628, KUN 1284088). **Markam (Mangkang Xian):** rd. betw. Banda & Zogang, 3800 m, 29 Aug. 1976, *s. coll. 76-565* (PE 01037103). **Qamdo (Changdu) Xian (now Karuo Qu):** Ritong Gou Kou, 3400 m, 25 Sep. 1976, *Qinghai-Xizang Team, Botany Group 10058* (PE 01037104). **Riwoqê (Leiwuqi) Xian:** Riwoqê, E river bank, 3900 m, 21 July 1976, *s. coll. 9230* (PE 01037095); Sangduo Tun, 3700 m, 27 Aug. 1976, *Qinghai-Xizang Exped. 12968* (HNWP 61610, KUN 0178239–40, PE 01037099–100).

115. Berberis bracteata (Ahrendt) Ahrendt, J. Linn. Soc., Bot. 57: 163. 1961. *Berberis dictyoneura* C. K. Schneid. var. *bracteata* Ahrendt, J. Bot. 80: 111. 1944. TYPE: China. NW Yunnan: [Gongshan Xian], upper Kiukiang (Dulong Jiang) Valley, (Clulung) Chöherton, 3200 m, 6 Aug. 1938, *T. T. Yu 19690* (holotype, E E00117365!; isotype, A 00038734!).

Shrubs, deciduous, to 1.5 m tall; mature stems pale yellow-brown, very sulcate; spines 3-fid, concolorous, 0.6–1.5 cm, terete, weak. Petiole almost absent; leaf blade abaxially pale yellow-green, adaxially deep green, shiny, obovate, 0.5–4 × (0.2–)0.7–1.7 cm, papery, midvein raised abaxially, venation and reticulation conspicuous and dense on both surfaces, base attenuate, margin spinose with 5 to 12 teeth on each side, apex rounded. Inflorescence a sub-raceme or sub-fascicle, 5- to 8-flowered, 1.5–2 cm overall including peduncle ca. 7 mm; bracts 4–5 mm; pedicel 2–5(–9) mm. Flowers 7 mm diam., otherwise unknown. Berry red, oblong, sometimes slightly bent at tip, 8–9 × 4–5 mm; style not persistent; seeds 2.

Phenology. *Berberis bracteata* has been collected in August with the remains of flowers and in September with fruit.

Distribution and habitat. *Berberis bracteata* is known from Gongshan Xian in northwest Yunnan. It has been collected on rocky slopes at ca. 3100–4265 m.

The protologue of *Berberis dictyoneura* var. *bracteata*, which cited only the holotype, described the flowers as being 7 mm in diameter but gave no further details. The holotype has one flower. This appears to have ovate outer sepals and possibly obovate inner sepals. The description of fruit given above comes from *Forrest 20295* (details below).

I have been unable to locate exactly where the type of *Berberis bracteata* was collected. The Dulong (Drung) Jiang rises in Zayü (Chayu) Xian in southeast Xizang, then flows into Gongshan Xian in Yunnan, and thence into Myanmar. Though the protologue states *T. T. Yu 19690* was collected in Zayü, the collection details on the sheet give Yunnan. The coordinates in the protologue appear to be speculation by Ahrendt since the collector's notes do not record this.

The collector's notes to the type describe it as 2 ft. (0.6 m) high, and those of *Forrest 20295* as 2–5 ft. (0.6–1.5 m).

Ying (2001: 200) described the mature stems of *Berberis bracteata* as sometimes dull purplish red. This appears to be based on *Beijing Hengduan Mountains Team 01448* (PE 01037117), identified here as *B. gongshanensis*.

Selected specimens.
NW Yunnan. [Gongshan Xian]: Mekong-Salwin divide, Sie-la, 28 28°N, 98.666667°E, 3960–4265 m, Sep. 1921, *G. Forrest 20295* (A 00280008. CAL, E E00395964, US 00945924).

116. Berberis brevipaniculata C. K. Schneid., Bull. Herb. Boissier, sér. 2, 8: 263. 1908. TYPE: China. W Hubei: Patung (Badong) Distr., May 1888, *A. Henry 4675* (holotype, K K000077377!; isotypes, K K00007737!, LE!).

Berberis oblanceifolia C. M. Hu, Bull. Bot. Res., Harbin 6(2): 12. 1986, syn. nov. TYPE: China. N Hunan: Shimen Xian, Dingping Shan, 1299 m, 21 July 1981, *H. S. Liao 15885* (holotype, CSFI CSFI004704 image!; isotypes, CSFI CSFI004711 image!, HNNU [2 sheets] 00002929–30 images!, IBSC 0000589 image!).

Shrubs, deciduous, to 1.5 m tall; mature stems purplish or reddish brown, sulcate; spines 3-fid, yellow, 0.6–1.5 cm, abaxially sulcate. Petiole almost absent or to 3 mm; leaf blade abaxially white pruinose, adaxially dull green, oblanceolate, 1–3 × 0.4–0.8 cm, papery, midvein and dense reticulation conspicuous and raised abaxially, conspicuous adaxially, base attenuate, margin entire, rarely spinose-serrulate with 1 or 2 teeth on each side, apex rounded. Inflorescence a loose, often umbellate panicle, 14- to 24-flowered, 2–9 cm overall; bracts triangular-ovate, 1–1.5 mm; pedicel 6–10 mm; bracteoles ovate, ca. 1 mm. Sepals in 2 whorls; outer sepals elliptic, 2–2.5 mm; inner sepals narrowly ovate, mostly 4.5–5 mm; petals narrowly ovate or obovate, ca. 3.5 mm, base clawed, glands separate, apex emarginate. Stamens ca. 2.5 mm; anther connective apiculate. Ovules 2. Fruit pink, ellipsoid, 4–5 × 3 mm; style persistent.

Phenology. *Berberis brevipaniculata* has been collected in flower from May to July and in fruit in September.

Distribution and habitat. *Berberis brevipaniculata* is known from southwest Hubei, north Hunan, and possibly northeast Jiangxi. It has been found in thickets on limestone hills and valley sides at ca. 800–1800 m.

The protologue of *Berberis brevipaniculata* stated that the type was collected at Ichang (Yichang) and this was repeated by Schneider (1918: 296) and Ahrendt (1961: 204). However, the collection details on the holotype state "Patung District, China. From Dr. A. Henry, Ichang, May 1888." Bretschneider (1898a: 774–775) noted that Patung was a general term used by Henry for Changyang and Badong districts and the mountainous area southwest of Yichang, and that between April and July 1888, Henry was on a collecting expedition travelling south of the Yangtze from Yichang to Wushan in Sichuan (Chongqing), the route of which is largely unknown (see Morley, 1979: 59–60). The "Yichang" on the specimen sheet of *Henry 4675* refers simply to where Henry was based from 1881–1888 (Morley op. cit. passim). As can be seen below, it appears the only areas in Hubei where *B. brevipaniculata* has been found subsequently are Hefeng Xian and Wufeng Xian, and it is possible it is in one of these that *Henry 4675* was collected. But interestingly, I am informed by S. O'Brien of National Botanic Gardens, Kilmacurragh, Ireland (pers. comm. 23 June 2010) that (unmentioned by either Bretschneider or Morley) in May during his 1888 expedition, either Henry or his Chinese collectors briefly visited Shimen Xian in Hunan, so it is possible that the type of *B. brevipaniculata* was collected in the same area as the type of *B. oblanceifolia* and was mislabelled as being collected in Hubei.

The protologue of *Berberis oblanceifolia* described the taxon as having very similar leaves to *B. brevipaniculata*, but with a very different inflorescence. However, a comparison of the two type gatherings confirms they are the same species. It is not just that the leaves are almost identical, but the color of the mature stems is similar (with *B. oblanceifolia*, this is apparent only from the isotypes) and the color of the spines is identical. The flower stalk of the two gatherings are different in length, those of *Henry 4675* being up to 2.8 cm whereas those of *H. S. Liao 15885* are up to 6 cm, but both are paniculate. However, it is clear that the length of the flower stalks is variable, those of *20100627004A* from Hefeng Xian, cited below, being between 2.5 and 9 cm.

The specimens from Jiangxi, cited below, are sterile, but the leaves appear to be the same as *Berberis brevipaniculata*.

Schneider (1916: 318) implied that *Berberis brevipaniculata* is deciduous, as did Ahrendt (1961: 201), though neither said so explicitly. From his key (1986: 3) it is clear that Hu determined *B. oblanceifolia* also to be deciduous. However, without specimens collected in the winter months, this is not fully proven.

There is a living plant in the Washington Park Arboretum, Seattle, U.S., grown from seed from central Hunan, Nanyue Qu, Mt. Heng, 27.25°N, 112.683333°E, s.d., provided by the Shanghai Botanic Garden and identified by them as *Berberis oblanceifolia*. This plant (which is deciduous) has leaves very similar to *B. oblanceifolia*, but a purely umbellate inflorescence. Its berries are shiny scarlet red, epruinose, ellipsoid, 6–7 × 3–4 mm; style not persistent; seeds one or two (information and images supplied by R. Hitchin of Washington Park Arboretum, 5 March and 7 May 2009). The relationship (if any) between this plant and *B. oblanceifolia* requires further investigation.

Both Fu (2001: 388) and Ying (2001: 205) treated *Berberis brevipaniculata* as a synonym of *B. aggregata*, but this is unsustainable since the leaves of the latter are not abaxially pruinose and are rarely entire and the inflorescence is a short, congested panicle. It should be noted that *B. aggregata* is recorded in neither Hubei nor Hunan. Ying (2001) did not notice *B. oblanceifolia*. The inclusion of *B. brevipaniculata* as a species found in the Hengduan mountain range by Ying (1993: 551) is likely to be the result of misidentification.

The relationship between *Berberis brevipaniculata* and *B. pingbaensis* (M. T. An, 2008) needs further investigation. Though the latter is found in central Guizhou, more than 600 km away from west Hubei and north Hunan, its leaves are almost identical as are its flowers which also have two ovules. However, its flower stalk, though paniculate, is up to 14 cm long and the species is semi-evergreen with globose pink fruit similar to that of *B. wilsoniae* (but see above in relation to whether *B. brevipaniculata* is deciduous). The type of *B. pingbaensis* is also from a limestone area.

Selected specimens.
SW Hubei. Hefeng Xian: Er Qu, Wuliping, 800 m, 14 Sep. 1958, *W. B. Lin 801* (HIB 0129627–29); 16 May 1959, *F. S. Peng 478* (HIB 0129619); Mulinzi, Tieluyidai, 22 May 1986, *Z. B. Huang 03864* (CCAU 0002131); Zouma Zhen, Wuli, Qingshan, 1127 m, 27 June 2010, *J. B. Zhang et al. 20100627004A* (TAIF 350885–86); Nanbei Zhen, Xiaojia Yu, 832 m, 27 June 2010, *J. B. Zhang et al. 20100627044B* (TAIF 354179–80). **Wufeng Xian:** Wantan Zhen, 1000 m, 29 Sep. 1990, *F. S. Peng 4648* (HIB 0129626); same location, 1010 m, 6 Oct. 1990, *F. S. Peng 4946* (HIB 0129625).

N Hunan. Dayong Xian: Tianmen Shan, 1400 m, 1 Aug. 1984, *Y. T. Xiao 40746* (CSFI CSFI004705–09, CSFI004712). **Shimen Xian:** Tuwan, 1350 m, 17 May 1987, *Hupingshan Exped. 87160*, (CSFI CSFI004710, PE 01376144); betw. Xiaoxi & Dingping, 1800 m, 12 July 1987, *Hupingshan Exped. 1386* (PE 01376141–3); Dingping, 1569 m, 4 July 2006, *06 Exped. B073* (HHNNR 1839).

Jiangxi. Wuning Xian: Zhujia Shan, 10 Aug. 1938, *Y. S. Xiong 01212* (LBG 00013821); **s. loc.**, s.d., *s. coll. 76-514* (LBG 00013788).

117. Berberis brevipedicellata Harber, Bot. Mag. (Kew Mag.) 33(1): 37. 2016. TYPE: China. NW Yunnan: Zhongdian (Xianggelila) Xian, slope on S side of Daxue Shan Yakou along hwy. S217, 28.562222°N, 99.830278°E, 4040 m, 2 Sep. 2013, *D. E. Boufford, J. F. Harber & X. H. Li 43402* (holotype, PE!; isotypes, A!, BM BM001121295!, CAS!, E [2 sheets] E00758241–42!, IBSC!, K!, KUN [2 sheets]!, MO!, PE!, TI!, US!).

Shrubs, deciduous, to 2.25 m tall; mature stems pale reddish brown, terete; spines 3-fid, pale yellow, 0.8–2.5 cm, slender, solitary or absent toward apex of stems. Petiole almost absent, rarely to 2 mm; leaf blade abaxially pale green, densely papillose, adaxially dull mid-green, oblanceolate or narrowly obovate, 2–4(–5.2) × 0.4–1(–1.3) cm, papery, with slightly raised midvein adaxially, lateral veins and reticulation inconspicuous abaxially, conspicuous adaxially, base narrowly attenuate, margin entire, apex obtuse or subacute. Inflorescence 1-flowered. Flowers unknown; fruiting pedicel 1–4 mm. Immature berry turning red, ovoid, rarely ellipsoid, 11–14 × 6–7 mm, apex sometimes bent; style persistent; seeds 2 to 4.

Phenology. Berberis brevipedicellata has been collected with immature fruit in September, otherwise its phenology is unknown.

Distribution and habitat. Berberis brevipedicellata is known from only the type collection from Zhongdian (Xianggelila) Xian in northwest Yunnan. This was found in roadside thickets on the edge of a deciduous forest on a mountain slope at 4040 m.

As noted in the protologue, only one small colony of this very distinctive species was found.

118. Berberis calcipratorum Ahrendt, J. Linn. Soc., Bot. 57: 130. 1961. TYPE: China. NW Yunnan: [Lijiang Shi], "Eastern flank of Lichiang range," 27.166667°N, 3350–3650 m, May 1910, *G. Forrest 5552* (holotype, K K000077417!; isotypes, BM BM001015565!, E E00217981!, P P00835675!, PE [2 sheets] 01432162–63!).

Shrubs, deciduous, to 2.5 m tall; mature stems dark purplish red, very sulcate, sometimes partially white pruinose; spines solitary or 3-fid, orange, 0.8–1.5 cm, terete or abaxially flat, stout. Petiole almost absent; leaf blade abaxially pale green, adaxially bluish green, narrowly or broadly obovate, 1–2.5 × 0.4–0.9 cm, thinly papery, white or glaucous pruinose abaxially, midvein, lateral veins, and reticulation conspicuous and raised abaxially, conspicuous adaxially, base cuneate, margin slightly thickened, entire, sometimes spinulose with 1 to 5 teeth on each side, apex rounded, sometimes retuse. Inflorescence a sub-umbel or sub-raceme, 2- to 12-flowered; bracts ca. 2 mm; pedicel 5–10 mm, white pruinose; bracteoles ovate, ca. 3 mm, apex acuminate. Flowers orange-yellow. Sepals in 2 whorls; outer sepals elliptic, ca. 5.5 × 4 mm; inner sepals obovate-elliptic, ca. 7.5 × 4 mm; petals oblong-obovate, ca. 5.5 × 3.5 mm, base clawed, glands separate, elliptic, apex emarginate. Stamens ca. 3.5 mm; anther connective slightly extended, apiculate. Ovules 2 to 4, shortly stipitate. Immature berry turning red, narrowly ellipsoid, ca. 7 × 4 mm, pruinose; style persistent.

Phenology. Berberis calcipratorum has been collected in flower in May and June and with immature fruit from July to September.

Distribution and habitat. Berberis calcipratorum is known from Yulong Xue Shan in Lijiang Shi, and Haba Yue Shan and Tianbao Xue Shan in Zhongdian (Xianggelila) Xian in northwest Yunnan. It has been collected from roadsides, forest margins, limestone meadows, and stony slopes at ca. 2900–3650 m.

The protologue of *Berberis calcipratorum* mistakenly gave the place of collection of the type as the Dali range.

Selected specimens.
NW Yunnan. Lijiang Shi: [Yulong Xue Shan], "in collibus prope pagum Ngu leh keh versus orient," ca. 3300 m, 25 July 1914, *C. K. Schneider 1993* (A 00279912); "Eastern slopes of Likiang Snow Range," May–Oct. 1922, *J. F. Rock 3476* (A 00279917, US 00945980); same details, *J. F. Rock 3842* (A 00279916, US 00945981); same details, but Lame-le, *J. F. Rock 3682* (A 00279915, US 00945978); Sandawan, 2900 m, 27 Apr. 1937, *T. T. Yu 5104* (KUN 0176462, PE 01033896); Yulong Shan, 3350 m, 22 June 1957, *Z. J. Gao 1332* (PEM 0002233–34); Yulong Xue Shan, Muzi Gou, Gan He Ba, 3000 m, 23 May 1985, *Kunming Edinburgh Yulong Shan Exped. 20* (E E00375995, KUN 0177176). **Zhongdian (Xianggelila) Xian:** "N flank of Haba Snow Range," 10 June 1939. *K. M. Feng 1274* (A 00279918, KUN 0177168–70); Haba Yue Shan, 3100 m, 30 June 1963, *Zhongdian Team 2948* (KUN 0176666, 0176673); Tinbao Xue Shan, 3400 m, 13 July 1963, *Zhongdian Team 3580* (KUN 0176667, 0176672); Tianbao Xue Shan, 3600–3650 m, 15 June 1981, *Qinghai-Xizang Team 1099* (CDBI CDBI0028170, 0028184, HITBC 003621, KUN 0176657, PE 01032922, 01032945); Xiao Zhongdian, Tianbao Shan, 27.605278°N, 99.885833°E, 4 Sep. 2013, *D. E. Boufford, J. F. Harber & X. H. Li 43439* (A 00914431, BM BM001190940, CAS, E E00833564. K, KUN, MO, PE).

119. Berberis calliobotrys Bien. ex Koehne, Deut. Dendrol., 168. 1893. TYPE: Afghanistan. Kurrum Valley, Drékalla, 17 Aug. 1879, *J. E. T. Aitchison 176/4* (neotype, designated here, K!, isoneotypes, G G00332688 image!, GH 01153598!).

Berberis aitchinsonii Ahrendt, J. Roy. Asiat. Soc. Bengal 11: 4. 1945. TYPE: Afghanistan. Kurrum Valley, Sergal to Gunht, 2745–3350 m, 10 Aug. 1879, *J. E. T. Aitchison 176/5* (K K000644868!).

Shrubs, deciduous, to 3 m tall; mature stems reddish purple or reddish brown, partially pruinose, sub-terete or sub-sulcate; spines 1- to 3-fid, yellowish brown, 1.7–3 cm, abaxially sulcate, stout. Petiole almost absent, sometimes to 2 mm; leaf blade abaxially pale green, papillose. adaxially green, narrowly obovate or obovate-elliptic, 1.5–2.3 × 0.5–0.8 cm, papery, midvein, venation, and reticulation conspicuous and raised abaxially, conspicuous adaxially, base narrowly attenuate, margin entire especially on upper part of twigs, sometimes spinose with 1 to 5 widely spaced teeth on each side, apex subacute or obtuse, mucronate. Flowers uncertain. Infructescence a fascicle, sub-fascicle, or short raceme, 3- to 8-fruited, to 1.7 cm overall; pedicel 6–8 mm. Berry black, globose, 6 × 6 mm; style persistent; seeds 2 or 3.

Phenology.   In China *Berberis calliobotrys* has been collected with immature fruit in June. Its phenology is otherwise unknown. The species is also found in Afghanistan and Pakistan.

Distribution and habitat.   *Berberis calliobotrys* is known in China from only one collection made in southwest Xinjiang. This was found among *Picea* and *Juniperus* on the slopes of a broad valley at 3000 m.

Shortly before completing this study, I was sent an image by Bruce Bartholomew of CAS of a *Berberis* specimen he had collected in southwest Xinjiang. I could see immediately that it was of not of any species previously recorded for Xinjiang or indeed anywhere else in China. It was, therefore, either of an unrecognized species or, more likely, of a species found in one or more of the neighboring areas of Tajikistan, Afghanistan, Pakistan, and India.

Information about the *Berberis* taxa of these neighboring areas is often confusing and contradictory and a substantial revision is needed. However, after a process of elimination, it became clear that the Xinjiang species is *B. calliobotrys*.

Berberis calliobotrys has had a somewhat complicated history. The name first appeared on annotations to various specimens in the Herb. Bunge collected by the Baltic German botanist Theophil Bienert on the Imperial Russian Geographical Society's expedition to Khorasan in 1858–1859 on what is now the Iranian border with Turkmenistan. These are at LE, but there are duplicates at G, K, and P.

The name first appeared in print in Aitchison (1881, 1882), a two-part account of the author's botanical col-

lections in 1879 in the Kuram (Kurrum) Valley, then in Afghanistan but now partially in the Federally Administered Tribal Areas of Pakistan. In the first part, Aitchison (1881: 31) identified as "*B. calliobotrys* Bienert" a collection numbered *726*, "Hills above Kaiwás, 11,000 feet; a large bush." Presumably he so identified it on the basis of seeing one of Bienert's specimens, probably the one at Kew. Since "bush" describes the vast majority of the species in the genus, this entry should be regarded as a nomen nudum. However, in the second part (Aitchison, 1882: 150–151), he did give a rudimentary description, but at the same time withdrew his previous identification (now given as *Berberis calliobotrys* Herb. Bunge) and stated his *726* as well as various others of his collection, *171, 176/1, 176/4, 352, 273,* and *490,* were one of three varieties of *B. vulgaris* he had collected on his expedition (these identifications were mistaken since *B. vulgaris* is not found in either Afghanistan or Pakistan).

The name was finally validly published by Koehne (see above), a professor of botany at Berlin, as "*B. calliobotrys* Bienert? im Herb. Bunge." The accompanying brief description cited no specimens, but from the information given by Koehne about the sources of his work, these were from that of Berlin-Dahlem and/or his personal herbarium (acquired by Berlin-Dahlem in 1919), specimens that will have been lost with the destruction of that herbarium in 1943. His description included information about the fruit of the taxon (described as almost as broad as long and having a conspicuous style) but none about its flowers except that the number of ovules was three (and contextually, this may just have been based on a seed count). Since the Bienert specimens have only floral material, at least one or more of the specimens Koehne was drawing on must have been different and it is reasonable to conclude that they must have included ones collected by Aitchison, since otherwise there would seem no reason for Koehne to use the name. This appears to be confirmed by the distribution which was given as "Afghan., Himal." Koehne's use of "?" suggests he had not actually seen any of Bienert's specimens at this time, but was taking his information from Aitchison's 1881 and 1882 entries. Had he done so, he would surely have included Persia in the list of where *Berberis calliobotrys* was found, since "Iter persicum [Persian journey]" is to be found on all of Bienert's specimens. That he had not seen Bienert's specimens, then, would seem to be confirmed subsequently by Koehne (1899: 20) where he states that Aitchison's *B. calliobotrys* was different from that of the "echten [true] *calliobotrys* Bienert."

Schneider (1905: 661–662) described a "*Berberis calliobotrys* Aich.," making it clear this was different from Bienert's use of the name. His description was based solely on fruiting material and was slightly different from that of Koehne and included giving the number of seeds as two. Only two specimens were cited: *Aitchison 176/4* and *Aitchison 171* (though the other numbers cited by Aitchison were listed). No herbarium was given for *176/4* and *171*, but there are specimens of both at G annotated on 28 August 1904 by Schneider as "Berberis calliobotrys Aitch. nec Bienert."

Ahrendt (1961: 225) reported the species as *"Berberis calliobotrys* Aitch . . . nomen" followed by citing Koehne and Schneider's publications dealing with the taxon. *Aitchison 490* at Kew (which has floral rather than fruiting material) was cited as the type. The number of ovules was given as two.

Both Jafri (1975) in the *Flora of West Pakistan* and Rao et al. (1998b: 134) in a revision of Indian *Berberis*, however, cited the taxon as Aich. ex Koehne, though they followed Ahrendt as giving the type as *Aitchison 490*.

A different approach was taken by Browicz and Zieliński in the Flora Iranica (1975). Here the taxon was described as "Koehne . . . non Bienert." The type was said to be from Afghanistan, but no actual specimen was given for this. Of the numbers listed by Aitchison in 1881 and 1882, the authors cited *176, 176/2, 176/3, 176/4, 273,* and *490*.

There is, therefore, an unresolved problem in that as the valid publisher of the name, Koehne made no reference to Aitchison and cited no specimens, but gave a passable description of fruit (though none of flowers), while Ahrendt credited Aitchison as the author of the name and cited as type a specimen with flowers. My solution to this is to designate *Aitchison 176/4* at Kew as a neotype. This corresponds largely to Koehne's admittedly brief description including his description of its fruit. Importantly, a dissection made of *176/4* at Kew produced two or three seeds per fruit (this is the same number as found in Bartholomew et al.'s Xinjiang collection).

It is possible that *Aitchison 490* is *Berberis calliobotrys*, in which case the description of flowers given by Ahrendt (1961: 225) would apply, but since he gives the number of ovules as being only two, it is not reproduced here.

Browicz and Zieliński (1975) synonymized *Berberis aitchisonii* (described in the protologue as having one to three seeds), *B. gambeliana* (Ahrendt, 1945b), and *B. wazaristanica* (Ahrendt, 1945b) though the types of the last two were not listed among the specimens they had seen. The type of *B. gambeliana* (K K000644933), the only specimen cited in the protologue, is a poor one, while I have been unable to locate the type of *B. wazaristanica.* I have not pursued these two taxa further.

The specimens annotated by Bienert as *Berberis calliobotrys* are, as Schneider and Koehne recognized,

clearly different from Aitchison's specimens, but I have not found it necessary to identify their species.

I am grateful to C. C. Yu of NTU who, while studying in the United States, undertook the dissection of fruit of the CAS specimen for me and thus avoided the necessity of it being sent to the U.K. and back.

Selected specimens.
SW Xinjiang. Kargilik (Yecheng) Xian, Paolong, N side of Kunlun Mtns., 37.088889°N, 77.121944°E, 3000 m, 30 June 2001, *B. Bartholomew, I. Al-Shehbaz, A. Abbas, A. Tumur, P. Bhati & Z. Zunus 8386* (CAS 1069009, MO 5724793, PE, XJU [2 sheets]).

120. Berberis campylotropa T. S. Ying, Fl. Xizang. 2: 152. 1985. TYPE: China. NE Xizang: Riwoqê (Leiwuqi) Xian, Sangduo Qu, 3700 m, 29 Aug. 1976, *Qinghai-Xizang Complex Exp. 12961* (holotype, PE 00935220!; isotypes, HNWP 61629 image!, KUN [2 sheets] 0178241–42!, PE 00935221!).

Shrubs, deciduous, to 1.5 m tall; mature stems reddish purple, terete; spines 3-fid or solitary, yellowish, 0.4–0.8 cm, terete, weak. Petiole almost absent; leaf blade abaxially gray-green, papillose, sometimes pruinose, adaxially dark green, narrowly elliptic, oblanceolate or narrowly or broadly obovate, 0.8–1.6(–2) × 0.3–0.7(–1) cm, papery, midvein raised abaxially, flat adaxially, lateral veins and reticulation conspicuous on both surfaces, base cuneate, margin entire or spinulose with 1 to 5 teeth on each side, apex subacute to obtuse. Inflorescence a fascicle, 1- to 3(to 4)-flowered, otherwise unknown; fruiting pedicel 6–12 mm. Berry red, slightly pruinose, oblong, 10–12 × 4–5 mm; style persistent, sometimes bent; seeds 2 or 3.

Phenology. *Berberis campylotropa* has been collected in fruit in July and August. Its flowering period is unknown.

Distribution and habitat. *Berberis campylotropa* is known from Dêngqên (Dinqing) and Riwoqê (Leiwuqi) Xian in northeast Xizang. It has been collected from mountain slopes, scrub in degraded shale areas, and roadsides at ca. 3700–3900 m.

The protologue cited *Qinghai-Xizang Complex Exp. 12916* as the type. This is clearly a typographical error, since the specimen of this at PE is not *Berberis*, but is identified as *Roegneria brevipes*, whereas *12961* at PE is annotated as the type.

Selected specimens.
NE Xizang. Dêngqên (Dinqing) Xian: Dangdui Qu, Baiguo Si, 3900 m, 15 July 1961, *s. coll. 1664* (KUN 0178407). **Riwoqê (Leiwuqi) Xian:** near Changmaoling, 25

July 1961, *s. coll. 1746* (KUN 0178792); WSW of the city of Riwoqê on hwy. 317 from Riwoqê to Dingqing (Tengchen), Logonda, 31.244444°N, 96.519167°E, 3900 m, 12 Aug. 2004, *D. E. Boufford et al. 32184* (A 00279937).

121. Berberis capillaris Cox ex Ahrendt, J. Bot. 79(Suppl.): 47. 1941. *Berberis ludlowii* Ahrendt var. *capillaris* (Cox ex Ahrendt) Ahrendt, J. Linn. Soc., Bot. 57: 115. 1961. TYPE: Myanmar (Burma). "Chimili Alps," 3950 m, 18 Oct. 1919, *R. Farrer 1399* (lectotype, designated here, E E00117388!).

Berberis muliensis Ahrendt var. *atuntzeana* Ahrendt, Bull. Misc. Inform. Kew 6: 269. 1939, syn. nov. TYPE: China. NW Yunnan: [Dêqên (Deqin) Xian], "mountains east of Atunze," 28.583333°N, 99.083333°E, 3650–3950 m, Sep. 1921, *G. Forrest 20713* (holotype, K K000077418!; isotypes, A 00058266!, CAL CAL0000027191 image!, E E00217959! US 00945915 image!, W 1925–0006272!).

Shrubs, deciduous, to 2 m tall; mature stems reddish purple, sulcate; spines 3- to 5(to 8)-fid, yellow-brown, 0.5–1.3 cm, terete or abaxially sulcate. Petiole almost absent; leaf blade abaxially pale green, adaxially dull olive green, obovate, obovate-elliptic, or oblong-obovate, 1.5–3 × 1–1.5 cm, papery, midvein raised abaxially, slightly impressed adaxially, lateral veins and reticulation conspicuous on both surfaces, base cuneate, margin spinose with 3 to 7 teeth on each side, occasionally entire, apex rounded. Inflorescence 1(or 2)-flowered; pedicel (15–)25–35 mm. Sepals in 2 whorls; outer sepals oblong, 5 × 2.5 mm; inner sepals broadly elliptic, 6 × 5 mm; petals obovate, 6 × 4 mm, base slightly clawed, glands separate, apex emarginate. Stamens ca. 3 mm; anther connective slightly extended, truncate. Ovules (5 to)8 or 9(or 10). Berry red, semi-globose, ellipsoid, or ovoid, (1–)1.5–1.8 × 1–1.4 cm; style not persistent.

Phenology. *Berberis capillaris* has been collected in flower from July to September and in fruit from August to November.

Distribution and habitat. In China *Berberis capillaris* is known from Dêqên (Deqin), Fugong, and Gongshan Xian in northwest Yunnan and Zayü (Chayu) Xian in southeast Xizang (Tibet). It is also known from Myanmar (Burma). It has been found among bamboo and *Rhododendron* scrub, and in grassy and rocky areas of open mountain tops and slopes at 3600–4100 m.

Berberis capillaris has had a checkered history. The protologue gave a brief description and cited *Farrar 1399*. Ahrendt (1941b: 47–48) was, therefore, wrong to describe the 1926 publication as being of a nomen nudum. The protologue did not indicate any herbarium

where *Farrer 1399* could be found and none was given by Ahrendt (1941b). What appears to be the only specimen has, therefore, been chosen as the lectotype. The indications are that Ahrendt never saw this since his collection details are limited to "Hpimaw," which is not to be found on the specimen sheet of no. *1399*. Indeed, it is likely that he did not know of its existence since later, Ahrendt (1961: 115) simply cited "Hipimaw, Farrer 1399 seed number."

Ahrendt (1961: 115) also decided that *Berberis capillaris* was, in fact, a variety of *B. ludlowii*, a species with narrowly obovate or oblong-obovate leaves up to 7 cm long otherwise known only from Mainling (Milin) and Nang (Lang) Xian in southeast Tibet. His justification appears to have been that *B. capillaris* and *B. ludlowii* have the same flower structure. No specimens of *B. capillaris* were cited to support this and, as noted under that species, the flower structure of *B. ludlowii* is unknown. The description of flowers here is taken from *Farrer 1916* (details below) which is described in the collector's notes as "= F 1399 in flower." It should also be noted that, had the infraspecificity claimed by Ahrendt been correct, *B. ludlowii* would have been a variety of *capillaris* since the latter was published first. Ying (2001: 79) treated *B. capillaris* as a synonym of *B. muliensis*, a semi-fascicled species from Muli Xian in Sichuan with three or four ovules. This is clearly unsustainable.

The protologue of *Berberis capillaris* described cultivated plants grown from *Farrer 1399* as having single flowers "the size of a shilling" (i.e., 2.3 cm diam.). This may reflect more favorable conditions in cultivation than in the wild in that the flowers of *Farrer 1916* and those of *Feng 5076*, *Feng 5326*, *Forrest 20264*, *J. F. Rock 22350*, and *Yu 22170*, listed below, are much smaller than this.

The type of *Berberis capillaris* has one set of paired fruit, while all of the specimens listed below are single-flowered or -fruited with the exception of *Feng 7670*, which has both single and paired fruit plus what appear to be three fruit at the tip of one of the branches of one of the specimens at KUN (barcode 0176549).

The height of *Farrer 1399* is given as 1–2 ft. (30–60 cm). Where given, the height of the specimens listed below varies between 20 cm and 1 m, the exceptions being *Forrest 20264* and *Gaolingshan Biodiversity Survey 2005 26558* which are recorded as 1.5 m, and *Qinghai-Xizang Team 10305* which is recorded as 2 m.

Berberis muliensis var. *atuntzeana* presents a possible problem. The protologue implied that it had two or three seeds per fruit. However, the specimens of the type gathering (recorded as being 1.2–1.8 m high) have fruit with up to 10 seeds and the leaves are more or less the same as those of the type of *B. capillaris*, as is the color of the mature stems. The specimen at E also has spines that are up to 8-fid, a rare characteristic of *Berberis* in China (though *Farrer 1916* has spines that are 3- to 5-fid). The difficulty is that, if the collection details are accurate, it was collected in an area that is significantly drier than where it is otherwise recorded. One possibility is that the collection details are inaccurate in that it is clear from other Forrest collections, including no. *20264* (details below) and the holotype of *Swertia forrestii, Forrest 20718* (E E00001933), that for at least some of the time in September 1921, Forrest and/or his collectors were in an area between the Mekong and the Salwin.

Ahrendt (1941b) cited as *Berberis capillaris, F. Ludlow & G. Sherriff 3647* (BM BM000939722) from the Pachakshiri district of Arunachal Pradesh in India, some 300 km to the west of where it is otherwise recorded. Though very similar, this is likely to be of another species.

Selected specimens.

SE Xizang. Zayü (Chayu) Xian: Tsarong, "On the Salwin-Kiu Chiang divide," 28.4°N, 98.4°E, 3660–3960 m, Sep. 1921, *G. Forrest 20264* (A 00279940, E E00117386); 28.416667°N, 97.916667°E, 3810 m, 2 Oct. 1931, *F. Kingdon-Ward 10006* (BM BM00794126); Tsarong, mtns. of Tjonatong, upper Salwin River, 3960 m, June–July 1932, *J. F. Rock 22350* (A 00279692, BM, E E00623036, K, UC UC516222, US 00945979); Ridong Qu, Shuzhuru, 3600 m, 9 Sep. 1982, *Qinghai-Xizang Team 10305* (KUN 0178893, PE 01037765).

NW Yunnan. Dêqên (Deqin) Xian: Doker-La, 3800 m, 11 Nov. 1937, *T. T. Yu 7916* (A 00279693, BM, E E00612566, KUN 0176531, PE 01037757); [Dêqên Xian–Gongshan Xian border], "Mekong-Salwin Divide, Sila," 3800 m, 5 Aug. 1938, *T. T. Yu 22170* (A 00279683, 00279686, KUN 0176552, PE 01037724–35); same details, but 3600 m, 10 Oct. 1938, *T. T. Yu 22790* (A 00279687, E E00612565, KUN 0176533, 0176551, PE 01037724–25); Cizhong, 3800–3850 m, 4 July 1940, *K. M. Feng 5076* (KUN 0176540, 0176557–58, PE 01037730). **Fugong Xian:** "Chih-tze-lo Hsien" (Bijiang Xian, now part of Fugong Xian), top of Pi-lo-shan (Biluo Shan), 4000 m, 26 Aug. 1934, *H. T. Tsai 58193* (A 00279697, IBSC 0092309, KUN 0176544, 0176555, LBG 00064189, NAS NAS00314080, PE 01037729, SZ 00291380, WUK 0039619); Bijiang Xian, Biluo Xue Shan, near Jiumingfang, 4100 m, 13 Sep. 1964, *S. G. Wu 8848* (KUN 0176529, 0176548, LBG 00064332); Lishadi Xiang, Yaduo Cun near Myanmar border, E side of Gaolingshan, 27.215833°N, 98.702222°E, 3650 m, 7 Aug. 2005, *Gaolingshan Biodiversity Survey 2005 26558* (CAS 00120166, GH 00247599, HAST 124225); same details, but 27.215278°N, 98.801389°E, 3640 m, 24 Aug. 2005, *Gaolingshan Biodiversity Survey 2005 28063* (CAS 00120167, GH 00247598, HAST 124224, MO 6408634). **Gongshan Xian:** Chiu-Kiang (Dulong Jiang), W of Champutung (Changputong), Oct. 1935, *C. W. Wang 67471* (A 00279714, KUN 0176541, PE 01037726–27); Changputong, 3600–3700 m, 10 Sep. 1940, *G. M. Feng 7670* (KUN 0176539, 0176549. 0176550, PE 01037733); Changputong, 3700–3800 m, 20 Sep. 1940, *G. M. Feng 7956* (KUN 0176534, 0176546–47); Salwin-Kiukiang (Nujiang-Dulong), Divide, Parolaka, 3300 m, 10 Oct. 1938, *T. T. Yu 20678* (A 00279688, E E00117387, KUN 0176542, PE 01037743–44); W side of divide above Dong Shao Fang Forest Station on Nan Mo Wang Shan,

27.691667°N, 98.456944°E, 3600 m, 22 Sep. 1997, *Gaoligong Shan Exped. 1997 9598* (E E00113404, MO 5198976); E Side of Gaoling Shan, W of Qiqi, betw. Dongshaofang & pass to Dulong Jiang valley, 27.696111°N, 98.455278°E, 3600–3680 m, 16 July 2000, *H. Li et al. 12690* (CAS 00238636, E E00238636, MO 5798166); Cikai, N of rd. from Gongshan to Kongdang, S face of Gongshan, 27.793083°N, 98.465917°E, 3429 m, 3 Oct, 2002, *Gaolingshan Biodiversity Survey 16942* (A 00352253).

MYANMAR (BURMA). Chawchi Pass, 3950 m, 7 Sep. 1920, *R. Farrer 1916* (E E00117385); "W flank of the Chimi-li, N'maikha-Salwin Divide," 26.4°N, 98.8°E, 3660–3960 m, June 1925, *G. Forrest 26945* (E E00117386, US 00967441); "Upper Burma," 1924–1925, *G. Forrest 27519* (WU 1933–3750).

122. Berberis caroli C. K. Schneid., Bull. Herb. Boissier, sér. 2, 5: 459. 1905. TYPE: China. Mongolia: [Nei Mongol], Ordos, [?21 Aug.], 1884, *G. N. Potanin s.n.* (lectotype, inadvertently designated by Mikhailova [2000: 60], LE!; isolectotype, G G00343716!).

Berberis integerrima Bunge var. *stenophylla* Maxim., Fl. Tangut. 31. 1889. TYPE: China. [Qinghai], Churmyn tributary of Khuane-khe, 2750–2900 m, 17 May 1880, *G. N. Potanin s.n.* (lectotype, designated here, LE!).
Berberis caroli C. K. Schneid. var. *hoanghensis* C. K Schneid., Bull. Herb. Boissier, sér. 2, 5: 459. 1905, syn. nov. TYPE: China. Gansu, 1880, *N. M. Przewalski s.n.* (lectotype, designated here, specimen with fruit on sheet W 1890–2027!; probable isolectotype E E00217982!).
Berberis vernae C. K. Schneid., Pl. Wilson. (Sargent) 1: 372–373. 1913, syn. nov. TYPE: China. S Gansu: Minchow (Min Xian), 2450–2750 m, s.d.[1911], *W. Purdom 1047* (lectotype, designated by Ahrendt [1961: 175], K K000077401!; isolectotypes, A 00038811!, 00038812!, US 00946049 image!).
Berberis jamesiana Forrest & W. W. Sm. var. *sepium* Ahrendt, J. Linn. Soc., Bot. 57: 180. 1961, syn. nov. TYPE: China. [N Sichuan]: Sep. 1903, *E. H. Wilson (Veitch) 3157* (holotype, K K000567820!; isotype, A 00280096!).
Berberis haoi T. S Ying, Acta Phytotax. Sin. 37(4): 339. 1999, syn. nov. TYPE: China. S Gansu: Minchow (Min Xian), Yench'eng, 1800 m, 26 June 1930, *K. S. Hao 533* (holotype, PE 00935140!; isotype, PE 00935141!).

Shrubs, deciduous, to 3 m tall; mature stems purplish, slender, sulcate, black verruculose; spines solitary, rarely 3-fid, yellow, 1–3 cm, abaxially sulcate, stout. Petiole 2–6 mm; leaf blade abaxially pale green, adaxially bright green, narrowly oblanceolate, spatulate-oblanceolate, obovate-lanceolate, or narrowly elliptic, 1.5–3.5 cm × 4–8 mm, papery, midvein raised abaxially, slightly impressed adaxially, lateral veins and reticulation raised on both surfaces, base attenuate, margin entire, apex obtuse, rarely acute, mucronate. Inflorescence a spikelike raceme, (6- to)15- to 35-flowered, 2–4 cm overall; peduncle 5–10 mm; bracts ca. 1.3 mm; pedicel 1–3 mm; bracteoles reddish, lanceolate, ca. 1 mm. Sepals in 2 whorls; outer sepals obovate, ca. 1.5 × 1 mm; inner sepals obovate, ca. 2.5 × 1.5 mm; petals obovate-elliptic, ca. 2.25 × 1.5 mm, base clawed, glands separate, apex subacute, entire. Stamens ca. 1.5 mm; anther connective not extended, truncate. Ovules 2, subsessile. Berry pale red, globose, 3.5–4.5 mm; style not persistent.

Phenology. *Berberis caroli* has been collected in flower between May and July and in fruit from August to September.

Distribution and habitat. *Berberis caroli* is known from Gansu, west Nei Mongol, Ningxia, Qinghai, north Sichuan, and probably north Shaanxi. It has been collected from forests on valley slopes, by roadsides, and on river beaches at 1270–3400 m.

The protologue of *Berberis caroli* cited only "Ordos [Potanin 1884]" but cited no herbarium. As noted in the introduction, at the time of his 1905 publication, Schneider had seen *Berberis* specimens from only B, G, M, W, and WU. The only specimen from these herbaria found matching the protologue description is at G and was annotated by Schneider as *B. caroli* on 28 October 1904. However, the statement by Mikhailova (2000: 60) that the specimen at LE is the holotype must be treated as an inadvertent lectotypification. The gathering is likely to be the *Potanin s.n.*, Ulan-Morin river valley, Ordos, 21 Aug. 1884, cited by Maximowicz (1889b: 32) as being an intermediate between *B. integerrima* and *B. integerrima* var. *stenophylla*.

The protologue of *Berberis integerrima* var. *stenophylla* did not use the word type, but listed five gatherings of Potanin and three of Przewalski, all from Tangut, a name for an area now partly in Qinghai and partly in Gansu. These should all be treated as syntypes. However, while Maximowicz gave details of exactly where and when these gatherings were made, the only Potanin and Przewalski specimens originally labelled *B. integerrima* var. *stenophylla* found with detailed collection information are at LE. The best of those I found there has been chosen as the lectotype. Specimens elsewhere with only the collector's name and year of collection are likely to be duplicates of ones listed in the protologue, but this cannot be conclusively proved. Borodina-Grabovskaya et al. (2001: 161, 168) synonymized *B. integerrima* var. *stenophylla* as *B. caroli*.

The protologue of *Berberis caroli* var. *hoanghensis* cited "Kansu, Tangut [Przewalski 1880]; Hoangho-Thai [Frutterer et Holderer n. 164 et 165]; ferner im westl. Teil des Gebietes [Potanin 1885]," but cited no herbaria. Subsequently, Schneider (1918: 218) cited as the type "Am hoangho-Tal, Mai und August 1880, Przewalski." Though no herbarium was cited, this will have referred to (or have included) a sheet with both floral and fruiting specimens at W which he annotated

as *B. caroli* var. *hoanghensis* on 12 November 1904, though the only collection details on the sheet are "Regio Tangut (prov. Kansu) 1880." The floral specimen is only a fragment, having no branches and only a few leaves; therefore, the specimen with fruit has been chosen as the lectotype.

Berberis vernae was published by Schneider (1913: 372–373) with the only collection cited being "West Kansu: Min-chou, alt. 3200–3600 m., W. Purdom," but without citing any herbarium. There are two specimens of a *W. Purdom 1047* at A annotated by Schneider in February 1913 as *B. vernae* (it is likely that the number was a later addition). Though the specimens are undated, it is clear from the catalogue of Purdom papers, held by the Arnold Arboretum, that they must have been collected in 1911. Ahrendt (1961: 175) cited as the type a further specimen (also numbered *1047*) at K. This is treated here as a lectotypification.

Berberis caroli has had an elusive history. Having published it together with *B. caroli* var. *hoanghensis*, describing the only differences to be that the latter had more flowers, pedicels that were 2 mm shorter, and somewhat smaller fruit, Schneider (1918: 218) reduced variety *hoanghensis* to a synonym of *B. vernae*, while only mentioning *B. caroli* itself in a footnote where he stated that, until there was more evidence, *B. caroli* was best kept separate.

Berberis caroli then largely disappeared from view. It was included by Ahrendt (1961: 176) with a two-line description with *Potanin s.n.* from Ordos cited as the type but with no herbarium cited. It and a separate *B. vernae* were reported from Nei Mongol by G. H. Liu and Zhou (1991: 580–582), but from the text and the accompanying line drawing, the entry for *B. caroli* appears actually to describe *B. dubia* (this is confirmed by specimens at HIMC identified as *B. caroli* but which are actually *B. dubia*). It was not included by Ying (2001). As noted above, it was listed by Borodina-Grabovskaya et al. (2010: 161, 168), but they treated it as separate from *B. vernae*.

Having examined the type specimens of *Berberis caroli*, *B. caroli* var. *hoanghensis*, and *B. vernae*, I can find no significant difference among them. It therefore follows from this that both *B. caroli* var. *hoanghensis* and *B. vernae* should be regarded as synonyms of *B. caroli*.

Schneider (1913: 368) cited *Wilson (Veitch) 3157* as an example of *Berberis caroli* var. *hoanghensis*. In publishing it as the type of *B. jamesiana* var. *sepium*, Ahrendt (1961: 180) made no reference to this. Where the gathering was made is recorded neither on the specimens at A and K, nor in Wilson's fieldbook (Wilson Papers, Arnold Arboretum); information as to where Wilson was in September 1903 when the gathering was made comes from Briggs (1993: 38).

The protologue of *Berberis haoi* cites only the type. This would seem indistinguishable from *B. caroli*. Previously, the collector himself, Hao (1938) had identified *K. S. Hao 553* as *B. sinensis*.

The leaves of *Berberis caroli* are unusually varied, often even on the same plant. Specimens with only narrowly oblanceolate leaves are sometimes confused with *B. chinensis*, but can be differentiated from the latter by the preponderance of stout, solitary spines and smaller flowers with two ovules (*B. chinensis* having only one).

Selected specimens.
Syntypes. Berberis integerrima var. *stenophylla.* [Gansu] S of Tetung river (Datong He), 2285 m, 10 Aug. 1880, *N. M. Przewalski s.n.* (LE); Qinghai, near confluence of Churmyn & Khuan-khe, 2750 m, 18 May 1880, *N. M. Przewalski s.n.* (LE); Qinghai, W of Dzhamba river, 15 Mar. 1885, *G. N. Potanin s.n.* (?LE not seen); Qinghai, Karysh valley, near Kazhyr, 5 May 1885, *G. N. Potanin s.n.* (?LE not seen); Qinghai, Khuan-khe valley, near Dunych, 6 May 1885, *G. N. Potanin s.n.* (?LE not seen); Qinghai, Nurun-Dzamba river, near Bou-nan, 2375 m, 12 May 1885, *G. N. Potanin s.n.* (?LE not seen); Qinghai, Lopsyr'gol river, 1 June 1886, *G. N. Potanin s.n.* (LE not seen). Probable syntypes. Gansu, 1880, *N. M. Przewalski s.n.* (fragment with flowers on sheet W 1890–2027); Gansu, 1886, *G. N. Potanin s.n.* (P P00580428).

Syntypes. Berberis caroli var. *hoanghensis.* Gansu Hoangho valley, *K. Futterer & J. Holderen 164* and *165* (not found ?B†); W Gansu, *G. N. Potanin s.n.* 1885 (K K000395212, LE 01041510, W 1890/2026).

Other specimens.
Gansu. Jonê (Zhuoni) Xian: vic. of Choni, 3100–3300 m, 15 Sep. 1923, *R. C. Ching 1001* (A 00280101, NAS NAS00314304, NAU 00024194, UC UC282913, US 00946046); Yan'ertan, 2500 m, 28 June 1987, *Meifujie Field Team, Group 1 s.n.* (LZU). **Kangle Xian:** Lianlu Shan, 2100 m, 20 Aug. 1996, *Y. S. Lian et al. 96846* (PE 01840139). **Lintan Xian:** Ligang Shan, 11 June 1957, *Tao He Team 3355* (IBK IBK00013168, IBSC 0092748, KUN 0177272, NAS NAS00314307, PE 01032528, WUK 0091304). **Lintao Xian:** Panjialiang, 2100 m, 17 June 1961, *X. X. Ma & Y. Q. Zhang s.n.* (LZU). **Su'nan Xian:** Xilian Shan, 2600 m, 30 June 1959, *Y. Q. He 3534* (PE 01032526, WUK 00158982). **Têwo (Diebu) Xian:** Dianga, 2450 m, 31 May 1999, *Bailongjiang Exped. 1541* (PE 01556131). **Tianzhu Xian:** Haixi Linchang, 2600 m, 18 July 1991, *T. N. He 2264* (CAS 796676, HNWP 167676). **Weiyuan Xian:** Majiaji Gongshe, Tianjia He Dadui, 2200 m, 1 July 1961, *Lanzhou University Biology Dept. 13* (WUK 00166873). **Xiahe Xian:** Labuleng Si, 22 Sep. 1934, *Z. W. Yao 473* (LZU, PE 01032521). **Yongdeng Xian:** Lian Cheng, Xing'er Gou, 3300 m, 3 Sep. 1960, *Chinese Academy of Science, Qinghai Gansu Survey Team 3509* (PE 01032533). **Zuanglang Xian:** 1950 m, 15 May 2014, *Y. He & J. C. Hao GSL2014050512* (BNU 001150).
Nei Mongol. Alashan You Qi: Taohua Shan, 11 Sep. 2006, *S. W. Zhao s.n.* (NMAC 00042657). **Otog (E'tuoke) Qi:** Zhaohuang Gongshe, 13 Aug. 1960, *s. coll. 15* (HIMC 0010775); A'erbasi, 12 July 2006, *Y. Tu 65* (NMAC 00053006). **Uxin (Wushen) Qi:** Wulantaolegai Gongshe, 12 July 1973, *Y. P. Xu 101* (HIMC 0010773).
Ningxia. Shizuishan [Pingluo Xian]: Helan Shan, Ruji Gou, Huangcao Tan, 2100 m, 11 Sep. 1981, *Z. Y. Yu & Y. P. Xu 2533* (WUK 00433576, 0449429); Helan Shan, Zhengy-

iguan Gou, 1300 m, 13 Sep. 1981, *Z. Y. Yu & Y. P. Xu 2558* (WUK 00433601, 0449422).
Qinghai. Datong Xian: Laoye Shan, 2700 m, 4 Aug. 1977, *K. J. Guan 77299* (PE 01032538). **Gonghe Xian:** Huang He Zhi, 12 Aug. 1956, *S. J. Ma et al. 00054* (PE 01032535). **Huangyuan Xian:** Daban Shan, 3400 m, 3 July 1990, *Z. H. Zhang et al. 5416* (HNWP 160486). [**Huangzhong Xian**]: Ba Valley, 2685 m, July 1926, *J. F. Rock 14362* (A 00280097, E E00623002, K, NAS NAS00313896). **Huzhu Xian:** Beishan Linchang, 12 June 1971, *B. Z. Guo 9103* (HNWP 0210210, KUN 0178141, PE 01032534). **Jainca (Jianzha) Xian:** near Angla Gongshe, 2100 m, 6 June 1970, *S. W. Liu & D. S. Luo 619* (HNWP 131567). **Jigzhi (Jiuzhi) Xian:** Yinshen Gou, 2300 m, 15 Sep. 1959, *Qinghai-Gansu Team 1796* (HNWP 006548, PE 01032524–25). **Ledu Xian:** Shanbei Shan, Linchang, 2600 m, 5 Aug. 1986, *Y. H. Wu et al. 4171* (HNWP 133900). **Menyuan Xian:** Xianmisi, 2400 m, 13 June 1960, *Chinese Academy of Science, Qinghai Gansu Survey Team 2403* (PE 01032547, WUK 00167756). **Minhe Xian:** Xing'er, 2100 m, 10 June 1970, *B. X. Guo 6876* (HNWP 21558). **Tongren Xian:** Zhamao Gongshe, Zhamao Gou, 2900 m, 29 July 1970, *L. H. Zhou & L. N. Sun 1623* (HNWP 22615). **Xinghai Xian:** Heka, Yangqu, 2800 m, 19 Sep. 1965, *T. N. He 475* (HNWP 12406). **Xining Shi:** 1965, *X. Z. Li 116* (HNWP 19149); Chengxi Qu, Huangshui Linchang, 2297 m, 21 July 1990, *Z. H. Zhang et al. 5415* (HNWP 160484–85). **Zêkog (Zeku) Xian:** Maixiu Linchang, 19 Aug. 1982, *B. Z. Guo 26431* (HNWP 102792, 102957).
N Sichuan. Songpan Xian: Songpan, 2750 m, Sep. 1910, *E. H. Wilson 4022* (A 00280095); near Wulishe, 2300 m, 30 May 1962, *Botany Team 0711* (CDBI CDBI0027762–63); Chuanli Si, 2900 m, 13 June 1983, *S. Z. Zhao & Z. J. Zhao 120450* (CQNM 0005137). **Zoigê (Ruo'ergai) Xian:** near Gouwa Cun, 33.701389°N, 103.403333°E, 22 Aug. 2007, *D. E. Boufford et al. 40311* (A, E).
Suiyuan (now part of Nei Mongol). Wufan Qi (now Otog [E'tuoke] Qian Qi): Maobulatan, 1290 m, 12 Aug. 1953, *K. J. Fu 7306* (IBSC 0091983, PE 01032610).

123. Berberis chinensis Poir., Encycl. 8: 617. 1808. TYPE: China. "Cette espèce, originaire de la Chine, est cultivée, depuis plusieurs années, au Jardin des Plantes de Paris," 1789, *s. coll. s.n.* (lectotype, designated here, P-LA!).

?Berberis sinensis Desf., Tabl. École Bot. 1: 150. 1804, nom. nud.
?Berberis sinensis Desf., Hist. Arbr. France 2: 27. 1809, nom. illeg., superfl. TYPE: Cultivated(?). S.d., *s. coll. s.n.* ex-Herb. Desfontaines (lectotype designated here, FI-W 005017 image!).
Berberis sinensis Desf. var. *angustifolia* Regel, Trudy Imp. S.-Peterburgsk. Bot. Sada 2(2): 416. 1873, syn. nov. *Berberis poiretii* C. K. Schneid. var. *angustifolia* (Regel) Nakai, Fl. Sylv. Kor. 21: 66. 1936. TYPE: China. Pekin, s.d., *K. Skatchkoff s.n.* (neotype, inadvertently designated by Imchanitzkaja [2005: 278], LE!; isoneotypes, BM BM000810308!, K ex HK, ex LE 000567994!, US 00945868 image!).
Berberis poiretii C. K. Schneid., Mitt. Deutsch. Dendrol. Ges. 15: 180. 1906, syn. nov. TYPE: China. "Route de Pekin à Lhe-hol," s.d. [1793], *Sir G. L. Staunton s.n.* (lectotype, designated here, W ex "Hrb. Meille, Dupl. Banks" 0032266!; isolectotype, "inter Peking & Jehol" BM BM000564432!).

Berberis poiretii f. *weichangensis* C. K. Schneid., Pl. Wilson.
(Sargent) 1: 372. 1913, syn. nov. TYPE: China. [Hebei,
Weichang Xian]: N Chili, E Wei-chang, 1909, *W. Pur-
dom 2* (holotype, A 00038781!).
Berberis beijingensis T. S. Ying, Acta Phytotax. Sin. 37(4):
324. 1999, syn. nov. TYPE: China. Hebei: Beijing,
Western Hills, 9 May 1930, *H. F. Chow 40202* (holotype,
PE 00935218!; isotypes, CQNM 0005154 image!, IFP
052010004w0006 image!, NAS [2 sheets] 00314137,
00314141 images!).

Shrubs, deciduous, to 2 m tall; mature stems slightly
shiny, stramineous to pale reddish brown, angled,
black verruculose; spines absent or solitary, sometimes
3-fid, concolorous, 0.4–0.9 cm, abaxially slightly sulcate. Petiole almost absent; leaf blade abaxially pale
green or grayish green, adaxially deep green, oblanceolate to narrowly oblanceolate, occasionally lanceolate-
spatulate, 1.5–4 × 0.5–1 cm, papery, midvein, lateral
veins, and reticulation raised on both surfaces, base
attenuate, margin entire, apex acuminate or acute,
mucronate. Inflorescence a raceme, sometimes sub-
umbellate at tip and or partly paniculate, 8- to 30-flow-
ered, 3–7 cm overall; peduncles ca. 1–2 cm; bracts
linear, 2–3 mm; pedicel 3–6 mm; bracteoles lanceolate, ca. 1.8 × 2 mm. Sepals in 2 whorls; outer sepals
elliptic or oblong-ovate, ca. 2 × 1.3–1.5 mm; inner sepals oblong-elliptic or obovate, ca. 3 × 2 mm; petals
obovate or elliptic, ca. 3 × 1.5 mm, base clawed, glands
separate, apex entire or incised. Stamens ca. 2 mm; anther connective not extended, truncate. Ovules 1. Berry
red, oblong, ca. 9 × 4–5 mm; style not persistent.

Phenology. *Berberis chinensis* has been collected
in flower in May and June and in fruit between July and
September.

Distribution and habitat. In China *Berberis chinensis* is known from Beijing, Hebei, Jilin, Liaoning, Nei
Mongol, Shandong, northeast Shanxi, and Tianjin. It
is also found in northern Korea. In China it has been
found in thickets and forests, and on riverbanks and
talus slopes at 300–2300 m.

There has been more confusion in relation to *Berberis chinensis* than any other *Berberis* species found
in China. Indeed, for more than a century following
Schneider (1906: 180), there has been widespread acceptance that the species is not found in China at all,
but is endemic to the Caucasus. This, however, is a
mistake.
Desfontaines (1804: 150) published the name *Berberis sinensis*. However, as a nomen nudum, it was not
validly published. Four years later, Poiret validly published the name *B. chinensis*. His protologue clearly
stated that the taxon was based on a specimen from a
plant of Chinese origin cultivated for several years in
the Jardin des Plantes, Paris. Given the French monar-
chy sent representatives to Beijing in the 18th century
and that, until 1790 the Jardin des Plantes was the Jardin du Roi, this origin seems perfectly credible. The
inflorescence was described as consisting of long, simple clusters, whereas that of the type itself (labelled
"Berberis Sinensis L. R. [?Louis Roi] 1789" and which,
though fragile, is an excellent one) is largely racemose
but also partially paniculate. No other specimens were
cited by Poiret and, although the type specimen has
only flowers, he also included a description of the fruit;
therefore, the specimen has been designated here as a
lectotype. There is, however, a further cultivated specimen that may have been taken from the same plant.
This is the ex-Herb. Lemonnier specimen at G, cited
below, Lemonnier (who died in 1799) being Desfontaines's predecessor as Professor of Botany at the Jardin
du Roi and whose herbarium was acquired by Delessert
in 1803. The specimen is labelled both as "*Berberis
sinensis*" and "*Berberis chinensis*" and also has a partially paniculate inflorescence.

In the year following Poiret's description of *Berberis
chinensis*, Desfontaines (1809: 27) again published his
name *B. sinensis*, this time with a description but with
no specimens cited. This description was even more
rudimentary than Poiret's and the flowers were described as being in numerous, small, pendent bunches.
The species was described as being from China. Under
Art. 53.3 of the Shenzhen Code (Turland et al., 2018),
B. sinensis must be regarded as a homonym of *B. chinensis* and hence illegitimate. There are no specimens of
B. sinensis in P-DESF and only one specimen in FI-W
annotated by Desfontaines. Although the name is illegitimate, it seems useful to designate this as the type of
B. sinensis.
Despite this history, the name *Berberis sinensis* came
into widespread use in the 19th century. This included
de Candolle (1821: 8), who mistakenly treated Poiret's
valid name *B. chinensis* (1808) as a synonym of Desfontaines's (1804) earlier but invalid *B. sinensis*. References to the latter name included Schrader (1838:
377–378) who, in a confused entry, appeared to differentiate between a *B. sinensis* Desf. and a separate
B. sinensis "H. Vindob et Desf.," as well as stating that
some plants identified as *B. sinensis* were, in fact, a new
species, *B. spathulata*, that Schrader published in the
same article. Nowhere did Schrader cite any actual
specimens of *B. sinensis*.
Equally relevant to subsequent consideration of the
taxon *Berberis sinensis* was Regel who stated (1873:
416) that *B. sinensis* Desf. was a deciduous species
found in south Europe, north Africa, middle and east
Asia, and North America. This included the variety
typica which he named "die ächte [the true] *B. sinensis*
Desf" found in north Africa, North America, the Orient,

and the Caucasus whose synonyms included *"Berberis sinensis* Schrad. herb. (ex horto Parisiensi) et Linnaea XII, 378." No details were given of the cultivated specimen in Schrader's herbarium except that it resembled one at P dating from Desfontaines's time. Regel's varieties of *B. sinensis* also included a variety *angustifolia* from Manchuria and north China but for which no actual specimens were cited.

Subsequently, Schneider (1905: 655) disregarded Regel and listed variety *angustifolia* as a synonym of *Berberis sinensis* Desf. citing six gatherings, all from China. Later, however, in a detailed commentary on unspecified *Berberis* specimens sent to Schneider from Schrader's Herbarium at LE, Schneider decided (1906: 179–180) that *B. sinensis* was not a Chinese species at all, but a Caucasian one. The relevant paragraph translated from the German is as follows:

Earlier I had noticed that *B. sinensis* [sic] Poir., in Lam Encycl. viii (1808) p 617 [Desfontaines only published this name with a poor description in 1809, before that it had just been a name, and one which Poiret made no reference to], did not match the description of wild-collected Chinese forms, nor did it match those described by Spach in his Hist. Veg viii (1839) p 42. However I continued to use the name *sinensis* Desf, because I wanted to clarify the situation in Paris first. Now it turns out that those specimens from the Hort. Paris described by Poiret as *B sinensis* [sic] and those described by Desfontaines, do not match the Chinese forms, but rather the Caucasian ones, which I have referred to under *B. spathulata* Schrad, see above. [*Berberis spathulata* was the prior entry in Schneider's article.] Therefore *B. sinensis* Poiret (and Desf.) is the oldest name for *B. spathulata* Schrad and *B. iberica* s.m. Thus the *sinensis* I described has to be renamed and I give it the name *B. Poireti* [sic].

It should be noted that in this paragraph Schneider cites no specific specimens and, indeed, provides no evidence that he had seen the specimen of *Berberis chinensis* in the Lamarck Herbarium (or the specimen at FI-W, or the cultivated ex Herb. Lemonnier one at G) as against other unspecified cultivated specimens from the Paris Botanic Garden (whether in the Paris herbarium, the Herb. Schrad., or elsewhere). Nor does he seek to explain why Poiret should describe a plant as being of Chinese origin if it was actually from the Caucasus. Subsequently, Schneider (1918: 226), in a footnote to an entry for *B. poiretii*, stated that he "would give the synonymy for the true west Asian *B. chinensis* (Poiret did not write *"sinensis"*) in another place."

Five years later, Schneider (1923: 216–219) returned to the subject. Here he gave a detailed description of *Berberis chinensis*, listed eight and possibly nine synonyms, seven of which were taxa based on cultivated specimens of unknown origin including *B. spathulata* and once again asserted that the species was of Caucasian origin. Before pursuing *B. chinensis* further, it is

therefore necessary to interrogate the only taxa of wild-collected origin that Schneider included in his synonyms, *B. iberica* and *B. vulgaris* var. *iberica*.

Berberis iberica was published by Sweet (1826: 13), referring to the earlier *B. vulgaris* var. *iberica* DC. de Candolle (1821: 6–7) described "var ? *iberica"* as a possible variety of *B. vulgaris* from Iberia (contextually this referred to the Caucasus rather than the Iberian Peninsula), the source being a specimen sent by Steven and Fisher. This is was a reference to the (poor) specimen in G-DC (00201881) annotated as *B. vulgaris* ? *iberica*, dated 1819 and recorded as having been received in correspondence from Steven and Fisher. There is no information on the sheet as to where the specimen was collected. Steven (1827: 138) gave an exceptionally brief description and stated that *B. iberica* was from the Caucasus (where he certainly collected) but cited no specimens. Schneider (1905: 656) later cited unnamed specimens he stated to be from the Caucasus as "ex Herb. Fuckel im Herb Barbey; ex Herb. Hort Petrop. in Herb Boisser," and appear to refer respectively to two sheets at Geneva (G G00096569, G00096568), both being annotated by Schneider as *B. iberica* in September 1904. However, the former is sterile and has no collection details, while the latter has floral material and, though a Herb. Fuckel label states "Berberis iberica Steven, B. vulgar v. iberica del Kuakasas," it is unclear whether this indicates its origin or is simply an attempt at identification.

The protologue of *Berberis spathulata* in Schrader (1838: 376–378) gave a variety of descriptions, but no actual specimens were identified, nor was any provenance given for the sources of these descriptions. Commenting later on the descriptions, Schneider maintained (1906: 179) that these showed that *B. spathulata* was the same as *B. iberica*, but provided no evidence that these descriptions related to specimens of Caucasian origin.

It is certainly true that there is a species in the Caucasus, which has some similarities with specimens from northeast China that Schneider listed in his entry for *Berberis sinensis* in 1905. But despite Schneider's conclusion in 1906 that this should properly be called *B. chinensis*, this name has been rarely used in botanical works covering the Caucasus. Fedtschenko (1937: 558) stated the species to be *B. crataegina* DC. and cited as synonyms *B. vulgaris* var. *iberica* and *B. iberica*. This synonymy was previously made by Grossheim (1930: 126). However, an editorial footnote to Fedtschenko's entry for *B. crataegina* noted that "Reports of the distribution of this species in the Caucasus are highly doubtful and should be referred to *B. orientalis* C. K. Schneid." This is a reference to Schneider (1905: 666) where he lists various syntypes and other specimens from Iran, Turkey, and the Caucasus as *B. orientalis*.

From the evidence of specimens of these at G-BOIS and at WU, these collections are not similar to the Chinese ones listed by Schneider for *B. sinensis* in 1905. However, various specimens from the Caucasus that I have seen in LE that *do* have some similarities with the specimens from northeast China listed by Schneider are annotated as *B. orientalis* while other similar specimens from the Caucasus in other herbaria are annotated as *B. iberica*. These latter are listed under selected specimens below.

But irrespective of which name it is given, it is clear that this species from the Caucasus, though somewhat similar, differs from the type of *Berberis chinensis* in a number of ways. These include sometimes having some leaves that are more spatulate than found in the type. But the most important difference is that the inflorescence is always racemose (as is the case with the type specimen of *B. vulgaris* ? *iberica*) and never paniculate, whereas, as noted above, the type of *B. chinensis* is partly paniculate (significantly, Schneider's 1923 description of the inflorescence of *B. chinensis* simply states it is racemose, thus confirming he had never seen the type). A similar partially paniculate inflorescence is present on two of the three specimens at P of *David 1723*, a Chinese gathering which Schneider listed in his entry for *B. sinensis* in 1905 and, indeed, these *David* specimens are remarkably similar to the type of *B. chinensis*. I conclude from this that there is no reason to doubt Poiret's statement that *B. chinensis* is a Chinese species, and it follows from this that *B. poiretii* is a synonym.

Whether the specimen at FI-W is *Berberis chinensis* is difficult to say. It has purely racemose fruit stalks and largely similar leaves, except that some of them have margins which are 3- to 5-spinose versus the entire leaves of *B. chinensis*. Given that it is likely to be a cultivated specimen, it is possible that it is a hybrid of some sort (the propensity of cultivated *Berberis* to hybridize was probably not known in Desfontaines's time). Interestingly, four other specimens at FI-W labelled as *B. sinensis*, ex "hort. Pari" and dated 1835, two of which have a partially paniculate inflorescence, are *B. chinensis*. This, however, is not the case for the various cultivated specimens at P cited below and which may be hybrids.

Whether there was a type of *Berberis poiretii* and, if so, where it could be found has perplexed successive authors. Fedtschenko (1937: 557), who confusedly cited *B. chinensis* Poir. as a synonym, stated "Type unknown," Ahrendt (1961: 174) did not cite a type, and Borodina-Grabovskaya et al. (2007: 166) wrote "Type ?," while Nakai (1936: 68), while stating the type of *B. poiretii* was "a peculiar form," did not refer to any particular specimen. In fact, it is clear that the six gatherings cited by Schneider (1905: 655) should all be regarded

as syntypes. Schneider cited no herbaria for these. At this date, he had seen *Berberis* specimens only at B, G, M, W, and WU, but only two of these are in W and none in G, M, and WU, which suggests the other four he saw were all in B and hence were lost in WWII. However, specimens of these four—*Bretschneider 49*, *David 1723*, and *Maximowicz s.n.* Amur s.d., and *Niederlein 210*—have been located elsewhere. Schneider did not give a collection date for *David 1723* and the specimens at P are from two dates. Those of May 1862 are designated here as *David 1723* "A" and those of April 1864 as *David 1723* "B." The specimen of *Neiderlein 210* at A is an exceptionally poor one.

The Staunton gathering listed by Schneider was undoubtedly the one referred to by de Candolle (1821: 8) as an example of *Berberis sinensis* in the herbaria of Banks and Lambert and was collected by Staunton between Beijing and Jehol (a province that has subsequently been divided between Hebei, Liaoning, and Nei Mongol) and is presumably the *Berberis* listed under "Plants collected in the Journey between Peking and Zhe-hol, in Tartary" in Staunton (1799: 97). The Staunton specimens at BM and W appear to be the oldest provable wild-collected *Berberis* specimens known from China and, as such, would seem the most appropriate collection for the type of *B. poiretii*. The W specimen has been chosen here as the lectotype because it is most likely to be the specimen that Schneider saw. It is possible that some or all of the Chinese specimens without collection details at G-DC are further duplicates of the lectotype since, according to Miller (1970: 543), Staunton's Chinese collection was given to Lambert who, in 1816, gave 300 of those specimens to de Candolle. This, however, cannot be proved.

Schneider (1918: 227) treated his *Berberis poiretii* f. *weichangensis* as a synonym of *B. poiretii*, and the forma name is synonymized herein.

No specimens were cited by Regel (1873) for *Berberis sinensis* var. *angustifolia*. Subsequently, Franchet (1883: 178) did cite *David 1723* at P as this taxon, but assigned no type status to it. Later, Imchanitzkaja (2005: 278) lectotypified *Skatchkoff s.n.* (LE) for the varietal name, but under the Shenzhen Code, Art. 9.10 (Turland et al., 2018), this is an inadvertent neotypification. The LE specimen was annotated as *B. poiretii* by Schneider and is a good example of *B. chinensis*.

Nakai (1936: 36) transferred Regel's variety *angustifolia* from *Berberis sinensis* to *B. poiretii*, citing as synonyms *B. chinensis* Desf. (non Poir[sic]), *B. sinensis* (non Desf.), *B. sinensis* f. *weichangensis*, and *B. sinensis* var. *weichangensis* (both of the latter mistakenly said to be published by Schneider). No types were cited and thus Imchanitzkaja's typification extends to this taxon.

Berberis beijingensis was differentiated by its author T. S. Ying from *B. poiretii* largely on the basis that it has a partly paniculate inflorescence. But, as noted above, the type of *B. chinensis* also exhibits this characteristic, and *B. beijingensis* is synonymized herein.

I have found no evidence to support the report of *Berberis chinensis* (as *B. poiretii*) in Heilongjiang by Lu et al. (1992: 123).

I am grateful to Porter P. Lowry of MO/P who enabled me to have access to the Lamarck Herbarium and thus to inspect the type of *Berberis chinensis* several years before an image of it appeared on the Paris Herbarium website.

Selected specimens.
Possible isolectotypes. Berberis poiretii. [?Staunton specimens], *s. coll.* (G-DC G00201766–67).

Syntypes. Berberis poiretii. Gehol, May 1862, *A. David 1723* "A" (K K000077405, P P00580425–26, P00580449, US 00945870, VNM); Gehol, Apr. 1864, *A. David 1723* "B" (K K000077405, P P00580424); Amur, s.d., *I. Maximowicz s.n.* (BM, K, P P06868502); Peking, s.d., *E. V. Bretschneider 49* (LE); Peking, Ku-pei-ku, s.d., *H. Wawra 969* (W 0032265); near the Great Wall, 25 Nov. 1898. *G. Niederlein 210* (A 00872733, ?B†).

Other specimens.
Beijing. Changping Qu: Shisan Ling Gongshe, Xiaozhu Yu, 21 Aug. 1970, *s. coll. 252* (PE 01197747). **Fangshan Qu:** Xiayungling, 18 May 1971, *Beijing Herbal Medicine Survey Team, Fangshan 91* (PE 01197753). **Haidian Qu:** Xishan, 30 Oct. 1956, *C. J. Liu et al. 742* (PE 01387620). **Huairou Qu:** Lulimiao Gongshe, Laogongying, Nanshan, 23 May 1964, 500 m, *s. coll. 0001* (PE 01236014). **Mentougou Qu:** Huang'antuo 920 m, 21 Aug. 1965, *s. coll. 1052* (PE 01236009). **Yanqing Xian:** Songshan Linchang, 750 m, 26 May 1984 *W. X. Zhang 159* (PE 01539748).
Hebei. Chengde Shi: Shuangfengso Gongshe, 600 m, 3 June 1964, *J. Zhang & J. J. Hu 122* (PE 01572050). **Chicheng Xian:** Dahaituoshan, 1250 m, 5 June 1985, *Y. Chen & S. Y. Song et al. 006* (PE 01236015). **Fuping Xian:** near Longquanguan, 19 Aug. 1934, *K. M. Liou L.3297* (PE 01032678, SYS SYS00052185, WUK 0004154). **Huai'an Xian:** Zuowei, Fenghuang Shan, 1 Sep. 1970, *S. R. Liu s.n.* (HIMC 0039331). **Huailai Hsien (Xian):** Sep. 1931, *H. F. Chow 41657* (PE 01032630). **Laishui Hsien (Xian):** near Jinshuikou, 770 m, 27 May 1934, *K. M. Liou L.2164* (N 093057081, PE 01032661, SYS SYS00052186, WUK 0004158). **Laiyuan Xian:** Dianziliang, 1200 m, 14 Aug. 1959, *s. coll. 2945* (HNWP 114092, PE 01032734). **Neiqiu Xian:** Xiolingdi Cun, Wuzhi Yao, 1600 m, 2 July 1950, *Y. Liu 12966* (PE 01032643, 01032653). **Pingshan Xian:** Jiaotanzhuang, 400 m, Sep. 1959, *s. coll. 1287* (HNWP 114086, PE 01032702). **Qianxi Xian:** 23 May 1972, *s. coll. Qianxi 100* (PE 01236022). **Qinglong Xian:** Niuxin Shan, 360 m, 3 Sep. 1971, *s. coll. Qinglong 863* (PE 01236011). **Weichang Xian:** 5 July 1971, *Chengde Team 249* (PE 01236021). **Xinglong Xian:** betw. Tao He & Meisijian, 1300 m, 2 Sep. 1959, *S. E. Liu et al. 4566* (NAS NAS00314133, PE 01032743). **Yi Xian:** Dayukou Cun, 23 July 1959, *s. coll. 3214* (PE 01032740–41). **Yu Xian:** near Xiao Wutai Shan Linchang, 1600 m, *Xiaowutai Team 74167* (PE 01032652). **Zhuolu Xian:** Taipingbao Gongshe, Baojiakou, 19 Aug. 1971, *261 Medical Treatment Team 064* (PE 01236010); Wutai Shan Natl. Nature Reserve, 1259 m, 4 Sep. 2008, *Z. G. Ma s.n.* (Wutai Shan Natl. Nature Reserve Herbarium 01225).

Jilin. Jilin Shi: Longtan Shan, 27 Aug. 1950, *Y. Y. Zhang & Z. C. Jiang 915* (FUS 00014351, IBK IBK00013137, IFP 05201004x0101–04, NEAU 052001004000001, PE 01032585). **Jiutai Shi:** Tumen Ling, June 1956, *North East Teachers University 8134* (NENU NENU00012557).
Liaoning. Anshan Shi: Qianshan, 28 Sep. 1950, *Q. T. Li & P. Y. Fu et al. 2582* (IFP 05201004x0028, PE 01032590). **Fencheng Xian:** Fenghuang Shan, 8 May 1950, *D. C. Zhao & L. N. Ba 331* (IBSC 0091933, IFP 05201004x0038–41, KUN 0178695, PE 01032589, 01032596). **Jianchang Xian:** Heishan Linchang, 9 May 1965, *S. C. Cui 528* (IBSC 0092788). **Jianping Xian:** Yebaishou Xiang, 13 Sep. 1958, *C. S. Wang 2814* (IFP 05201004x0092–94, PE 01032605, 01032608). **Jinzhou Shi:** Nan Shan, 13 Sep. 1951, *G. Z. Wang et al. 71* (IFP 05201004x0015, PE 01032602). **Lianshan Qu:** Guanmotian Ling, 24 May 1950, *D. C. Zhao et al. 747* (IFP 05201004x0004, PE 01032591). **Lingyuan Xian:** Dahe Bei Gongshe, Laodong Gou, 16 May 1965, *S. C. Cui & W. J. Ju 620* (PE 01032588). **Shenyang Shi:** Donglin, 24 May 1964, *W. Wang et al. 2690* (IBK IBK00013128–29, PE 01032594). **Tiehling (Tieling Xian):** Chikuanshan (Jiguanshan), Fengtien, 280 m, 18 July 1930, *H. W. Kung 652* (PE 01032595, WUK 0004153).
Nei Mongol. Duolun Xian: 19 July 2006, *D. M. Lan s.n.* (NMAC 00027099). **Kelaqin Qi:** Wangyedian, Datou Shan, 6 June 1989, *G. H. Liu s.n.* (NMAC 00005173). **Ningcheng Xian:** Dayinghzi, 13 Aug. 1985, *C. P. Wang 159* (PE 01032612). **Tumote You Qi:** 5 July 1972, *s. coll. s.n.* (HIMC 0010750). **Xilinhaote Shi:** Bailin Xile, Muchang, 25 Apr. 1979, *Y. M. Ma s.n.* (NMAC 00005231). **Xinghe Xian:** 9 Sep. 1972, *Q. R. Wu 315* (HIMC 0010763). **Xiwuzhumuqin Qi:** Di'anmiao, 1100 m, *Nei Mongol Medicinal Plant Survey Team 461* (HIMC 0010765). **Zhenglan Qi:** 7 Sep. 1952, *Y. S. Liu 021* (HIMC 0010767).
Shanxi. Qin Xian: Funiu Shan, 11 May 1959, *K. J. Guan & Y. L. Chen 124* (HSIB HSIB003773, WUK 0323119). **Wutai Xian:** 9 Sep. 1934, *K. M. Liou 3652* (IBSC 0091930, N 093057085, WUK 0004146); Gengzhen Cun, 1800 m, 11 July 1959, *Y. L. Chen & K. J. Guan 2252* (HSIB HSIB003776, PE 01032750).
Tianjin. Ji Xian: Xiayin Gongshe, 300 m, 20 July 1983, *82 Grade Students 037* (NKU NK013495); Heishui He, Baxian Shan Nature Reserve, 540 m, 15 May 2008, *J. T. Zhao Z0175-H* (Baxian Shan Nature Reserve Herbarium BXS0227–1).
NORTH KOREA. Igen (Weewon), 17 Aug. 1912, *H. Imai 250* (TI); Sozan (Chosan), 17 June 1912, *T. Nakai 2014* (TI); Unzan, Hakukekizan (Unsan, Baekbeoksan), 28 Sep. 1912, *T. Ishidoya 48* (TI).

Specimens, images not seen.
Liaoning. Benxi Xian: 24 May 1950, *D. C. Zhao 747* (IFP 05201004x0004, x0060). **Changtu Xian:** 12 June 1949, *Y. L. Zhang s.n.* (IFP 05201004x0011, 05201004x0011). **Chaoyang Xian:** 29 May 1961, *P. Y. Fu 241* (IFP 05201004x0090). **Gaizhou Shi:** 22 May 1950, *S. E. Liu 337* (IFP 05201004x0070). **Qingyuan Xian:** 25 May 1981, *Z. F. Fang 2433* (IFP 05201004x0003). **Xifeng Xian:** 6 June 1961, *Y. C. Deng 181* (IFP 05201004x0074). **Xinbin Xian:** 24 May 1981, *Z. F. Fang 2410* (IFP 05201004x0095). **Xingcheng Shi:** 8 June 1987, *Z. S. Qin 189* (IFP 05201004x0010). **Zhuanghe Shi:** 5 Sep. 1959, *J. Y. Li 97* (IFP 05201004x0066).

Cultivated specimens.
"Ex-horto Paris 1819 China bor.," *s. coll. s.n.* (pro parte) (W ex Herb. Boos); "Hort. Mus. Par." 1831, *s. coll. s.n.* (P ex Herb. Spach P02559831, P02559835); "H[orto] P[aris]," 1832, *s. coll. s.n.* (P ex Herb. Spach P02559833); "Hortus pa-

risiensis," s.d., *s. coll. s.n.* (P P02559832); "Culta in Hort. Paris," s.d., *s. coll. s.n.* (G ex Herb. Lemonnier, ex Herb. Delessert); "*Berberis sinensis*, hort. Pari," *s. coll. s.n.*, 1835 (FI-W [4]).

Selected non-Berberis chinensis specimens from the Caucasus referred to above.
AZERBAIJAN. S. loc., *Z. Novruzova & L. Prilipko s.n.* (BM ex BAK); NE Azerbaijan, SE of Quba, 13 May 1999, *J. F. Gaskin 212* (MO 55578140); SW Azerbaijan, Nakhichevan, Norashensky Distr. 5 June 1999, *J. F. Gaskin 546* (MO 5619473).
GEORGIA. Borjomi Distr., Mtkvan River, 9 km SW of Borjomi, ca. 41.786944°N, 43.296111°E, ca. 840 m, 7 Sep. 2007, *D. E. Altha et al. 5933* (NY); Dedoplistskaro Distr., Vashlovani Reserve, 360 m, 20 July 2002, *K. Iashagashvali & N. Lachashvili 000019* (NY); Pantishara Gorge, ca. 267 m, 17 May 2006, *N. Lachashvili & M. Khutsishvili 387* (NY, W 2008–03097); Dedoplistskaro Distr., Vashlovani Natl. Park, *D. E. Atha 4551* (K, NY).
RUSSIA. Dagestan, near Achty, 1125 m, 5–7 Aug. 1898, *M. A. Alexeenko s.n.* (B); Dagestan, 15 km NW of Gerebil, shore of Avarskoe-Koisu River, 30 July 1987, *J. Maituluna et al. 135* (NA 0065149, W 2011–0299).

124. Berberis circumserrata (C. K. Schneid.) C. K. Schneid., Pl. Wilson. (Sargent) 3: 435. 1917. *Berberis diaphana* Maxim. var. *circumserrata* C. K. Schneid., Pl. Wilson. (Sargent) 1: 354. 1913. TYPE: China. S Shaanxi: Tai-pai-shan (Taibai Shan), 1910, *W. Purdom 4* (holotype, A 00038725!).

Shrubs, deciduous, to 2 m tall; mature stems yellow, sulcate, sparsely black verruculose; spines 3-fid, concolorous, 2–4 cm, terete or abaxially sulcate, stout. Petiole almost absent; leaf blade abaxially gray-green, adaxially dull green, obovate, obovate-orbicular, or obovate-elliptic, rarely elliptic, 1.4–2.8(–3.4) × 0.5–2.5 cm, papery, midvein raised abaxially, venation indistinct or obscure on both surfaces, base attenuate, margin spinose with 15 to 40 closely spaced teeth on each side, apex rounded. Inflorescence 1(to 3)-flowered; pedicel 15–45 mm. Sepals in 2 whorls; outer sepals elliptic-ovate, 6 × 4 mm; inner sepals obovate, 7 × 4 mm; petals orbicular-obovate, 5.5 × 4 mm, base clawed, glands separate, apex entire. Stamens ca. 4 mm; anther connective scarcely extended, rounded or truncate. Ovules 4. Berry red, slightly or not pruinose, ellipsoid to oblong or ovoid, bent at apex, 13–15 × 5–6 mm; style persistent.

Phenology. *Berberis circumserrata* has been collected in flower in June and July and in fruit from August and October.

Distribution and habitat. *Berberis circumserrata* is known from Taibai Shan and Taibai Xian in south Shaanxi, and Lingbao and Lushi Xian in northeast Henan. On Taibai Shan it has been found by roadsides, in thickets, and on steep mountain slopes at ca. 2350–2900 m. In Henan it has been found in scrub at 1450–2000 m.

The protologue of *Berberis diaphana* var. *circumserrata* cited only *Purdom 4* and included the statement "The specimen before me." Given that Schneider was sent specimens from A to identify and no other specimens of *Purdom 4* are known, this statement would seem sufficient evidence to treat the specimen at A as a holotype. In raising it to the status of a species, Schneider cited as "co-types" a number of living plants at the Arnold Arboretum, U.S., grown from seeds of *Purdom 604, 604a,* and *608,* Taibai Shan, 8 February 1911. These should be regarded as simply providing additional information.

The description of flowers used here is from *Chen et al. 2043* (details below). As such, it differs somewhat from that given by Ahrendt (1961: 122) who based it on a cultivated plant reputedly grown from *Purdom 604a.*

Berberis circumserrata appears to have a very limited range in Shaanxi, being restricted to the middle elevations of one mountain.

Berberis circumserrata was reported from Ningxia by D. Z. Ma (2007: 250), from Qinghai by Zhou (1997: 373), and from Gansu by Ying (2001: 97). No evidence has been found to support any of these reports. Specimens located from there and identified as such are mostly *B. diaphana.*

Selected specimens.
NE Henan. Lingbao Xian: Laoyachashan, 1600 m, 26 June 2005, *M. Liu et al. H10075* (PE 01840122); Laoyacha, 7 Aug. 2014, *J. M. Li et al. 140807156* (HEAC 0000253). **Lushi Xian:** Lao Kiün Shan (Laojunshan), 2000 m, 21 Sep. 1919, *J. Hers 1202* (A 00280188, AU 038760, BR, K, P P02313066); Lao-kiun-shan, 1450 m, 26 Aug. 1935, *K. M. Liao 5270* (PE 01030915, WUK 0004087). **S Shaanxi. Taibai Xian:** Taibai Shan, Tomukung, 2350 m, 17 Sep. 1932, *K. S. Hao 4263* (PE 01030890); P'ingansze, 2700–2900 m, 3 Aug. 1933, *T. P. Wang 1692* (PE 01030889, WUK 0004102), "Taipei-schan," 2500 m, 22 July–5 Sep. 1934, *G. Fenzel 763* (A 00280187, W, WUK 0072636, 0170978) and *750* (W, WUK 0072621, 0170976); Wu-kwan-T'ai, 4 Oct. 1934, *Y. Y. Pai 1516* (PE 01030884, WUK 0004088); on way from Ta-tien to Pinganszu, 2800 m, 6 Sep. 1937, *T. N. Liou & P. C. Tsoong 685* (PE 01030886, WUK 0004090, 0004092, 0094173); Yuhuang Shan, 2840 m, 22 June 1950, *J. M. Liu 11244* (PE 01033696, WUK 0020766); Doumugong, 2800 m, 29 Sep. 1955, *K. J. Fu et al. 0154* (PE 01030895); Ping'an Si, 2600 m, 6 Oct. 1956, *Seed Collection Team 350* (KUN 0178643, PE 01811887, WUK 0080821, 0093975); Doumugong, 2800 m, 2 July 1958, *58 Qinling Team 185* (PE 01387589); Ping'an Si, 9 Sep. 1958, *58 Qinling Team 322* (PE 01387586, 01387588); Baihe'an, 2540 m, 7 Aug. 1957, *K. J. Fu 10239* (WUK 00172359, 0090867); Ping'an Si, 2660 m, 27 Sep. 1959, *59 Qinling Mountain Team 1019* (PE 01387594–95); Tangyu, Shangbang Si, 2 July 1999, *Zhu et al. 1433* (KUN 0179273); Tangyu, Shangbang Si, 34°N, 109.12°E, 2400–2700 m, 2 July 1999, *G. H. Chen et al. 2043* (MO 5680631, 6408644).

125. Berberis concolor W. W. Sm., Notes Roy. Bot. Gard. Edinburgh 11: 199. 1920. TYPE: China. NW Yunnan: near Atunze (Dêqên [Deqin] Xian), Tung-chuling (Dongzhulin), 3050 m, 31 May 1913, *F. Kingdon-Ward 315* (holotype, E E00217986!).

Berberis solutiflora Ahrendt. J. Linn. Soc., Bot. 57: 205. 1961, syn. nov. TYPE: China. NW Yunnan: Mekong-Yangtze divide, SE of Dêqên, Pei-ma Shan (Beima Shan), 3650 m, 5 June 1932, *J. F. Rock 22835* (holotype, K K000077376!; isotypes, A 00279986!, UC UC516221!).

Berberis spraguei Ahrendt var. *pedunculata* Ahrendt, J. Linn. Soc., Bot. 57: 162. 1961, syn. nov. TYPE: Cultivated by Ahrendt, 6 May 1943, mistakenly reported as being from seed of *G. Forrest 30717* (lectotype, designated here, OXF!).

Berberis trichiata T. S. Ying, Fl. Xizang. 2: 125. 1985, syn. nov. TYPE: China. E Xizang (Tibet): Markam (Mangkang) Xian, Yanjing, Jiaolong Gongshe, 3500 m, 9 June 1976, *Qinghai-Xizang Complex Exped. 11833* (holotype, PE 00935246!; isotypes, HNWP 61920 image!, KUN 0179149!, PE 00935247!).

Berberis contracta T. S. Ying, Acta Phytotax. Sin. 37(4): 322. 1999, syn. nov. TYPE: China. NW Yunnan: Dêqên Xian, Benzi Lan Gongshe, Yeri Da, 122 Rd. Maintenance Station, 3000 m, 9 July 1981, *Qinghai-Xizang Exped. 2530* (holotype, PE 00935132!; isotypes, CDBI [2 sheets] 0027149–50 images!, HITBC 003592 image!, KUN [2 sheets] 0175828–9!, PE 00935133!).

Shrubs, deciduous, to 2 m tall; mature stems pale yellow or pale brownish yellow, subterete, verruculose; spines 3-fid, solitary or absent toward apex of stems, concolorous, 0.6–1.2(–1.5) cm, terete. Petiole almost absent; leaf blade concolorous green on both surfaces, abaxially sparsely finely papillose, narrowly obovate, obovate-oblong or obovate, 1–2.7(–3.5) × 0.3–0.9(–1.5) cm, thickly papery, midvein, lateral veins, and much-branched reticulation raised and conspicuous on both surfaces, base cuneate, margin entire, occasionally spinulose with 1 to 3 teeth on each side, apex obtuse. Inflorescence a panicle, sometimes compound at base, 10- to 40-flowered, 3–7 cm overall; bracts 1–2.5 mm; pedicel yellow, 2–3 mm in compound parts of inflorescence, otherwise 5–7(–10) mm. Sepals in 2 whorls; outer sepals broadly ovate, 2.5–3 × 2–2.5 mm; inner sepals obovate, 4–4.5 × 3–3.5 mm; petals obovate, ca. 3.5 × 2.5 mm, base shortly clawed, glands separate, apex slightly incised. Stamens ca. 3 mm; anther connective slightly extended, truncate. Ovules 1 or 2. Berry dark red to red-purple, ellipsoid, 7–8 × 5–6 mm; style persistent, short.

Phenology. *Berberis concolor* has been collected in flower in May and June and in fruit from July to September.

Distribution and habitat. *Berberis concolor* appears to be endemic to Dêqên Xian in northwest Yunnan and the neighboring Markam (Mangkang) Xian in Xizang (Tibet). It has been collected from dry, exposed mountainsides, alpine meadows, forest margins, and thickets by streams and roadsides at 2300–3650 m.

Berberis concolor can be distinguished from other species from the same area with similar leaves by its paniculate inflorescence. This characteristic is present in the type specimens of all the synonyms listed above.

Ahrendt cited unspecified specimens at BM and OXF as the type of *Berberis spraguei* var. *pedunculata*. One at OXF with extensive floral material has been chosen as the lectotype. Ahrendt's claim that this was grown from seed of *G. Forrest 30717* is unsustainable in that, not only do wild-collected specimens with this number (BM, E E00351976, HBG HBG-506733, PE 01032928) have floral material and were clearly collected at a time of year when seeds were not available, but in any case, appear to be of another species.

Kingdon-Ward (1923: 31) noted the abundance of *Berberis concolor* in the area where the type was collected. It would be interesting to know if this is still the case.

Ying (2001: 170) treated *Berberis spraguei* var. *pedunculata* as a synonym of *B. virescens*, a species from Nepal, Sikkim, and south Xizang. His treatment did not notice *B. solutiflora*.

Selected specimens.
NW Yunnan. Dêqên Xian: Doker-la, 28.333°N, 98.667°E, 3650 m, Sep. 1921, *G. Forrest 20292* (A 00279985, E E00612433, K, P P02313060, US 00945864); A-tun-tze, 2700 m, Sep. 1935, *C. W. Wang 70363* (A 00279396, KUN, 017583, NAS NAS00313877, PE 01037546, 01037548); near Dêqên, 2850 m, 14 Sep. 1937, *T. T. Yu 10207* (E E00612429, KUN 0175837, PE 01037545); Cang Jiang, Yongzi, 2300 m, 25 July 1940, *G. M. Feng 5801* (KUN 0175838–40, PE 01037547); side of Lan Cang Jiang, 2500 m, 17 July 1981, *Qinghai-Xizang Exped. 2879* (CDBI CDBI0028054, HITBC 003660, KUN 0175832–33, PE 00935134); Ba-li-da, Guonian, 28.300556°N, 98.830833°E, 2975 m, 20 May 2004, *S. G. Wu et al. WB-042* (MO 6173045); by hwy. G214, 28.473889°N, 98.912778°E, 3350 m, 20 Aug. 2013, *D. E. Boufford, J. F. Harber & Q. Wang 43135* (A 00914428, CAS, E E00833555, K, KUN 1278354, PE, TI).

126. Berberis cooperi Ahrendt, Gard. Chron. ser. 3, 109: 100. 1941. TYPE. Cultivated. Royal Botanic Gardens, Kew, 1926 "no 26 . . . 12–15 Bees," from seeds of *R. E. Cooper 2979*, W Bhutan. Thimphu, Parshong (Barshong), 26 Sep. 1914 (lectotype, designated here, K K000395254!).

?*Berberis angulosa* Wall. ex Hook. f. & Thomson var. *fasciculata* Ahrendt, J. Bot. 79(Suppl.): 42. 1941. *Berberis angulosa β* Hook. f. & Thomson, Fl. Ind. 227, 1855, syn. nov. TYPE: India. Sikkim, "10,700 ped.," s.d., *J. D. Hooker s.n.* (lectotype, designated by Ahrendt [1961: 114], K K000077363!).

Berberis lasioclema Ahrendt, J. Bot. 79(Suppl.): 57–58. 1941. TYPE: W Bhutan. Charithang, [27.416667°N, 89.033333°E], 3505 m, 31 May 1933, *F. Ludlow & G. Sherriff 41* (holotype, BM BM000551292!; isotype, E E00170012!).

?*Berberis hobsonii* Ahrendt, J. Linn. Soc., Bot. 57: 137. 1961. TYPE: China. S Xizang (Tibet): Yatung (Yadong), 27.85°N, 88.5833331897°E, *H. E. Hobson s.n.* (holotype, K K000395223!).

Shrubs, deciduous, to 2 m tall; mature stems pale yellow, very sulcate; spines solitary or 3-fid, concolorous, 1–1.5 cm, terete or abaxially sulcate, weak. Petiole almost absent; leaf blade abaxially pale green, shiny, adaxially deep green, shiny, obovate, 1.5–3.2 × 0.6–1 cm, papery, midvein raised abaxially, slightly impressed adaxially, lateral veins and reticulation indistinct abaxially, inconspicuous adaxially, base cuneate, margin entire, apex obtuse. Inflorescence a fascicle or sub-fascicle, 2- to 7-flowered; pedicel 6–12(–16) mm. Flowers bright yellow. Sepals in 3 whorls; outer sepals oblong-ovate, 6.5 × 4 mm; median sepals oblong-elliptic, 8 × 5 mm; inner sepals broadly obovate, 7 × 5 mm; petals orbicular-elliptic, 5.5 × 4 mm, base slightly clawed, glands separate, ovate, ca. 1 mm, apex rounded, entire. Stamens ca. 3 mm; anther connective extended, obtuse. Ovules 4 to 8. Berry red, globose, 6–8 × 7–8 mm; style persistent; seeds black.

Phenology. *Berberis cooperi* has been collected in fruit in China in September. In China its flowering season is unknown.

Distribution and habitat. In China *Berberis cooperi* is known from Yadong Xian in south Xizang. It is also known from west Bhutan and possibly from Sikkim, India. In Yadong it has been found in mountainside thickets at 3800 m.

Determining a lectotype of *Berberis cooperi* is somewhat complex. In the protologue, Ahrendt did not use the word type but in citing *Cooper 2979* stated, "I am indebted for material of this to the Royal Botanic Gardens at Kew and Edinburgh." Later that year (1941b: 61), Ahrendt did cite *Cooper 2979* as the type explaining, "This description is compiled from material sent from Kew of the well grown bush there raised from seed of Cooper's number 2979 which was received from Mr A. K. Bulley of Cheshire." Subsequently, Ahrendt (1961: 142) stated, "Cultivated L. A. 491, from seed of *Cooper 2979*, fl June 1938; fl. June 1938, fr. 26 Oct. 1938 (both Kew 74) (Type BM)."

There are four cultivated specimens of *Cooper 2979* at BM, none of which have any reference to "L.A. 491" or "Kew 74." Three dated 26 October 1938 with floral material are labelled as being from Kew and one dated

9 October 1940 with fruit is labelled as being from Edinburgh. All are labelled as being "Communicated" by Ahrendt and all are annotated by Ahrendt as "type," one of annotations to the Kew specimen being dated 21 February 1943, one 25 February 1943, and one being undated. The annotation to the Edinburgh specimen is dated 21 January 1942. In addition, there are two further cultivated specimens at Kew, one (K K000395254) is dated 1926 and labelled "no 74, Bees 12–15" and is annotated by Ahrendt as type with the date of 4 July 1940, the second "no. 74" is also annotated by Ahrendt as type with the date 11 March 1942. K000395254 at K has been chosen as the lectotype because Bees was the seed company of A. K. Bulley and the sheet provides clear evidence that Ahrendt saw it before the protologue of *Berberis cooperi* was published.

It should also be noted there is a wild-collected specimen of *Cooper 2979* at Edinburgh ex Cooper's herbarium (according to Long [1979], this herbarium only became available in 1952). This wild-collected specimen is the source of the collection details of the seeds of *Cooper 2979*.

The description of flowers used here is from a plant in my own collection grown from seeds of *R. Liddington s.n.*, yak pastures below Phajodhing, Thimphu, Bhutan, 1999 (where I saw the species growing in 1997). As such, it differs slightly from Ahrendt's descriptions of *Berberis cooperi* and *B. lasioclema*. Ahrendt's description of the flowers of *B. hobsonii* is somewhat different again, but K's policy in relation to type specimens has precluded me verifying his description. In my opinion, the similarities among the types of these three taxa are such that these are unlikely to represent different species, especially as species diversity of *Berberis* in this part of the Himalayas appears to be limited; therefore, the synonymy made by Grierson (1984: 327) is followed here. It should be noted that, though the collector's notes to the type of *B. hobsonii* say, "Yatung, Tibet," the coordinates given situate the place of collection some 20 km south in Sikkim.

Hooker and Thomson (1855: 227) listed a "*Berberis angulosa β*" which was described as having the same characteristics as *B. angulosa* but was fascicled rather than single-flowered. The evidence cited was "Sikkim, alt 10,000 ped." No collector for this was given, but from contextual evidence, this was Hooker. The lectotype of *B. angulosa* var. *fasciculata* has a label in Hooker's hand describing it as *B. angulosa* var. *β*, so it can be assumed this was one of the specimens Hooker and Thomson were referring to. Unfortunately, it is a very poor one with no leaves and only a few fruit, so the synonymy with *B. cooperi* is possible rather than definite. The lectotypification was by Ahrendt (1961: 114).

Besides the type of *Berberis angulosa* var. *fasciculata*, Ahrendt also cited *F. Kingdon-Ward 11653* (BM

BM001010603), Assam Himalaya, Mago, 9 June 1935. Mago is near the border with east Bhutan and is in that part of Arunachal Pradesh where there is a conflicting territorial claim by China. The specimen suffered much damage while on loan to an Indian herbarium in 1985–1986 and now has only a few leaves and the remnants of floral material. Hence, its determination is impossible to verify (unfortunately Ahrendt gives only a sketchy description of its flowers), but from the evidence of what remains of the specimen, it appears to be similar to various specimens collected nearby in Cona (Cuona) Xian in southeast Xizang. These are *Qinghai-Xizang Team Botany Group 2619* (PE 00049418, 01030773), south of Boshan Kou, 3800 m, 10 August 1974; *Qinghai-Xizang Supplementary Collections 75-1605* (PE 01030774), same location, 4300 m, 7 September 1975; *Qinghai-Xizang Supplementary Team 75-1742* (PE 01030775), same location, 3500 m, 10 September 1975; and *Qinghai-Xizang Supplementary Team 751894* (HNWP 92506, PE 01030776, 01840028), Boshan, south slope, 3800 m, 10 September 1975 (these are all annotated as *B. hobsonii*). The relationship (if any) of these specimens to *B. cooperi*, whose distribution in Bhutan appears to be unknown, requires further investigation. If they are not *B. cooperi*, they are likely to be of an undescribed species.

Selected specimens.
S Xizang (Tibet). Gyirong (Jilong) Xian: "Yatung Canyon (Jilung Canyon), 300–3100 m, 14 Aug. 2012, *Phylogeny of Chinese Land Plants, National Taiwan University Expedition to Tibet 2012 1080* (E E00662306). **Yadong Xian:** Chunpikuang, 3800 m, 14 Sep. 1974, *Qinghai-Xizang Team 74-2463* (HNWP 89526, KUN 0178676, PE 01030769–70).
BHUTAN. Thimphu, Parshong (Barshong), [27.7°N, 89.55°E], 3505 m, 26 Sep. 1914, *R. E. Cooper 2979* (E E00170023).

Cultivated material:
Living cultivated plants.
Foster Clough, Mytholmroyd, West Yorkshire, U.K., from *R. Liddington s.n.*, yak pastures below Phajodhing, Thimphu, Bhutan, 1999.

127. Berberis cornuta Harber, sp. nov. TYPE: China. SE Xizang (Tibet): [Lhünzê (Longzi) Xian]: Podzö Sumdo, Tsari Chu, 28.683333°N, 93.466667°E, 3350–3960 m, 21 May 1936, *F. Ludlow & G. Sherriff 1626* (holotype, BM BM000939688).

Diagnosis. Berberis cornuta has mature stems with a similar color and leaves somewhat similar in shape and size to *B. nambuensis*, but a very different flower structure with three whorls of sepals and stamens with a horned apex versus the two whorls of sepals and truncate apex of *B. nambuensis*.

Shrubs, deciduous, to 3 m tall; mature stems pale yellowish brown, sulcate; spines 3-fid, solitary toward apex of stems, concolorous, 0.6–1.6 cm. Petiole almost absent; leaf blade abaxially pale green, papillose, adaxially green, narrowly obovate or obovate-elliptic, (1.2–)2–2.8 × 0.6–0.9 cm, papery, venation inconspicuous on both surfaces, base attenuate, margin entire, apex subobtuse to subacute. Inflorescence 1-flowered; pedicel 18–28 mm. Flowers orange-yellow. Sepals in 3 whorls; outer sepals oblong-ovate, 9 × 3.5 mm; median sepals ovate, 8 × 5 mm; inner sepals obovate-elliptic, 8–9 × 6 mm; petals obovate, 7 × 4.5 mm, base clawed, glands separate, apex slightly emarginate or entire. Stamens 5.5–6 mm; anther connective extended, horned. Ovules unknown. Berry unknown.

Phenology. Berberis cornuta has been collected in flower in May. Its fruiting season is unknown.

Distribution and habitat. Berberis cornuta is known from only the type collected among scrub jungle in the bed of a valley at 3350–3960 m in Lhünzê (Longzi) Xian in southeast Xizang (Tibet).

Etymology. Berberis cornuta is from cornutus, Latin for horned, and has been chosen because of the shape of the apex of the species' stamens.

IUCN Red List category. Berberis cornuta is assessed as DD or Data Deficient, according to IUCN (2001) criteria.

F. Ludlow & G. Sherriff 1626 was identified as *Berberis ludlowii* by Ahrendt (1941b: 43). This, despite the fact that *B. ludlowii* has mature stems that are dark purple, leaves up to 7 cm long, and pedicels up to 55 mm long.

As far as I can ascertain, the horned apex of the stamens of *Berberis cornuta* is not recorded for any other *Berberis* species in the genus. This is certainly the case in relation to species found in China.

Unfortunately, the pistil of the flower chosen for dissection from the type specimen proved to be very damaged and so the number of ovules could not be ascertained.

128. Berberis crassilimba C. Y. Wu ex S. Y. Bao, Bull. Bot. Res., Harbin 5(3): 2. 1985. TYPE: China. NW Yunnan: Zhongdian (Xianggelila) Xian, N Flank of Haba Yueshan, 9 June 1939, *K. M. Feng 1252* (holotype, KUN 0160526!; isotypes, A 00038727!, KUN [2 sheets] 0175850–51).

Shrubs, deciduous, to 80 cm tall; mature stems purplish red, sub-sulcate; spines 3-fid, pale yellowish brown, 1–1.5 cm, abaxially sulcate. Petiole almost absent; leaf blade abaxially pale green, papillose, adaxially green, obovate or obovate-elliptic, ca. 0.6–1.2 ×

0.5–0.6 cm, papery, midvein and lateral venation inconspicuous or indistinct on both surfaces, base cuneate, margin entire, apex rounded, mucronate. Inflorescence 1-flowered; pedicel 3–6 mm. Flowers unknown. Berry globose or subglobose, red, 5–7 × 5–6 mm; style not persistent; seeds 2 to 5.

Phenology. *Berberis crassilimba* has been recorded as being collected in fruit in June. Its phenology is otherwise unknown.

Distribution and habitat. *Berberis crassilimba* is reliably known from only the type collection from Haba Shan in Zhongdian (Xianggelila) Xian in northwest Yunnan. This was found on a open stony slope at an unrecorded elevation.

The holotype of *Berberis crassilimba* has only mature fruit (because of this, the June collection date should be treated as doubtful). However, the protologue gave a description of flowers. This must have come from one or both of the only other specimens cited in the protologue: *K. M. Feng 1147*, also from Haba Shan, and *Feng 657* from Yulongshan (details of both below). This included giving the number of ovules as three. However, fruit from the isotype of *B. crassilimba* at A produced two to five seeds. Moreover, the description of flowers in the protologue is more or less the same as that of *B. exigua*, whose type is from seeds collected on Yulongshan and which also has three ovules but narrowly ovoid fruit versus the globose or semi-globose fruit of *B. crassilimba*. This suggests that either or both *Feng 1147* and *Feng 657* are, in fact, *B. exigua*.

Selected specimens.
NW Yunnan. Lijiang Xian: [that part now in Yulong Xian], Pai-shu-ho (Baishui He), 3 Apr. 1939, *K. N. Feng 657* (A 00279397, KUN 0175852–53). **Zhongdian (Xianggelila) Xian:** N flank of Haba Snow Range, 3 May 1939, *K. M. Feng 1147* (A 00279398, KUN 0175847–48).

129. Berberis dahaiensis Harber, sp. nov. TYPE: China. NE Yunnan: Huize Xian, Dahai Xiang, Dahai Caoshan, near Dishuiyan, 26.190278°N, 103.269167°E, 3600 m, 10 Sep. 2013, *D. E. Boufford, J. F. Harber & X. H. Li 43529* (holotype, PE!; isotypes, A!, BM!, CAS!, E!, K!, KUN!, TI!).

Diagnosis. *Berberis dahaiensis* has leaves that are somewhat similar to *B. aggregata*, (though with coarser marginal spines), but has purple mature stems and a very different inflorescence.

Shrubs, deciduous, to 1 m tall; mature stems purple, sulcate; spines 3-fid, pale brownish yellow, 1–2 cm, abaxially sulcate. Petiole almost absent; leaf blade abaxially pale green, papillose, sometimes pruinose,

adaxially green, narrowly obovate or narrowly elliptic-obovate, 1.5–2.8 × 0.6–1 cm, thinly papery, midvein raised abaxially, lateral veins and reticulation conspicuous on both surfaces, base attenuate, margin spinose with 4 to 8 coarse teeth on each side, spines to 2 mm, sometimes entire on young stems, apex acute, subacute, or obtuse, rarely mucronate. Inflorescence a fascicle or a sub-umbel, 2- to 5-flowered, otherwise unknown; fruiting pedicel 5–8 mm; peduncle (when present) to 1 cm. Berry red, oblong or ellipsoid, 4–5 × 8 mm; style short or absent; seeds 2.

Phenology. *Berberis dahaiensis* has been collected in fruit from July to September. Its flowering period is unknown.

Distribution and habitat. *Berberis dahaiensis* is known from only the open alpine grasslands of Dahai Caoshan, Huize Xian, in northeast Yunnan, where it has been found at 3520–3600 m.

IUCN Red List category. *Berberis dahaiensis* is assessed as DD or Data Deficient, according to IUCN (2001) criteria.

I made a brief visit to Dahai Caoshan (grass mountain) in September 2013 to look for this species, having seen images of the specimens at KUN cited below. I found only one small colony, though given the vast expanse of grassland (some 120 square kilometers), there are likely to be others. The only other *Berberis* I found here was *B. qiaojiaensis*, which was much commoner.

The coarse marginal spines of the leaves of *Berberis dahaiensis*, which are particularly large in relation to the size of the leaves, make it a very distinct species. Both of the collections cited below have leaves that are markedly abaxially pruinose, but none of the plants in the colony I found exhibited this characteristic.

Selected specimens.
NE Yunnan. Huize Xian: Dahai Muchang, 3520 m, 28 Sep. 1963, *Dongchuan Team 63-141* (KUN 0175726, 0177408, LBG 00064322); Dahai, betw. Dishuiyan & Hongzikou, 3600 m, 21 July 1964, *Northeast Yunnan Team 270* (KUN 0175724, 0177409, LBG 00064323).

130. Berberis daiana T. S. Ying, Acta Phytotax. Sin. 37: 345. 1999. TYPE: China. NE Sichuan (Chongqing): Chengkou, Yiziliang, Fangya Kou, 2550 m, 20 Sep. 1958, *T. L. Dai 107268* (holotype, PE 00935127!; isotypes, PE [8 sheets] 00935128–29!, 01037407–12!), SZ [2 sheets] 00289755, 00287775 images!).

Shrubs, deciduous, to 1 m tall; mature stems pale brownish yellow; sulcate; spines 3(or 5)-fid, concolorous, 0.5–1.2 cm, terete, slender. Petiole almost absent or to 5 mm; leaf blade abaxially pale yellow-green, adaxially deep green, elliptic-obovate or obovate, 2–5 × 1.2–2.2 cm, papery; midvein and lateral veins raised abaxially, dense reticulation inconspicuous abaxially, conspicuous adaxially, base cuneate, margin entire or spinose with 3 to 8(or 10 to 20) inconspicuous teeth on each side, apex rounded. Inflorescence 1-flowered, sometimes paired at the apex of stems; pedicel 15–25 mm, stout, tip sometimes bent. Sepals in 2 whorls; outer sepals elliptic, 7.5–8 × 4.5–5 mm; inner sepals broadly obovate-elliptic, 7–7.2 × ca. 5 mm; petals obovate, ca. 6 × 4 mm, base clawed, glands separate, elliptic, apex narrowly incised with acute lobes. Stamens ca. 3 mm; anther connective extended, rounded. Ovules 11. Berry red, slightly pruinose, subglobose, 1.3–1.5 × 1.2–1.3 cm; style not persistent; seeds 5 to 8.

Phenology. *Berberis daiana* has been collected in flower in June and in fruit from August to September.

Distribution and habitat. *Berberis daiana* is known from Chengkou and Wuxi Xian in Chongqing (northeast Sichuan) and Shennongjia Lin Qu in west Hubei. It has been found in thickets and forested areas on mountain slopes and above cliffs at 2000–2830 m.

The protologue describes *Berberis daiana* as evergreen, but all the evidence suggests it is deciduous.
Farges 1245 (details below) was probably collected between 1892 and 1900. There are three specimens with this number at P, two with flowers which are designated here as *Farges 1245* "A," and one with fruit which is designated *Farges 1245* "B." As can be seen from the above, it was not until a century after this collection that *Berberis daiana* was recognized.
The specimens below from Shennongjia Lin Qu have mostly been identified as *Berberis circumserrata*, which has four ovules. This was an identification I was initially inclined to accept; however, a dissection kindly made on my behalf by Bonnie Isaac of the Carnegie Museum of Natural History of CM's specimen of *Sino-Amer. Bot. Exped. 255* produced eight seeds.

Selected specimens.
Chongqing (NE Sichuan). Chengkou Xian: Tschen-ke-ou-tin (Chengkou), s.d., *P. G. Farges 1245* "A" (P P02682018, P06868329), *P. G. Farges 1245* "B" (P P06868330); Yizi-liang, Dayakou, 2500 m, 11 June 1958, *T. L. Dai 100826* (MO 04533964, PE 00935130, SZ 00289758); Yiziliang, Daya Kou, 2350 m, 11 Aug. 1958, *T. L. Dai 101835* (CDBI CDBI0027153, MO 9470954, PE 00935131, SZ 00289756–57); Yiziliang, Daya Kou, 2500 m, 20 Sep. 1958, *T. L. Dai 100826* (PE 00935130, SZ 00289758). **Wuxi Xian:** Shuanghe Xiang, Lanmei Zhai, 2600 m, 14 July 1958, *G. H. Yang 58758* (IBSC 0092618, PE 01037415–16, SZ 00289759–60); W of Hongchi Ba, 2000 m, 31 Aug. 1958, *G. H. Yang 59474* (CDBI

CDBI0027154, FUS 00014399, IBSC 0092632, PE 01037417); betw. Xiliu & Waituancheng, 2500 m, 20 July 1962, *B. C. Ni 00527* (CDBI CDBI0027155).
W Hubei. Shennongjia Lin Qu: S slope of Dashenong-jia, 2700 m, 5 July 1976, *Shennongjia Forestry Team 10711* (HIB 0129283–84, PE 01030911–12); Xiao Shennongjia, 2800 m, 9 July 1976, *Shenongjia Forestry Team 10747* (PE 01030913–14); Chuifeng pass, 2760 m, 31.5°N, 110.5°E, 26 Aug. 1980, *Sino-Amer. Bot. Exped. 255* (A 0280191, CM 274234, HIB 0129285–86, KUN 0175825, NA 0017420, NY, PE 00996104, UC); Shennong Gu, 31.443055°N, 110.2730556°E, 2830 m, 19 Sep. 2011, *Wuling Shan Plant Research Institute 20110919009-1-5* (Wuling Shan Research Institute Herbarium); Jiuhu Zhen, Pingqian Cun, Houzishi Shan, 18 July 2011, *D. G. Zhang zdg1885* (Shennongjia Forest Herbarium); s. loc., May 2014, *W. Du 14092* (WH 15074081).

131. Berberis daochengensis T. S. Ying, Acta Phytotax. Sin. 37: 336–338. 1999. TYPE: China. W Sichuan: Daocheng Xian, Gongling Qu, Lamuge, 3380 m, 31 July 1973, *Sichuan Botany Team 2426* (holotype, PE 00935222!; isotypes, CDBI CDBI0027982 image!, KUN 0178743!, SITC 00012147 image!, SWCTU 00006289 image!).

Shrubs, deciduous, to 2 m tall; mature stems pale brownish yellow, terete; spines absent, rarely solitary or paired, 1–1.4 cm, terete. Petiole almost absent or to 4 mm; leaf blade abaxially pale green, adaxially green, obovate or narrowly obovate, rarely elliptic, 1.5–3.8 × 0.6–1.5 cm, papery, midvein raised abaxially, slightly impressed adaxially, lateral veins and reticulation inconspicuous on both surfaces, base cuneate, margin entire or spinulose with (1 to)10 to 20 teeth on each side, apex rounded, obtuse, or subacute. Inflorescence a raceme, rarely partially paniculate, 5- to 15-flowered, 2.5–4.5 cm including peduncle 1–2.5 cm; pedicel 3–6 mm; bracts ovate, 1–1.5 mm, apex acuminate. Flowers pale yellow. Sepals in 3 whorls; outer sepals oblong-ovate, 6 × 3.5 mm; median sepals obovate-elliptic, 7 × 5 mm; inner sepals obovate-elliptic or obovate-orbicular, 7–9 × 6–8 mm; petals obovate-elliptic, 7.5 × 5.5 mm, base clawed, glands separate, ca. 1 mm, apex entire or notched. Stamens 5 mm; anther connective distinctly apiculate. Pistil 5 mm; ovules 2, sessile. Berry red, oblong, 8–9 × 3–4 mm; style not persistent or persistent and short.

Phenology. *Berberis daochengensis* has been collected in fruit from July to September. In cultivation it has flowered in June.

Distribution and habitat. *Berberis daochengensis* is known from Daocheng Xian in west Sichuan. It has been found by road and river sides and on an open, mossy forest floor at 3380–3800 m.

Berberis daochengensis was published solely on the basis of the holotype and was described in the protologue as having spineless stems and leaves that were entire or occasionally spinulose with 1 or 2 teeth on either side. The isotypes, however, evidence of solitary spines and leaves that are spinulose with up to 20 teeth. These characteristics are confirmed by the living cultivated plant cited below.

The plant of *SICH 1347* is the source of the color of the mature stems and the flower structure given above.

Selected specimens.
W Sichuan. Daocheng Xian: Riwa Xiang, Rencunshe, 3800 m, 26 Sep. 1971, *Sichuan Economic Plant Exped. 0763* (CDBI CDBI0027156); betw. Mula Xiang & Kalong, 3940 m, 20 June 1973, *Sichuan Botany Team 1863* (CBDI 0027191, 0027563, 0028047–48, SITC 00012132, SWCTU 00006215); Yading Cun, 28.467639°E, 100.351194°N, 2690 m, 19 Aug. 2007, *Z. K. Zha & G. Y. Liu YA274* (Ganzi Institute of Forestry Herbarium 0027944–47).

Living cultivated plants.
Royal Botanic Gardens, Kew, and Howick Hall, Northumberland, U.K., from seeds of *C. Erskine, H. Fliegner, C. Howick & W. McNamara (SICH) 1347*, Daocheng Xian, above Camp 1, Yading farmsteads, following Yading River upstream on NE side of Xiao Gonga Shan, ca. 3700 m, 23 Sep. 1994.

132. Berberis dasystachya Maxim., Bull. Acad. Imp. Sci. Saint-Pétersbourg, sér. 3, 23: 308. 1877. TYPE: China. Terra Tangutorum (Province Kansu), "Jugum a fl. Tetung S.," "19/31 V 1873," *N. M. Przewalski s.n.* (lectotype, designated by Mikhailova [2000: 60], LE!; isolectotypes, LE! [3 sheets]; possible isolectotypes, Terra Tangutorum (Province Kansu), May 1873, *N. M. Przewalski s.n.* (E E00217987!, K K000077395!).

Berberis dielsiana Fedde, Bot. Jahrb. Syst. 36(Beibl. 82): 41. 1905, syn. nov. TYPE: China. S Shaanxi: [Hu Xian], Sciu-jan-shan (Shouyang Shan), beside river Kan y huo, 15 May 1899, *G. Giraldi 2298* (lectotype, designated here, FI!; isolectotype, B†, photograph, and fragments at A 00038740!).
Berberis dolichobotrys Fedde, Bot. Jahrb. Syst. 36(Beibl. 82): 41–42. 1905. TYPE: China. S Shaanxi: [Taibai Xian], Thae-pei-san (Taibai Shan), 10–20 Aug. 1894, *G. Giraldi 51* (lectotype, designated here, FI; isolectotype, B†).
Berberis kansuensis C. K. Schneid., Oesterr. Bot. Z. 67: 288. 1918, syn. nov. TYPE: China. SW Gansu: Minchow-Choni (Min Xian-Jonê) distr., 1911, *W. Purdom 1014* (holotype, A 00038770!; isotypes, E E00217950!, K K000077398!, US 00946011 image!).
Berberis kansuensis var. *procera* Ahrendt, J. Linn. Soc., Bot. 57: 183. 1961. TYPE: China. SW Gansu: Komang ssu, 2990 m, Oct. 1925, *J. F. Rock 13290* (holotype, K K000077397!; isotypes, A 00038771!, E E00217949!, G G00226030!, P P00716530!, S 08–796 image!, UC UC381628!).
Berberis dasystachya Maxim. var. *pluriflora* P. Y. Li, Acta Phytotax. Sin. 10(3): 213. 1965, syn. nov. TYPE: China. Qinghai: Mengyuan Xian, Chu-ku (Kyikug), 2500 m, 18

June 1960, *Qinghai-Gansu Exped. 2607* (lectotype, designated here, WUK 00167536 image!; isotypes, PE 01033419!, WUK 00412542 image!).

Shrubs, deciduous, to 4.5 m tall; mature stems shiny, reddish or purplish brown or reddish purple, terete, very sparsely verruculose; spines 3-fid, but often solitary or absent toward apex of stems, concolorous or slightly lighter, 0.2–4 cm, terete or abaxially sulcate. Petiole almost absent or to 50 mm; leaf blade abaxially yellow-green, papillose, adaxially dark yellow-green, obovate, suborbicular, oblong-elliptic, or broadly elliptic, 3–6 × 2.5–4 cm, papery, midvein raised abaxially, slightly impressed adaxially, lateral veins and reticulation indistinct or obscure on both surfaces, inconspicuous or conspicuous on both surfaces when dry, base cuneate, rounded, or cordate, margin spinulose with 25 to 50 teeth on each side, sometimes entire, apex rounded or subacute. Inflorescence a raceme or spike-like raceme, often pointing upward, (10- to)15- to 50(to 75)-flowered, rarely compound at base, 3–20 cm overall; peduncle 1–3 cm; pedicel 4–7 mm; bracteoles lanceolate or triangular, ca. 2 × 0.5 mm. Sepals in 2 whorls; outer sepals lanceolate, ca. 3.5 × 2 mm; inner sepals obovate or obovate-oblong, ca. 5 × 3 mm, base slightly clawed; petals obovate or elliptic, ca. 4 × 2.5 mm, base clawed, glands separate, oblong-elliptic, apex entire or emarginate. Stamens ca. 2.5 mm; anther connective not extended, truncate. Ovules 1 or 2. Berry red, ellipsoid or oblong, 6–7 × 5–5.5 mm; style not persistent.

Phenology. *Berberis dasystachya* has been collected in flower between April and June and in fruit between June and September.

Distribution and habitat. *Berberis dasystachya* is known from Gansu, west Hubei, Qinghai, south Shaanxi, south Shanxi, north Sichuan, and possibly northwest Henan. It has been found in thickets, forests, and on forest margins, mountain peaks and slopes, and streamsides at ca. 1200–3800 m.

The protologue of *Berberis dasystachya* cited only "Kansu (Przewalski, 1873)." Subsequently, Schneider (1918: 287) cited as "Typ" Przewalski specimens from the Tetung Mountains of 31 May 1872 and the upper Huangho of 5 August 1880, both of these specimens having been cited by Maximowicz (1889a: 30), though not as types. The latter cannot be a type, but the former can be provided "1872" is accepted as a misreporting of 1873 on the basis that Bretschneider (1898a: 961–962) records that, in May and June of 1872, Przewalski was in what is now Nei Mongol, whereas in May 1873, he was in the Tetung Mountains. If this is accepted as misreporting, then Przewalski specimens of 31 May 1873 should be regarded as types. There are four of these at LE, only one of which has the collection details "Jugum a fl. Tetung, S." This was lectotypified by Mikhailova (2000: 60). Ahrendt (1961: 184) cited *Przewalski s.n.*, 1873 at K as type. There are no collection details for this other than it was collected in May in Gansu. This and a specimen at E with the same collection details are treated here as possible isolectotypes.

Berberis dielsiana and *B. dolichobotrys* were published by Fedde on the basis of specimens sent by FI to Berlin for identification. The protologue of *B. dielsiana* cited only *Giraldi 2298* and that of *B. dolichobotrys Giraldi 51* and *49*, but neither case cited any herbarium nor used the word type. Schneider cited *Giraldi 51* as the type of *B. dolichobotrys*, but again without citing any herbarium. From the evidence of the photographs at A cited above and below, it is certain that specimens of *Giraldi 2298* and *49* were retained at Berlin. The same is likely for *Giraldi 51*. Any such specimens were destroyed in WWII; therefore, specimens of *2298* and *51* at FI have been designated here as lectotypes.

The protologue of *Berberis kansuensis* cited as type *Purdom s.n.* at A with the collection details "Minchow and Choni" districts, probably of 1910. There is no *Purdom s.n.* at A corresponding to this description, but there is a *Purdom 1014* that does and which was annotated as type by Schneider on 3 October 1916. It is possible the (separate) label with the number was added later. The date of 1911 comes from the isotype at Kew.

The protologue of *Berberis dasystachya* var. *pluriflora* cited *Qinghai-Gansu Exped. 2607* as "Typus" but did not designate a herbarium. The specimen at WUK with the sheet number 00167536 has been chosen as the lectotype because WUK was where the author was based and the specimen is annotated as "Typus." The isolectotype at WUK is not so annotated.

There has been substantial confusion in relation to *Berberis dasystachya*. This seems to stem primarily from the very variable shape of the leaves, which can be obovate, suborbicular, oblong-elliptic, or broadly elliptic, and the length of the petiole, which can range from being almost absent to 50 mm. This variability can frequently be found in leaves on the same plant. It is in this context that the status of *B. dolichobotrys* and *B. dielsiana* need to be considered. In the protologue of *B. dolichobotrys*, Fedde acknowledged that the fruit stalk showed it was closely related to that of *B. dielsiana* but differed from it in its "broader, very rounded leaves that are slightly cordate at the base, the considerably produced petioles and also by the narrow triangular-lanceolate bracts and the cylindrical filaments" (in the protologue of *B. dielsiana*, Fedde described the bracts as narrowly triangular and the filaments as being obclavate, which seems a trivial difference). Subse-

quently, Schneider (1918: 287) stated he could see no difference between *B. dolichobotrys* and *B. dasystachya* except in the length of the flower stalk which Fedde reported in the former as being up to 10 cm long, whereas Schneider had seen examples for only the latter of up to 7 cm and he therefore treated *B. dolichobotrys* as a likely synonym. This synonymy was ignored by Ahrendt (1961: 183–184) who clearly had not seen the type specimens, but was followed by Ying (2001: 189). It is accepted here.

The trajectory of *Berberis dielsiana* has been quite different. Schneider (1908: 261) stated he believed *B. dielsiana* to be either identical to or hardly separable from *B. feddeana* (treated here as a synonym of *B. henryana*). Subsequently (1917: 441), he abandoned this idea and cited as examples of *B. dielsiana*, *Wilson 2863* (A, BM, E, HBG, K, US), west Sichuan, Ebian Xian, 1908, and *Purdom 341* (A, K, US), north Shaanxi, south of Yenan (Yan'an), as well as living plants at the Arnold Arboretum grown from seeds of *Purdom 543, 549*, and *605*. But he conceded that, not having seen *Giraldi 2298*, he was "not entirely convinced that Wilson's and Purdom's plants belong to the same species" as *B. dielsiana* (the type of which he mistakenly stated was also collected near Yan'an). By the following year (1918: 292), while still citing *Wilson 2863*, he seems to have decided that the Purdom collections were not *B. dielsiana*. The situation was further complicated by Ahrendt (1961: 183) who, as well as *Giraldi 2998* (which he also had clearly not seen), cited *Purdom 341* and *Hers 598* (A, BR, K, P) from north Henan while giving a full description of flowers based on a cultivated plant whose provenance was not given (there are specimens from this plant at BM annotated by him as *B. dielsiana*). However, not only are *Wilson 2863, Purdom 341*, and *Hers 598* different from each other, and the cultivated specimens at the BM different again, but none resemble *Giraldi 2298*. *Purdom 341* is *B. purdomii* as are cultivated specimens of *Purdom 543* and *549* at A, while *Wilson 2863* appears to be *B. francisci-ferdinandi* (I have been unable to identify *Hers 598*). Moreover, thanks to a flower from the type specimen kindly provided by FI, the flower structure of *Giraldi 2298* can be reported as follows: "bracteoles triangular, 1.3–2 × 0.8–1 mm; outer sepals oblong-ovate, 2.5–3 × 1.5–2 mm; inner sepals obovate, 3.5–4 × 2.5–3 mm; petals obovate, base very distinctly clawed, 3.3–3.8 × 2.3–2.7 mm, glands separate, ellipsoid, 5–7 mm. Stamens 2–2.5 mm, filament cylindrical or obclavate; anther connective truncate or shortly apiculate. Pistil 2–2.5 mm, ovules 2." It should also be reported that the leaves of *Giraldi 2298* are obovate, oblong-elliptic, or broadly elliptic, 3.5–6.5 × 1.5–3.5 cm, whereas in the description given by Ying (2001: 198) they are said to be elliptic or elliptic-lanceolate, 4–9 × 1–2 cm, or the leaves of Ahrendt's cultivated plant which are narrowly elliptic or lanceolate. As such, the flower structure differs little from the standard account given above and, when taken with the leaves, I conclude that *B. dielsiana* is a synonym of *B. dasystachya*.

Finally, it needs to be noted that Ying's (2001) mistaken description of the shape of the leaves of *Berberis dielsiana* may be why numerous specimens in Chinese herbaria have been misidentified as this taxon, most being *B. gilgiana*, *B. purdomii*, or *B. salicaria*. The original author of this mistake, however, appears to be Alfred Rehder, who identified various cultivated plants of *B. gilgiana* in the Arnold Arboretum as *B. dielsiana*, specimens of which were distributed to various Chinese herbaria.

Schneider (1918: 289), while noting the flowers of his *Berberis kansuensis* were "ganz auffallend" (quite strikingly) similar to those of *B. dasystachya*, claimed that the purple color of the mature stems of *B. kansuensis* was quite different from that of *B. dasystachya*. I have been unable to discover what Schneider believed this color to be, but the color of the type collection of *B. dasystachya* is purplish brown and, as such, is little different from that of some of the stems of the type collection of *B. kansuensis*; hence my synonymy.

Ahrendt (1961: 183) differentiated *Berberis kansuensis* var. *procera* from *B. kansuensis* solely on the basis that the former was 12–15 ft. (3.6–4.5 m) high versus the 3–5 ft. (0.9–1.5 m) of the latter. As it happens, even on the basis of the type collections of the two taxa, this was inaccurate in that the collector's notes of the holotype of *B. kansuensis* describes it as 6 ft. (1.8 m) high. In any case, from the evidence of numerous specimens, the height of *B. dasystachya* is very variable, no doubt reflecting age, climatic conditions, etc. Ying (2001: 189) synonymized variety *procera* but treated *B. kansuensis* as a separate species.

According to the protologue, *Berberis dasystachya* var. *pluriflora* differs from *B. dasystachya* only by having up to 75 flowers and stamens ca. 1.5 mm. This seems an insufficient basis for recognizing a separate variety. *Berberis dasystachya* var. *pluriflora* was not noticed by Ying (2001).

There are unresolved issues in relation to *Berberis dasystachya* in Shanxi and northwest Henan. In the former, *B. amurensis* is also found. Though this differs from *B. dasystachya* in having mature stems that are pale yellow or yellow-brown and a different flower structure, in the case of herbarium specimens from Shanxi that have neither, it is sometimes difficult to distinguish between the two species. It would seem that *B. amurensis* is found in the north and *B. dasystachya* in the south of Shanxi, but where the dividing line is and whether there are areas where both are found remains to be discovered. In northwest Henan, matters appear to be complicated by *B. honanensis* published by Ahrendt (1944b) on the basis of cultivated material reputedly from seed from Lushan in Henan. This is described as differing from *B. dasystachya* primarily by having yellow mature stems and flowers with three whorls of sepals (for further details of the taxon, see under Taxa Incompletely Known). None of the wild-collected possible candidates for this taxon from northwest Henan located (including those listed below) have either mature stems or floral material, and hence I have found it impossible to rule out the possibility that they are *B. honanensis* (assuming it exists in the wild and that the seeds of *B. honanensis* were collected in Lushan, neither of which may be true; for further details, see under Taxa Incompletely Known).

Selected specimens.
Syntypes. Berberis dasystachya. "Jugum a fl. Tetung, 5 versus," June 1873, *N. M. Przewalski s.n.* (LE, PE 01896189); Gansu, Terra Tangutum, June 1873, *N. M. Przewalski s.n.* (W 1890–2025); Gansu, Terra Tangutum,1873, *N. M. Przewalski s.n.* (P P00716592).

Syntype. Berberis dolichobotrys. Shaanxi. "Piccolo," Monte Hua tzo pin," 10 km from "Monte Tun u sse," 25 km from "Han kiun," 20 June 1984, *G. Giraldi 49* (B†, photograph and fragm. A 00038741, FI).

Other specimens.
Chongqing (NE Sichuan). Chengkou Xian: "District de Tchen-kéou tin," s.d., *P. G. Farges 398* (P 02682020), *P. G. Farges 742* (P P02682024), *P. G. Farges s.n.* (P P02682019). **Wuxi Xian:** Shuanghe Xiang, Lanying Zhai, 2600 m, 16 July 1958, *G. H. Yang 58834* (PE 01033003, 01033512).

Gansu. Jonê (Zhuoni) Xian: "Chuounihsien," La-Ki-Ko, 2700 m, 31 Aug. 1940, *W. Y. Hsia 8382* (WUK 0067946). **Lanzhou Shi:** Xinglong Shan, 2100 m, 2 Aug. 1959, *Y. Q. He 5352* (PE 01033371, WUK 00391524). **Li Hsien (Xian):** 1800 m, 21 June 1936, *T. P. Wang 4432* (KUN 0178636, PE 01032488, WUK 0066929). **Lintan Xian:** Yangsha, 2400 m, 15 June 1957, *Tao He Team 3384* (IBK IBK00013169, IBSC 0092743, KUN 0176442, 0178662, NAS NAS00313906, PE 01033407, WUK 0091308). **Min Xian:** Baipo Bao, 2300 m, 29 May 1951, *Z. B. Wang 14045* (PE 01033400, WUK 0020782). **Têwo (Diebu) Xian:** Lazikou Linchang, Meilong Gou, 34.171944°N, 103.8°E, 2700 m, 26 July 1998, *Bailongjiang Exped. 830* (PE 01556137). **Tianshui Shi:** Jiaochuan, Baojia Gou, Guanyin Ya, 1850 m, 23 May 1963, *Q. X. Li 208* (PE 01033409, 01033414). **Tianzhu Xian:** Zhu Xiang, 2500 m, 14 July 1959, *Y. Q. He 4769* (PE 01033396, WUK 00156639). **Weiyuan Xian:** Dongfeng Qu, Laojun Shan, 14 July 1956, *Chinese Academy of Science, Huang He Investigation Team, Gansu 1st division Team 1450* (WUK 0082571). **Wen Xian:** Motiangling Shan, Baishui Jiang Nature Reserve, Qiujiaba, ravine along NW branch of Baima He, 32.9225°N, 104.321667°E, 2320–2725 m, 19 May 2007, *D. E. Boufford & Y Jia 37772* (A 00285876, PE 01814738). **Xiahe Xian:** Qingshui, 2400 m, 2 July 1937, *K. T. Fu 934* (PE 01033451, WUK 0004192). **Xigu Qu:** Baicangzi, 3300 m, 19 June 1951, *Z. B. Wang 14192* (KUN 0179159, PE 01033398, WUK 0020772). **Yongdeng Xian:** Liancheng, Tulu Gou, 3000 m, 8 July 1991, *R. F. Huang 2155* (HNWP 0231962, 167408).

Yuzhong Xian: Mayu Shan, 7 Sep. 1956, *Huang He team 3060* (KUN 0179190, PE 01033395, 01601585). **Zhang Xian:** Hedi Cun, Songhu Shan, 11 June 1956, *Huang He Team 4685* (KUN 0178353, PE 01033444, WUK 0084598). **Zhangjiachuan Xian:** 1807 m, 15 May 2014, *Y. He & J. C. Hao GSL2014050537* (BNU 001352). **Zhuanglang Xian:** en route to Yunya Si, 1 June 2015, *Y. He & J. C. Hao GSL2015050976* (BNU 0020857). **Zhugqu (Zhouqu) Xian:** Dahaigou, 33.626389°N, 104.241111°E, 2000 m, 18 July 1998, *Bailongjiang Exped. 287* (PE 01556139).

W Hubei. Shennongjia Lin Qu: vic. of Honghe, 31.5°N, 110.5°E, 2200 m, 24 Sep. 1980, *Sino-Amer. Bot. Exped 1782* (A 280197–98, BISH, E E00392871, HIB 0129343–44, KUN 0179069, NAS NAS00313973–74, PE 00996109–10, UC UC1491775–6, WH 17086975).

S Ningxia. Jingyuan Xian: Wanhua Nan, 2230 m, 17 Aug. 1985, *Huangtu Team 4702* (WUK 00463363–64).

Qinghai. Datong Xian: Laoye Shan, 8 June 1980, *s. coll. 386* (HNWP 006528). **Huangzhong Xian:** near Sanhe Linchang, 21 Aug. 1972, *S. W. Liu 2120* (HNWP 28663). **Hua Xian:** Jiading, Zhalong Gou, 21 June 1971, *B. Z. Guo 9229* (HNWP 82908, KUN 0178200, PE 01033372). **Jainca (Jianzha) Xian:** Dangshun Gongshe, Bailang Gou, 2540 m, 10 June 1970, *L. H. Zhou & L. N. Sun 649* (HNWP 131568). **Ledu Xian:** Shangbei Shan Linchang, 3000 m, 3 Aug. 1986, *Y. H. Wu et al. 4108* (HNWP 134498). **Menyuan Xian:** Xianmi Linchang, 3200 m, 13 Sep. 1975, *B. Z. Guo & W. Y. Wang 12645* (HNWP 48257, PE 01033421). **Minhe Xian:** Xing'er, Beishan, 2500 m, 11 June 1970, *B. Z. Guo 7006* (HNWP 21519, WUK 0308131). **Ping'an Xian:** Gucheng Xiang, 3800 m, 5 Aug. 1986, *Y. H. Wu et al. 4193* (HNWP 133779, 0210200, 0210200). **Tongren Xian:** Xiabulong, Qianquang, 3000 m, 9 June 1972, *B. Z. Guo 10394* (HNWP 32474). **Xunhua Xian:** Mengda, 2500 m, 11 June 1981, *B. Z. Guo & W. Y. Wang 5262* (HNWP 98774). **Zêkog (Zeku) Xian:** Maixiu Linqu, Laocan Gou, 3060 m, 26 Aug. 1970, *L. H. Zhou & L. N. Sun 1899* (CAS 897415, HNWP 23041).

C & S Shaanxi. Chang'an Xian: Taiyigong, Cuihua Shan, 7 May 1956, *Chinese Academy of Science, Huang He Investigation Team 7* (KUN 0178805, PE 01033133, WUK 0081542). **Danfeng Xian:** Yaozhuang Xiang, 1500 m, 23 Sep. 1958, *S. B. He 422* (WUK 00102074). **Foping Xian:** Hetaoping, Shanren Gou, 1800, 25 June 1952, *K. J. Fu 4791* (IBK IBK00013113, PE 01033503, WUK 0059756). **Hu Xian:** Guangtou Shan, Gaojialing, 2000 m, 9 Sep. 1962, *K. J. Fu 14501* (WUK 00185672). **Lan'gao Xian:** Sijihe, 2200 m, 17 July 1959, *Y. L. Qiao 1509* (KUN 0178685). **Lung Hsien (Long Xian):** Kwanshan, 2100 m, 2 June 1936, *T. P. Wang 4084* (CDBI CDBI0027332, PE 01033649, WUK 0066943). **Mei Xian:** Tangyu, Jiaokou, 11 Oct. 1958, *X. M. Zhang 736* (KUN 0175860). **Ningshan Xian:** Xunyangba, Dagou, 1600 m, 7 June 1959, *J. Q. Xing 6241* (IBK IBK00013110, WUK 00128199). **Pingli Xian:** Hualong Shan, 2710 m, 1 July 1986, *K. L Yuan, Pingli 1-0534* (PE 01840084). **Shang Xian (now Shangzhou Qu):** Niutouya, 1200 m, 20 June 1952, *Z. B. Wang 15419* (KUN 0178829, PE 01033653, WUK 0062120). **Shiquan Xian:** Yangba, Zhifang Gou, 1700 m, 20 June 1960, *s. coll. 434* (KUN 0175863). **Taibai Xian:** Dadian, 5 Sep. 1956, *Seed Collection Team 233* (KUN 0175859, PE 01811888). **Yang Xian:** ironworks on way to Yuanbazi, 6 Sep. 1959, *Qingling Team 946* (PE 01387603). **Zhashui Xian:** Yingpan Linchang, 1800 m, 21 June 1973, *X. X. Hou et al. 987* (FJSI 008977, IBSC 0091695, WUK 00295725). **Zhen'an Xian:** Heigou Linchang, Mihunzhen Shan, 2440 m, 5 June 1973, *X. X. Hou 794* (FJSI 008976, IBSC 0091692). **Zhouzhi Xian:** Chengouwan, 1600 m, 23 Aug. 1959, *Qinling Team 612* (PE 01387590–91).

S Shanxi. Yangcheng Xian: Yunmeng Shan, 18 May 1959, *S. Y. Bao & S. J. Yan 161* (HNWP 114134, PE 01033477, 01033483). **Yicheng Xian:** Shihe Gongshe, Liujiaqu, Dacao Gou, 1480 m, 30 May 1960, *X. Y. Liu 20304* (HNWP 114064, PE 01033481). **Yuanqu Xian:** Meigu, Manwo cattle farm, Laojun, 2800 m, 29 May 1983, *Z. Liu 1009* (HSIB 003767, PE 01840067).

NC, N & NE Sichuan. Li Xian: Hwy. 213, N of Miyaluo, 31.725556°N, 102.744167°E, 3000–3200 m, 8 Sep. 1997, *D. E. Boufford et al. 27985* (A 00236480, HAST 115102). **Ma'erkang (Barkam) Xian:** Mozi Gou, 2850 m, 26 May 1957, *X. Li 71188* (IBSC 0091774, 0092637, IFP 05201999w0012, KUN 0175855, NAS NAS00313920–21, PE 01033687, WUK 00253444). **Nanjiang Xian:** Guanwu Shan Natl. Forest, 10 Aug. 2011, *J. X. Li et al. 1340* (CSFI I050932–34). **Nanping (now Jiuzhaigou) Xian:** Jiuzhaigou, 2800 m, 7 June 1983, *K. Y. Lang et al. 1610* (KUN 0179142, PE 01033522, 01033528). **Sungpan Hsien (Songpan Xian):** Kungkangling, 3000 m, 19 Oct. 1937, *K. T. Fu 2061* (PE 01033529, WUK 0004124). **Xiaojin Xian:** (A, E), NE of the city of Xiaojin near Muer Zhai, 30.997222°N, 102.693611°E, 2900–3000 m, 28 July 2007, *D. E. Boufford et al. 38456* (A, E). **Zoigê (Ruo'ergai) Xian:** Ri'er Gongshi, 2900 m, 21 June 1983, *K. Y. Lang et al. 1918* (KUN 0179136, PE 01033532, 01033535).

Possible specimens.

NW Henan. Lingbao Xian: Laoyacha Shan, 1700 m, 26 June 2005, *M. Liu et al. H10073* (PE 01840123). **Lushi Xian:** Laochünshan (Laojunshan), 1450 m, 13 Aug. 1935, *J. K. Liou L.5104* (K, PE 01033486, 01033490, WUK 0004095); Dakuai Di, Yuhuangjian, 2030 m, 8 July 1959, *Henan Team 3448* (PE 01033488–89); Shangpen Zhen, Guanyun Shan, 1700 m, 4 July 2005, *M. Liu et al. H20107* (PE 01840124). **Song Xian:** Shiren Shan, 2000 m, 21 Sep. 1983, *Botanical Resources Investigation Team L0524* (PE 01033492). **Xixia Xian:** Laojun Shan, 2100 m, 18 July 1960, *Henan Team 1212* (PE 01033487, 01033494).

133. Berberis dawoensis K. Mey., Repert. Spec. Nov. Regni Veg. 12: 379. 1922. TYPE: China. NW Sichuan: Dawo (Dawu, [Daofu] Xian), "Am Lamatempel Belo retscho, westlich Tschi sse tsung," 4000 m, 6 July 1914, *W. Limpricht 1936*, (lectotype, designated here, WU 039238!; isolectotype, WRSL†, photograph and twig from WRSL specimen, A 00038728!, photograph of WRSL specimen, SYS SYS00052310!)

Shrubs, deciduous, to 2.5 m tall; mature stems pale yellowish brown, angled; spines 3-fid, pale yellow, 0.6–1.5 cm, abaxially slightly sulcate, weak. Petiole almost absent; leaf blade abaxially grayish green, adaxially green, narrowly obovate to narrowly oblong-obovate or obovate-spatulate, 1–3.5 × 0.7–1.1 cm, papery, midvein slightly raised abaxially, venation inconspicuous on both surfaces, base cuneate, margin spinulose with 2 to 10 teeth on each side, sometimes entire, apex rounded. Inflorescence a fascicle, sub-fascicle, sub-umbel, or raceme, (1- to)3- to 5(to 10)-flowered, to 3 cm overall; pedicel 8–10 mm, slender. Sepals in 2 whorls;

outer sepals oblong-ovate, ca. 2.5 × 0.8 mm; inner sepals narrowly elliptic, 4 × 1.75 mm; petals obovate-elliptic, 3.5 × 1.75–2.25 mm, base slightly clawed, glands separate, apex distinctly incised, lobes acute. Stamens ca. 2 mm; anther connective not extended. Ovules 2. Berry red, oblong-ovoid, ca. 8 × 4 mm; style persistent, short.

Phenology. *Berberis dawoensis* has been collected in flower in June and July and in fruit in August.

Distribution and habitat. *Berberis dawoensis* is known from Dawu (Daofu) Xian in northwest Sichuan and Baima (Banma) Xian in south Qinghai. It has been collected from *Pinus* forest, alpine valleys, hillsides, and roadsides at ca. 3350–4000 m.

The protologue of *Berberis dawoensis* cited only *Limpricht 1936* but without citing a herbarium. From the evidence of the black-and-white photographs at A and SYS, the best specimen was that at WRSL, but this was apparently destroyed in WWII; therefore, the duplicate at WU has been designated the lectotype. The twig at A includes some floral material.

From Limpricht (1922: 165–166, Map 6), it is possible to locate the place of the collection of the type as being between Bamah (Bamai Zhen) and Dawo. This information was not noticed by Ahrendt (1961: 161), who both misreported the collection as being from Tsarong in Zayü (Chayu) Xian in southeast Xizang and, on the basis of *T. T. Yu 10560* (E E00623042) from northwest Yunnan and a cultivated plant apparently grown from this number (see BM BM000895027 for a specimen of this), substantially amended and augmented Meyer's protologue description. I have been unable to identify *Yu 10560*, but it is certainly not *Berberis dawoensis*.

The protologue of *Berberis dawoensis* gave only a partial description of the flower structure. The description used here is taken from a line drawing by Schneider on the sheet of the lectotype. The protologue also stated the inflorescence was mostly fascicled and, as such, this reported the evidence of the type specimens (though the description of the pedicels as being 1 mm was clearly a misprint for 1 cm). The wild-collected specimens of *H. Smith 12532* cited below are most fascicled, sub-fascicled, or sub-umbellate though the UPS specimen has evidence of a racemose fruit stalk. The cultivated specimens from this number, however, have inflorescences that are largely racemose. Investigation of the flower structure of UPS BOT:V-159617 showed it to be the same as in Schneider's line drawing; hence my amendments to the protologue description. I have not investigated the flower structure of *Sichuan Botany Team 05386* which is largely fascicled.

Sino-British Qinghai Alpine Garden Society Expedition 799 (details below) has an inflorescence as follows: a sub-raceme or sub-umbel, sometimes fascicled at base, 4- to 8(to 13)-flowered; flowers greenish yellow, 2.5–4 cm overall; peduncle 0.5–2.5 cm; pedicel 4–15 mm (to 20 mm at base). An investigation of the structure of its flowers by myself and David Chamberlain of E revealed them to be almost identical to those of the type of *Berberis dawoensis*. Found some 250 km to the north of where *Limpricht 1936* was collected, it may be evidence of an interesting disjunct, but given that the *Berberis* in the area between the two collections is in need of much more research, it is too early to definitely conclude this. There appear to be only a handful of other *Berberis* specimens of any description from Baima Xian itself.

The reports of *Berberis dawoensis* in Yunnan by Ying (2001: 197) and Bao (1997: 82) were no doubt based on Ahrendt's misreporting of the collection area of the type.

Selected specimens.
S Qinghai. Baima (Banma) Xian: valley S of Baima along Make He, 32.790278°N, 100.799444°E, 3350 m, 8 July 2000, *Sino-British Qinghai Alpine Garden Society Expedition 799* (E E00274455).
NW Sichuan. Dawu (Daofu) Xian: Lhamo Mondel La, ca. 3500 m, 20 Sep. 1934, *H. Smith 12352* (MO 4367267, PE 01037865, UPS BOT:V-040916); Qianning, Zha'e, 3920 m, 26 Aug. 1961, *s. coll. 61043* (CDBI CDBI0028207–08); Qianning Qu, 4000 m, 14 June 1974, *Sichuan Botany Team 05386* (CDBI CDBI0027464–66).

Cultivated material:
Cultivated specimens.
Uppsala Botanic Garden, from *H. Smith 12352*, 8 June 1945 (PE 01032919, UPS BOT:V-159617); 2 June 1946 (MO 6408631); 11 June 1946 (MO 6408635, UPS BOT:V-159632).

134. Berberis deqenensis Harber, sp. nov. TYPE: China. NW Yunnan: Dêqên (Deqin) Xian, side valley betw. Dêqên & Mekong River, 28.466667°N, 98.85°E, 3100 m, 2 June 1993, *Kunming, Edinburgh, Gothenburg Exped. 683* (holotype, E E00125284!; isotype, GB 012 8809 image!).

Diagnosis. *Berberis deqenensis* has leaves that are somewhat similar in shape and color to those of *B. polyantha*, but has a very different inflorescence. The leaves are also somewhat similar in shape to those of *B. bracteata*, but are lighter in color and with far fewer or no marginal spines.

Shrubs, deciduous, to 2.5 m tall; mature stems pale brown, sulcate; spines 3-fid, concolorous, (0.5–)1–2.5 cm, terete. Petiole almost absent; leaf blade abaxially pale green, papillose, adaxially green, obovate, mostly broadly so, 1.4–2.5 × 1–1.4 cm, papery, midvein, lateral veins, and dense reticulation raised and conspicuous on both surfaces, base attenuate, margin spinose with 2 to 5 widely spaced teeth on each side or entire, apex obtuse or rounded, sometimes mucronate. Inflorescence a fascicle, sub-fascicle, or sub-raceme, to 3.5 cm overall, 4- to 10-flowered; pedicel (3–)9–15(–20) mm. Sepals in 2 whorls; outer sepals triangular-ovate, 2 × 1.5 mm; inner sepals obovate-elliptic, 3.5 × 3 mm; petals obovate, 3 × 2 mm, base attenuate, glands oblong, 0.75 mm, contiguous to margins, apex entire. Stamens ca. 2 mm; anther connective slightly extended, obtuse. Pistil ca. 2 mm; ovules 2. Berry unknown.

Phenology. *Berberis deqenensis* has been collected in flower in June. Its fruiting season is unknown.

Distribution and habitat. *Berberis deqenensis* is known from Dêqên (Deqin) Xian in northwest Yunnan. It has been found in a degraded mixed forest at 3100 m.

IUCN Red List category. *Berberis deqenensis* is assessed as DD or Data Deficient, according to IUCN (2001) criteria.

The evidence of the color of the mature stems of *Berberis deqenensis* comes not from the type specimens but from the collectors' notes which state, "flowering branches pale brown."

The relationship between *Berberis deqenensis* and *B. bracteata* needs further investigation, but is hampered by the lack of any account of the flower structure of the latter.

135. Berberis derongensis T. S. Ying, Acta Phytotax. Sin. 37: 333. 1999. TYPE: China. W Sichuan: Derong Xian, betw. Derong & Riyu, 3200 m, 5 Aug. 1981, *Qinghai-Xizang Exped. 003305* (holotype, PE 00935136!; isotypes, CDBI [2 sheets] 0027181–82 images!, HITBC 003605, image!, KUN 0178920!, PE 00935137!).

Shrubs, deciduous, to 1.75 m tall; mature stems dark purplish red, sometimes partially pruinose, angled, not verruculose; spines 3(or 5)-fid, solitary toward apex of stems, pale bright yellow, 0.6–1.2 cm, terete. Petiole almost absent; leaf blade abaxially pale green, papillose, adaxially dull mid-green, narrowly elliptic or obovate, 0.5–1.5 × 0.3–0.5 cm, thickly papery, midvein and lateral venation raised abaxially, venation inconspicuous adaxially, base attenuate, margin entire, rarely spinulose with 1 to 5 teeth on each side, apex acute or obtuse. Inflorescence a sub-fascicle, sub-raceme, or sub-umbel, 4- to 7-flowered, to 2.5 cm overall; peduncle to 1.5 cm; pedicel 2–7 mm. Flowers unknown. Berry black, obovoid-oblong, 7–8 × 3–4 mm, slightly pruinose; style 1–2 mm; seeds 2.

Phenology. *Berberis derongensis* has been collected in fruit in July and August. Its flowering season is unknown.

Distribution and habitat. *Berberis derongensis* is known from Derong Xian in west Sichuan. It has been found in thickets on slopes in open places at 2920–3325 m.

The protologue of *Berberis derongensis* cited only the holotype and described the species as having an umbellate raceme with up to seven black fruit with two seeds. However, the isotypes have evidence of sub-fascicles and sub-racemes. *Qinghai-Xizang Team 1674* (details below) appears to be *B. derongensis* and has floral material, but I have not had an opportunity to examine this. The protologue also described the species as evergreen, but it appears to be deciduous.

It is possible that *D. E. Boufford, J. F. Harber & Q. Wang 43164* (details below) is *Berberis derongensis.* This is fascicled, sub-fascicled, or sub-racemose with up to seven immature fruit turning purple with two or three seeds. The mature stem color and leaf size and shape are the same as the type.

Selected specimens.
W Sichuan. Derong Xian: Zigen, 2920 m, 1 July 1981, *Qinghai-Xizang Team 1674* (CDBI CDBI0028055–56, HITBC 003598, KUN 0178919, 0179249); Zigen, 3250 m, 2 July 1981, *Qinghai-Xizang Team 1747* (CDBI CDBI0028061–62, HITBC 003600, KUN 0178314, 0178862, PE 01037163–64); rd. from Derong to Batang, betw. Gangran Cun & Yangedi, 29.092778°N, 99.375556°E, 3325 m, 22 Aug. 2013, *D. E. Boufford, J. F. Harber & Q. Wang 43164* (A, BM, CAS, E, K, KUN, PE, TI).

136. Berberis diaphana Maxim., Bull. Acad. Imp. Sci. Saint-Pétersbourg, sér. 3, 23: 309. 1877. TYPE: China. SW Gansu: "Jugum a fl. Tetung [Tatung]," 27 June 1872 [Gregorian calendar], 9 July 1872, *N. M. Przewalski s.n.* (lectotype, designated by Mikhailova [2000: 60], LE!).

Berberis circumserrata (C. K. Schneid.) C. K. Schneid. var. *occidentalior* Ahrendt, J. Linn. Soc., Bot. 57: 122. 1961, syn. nov. TYPE: China. E Qinghai: Koko Gorge, Sep. 1925, *J. F. Rock 13272* (holotype, K K000077358!; isotypes, A 00038726!, E E00259016!, P P02313612!, S S08–806 image!, UC UC381473!).

Shrubs, deciduous, to 2.5 m tall; mature stems pale yellowish brown, angled or sulcate; spines 3-fid, semi-concolorous, 1–2 cm, terete or abaxially slightly sulcate, stout. Petiole almost absent; leaf blade abaxially pale green, papillose, adaxially dark green, oblong, obovate-oblong, or obovate, 1.5–4 × 0.5–1.6 cm, papery, midvein and lateral venation raised abaxially, slightly impressed adaxially, reticulation conspicuous abaxially, inconspicuous adaxially, base attenuate, margin spinulose with 2 to 14 teeth on each side or entire, apex obtuse. Inflorescence a fascicle, 1- to 5-flowered, with single or paired flowers frequently found toward the end of branches; pedicel 12–22 mm. Sepals in 2 whorls; outer sepals ovate, 6 × 4 mm, inner sepals obovate-elliptic, 5 × 4 mm; petals obovate, base clawed, glands separate, apex obtuse, incised. Anther connective slightly extended, obtuse. Ovules 4 to 6. Berry red, ovoid-oblong, 10–12 × 6–7 mm; style persistent, apex slightly bent.

Phenology. Berberis diaphana has been collected in flower in May and June and in fruit between July and October.

Distribution and habitat. Berberis diaphana is known from Gansu, Ningxia, Qinghai, and north Sichuan. It has been collected in alpine meadows, forests, and forest margins at ca. 1600–3700 m.

In the protologue of *Berberis diaphana*, Maximowicz cited simply "*Przewalski* 1872" from Gansu. Since he included a description of both flowers and fruit, it is reasonable to assume this was based on more than one gathering and that specimens of these were at LE where Maximowicz was based, and, indeed, there are three *Przewalski* sheets at LE from Gansu dated 1872, one "Jugum a fl. Tetung," 27 June 1872 [Gregorian calendar], 9 July 1872, and two of 25 August 1872 [Gregorian calendar], 6 September 1872, one of which is also labelled as being "Jugum a fl. Tetung." Schneider (1916: 321) designated "Tetung Gerbirge" *Przewalski s.n.* 9 July 1872 as the type, but without citing a herbarium. This is likely to have been a duplicate of the LE specimen since Schneider cites Bretschneider (1898a: 971) as evidence of the place of collection rather than the sheet itself. If so, this duplicate is likely to have been at B and, as such, was destroyed in WWII. Mikhailova (2000: 60) lectotypified the LE specimen.

Maximowicz described the number of ovules of *Berberis diaphana* as six, while dissection of two of the fruit from one of the syntype gatherings of 25 August/6 September at LE produced three and five seeds. Schneider's assertion (1913: 353) that the species has six to eight ovules appears to be based on specimens from west Sichuan, one of which, *E. H. Wilson 930* (A, BM, P, US) from Wa Shan, he later (1917: 434) designated as the type of *B. aemulans*, which does have this number of ovules; but at the same time, he then appears to have increased the number of ovules of *B. diaphana* to six to 12 without citing any additional evidence. Ahrendt (1961: 123) described the number of ovules as six to 10, but this appears to be based on a cultivated plant of unknown origin.

The type of *Berberis circumserrata* var. *occidentalior*, *J. F. Rock 13272*, has only fruit. From the evidence of the isotype at A, the gathering was determined by Schneider to be *B. diaphana*.

The report by Bao (1997: 67) of *Berberis diaphana* being found in Yunnan is mistaken.

Selected specimens.

Syntypes. Berberis diaphana. Gansu: 25 Aug. 1872 [Gregorian calendar], 6 Sep. 1872, *Przewalski s.n.* (LE [2 sheets], PE 01896191); Gansu: "Terra Tangutorum," 1872, *Przewalski s.n.* (P P00716589).

Other specimens.

Gansu. Vic. of Choni (Jonê [Zhuoni] Xian), 3100–3300 m, 7–15 Sep. 1923, *R. C. Ching 961* (A 00279393, NAS NAS00313963, US 00945854). **[Kangle Xian]:** Lien hoa shan (Lianhuashan), Oct. 1925, *J. F. Rock 13474* (A 0027939, NAS NAS00313961). **Lintan Xian:** Heisong Ling, 2800 m, 13 Sep. 1982, *Q. R. Wang & M. S. Yan 11047* (PE 01030910). **Linxia Xian:** Minglu Xiang, Wanzhuzi Gou, 2550 m, 21 Aug. 2001, *X. G. Sun et al. 20010485* (PE 01801146). **Luqu Xian:** Zecha Gou, 3200 m, 7 July 1996, *G. L. Zhang & J. Z. Sun s.n.* (LZU 00008884–85). **Shandan Xian:** Zhongsan Si, 2400 m, 23 July 1991, *T. N. He 2603* (CAS 928715, HNWP 165681). **Su'nan Xian:** Mati Si, 2800 m, 28 July 1981, *T. N. He 2898* (HNWP 165607). **[Têwo (Diebu) Xian]:** Lanzi Gou, Tiechliang, 3150 m, 28 July 1998, *Bailongjiang Exped. 935* (PE 01556143). **Tianzhu Xian:** Haxi Linchang, 2600 m, 18 July 1991, *T. N. He 2275* (CAS 796674, HNWP 167069). **Wen Xian:** Baishuijiang Nature Reserve, 3148 m, 27 June 2006, *Baishuijiang Investigation Team 0361* (PE 01359664). **Xiahe Xian:** on the way to Hezuo, 26 Sep. 1991, *s. coll. 917* (KUN 0178579). **Yongdeng Xian:** Liancheng, Tulugou, 3000 m, 23 July 1991, *T. N. He 1795* (HNWP 165848). **Yuzhong Xian:** Xinglong Shan, Dahui Gou, 1 Sep. 1956, *Huang He Team 3438* (PE 01037961). **Zhugqu (Zhouqu) Xian:** Yangbu Gou, Shatan Linchang, 3300 m, 1 Aug. 1964, *B. Z. Guo 5261* (WUK 0231675).

Ningxia. Jingyuan Xian: Xixia, 2140 m. 7 July 1984, *J. W. Zhang s.n.* (LZU 00009184).

Qinghai. Koko Gorge, Sep. 1925, *J. F. Rock 13263* (A 00279386, NTUF). **Datong Xian:** Laoye Shan, 2935 m, 22 June 1962, *s. coll. 394* (HNWP 006555). **Guide Xian:** Galang Xiang, Qianhu Cun, 3400 m, 24 July 1989, *S. W. Liu 3362* (HNWP 152986, PE 01037946). **Henan Xian:** Ningmute Xiang, 3260 m, 12 July 1967, *s. coll. 1705* (HNWP 18675). **Huangyuan Xian:** Xianmi Si, Qihanke Gou, 1620 m, 31 July 1958, *B. Q. Zhong 10030* (HNWP 006554, PE 01037955, 01037958, WUK 0100385). **Huangzhong Xian:** near Sanhe Linchang, 16 Aug. 1972, *S. W. Liu 2070* (HNWP 28613). **Huzhu Xian:** Beishan Linchang, Zhong Gou, 3200 m, 15 July 1986, *Y. H. Wu et al. 2969* (HNWP 130786). **Jainca (Jianzha) Xian:** Dangshun Gongshe, Si'erdong Gou, 2900 m, 6 June 1970, *L. H. Zhou & L. N. Sun 588* (HNWP 21864). **Ledu Xian:** Yaocaotai Linchang, 3000 m, 21 July 1986, *Y. H. Wu et al. 3212* (HNWP 131087). **Maqên (Maqin) Xian:** Xihalong He Gu, 3550 m, 8 Aug. 1990, *Y. H. Wu et al. 5713* (HNWP 163800). **Menyuan Xian:** Huangchang Qu, Liu Gou, 2500 m, 14 July 1960, *Chinese Academy of Science, Qinghai Gansu Survey Team 2514* (PE 01033127). **Minhe Xian:** Xing'er, 2500 m, 11 June 1970, *B. Z. Guo 7008* (HNWP 21517). **Tongde Xian:** Hebei Xiang, 3800 m, 27 July 1990, *Y. H. Wu et al. 5042* (HNWP 163974). **Tongren Xian:** near Hebei Xiang, Saiqian Gou, 3200 m, 11 Aug. 1990, *Y. H. Wu et al.*

6482 (HNWP 162854). **Xunhua Xian:** Mengda, 2500 m, 12 June 1981, *B. Z. Guo & W. Y. Wang 25264* (HNWP 98776). **Zêkog (Zeku) Xian:** Maixiulin Qu, 3500 m, 27 Aug. 1970, *L. H. Zhou & L. N. Sun 1987* (HNWP 23178).

N Sichuan. Ngawa (Aba) Xian: Sep. 1977, *S. X. Tan 1329* (PE 01033805). **Pingwu Xian:** Lanping, toward Yakou, 1650 m, 21 Sep. 1986, *Z. Wu et al. 257* (KUN 0178510). **Songpan Xian:** Zhangla Qu, Zhangjin Xiang, 3170 m, 25 Sep. 1961, *X. N. Tang et al. 00931* (CDBI CDBI0027189).

137. Berberis dictyoneura C. K. Schneid., Pl. Wilson. (Sargent) 1: 374. 1913. TYPE: China. N Sichuan: Min valley, near Songpan, 2450–2750 m, Aug. 1910, *E. H. Wilson 4633* (lectotype, designated by Ahrendt [1961: 161], K missing, new lectotype, designated here, A 00038738!).

Shrubs, deciduous, to 1.8 m tall; mature stems dark reddish purple, angled; spines 3-fid, pale yellowish brown, 0.5–1 cm, solitary or absent toward apex of stems, abaxially sulcate, weak. Petiole almost absent or to 6 mm; leaf blade abaxially and adaxially semiconcolorous green, slightly shiny, elliptic or elliptic-obovate., 1–3 × 0.5–1.4 cm, papery, midvein, lateral veins, and dense reticulation raised and conspicuous abaxially, especially when dry, less conspicuous adaxially, base attenuate, margin spinose with very conspicuous 10 to 20 teeth on each side, apex subacute or obtuse. Inflorescence a sub-fascicled raceme, 3- to 6-flowered, 2–3 cm overall; bracts triangular, 1.5 mm; pedicel 8–14 mm. Flowers unknown. Immature berry obovoid-ovoid, 7–8 × 4.6 mm; style persistent, short; seeds 2.

Phenology. *Berberis dictyoneura* has been collected in fruit in August and possibly in October. Its flowering season is unknown.

Distribution and habitat. *Berberis dictyoneura* is known from only two collections from north Sichuan— the type collection made on a roadside in the Min valley near Songpan at 2450–2750 m, and a collection from a trailside in a narrow ravine in Ngawa (Aba) Xian at 3100 m—and possibly one from Wen Xian in south Gansu collected on a forest margin at 1900 m.

The protologue of *Berberis dictyoneura* cited only the type, *Wilson 4633*, but without citing any herbarium. Ahrendt's (1961: 161) citing of a specimen at K was an effective lectotypification, but this is missing. The specimen at A which is annotated as type in Wilson's hand is therefore designated here as a new lectotype.

Schneider's description in the protologue of *Berberis dictyoneura* that the leaves of the type are ovate is inaccurate; they are either elliptic or elliptic-obovate. The type of *B. dictyoneura* has no mature stems; the

description of their color comes from *D. E. Boufford, B. Bartholomew, J. F. Harber, Q. Li & J. P. Yue 44503* (details below).

Ahrendt (1961: 161) gave a description of *Berberis dictyoneura* which included detailing the structure of the flowers. This was based on material from a cultivated plant in his own collection which he stated was grown from seeds of the type. There are specimens of this at BM and OXF. Ahrendt gave no information as to the provenance of his plant, but there is a plant identified as *B. dictyoneura* growing at the Sir Harold Hillier Gardens, Hampshire, U.K., (herbarium specimen from the plant: HILL 6576) which is very similar to Ahrendt's material. This, however, has no provenance details either. Neither Ahrendt's material nor the Hillier plant completely resemble the type collection, nor is there any evidence in the Wilson papers at the Arnold Arboretum that seeds from the type were collected, and so these have not been drawn on for the above description.

Ying (2001: 199) synonymized as *Berberis dictyoneura*, *B. brachystachys* whose type is from northeast Xizang. However, this synonymy is mistaken and *B. brachystachys* is treated here as a distinct species. No evidence has been found for the report by Ying (2001) of *B. dictyoneura* being found also in Qinghai and Shanxi nor the report by D. Z. Ma (2007: 252) of it being found in Ningxia.

Selected specimens.
S Gansu. Wen Xian: Tielou, 1900 m, 7 Oct. 1958, *Y. Q. He 1020* (PE 01037072, WUK 0166968).

N Sichuan. Ngawa (Aba) Xian: He Dongqiongdang along Aliang Rd. betw. the city of Ngawa (Aba) & Zamtang (Rangtang), 32.66305°N, 101.584444°E, 3100 m, 8 Aug. 2018, *D. E. Boufford, B. Bartholomew, J. F. Harber, Q. Li & J. P. Yue 44503* (A, CAS, E, KUN, PE).

138. Berberis dictyophylla Franch., Pl. Delavay. 39. 1889. TYPE: China. NW Yunnan: [Heqing Xian], "Au col de Yen tze hay au dessus de Mo-so-yn," 3200 m, 14 Oct. 1887, *J. M. Delavay s.n.* (lectotype, designated by Imchanitzkaja [2005: 270–271], P P00716582!; isolectotype, LE fragm.!; possible isolectotype, specimen with fruit K K000077420).

Berberis stiebritziana C. K. Schneid., Oesterr. Bot. Z. 66: 320. 1916, syn. nov. TYPE: China. NW Yunnan: Lijiang, foot of Snow Mt., ca. 3000 m, 16 Sep. 1914, *C. K. Schneider 2908* (lectotype, designated by Ahrendt [1961: 129], K K000077423!; isolectotypes, A 00038797!, G G00226032!).

Shrubs, deciduous, to 2 m tall; mature stems purplish red or brownish red, subterete, sometimes partially pruinose, young shoots densely white pruinose; spines 3-fid, sometimes solitary toward apex of stems, pale yellow or yellow-brown, 1–3 cm, mostly abaxially sulcate. Petiole almost absent; leaf blade abaxially gray-green, densely white or glaucous pruinose, adaxially dull green or bluish green, obovate or obovate-elliptic, very rarely obovate-orbicular, 1–2.5(–5) × 0.4–1(–2) cm, papery, midvein slightly raised abaxially, lateral veins and reticulation indistinct or obscure on both surfaces, base cuneate, margin entire, rarely spinulose with 1 to 3 teeth on each side, apex rounded, often mucronate. Inflorescence 1-flowered, very rarely paired; pedicel (2–)6–15 mm, pruinose at first. Flowers pale yellow. Sepals in 2 whorls; outer sepals initially orangish, oblong-elliptic, ca. 6 × 2.5 mm; inner sepals oblong-elliptic, 8–9 × ca. 4 mm; petals elliptic, ca. 8 × 3–6 mm, base clawed, glands widely separate, apex entire. Stamens 4.5–5 mm; anther connective slightly extended, shortly apiculate. Ovules 3 to 5. Berry red, pruinose, ovoid, 9–14 × 6–8 mm; style persistent.

Phenology. *Berberis dictyophylla* has been collected in flower in May and June and in fruit between July and December.

Distribution and habitat. *Berberis dictyophylla* is known from northwest and northeast Yunnan, southwest Sichuan, and Zayü (Chayu) Xian in southeast Xizang (Tibet). It has been found in thickets, on mountain slopes, forest margins, and trail and streamsides at ca. 2700–3700 m.

Franchet's protologue of *Berberis dictyophylla* cited Delavay collections of 31 May 1886 and 14 October 1887 from Yen-tze-hay, and of 9 May 1887 from Fang-yang-tchang. Later, Schneider (1916: 319) described the first two as "Typ" without indicating any herbarium. Ahrendt (1961: 127) cited "Yunnan, Yen-tze-hay, above Lankong, and above Mo-so-yn, fl 31 May 1885; fr 14 Oct. 1887, *Delavay s.n.* (Type, K)." This was presumably a reference to a sheet 000077420 at K labelled "ad collum Yen-tze-hay, supra Mos-so-yn 31 May 1885" and which has various specimens with flowers and one with fruit. This latter may be of 14 October 1887, but this is unprovable. Imchanitzkaja (2005: 270–271) lectotypified the specimen of 14 October 1887 at P, though this appears to have been on the basis of an ex P fragment at LE rather than having actually seen the specimen. Nevertheless, in accordance with Art. 9.19 of the Shenzhen Code (Turland et al., 2018), this lectotypification stands.

The protologue of *Berberis stiebritziana* cited the type as *Schneider 2908*, but without citing a herbarium. Ahrendt's citation (1961: 129) of the specimen at K as type constituted an effective lectotypification. In the protologue, Schneider differentiated *B. stiebritziana* from *B. dictyophylla* solely on the basis that its leaves had no or little reticulation and that its fruit stalks were,

on average, longer. This would seem a trivial difference. *C. K. Schneider 1922* (A 00279888, GH 00279887), northwest Yunnan, "ad Latora orient. mont. niveor. prope Lichiang," ca. 3300 m, 19 July 1914, also cited in the protologue as an example of *B. stiebritziana*, is not *B. dictyophylla* and is probably *B. minutiflora*. Ahrendt's (1961: 129) statement that the pedicels of the type of *B. stiebritziana* are 12–25 mm long was mistaken. As Schneider noted in the protologue, they are 8–15 mm. Ying treated *B. stiebritziana* as a synonym of *B. approximata* (listed here under Excluded Taxa).

For details of *Berberis dictyophylla* var. *epruinosa*, see under *B. ambrozyana*; for details of *B. dictyophylla* var. *campylogyna*, see under Excluded Taxa.

Berberis dictyophylla has one of the most extensive distributions of any deciduous *Berberis* of Yunnan and neighboring areas of southwest Sichuan. It has, however, also become a frequent taxonomic "dustbin" for 1-flowered specimens from west and north Sichuan, northeast Xizang, and south Qinghai. In fact, the farthest north I have found a collection is *K. Y. Lang et al. 2327* from south Batang Xian (details below). For 1-flowered species north of here besides *B. ambrozyana*, see under *B. baiyuensis*, *B. barkamensis*, *B. bouffordii*, *B. epedicellata*, *B. kangdingensis*, *B. markamensis*, and *B. qinghaiensis*. There are likely to be undescribed species in addition to these.

While collecting in Daocheng, Derong, Xiangcheng, and Muli Xian in August and September 2013, I found *Berberis dictyophylla* to be the most common *Berberis* species. Many of these plants had significantly shorter pedicels (sometimes only 2–3 mm long) than plants found farther south in northwestern Yunnan. Very short (or almost non-existent) pedicels are also a feature of the very similar *B. epedicellata* found farther north in Kangding Xian, but the two taxa have different flower structures.

The densely pruinose leaves and young shoots of *Berberis dictyophylla* make it an extremely attractive plant.

D. E. Boufford, J. F. Harber & Q. Wang 43166 (A, E, KUN, PE) found growing among no. *43165* (details below) and no. *43164* (for which, see under *Berberis derongensis*, where it is tentatively identified as such) appears to be a natural hybrid between these two species.

Selected specimens.
Syntypes. Berberis dictyophylla. [Heqing Xian], "Au col de Yen tze hay (Lan kong)," 3200 m, 31 May 1886, *J. M. Delavay s.n.* (B B10 0250731, P [2 sheets] P00716580–81, P06868526, K K000077420, specimens with flowers, LE fragm.); [Heqing Xian], "les pasturages du Fang-yang-tchang au dessus de Mo-so-yn," 3000 m, 9 May 1887, *J. M. Delavay s.n.* (LE fragm., P [4 sheets] P00716583–85, P02682377, UC UC1038328,).

Other specimens.

SW & W Sichuan. Batang Xian: Zhongzan Qu, Lifu, 3400 m, 24 July 1983, *K. Y. Lang et al. 2327* (KUN 0179040, PE 01037641–42). **Daocheng Xian:** above Axi village off hwy. S216, S of Daocheng at Haiyi (locally called Tedu), 28.69°N, 100.233056°E, 3665 m, 28 Aug. 2013, *D. E. Boufford, J. F. Harber & X. H. Li 43286* (A 00914598, E E00663242, K, KUN, PE) and *43287* (A 00619091, E, KUN). **Derong Xian:** Zigen, 3440 m, 2 July 1981, *Qinghai-Xizang Team 1720* (CBDI 0027229–30, HITBC 003643, KUN 0175996–97, PE 01037639–40); along rd. from Derong to Batang WNW of the village of Lanzhu, 29.092778°N, 99.375556°E, 3325 m, 22 Aug. 2013, *D. E. Boufford, J. F. Harber & Q. Wang 43165* (A 00914593, E E00663245, K, KUN). **Muli Xian:** Wachin, near Lamasary, 3100 m, 2 Oct. 1937, *T. T. Yu 14412* (A 00279445, BM, E E00612470, KUN 0176090); Mianbu Cun, S of Muli city along hwy. S216, Mian Yakou, 27.685°N, 101.223333°E, 3240 m, 6 Sep. 2013, *D. E. Boufford, J. F. Harber & X. H. Li 43472* (A 00914600, CAS, E E00663250, K, KUN, PE, TI); N of the city of Muli, on hwy. Z020, near Kangwu Dasi, 28.106944°N, 101.200278°E, 3710 m, 7 Sep. 2013, *D. E. Boufford, J. F. Harber & X. H. Li 43499* (A 00619090, CAS, E E00663251, K, KUN, PE, TI). **Xiangcheng Xian:** W of Reda Xiang along Z006 to Muyucun, 29.125°N, 99.646389°E, 3445 m, 23 Aug. 2013, *D. E. Boufford, J. F. Harber & X. H. Li 43182* (A 00914597, CAS, E E00826404, K, KUN, PE). **Yanyuan Xian:** Guabie Qu, Lianhe Gongshe, Yida Dui, Huolu Shan, 3400 m, 26 July 1983, *Qinghai-Xizang Team 12492* (KUN 0179038, PE 01037637–38, 01037649).

SE Xizang (Tibet). Zayü (Chayu) Xian: Tsa-wa-rung, Nar-jou, 3400 m, Sep. 1934, *C. W. Wang 66462* (A 00279447, IBSC 0092348, KUN 0176058, LBG 00064311, NAS NAS00313985, PE 01037629–30, WUK 0035403); same details, *C. W. Wang 66566* (A 00279482, IBSC 0092349, KUN 0176059, LBG 00064314, NAS NAS00313983, PE 01037631–32, WUK 0035584); Chawalong, Songta Xueshan, 3200 m, 29 June 1982, *Qinghai-Xizang Team 7768* (KUN 0175993–94, PE 01037635–36).

NE Yunnan. Qiaojia Xian: Wuming Shan on way to Yao Shan along Xiaoshui section of hwy. X250, 27.091389°N, 102.99°E, 2850 m, 9 Sep. 2013, *D. E. Boufford, J. F. Harber & X. H. Li 43506* (A 00914594, CAS, E E00833565, K, KUN, PE, TI).

NW Yunnan. Binchuan Xian: Jizhu Shan, betw. Zhusheng Si & Jinding Si, 21 Dec. 1946, *S. E. Liu 22095* (IBSC 0092323, 0093705, KUN 0176046, PE 01037684). **Dali Shi:** "West of Talifu, Mekong watershed, en route to Youngchang and Tengyueh," Sep.–Oct. 1922, *J. F. Rock 6804* (A 00279470, US 0945920); Cangshan, 3400 m, 2 Aug. 1963, *Zhongdian Team 3742* (KUN 0175958, 0176072). **Dayao Xian:** Santai Qu, Bakou Cun, Duodihe, Bizhayang Wozi, 3300 m, 12 July 1965, *Woody Oil Investigation Team 468* (KUN 0176039, 0176051). **Dêqên (Deqin) Xian:** A-tun-tze, 2700 m, Sep. 1934, *C. W. Wang 70033* (A 00279465, IBSC 0092350, KUN 01037817, LBG 00064359, NAS NAS00313984, PE 01037817–18); Beima Shan, W flank, 3600 m, 23 July 1981, *Qinghai-Xizang Team 3628* (CDBI CDBI0027234, 0027241, HITBC 003582, KUN 0175991–92, PE 01037627). **Heqing Xian:** "Au col de Yen-tze-hay (Lankong)," 3200 m, 1 June 1886, *J. M. Delavay s.n.* (KUN 1206512, P P02682375); "In dumetis montis Fang-yang-tchang supra Lankong," June 1887, *J. M. Delavay s.n.* (VNM); Songgui Qu, Ma'ershan, Chamu Jing, 3200 m, 21 July 1963, *Zhongdian Team/Northwest Yunnan Jinsha River Team 4771* (KUN 0175978, 0176073, PE 01037626). **Lijiang Xian:** Lijiang Snow Range, E flank,

2745–3050 m, May–Oct. 1922, *J. F. Rock 3388* (A 00279437, E E00612470, US 00945918); Yulong Shan, 3200 m, 26 July 1980, *S. T. Li 262* (KUN 0176062–63). **Zhongdian (Xianggelila) Xian:** "SE Chungtien: betw. Chiao-tou (On Yangtze bank) to Hsiao Chungtien [Xiao Zhongdian]," 13 May 1939, *K. M. Feng 889* (A 00279448, KUN 0176036–37, 0176048); Nixi Gongshe, 3200 m, 1 July 1981, *Qinghai-Xizang Team 2029* (CDBI CDBI0027235–36, HITBC 003601, KUN 0175986, 0179248, PE 01037597–98).

139. Berberis difficilis Harber, sp. nov. TYPE: China. W Sichuan: Daocheng Xian, NNW of Sandui Xiang and N of Bengpugunba on hwy. S217, 29.2275°N, 100.089722°E, 3990 m, 26 Aug. 2013, *D. E. Boufford, J. F. Harber & Q. Wang 43269* (holotype, PE!; isotypes, A!, BM!, CAS!, E!, K!, KUN!, MO!, TI!).

Diagnosis. *Berberis difficilis* is similar to various mainly fascicled and sub-fascicled species with two ovules from northwest Yunnan and west and southwest Sichuan, but with a different flower structure; for further details, see the multi-access key (Key 9). *Berberis difficilis* also has distinctive bluish green leaves.

Shrubs, deciduous, to 3 m tall; branches spreading, mature stems pale reddish brown, sulcate; spines 3-fid, solitary or absent toward apex of stems, pale yellow, 0.8–1.5(–2) cm, abaxially sulcate. Petiole almost absent; leaf blade abaxially pale green, adaxially dull bluish green, narrowly obovate or oblanceolate, (1.4–)2–2.5(–3.5) × 0.5–0.8(–1) cm, papery, abaxially lightly papillose, midvein raised abaxially, flat adaxially, venation inconspicuous on both surfaces, base narrowly attenuate, margin entire, very rarely spinose with 1 or 2 teeth on each side, apex acute or subacute, sometimes obtuse, mucronate. Inflorescence a fascicle or subfascicle, 1- to 6-flowered; pedicel 5–15(–22) mm; bracteoles triangular, pinkish, 2–2.5 × 1 mm. Flowers lemon yellow. Sepals in 2 whorls; outer sepals elliptic-ovate, 5–6 × 2.5 mm; inner sepals oblong-elliptic, 6 × 3.5 mm; petals obovate, 4.5 × 3 mm, base slightly clawed, glands separate, 0.75 mm, apex entire. Stamens ca. 2.5 mm; anther connective not or slightly extended, truncate. Pistil 3.5 mm; ovules 2. Immature berry turning red, ellipsoid or oblong, 7–8 × 3–4 mm; style persistent.

Phenology. *Berberis difficilis* has been collected in flower and with immature fruit in August. Its phenology is otherwise unknown.

Distribution and habitat. *Berberis difficilis* is known from Daocheng and Xiangcheng Xian in west Sichuan. It has been found on riverside and roadside meadows and on a barren former road construction site at 3950–4125 m.

Though the description above gives the color of the mature stems as pale reddish brown, it should be noted that this was rarely in evidence on any of the plants I saw or collected from.

On our 2013 National Geographic Society (NGS) expedition, three collections, nos. *43170*, *43203*, and *43205* (details below), were made from plants whose leaf shape and size and number and length of pedicels were almost identical to those of *Berberis difficilis*, but whose leaves were green rather than blue-green and whose habit was upright rather than spreading. The flowers were bright yellow and a dissection of those of no. *43170* produced the following: "Sepals in 2 whorls; outer sepals ovate, 3.5–4 × 2–2.5 mm; inner sepals broadly obovate, 5 × 3.75 mm; petals obovate, 3.5–4 × 3–4 mm, base slightly clawed, glands separate, ca. 1 mm, apex entire. Stamens ca. 2.5 mm; anther connective not or slightly extended, truncate." This is so similar to that of *B. difficilis* that it is not published here as a separate species, though further investigation might conclude it should be.

IUCN Red List category. *Berberis difficilis* is assessed as DD or Data Deficient, according to IUCN (2001) criteria.

Selected specimens.
W Sichuan. Daocheng Xian: near Laolinkou, 3700–4000 m, 3 Aug. 1983, *K. Y. Lang et al. 2729* (KUN 0179043, PE 01033837, 01033844); betw. Sandui & Litang, SW of Haizi Shan, 29.271111°N, 100.082222°E, 3980 m, 1 July 1998, *D. E. Boufford et al. 28085* (A 00280068); hwy. 217 from Litang to Sandui, E side of Haizi Shan, NNE of Sandui, 4055 m, 29.280278°N, 100.082222°E, 27 Aug. 2006, *D. E. Boufford et al. 37359* (A, E, KUN). **Xiangcheng Xian:** W of the city of Xiangcheng on hwy. XV09 to Reda Xiang, near pass on Ma'an Shan, 28.996389°N, 99.744722°E, 4125 m, 24 Aug. 2013, *D. E. Boufford, J. F. Harber & Q. Wang 43201* (A, BM, CAS, E, K, KUN, PE, TI); along hwy. S217 from Xiangcheng to Sandui, S of Wuming Shan near Chaoyanggou Daoban, 3950 m, 26 Aug. 2013, 29.123611°N, 99.99°E, *D. E. Boufford, J. F. Harber & Q. Wang 43278* (A, CAS, E, K, KUN, PE) and *43279* (A, E, KUN).

Specimens very similar to Berberis difficilis.
W Sichuan. Batang Xian, border with Derong Xian: N of Ciwu Xiang, 29.138889°N, 99.335°E, 4100 m, 22 Aug. 2013, *D. E. Boufford, J. F. Harber & Q. Wang 43170* (A, BM, CAS, E, K, KUN, T, TI). **Xiangcheng Xian:** W of Reda Xiang, along rd. Z006 to Muyucun, 29.145°N, 99.671667°E, 3475 m, 24 Aug. 2013, *D. E. Boufford, J. F. Harber & Q. Wang 43203* (A, E, KUN) and *43205* (A, CAS, E, K, KUN, PE).

140. Berberis dispersa (Ahrendt) Harber, comb. et stat. nov. Basionym: *Berberis humidoumbrosa* Ahrendt var. *dispersa* Ahrendt, J. Linn. Soc., Bot. 57: 160. 1961, as "*humido-umbrosa* var. *dispersa*." TYPE: China. SE Xizang: [Cona (Cuona) Xian], Trimo, Nyam Jang Chu, 27.916667°N, 91.9°E, 3200 m, 23 May 1947, *F. Ludlow, G. Sherriff & H. H. Elliot 12532* (holotype, BM BM000564308!; isotype, E E00162254!).

Shrubs, deciduous, to 3 m tall; mature stems reddish purple, sulcate, shiny, very sparsely verruculose; spines 3-fid, solitary or absent toward apex of stems, semi-concolorous, 0.4–1 cm, terete, weak. Petiole 1–5 mm or almost absent; leaf blade abaxially green, adaxially dark green, obovate, obovate-elliptic, or obovate-oblanceolate, (1.5–)2–3 × 0.5–0.8(–1) cm, papery, midvein slightly raised abaxially, slightly impressed adaxially, lateral veins and reticulation inconspicuous abaxially, inconspicuous or indistinct adaxially, base narrowly attenuate, margin entire, rarely spinulose with 1 or 2 teeth on each side, apex acute, mucronate. Inflorescence a loose raceme, rarely partially paniculate, (3-to)7- to 12-flowered, 2.5–4 cm overall; peduncle 1–2.5 cm; bracteoles 1.5 mm, apex acuminate; pedicel 7–14 mm. Flowers lemon yellow, slender, 3–5 mm diam. Sepals in 2 whorls; outer sepals oblong-ovate, 3–5.5 × 1.5–2 mm, apex subacute; inner sepals obovate, ca. 5 × 3.5 mm, apex rounded; petals obovate, ca. 3.5 × 2.5 mm, apex entire. Stamens 2.75 mm; anther connective extended, obtuse. Ovules 3. Berry red, oblong-ovoid or ellipsoid, 11–12 × 4–5 mm; style persistent.

Phenology. *Berberis dispersa* has been collected in flower in May and in fruit in September.

Distribution and habitat. *Berberis dispersa* is known from [Cona (Cuona) Xian] in southeast Xizang. It has been collected in shrub forest and forest glades at ca. 3200–3800 m.

On current evidence, there would seem to be no reason for associating *F. Ludlow, G. Sherriff & H. H. Elliot 12532* with *Berberis humidoumbrosa*, which has yellow mature stems and two ovules.

The protologue of *Berberis humidoumbrosa* var. *dispersa* cites only the type. The description of fruit used here is from the three *Qinghai-Xizang Supplementary Team* gatherings cited below. All these gatherings have reddish purple mature stems. An annotation to no. *75-1898* notes the fruit have three seeds.

Selected specimens.
SE Xizang (Tibet). Cona (Cuona) Xian: Mama, 2900–3100 m, 2 Sep. 1975, *Qinghai-Xizang Supplementary Team 75-1809* (HNWP 52423, KUN 0178134–35, PE 01037085, 01037090); S side of Bu Shan, 3800 m, 10 Sep. 1975, *Qinghai-Xizang Supplementary Team 75-1898* (HNWP P0226924, KUN 0178679–80, PE 01037086, 01840024); Bu Shan, S of river, 3500 m, 10 Sep. 1975, *Qinghai-Xizang Supplementary Team 75-1740* (PE 01037094).

141. Berberis dokerlaica Harber, Curtis's Bot. Mag. 33(1): 26. 2016. TYPE: Cultivated. Dawyck Botanic Garden, Stobo, Scotland, 30 May 2013, *P. Brownless 431* from *P. A. Cox 6110*, China. NW Yunnan: Dêqên (Deqin) Xian, W of Mekong River, above Camp 6 on path to Doker La, 3600 m, 24 Sep. 1992 (holotype, E E00668752!).

Shrubs, deciduous, to 3 m tall; mature stems dark purple turning reddish brown, sulcate; young shoots pinkish purple, pruinose; spines 3-fid, solitary or absent toward apex of stems, pale yellow, 0.6–2 cm, abaxially sulcate. Petiole almost absent; leaf blade abaxially and adaxially glaucous, abaxially densely papillose, obovate or oblong-obovate, 2–4.3 × 1–1.8 cm, papery, midvein raised abaxially, flat adaxially, lateral veins and reticulation inconspicuous abaxially (though conspicuous when dry), conspicuous adaxially, base attenuate, margin entire, apex obtuse or rounded, sometimes subacute, sometimes mucronate. Inflorescence a subraceme, sometimes fascicled at base or a fascicle toward apex of stems, (2- to)4- to 10-flowered, 3–4 cm overall including peduncle 1–1.8 cm; pedicel 6–8 mm but to 15 mm when from base. Sepals in 3 whorls; outer sepals oblong-ovate, 5–6 × 2.5–3.5 mm; median sepals oblong-elliptic, 7 × 4 mm; inner sepals broadly elliptic, 7–8 × 4–5 mm; petals obovate, 4.5 × 3 mm, base slightly clawed, glands widely separate, ca. 1 mm, apex entire. Stamens ca. 3.5 mm; anther connective distinctly extended, truncate or rounded. Pistil ca. 3 mm; ovules 3 to 5. Berry bright red, ovoid or oblong, 12–13 × 5–7 mm; style persistent; seeds reddish brown.

Phenology. *Berberis dokerlaica* has been collected in fruit in August and September. Its flowering period in the wild is unknown.

Distribution and habitat. *Berberis dokerlaica* is known from only Doker La in Dêqên (Deqin) Xian in northwest Yunnan. The seeds of the type were collected from the side of a path in mixed forest at 3600 m; *G. M. Feng 5869* and *5877* (see below) were collected from mixed forest at ca. 3000–3400 m.

As noted in the protologue of *B. dokerlaica*, the living plants at Dawyck were first identified as *Berberis franchetiana* var. *glabripes* whose type, *T. T. Yu 7864*, is also from Doker La. However, the collection details of this record have it with two yellow seeds per fruit and dissection of fruits of the specimen at KUN confirmed this (for more details of this taxon, see under Taxa Incompletely Known). But fruit from *G. M. Feng 5877*, listed below, produced two to four seeds with the same reddish color as seeds of the type. Specimens of *G. M. Feng 5869* are identical to those of no. *5877*.

Selected specimens.
NW Yunnan. Dêqên (Deqin) Xian: Lancanjiang-Nujiang divide, Dokerla, 3000–3400 m, 3 Aug. 1940, *G. M. Feng 5869* (KUN 0176201–3); same details, but 3100–3400 m, *G. M. Feng 5877* (KUN 0176184, 0176204–5).

Cultivated material:
Dawyck Botanic Garden, Stobo, Scotland, 27 Sep. 1999, *D. G. Knott. s.n.* from *P. A. Cox 6110* (E E00269076); same details, but 20 May 2014, *P. Brownless 493* (E E00705629).

142. Berberis dubia C. K. Schneid., Bull. Herb. Boissier, sér. 2, 5: 663. 1905. TYPE: China. E Gansu: 1885, *G. N. Potanin s.n.* (lectotype, designated here, W 1890/2029!); possible isolectotypes, one of the following: "Valle fl. Karyn," 4 May 1885, *G. N. Potanin s.n.* (LE!), "Valle fl. Labran," 21 May 1885, *G. N. Potanin s.n.* (LE!); "Inter Mörpin et U pin," 28 June 1885, *G. N. Potanin s.n.* (?B†).

Shrubs, deciduous, to 3 m tall; mature stems purplish red, shiny, sub-sulcate; spines solitary or 3(to 5)-fid, concolorous, 0.7–2 cm, mostly terete. Petiole 1–3 mm; leaf blade abaxially pale yellow-green, adaxially green, narrowly obovate, 1.5–3 × 0.5–1.8 cm, papery, midvein, lateral veins, and dense reticulation raised and conspicuous abaxially, conspicuous adaxially, base attenuate, margin spinulose with 6 to 14 teeth on each side, but leaves at apex of stems often entire, apex subacuminate. Inflorescence a raceme, 5- to 10-flowered, 1–3 cm overall; peduncle 0.5–1 cm; pedicel weak, 3–6 mm; bracteoles lanceolate, ca. 1.5 mm, apex acute. Sepals in 2 whorls; outer sepals ovate, ca. 2.5 × 1.5 mm; inner sepals broadly obovate, ca. 4.5 × 3.5 mm; petals elliptic, ca. 3.5 × 2.5 mm, base cuneate, glands separate, apex shortly emarginate. Stamens ca. 2.5 mm; anther connective extended, shortly apiculate. Ovules 2. Berry red, obovoid-ellipsoid, ca. 8 × 4 mm; style not persistent.

Phenology. *Berberis dubia* has been collected in flower in May and June and in fruit from July to September.

Distribution and habitat. *Berberis dubia* is known from Gansu, Nei Mongol, Ningxia, Qinghai, and north Sichuan. It has been found on stony and mudstone mountain and valley sides, open grasslands, and forest understories at ca. 1350–3700 m.

The protologue of *Berberis dubia* cited simply "O – Kansu [Potanin 1885]" and "O – Mongolei: Mont Alashau [Przewalski 1873]" and stated "Typus in Herb. Berlin etc." Any specimens in Berlin were destroyed in WWII. However, there are specimens of both at W whose labels have exactly these collection details, the former annotated by Schneider on 26 October 1904 and the latter on 13 August 1904. The former has been chosen here as the lectotype because it has floral material, which is a more useful tool for identification than fruit. Subsequently, Schneider (1918: 228) cited *Potanin* "Inter Mörpin et U pin," 28 June 1885, as type but cited no herbarium. No specimen with this description has been located; it may also have been at Berlin and hence also destroyed in WWII. In any case, this may have been a duplicate of the specimen at W, though this is unprovable. Mikhailova (2000: 60) lectotypified "Valle fl. Labran," 21 May 1885, *Potanin s.n.* at LE. However, though from an annotation it is clear that Schneider saw this specimen, from contextual evidence this was not until October 1906, i.e., after he had published *B. dubia*. This specimen (which was annotated by Schneider not as type but only as "formiae dubiae") may be a duplicate of the Vienna specimen, but as with the specimen cited by Schneider in 1918, this is unprovable. Moreover, there is yet another candidate to consider as an isolectotype, "Valle fl. Karyn" 4 May 1885, *Potanin s.n.* (LE) which, though labelled as being from Gansu, from the evidence of Bretschneider (1898a: 1013), appears to have been collected in Guide Xian in Qinghai. It is, of course, possible than none of the three specimens are (or were) duplicates and that the Vienna specimen is from a fourth gathering; therefore, Mikhailova's lectotypification is not accepted here.

From the evidence of Bretschneider (1898a), it appears that the Potanin specimen of 21 May 1885 was collected in Xiahe Xian, Gansu.

Many of the specimens at NMAC from Nei Mongol identified as *Berberis vernae* are, in fact, *B. dubia*.

Selected specimens.

Syntypes. W Nei Mongol, Mt. Alashan, 1873, *N. M. Przewalski s.n.* (W); possible syntypes, Mt. Alashan, 2 July 1873 (Gregorian calendar), 14 July 1873 (modern calendar), *N. M. Przewalski s.n.* (LE); W Nei Mongol, Mt. Alashan, July 1873, *N. M. Przewalski s.n.* (LE [2 sheets]).

Other specimens.

Gansu. "Upper Tebbu Country," S slopes of Minshan, 2925–3200 m, June 1925, *J. F. Rock 12513* (A 00279332, E E00395996, K). **Huining Xian:** Shoacha Gou, 10 Aug. 1956, *Huang He Team 5789* (KUN 0178438, PE 01037077). **Jiuquan Shi:** Jiayuguan, 2800 m, 28 July 1960, *Chinese Academy of Science, Qingai-Gansu Team 3053* (PE 01033770, WUK 00167579). **Linxia Xian:** betw. Hezheng & Minglu Xiang, Wanzhuzi Gou, 2500 m, 21 Aug. 2001, *X. G. Sun et al. 20010486* (PE 01801148). **[Subei Xian]:** Danghe Nanshan, Taben-buluk, ca. 700 m above BB Camp 68, 5 July 1931, *B. Bohlin 2158* (S 05–4896). **Su'nan Xian:** Haiya Gou, 2500 m, 13 July 1960, *Chinese Academy of Science, Qinghai-Gansu Investigation Team 3228* (PE 01033769, WUK 00167902). **[Suzhou Qu]:** Xigou, 2800 m, 28 July 1960, *Chinese Academy of Science, Qinghai-Gansu Investigation Team 3053* (PE 01033770, WUK 0167579). **Yuzhong Xian:** Xinlong Shan, 2500 m, 15 July 1993, *Z. Y. Zhong 508* (MO 04751423, PE

01839970). **Zhangye Shi:** Nanshan gypsum mine, 30 Aug. 1958, *B. Q. Zhong 9065* (HNWP 006551, KUN 0179014, PE 01037078–79).

Nei Mongol. "Montes Suma-hada," 300 versts W of Kalgan (Zhangjiakou), 1871, *N. M. Przewalski s.n.* (K K00395216); Wulashan, 1400 m, 20 Aug. 1931, *W. Y. Hsia 3076* (NKU NK 003823, PE 01033771–72); Ahuikou, ca. 100 km to SW of town of Paotou (Baotou), 23 Aug. 1937, *E. Licent 13814* (TIE 00048686, as *E. Licent s.n.* W 1940/12092). **Aixa (Alanshan)You Qi:** Longshou Shan, 21 July 1982, *X. T. Lei 45* (NMAC 0034388). **Aixa (Alanshan) Zuo Qi:** Helan Shan, Halawu Gou, 2050 m, 1 June 1959, *Y. Q. He 2824* (PE 01033776); Helan Shan, Bei Si, 5 Aug. 1962, *S. T. Zeng & Y. Q. Ma 15* (HIMC 0010758, PE 01033773); Helan Shan, 27 July 1963, *L. Q. Ma et al. 67* (N 093057129); Helan Shan, 2861 m, 9 Sep. 2006, *Z. B. Wu 195* (NMAC 00029219); Helan Shan, Nansi Gou, 31 July 2014, *Y. S. Cheng et al. 141138* (PE 02031092). **Baotou Shi:** [Tümed Baraɣun qosiɣu], Daqing Shan, 5 July 1993, *s. coll. s.n.* (TNM S60322). **Jungar (Zhunge'er) Qí:** Egui Miao, 25 May 2002, *Y. Z. Zhao 012* (HIMC 0010803). **Tumote You Qi:** Jiufeng Shan, Zao Gou, 4 Aug. 2014, *Y. S. Chen et al. 141309* (PE 02036601). **Urad (Wulate) Qian Qi:** Wula Shan, Dahuabei, 14 July 1960, *Bameng Wild Plant Survey Team 165* (HIMC 0010757).

Ningxia. Haiyuan Xian: Hongyang Qu, Yuliang Shan, 7 July 1956, *Huang He Team 5593* (PE 01037075, WUK 0085873). **Pingluo Xian:** Changgong, Dashui Gou, 1350 m, 13 Aug. 1980, *Z. Y. Yu 247* (WUK 00449420). **[Tongxin Xian]:** Daluo Shan, 2300–2700 m, 8 Sep. 1956, *Huang He Team 8729* (KUN 0178620, 0178341, PE 01037080, WUK 0087515). **Yinchuan Shi:** Ho Lan Shan (Helan Shan), 1375–2400 m, 10–25 May 1923, *R. C. Ching 167* (A 00279773, GH 00279774, NAU 00024190, PE 01032570, US 00945948); Helan Shan, 2 June 1973, *Ningxia Institute for Drug Control 4* (PE 01030745–6); Xiaokuzi, 1600 m, 17 Aug. 1980, *Z. Y. Yu 395* (WUK 00449416).

Qinghai. [Da Quaidam (Dachaidam)]: Xiariha, 3300 m, 28 Aug. 1983, *Botanic Geography Group 325* (HNWP 105658). **Delhi (Delingha) Shi:** Baishu Shan, 3800 m, 23 July 1975, *B. Z. Guo & W. Y. Wang 11647* (HNWP 47270, PE 01033785). **Dulan Xian:** Yingdebu Yangchang, 3550 m, 2 Sep. 1981, *Q. Du 407* (HNWP 99936). **Guide Xian:** Galang Xiang, Qianhu Cun, 2800 m, 24 July 1989, *S. W. Liu 3277* (HNWP 152901, PE 01033813). **Guinan Xian:** Nianma Chang, 3200 m, 23 July 1967, *B. W. Li & H. Z. Zhang 47* (HNWP 33497). **Huzhu Xian:** Jiading Gongshe, 19 May 1975, *G. R. Wu 75-21* (HIMC 0010756). **Jainca (Jianzha) Xian:** Dangshun Gongeshe, Si'erdong Gou, 2700 m, 9 June 1970, *L. H. Zhou & L. N. Sun 631* (HNWP 21907). **Menyuan Xian:** Xianmi Qi, Hanke, 2450 m, 3 June 1960, *Chinese Academy of Science, Qinghai and Gansu Investigation Team 2430* (PE 01033787, WUK 00167963). **Qilian Xian:** Babaoshan, 2800 m, 27 June 1960, *Chinese Academy of Science, Qinghai and Gansu Investigation Team 2752* (PE 01033791, WUK 0167791). **Tongde Xian:** Lajiaxiang, 34.683333°N, 100.683333°E, 3100–3200 m, 21 July 1993, *T. N. Ho et al. 102* (BM BM001010957, CAS 914844, HNWP 171378, PE 01840127). **Tongren Xian:** Lancai Gongshe, Yangzhi Gou, 3000 m, 25 July 1970, *S. W. Liu & D. S. Luo 1468* (CAS 898650, HNWP 26318). **Ulan (Wulan) Xian:** Chaidamu, Xili Gou, Dulan Si, 3700 m, 25 July 1959, *Qinghai-Gansu Team 619* (PE 01033798, WUK 00169366). **Zeku Xian:** Duofutun Gongshe, Maixiu, Laozang Gou, 3090 m, 26 Aug. 1970, *L. H. Zhou & L. N. Sun 1862* (HNWP 23025).

N Sichuan. Zoigê [Ruo'ergai] Xian: S313, Xiashan Cun, 34.1675°N, 102.88611°E, 2810 m, 23 Aug. 2007, *D. E. Boufford et al. 40357* (A, E).

143. Berberis dulongjiangensis Harber, sp. nov.
TYPE: Cultivated. Foster Clough, Mytholmroyd,
West Yorkshire, U.K., 30 Apr. 2015, *J. F. Harber
2015-21* from seeds of *Z. L. Dao, M. Lear, Y. H. Li
& M. Wickenden (KWL) 63*, China. NW Yunnan:
Gongshan Xian, Baimawang (tributary of Dulong
Jiang), 28.1388°N, 98.271917°E, 2590 m, 14
Sep. 2008 (holotype, PE!; isotypes, A!, CAS!, E!,
K!, KUN!, TI!).

Diagnosis. *Berberis dulongjiangensis* has leaves that are
somewhat similar to those of *B. bracteata* and *B. gongshanensis* (both also from Gongshan Xian) but differs from the former
in having dark reddish purple mature stems and from the latter by its flower structure with three ovules rather than six.

Shrubs, deciduous, height unrecorded in wild; mature stems dark reddish purple, very sulcate; spines
1- to 3-fid, concolorous, 0.8–1.2(–1.5) cm, weak. Petiole 2–7(–10) mm, sometimes almost absent; leaf blade
abaxially pale green, papillose, adaxially slightly yellowish dull green, obovate or obovate-elliptic, 2.5–3.5
× 1.8–2.2 cm, papery, midvein raised abaxially, impressed adaxially, lateral veins and reticulation indistinct on both surfaces, base attenuate, margin spinulose
with 12 to 15 teeth on each side, reddish at base of
teeth at first, apex obtuse, mucronate. Inflorescence a
fascicle, sub-fascicle, or sub-umbel, 3- to 6-flowered;
pedicel 5–15 mm. Sepals in 3 whorls; outer sepals
ovate, 5 × 3–3.5 mm; median sepals elliptic-ovate,
5.5–6 × 4–4.5 mm; inner sepals broadly obovate or
obovate-orbicular, 5–5.5 × 4–4.5 mm; petals broadly
obovate, 3–3.5 × 3–3.5 mm, base distinctly clawed,
glands close together, ca. 1 mm, apex slightly notched.
Stamens ca. 3 mm; anther connective not extended,
truncate. Pistil ca. 3 mm; ovules 3, stipitate. Berry red,
obovoid, oblong, or ellipsoid, 11 × 6 mm; style absent
or persistent and short.

Phenology. *Berberis dulongjiangensis* has been collected in fruit in July and September. Its flowering season in the wild is unknown.

Distribution and habitat. *Berberis dulongjiangensis*
is known from only two collections in Gongshan Xian
in northwest Yunnan. *KWL 63* was made at 2590 m and
ST1150 at 2400–3300 m. In neither case was the habitat recorded.

IUCN Red List category. *Berberis dulongjiangensis*
is assessed as DD or Data Deficient, according to IUCN
(2001) criteria.

KWL 63 and *ST1150* were collected some 16 km
from each other.

My plants of *Berberis dulongjiangensis* were supplied by the late Michael Wickenden of Cally Gardens,

Castle Douglas, Scotland. For an account of the expedition that collected the seeds of *KWL 63*, see Wickenden and Lear (2010). At the time of writing, the largest
plant was 1.75 m high.

Selected specimens.
NW Yunnan. Gongshan Xian: Baimawang (tributary of
Dulong Jiang), 28.1388°N, 98.271917°E, 2590 m, 14 Sep.
2008, *Z. L. Dao, M. Lear, Y. H. Li & M. Wickenden (KWL) 63*
(KUN); mtn. of Nandai, Dulongjiang region, 2400–3300 m, 27
July 2013, *X. H. Jin et al. ST1150* (PE 01977427–28).

Cultivated material:
Cultivated plants.
Foster Clough, Mytholmroyd, West Yorkshire, U.K., from
Z. L. Dao, M. Lear, Y. H. Li & M. Wickenden (KWL) 63.

Cultivated specimens.
From above plant, *J. F. Harber 2015-24*, 18 Aug. 2015 (A,
E, K, KUN, PE, TI).

144. Berberis elliptifolia Harber, sp. nov. TYPE:
China. Qinghai: Nangqên (Nanqian) Xian, ca.
15–20 km E of the town of Nangqen along the
Za-Qu (Za River) in Shixia Gou, 31.893611°N,
96.596111°E, 4100–4200 m, 22 June 1995, *D. E.
Boufford, M. J. Donoghue, X. F. Lu & T. S. Ying
26652* (holotype, A 00279960!; isotype, MO
4920164 image!).

Diagnosis. The leaves of *Berberis elliptifolia* are similar
to *B. heteropoda*, but the color of mature stems and the flower
structure color are different. It is somewhat similar to *B. francisci-ferdinandi*, but its leaf shape is less varied, and its inflorescence much shorter and never paniculate and with a different flower structure.

Shrubs, deciduous, to 3 m tall; mature stems purple
or brownish purple, subterete, densely verruculose;
spines solitary, rarely 2- or 3-fid, pale yellowish brown,
0.6–1.6 cm, verruculose, abaxially sulcate. Petiole almost absent or to 7(–9) mm; leaf blade abaxially pale
green, not papillose, adaxially mid-green, slightly shiny,
elliptic, sometimes elliptic-lanceolate, 2–5.5 × 1–2.2
cm, papery, midvein, lateral veins, and reticulation
raised and conspicuous on both surfaces, especially
when dry, base attenuate, margin spinulose with 5 to 12
teeth on each side, sometimes entire, apex acute or
subacute. Inflorescence a raceme, sometimes fascicled
at base, 5- to 12-flowered, 1–4 cm overall; peduncle
0.3–0.5 cm; pedicel 5–12 mm. Sepals in 3 whorls;
outer sepals oblong-ovate, ca. 1.25–1.5 × 0.75 mm,
apex acute; median sepals ovate, 1.5–2 × 1.25 mm;
inner sepals broadly elliptic, 3.25–3.5 × 2–2.5 mm;
petals obovate, 2.75 × 2.5 mm, base slightly clawed,
glands separate, ca. 0.6 mm, apex rounded. Stamens
ca. 2 mm; anther connective not extended, truncate.
Ovules 2. Berry red, oblong-ellipsoid, 9–11 × 5–6 mm;
style not or slightly persistent.

Phenology. Berberis elliptifolia has been collected in flower from May to June and in fruit in July and August.

Distribution and habitat. Berberis elliptifolia is known from south Qinghai, northwest Sichuan, and northeast Xizang. It has been found by stream and river sides, ravine bottoms, and on open hillsides and forest margins at ca. 3480–4300 m.

Selected specimens.
Qinghai. Chindu (Chengduo) Xian: Gaduo Gongshe, 3950 m, 25 Aug. 1983, *X. J. Gou & Y. Y. Liu 85333* (HNWP 106794). **Nangqên (Nangqian) Xian:** 15–20 km from the town of Nangqen along the Za-Qu in Shixia Gou, 32.29972222°N, 96.54138889°E, 3700–3800 m, 21 June 1995, *D. E. Boufford et al. 26625* (A 00279961); 15–20 km E of the town of Nangqen along the Za-Qu (Za River) in Shixia Gou, 31.89361111°N, 96.59611111°E, 4100–4200 m, 14 Aug. 2018, *D. E. Boufford, B. Bartholomew, J. L. Guo, J. F. Harber, Q. Li & J. P. Yue 44631* (A, CAS, E, KUN, PE). **Qumarlêb (Qumalai) Xian:** Dongfeng Xian, Jianglang Si, 4100 m, 22 Aug. 1966, *S. W. Liu 860* (HNWP 17516). **Yushu Xian:** SE of the city of Yushu, S of hwy. G214 along rd. on E side of the Tongtian He, ca. halfway betw. Tangka & Wuwa, 32.88305556°N, 97.2575°E; 3545 m, 12 Aug. 2018, *D. E. Boufford, B. Bartholomew, J. L. Guo, J. F. Harber, Q. Li & J. P. Yue 44580* (A, CAS, E, KUN); along Xian rd. X817 betw. km markers 95 & 96, ravine flowing into Za Qu (Za River), 32.14083333°N, 97.035°E, 3700 m, 17 Aug. 2018, *D. E. Boufford, B. Bartholomew, J. L. Guo, J. F. Harber, Q. Li & J. P. Yue 44709* (A, CAS, E, KUN, PE). **Zadoi (Zaduo) Xian:** Angsai Xian, 4000 m, 9 July 2005, *Y. H. Wu 32979* (HNWP 0214098, 0214100); Angsai Xian, Guodao Gou, 4100 m, 13 July 2005, *Y. H. Wu 33660* (HNWP 0215658) and *33756* (HNWP 0215260, 0215262); along Xian rd. X810 S of the confluence of the Sha Qu and Za Qu on E side of the Za Qu, N of Angsai-Xiang, 32.83611111°N, 95.5625°E, 3980–4125 m, 19 Aug. 2018, *D. E. Boufford, B. Bartholomew, J. L. Guo, J. F. Harber, Q. Li & J. P. Yue 44760* (A, CAS, E, KUN, PE). **Zhidoi (Zhiduo) Xian:** Gangcha Xiang, behind Gangcha Si, 4040 m, 20 June 1966, *L. H. Zhou 85* (HNWP 18070).
NW Sichuan. Sêrxü (Shiqu) Xian: Luozu Qu, Xinzhonglong Gou, 4 July 1988, *L. C. He: Shi 24* (Ganzi Institute of Forestry Herbarium 00008622–24).
NE Xizang (Tibet). Jomda (Jiangda) Xian: rd. from Jiangda to Dêgê, Sichuan (hwy. 317), W side of Leijila Shan, 31.64666667°N, 98.39055556°E, 3485–3500 m, 2 Aug. 2004, *D. E. Boufford et al. 31559* (A 00279932). **Qamdo (Changdu) Xian (now Karuo Qu):** hwy. 317 from Changdu to Jiangda, along Zha Qu, W of Toba, 31.507222°N, 97.334444°E, 3480–3550 m, 17 Aug. 2004, *D. E. Boufford et al. 32515* (A 00279530, CAS 1078921). **Riwoqê (Leiwuqi) Xian:** E of the city of Riwoqê on hwy. 317 + 214, from Riwoqê to Changdu (Chamdo) along Gei Qu (river), 31.184167°N, 96.605278°E, 4300 m, 12 Aug. 2004, *D. E. Boufford et al. 32191* (A 00279935, F 2307936); Guowa-Cun, betw. hwy. G214 & the Ji Qu, 32.29972222°N, 96 54138889°E, 3700–3800 m, 15 Aug. 2018, *D. E. Boufford, B. Bartholomew, J. L. Guo, J. F. Harber, Q. Li & J. P. Yue 44668* (A, CAS, E, K, KUN, PE).

145. Berberis emeishanensis Harber, sp. nov. TYPE: China. WC Sichuan: [Emeishan Shi], Mt.

Omei, June 1904, *E. H. Wilson (Veitch) 4726* "A" (holotype, A 00279867!; isotypes, A 00279868!, IBSC 0000591 image!, K[2 sheets], P P02313289!).

Diagnosis. Berberis emeishanensis is somewhat similar to *B. silva-taroucana,* but with a different inflorescence and flower structure. It is somewhat similar to *B. henryana* but with a different mature stem color, mostly absent spines, fewer flowers, longer pedicels, and a different flower structure.

Shrubs, deciduous, to 3 m tall; mature stems brownish or grayish purple, sulcate; spines absent, rarely 1- to 3-fid, semi-concolorous, 0.3–1.3 cm, terete, weak. Petiole almost absent to 12(–22) mm; leaf blade abaxially gray-green, papillose, adaxially green, elliptic or obovate-elliptic, 1.5–4(–5.4) × (0.6–)1.5–2(–2.6) cm (excluding petiole), papery, midvein raised abaxially, lateral veins and reticulation conspicuous abaxially, inconspicuous adaxially, base attenuate, margin entire, rarely spinulose with 1 to 7 teeth on each side, apex obtuse or subacute, sometimes retuse. Inflorescence a raceme, 4- to 12-flowered, 1.5–4.5 cm overall including peduncle 0.5–2.5 cm; pedicel 8–15 mm; bracteoles triangular, 2 × 1.25 mm. Sepals in 2 whorls; outer sepals broadly ovate, 3 × 2.5 mm; inner sepals elliptic-orbicular, 4 × 4 mm; petals obovate, 4 × 2.5 mm, base cuneate, glands approximate, apex entire. Stamens ca. 2.75 mm; anther connective not extended, truncate. Pistil ca. 3 mm; ovules 2. Berry red, ellipsoid, 7–10 × 5–6 mm; style not persistent.

Phenology. Berberis emeishanensis has been collected in flower in May and June and in fruit between August and October.

Distribution and habitat. Berberis emeishanensis is known from Emei Shi and Boaxing, Ebian, Hongya, and Luding Xian in west-central Sichuan. It has been found on wooded mountain slopes at ca. 1500–2950 m.

IUCN Red List category. Berberis emeishanensis is assessed as DD or Data Deficient, according to IUCN (2001) criteria.

Wilson 4726 consists of two gatherings. That of June 1904 is designated here as *4736* "A" and that of October 1904 as *4736* "B." *Wilson 1012a* (details below) also consists of two gatherings, and that of July 1908 is designated here as *1012a* "A" and that of October 1908 as *1012a* "B."

Wilson 4726 from Emei Shan was included as an example of *Berberis silva-taroucana* by Schneider (1913: 371), as were *Wilson 2861* and *1012a* from Mupin (Baoxing), as well as *Wilson 2858* and *955* from Wa Shan. However, all of these are racemose whereas the flower stalk of the type of *B. silva-taroucana* is a fascicle, sub-fascicle, umbel, sub-umbel, or sub-raceme. More-over, the flowers of *Wilson 4736* "A" and *Wilson 2861* (which are the same) have a different structure from *B. silva-taroucana*. (I have not investigated the flowers of *Wilson 2858* from Wa Shan, but four other *Berberis* species found on Emei Shan—*B. aemulans*, *B. simu-lans*, *B. verruculosa*, and *B. wilsoniae*—are also found on Wa Shan, which suggests the vegetation of the two areas is similar. For a comparison of the two areas, see Callaghan (2011), though his identification of the *Ber-beris* species of Wa Shan contains various errors.

The report of *Berberis silva-taroucana* in Baoxing by X. H. Li et al. (2017) is probably referable to *B. emeishanensis*.

Selected specimens.
WC Sichuan. Baoxing Xian: Mupin, 2135–2590 m, June 1908, *E. H. Wilson 2861* (A 00279854, E E00612584, HBG-506685, K, LE, US 00945938); Mupin, 1525–2285 m, July 1908, *E. H. Wilson 1012a* "A" (A 00279869, BM BM000810088, E E00612586, HBG-506687, US 00945939); Mupin, 1525–2285 m, Oct. 1908, *E. H. Wilson 1012a* "B" (A 00279869, US 00945939); Dengchi Gou, 2960 m, 25 June 1933, *T. H. Tu 4278* (CQNM 0005203, PE 01032883); above Lengpu Gou, 2400 m, 5 Sep. 1933, *T. T. Yu 1936* (CQNM 005204, PE 01032891); Xinzhaizi Gou, 2700 m, 5 Sep. 1938, *K. L. Chu 6336* (FUS 00014214, IBSC 0092591, PE 01032890). **Ebian Xian:** Wa shan, 2440 m, June 1908, *E. H. Wilson 2858* (A 00279861, BM BM000810085, US 00945937); Wa-shan, 1830 m, Sep. 1908, *E. H. Wilson 955* (A 00279856, US 0945942); Wa Shan, 1 Oct. 1938, *Z. W. Yao 3097* (PE 01032950); Wa Shan, 2 Oct. 1938, *Z. W. Yao 3098* (KUN 0177084, PE 01032951, 01033328). **Emei Shi, Mt. Omei (Emei Shan):** Oct. 1904, *E. H. Wilson (Veitch) 4726* "B" (A 00279867, K); above Dacheng Si, 2800 m, 15 June 1935, *F. H. Tu 296* (IBK IBK00013093, IBSC 0092435, PE 01032852, 01032855, 01032965); Hungchunp'ing, 19 July 1938, *H. C. Chow 7848* (A 00279571, KUN 0177083); 1800 m, 4 Aug. 1938, *W. P. Fang 12935* (A 00279570); below Hsi-Hsiang-Chih, 10 June 1939, *S. C. Sun & K Chang 137* (A 00279568, WH 06008631); Lei-tung-ping, 2500 m, 1939, *S. C. Sun & K. Chang 1152* (A 00279562); 14 Oct. 1940, *T. C. Lee 3904* (A 00279569, KUN 0177082); Dacheng Si, 2400 m, 26 May 1956, *S. Z. Yu 49489* (SZ 00290613–15); 18 Aug. 1956, *Y. H. Tao 51579* (SZ 00290612); Gongbei Shan, 2300 m, 7 June 1957, *G. H. Yang 55191* (IBSC 0091744, KUN 0177092, PE 01032857, SZ 00290599); above Dachang Si, 2300 m, 9 June 1957, *G. H. Yang 55213* (IBSC 0092675, PE 01032856, SZ 00290603); below Jingangzui, 2300 m, 9 June 1957, *G. H. Yang 55258* (IBSC 0092448, KUN 0177091, PE 01032859, SZ 00290591); Dacheng Si, Baiyanwan, 2400 m, 26 Sep. 1957, *G. H. Yang 57411* (IBSC 0092077, KUN 0177090, PE 01032851, SZ 00290602). **Hongya Xian:** Linchang Heishan, 3 July 1974, *W. K. Bao et al. 2834* (CDBI CDBI0028422–23). **Luding Xian:** Moxi Gongshe, Hailuogou, 2950 m, 8 June 1980, *Z. A. Liu & Q. Q. Wang 22205* (CDBI CDBI0027429–30, IBSC0091777–78); Heishui Gou, 2400 m, 1 June 1983, *Botany Team 30866* (CDBI CDBI0027433–34).

146. Berberis epedicellata Harber, sp. nov. TYPE: Cultivated. Royal Botanic Gardens, Kew, 5 May 1995, *s. coll. s.n.*, from seeds of *J. Simmons, C. Er-skine, C. Howick & W. McNamara (SICH) 538,* China. Kangding Xian, Paoma Shan, ca. 2820 m, 20 Sep. 1991 (holotype, K!).

Diagnosis. Berberis epedicellata is very similar to *B. dic-tyophylla*, but has smaller and often slightly broader leaves. Most importantly, its flower structure is also different. *Berberis epedicellata* is also somewhat similar to *B. baiyuensis*, which is also epedicellate.

Shrubs, deciduous, to 2.5 m tall; mature stems pale or dark purple, mostly pruinose, subterete; spines 3-fid, sometimes solitary toward apex of stems, pale yellow, 1–2 cm, abaxially sulcate. Petiole almost absent; leaf blade abaxially pale green, mostly white pruinose, oth-erwise papillose, adaxially dull mid-green, obovate or obovate-elliptic, 1–1.6 × 0.8–1.2(–2) cm, papery, mid-vein slightly raised abaxially, lateral veins and reticu-lation inconspicuous or indistinct on both surfaces, base attenuate, margin entire or spinulose with 1 to 10 teeth on each side, apex rounded or obtuse, sometimes retuse, rarely mucronate. Inflorescence 1-flowered; ped-icel absent or to ca. 2 mm. Flowers yellow. Sepals in 2 whorls; outer sepals obovate, 3.5–4 × 2–2.5 mm; inner sepals broadly obovate, 5–6 × 3–4 mm; petals obovate, 4–4.5 × 3 mm, base cuneate, glands separate. Stamens 4 mm; anther connective extended, truncate. Pistil ca. 4 mm; ovules 3 to 4. Berry dull red, pruinose, subglo-bose, 9 × 8 mm; style persistent, short.

Phenology. Berberis epedicellata has been collected in fruit between May and September. Its flowering sea-son in the wild is unknown.

Distribution and habitat. Berberis epedicellata is known from Kangding Xian in west-central Sichuan. It has been found on mountain sides, in thickets, among dense secondary shrubs, and in forests at 2600–3070 m.

IUCN Red List category. Berberis epedicellata is assessed as DD or Data Deficient, according to IUCN (2001) criteria.

Berberis epedicellata is one of at least three 1-flow-ered species found in Kangding Xian, the other two being *B. ambrozyana* and the dwarf *B. kangdingensis*. Specimens of *B. epedicellata* from Kangding have vari-ously been identified as *B. ambrozyana*, *B. dictyo-phylla*, *B. dictyophylla* var. *epruinosa* (treated here as a synonym of *B. ambrozyana*), *B. dictyophylla* var. *campylogyna*, and *B. approximata* (for these latter two, see under Excluded Taxa), or even, on occasion, *B. wilsoniae*.

Fang 3645 (details below) was mistakenly identified as *Berberis ambrozyana* by Ahrendt (1961: 128) and was the source of his description of the berry of that species.

A cultivated specimen of *Berberis epedicellata* has been selected as the holotype because no wild-collected specimens with floral material have been located.

Some of the habitat of *Berberis epedicellata* is likely to have been destroyed by the recent rapid urbanization of the area south of Kangding town (the once-separate village of Yulin is now no longer recorded on maps).

Selected specimens.
Sichuan. Kangding Xian: Kangtin Hsien, Tachienlu, 2745–2895 m, 27 Sep. 1928, *W. P. Fang 3645* (A 00279435, E E00612468, IBSC 0092218, K001273122, N 093057107, NAS NAS00314434, P P02313566, PE 01030795, 01037652, 01037667); near Yulin Gong, 7 Sep. 1951, 2600 m, *W. G. Hu & Z. He 11170* (HGAS 013676, PE 01037674); Dapaoshan, Daya Gou, 3070 m, 20 Sep. 1953, *X. L. Jiang 36959* (IBK IBK00012801, PE 01037680); Lijia Gou, 3000 m, 6 June 1973, *Sichuan Vegetation Team 4212* (CDBI CDBI0027112–14); near Kangding Town, Lijia Gou, 3000 m, 6 June 1973, *Sichuan Vegetation Team 4213* (CDBI CDBI0027248–50); Zheduo Tang, 2800 m, 31 May 1981, *Z. J. Zhao et al. 114301* (CDBI CDBI0027226); Paoma Shan, 2860 m, 2 Oct. 1988, *C. Erskine, H. Fliegner, C. Howick & W. McNamara (SICH) 340* (K 001273117); Paoma Shan, ca. 2820 m, 20 Sep. 1991, *J. Simmons, C. Erskine, C. Howick & W. McNamara (SICH) 538* (K 001273116).

Cultivated material:
Living cultivated plants.
Royal Botanic Gardens, Kew, and Howick Hall, Northumberland, U.K., Quarryhill Botanical Garden California, U.S. from seeds of *SICH 538*.

147. Berberis erythroclada Ahrendt, J. Bot. 79 (Suppl.): 49. 1941. TYPE: China. SE Xizang (Tibet): Zayü (Chayu) Xian, Ata Lang La, Chutong Camp, ca. 29.069374°N, 96.813583°E, 3950 m, 6 Aug. 1933, *F. Kingdon-Ward 10562* (holotype, BM BM000559583!).

Berberis ludlowii Ahrendt var. *saxiclivicola* Ahrendt, J. Linn. Soc., Bot. 57: 115. 1961. syn. nov. TYPE: China. SE Xizang: [Mainling (Milin) Xian], Takpo province, Chubumbu La near Langong, 28.78°N, 93.73°E, ca. 3950 m, 6 June 1938, *F. Ludlow, G. Sherriff & G. Taylor 3968* (holotype, BM [2 sheets] BM000559580–81!).

Shrubs, deciduous, to 30 cm tall; mature stems dark red, sulcate, shiny; spines 5- to 7(to 9)-fid, orange-red, 0.5–1 cm, terete or abaxially sulcate. Petiole almost absent; leaf blade abaxially pale or gray-green, ultimately pale green, slightly pruinose, adaxially dark green, obovate or obovate-elliptic, 1–2.8 × 0.3–0.7 cm, papery, midvein raised abaxially, lateral veins and reticulation inconspicuous on both surfaces, base cuneate, margin spinose with (5 to)7 to 15 teeth on each side, apex rounded, rarely acute. Inflorescence 1-flowered; pedicel 15–25 mm, very slender. Flowers 11–14 mm diam. Sepals in 2 whorls; outer sepals oblong-elliptic, ca. 5.5 × 3 mm; inner sepals obovate, ca. 7 × 5 mm;

petals obovate, ca. 4.5 × 2.5 mm, apex subentire or sub-emarginate. Stamens ca. 3 mm; anther connective slightly extended, subretuse. Ovules 6 to 9. Fruit bright coral pink, otherwise unknown.

Phenology. In China *Berberis erythroclada* has been collected in flower from June to August. Its fruiting season is unknown.

Distribution and habitat. In China *Berberis erythroclada* has been found in Mainling (Minlin), Mêdog (Motuo), and Zayü (Chayu) Xian in Xizang (Tibet) in scrub (including among *Rhododendron*) on open, stony mountain slopes at ca. 3050–4000 m. It is also known from northeast Arunachal Pradesh in India and northern Myanmar (Burma).

The source of the color of fruit used above is from the collector's notes to *Kingdon-Ward 6233* (details below). The specimen itself is, however, sterile.

The spines of *Berberis erythroclada* are highly unusual in that they are sometimes branched. This is very evident on the type specimen, but this feature is also found to a lesser extent on the type of *B. ludlowii* var. *saxiclivicola*.

It is unclear why Ahrendt determined *F. Ludlow, G. Sherriff & G. Taylor 3968*, which the collectors' notes record as being from a plant 30 cm high, as a variety of *Berberis ludlowii*, which has completely different leaves and is up to 4 m tall, except both are 1-flowered and have a large number of ovules (the protologue stated these to be seven).

Ying (2001: 91) treated *Berberis erythroclada* var. *trulungensis* Ahrendt (1961: 118) as a synonym of *B. erythroclada*. This taxon, which is up to 1.8 m tall versus the 30 cm of *B. erythroclada*, is treated here as a synonym of *B. temolaica*.

In the U.K. at least, *Berberis erythroclada* has sometimes been confused with *B. concinna* from Nepal and India (Sikkim). However, as well as being found in a different part of the Himalayas, the latter is semi-evergreen and has 3-fid spines and obovate-rhombic leaves usually with only one or two marginal spines.

Selected specimens.
SE Xizang (Tibet). Mainling (Milin) Xian: W of Dongxiongla, 4000 m, 19 Aug. 1982, *Z. C. Ni et al. 3019* (PE 01037768). **[Mêdog (Motuo) Xian]:** Pemako, Doshong La, 3050–3650 m, 20 Oct. 1924, *F. Kingdon-Ward 6233* (K); Nage to Dongxiongla, 3800 m, 31 July 1974, *Qinghai-Xizang Team 74-3864* (KUN 0178675, PE 01037766–67).
INDIA. NE Arunachal Pradesh: Kaso, Delai valley, 28.35°N, 96.613337°E, 3650–3960 m, 4 July 1928, *F. Kingdon-Ward 8427* (K).
N MYANMAR (BURMA). Seinghku Wang, 28.133333°N, 97.4°E, 3350–3650 m, 22 June 1926, *F. Kingdon-Ward 6969* (K).

148. Berberis everestiana Ahrendt, J. Linn. Soc.,
Bot. 57: 116. 1961. TYPE: China. S Xizang
(Tibet): [Tingri (Dingri) Xian], N of Karta Chu,
3655 m, 11 June 1922, *Mt. Everest 1922 Expedition, E. F. Norton 77* (holotype, K K000644935!).

Berberis multicaulis T. S. Ying, Fl. Xizang. 2: 147. 1985, syn.
nov. TYPE: China. S Xizang (Tibet): Tingri (Dingri) Xian,
Kada Qu, 3600 m, 20 June 1975, *Qinghai-Xizang Complex Exped. 4098* (holotype, PE 00935171!; isotype, PE
00935172!).

Shrubs, deciduous, to 30 cm tall. Mature stem pale
yellow, terete to sub-sulcate, sparsely verruculose;
spines 3-fid, semi-concolorous, 0.5–1.4 cm, terete,
slender. Petiole almost absent; leaf blade abaxially
pale yellow-green, adaxially green, obovate or elliptic,
0.7–1.6 × 0.4–0.6 cm, papery, midvein raised abaxially, lateral veins conspicuous abaxially, inconspicuous adaxially, base cuneate, margin entire or spinulose
with 1 to 3 teeth on each side, apex obtuse or acute,
mucronate. Inflorescence 1-flowered; pedicel 2–3 mm;
bracteoles oblong-lanceolate, ca. 4 × 1.8 mm. Sepals in
2 whorls; outer sepals oblong, ca. 5–7 × 3–4.5 mm,
apex acute; inner sepals obovate-oblong, 7–8 × 4–6 mm;
petals elliptic, 6–6.5 × 3–3.5 mm, base clawed, glands
separate, apex emarginate. Stamens 3–4 mm; anther
connective extended. Ovules 3 or 4. Berry red, subglobose or oblong-ovoid, 10 × 6 mm; style persistent.

Phenology.   *Berberis everestiana* has been collected
in flower in June. Its fruiting season is unknown.

Distribution and habitat.   *Berberis everestiana* is
known from Tingri (Dingri), Gyrong (Jilong), and possibly Nyalam (Nielamu) Xian in south Xizang (Tibet). It
has been found in thickets and on stony ground at ca.
3600–4200 m.

Herbarium specimens of *Berberis everestiana* are
sometimes confused with *B. tsarica*. However, *B. tsarica* has purple mature stems and it often has 5-fid
spines, whereas the mature stems of *B. everestiana* are
pale yellow and the spines appear to be always 3-fid.
Besides the type, the protologue cited various other
specimens from Xizang and Nepal. Not all of these
have been found, but the three I have seen are not
B. everestiana; *Mt. Everest 1922 Expedition, Norton 90*
(K K000395221), Kharta, 3660 m, 11 June 1922, and
F. Kingdon-Ward 5811 (E E00395954, K), Pome,
Nyuna La, 4370–4570 m, 20 June 1924, are *B. tsarica*.
Bailey 46 (BM), Nepal, Chilang Pati, 25 October 1935
is *B. angulosa*.

The type of *Berberis multicaulis* was collected in the
same area and at a similar elevation as the type of *B.
everestiana*.

Ahrendt (1961: 117) recognized a *Berberis everestiana* var. *nambuensis* with a type from southeast Xizang
and a variety *ventosa* from Nepal. The former is recognized here as a separate species, *B. nambuensis*. According to Adhikari et al. (2012: 464–466), who record
no variety *everestiana* from Nepal, variety *ventosa* differs from variety *everestiana* by its distinctly extended
anther connective. But this distinction appears to rely
on Ahrendt's protologue, which describes the anther
connective of variety *everestiana* as "scarcely extended,"
the accuracy of which I have been unable to check.
They also describe variety *ventosa* as having five to
seven ovules versus the three or four given by Ahrendt
for variety *everestiana*. Interestingly, the protologue of
B. multicaulis describes the anther connective as "prolonged, shortly apiculate and with 1 tooth," but the
number of ovules as four. Clearly more research is
needed here.

Besides the collection cited below, it is possible that
Z. C. Ni & D. D. Ci 2055 (PE 01037423), south Xizang,
Nyalam (Nielamu) Xian, near Xiancheng, 4100 m, 7
September 1981, is *Berberis everestiana*. This, however,
has no mature stems and is sterile.

Selected specimens.
S Xizang. Gyrong (Jilong) Xian: Gongdang Qu, Shala
Shan, 4200 m, 4 Aug. 1974, *Qinghai-Xizang Botanic Team
5734* (PE 01037418–20).

149. Berberis exigua Harber, sp. nov. TYPE: Cultivated. Foster Clough, Mytholmroyd, W. Yorkshire,
U.K., 12 May 2016, *J. F. Harber 2016-04*, from
S & D. Rankin 4219, China. NW Yunnan: Yulongshan, near Lijiang Field Station, 3250 m, 2 Sep.
2005 (holotype, PE!; isotypes, A!, BM!, KUN!,
MO!, P!, TI!).

Diagnosis.   *Berberis exigua* is very similar to *B. minutiflora*, but has a different flower structure including having
three ovules versus the two of *B. minutiflora*. The shape of its
fruit is narrowly ovoid versus the ellipsoid or ovoid-ellipsoid
fruit of *B. minutiflora*.

Shrubs, deciduous, to 80 cm tall in cultivation; mature stems reddish purple, sulcate; spines 3-fid, pale
yellowish brown, 0.6–1 cm, terete or abaxially subsulcate. Petiole almost absent; leaf blade abaxially pale
gray-green, papillose, adaxially green, narrowly obovate
or oblanceolate, 0.6–1.7 × 0.3–0.4 cm, papery, midvein raised abaxially, slightly impressed adaxially, venation indistinct or inconspicuous on both surfaces, base
attenuate, margin entire, apex obtuse, rarely subacute
and minutely mucronate. Inflorescence 1-flowered, ca.
10 mm diam.; pedicel 5–7 mm. Sepals in 2 whorls;
outer sepals oblong-elliptic, 6.5 × 3.5 mm; inner sepals
broadly obovate, 6 × 4.5–5 mm; petals obovate, 5 × 3.5
mm, base clawed, glands very close together, ca. 1 mm,

apex emarginate. Stamens ca. 3 mm; anther connective distinctly extended, obtuse or sub-apiculate. Pistil 2.75 mm; ovules 3. Berry red, narrowly ovoid, 12 × 6 mm; style not persistent.

Phenology. *Berberis exigua* has been recorded as being collected in fruit in September. Its flowering season outside cultivation is uncertain.

Distribution and habitat. *Berberis exigua* is reliably known from only the type collection from seed collected on Yulongshan. This was collected from a plant in an unrecorded habitat at 3250 m.

As noted under *Berberis minutiflora*, there are complex and unresolved issues in relation to dwarf *Berberis* species with reddish brown or reddish purple mature stems from northwest Yunnan and Muli Xian in Sichuan. This treatment is based on the premise that *B. minutiflora* has flowers with two ovules and that specimens with more than this are of other species.

The flower structure given above (including the number of ovules as three) is little different for that given in the protologue of *Berberis crassilimba*. But as noted in the discussion of that species above, this was based not on the type collection, which has fruit with up to five seeds, but either or both *K. M. Feng 1147* from Haba Shan and *Feng 657* from Yulongshan (details below). This would seem to show that either or both are actually *B. exigua*.

Both *Berberis minutiflora* and *B. exigua* are found in Yulongshan. Given this, it is likely that some specimens from this area listed here under *B. minutiflora* are, in fact, *B. exigua*. In each case, only dissections of their flowers or fruit would show this or otherwise.

Selected specimens.
NW Yunnan. Zhongdian (Xianggelila) Xian: N flank of Haba Snow Range, 3 May 1939, *K. M. Feng 1147* (A 00279398, KUN 0175847–48). **Lijiang Xian:** [that part now in Yulong Xian], Pai-shu-ho (Baishui He), 3 Apr. 1939, *K. N. Feng 657* (A 00279397, KUN 0175852–53).

150. Berberis fengii S. Y. Bao, Bull. Bot. Res., Harbin 5(3): 3. 1985. TYPE: China. NW Yunnan: Zhongdian (Xianggelila) Xian, Haba Yue Shan, 3700 m, 12 Oct. 1955, *K. M. Feng 21044* (holotype, KUN 0160894!; isotypes, KUN 0160895!, SZ 00290949 image!).

Shrubs, deciduous, to 1 m tall; mature stems purple, sulcate, black verruculose; spines 3-fid, pale brown, 1.5–2.5 cm, slender, abaxially slightly sulcate. Petiole 2–3 mm; leaf blade abaxially pale green, sometimes slightly pruinose, adaxially olivaceous, elliptic or obovate, 1.5–2.5 × 0.5–1.5 cm, papery, midvein, lateral

veins, and reticulation raised and conspicuous abaxially, conspicuous adaxially, base cuneate, margin entire, rarely spinulose with 3 to 6 teeth on each side, apex rounded, mucronate. Inflorescence 1(or 2)-flowered, otherwise unknown; fruit stalk ca. 10 mm, stout. Berry red, broadly ellipsoid, ca. 20 × 12–14 mm, blue pruinose; style not persistent; seeds 6 or 7.

Phenology. *Berberis fengii* has been collected in fruit in October. Its flowering season is unknown.

Distribution and habitat. *Berberis fengii* is known from only the type collection made at 3700 m on a dry, grassy slope of the Haba Yue Shan in Zhongdian (Xianggelila) Xian in northwest Yunnan.

The protologue of *Berberis fengii*, which cited only the holotype, described the species as having densely spinose leaves and solitary flowers. However, both the holotype and isotypes have largely entire leaves, and the holotype and the isotype at SZ have some fruit in pairs. Evidence for the color of the mature stems comes from the isotype at SZ. The exceptionally large fruit and large number of seeds of *B. fengii* are somewhat similar to those of *B. capillaris*, but that species has leaves that are spinose and much longer pedicels.

151. Berberis forrestii Ahrendt, Gard. Chron., ser. 3, 109: 101. 1941. TYPE: China. NW Yunnan: [Lijiang Xian, now Yulong Xian], W flank of Lichiang (Lijiang) range, 27.3333°N, 3660–3960 m, 15 Oct. 1918, *G. Forrest 17143* (holotype, K K000077406!; isotypes, E E00217995!, WSY 0057599!).

Shrubs, deciduous, to 2.75 m tall; mature stems pale yellowish brown, sulcate, young shoots pink; spines mostly absent, sometimes solitary, rarely 3-fid, concolorous, (0.6–)1.2–2.5 cm, abaxially sulcate, weak. Petiole almost absent or to 4 mm; leaf blade abaxially pale green, papillose, adaxially mid-green, obovate, 2.4–4 × 0.8–1.6 cm, papery, midvein raised abaxially, slightly impressed adaxially, lateral veins and reticulation inconspicuous on both sides, base attenuate, margin entire, rarely spinose-serrulate with 1 or 2 teeth on each side, apex rounded or obtuse, rarely subacute, sometimes mucronate. Inflorescence a raceme, sometimes compound or fascicled at base (5- to)8- to 14-flowered, 6–9 cm overall, including peduncle to 3.5 cm; pedicel 7–10 mm, but to 20 mm from base. Sepals in 3 whorls; outer sepals lanceolate, 4.5 × 1.5 mm; median and inner sepals oblong-obovate, 5–6 × 3–4 mm; petals obovate-elliptic, 4–5 × 2–3 mm, base cuneate, not clawed, glands approximate, apex distinctly emarginate or incised. Stamens ca. 2.5–3 mm; anther connective extended, slightly apiculate. Ovules 1 to 3. Berry scarlet red, oblong-ovoid, 10–12 × 7–8 mm; style not persistent or persistent and short; seeds 3.

Phenology. *Berberis forrestii* has been collected in fruit in October. Its flowering season in the wild is unknown.

Distribution and habitat. *Berberis forrestii* is reliably known from only the type collection from the west flank of the Lijiang range in northwest Yunnan. The collector's notes record the plants growing in open meadows and pine forests at 3660–3960 m.

The protologue of *Berberis forrestii* gave only a cursory description and stated, "I am grateful for material of number 17141, B. Forrestii from Wisley [the Botanic Garden of the Royal Horticultural Society in Surrey, U.K.] and facilities for examining the type herbarium sheet at Kew." The description was substantially expanded by Ahrendt (1944a: 108–109), which described the young shoots as red and the mature stems as reddish brown, as well as including the description of the flowers given above, except that the number of ovules was described as one or two. In the discussion, Ahrendt again referred to cultivated material supplied by Wisley, but also to material from a Mr. R. D. Trotter of Leith Vale, Ockley, Surrey, though the provenance of this latter material was not given. The description of both the stem color and the flowers had to have come from this cultivated material since the holotype at K has neither, though whether these additional details came from the Wisley or the Trotter material or from both was not made clear. It is possible a sheet of cultivated material at BM (details below) annotated by Ahrendt as "Forrest 17141 Cult. Type" on 29 September 1943 is this material, though its provenance is unclear. The sheet has four specimens, all poor ones. Three have fruit or a fruit stalk, and one has a few flowers (whose structure I have not investigated). None of these have evidence of mature stem color. There is a cultivated specimen at WSY grown from seeds of *Forrest 17141*. This has no floral material but does, however, have mature stems which are pale yellowish brown rather than reddish brown, and this is the source of the stem description given above.

Uncited by Ahrendt, there are two further wild-collected specimens of *Forrest 17141* at E and WSY. These do not have mature stems either but, importantly, a line drawing of a fruit by Schneider attached to the specimen at E shows it has three seeds. Ahrendt's description of the number of ovules has, therefore, been amended accordingly.

Subsequently, Ahrendt (1961: 172–173) reproduced his earlier description of the flowers but, without explanation, ignored his earlier account of stems in favor of

"young shoots bright red, becoming yellowish red, finally yellow when mature." He now cited not just the type and cultivated material from Wisley and from his own living collection, but also wild-collected specimens from the east flank of the Lichiang range: *Forrest 2271, 5553,* and *5554. Forrest 5553* has a flower structure different from that given by Ahrendt for *Berberis forrestii*. It also has mature stems which are dark reddish purple. It is treated here as the type of a new species, *B. zhaoi*, with *Forrest 2271* and *5554* as examples of the same species.

Additionally, Ahrendt, in 1961, cited a plant in his own living collection "labelled *Yu 10819*." Why he described this plant as "labelled *Yu 10819*" is apparent from his 1961 entry for his *Berberis franchetiana* var. *glabripes* (1961: 153) where he cited a wild-collected specimen of *Yu 10819* as this taxon while stating "Plants distributed as *Yu 10819* were *B. forrestii*." No cultivated specimens of *Yu 10819* have been found, while wild-collected specimens with the same number—Atunze (Dêqên [Deqin]) Xian, Bai-ma Shan, 3600 m, 28 November 1937 (BM, E E00612477, KUN 0176195, PE 01033851)—are extremely poor ones with no leaf materials, fruit stalks much shorter than those of *Forrest 17141*, and with smaller fruit with two seeds. It is, therefore, a possibility that Ahrendt's plant was from the wild-collected seed of *Yu 10919*. In any event, it is likely the plant was not *B. forrestii* (for further information on *B. franchetiana* var. *glabripes*, see under Taxa Incompletely Known below and under *B. dokerlaica* above).

As noted above, the type of *Berberis forrestii* was collected on the western flank of the Lijiang range. However, it is unclear whether Forrest was referring to the west side of Yulongshan or to Haba Shan on the other side of the Yangtze Gorge, the collection details giving only a northern co-ordinate. In this context, it is perhaps worth noting *J. F. Rock 25310* (A 00279886) collected in 1932. This has leaves and an infructescence very similar to *Forrest 17143* (though three of its fruit all produced two rather than three seeds), its collection details being given as "Haba shan, north of the Yangtze loop, third peak of the Likiang snow range, Bardar," thus showing that, for Rock at least, the Lijiang range was more than Yulongshan. Where the location is given, most *Berberis* specimens (including those of *B. zhaoi*) are from the east flank of the Yulongshan. Clearly, more needs to be known about *Berberis* species on the west flank and, in particular, their relationship to species on Haba Shan.

Cultivated material:
Cultivated specimens.
Royal Horticultural Society Gardens, Wisley, Surrey, Oct. 1936, from seeds of *Forrest 17141, s. coll. s.n.* (WYS

WSY0057600); "Stonefield, Watlington, Oxfordshire?, 29. 9. 1943 . . . Communicated by L. W. A. Ahrendt" (BM).

152. Berberis francisci-ferdinandi C. K. Schneid., Pl. Wilson. (Sargent) 1: 367–368. 1913. TYPE: China. C Sichuan: Mao-chou (Mao Xian), 1200–2150 m, Oct. 1908, *E. H. Wilson 1180* "B" (lectotype, designated here, specimen with fruit A 00038753!; isolectotypes, BM BM001050025!, E E00217935!, HBG HBG-506678 image!, K K000644923!, LE!, US 00103879 image!, W 1914–0004798!).

Berberis pingwuensis T. S. Ying, Acta Phytotax. Sin. 37(4): 339. 1999, syn. nov. TYPE: China. N Sichuan: Pingwu Xian, Zhenliu Gongshe, Moyu Gou, 1800 m, 10 Sep. 1971, *236 Coll. Team 0882* (holotype, PE 00935180!; isotype, PE 00935181!).

Shrubs, deciduous, to 3 m tall; mature stems reddish brown, subterete, scarcely verruculose; spines 3-fid, sometimes solitary or absent toward apex of stems, concolorous, 0.5–2 cm, abaxially sulcate. Petiole 5–15 mm; leaf blade abaxially pale yellow-green, adaxially green, slightly shiny, ovate to elliptic, sometimes oblong-lanceolate, 2–7 × 1–3 cm, papery, midvein, lateral veins, and reticulation raised and conspicuous abaxially, midvein slightly impressed adaxially, lateral veins and reticulation inconspicuous adaxially, base narrowly attenuate, margin spinulose with 15 to 30 teeth on each side, apex acute or subacute. Inflorescence a loose raceme, often partially paniculate, (9- to)20- to 40-flowered, 5–14 cm overall; peduncle 1–3 cm; bracts triangular, 1.5–2 mm, acuminate; pedicel 4–8 mm, slender; bracteoles reddish, 1.5–2 mm, apex acute. Sepals in 3 whorls; outer sepals ovate, ca. 2.4 × 1.5 mm, apex acute; median sepals ovate, ca. 3 × 2 mm; inner sepals obovate, 3.3–4.3 × 2–2.5 mm; petals oblong, 3.5–4.5 × 2.5–3 mm, base cuneate, glands separate, oblong, apex acute, slightly incised. Stamens 2.5–3.5 mm; anther connective slightly extended, rounded. Ovules 2, sessile. Berry scarlet, obovoid-ellipsoid, 11–12 cm × 4–6 mm; style ± persistent.

Phenology. *Berberis francisci-ferdinandi* has been collected in flower in May and June and in fruit between July and October.

Distribution and habitat. *Berberis francisci-ferdinandi* has been found in south Gansu and Sichuan in thickets, forest margins, and understories, mountain slopes, and ravine sides at ca. 1500–3350 m.

In the protologue, Schneider cited *Wilson 1180* as the type of *Berberis francisci-ferdinandi* with dates of June and October 1908. In fact, there is also a specimen at

A with the same number but with the date 25 May 1908. Specimens with these three dates are designated here as "A," "B," and "C," respectively. Given Wilson's at times somewhat haphazard approach to collection details, it is possible that "A" and "C" are, in fact, from the same gathering. Later, Schneider (1918: 213) designated *Wilson 1180*, October 1908 as the type, but without citing any herbarium. His lectotypification has been completed by choosing the specimen at A, because Schneider's protologue was based on material sent to him for identification by the Arnold Arboretum.

The description of *Berberis pingwuensis* in the protologue differs little from that of *B. francisci-ferdinandi*, except the flowers are described as having four whorls of sepals and the inner whorls of petals was described as being different from the outer. This is not based on the type gathering, but presumably on another of the gatherings citing *X. L. Jiang 10241* (details below) which has floral material. Differing inner and outer petals appear not to have been recorded for any other taxon in the genus, so, if correct, it may be just an unusual form.

Schneider (1917: 442) reported that he believed *Wilson 2862*, cited below, from northeast of Kangding to be *Berberis feddeana*. However, subsequently (1918: 290), he changed his opinion and left it an open question as to whether it was related to *B. feddeana* or to *B. henryana*. In fact, from the evidence of the flower structure, it appears to be *B. francisci-ferdinandi*. Ahrendt (1961: 184) mistakenly reported *Wilson 2862* as the type of *B. feddeana*.

The inflorescences of the specimens from Leibo Xian in south Sichuan, cited below, are longer and more paniculate than any specimens of *Berberis francisci-ferdinandi* located from elsewhere. Further investigation is needed to confirm or otherwise that they are this species.

No evidence has been found to support the report by Ying (2001: 207–208) of *Berberis francisci-ferdinandi* being found in Shanxi and Xizang. In the case of Xizang, the report is likely to be based on specimens that may be either *B. brachystachys* or *B. elliptifolia* (for further details, see under those species).

Syntype. *Berberis francisci-ferdinandi*. Sichuan. Mao-chou (Mao Xian), 1200–2150 m, Oct. 1908, *E. H. Wilson 1180* "A" (A 00038753, 0038754, BM BM000564430, E E00217935, HBG HBG-506678, K K000077399, LE, US 00103879, W 1914–0004798). Possible syntype, Mao-chou (Mao Xian), 1200–2150 m, 24 May 1908, *E. H. Wilson 1180* "C" (A 00038754).

Selected specimens.
S Gansu. Kang Xian: Anmenkou, 23 June 1959, *Survey Team 0544* (LZU 009206–7). **Wudu Xian:** Luotang, Majialeng, 2200 m, 19 May 1959, *Z. Y. Zhang 2105* (LBG 00064027, WUK 0145467). **Zhugqu (Zhouqu) Xian:** Jiao'er

Qiao, Ge'er Gou, 2100 m, 19 Sep. 1958, *Z. P. Wei 2598* (HHBG HZ009010, WUK 0107224).

Sichuan. 2440–3050 m, May 1904, *E. H. Wilson (Veitch) 3151* (A 00279512–13, K, P P02313116); NE of Tachien-lu (Kangding), 2740–3050 m, 9 July 1908, *E. H. Wilson 2862* (A 00038744, 00306058, BM, K K00007396, US 00945903); W of Wenchuan Xian, 2400 m, 29 May 1930, *F. T. Wang 21041* (A 00279515, IBSC 0091725, KUN 0176229, NAS NAS00313970, PE 01037221, TAI 047307). **Baoxing Xian:** "Pao-Hsing Hsien," Xinjia Gou, 2500 m, 22 June 1936, *K. L. Chu 2908* (E E00612489, IBSC 0091721, 0092644, P P02313117, PE 01037239–41); Mahuang Gou, 2400 m, 29 June 1958, *X. S. Zhang & Y. X. Ren 5707* (GF 09001080-81, PE 01037234). [**Ebian Xian**]: Wa-shan, 1525–1825 m, May & Oct. 1908, *E. H. Wilson 2863* "A" & "B" (A 00274275, 00279852, BM, E E00392870, K, US 00945912); *E. H. Wilson 2863* "A" (HBG HBG-506756). **Ganluo Xian:** Pingba, 25 July 1959, *s. coll. 4321* (PE 01037249–50). **Hanyuan Xian:** "Ching-chi Hsien," Fei-yueh-ling, 2440 m, May 1908, *E. H. Wilson 2869* (A 00279516, E E00612487, K, US 00945900); [Hanyuan Xian], Feiyoling, 8 Oct. 1930, *W. C. Cheng 1994* (FUS 00014180, LBG 00064347, NAS NAS00313969, PE 01037219). **Jinchuan Xian:** Xilizhai Gou, 3000 m, 7 June 1983, *Z. X. Tang 951* (PE 01605710). **Jiulong Xian:** Naiqu Gongshe, Biri Production Team, 2500 m, 22 June 1974, *F. Y. Qiu 4659* (CDBI CDBI0027448–51, PE 01033140). **Kangding Xian:** Dishui Yan, 2400 m, 9 May 1974, *Q. S. Zhao & Y. T. Bai 110918* (PE 01037214). **Leibo Xian:** Gudui Xiang, 2100 m, 17 June 1959, *Sichuan Economic Plants Exp. Liangshan Group 1959* (CDBI CDBI0027374, KUN 0176233, PE 01037242–3); Ahe Gou, Linchang, 2400 m, 12 Aug. 1972, *236 Taskforce 0718* (PE 01037244, 01037246); Ahe Gou, Linchang, 3000 m, 14 Aug. 1972, *236 Taskforce 0798* (PE 01037245). **Li Xian:** Laisou Gou, Da Lasng Ba, 29 Gou, 3350 m, 4 Sep. 1956, *H. P. He 46167* (IBSC 0091732). **Luding Xian:** Erlang Shan, Machang, 2700 m, 14 June 1959, *Sichuan Agricultural College 0026* (CDBI CDBI0027375, KUN 0176231, PE 01037211). **Mao Xian:** 2100 m, 30 July 1930, *F. T. Wang 21972* (PE 01037229–30); Fengyi Qu, Machang, 15 June 1959, *Maowen Group 2710* (CDBI CDBI0027376–77, KUN 0176232, SM SM704800262). **Nanping (now Jiuzhaigou) Xian:** Jiuzhai Gou, 2650 m, 25 July 1984, *s. coll. 8515* (PE 01037225, 02046858). **Pingwu Xian:** Lijia Xiang, Yangliuping Cave, 2050 m, 15 May 1958, *X. L. Jiang 10241* (PE 00935185, 00935187–89); same details, but 1830 m, 16 May 1958, *X. L. Jiang 10236* (PE 01037207–09, SZ 00291487); Wangbameng, Luotongba, 2160 m, 11 July 1958, *X. L. Jiang 10783* (IBSC 0091731, PE 00935182–84, 00935186); Wangba Dam, 2120 m, 23 May 1961, *X. N. Teng et al. 00232* (CDBI CDBI0027370). **Wenchuan Xian:** Wolong Natl. Nature Reserve, Balang, 2100 m, 12 Aug. 1982, *K. Y. Lang et al. 1332* (IBSC 0091638, KUN 0178903, PE 01037222–23). **Xiaojin Xian:** near Shanong Gongshe, 2400 m, 12 Aug. 1975, *Sichuan Botany Team 9705* (CDBI CDBI0027339–40, PE 01037263). **Yuexi Xian:** "Juei-she-Hsien," 2300 m, 26 May 1932, *T. T. Yu 910* (A 00279957, CQNM 0005212, PE 01037217). **Zoigê (Ruo'ergai) Xian:** S313 to Jiangzha Xiang, Xiarelong Cun., 34.161667°N, 102.912500°E, 2715 m, 23 Aug. 2007, *D. E. Boufford et al. 40338* (A, E).

153. Berberis gaoshanensis Harber, sp. nov. TYPE: China. W Sichuan: Daocheng Xian, Haizi Shan, NNE of Sandui Xiang & NNE of Haizi Shan Daoban on hwy. S217 to Litang, 29.415278°N, 100.176389°E, 4602 m, 27 Aug. 2013, *D. E. Boufford, J. F. Harber & Q. Wang 43280* (holotype, PE!; isotypes, A!, CAS!, E!, K!, KUN!, TI!).

Diagnosis. Berberis gaoshanensis has leaves that are somewhat similar in shape to *B. dictyophylla*. However, unlike *B. dictyophylla*, its leaves are not abaxially pruinose, nor does it have pruinose mature stems. Like *B. dictyophylla*, it is mainly single-flowered on short pedicels, but its flower structure and berry shape and color are different.

Shrubs, deciduous, to 1.5 m tall; mature stems reddish brown, sulcate; spines 3-fid, pale, solitary toward apex of stems, brownish or orangish yellow, (0.5–)0.8–1.5(–2) cm, abaxially not sulcate. Petiole almost absent; leaf blade abaxially pale green, papillose, adaxially mid-green, oblanceolate, narrowly obovate, or narrowly elliptic, (1–)1.5–2 × 0.3–0.5 cm, papery, midvein raised abaxially, lateral veins indistinct on both surfaces, base attenuate or cuneate, margin entire, sometimes spinulose with 3 to 6 teeth on each side on young stems, apex subacute or acute, rarely obtuse, mucronate. Inflorescence 1-flowered, rarely paired; pedicel 3–5(–10) mm. Sepals in 2 whorls; outer sepals oblong-elliptic, with orange tinge, 4.5 × 3 mm; inner sepals broadly obovate, 5.5–6 × 4.5–5 mm; petals obovate or orbicular-obovate, 4–5 × 3–4.5 mm, base distinctly clawed, glands separate, elliptic or obovate, 0.75–1 mm, apex entire. Stamens 2.5–3 mm; anther connective extended, obtuse. Ovules 4. Berry bright red, ellipsoid, obovoid-ellipsoid, or oblong, 8–10 × 4–6 mm; style persistent, short.

Phenology. Berberis gaoshanensis has been collected in flower and fruit in August. Its phenology is otherwise unknown.

Distribution and habitat. Berberis gaoshanensis is known from Daocheng, Derong, and Xiangcheng Xian in west Sichuan. It has been found on glacier-strewn, granite boulder fields, gravelly slopes, and by roadsides at 3990–4602 m.

Etymology. Berberis gaoshanensis is derived from the Chinese for high mountains, 高山—gaoshan.

IUCN Red List category. Berberis gaoshanensis is assessed as DD or Data Deficient, according to IUCN (2001) criteria.

Berberis gaoshanensis was the only *Berberis* species found on our NGS visit to the desolate boulder-strewn Haizi Shan and on a revisit in 2018. Outside Tibet, few other *Berberis* specimens in China have been collected at elevations as high as 4600 m.

Selected specimens.

W Sichuan. Daocheng Xian: NNW of Sandui Xiang and N of Bengpugunba on hwy. S217 to Haizi Shan and Litang,

29.2275°N, 100.089722°E, 3990 m, 26 Aug. 2013, *D. E. Boufford, J. F. Harber & Q. Wang 43270* (A, CAS, E, K, KUN, PE, TI); Haizi Shan, NNE of Sandui Xiang and N of Xiaohekou Daoban on hwy. S217 to Litang, 29.513889°N, 100.246667°E, 4430 m, 27 Aug. 2013, *D. E. Boufford, J. F. Harber & Q. Wang 43281* (A, BM, CAS, E, HAST, IBSC, K, KUN, MO, PE, TI); same details, but 27 Aug. 2018, *D. E. Boufford, B. Bartholomew, J. L. Guo & J. F. Harber 44987* (A, CAS, E, K, KUN, PE). **Derong Xian:** Gajingxue Shan, NNW of city of Derong, 28.816389°N, 99.215833°E, 4200–4400 m, 19 July 2004, *D. E. Boufford et al. 30857* (A 00279961). **Xiangcheng Xian:** Disi Qu, E side of Rizhao Shan, 4300 m, 12 Aug. 1981, *Qinghai-Xizang Team 4685* (CDBI CDBI0027304, HITBC 003573, KUN 0178927, PE 01037791); along province rd. S217 N of Maxiong Gou on W flank of Kuluke (Wuming) Shan, 29.14°N, 100.0386°E, 4600–4610 m, 28 Aug, 2018, *D. E. Boufford, B. Bartholomew, J. L. Guo & J. F. Harber 44988* (A, CAS, E, KUN).

154. Berberis gilgiana Fedde, Bot. Jahrb. Syst. 36 (Beibl. 82): 43. 1905. TYPE: China. C Shaanxi: "In-kia-po (Lao y san)," May 1899, *G. Giraldi 2307* (lectotype, designated by Ahrendt [1961: 197], K K000077385!; isolectotypes, B†, FI image!).

Berberis pubescens Pamp., Nuovo Giorn. Bot. Ital. 17: 273. 1910. TYPE: China. N Hubei: [Yunxi Xian], "Niang-Niang monte," 1850 m, July 1907, *C. Silvestri 718*, (lectotype, designated here, FI image!, fragments and photograph of FI specimen A 00280172!; possible isolectotype, specimen with no collection details in same folder as lectotype FI image!).

?*Berberis oritrepha* C. K. Schneid., Oesterr. Bot. Z. 67: 293. 1918, syn. nov. TYPE: Cultivated. Golden Gate Park, San Francisco, U.S., 3 Oct. 1916, from *W. Purdom 592*, Taibai Shan, Shaanxi, 8 Feb. 1911 (lectotype, designated here, A 00038678!).

Berberis poiretii C. K. Schneid. var. *biseminalis* P. Y. Li, Acta Phytotax. Sin. 10: 212. 1965, syn. nov. TYPE: China. E Shaanxi: Weinan Shi, [Luonan Xian], Beishuan, Qinggangping, 1500 m, 29 June 1952, *T. P. Wang 15554* (lectotype, designated here, WUK 0062117 image!; isolectotypes, KUN 0179012!, PE 01032566!).

Shrubs, deciduous, to 1.5 m tall, mature stems reddish or purplish brown, terete; spines solitary or 3-fid, often absent toward apex of stems, semi-concolorous, 0.5–1.5 cm, terete, weak. Petiole almost absent or to 5 mm; leaf blade abaxially pale green, pubescent, adaxially mid-green, lanceolate, obovate-lanceolate, or narrowly elliptic, (1.5–)2.5–4(–6) × 0.3–1.5 cm, reticulations inconspicuous on both surfaces, base attenuate, margin entire, rarely spinulose with 1 to 4 teeth on each side, apex acute. Inflorescence a spikelike raceme, 10- to 25-flowered, 3–6 cm including pubescent peduncle 1–3 cm; bracts triangular-lanceolate, apex acute; pedicel 2–4 mm; bracteoles triangular. Sepals in 2 whorls, both obovate-orbicular; outer sepals much smaller than petals; inner sepals a little larger than petals. Ovules 2. Berry red, ellipsoid, 8–9 × ca. 6 mm; style not persistent.

Phenology. *Berberis gilgiana* has been collected in flower in April and May and in fruit between July and October.

Distribution and habitat. *Berberis gilgiana* is known from central Shaanxi, northwest Hubei, Chengkou Xian in Chongqing, and possibly northwest Henan and south Shanxi. It has been collected from thickets and road, stream, and valley sides at ca. 600–2250 m.

The protologue of *Berberis gilgiana* cited *Giraldi 2307, 2308,* and *2309,* but did not use the word type. These were specimens sent by FI to Berlin for identification, and it is clear that specimens of at least *2307* and *2309* were retained by Berlin since the Kew specimens are ex-Berlin. These retained specimens were destroyed in WWII. Schneider (1918: 285) designated *2307* as type, but did not cite a herbarium. Ahrendt (1961: 197) lectotypified the specimen at K. This, however, consists of only a flower stem and one complete leaf and a leaf fragment. The specimen at FI is a much better one.

The protologue of *Berberis pubescens* was in a work listing specimens of many genera collected by Silvestri, the introduction to which Pampanini (1910: 223) implied that specimens of each were at FI. Though the protologue only cited *Silvestri 718*, it seems unwise to treat the specimen of this at FI as a holotype since it cannot be ruled out that duplicates were sent elsewhere (Silvestri specimens were fairly widely distributed); hence it has been lectotypified.

The protologue of *Berberis oritrepha* cited only "Tai pei shan, 8 Februar 1911, W. Purdom (Samen no 592)" without citing any herbarium, stating, "So far from plants from Purdom's seed I have only seen fruiting branches from the Golden Gate Park in San Francisco." The specimen at A cited above was annotated as type by Schneider on 3 September 1916. The sheet includes line drawings of fruit stalks other than ones that are on the sheet. It is likely these were from material Schneider retained for his personal herbarium which was subsequently destroyed in WWII. Because of this, the specimen at A cannot be treated as a holotype; hence it has been lectotypified. The specimen at CAS (which is sterile) was almost certainly from the same plant and may have been collected at the same time, but this cannot be proved.

The protologue of *Berberis poiretii* var. *biseminalis* cited *Wang 15554* as "Typus," but did not designate a herbarium. The specimen at WUK has been chosen as the lectotype because WUK was where the author was based and the specimen is annotated as "Typus." This specimen is a poor one, but from the better specimens at KUN and PE, the taxon would appear to be *B. gilgiana*. This is also case with the only other specimen cited in the protologue, *P. Y. Li 89* from Taibai Shan (details below).

There are complications in relation to where most of the type specimens listed above and below were collected. As noted in the introduction, the transliteration of Chinese place names of specimens collected by Giraldi and Silvestri is eclectic. As a consequence, I have been unable to locate where *Giraldi 2307, 2308,* or *2309* were collected. The specimen sheet of the type of *Berberis pubescens* states, " Hupei Niang Niang Monte." This is shown on a map in Pampanini (1910: 279). Though the map does not indicate administrative divisions, this can be located in Yunxi Xian. The collection details on the sheet of the type of *B. poiretii* var. *biseminalis* record "Weinan, Beichuan," but Beichuan appears to be a little farther south in Luonan Xian in Shangluo Shi.

Though he had not seen the type, Schneider treated *Berberis pubescens* as a synonym of *B. gilgiana*, noting that Pampanini had overlooked the latter's publication and that his description of *B. pubescens* correlated well with that of *B. gilgiana*. However, both the type and only specimen cited of *B. pubescens* and the specimen in the same folder at FI have only immature fruit, so this cannot be conclusively proved, though its abaxially pubescent leaves makes this likely. The presence of *B. gilgiana* in Yunxi Xian is confirmed by *J. X. Yang & Y. M. Liang 2533* (details below). Schneider's synonymy was overlooked by Ahrendt (1961: 198) and Ying (2001: 175).

In the protologue to *Berberis oritrepha*, Schneider stated that the leaves of the type specimen had similarities with *B. purdomii*, while the fruit stalk was similar to that of *B. dictyoneura*. In fact, they resemble *B. gilgiana* which, as can be seen from the specimens cited below, is certainly found on the lower slopes of Taibaishan.

The specimens cited below from northwest Henan and south Shanxi appear to resemble *Berberis gilgiana*, though none have any floral material. If they are not this taxon, then they are of an undescribed species. *H. Smith 6026* is the only *Berberis* collection of any description located from Zhongtiao Shan.

Selected specimens.

Syntypes. Berberis gilgiana. Shaanxi: "Kan-y san, Cantena del Lao-y san," 2 May 1899, *Giraldi 2308* (FI); Shaanxi: "Alle falde del Monte Quan-tou-san," 5 May 1899, *Giraldi 2309* (FI, K fragm.).

Other specimens.
Chongqing (Sichuan). Chengkou Xian: Beiping Zhen, Shentian Cun, 2000 m, 1 July 2004, *Y. S. Chen et al. 608* (WUK 0488459–61).

NW Henan. Lingbao Xian: 12 June 1935, *K. M. Liou L.4264* (NKU NK003805, PE 01036949–50, WUK 0004121); Funiu Shan, Xiaoganjia Kou, 15 Aug. 1958, *J. Q. Fu 89* (KUN

0176741, NAS NAS00314168); Funiu Shan, S flank, near Laoya Kou, 1240 m, 22 Aug. 1958, *J. Q. Fu 278* (IBSC 0092266, NAS NAS00314123); Laoyacha Shan, 750 m, 29 July 2005, *M. Liu et al. H100091* (PE 01840126). **Luoning Xian:** Quanbao Shan, 1400 m, 15 July 2005, *M. Liu et al. H30180* (PE 01840125, TAIF 332219). **Lushi Xian:** Hsiungerhling, 650 m, 25 June 1935, *L. M. Liou L.4594* (PE 01036962, WUK 0004123); 1000 m, 29 June 1935, *K. M. Liou L.4694* (PE 01036960, WUK 0004122); 600 m, 3 Aug. 1935, *K. M. Liou L.4887* (PE 01036958, WUK 0004120); Panhe, Sanguanmiao, 1100 m, 1 Aug. 1959, *Henan Team 34362* (PE 01036956–57); en route from Huantong Si to Shiziping, 1140 m, 17 Oct. 1958, *J. Q. Fu 2424* (KUN 0176678, NAS NAS00314118).

N & NC Hubei. Baokang Xian: Tongsheng Qu, Longping, Huanglianpeng, 700 m, 13 Aug. 1974, *R. H. Huang 3038* (HIB 0129536, 0129605). **Suizhou Shi:** Dahong Shan, Bailong Chi, 800 m, 21 June 1974, *R. H. Huang 2894* (HIB 0129318). **Yunxi Xian:** Sunjia Gongshe, Tianfeng Shan, 700 m, 5 May 1964, *J. X. Yang & Y. M. Liang 2533* (PEM 0002332). **Zhongxiang Xian:** Kedian, Heihu Miao, 1000 m, 24 Apr. 1973, *Q. H. Liu 139* (HIB 0129321).

C & E Shaanxi. Huayin Xian: Huashan, 1000–1500 m, 31 Oct. 1924, *J. Hers 3096* (A 00280217); Wulihuan to Chengkoping (Qingkeping), 30 Apr. 1937, *W. Y. Hsia & C. H. Wang 64* (KUN 0178806–07, PE 01032783, WUK 0053829); Shadyuantung, 10 June 1939, *T. N. Liou 10752* (PE 01032781, WUK 0051492); Chingkehping (Qingkeping), 9 Sep. 1939, *T. N. Liou 10672* (PE 01032780, WUK 0051491); SW of Qingkeping, 1200 m, 11 May 1961, *Botany Dept. 212* (KUN 0178251, WUK 00483128); N flank of South Peak, 2100 m, 13 May 1961, *Botany Dept. 295* (KUN 0178252, WUK 00483126); Yuwang Gongshe, Yuwang Dadui, Huangni Gou, 5 June 1964, *J. X. Yang & Y. M. Liang 2848* (PE 01032784, WUK 00227980). **Mei Xian:** Taibai Shan, Hoaping Si, 19 June 1955, *P. Y. Li 89* (WUK 0166877); Tangyu Gou, 20 Apr. 1957, *Z. B. Wang 17647* (KUN 0178688, WUK 0090925, 0170638); Tangyu Gou, 1000 m, 25 Apr. 1957, *Z. B. Wang 17744* (KUN 0178253, WUK 0090993, 0170534); Tangyu, Xiao Taibai Si, 11 Oct. 1958, *X. M. Zhang 711* (IBK IBK00013107, IBSC 0092764, KUN 0176239, 0178701, NAS NAS00314029, WUK 00103608); Tangyu, 750 m, 15 June 1988, *Z. H. Wu 88-529* (FI, PE 01839987); Tangyu Zhen, Taibai Shan, 1100 m, 3 Sep. 2006, *H. N. Tan et al. 19297* (PE 01037589–90). **Ningshan Xian:** Guankou new rd., 2 June 1957, *J. Q. Xing 4291* (IBSC 0092779, NAS NAS00314015, WUK 0125098). **Shang Xian (now Shangzhou Qu):** Jieshuijin Gou, 1300 m, 13 Aug. 1952, *Z. B. Wang 16104* (PE 01032762, WUK 0062040). **Shangnang Xian:** Kai He, 800 m, 17 Oct. 1958, *B. Z. Guo 4400* (WUK 00112382). **Shanyang Xian:** Yugang Gongshe, Yugang He Dadui, Huangni Gou, 1100 m, 5 June 1964, *J. X. Yang & Y. M. Liang 2848* (WUK 00227980). **Zhen'an Xian:** Heiyao Gou, Linchang, 1250 m, 4 June 1973, *X. X. Hou 782* (IBSC 0091588). **Zhouzhi Xian:** Taibai Shan, 1910 *W. Purdom 6* (A 00280219, K, US 00945894), *W. Purdom 8* (A 00280218); Houzhingzhi, 1400 m, 24 July 1999, *Zhu et al. 1579* (KUN 0179272); Wutaiwan, 2250 m, 4 Oct. 1959, *Qingling Team 1107* (PE 01387615).

Cultivated specimens.
Golden Gate Park, San Francisco, U.S., 11 Oct. 1916, from *W. Purdom 592* (CAS).

Possible specimens.
S Shanxi. Yün-cheng (Yuncheng Shi): Chuntiaoshan (Zongtiao Shan), 1000 m, 2 July 1924, *H. Smith 6026* (A

00280206, MO 4367265, PE 01037011, UPS BOT:V-040891).

155. Berberis gilungensis T. S. Ying, Fl. Xizang. 2: 134. 1985. TYPE: China. S Xizang (Tibet): Gyirong (Jilong) Xian, Jilong Zhen, 3200 m, 5 June 1972, *Xizang Exped. Chin. Herb. Medic. 135* (holotype, PE 00935223!; isotypes, HNWP 29138 image!, PE 00935224!).

Shrubs, deciduous, to 2 m tall; mature stems reddish brown or reddish purple, subterete, not verruculose; spines solitary or 3-fid, 0.2–2 cm, mostly terete. Petiole almost absent; leaf blade abaxially grayish green, adaxially dark green, obovate to obovate-elliptic, 1.5–5 × 1–2.2 cm, papery, midvein raised abaxially, lateral veins and reticulation largely inconspicuous abaxially (more conspicuous when dry), conspicuous adaxially, base attenuate, margin entire, rarely spinulose with 1 to 5 teeth on either side, apex obtuse. Inflorescence a sub-umbel, 4- or 5-flowered, 2–3.5 cm overall; pedicel 5–9 mm; bracteoles oblong, ca. 3.2 × 1.2 mm. Sepals in 2 whorls; outer sepals oblong-elliptic, ca. 5.5 × 3 mm; inner sepals oblong-elliptic, ca. 8 × 6 mm; petals obovate, ca. 5.2 × 4 mm, glands separate, apex entire. Stamens ca. 3 mm; anther connective slightly extended. Ovules 4. Berry red, oblong, 10–12 × 4–6 mm; style persistent, short.

Phenology. *Berberis gilungensis* has been collected in flower and in fruit in August.

Distribution and habitat. *Berberis gilungensis* is known from Gyirong (Jilong) Xian in south Xizang. It has been collected in forests and forest margins at 3200 and 3400 m.

Selected specimens.
S Xizang (Tibet). Gyirong Xian: Rema Gou, 3400 m, 27 June 1975, *Integrated Research Team 75-399* (PE 00935225); Yatung Canyon (Jilung Canyon), 3100 m, 13 Aug. 2012, *Phylogeny of Chinese Land Plants, Academia Sinica Xizang Exped. 2012 1041* (E E00662307).

156. Berberis glabramea (Ahrendt) Harber, comb. et stat. nov. Basionym: *Berberis minutiflora* C. K. Schneid. var. *glabramea* Ahrendt, J. Linn. Soc., Bot. 57: 152. 1961. TYPE: China. SE Xizang (Tibet): [Gyaca (Jiacha) Xian], Lhapso Dzong, 29.116667°N, 92.533333°E, ca. 3500 m, 5 Apr. 1947, *F. Ludlow, G. Sherriff & H. H. Elliot 12437* (holotype, BM BM000559600!; isotype, E E00623202!).

Shrubs, deciduous, to 3.5 m tall; mature stems pale reddish yellow, sulcate; spines 3-fid, concolorous, 0.6– 1.5 cm, terete or abaxially flat. Petiole almost absent; leaf blade abaxially pale green, adaxially green, narrowly obovate or oblanceolate, 0.6–1.2 × 0.2–0.4 cm, papery, midvein raised abaxially, reticulate veins inconspicuous on both surfaces, base cuneate, margin slightly revolute when dry, spinulose with 1 or 2 teeth on each side or entire, apex acute, mucronate. Inflorescence a fascicle, sub-fascicle, or sub-umbel, 1- to 4(to 6)-flowered; pedicel 2–5 mm, slender. Sepals in 2 whorls; outer sepals oblong-ovate, ca. 3 × 2 mm; inner sepals ovate-elliptic, ca. 4 × 3 mm; petals obovate-elliptic, ca. 3 × 2 mm, base broadly cuneate, slightly clawed, glands separate, apex emarginate. Stamens 2.5–3 mm; anther connective slightly extended, rounded. Ovules 2 to 4. Berry unknown.

Phenology. *Berberis glabramea* has been collected in flower in April and May; its fruiting season is unknown.

Distribution and habitat. *Berberis glabramea* is known from Gyaca (Jiacha) and Nang (Lang) Xian in southeast Xizang (Tibet). It has been found on steep mountain sides among *Rosa* and other shrubs, in thickets on shaded slopes, by streams, and in *Betula* forest at ca. 3050–3500 m.

It is unclear why Ahrendt determined that *F. Ludlow, G. Sherriff & H. H. Elliot 12437* was a variety of the 1-flowered dwarf species, *Berberis minutiflora*, endemic to northwest Yunnan and southwest Sichuan, given that it differs from that species by being much taller, having reddish yellow mature stems, larger leaves, being 1- to 6-flowered, and having a different flower structure.

Ahrendt (1941b: 60, 66) identified *Kingdon-Ward 5632* and *5633* as *Berberis virescens* var. *ignorata*, and *F. Ludlow, G. Sherriff & G. Taylor 4239* as *B. jaeschkaena* var. *bimbilaica*, while later (1961: 145, 160) determining the first two to be *B. ignorata* and identifying *F. Ludlow, G. Sherriff & H. H. Elliot 12425* as *B. humidoumbrosa*. However, all are *B. glabramea*.

Ying (2001: 86) treated *Berberis minutiflora* var. *glabramea* as a synonym of *B. minutiflora*.

It is worth noting here *F. Ludlow, G. Sherriff & H. H. Elliot 12414* (BM BM00794127, E E00351614), Tsangpo Valley, Kamchang, 29.083333°N, 93.5°E, 3050 m, 27 April 1947. Collected some 100 km to the east of the type of *Berberis glabramea*, this has very similar leaves and flowers with three or four ovules, but has bright red mature stems and is 3- to 7-fascicled. This was also misidentified by Ahrendt (1961: 145) as *B. ignorata* (treated here as a synonym of *B. virescens*) and may be of an undescribed species.

Selected specimens.
SE Xizang (Tibet). Gyaca (Jiacha) Xian: Tsangpo Valley, below Tsetang, ca. 3050–3350 m, 29 Apr. 1924, *F. Kingdon-Ward 5632* (K K000568153–54); same details, *F. Kingdon-Ward 5633* (E E00351616, K K000568152). **[Nang (Lang) Xian]:** Nang Zhong, 29.05°N, 93.166667°E, 3350 m, May 13 1938, *F. Ludlow, G. Sherriff & G. Taylor 4239* (BM BM000895070); Kongbo, Takpo, Nga La, 29.016667°N, 93.20°E, 3505 m, 30 Apr. 1947, *F. Ludlow, G. Sherriff & H. H. Elliot 12425* (BM BM000939714).

157. Berberis gongshanensis Harber, sp. nov. TYPE: China. NW Yunnan: Gongshan Xian, E side of Gaoligong Shan, W of Gongshan, on trail from Qiqi to Dongshao Fang & pass to Dulong Jiang valley, 3250 m, 27.713056°N, 98.491667°E, 17 July 2000, *H. Li, B. Bartholomew, P. Thomas, P. W. Fritsch, Z. L. Dao, Z. L. Wang & R. Li 12776* (holotype, E E00238635!; isotypes, CAS 00120137!, GH 00280003!, HAST 89937 image!, KUN 1410011 image!, MO 5754350!).

Diagnosis. *Berberis gongshanensis* has similar leaves to both *B. bracteata* and *B. multiserrata*, though unlike the latter, these are never entire. The mature stems are a different color from *B. bracteata*. They are the same color as *B. multiserrata* but are epruinose. Unlike *B. bracteata* and *B. multiserrata*, *B. gongshanensis* is consistently fascicled. Its flower structure is also different from *B. multiserrata* (the flower structure of *B. bracteata* is unknown, but it has only two ovules).

Shrubs, deciduous, to 1.2 m tall; mature stems dark purple or reddish purple, subterete; spines 1- to 3-fid, pale brownish yellow, 0.5–1 cm, abaxially sulcate, weak. Petiole almost absent or to 2 mm; leaf blade abaxially grayish green, papillose, adaxially dark green, obovate, (2–)2.7–6 × (1.2–)1.5–2.5 cm, thickly papery, midvein, lateral veins, and reticulation raised on both surfaces, base narrowly attenuate, margin dentate with 2 to 10 widely spaced coarse teeth on each side, apex rounded or obtuse, sometimes retuse. Inflorescence a fascicle, 4- to 9-flowered; pedicel 12–17 mm. Sepals in 3 whorls; outer sepals narrowly ovate, 5.5–6 × 2.5–3 mm, apex acute; median sepals elliptic, 6–6.5 × 4–4.5 mm, apex rounded; inner sepals oblong-elliptic, 6.5–7 × 4–4.5 mm, apex subacute; petals obovate, 5–5.5 × 3.5–4 mm, glands separate, obovate, base slightly clawed, apex emarginate. Stamens 3–3.5 mm; anther connective not extended, rounded. Ovules 6. Berry unknown.

Phenology. *Berberis gongshanensis* has been collected in flower in July. Its fruiting period is unknown.

Distribution and habitat. *Berberis gongshanensis* is known from Gongshan Xian and the Gongshan Xian–Weixi Xian border area in northwest Yunnan. It has been found in open mixed and evergreen woodland at 3250–3300 m.

IUCN Red List category. *Berberis gongshanensis* is assessed as DD or Data Deficient, according to IUCN (2001) criteria.

The MO specimen records there were eight duplicates of *Li et al. 12776*. Six are recorded above. I have been unable to establish the whereabouts of the other two.

Selected specimens.
NW Yunnan. Gongshan Xian/Weixi Xian border: Biluo Xue Shan, Dongpo Qu, E Guardhouse, 3300 m, 13 July 1981, *Beijing Hengduan Mt Botany Team 01448* (PE 01037117, 02046550-51).

158. Berberis gyaitangensis Harber, sp. nov. TYPE: Cultivated. Royal Botanic Garden Edinburgh, 15 June 2016, *P. Brownless 1049* grown from *Alpine Garden Society China Exped. (ACE) 2237*, China. NW Yunnan: Zhongdian (Xianggelila) Xian, above Napa Hai, 3960 m, 17 Oct. 1994 (holotype, E E00831857!; isotype, A!)

Diagnosis. *Berberis gyaitangensis* is somewhat similar to *B. polybotrys* but with leaves that are more broadly obovate, a flower stalk that is more varied, and with a different flower structure.

Shrubs, deciduous, to 1.5 m tall in cultivation; mature stems pale reddish brown, subterete; spines largely absent, when present 1- to 3-fid, pale yellowish brown, 0.8–3 cm, terete. Petiole almost absent, rarely to 15 mm; leaf blade abaxially pale gray-green, papillose, adaxially mid-green, obovate, 2–3(–4.2) × 0.8–2(–3) cm, papery, midvein raised abaxially, lateral veins and reticulation inconspicuous on both surfaces, base attenuate, margin entire, apex obtuse, rounded, or subacute. Inflorescence a sub-raceme, sub-umbel, or umbel, sometimes fascicled at base, 3- to 12-flowered, 1.5–6 cm overall; peduncle 0.5–1.5 cm; pedicel 4–8 mm, to 18 mm when from base. Sepals in 2 whorls; outer sepals oblong-ovate, 4 × 2 mm; inner sepals obovate or obovate-orbicular, 4.5 × 3.5–4 mm; petals broadly obovate or orbicular-obovate, 3 × 3 mm, base clawed, glands widely separated, elliptic or elliptic-obovate, ca. 0.5 mm, apex obtuse, entire or slightly emarginate. Stamens ca. 2.5 mm; anther connective slightly extended, obtuse. Pistil 2.75 mm; ovules 2. Berry red, oblong, 8 × 4 mm; style persistent; seeds 2.

Phenology. *Berberis gyaitangensis* has been collected in fruit in October. Its flowering season in the wild is unknown.

Distribution and habitat. *Berberis gyaitangensis* is known from only the type collection from a cultivated plant grown from seed from above Zhongdian (Xianggelila) Xian in northwest Yunnan. This was collected on a high ridge in open moorland at 3960 m.

Etymology. *Berberis gyaitangensis* is named after Gyaitang, the ancient Tibetan name for the Zhongdian plateau.

IUCN Red List category. *Berberis gyaitangensis* is assessed as DD or Data Deficient, according to IUCN (2001) criteria.

Unlike most other *Berberis* collections made by the U.K.'s Alpine Garden Society's 1994 expedition to northwest Yunnan, it appears that no herbarium specimen was collected of *ACE 2237*, only seeds.

Cultivated material:
Living cultivated plants.
Royal Botanic Garden Edinburgh, Howick Hall, Northumberland, Ness Botanic Garden, Cheshire, and Foster Clough, Mytholmroyd, West Yorkshire, U.K., all from *ACE 2237*.

159. Berberis gyalaica Ahrendt, Gard. Chron., ser. 3, 109: 101. 1941. TYPE: Cultivated. S.d., *s. coll. s.n.*, from plant grown at Exbury Gardens, Hampshire, U.K., apparently from seed from the same plant as *F. Kingdon-Ward 5962*, China. SE Xizang (Tibet): [Mainling (Milin) Xian], gorge of the Tsangpo, Gyala, 2740–3050 m, 19 July 1924 (lectotype, designated here, BM BM000559170!).

Berberis taylorii Ahrendt, J. Bot. 79(Suppl.): 71. 1942. TYPE: China. SE Xizang (Tibet): [Mainling Xian], Gongbo Province, Molo, Lilung Chu, 29.066667°N, 93.933333°E, 2980 m, 4 Oct. 1938, *F. Ludlow, G. Sherriff & H. H. Elliot 7163* (holotype, BM [3 sheets] BM00571166–68!).
Berberis gyalaica var. *maximiflora* Ahrendt, J. Linn. Soc., Bot. 57: 218. 1961. TYPE: China. SE Xizang (Tibet): [Mainling Xian], Gongbo, Temo La, 29.583333°N, 94.633333°E, 3350 m, 4 Oct. 1947, *F. Ludlow, G. Sherriff & H. H. Elliot 15826* (holotype, BM BM000571169!).
Berberis gyalaica var. *minuata* Ahrendt, J. Linn. Soc., Bot. 57: 218. 1961. TYPE: China. SE Xizang (Tibet): [Mainling Xian], Gongbo, Tsangpo Valley, Tse, 29.40°N, 94.366667°E, 2900 m, 10 Oct. 1947, *F. Ludlow, G. Sherriff & H. H. Elliot 13304* (holotype, BM BM000573790!).

Shrubs, deciduous, to 3 m tall; mature stems purplish brown, sulcate; spines 1- to 3-fid, pale yellow, 0.4–1 cm, terete, weak. Petiole almost absent, rarely to 4 mm; leaf blade abaxially grayish green, adaxially dark green, obovate or elliptic, 1.2–3.2 × 0.7–1.7 cm, papery, midvein raised abaxially, lateral veins and reticulation indistinct on both surfaces (distinct but inconspicuous when dry), base cuneate, margin entire, occasionally spinulose with 2 to 4 teeth on each side, apex acute or rounded. Inflorescence a panicle, 15- to 30(to 50)-flowered, 4.5–7.5(–16) cm overall; peduncle 1–2 cm; bracts 2–5 mm; pedicel 2–5 mm; bracteoles oblong, 2–3 × ca. 2 mm, apex acute. Sepals in 2 whorls; outer sepals oblong-obovate, 4–5 × 3–4.5 mm, apex

acute; inner sepals oblong-suborbicular, ca. 6 × 4–6 mm; petals obovate, ca. 4 × 3–4 mm, base clawed, glands widely separated, submarginal, apex incised. Stamens ca. 3 mm; anther connective not extended, truncate. Ovules 3 to 5. Berry purplish black, slightly pruinose, oblong-ovoid, 9–10 × 4–5 mm; style not persistent; seeds black.

Phenology. *Berberis gyalaica* has been collected in flower between July and September and in fruit in October.

Distribution and habitat. *Berberis gyalaica* is known from Bomê (Bomi), Mêdog (Motuo), Mainling (Milin), Bayi Qu (formerly Nyingchi [Linzhi] Xian), and possibly Zayü (Chayu) Xian in southeast Xizang (Tibet). It has been collected from thickets, river banks, and roadsides at 2030–3050 m.

Berberis gyalaica is one of three species from Xizang whose types, designated by Ahrendt, are cultivated specimens grown from seeds collected by Kingdon-Ward in the autumn of his 1924 expedition to the Tsangpo valley, the others being *B. johannis* and *B. temolaica*. In each of these cases, there are wild-collected specimens at K with the same number collected in the summer of 1924, the explanation being that it was Kingdon-Ward's practice on this expedition to collect herbarium specimens in the summer months, mark the plants, and return to collect seeds in the autumn.

As with *Berberis johannis* and *B. temolaica*, determining which cultivated specimen is the type of *B. gyalaica* is not a simple exercise. The protologue simply cites, "Gyala 1924, *Kingdon-Ward 5962*." Subsequently, Ahrendt (1941b: 78) expanded this, stating that *Kingdon-Ward 5962* had been collected on 19 July 1924, "But I can find no corresponding herbarium sheet. Consequently, the type is taken from a cultivated plant at Exbury from which Mr. Lionel de Rothschild kindly sent me copious material." This type, cited as in the "Herb. Ahrendt," almost certainly referred to more than one specimen. Later, Ahrendt (1961: 218) cited the type as "Cultivated; fl., fr 1939 (seed of *KW 5962*), at Exbury . . . (Type BM)." There are four cultivated specimens of *Kingdon-Ward 5962* at BM identified as ex-Exbury, and three at OXF and two at K similarly identified. Some or all may be ex Herb. Ahrendt. The specimen at BM dated July 1924 has been chosen as the lectotype.

Seeds from OXF and K specimens of *5962* are black versus the pale yellow given by Ahrendt in 1941 and 1961, as are the seeds of the only other specimen cited by Ahrendt (1961: 218) as *Berberis gyalaica*—*F. Ludlow, G. Sherriff & H. H. Elliot 15825* (details below)—and his description has been amended accordingly. This latter specimen was collected in the same place and on the same date as the type of *B. gyalaica* var. *maximiflora*, but at a slightly lower elevation.

In the protologue of *Berberis taylorii*, Ahrendt stated that the taxon had only one ovule and the fruit only one pale brown seed. However, fruit of the type have black seeds and evidence of up to four ovules. Ahrendt also cited as *B. taylori, F. Ludlow, G. Sherriff & G. Taylor 5180, 5821*, and *5686*. The first two are *B. gyalaica*. Ying (2001: 209) synonymized *B. taylorii* and all the varieties of *B. gyalaica* listed above.

It is possible that *F. Kingdon-Ward 10988* (BM), Zayü (Chayu) Xian, Rongtö Valley, north of Rima, 1220–2400 m, 1 December 1933, is *Berberis gyalaica*. The colors of the mature stems are consistent with the type, the leaves are very similar, and it has a paniculate inflorescence with violet-blue fruit with three seeds. The collector's notes describe it as "common in thickets all up the valley." If it is *B. gyalaica*, then it is a considerable distance from where it is otherwise recorded and at a lower elevation. Only specimens with floral material from the same area would confirm this identification or otherwise.

Selected specimens.
SE Xizang. Bomê (Bomi) Xian: Zhamu, 2700 m, 9 July 1965, *Y. T. Zhang & K. Y Lang 502* (PE 01037156); Tongmai, near military depot, 2030 m, 19 July 1965, *Y. T. Zhang & K. Y Lang 684* (PE 01037271–72); Tongmai, behind military depot, 2030 m, 21 July 1965, *Y. T. Zhang & K. Y Lang 792* (PE 01037269); Yuren Qu, Derong Gongshe, 3000 m, 15 Sep. 1975, *S. Q. Sun & S. G. Tang 0172* (PE 01037158); W of city of Bomi on hwy. 318 along N side of Palongzangbu, 2510 m, 31 July 2000, *D. E. Boufford et al. 29817* (A 00279639, HAST 115067, P P02797877). **Mainling (Milin) Xian:** gorge of the Tsangpo, Gyala, 2740–3050 m, 19 July 1924, *F. Kingdon-Ward 5962* (K K000568156); Lilung Chu, near Molo, 29.066667°N, 93.933333°E, 3050 m, 13 July 1938, *F. Ludlow, G. Sherriff & G. Taylor 5686* (BM BM000939696); Kongbo Province, Tsangpo Valley, Lamdo, 29.333333°N, 94.283333°E, 3050 m, 13 July 1938, *F. Ludlow, G. Sherriff & G. Taylor 5821* (BM BM000939700, E E00351621); Pome, Showa Dzong, 29.866667°N, 95.416667°E, 2285 m, 19 June 1947, *F. Ludlow, G. Sherriff & H. H. Elliot 13188* (BM BM000939683, E E00351619, G, P P02682338); Mainling Xian, 3000 m, 14 July 1972, *Tibetan Herbal Medicine Survey Team 3943* (HNWP 30633, 32989, PE 00049414, 00049414); Wolong Gou, 3100 m, 21 July 1972, *Tibetan Herbal Medicine Survey Team 3836* (HNWP 30622, 32889, PE 00049415, 01037287); near Pai Qu, 3100 m, 12 Sep. 1974, *Qinghai-Xizang Team 74-4659* (KUN 0178142, PE 01037280–81); Jiage Shan, Micun, 3200 m, 27 July 1975, *Qinghai-Xizang Supplementary collections 750868* (HNWP 0226278, PE 01037159–60). **Mêdog (Motuo) Xian:** Kongbo, Peru La, 29.5°N, 95°E, 3200 m, 8 July 1938, *F. Ludlow, G. Sherriff & G. Taylor 5180* (BM BM000939697). **Nyingchi (Linzhi) Xian (now Bayi Qu):** Nambu La, 29.983333°N, 94.31667°E, 3200 m, 9 July 1947, *F. Ludlow, G. Sherriff & H. H. Elliot 15351* (A 00279908, BM BM000939695, E E00351622) and *15351a* (BM BM000939698); Chab, left bank Gyanda Chu, 29.5°N, 94.183333°E, 2990 m, 29 July 1947, *F. Ludlow, G. Sherriff*

& *H. H. Elliot 14204* (A 00279909, BM BM000939699, E E00351620); Kongbo, Temo La, 3200 m, 4 Oct. 1947, *F. Ludlow, G. Sherriff & H. H. Elliot 15825* (BM BM000939706); Nixi, 3040 m, 29 July 1965, *Y. T. Zhang & K. Y. Lang 1107* (PE 01037291–92); Aixin, 3200 m, 25 July 1974, *s. coll. 2013* (PE 01037295); Mafeng Gou, 3100 m, 17 Aug. 1974, *s. coll. 2159* (PE 01037293); betw. Dongjiu & Tongmai, 2480 m, 3 Aug. 1975, *Qinghai-Xizang Supplementary Collections 751239* (HNWP 180607, KUN 0178770, 0179340, PE 01037296–97).

160. Berberis heishuiensis Harber, sp. nov. TYPE: China. NC Sichuan: Heishui Xian, W of the city of Heishui on provincial rd. S302, betw. Ganshiba Cun & Shashiduo Xiang, 32.089558°N, 102.785008°E, 2700 m, 8 May 2014, *D. E. Boufford, S. Cristoph, C. Davidson, Y. D. Gao & Q. Y. Xiang 43610* (holotype, PE image!; isotypes, A!, CDBI image!, E!, MO image!, TI image!).

Diagnosis. *Berberis heishuiensis* has leaves that are very similar in shape to those of *B. dasystachya* but smaller. Its inflorescence, however, is very different to the spike of *B. dasystachya*.

Shrubs, deciduous, to 2.5 m tall; mature stems very pale brownish yellow, sulcate; spines 3-fid, solitary or absent toward apex of stems, concolorous, (0.6–)1.2–1.5 cm, terete or abaxially slightly sulcate, weak. Petiole sometimes almost absent, but mostly to 11 mm; leaf blade abaxially pale gray-green, papillose, adaxially dull green, broadly obovate, broadly obovate-elliptic, or obovate-orbicular, (0.5–)0.8–1.4(–2) × (0.4–)0.6–1.2(–1.8) cm, papery, midvein slightly raised abaxially, slightly impressed adaxially, lateral veins and reticulation inconspicuous on both surfaces, base attenuate, sometimes narrowly so, margin entire, rarely spinulose with 1 to 6 teeth on each side, apex rounded, obtuse, or subacute, rarely mucronate. Inflorescence a fascicle, sub-fascicle, sub-umbel, or rarely a sub-raceme, 4- to 8-flowered, to 2.5 cm overall; pedicel 8–15 mm; bracteoles triangular, 1.5–2 × 0.75 mm. Sepals in 2 whorls; outer sepals narrowly ovate, apex acute, 2.5–2.75 × 1.25–1.75 mm; inner sepals obovate, 4–4.5 × 2.5–3 mm; petals narrowly obovate, 4 × 1.75 mm, base clawed, glands separate, ca. 0.5 mm, apex entire. Stamens ca. 2.5 mm; anther connective not extended, truncate. Pistil ca. 3 mm; ovules 2. Immature berry purplish red, ellipsoid, 7 × 4 mm; style not persistent.

Phenology. *Berberis heishuiensis* has been collected in flower in May and with immature fruit in July.

Distribution and habitat. *Berberis heishuiensis* is known from Heishui Xian in north-central Sichuan. It has been found on forest margins, stream sides, and mountain slopes at 2450–2700 m.

IUCN Red List category. *Berberis heishuiensis* is assessed as DD or Data Deficient, according to IUCN (2001) criteria.

Selected specimens.
NC Sichuan. Heishui Xian: Luhua Gu, Ru Gou, 2600 m, 24 July 1957, *X. Li & J. X. Zhou 73790* (IBSC 0092609, PE 01033025, SZ 0029063); [unrecorded], Linchang, 2450 m, 2 July 1959, *Nanshuibeidiao Exp.- S. Yang et al. 01522* (KUN 0178626, PE 01033031, SZ 00290900); Zhimulin, Haibishi Gou, 2600 m, 4 July 1982, *C. R. Sun 178* (IBSC 0091659–60).

161. Berberis hemsleyana Ahrendt, J. Linn. Soc., Bot. 57: 213. 1961. TYPE: China. SC Xizang (Tibet): [Dagzê (Dazi) Xian], Ganden, 25 mi. (40 km) E of Lhasa, 29.683333°N, 91.45°E, 4100 m, 22 Sep. 1943, *F. Ludlow & G. Sherriff 9948* (holotype, BM BM000571165!).

Berberis nullinervis T. S. Ying, Fl. Xizang. 2: 141. 1985, syn. nov. TYPE: China. SC Xizang (Tibet): Namling (Nanmulin) Xian, Rendui Qu, near Jiao Gongshe, 4200–4300 m, 11 Sep. 1975, *Qinghai-Xizang Complex Exp. 7535* (holotype, PE 00935177!; isotypes, HNWP 49915 image!, KUN [2 sheets] 0178306!, 0179353!, PE 00935178!).

Shrubs, deciduous, to 2 m tall; mature stems dark red or reddish purple, sulcate; spines 3-fid, pale brownish or orangish yellow, 0.8–2.5 cm, stout, abaxially sulcate. Petiole almost absent; leaf blade oblanceolate or very narrowly elliptic, 1–2.5 × 0.5–0.7 cm, abaxially pale gray-green, papillose, adaxially dull dark green, papery, midvein and lateral veins slightly raised abaxially, adaxially with flat midvein, lateral veins slightly impressed adaxially, reticulation inconspicuous abaxially, indistinct adaxially, base attenuate, margin entire, occasionally spinulose with 1 or 2 teeth on each side, apex acute, mucronate. Inflorescence a fascicle, subfascicle, or umbellate sub-raceme, 1- to 4(or 5)-flowered, 1–2.5 cm overall; pedicel 7–16 mm, stout. Sepals in 2 whorls; outer sepals ovate, ca. 3.5 × 2.5 mm; inner sepals obovate, ca. 5.5 × 3.5 mm; petals obovate, ca. 3.5 × 2 mm, base clawed, glands separate, apex entire. Stamens ca. 3 mm; anther connective extended, truncate. Ovules 2(or 3). Berry red, oblong, ca. 10 × 5 mm; style not persistent.

Phenology. *Berberis hemsleyana* has been collected in flower between April and June and in fruit in August and September.

Distribution and habitat. *Berberis hemsleyana* is known from a 350-km stretch of the Yarlung Zangbo valley from Lhazê (Lazi) Xian in the west to Sangri Xian in the east, and from its northern tributary, the Lhasa river, and from Gyangzê (Jiangzi) Xian. It has

been found near river banks, open grassy places, and rocky hillsides at ca. 3000–4300 m.

The protologue of *Berberis nullinervis* cited only the type and *Qinghai-Xizang Complex Exp. 6112* (details below). These specimens appear to be indistinguishable from *B. hemsleyana*. No other *Berberis* specimens of any description from Namling Xian have been located.

Selected specimens.
Xizang (Tibet). Gyangzê (Jiangzi Xian): betw. Gyangzê & Khangma, 4150 m, 20 May 1961, *Xizang Team 1441* (PE 00049423, 01032827). **Lhasa (Lasa) Shi:** "Lhasa and surrounding hills, also 40 mi. [64 km], north of Lhasa," to 4570 m, Apr.–June 1939, *H. E. Richardson 4* (BM BM000794133); Netang (Niedang), near Lhasa, 3505 m, 21 Apr. 1942, *F. Ludlow & G. Sherriff 8589* (BM BM000939710); hills N of Lhasa, 4115 m, 6 June 1943, *F. Ludlow & G. Sherriff 9556* (BM BM000794135, E E00351615); 3960 m, 17 May 1943, *F. Ludlow & G. Sherriff 9504* (BM BM000939710), 3750 m, 27 May 1975, *s. coll. 75-73* (PE 01032835); 3960 m, 5 June 1980, *J. D. A. Stainton 8167* (E E00351617). **Lhazê (Lazi) Xian:** betw. Jiadang & Pengcuolin, 24 Aug. 1953, *B. Q. Zhong 6119* (KUN 0178822, PE 01032836, 01032841). **Maizhokunggar (Mozhugongka) Xian:** 4100 m, 28 May 1960, *G. X. Fu 128* (PE 01032840). **Namling (Nanmulin) Xian:** Rendui Qu, near Jiao Gongshe, 4200–4300 m, 11 Sep. 1975, *Qinghai-Xizang Complex Exp. 6112* (PE 01037974, 01037977). **Qüxü (Qushui) Xian:** near Niedang, 15 May 1953, *B. Q. Zhong 5601a* (PE 01032839); Qüxü Xian, near ferry crossing, 3700 m, 25 May 1960, *G. X. Fu 87* (PE 01032826); Qüxü, 3550 m, 26 May 1986, *T. Naito el al. 1404* (PE 01840042, TUS). **Rinbung (Renbu) Xian:** Qianjigunba, 3860 m, 4 June 1961, *Xizang-Qinghai Team 1500* (PE 00049424). **[Sangri Xian]:** Tsetang, 29.25°N, 91.833333°E, 3505 m, 3 May 1938, *F. Ludlow, G. Sherriff & G. Taylor 4110* (BM). **Xigazê (Rikase) Shi:** Dazhuka, 4300 m, 24 Aug. 1963, *J. X. Yang 2305* (HNWP 006537, PE 01032832, WUK 00215482).

162. Berberis henryana C. K. Schneid., Bull. Herb. Boissier, sér. 2, 5: 664. 1905. TYPE: China. W Hubei: Hsingshau (Xingshan), betw. 1885 & 1888, *A. Henry 5470B*, (holotype, G ex Herb. Barbey-Boissier 0096213!; isotypes, K K00077389!, LE!).

?*Berberis feddeana* C. K. Schneid., Bull. Herb. Boissier, sér. 2, 5: 665. 1905, syn. nov. TYPE: China. SE Sichuan (Chongqing): Nanchuan, ca. 1894, *A. von Rosthorn 2044* (holotype, B†).

Shrubs, deciduous, to 3 m tall; mature stems reddish brown, subterete to sub-sulcate; spines 3-fid, solitary or absent toward apex of stems, pale yellowish brown, 1–3 cm, abaxially sulcate. Petiole 4–15 mm; leaf blade abaxially gray-green, adaxially dark green, elliptic or obovate-elliptic, 1.5–3(–6) × 0.8–1.8(–3) cm, papery, midvein raised abaxially, slightly impressed adaxially, lateral veins and reticulation conspicuous abaxially, inconspicuous adaxially, base cuneate, margin spinulose with 10 to 20 inconspicuous teeth on each side, apex rounded. Inflorescence a raceme, 10- to 20-flowered,

2–6 cm overall including peduncle 1–2 cm; bracts 1–1.5 mm; pedicel 5–10 mm; bracteoles lanceolate, 1–1.5 mm, apex acuminate. Sepals in 2 whorls; outer sepals oblong-obovate, 2.5–3.5 × 1.5–2 mm; inner sepals obovate, 5.5–6.5 × 4–4.5 mm; petals oblong-obovate, 5–6 × 4–5 mm, glands separate, apex incised. Stamens 3.5–4.5 mm; anther connective not extended, truncate. Ovules 2. Berry red, ellipsoid, ca. 9 × 6 mm; style persistent, short.

Phenology. *Berberis henryana* has been collected in flower in May and June and in fruit between July and September.

Distribution and habitat. *Berberis henryana* is known from Chongqing, northeast Guizhou, west Hubei, northeast Sichuan, and southeast Shaanxi. It has been found in thickets, roadsides, meadows, scrub, forests, and forest margins at ca. 1300–2500 m.

The protologue of *Berberis henryana* stated, "Typus in Herb. Barbey-Boissier." This was subsequently incorporated into the Geneva Herbarium. Schneider's later statements (1917: 441, 1918: 292) that the type was in the Vienna herbarium were clearly a memory lapse.

Schneider's protologue of *Berberis feddeana* cited only *von Rosthorn 2044* at B. This was originally at O and was sent with other von Rosthorn collections to Berlin for identification. However, it was never returned and was subsequently destroyed during WWII. The only herbarium besides O that holds von Rosthorn duplicates is GZU, but *von Rosthorn 2044* is not among them. It appears that, subsequently, Schneider (1908: 261–262, 1913: 367–368, 1917: 441–442, 1918: 289–290, 292) was uncertain about the characteristics and status of *B. feddeana*. In 1908, he stated that it seemed to him that it was identical to or hardly separable from *B. dielsiana*, whose type is from Shaanxi, and cited *Giraldi 2295* from Shaanxi and *Henry 4937* from Hubei (Patung) as possible specimens, while identifying *Giraldi 2303* and *2310* from Shaanxi, *von Rosthorn 2037* from Nanchuan, and *Henry 5470* from Hubei as *B. henryana*. At the same time, *Wilson 3151* from "western China" was cited as having an inflorescence similar to *B. feddeana*. In 1913, however, he identified this latter as *B. francisci-ferdinandi*, while in 1917, he restored *B. dielsiana* as a separate species and now identified the Giraldi specimens as being of that taxon rather than *B. feddeana*. He now listed *Wilson 2862* from northeast of Kangding as an example of *B. feddeana*. In 1918, stating currently he had no access to the type of *B. feddeana*, he was extremely cautious; now no specimens beyond the type were cited except for *Wilson 2862*, which he left as an open question as to whether it was

related to *B. feddeana* or to *B. henryana*. All of this was largely ignored by Ahrendt (1961: 184) who mistakenly reported *Wilson 2862* as the type of *B. feddeana* (a report followed by Ying [2001: 181]) while also citing *Wilson 2863* from Washan, Sichuan, and *Henry 4937* from Hubei as examples.

All of this needs to be unpicked. *Berberis dielsiana* is treated here as a synonym of *B. dasystachya* and, while I have not seen *Giraldi 2295*, both *Giraldi 2303* and *2310* (both FI) are also *B. dasystachya*. *Henry 5470* and *Henry 4937* are *B. henryana* (see selected specimens below). *Wilson (Veitch) 3151* (A 00279512–13) is indeed *B. francisci-ferdinandi* as it appears in *Wilson 2862* (A 00038744, 00306058, BM, K K000077396, US 00945903) and *Wilson 2863* (for further details, see under *B. francisci-ferdinandi*).

As with the type of *Berberis feddeana*, the other specimen from Nanchuan cited by Schneider, *von Rosthorn 2037*, is missing and presumably was also destroyed in WWII. There are, however, various deciduous *Berberis* specimens from Nanchuan at A, CDBI, IBSC, KUN, NAS, and PE. Where a specific location is given, these are all from Jinfo Shan, so it is likely this is where the type of *B. feddeana* was collected. These specimens have been variously identified as *B. francisci-ferdinandi*, *B. feddeana* (rarely), *B. henryana*, *B. mouillacana*, and *B. silva-taroucana*, but all appear to be of the same species. This would seem to be *B. henryana*. The evidence for this is not just the leaf shape and mature stem color, but also the flower structure. For flowers from the specimen of *J. H. Xiong & Z. L. Zhou 90823* at PE, cited below, characters can be reported as follows: bracteoles triangular-lanceolate, 1.5–2 × 0.8–1.2 mm; outer sepals oblong-ovate, 2.8–3.2 × 2.3–2.6 mm; inner sepals obovate, 3–3.5 × 3–3.5 mm; petals obovate, 3–3.2 × 1.8–2 mm, glands separate. Stamens 2.5–2.7 mm; anther connective not extended, truncate. Pistil 2–2.3 mm; ovules 2 (dissection made on my behalf by B. Adhikari). This differs from the account of the flowers of *B. henryana* used above (which is that given by Ahrendt [1961: 192] based on specimens from Hubei) only in having the inner sepals of a similar size to the outer sepals rather than being twice the size. Even if such a difference between flowers from Hubei and Nanchuan was generally in evidence, it would not in itself seem a sufficient basis even for a variety of *B. henryana*. I conclude from this that *B. feddeana* should be treated as a probable synonym. However, the issue would benefit from more study of specimens from the two areas.

No evidence has been found to support the report of *Berberis henryana* in Henan by B. Z. Ding et al. (1981: 492) or in Hunan by L. H. Liu (2000: 715). The report of *B. henryana* in Gansu by Ying (2001: 187) is likely to be referable to *B. dasystachya*. Reports of the species in Anhui and Zhejiang are referable to the very similar *B. anhweiensis*. Specimens from Emei Shan in Sichuan are referable to *B. emeishanensis*.

Selected specimens.
Chongqing (Sichuan). betw. 1885 & 1888, *A. Henry 5470D* (E E00612501, GH 00279567). **Chengkou Xian:** Yizigenbai, Muping, 1300 m, 19 Apr. 1958, *G. H. Yang 100129* (HNWP 34230, PE 01033001); Beiping Xiang, Jieliang, 2175 m, 19 July 2008, *Ba Shan Collection Team 2059* (PE 01861159). **Fengjie Xian:** Guanlong Qu, behind Hongyanwan, 2000 m, 19 June 1958, *Z. R. Zhang 25309* (PE 01033005); near Huanglianba, 1800 m, 30 May 1964, *Sichuan University–East Sichuan Botanical Investigation Team 108357* (IBSC 0092641, PE 01033006, SZ 00290656). **Jinfo Shan:** "Chinfu Shan," 1830–2135 m, 18 May 1928, *W. P. Fang 970* (A 00279566, E E00612497, 00612499, PE 01033332); betw. Fenhuang Si & Touwan, 1730 m, 17 May 1957, *J. H. Xiong & Z. L. Zhou 90823* (HIB 0129386, IBSC 0091736, KUN 0179228, PE 01032902); Gufedong, 1700 m, 20 May 1957, *G. F. Li 61546* (IBSC 0091902, KUN 0178451, NAS NAS00314116, PE 01032908); Fenhuang Si, Qincai Ba, 2060 m, 10 Aug. 1957, *J. H. Xiong & Z. L. Zhou 92538* (HIB 0129386, IBSC 009174, KUN 0179254, PE 01033125); betw. Shizikou & Qinyuping, 1800 m, 28 July 1978, *Phytogeographic Team 810* (CDBI CDBI0027336, PE 01033323, 01033326). **Kai Xian:** Xuebaoshan, Mayun Linchang, 31.64675°N, 108.741361°E, 2140 m, 27 July 2008, *Ba Shan Collection Team 2289* (PE 01861162); Xuebaoshan, 31.664611°N, 108.780028°E, 27 Aug. 2008, *Ba Shan Collection Team 2446* (PE 01861161). **Nanchuan Xian:** 2440–2740 m, 24 May 1928, *W. P. Fang 890* (A 00279560, E E00612498, 00612500, LBG 00064042). **Pengshui Xian:** Sanchahe, Baishui He Nature Reserve, 2000 m, 17 May 2007, *Z. B. Feng et al. 20070339* (PE 01032999–3000, WCSBG 57966). **Wushan Xian:** Liaowangtai, Liziping Linchang, 1900–2310 m, 4 June 1996, *Z. D. Chen et al. 960963* (MO 5062454, PE 01604646). **Wuxi Xian:** Xining Xiang, Mao'erbei, Changcao Shan, 2300 m, 19 Sep. 1958, *G. H. Yang 65441* (CDBI CDBI0028121, KUN 0179225, PE 01033004); Tongcheng, Baiguo Linchang, 2200 m, 21 June 1962, *B. C. Ni 00326* (CDBI CDBI0027157–8); Zhuan Ping, Baiguo Linchang, 1610–1710 m, May 29 1996, *Z. D. Chen et al. 960701* (MO 5062453, PE 01840068). **Youyang Xian:** Xiaoxian, Daban, Yingdao Gou, 1500 m, 24 Aug. 1984, *Z. Y. Liu 6882* (IMC 0053578, PE 01033518). **Yunyang Xian:** Nongba Zhen, Yunfeng Cun, 31.43°N, 108.76°E, 1705 m, 5 May 2008, *Three Gorges Plant Exped. 1084* (PE 01788891).
N & NE Guizhou. Jiangkou Xian: Kaitu He tributary, 1700 m, 3 June 1964, *Z. S. Zhang 402237* (HGAS 013646, PE 01033333–34). **Suiyang Xian:** 12 May 1989, *K. M. Lan SY-890024* (GZAC 0030125–26).
W Hubei. betw. 1885 & 1888, *A. Henry 5470* (GH 00038763). **Badong Xian:** Patung (Badong), betw. 1885 & 1888, *A. Henry 4937* (K); So-Patung, betw. 1885 & 1888, *A. Henry 5470A* (A 00274274, E E00612503, GH 00038762, K K000077388); Niudongwan, 2500 m, 24 July 1957, *M. X. Nie & Q. H. Li 1047* (HIB 0129379, IBSC 0092252, LBG 00064061, PE 01033305–07). **Changyang (Xian):** May 1900, *E. H. Wilson (Veitch) 645* (A 00279863, 00280223, E E00612502, K, P P02313106). **Hsing-shan Hsien (Xingshan Xian):** Wan tiao Shan, 5 June 1907, 2440 m, *E. H. Wilson 2864* (A 00280226). **Shennongjia Lin Qu:** Xingshan Xian (that part now in Shennongjia Lin Qu), Laojunshan, 1900 m, 28 May 1957, *Y. Liu 547* (HIB 0129236, PE 01033308–10); Honghe, 2240 m, 19 June 1976, *Hubei Shennongjia Botanic Exped. 10264* (HIB 0129366–677, PE

01033303–4); Hongping Linchang, 1530 m, 24 Aug. 1976, *Hubei Shennongjia Botanic Exped. 32685* (HIB 0129351–522, PE 01033311); Qiaodonggou canyon, W of the rd. betw. Jiuhuping Linchang & Bancang, 31.50°N, 110.50°E, 1900 m, 27 Aug. 1980, *Sino-Amer. Exped 373* (A 00280228, HIB 0129375, PE 00996114, UC UC1492057).
SE Shaanxi. Lan'gao Xian: Nangongshan, 2000 m, 12 June 2005, *Y. S. Chen et al. 2498* (WUK 0493969–72). **Pingli Xian:** Hualong Shan, 2500 m, 12 Aug. 1988, *G. Y. Xu 4460* (WUK 0474858–59). **Zhenping Xian:** Niutoudian, Hualong Shan, Taiping He, 2250 m, 12 Aug. 1991, *J. S. Ying et al. 599* (WUK 0478537).
NE Sichuan. Wanyuan Xian: Hua'ershan, Yezhu Cao, 1850 m, 3 June 1959, *X. Z. Wang 2272* (FUS 00014197, KUN 0177087, SWCTU 00014197).

163. Berberis heteropoda Schrenk, Enum. Pl. Nov. 1: 102. 1841. TYPE: China–Kazakhstan border area, "Norberge des Alatau," 19–24 June 1840, *A. G. Schrenk s.n.* (lectotype, designated here, sheet stamped "Herbarium Academiae Scientiarum Petropol" LE!; isolectotypes LE 2 sheets!, possible isolectotype OXF!).

Berberis sphaerocarpa Kar. & Kir., Bull. Soc. Imp. Naturalistes Moscou 14: 376. 1841. *Berberis heteropoda* var. *sphaerocarpa* (Kar. & Kir.) Ahrendt, J. Linn. Soc., Bot. 57: 227. 1961. TYPE: E Kazakhstan. Uldschar River, near Tarbagatai Mtns., 1840, *G. S. Karelin & I. P. Kirilov 61* (lectotype, designated by Gubanov et al. [1998: 23], MW MW0592463 image!).

Shrubs, deciduous, to 3 m tall; mature stems dark red or brownish red, shiny, subterete, not verruculose; spines 3-fid, solitary or absent toward apex of stems, red, 0.5–1 cm, abaxially flat to subterete. Petiole 3–10 mm; leaf blade abaxially slightly shiny, pale green, adaxially green, obovate-elliptic or elliptic-orbicular, 2–6 × 1–4 cm, thickly papery, midvein, lateral veins, and reticulation slightly raised abaxially, inconspicuous adaxially, base broadly cuneate, margin entire, rarely sub-spinulose with 1 to 5 indistinct teeth on each side, apex rounded. Inflorescence a fascicle, subfascicle, or umbellate raceme, 4- to 10-flowered, 1.5–3(–5) cm; bracts ovate-lanceolate, 1.5–3 mm; pedicel 9–17 mm. Sepals in 2 whorls; outer sepals elliptic, ca. 5 × 4 mm, apex rounded; inner sepals obovate, ca. 7 × 5 mm; petals obovate-spatulate, ca. 6 × 4 mm, base cuneate, glands separate, apex rounded, entire. Stamens ca. 4.5 mm; anther connective extended, apiculate. Ovules 4 to 6, stipitate. Berry black, slightly pruinose, oblong or globose, 11–12 × 9–10 mm; style persistent.

Phenology. In China *Berberis heteropoda* has been collected in flower in April and May and in fruit from June to October.

Distribution and habitat. In China *Berberis heteropoda* is known from north and northwest Xinjiang. It has been collected from mountain slopes, river valleys, and steppe lands at ca. 575–3200 m. It is also known from Kazakhstan and Kyrgyzstan.

The protologue of *Berberis heteropoda* gave no collection details beyond Alatau, June, though from elsewhere in Schrenk (1841), it is clear this is a reference to one or more Schrenk specimens collected in 1840. Fedtschenko (1937: 557) stated the type of *B. heteropoda* to be at LE, but gave no further details. There are three specimens at LE labelled as being collected 19–24 June 1840 at "Norberge von Alatau." Fedtschenko's statement has, therefore, been treated as a first-step lectotypification and is completed here by selecting the best of the specimens. The label on the Schrenk specimen at OXF, cited above as a possible isolectotype, states it to be from "Norberge von Alatau," but is undated.

Ahrendt (1961: 227) cited the type of *Berberis heteropoda* to be "Sinkiang: Dzungartia, *Schrenk s.n.*" at K. This was a reference to one or more of the four specimens at K, all ex. LE and labelled "Songarei Schrenk," but with no further collection details. These could be further duplicates of the lectotype at LE, but they are more likely to have been collected by Schrenk on a later expedition. However, because of the uncertainty about this, I have listed them together with the many similarly labelled specimens elsewhere as possible syntypes.

The protologue of *Berberis sphaerocarpa* cited a specimen collected "ad Fl. Uldschar circa montes Targatabai." No herbarium was cited for this, but *G. S. Karelin & I. P. Kirilov 61* at MW labelled as being collected in 1840 and "Soc. Imp. Nat. Cur. Mosqu." would seem to be almost certainly the specimen referred to. This was lectotypified by Gubanov et al. (1998: 23).

Karelin and Kirilov (1842: 12), in describing specimens of *Berberis heteropoda* they had collected in the Alatau mountains in 1841, treated their *Berberis sphaerocarpa* as a synonym. It is almost certain that these specimens were the widely distributed *G. S. Karelin & I. P. Kirilov 1176* labelled as *B. sphaerocarpa* and "In montibus Alatau frequentissima" and as collected in 1841. Ahrendt (who made no reference to Karelin and Kirilov's synonymy nor that subsequently made by Regel [1877: 226]) determined *G. S. Karelin & I. P. Kirilov 1176* at K to be the type of a stat. nov. *B. heteropoda* var. *sphaerocarpa* (Kar. & Kir.). However, in accordance with Art. 7.3 of the Shenzhen Code (Turland et al., 2018), the type of this taxon can only be *G. S. Karelin & I. P. Kirilov 61*.

Schneider (1905: 457) treated *Berberis heteropoda* var. *caerulea* (Regel, 1877: 227) as a synonym of *B. heteropoda*. Given that Regel's rationale for this variety was simply that it had blue-colored fruit, this synonymy was unsurprising. The protologue of *B. heteropoda* var.

caerulea, however, did not specify a type, stating only "Hab. in montibus alatavicis prope Wernoje (A. Regel, Kuschakewicz) et in Kokania (O. Fedtschenko)." Wernoje is in Kazakhstan and Kokania [Kokand] is in east Uzbekistan. I have not investigated the matter further.

Possible syntypes. *Berberis heteropoda.* "In saltibus mont. Tarbagatai. Jul. Aug. 1840," *s. coll. s.n.* (MANCH); [W Xinjiang], Songaria, s.d., *A. G. Shrenk s.n.* (BM BM001015555, G [3 sheets] 00226027–29, GH 00038807, H H1347839, K [4 sheets] 00395202, 00395205–07, M 0164238, MO, NY 00000040, OXF, PR 360871, W [2 sheets] 1889–0322972, 0015486).

Selected specimens.
N, NW & W Xinjiang. Akqi (Aheqi) Xian: Ta'erdibulake, 6 Sep. 1992, *B. Wang 92-1931* (XJA 00063917). **Aksu (Akesu) Xian:** Tagelake, Meikuang, 574 m, 18 June 1978, *Xinjiang Collection Team 574* (PE 01033959, XJBI 00036170). **Akto (Aketao) Xian:** Ayisu, Linchang, 2800 m, 22 June 1987, *S. B. Li et al. 10165* (PE 01352397). **Altay (Aletai) Shi:** bank of River Kran, 47.833333°N, 88.2°E, 23 May 2004, *Sino-Russian Altai Exped. SRAE2004044* (PE 01766782). **Baicheng Xian:** Kuangchan Gou, 29 July 1993, *C. Y. Yang 1968* (XJA 00063906). **Bole Shi:** 1700 m, 13 Sep. 1978, *NW Institute of Botany Xinjiang Exped. 3793* (PE 01033960, XJBI 00036171). **Burqin (Bu'erjin) Xian:** 26 Aug. 1996, *s. coll. s.n.* (XJA 00064083). **Emin Xian:** Shiyue Gongshe, Luerke He, Ershan, 2 May 1964, *s. coll. 011* (XJBI 00036141). **Fukang Xian:** Tianchi Qu, en route to Tianchi, 18 Sep. 1957, *K. J. Guan 4272* (PE 01033970, XJBI 00036239–40). **Fuyun Xian:** Ku'erte, 1100 m, 23 Aug. 1987, *F. Konta et al. 232* (PE 01196932). **Gongliu Xian:** Mohe Gongshe, Ku'erdun, 21 Sep. 1978, *West Xinjiang Botany Team 2607* (PE 01036872). **Habahe Xian:** Fanxiu Gongshe, 2 Dadui, 22 Sep. 1976, *s. coll. 11120* (XJBI 00036151). **Hoboksar (Hebukesai'er) Xian:** Awusiqiwenge, 31 July 1983, *C. Y. Yang & B. Wang B 83-1382* (XJA 00063816–18). **Huocheng Xian:** en route from Ertai Rd. Maintenance Station to Yining, 1300 m, 14 July 1974, *Z. W. Zhang et al. 3304* (XJBI 00036205). **Hutubi Xian:** Shihezi, *s. coll. 92340* (XJBI 00036255). **Jimsar (Jimusa'er) Xian:** Laolong Gou, 22 July 1976, *Y. L. Han & N. S. You 760476* (XJA 00063876–77). **Jinghe Xian:** Nanshan, Kusongtai, 11 July 1973, *s. coll. 730764* (XJA 00063835–36). **Kargilik (Yecheng) Xian:** Supikeya, 3100 m, 15 Aug. 1987, *Qinghai-Xizang Team 1080* (PE 01033988). **Luntai Xian:** Baxi, Daban He, 1 Sep. 1983, *s. coll. 06–101* (XJA 00063899). **Manas (Manasi) Xian:** Shan Qu Linchang, Changbu, 13 May 1991, *B. Wang 93-034* (XJA 00063838–40). **Nilka (Nileke) Xian:** timber inspection yard, 1280 m, 13 Apr. 1982, *L. M. Ke 4153* (XJBI 00036224). **Qapqal (Chabucha'er) Xian:** near Balian Shan, 1300 m, 29 Aug. 1978, *Xinjiang Investigation Team 3028* (PE 01036837). **Qitai Xian:** Linchang, 28 July 1978, *s. coll. Ch7870097* (XJA 0006383879–80). **Shawan Xian:** Jiang Gou coal mine, 1150 m, 2 Oct. 1956, *R. C. Chang 3698* (PE 01033936, 01033941). **Tacheng Shi:** Bei Shan, 5 June 1997, *X. M. Guo 97-020* (XJA 00063809–10). **Tekes (Tekesi) Xian:** Si Muchang, Tekesi He ferry crossing, 1200 m, 29 June 1959, *T. Y. Zhou et al. 650603* (IBSC 0092812, KUN 0178751, PE 01033948, XJBI 00036192, 10019969). **Toli (Tuoli) Xian:** 10 km E of Ha'ersu, 20 July 1996, *B. Wang 96-203* (XJA 00063820). **Ulugqat (Wuqia) Xian:** Bositanteilieke, 17 Aug. 1993, *B. Wang 93-1732* (XJA 00063920–21). **Uqturpan (Wushi) Xian:** Mengkuzi, 6 Aug. 1993, *B. Wang 93-957* (XJA 00063900). **Urumqi (Wulumuqi) Shi:** Baiyanggou, Nanshan, 25 May

1994, *S. Y. Liang 9425* (PE 01839988). **Wenquan Xian:** Mi'erjike, near Fangzhan, 9 Sep. 1978, *Xinjiang Investigation Team 3649* (PE 010368670, XJBI 00036172). **Wensu Xian:** Pochengzi, Kukanke He, 2100 m, 13 May 1978, *s. coll. 780056* (XJBI 00036169). **Xinyuan Xian:** Yeguolin, Gailiangchang, 1000 m, 18 Aug. 1968, *Northwest Institute of Botany Xinjiang Exp. 2589* (PE 01033947). **Yarkant (Shache) Xian:** Kalatuzi Quan Qu, 3200 m, 22 July 1987, *Qinghai Xizang Team 870661* (PE 01033987). **Yumin Xian:** Ku'er Sayi, 1700 m, 31 July 1972, *G. J. Liu 443* (XJBI 00036182). **Zhaosu Xian:** en route to Qapqal Xian, 2000 m, 28 Aug. 1978, *West Xinjiang Investigation Team 2999* (PE 01033968). KAZAKHSTAN–CHINA BORDER AREA. Alatau Mtns., 1841, *G. S. Karelin & I. P. Kirilov 1176* (BM 000939606, 001015550, BR, K K00395203–4, M 0164236, NY, OXF, P P00716590–91, 02313101, W 040040–42).

KAZAKHSTAN. Tian Shan, Transiliense (Zailijski) Range, above Alma-Alta, 1500 m, 15 Sep. 1963, *A. K. Skvorsov s.n.* (A).

KYRGYZSTAN. Tian Shan, Terskej Range, Alatau, Aksu valley, 28 Aug. 1931, *M. Petrov et al. 6463* (A, LE).

164. Berberis hubianensis Harber, sp. nov. TYPE: China. NW Yunnan: Zhongdian (Xianggelila) Xian, Bigu Tianchi, 27.525556°N, 99.6425°E, 3890 m, 31 Aug. 2013, *D. E. Boufford, J. F. Harber & X. H. Li 43336* (holotype, PE!; isotypes, A!, E!, KUN!).

Diagnosis. *Berberis hubianensis* is somewhat similar to *B. mekongensis*, but has narrower leaves and up to three seeds per fruit.

Shrubs, deciduous, to 1.5 m tall; mature stems yellow-gray, sulcate; spines 3-fid, pale brownish yellow, 1–2.5 cm, terete. Petiole almost absent; leaf blade abaxially pale green, densely papillose, adaxially dark green, slightly shiny, obovate, (2.5–)3–4.5 × 1.2–1.7 cm, papery, midvein raised abaxially, slightly impressed adaxially, lateral veins and reticulation inconspicuous abaxially in living plants, conspicuous when dry, conspicuous adaxially, base attenuate, spinose with 3 to 15 teeth on each side or entire particularly toward apex of stems, apex subacute, rarely obtuse. Inflorescence a loose raceme or sub-raceme, sometimes with a few fascicled flowers at base, sometimes a sub-fascicle or sub-umbel toward apex of stems, 7- to 12-flowered, to 4 cm overall; peduncle (when present) to 2 cm; pedicel 6–12 mm, to 25 mm when from base. Flowers ca. 4 mm diam. Sepals in 2 whorls; outer sepals oblong-ovate, 3 × 1.75 mm; inner sepals obovate-elliptic, 4 × 2.5–3 mm; petals broadly obovate or orbicular-obovate, 2.5–3 × 2–2.5 mm, base slightly clawed, glands widely separate, 0.5 mm, ovoid, apex incised. Stamens 2.25–2.5 mm; anther connective slightly extended, obtuse. Pistil 2.25 mm; ovules 2 or 3. Berry orange-red, oblong, 8–9 × 3–4 mm; style persistent.

Phenology. *Berberis hubianensis* has been collected in fruit in August and September. Its flowering period in the wild is unknown.

Distribution and habitat. *Berberis hubianensis* is known only from the Bigu Tianchi lake in Zhongdian (Xianggelila) Xian in northwest Yunnan. It has been found in open areas on the slopes surrounding the lake at ca. 3800–3890 m.

Etymology. *Berberis hubianensis* is derived from the Chinese for lakeside, 湖边—hubian.

IUCN Red List category. *Berberis hubianensis* is assessed as DD or Data Deficient, according to IUCN (2001) criteria.

The description of flowers used above is from my plant of *ACE 1207* (details below). This flowered between the end of June and the beginning of July 2014.

Seeds from the original *ACE 1207* collection (details below) were germinated in 2013 at the Royal Botanic Garden Edinburgh (P. Brownless of RBGE, pers. comm. 9 Sep. 2013) and should provide additional cultivated plants.

Selected specimens.
NW Yunnan. Zhongdian (Xianggelila) Xian: Xiao Zhongdian, Tianchi Lake, 27.631667°N, 99.648611°E, 3825 m, 23 Sep. 1994, *Alpine Garden Society China Exped. (ACE) 1207* (E E00039599, LIV, WSY WSY0057769).

Cultivated material:
Living cultivated plants.
Foster Clough, Mytholmroyd, West Yorkshire, U.K., from *ACE 1207* and from *B. & S. Wynn-Jones 7674*, Zhongdian, Tianchi Lake, 3800 m, 28 Sep. 2000.

165. Berberis humidoumbrosa Ahrendt, J. Bot. 80(Suppl.): 115. 1945, as "*humido-umbrosa.*" TYPE: China. SE Xizang (Tibet): [Lhünzê (Longzi) Xian], Chayul, Sanga Chöling, 28.55°N, 93°E, 3350–3650 m, 18 Sep. 1935, *F. Kingdon-Ward 12357* "A" (holotype, BM BM000564307!).

Berberis racemulosa T. S. Ying, Fl. Xizang. 2: 129. 1985, syn. nov. TYPE: China. SE Xizang (Tibet): Lhünzê (Longzi), Xian, near Jiayu, 3600 m, 1 July 1975, *Qinghai-Xizang Complex Exp. 750430* (holotype, PE 00935192!; isotypes, HNWP 51066 image!, KUN [2 sheets] 0179325–26!, PE 00935193!).

Shrubs, deciduous, to 2.4 m tall; mature stems pale yellowish or reddish brown, sub-sulcate; spines 3-fid, solitary toward apex of stems, semi-concolorous, 0.7–2.5 cm, terete or abaxially sub-sulcate. Petiole almost absent; leaf blade abaxially pale green, papillose, adaxially green, narrowly obovate or elliptic-obovate, 0.6–2(–2.5) × 0.4–0.8 cm, papery, midvein raised abaxially, lateral veins and reticulation inconspicuous on both surfaces, base attenuate, margin entire, apex acute, subacute, or obtuse. Inflorescence a raceme, 4-

to 9-flowered, 1–1.5 cm overall; pedicel 3–6 mm. Sepals in 2 whorls; outer sepals ovate, ca. 2.8 × 1.8 mm, apex obtuse; inner sepals obovate, ca. 3 × 2.5 mm; petals obovate, ca. 3.3 × 2.5 mm, glands separate, elliptic, apex entire. Stamens ca. 2.3 mm; anther connective extended, truncate. Ovules 2 to 4. Berry red, densely blue or violet pruinose, obovoid, 7–8 × 4.5–5 mm; style persistent, conspicuous; seeds 2 to 4.

Phenology. *Berberis humidoumbrosa* has been collected in flower in April and May and in fruit from July to September.

Distribution and habitat. *Berberis humidoumbrosa* is known from Lhünzê (Longzi) Xian in southeast Xizang. It has been found in ravines, shady places, and field and streamsides at ca. 3350–3650 m.

The sheet of *F. Kingdon-Ward 12357* has two specimens. The collector's notes state, "Fruit either red or blue-violet. These two bushes were growing side by side." Ahrendt's commentary in the protologue stated, "The difference in color of the fruit appears to be due to the varying density of the mauve bloom evident in both specimens of no. *12357*. I have taken as the type that with the more distinct bloom." It is easy to identify which of the two specimens Ahrendt is referring to here but, to clarify matters, this is designated here as *12357* "A" and the other specimen as *12357* "B."

The type of *Berberis racemulosa* was collected in the same area and at more or less the same elevation as *B. humidoumbrosa*. The collection notes of the latter report "common in ravines, or wherever there is water or shade," while those of the former note it was growing on a field side by a river beach. The two collections have leaves of the same size and shape and the same fruit stalk. Nevertheless, there are unresolved issues here relating to the flower structure. The protologue of *B. humidoumbrosa* gave a description which contextually appears to have been based on *F. Ludlow & G. Sherriff 1547* (details below) from the same area of the two type collections. This differs from that given in the protologue of *B. racemulosa*, though both agree that the number of ovules is two, but the source of Ying's description is not given. It could have come from the few flowers at the apex of the holotype (all the isotypes having only immature fruit), but it could be from one or both of two collections cited in the protologue from Lhozhag (Luozha) Xian—*Qinghai-Xizang Complex Exp. 6879* (PE 01033060, 01840005), between Gabo and Zazhong, 3400 m, 4 June 1975; and *Qinghai-Xizang Complex Exp. 7006* (PE 01033055–56, 01033059), 3200 m, Lakang, 5 June 1975—both of which have floral material. However, these gatherings, made some 160 km west of where the specimens of various speci-

mens of *Berberis humidoumbrosa* were collected, have larger obovate leaves and, from the evidence of an annotation to one of the sheets of *7006*, have five or six ovules. A further complication is that dissection of various fruits of one of the isotypes of *B. racemulosa* at KUN produced two to four seeds per fruit. Given all of this, the account of flowers given above follows Ahrendt, but the number of ovules is given as two to four.

The protologue of *Berberis humidoumbrosa* also cited *F. Ludlow & G. Sherriff 1325* and *T. T. Yu 10560* (E E00623042) from northwest Yunnan, but subsequently Ahrendt (1961: 161) identified *T. T. Yu 10560* as *B. dawoensis*. In fact, as noted under that species, it is neither. Subsequently, Ahrendt also (1961: 60) cited a number of other specimens from elsewhere in southeast Xizang as *B. humidoumbrosa*. None of these are this species.

Ahrendt (1945a: 116) published a *Berberis humidoumbrosa* var. *inornata* with the type from Zhongdian (Xianggelila) Xian, northwest Yunnan, and subsequently (1961: 160) a *B. humidoumbrosa* var. *dispersa* with a type from Cona Xian in southeast Xizang. Neither have any relation with *B. humidoumbrosa*. For the former, see under Taxa Incompletely Known below. The latter is treated here as a new species, *B. dispersa*.

Syntype. *Berberis humidoumbrosa.* SE Xizang, [Lhünzê (Longzi) Xian], Chayul, Sanga Chöling, 28.55°N, 93°E, 3350–3650 m, 18 Sep. 1935, *Kingdon-Ward 12357* "B" (BM 000564307).

Selected specimens.
SE Xizang. Lhünzê Xian (Longzi) Xian: Chayul Dzong, Loro Chu, 3500 m, 22 Apr. 1936, *F. Ludlow & G. Sherriff 1325* (BM 000939712, MO 1618677); Chayul Dzong (Jiayu), Chayul Chu, 28.3°N, 92.8°E, 3500 m, 5 May 1936, *F. Ludlow & G. Sherriff 1547* (BM 000939712); near Jiayu, 4000 m, 1 July 1975, *Qinghai-Xizang Complex Exp. 750420* (HNWP 51056, 96822, KUN 0179324, 0179332, PE 01033057–58).

166. Berberis hypericifolia T. S. Ying, Fl. Xizang. 2: 140. 1985. TYPE: China. S Xizang (Tibet): Yadong Xian, Tangapu Gou, 3400 m, 10 Aug. 1975, *Qinghai-Xizang Complex Exped. 75-945* (holotype, PE 00935142!).

Shrubs, deciduous, to 2 m tall; mature stems purplish red, shiny, subterete, sparsely black verruculose; spines solitary or absent at apex of stems, concolorous, 0.6–1 cm, terete, weak. Petiole 2–5 mm; leaf blade abaxially pale yellow-green, adaxially dark green, obovate, occasionally suborbicular, 0.7–2.5 × 0.4–1.7 cm, papery, midvein raised abaxially, slightly impressed adaxially, lateral veins and reticulation inconspicuous or indistinct on both surfaces, base cuneate, margin entire, apex rounded, rarely minutely mucronate. Inflorescence a fascicle, 3- to 6-flowered, otherwise unknown;

fruiting pedicel 8–14 mm. Berry red, oblong, ca. 10 × 5 mm; style not persistent; seeds 3.

Phenology. *Berberis hypericifolia* has been collected in fruit in August. Its flowering season is unknown.

Distribution and habitat. *Berberis hypericifolia* is known only from the type from Yadong Xian, south Xizang (Tibet). This was found in slash in a recently logged area at 3400 m.

167. Berberis integerrima Bunge, Index Seminum [Tartu], 6. 1843. TYPE: Tajikistan. Upper Sarafschan (Zeravshan) River, [betw. Pendshukend (Panjakent) & Jar-Kischlak], 6 Sep. 1841, *A. Lehmann 44* (lectotype, designated here, specimen annotated by Schneider on 3 Oct. 1906, LE!; isolectotypes, K K000395215!, LE! [2 sheets], P [2 sheets] P00716587!, P00580435!, W 0015482!).

Berberis nummularia Bunge var. *schrenkiana* C. K. Schneid., Bull. Herb. Boissier, sér. 2, 5: 460. 1905, syn. nov. TYPE: China. [W Xinjiang]: Songaria, *C. A. Mayer s.n.* (lectotype, designated by Ahrendt [1961:178], OXF missing, new lectotype designated here, Songaria, 10 Aug. 1843, *C. A. Mayer s.n.*, G G00418039!).
Berberis iliensis Popov, Ind. Sem. Hort. Bot. Almaat. Acad. Sci. URSS 3: 3–4. 1936, syn. nov. TYPE: none cited.

Shrubs, deciduous, to 2.5 m tall; mature stems shiny, reddish brown, angled; spines mostly solitary, sometimes 2- or 3-fid, brownish yellow, shiny, 1–4.75 cm, terete, stout. Petiole almost absent or to 10 mm; leaf blade green on both surfaces, obovate or oblong, sometimes oblanceolate especially on young shoots, 2–5 × 1.3–1.8 cm, thickly papery, midvein, lateral veins, and reticulation raised abaxially, inconspicuous adaxially, base cuneate, margin entire, apex acute or rounded. Inflorescence a raceme, rarely compound at base, 10- to 20(to 25)-flowered, to ca. 5 cm overall; peduncle short; pedicel 5–8 mm. Sepals in 2 whorls; outer sepals elliptic, ca. 2.8 × 1.8 mm; inner sepals obovate, 3.8–4 × ca. 2.5 mm; petals obovate, ca. 3.5 × 2 mm. Ovules 2 or 3. Berry red, sometimes slightly pruinose at first, obovoid-ellipsoid, 7–8 × 2–3 mm; style not persistent.

Phenology. In China *Berberis integerrima* has been collected in flower in May and in fruit between June and September.

Distribution and habitat. In China *Berberis integerrima* is known from the Ili River basin in northwest and the Tarim basin and Kashgar oasis in west Xinjiang. It has been found on stony mountain slopes, river beaches, in dry river valleys, and at oases at ca. 500–2100 m. It is also known from the Ili River area of the Almaty Province of southeast Kazakhstan, and the upper Zeravshan River valley of northeast Tajikistan. It is possible that it is found in other areas outside China (see discussion below).

The protologue of *Berberis integerrima* gave a sparse description, the only reference to any specimen being to "In saxosis montium Karatau prope Samarkand (A. Lehm.)." The species was published simultaneously with *B. nummularia*, identified as being from the same area, the latter being described as also having a racemose inflorescence, but being differentiated from *B. integerrima* largely on the basis of having obovate-orbicular leaves versus the obovate-oblong leaves of *B. integerrima*. Subsequently, Bunge (1847: 129–130) identified the specimens as being respectively *Lehman 44* and *Lehmann 45*. The habitat of the former was described as being "Frequent on the stony banks of the upper Sarafschan and on damp slopes of the Karatau mountains" and the collection date was given as 6 September 1841. These two locations are quite separate, the Karatau Mountains being in south Kazakhstan, while the Sarafschan River rises in Tajikistan and flows westward into Uzbekistan. From Helmerson (1852: 112), it is clear that, on 6 September 1841, Lehmann was in what is now Tajikistan travelling up the Sarafschan valley from Pendshukend (Panjakent) to Jar-Kischlak, Helmerson specifically noting that among the plants found on the stony valley floor on that date was "*Berberis vulgaris.*"

No herbarium was given for any specimen of the type of *Berberis integerrima* until Fedtschenko (1937: 558), who cited LE though without specifying that this was *Lehman 44*, only that it was from Zeravshan. Subsequently, Browicz and Zielinski (1975: 10) stated that *Lehmann 44* at LE was the type, but omitted to note that LE holds three specimens of this. The best of these has been chosen here as the lectotype. The isolectotypes at K and P, however, are much better specimens than any of these. There is a photograph of one of the LE isolectotypes at E (E00326964).

The protologue of *Berberis nummularia* var. *schrenkiana* did not use the word type and cited six gatherings without citing a herbarium for any of them except in a footnote drawing attention to one of a number of Schrenk specimens at B (specimens which will have been destroyed in WWII). Ahrendt (1961: 178) cited as type a specimen at OXF of one of the other gatherings cited by Schneider, *C. A. Mayer s.n.*, but this is missing. A C. A. Mayer specimen at G annotated by Schneider on 28 August 1905 as *B. nummularia* var. *schrenkiana* has therefore been chosen as the lectotype. Specimens from four of the other five gatherings cited in the protologue have been located, all of which are annotated as *B. nummularia* var. *schrenkiana* by Schneider. These are *A. G. Schrenk s.n.* at G, M, and W; *A. Kuschakewitz s.n.* at M; *A. P. Fedjenko* [Fedtschenko] *s.n.* at G; and *P. P. Semenow s.n.* at G and K.

All of the specimens of *Berberis nummularia* var. *schrenkiana* annotated by Schneider cited in the previous paragraph are ex LE; all were labelled by LE as *B. integerrima*, and all appear to be indistinguishable from *Lehman 44*. It is, therefore, difficult to understand why Schneider determined them to be a different new taxon, especially as his description differs little from that of Bunge's of *B. integerrima*, which Schneider (1905: 461) maintained as a separate species. The explanation may be that, at the time, Schneider appeared to believe there was a *Bunge 44* from Karatau which was different from a Lehmann specimen from Sarafschan (this being confirmed by an annotation dated 3 August 1903 to the specimen of *Lehmann 44* at W, which is annotated as being "Ad montes Karatau," and may actually be of a different collection from the lectotype). In any event, his justification for the specimens being a different variety of *B. nummularia* was that they had obovate-oblong or oblanceolate leaves versus the obovate-suborbicular leaves of *B. nummularia*, i.e., making the same distinction that Bunge had made between *B. integerrima* and *B. nummularia*.

Berberis iliensis was published by Popov as a species from east Tianshan, but with no specimens cited. Given that the protologue was in an "Index Seminum," it is possible that the source was a cultivated plant or, if wild-collected, perhaps no plant material was conserved. In either case, given Popov's name for the taxon, the likely geographical origin of *B. iliensis* is that part of the Ili River which is in Kazakhstan. Popov's main rationale for publishing it was that it was a different species from *B. nummularia* in that it had oblong or oblanceolate-spatulate leaves versus the suborbicular leaves of *B. nummularia*, i.e., he reproduced the differences made by Bunge between *B. integerrima* (which he did not mention) and *B. nummularia*, and those made by Schneider in relation to the differences between *B. nummularia* var. *schrenkiana* (also not mentioned by Popov) and *B. nummularia*. I, therefore, deduce that what Popov was describing was *B. integerrima*. In these circumstances, it seems pointless to designate a neotype of *B. iliensis*. The similarities of the description of *B. iliensis* and *B. nummularia* var. *schrenkiana* were noticed by Ahrendt (1961: 178), who treated the latter as a synonym of the former. This was followed by Ying (2001: 182) who did not notice *B. integerrima*.

Berberis iliensis in Kazakhstan and China is on the current IUCN Red List of Threatened Species (Participants of the FFI/IUCN SSC Central Asian regional tree Red Listing workshop, 2007a).

Delavay gatherings from northwest Yunnan identified as *Berberis integerrima* by Franchet (1887: 386) are *B. jamesiana*. For further details, see discussion under that species.

Berberis integerrima has been variously reported over an extensive geographical range including Afghanistan, Pakistan, Turkmenistan, east Anatolia, the Caucasus, Iraq, and Iran. These reports include Browicz and Zielinski (1975: 9–10) who listed no fewer than 17 species and varieties as synonyms. This was reduced to four synonyms by Browicz (1988: 11–12), but now included *B. iberica* (Steven, 1827) which, if correct (and in my opinion it is not; see discussion under *Berberis chinensis*), would have important nomenclatural consequences since *B. iberica* was published first. In my view, such an extensive geographical range would seem inherently unlikely, but I have not pursued this matter further and the issues it raises can probably only be resolved through the much needed revision of *Berberis* species from central and southwest Asia.

Syntypes. Berberis nummularia var. *schrenkiana*. China. Xinjiang: Tianshan, Songaria, s.d., *A. G. Schrenk s.n.* (B†, G G00226031, K K000395211, M 0164255, MPU [2 sheets] 017787–88, OXF [2 sheets], P P00580430, P00580434, PR [2 sheets] 72610–11, PRC [2 sheets], W [3 sheets] 1889–0322973, 0015477, 1944–629). Turkestan. S.d., *A. P. Fedjenko* [Fedtschenko] *s.n.* (G G00343559); Tianshan, Ili River, s.d., *A. Kuschakewitz s.n.* (M M0164256); Songaria, Tianshan, 1887, *N. M. Przewalski s.n.* (not found); Ilitcol, s.d., *P. P. Semenow s.n.* (G G00343556, K K000395210).

Selected specimens.
NW & W Xinjiang. "Turkestanica Chinensis," 28 Aug. 1876, *Przewalski s.n.* (LE); 5 July 1885, *N. M. Przewalski s.n.* (LE); Abad, "Ausgang des südlichen Musart Tales, 30 verst," 1903, *G. Merzbacher 443* (M). **Akqi (Aheqi) Xian:** outskirts of Akqi town, 1950 m, 30 Aug. 1975, *C. Y. Yang 750937* (XJA 00063998–64001). **Aksu (Akesu) Shi:** en route to Tagelake, 1960 m, 18 June 1976, *Xinjiang Collection Team 519* (PE 01033236, XJBI 00036274). **Akto (Aketao) Xian:** Wuyitake Linchang, 18 Aug. 1975, *C. Y. Yang 750769* (XJA 00063997). **Gongliu Xian:** Mo He Kou, 770 m, 5 Sep. 1989, *L. X. Gong 271* (PE 01033242–43). **Hejing Xian:** Baluntai, 9 Aug. 1958, *Xinjiang Team 9024* (KUN 0178837, PE 01033232). **Hoxuci (Heshuo) Xian:** Qingshui He, 28 July 1976, *Y. L. Han & N. S. You 760524* (XJA 0063980–82). **Huocheng Xian:** Santai Linchang, 30 July 1997, *C. Y. Yang 163* (XJA 00063965). **Jinghe Xian:** 28 July 1997, *C. Y. Yang 065* (XJA 00063964). **Kalpin (Keping) Xian:** Jige Daike, Yushuikeng, 2000 m, 9 Sep. 1958, *Xinjiang Comprehensive Survey Team 7472* (PE 01033225). **Kuqa (Kuche) Xian:** Kezi'er Qinanfodong, 1080 m, 6 Sep. 1987, *K. Y. Lang et al. 424* (PE 01196940–41). **Nilka (Nileke) Xian:** to Kalasu, 30 Aug. 1957, *K. J. Guan 3888* (PE 01033230). **Pishan Xian:** Yujinka'ersu, 27 July 1983, *s. coll. 102* (XJA 00064005). **Qapqal (Chabucha'er) Xian:** near Touhu, 530 m, 30 Aug. 1978, *Northwest Institute of Botany Xinjiang Exp. 3405* (HNWP 63284, PE 01033231). **Qira (Cele) Xin:** Qiaha, Maku, 15 Aug. 1989, *s. coll. 265* (XJA 00064011–12). **Shihezi Shi:** Sep. 1985, *s. coll. s.n.* (XJA 00063849). **Shule Xian:** Yapu Quan, 1250 m, 9 June 1959, *K. Xin 0081* (PE 01033238, XJBI 00036278). **Tekes (Tekesi) Xian:** near Linchang, 8 Aug 1974, *S. Q. Huang et al. T7400636* (XJA 000638). **Ulugqat (Wuqia) Xian:** 2100 m, 20 June 1987, *B. S. Li et al.*

10100 (PE 01352400–02). **Xinyuan Xian:** Nongsishijing Sheep Farm en route to Nilka (Nileke) Xian, 1550 m, 24 Aug. 1957, *Xinjiang Comprehensive Survey Team 3784* (PE 01033228, XJBI 00036282). **Yanqi Xian:** 1500 m, 10 Aug. 2007, *T. J. Mai 07-010* (XJA 00063978). **Yengisar (Yingjisha) Xian:** Linchang, 20 Aug. 1981, *H. Tang 英0068* (XJA 00063990). **Yining Shi:** 630 m, 22 May 1959, *Xinjiang Comprehensive Survey Team 10446* (KUN 0179016, PE 01033226–27, XJBI 10020063, 00036286). **Yining Xian:** en route to Gongliu Xian, bank of Ili River, near Yemadu, 700 m, 18 July 1974, *Y. R. Lin 74260* (IBSC 0092806, PE 01033241).

KAZAKHSTAN. Ili River valley, "20 km supra pagum Ilijsk," 6 June 1955, *4113 V. Goloskokov s.n.* (LE, US 01121945, W 1958–17413); left bank of Ili River, near train station, 6 Sep. 1959, *L. P. Velikanov s.n.* (LE, PE 01806106, W 1963–14329); left bank of Ili River, Kapczegaj, 4 Sep. 1962, *L. Slizik s.n.* (LE, NY, PE 01033246, W 1967–19784).

168. Berberis integripetala T. S. Ying, Acta Phytotax. Sin. 37: 334. 1999. TYPE: China. SE Gansu: Cheng Xian, Zhaoba Linchang, 1800 m, 9 May 1984, *Q. R. Wang 11510* (holotype, PE 00935143!).

Shrubs, deciduous, to 1.5 m tall; mature stems purple, terete, scarcely black verruculose; spines solitary, concolorous, 1–2 cm, abaxially sulcate. Petiole almost absent; leaf blade abaxially pale green, adaxially green, elliptic or obovate-elliptic, 0.8–2 × 0.4–1 cm, papery, midvein, lateral veins, and reticulation raised abaxially, conspicuous adaxially, base cuneate, margin entire, apex obtuse or subacute. Inflorescence a sub-umbel or sub-raceme, 3- to 7-flowered, 3–4 cm overall; peduncle 1–2.2 cm; pedicel 8–12 mm. Sepals in 2 whorls; outer sepals elliptic, 3.8–4 × 2–2.1 mm; inner sepals broadly elliptic, 4–4.2 × 3–3.2 mm; petals elliptic, 5–5.5 × 3–3.4 mm, base cuneate, glands separate, elliptic, apex entire. Stamens ca. 3 mm; anther connective not extended, truncate. Ovules 3, shortly stipitate. Immature berry ellipsoid, style not persistent.

Phenology. *Berberis integripetala* has been collected in flower in May and photographed in fruit in July.

Distribution and habitat. *Berberis integripetala* is known from only the type collected from a mountainside forest understory at 1800 m in Cheng Xian in Gansu and from a photograph taken also in Cheng Xian.

The information above on the fruit of *Berberis integripetala* comes from photographs emailed to me for identification by Zhejiang University PhD student Yichao Gan on 24 July 2019. These were taken the previous day at Jifengshan Forest, Park, Cheng Xian, Gansu at 1300–1600 m.

169. Berberis jaeschkeana C. K. Schneid. var. **usteriana** C. K Schneid., Bull. Herb. Boissier, sér. 2, 5: 399. 1905. *Berberis usteriana* (C. K. Schneid.) R. Parker, Indian Forester 50: 399. 1924. TYPE: India. Kumaon, Chalek, Byans, 3660–3960 m, 23 July 1886, *J. F. Duthie 5307* (lectotype, designated by Ahrendt [1961: 139], K K00064942!; isolectotypes, CAL CAL0000004808 image!, DD image!, WU 0040034!).

Berberis pulangensis T. S. Ying, Fl. Xizang. 2: 133. 1985, syn. nov. TYPE: China. SW Xizang (Tibet): border with NW Nepal and India, Burang (Pulan) Xian, Kejia Xiang, 3700 m, 16 July 1976, *Qinghai-Xizang Complex Exp. 76-8494* (holotype, PE 00935236!; isotypes, HNWP 60441 image!, KUN [2 sheets] 0178243–44!, PE 00935237!).

Shrubs, deciduous, to 2 m tall; mature stem pale yellow or brownish yellow, terete or subterete; spines 3-fid, concolorous, 1–2 cm, abaxially sulcate. Petiole almost absent; leaf blade abaxially grayish green, papillose, adaxially dark green, obovate, 1–1.5 × 0.5–1 cm, thinly leathery, midvein, lateral venation, and reticulation raised abaxially, conspicuous adaxially, base cuneate, margin spinose with 1 to 4 teeth on each side, sometimes entire, apex subacute or obtuse, sometimes mucronate. Inflorescence a sub-umbel or sub-raceme, 2- to 5-flowered, 1.5–2.5 cm overall; bracts ovate-triangular, 1–2 mm; pedicel 5–10 mm. Sepals in 3 whorls; outer sepals ovate or oblong-ovate, 3–5.5 × 1–3 mm; median sepals obovate-elliptic, 4–7.5 × 2–4 mm; inner sepals broadly obovate, 6–8.5 × 4–6 mm; petals obovate, 5–6 × 3–3.5 mm, base cuneate, glands elliptic or obovate-elliptic, 0.8–1 mm, apex emarginate. Stamens 3.5–4 mm; anther connective scarcely or not extended. Pistil 3–4 mm; ovules 2 to 5. Berry red, oblong-ellipsoid, 8–9 × ca. 4 mm; style persistent, short.

Phenology. In China, *Berberis jaeschkeana* var. *usteriana* has been collected in fruit from July to September. Its flowering period is unknown. In India it has been collected in flower in July, and in Nepal in June.

Distribution and habitat. In China *Berberis jaeschkeana* var. *usteriana* is known from only the type of *B. pulangensis* and two other collections, all from Burang (Pulan) Xian, Xizang (Tibet). The type was collected on mountain slopes at 3700 m, *Xizang Institute of Biological Research 4097* on a cliffside at 3750 m, and *FLPH Tibet Exped. 12-0144* from a grassy place in a valley at 3671 m. *Berberis jaeschkeana* var. *usteriana* is also known from India and Nepal.

The protologue of *Berberis jaeschkeana* var. *usteriana* cited *Duthie 5306* and *5307* without citing any herbarium. The citing by Ahrendt (1961: 139) of *Duthie 5307* at K as type was an effective lectotypification. The designation of the specimen at WU as the lectotype by

Adhikari et al. (2012: 497–500) is, therefore, unsustainable.

Berberis jaeschkeana var. *usteriana* was published simultaneously with a *B. jaeschkeana* var. "typica" whose type was *H. Falconer 97* from Kashmir (holotype, W 001548; isotypes, M M0164257, S S12–25269). The main basis for differentiating between the two taxa was the shape of the leaves and the number of marginal spines. No account of the flowers of either taxa was given and, whereas variety *usteriana* has floral material, the three specimens of *Falconer 97* have only fruit. Subsequently, Ahrendt (1961: 138) differentiated the two varieties on the basis that variety *usteriana* had emarginate petals whereas those of variety *jaeschkeana* were entire, and this was followed by Rao et al. (1998b). However, no indication was given by Ahrendt as to which specimen or specimens of variety *jaeschkeana* this was based on. Complicating matters further are three further specimens of *Falconer 97* (GH 00038829, L L.174243, P P03645695) which do have flowers. The P specimen was identified by Schneider in June 1906 as *B. jaeschkeana*, while the one at GH was annotated by Schneider in June 1916 as "seem[s] to belong to *B. orthobotrys* Bienert." All three appear to be the same as a *Falconer 98* (K) also from Kashmir and also identified by Schneider as *B. orthobotrys*. I do not accept Schneider's identifications of these as *B. orthobotrys* (whose type is from northeast Iran) but, nevertheless, if the GH and P specimens of *Falconer 97* are not *B. jaeschkeana*, then it is possible that variety *usteriana* should be treated as a synonym of *B. jaeschkeana*. I have not investigated the matter further, but it appears that both the type of *B. pulangensis*, *Xizang Institute of Biological Research 4097* and *FLPH Tibet Exped. 12-0144*, more resemble the lectotype and syntype of variety *usteriana* than the holotype of variety *jaeschkeana*, and Kumaon (now a district of Uttarakhand) is much nearer to where the two Tibetan collections were made than Kashmir.

The account of flowers used here is based on that given by Adhikari et al. (2012: 497–500), which noted that all Nepalese specimens of variety *usteriana* they had examined had emarginate petals.

The protologue of *Berberis pulangensis* mistakenly described the taxa as evergreen.

Selected specimens.
Syntypes. Berberis jaeschkeana var. *usteriana*. India. Kumaon, Chalek, Byans, 3660–3960 m, *J. F. Duthie 5306* (CAL CAL0000004809, DD, WU 00400350).

Other specimens.
SW Xizang (Tibet). Burang (Pulan) Xian: Kejia, 3750 m, 20 Aug. 1974, *Xizang Institute of Biological Research 4097* (HNWP 40635, 88765, 776477, XJBI 00036294, 10020071); near river port, 30.153572°N, 81.323525°E,

3671 m, 4 Sep. 2012, *FLPH Tibet Exped. 12-0144* (PE 01957660, 01961392–93).
NEPAL. Muktinath, Mustang Distr.; (Dhawalagiri), 28.8167°N, 83.8708°E, 3810 m, 9 June 1954, *J. D. A. Stainton, W. R. Sykes & L. H. J. Williams 5694* (BM BM000897135).

170. Berberis jamesiana Forrest & W. W. Sm., Notes Roy. Bot. Gard. Edinburgh 9: 81–82. 1916. TYPE: China. NW Yunnan: NE of Yangtze bend, 27.75°N, 3350 m, Oct. 1913, *G. Forrest 11474* (lectotype, designated here, E E00220925!; isolectotype, K missing).

Berberis jamesiana var. *leucocarpa* (W. W. Sm.) Ahrendt, J. Linn. Soc., Bot. 57: 180. 1961. *Berberis leucocarpa* W. W. Sm., Notes Roy. Bot. Gard. Edinburgh 9: 82. 1916. TYPE: China. NW Yunnan: Mekong-Yangtze divide, 27.75°N, 3350 m, July 1914, *G. Forrest 12855* (holotype, E E00217946; isotype, K missing).
Berberis nummularia Bunge var. *sinica* C. K. Schneid., Bull. Herb. Boissier, sér. 2, 8: 202. 1908. TYPE: China. NW Yunnan, [Heqing Xian], "in monte Hee-chan-men," 25 May 1883, *J. M. Delavay 77* (lectotype, designated here, P P00640986!).

Shrubs, deciduous, to 3.5 m tall; mature stems dark red or purple, shiny, verruculose, subterete or sulcate; spines solitary or 3-fid, semi-concolorous, 1.5–5 cm, stout, abaxially slightly sulcate. Petiole 2–8(–15) mm; leaf blade abaxially pale green, adaxially blue-green, young leaves turquoise, elliptic or oblong-obovate, 2.5–8 × 1–4 cm, subleathery, midvein raised adaxially, slightly impressed adaxially, lateral veins and reticulation indistinct abaxially (conspicuous when dry), inconspicuous adaxially, base cuneate, margin entire or closely spinulose with 20 to 40 teeth on each side, apex rounded or retuse. Inflorescence a raceme, sometimes compound at base, (4- to)20- to 40-flowered, (1.8–)7–10 cm; pedicel 7–10 mm, slender; bracteoles ovate, 2–2.5 × ca. 1.5 mm, apex acute. Sepals in 2 whorls; outer sepals oblong-obovate, ca. 3 × 2 mm; inner sepals narrowly obovate, ca. 4.5 × 2.5 mm; petals obovate or narrowly oblong-elliptic, ca. 4.5 × 2 mm, base clawed, glands separate, apex emarginate, lobes acute. Stamens ca. 3 mm; anther connective extended, slightly apiculate. Ovules 2 or 3. Berry initially creamy white, ultimately pale red or pink, translucent, subglobose, ca. 10 × 7–8 mm; style not persistent.

Phenology. Berberis jamesiana has been collected in flower in April and May and in fruit between June and October.

Distribution and habitat. Berberis jamesiana is known from southwest Sichuan, central and northwest Yunnan, and Zayü (Chayu) Xian in southeast Xizang (Tibet). It has been found in thickets and on forest margins and mountain slopes, especially on limestone, at ca. 2100–3350 m.

The protologue of *Berberis jamesiana* cited *G. Forrest 11474, 10633,* and *13566,* referring in the discussion to *11474* as "the type-plant" (Smith, 1916). Schneider (1918: 218) ignored this, describing *10633* as "Typ. der *Jamesiana*" and *11474* as "Typ. Ex auct." I have followed Ahrendt (1961: 179–180) in treating *11474* as the type. The protologue did not cite any herbarium and Ahrendt (1961) cited specimens at E and K. Given that the protologue was published as one of the "specierum novarum in herbario Horti Regii Botanici Edinburgensis," the specimen at E has been chosen as the lectotype. In any case, the specimen at K is missing.

W. W. Smith differentiated *Berberis leucocarpa* from *B. jamesiana* largely on the basis that the fruit of the former was whitish while the latter was red. In fact, an unusual feature of *B. jamesiana* is that the immature fruit are white, turning red on maturity. The synonymy was made by Ying (2001: 182).

Franchet (1886: 386, 1889: 36) cited two Delavay collections which he identified as *Berberis integerrima*: "Yun-nan ad vicum kiao-che-tong, in monte Hee-chan-men orientem versus, prope Ho-kin, alt 2500 m; fl. 2 maj; fr. 9 oct 1884 (Delav. N. 881 et 77)." Subsequently, Schneider (1908: 202) decided these were of a new taxon, *B. nummularia* var. *sinica,* his protologue citing "Delavay Nr. 77 in monte Hee-chan-men 25 Mai 1883 (flor), Octob. 1885 et 10 Sept. 1885 (fruct)," "Orig. in Herb. Paris." This information was repeated by Schneider (1918: 218) when he synonymized the taxon as *B. jamesiana,* but with *Delavay 77* now described as "Typ der var. sinica." There are 13 Delavay sheets of *B. jamesiana* at P, all but two of them identified as *B. integerrima.* However, there is only one specimen unambiguously numbered *77* (P P00640986). The collection details are "monte Hee-chan-men 25 Mai 1883" and, as such, are consistent with the first collection cited in the protologue and, though it has immature fruit rather than flowers, it has been chosen as the lectotype. There is a further sheet with *881* crossed out and replaced by *77.* This has no collection details and consists of three specimens—one with mature fruit (P P00640989), one with immature fruit and one flower, and one with semi-mature fruit (both P P00640988). This is the only sheet that has an annotation by Schneider. Dated 16 June 1906, it reads "No 77 et 881 B. integerrima non est . . . spec nova." It is possible that the specimen with a flower on this sheet is an isolectotype, but this cannot be proved. No specimen of the syntype of October 1885 cited in the protologue has been found, although there is one of 10 September 1885 (P P00640987) from "Kiao Che Ton, Hee Chan-men," though it is numbered

881 rather than *77.* Despite this, it is treated here as a syntype.

There are other specimens of *Delavay 881* at P with other dates. Finding only ones of 15 May 1884 and 9 October 1884 (collection details below) on two visits to the herbarium before its reorganization, I annotated them respectively as *Delavay 881* "A" and *Delavay 881* "B." These designations are confirmed here while that of 2 May 1884 cited by Delavay (P02682360) is designated here as *Delavay 881* "C," and the syntype of 10 September 1885 is designated here as *Delavay 881* "D." The specimen with mature fruit (P P00640989) referred to above may be *Delavay 881* "B," and that with semi-mature fruit on the same sheet *Delavay 881* "D." A further undated *Delavay 881* with flowers (P P06868282) may be *Delavay 881* "C" as may a *Delavay 881* with flowers at Harvard (A 00279579), "in monte Hee-chan-mon, prope Ho-kin," dated 1883–1885. But none of this can be proved.

Besides the color of its fruit *Berberis jamesiana* is unusual in that it can have an exceptionally large variation of leaf shape and size on the same plant. This is apparent from living plants, but can cause confusion with herbarium specimens which may have only very large leaves or only small ones.

Ying (2001) treated *Berberis jamesiana* var. *sepium* as a synonym of *B. jamesiana.* This is treated here as a synonym of *B. caroli.*

The report of *Berberis jamesiana* in Qinghai by Zhou (1997: 376) is the result of misidentification and may be referable to *B. elliptifolia.*

Syntype. Berberis nummularia var. *sinica.* NW Yunnan [Heqing Xian]: "Kiao che ton, Hee chan men," 10 Sep. 1885, *J. M. Delavay 881* "D" (P P00640987).

Selected specimens.

SW Sichuan. Muli Xian: Yi Qu, Shuiluo Xiang, 29 Oct. 1959, *S. G. Wu 3479* (KUN 0176411–12). **Yanyuan Xian:** "in viculos Kalapa et Liuku," 3000 m, 17 May 1914, *C. K. Schneider 1260* (A 00279615, E E00612519, GH 00279614, K); "inter vicos Dubrilliangdse et Hungga ubique," 2750–2950 m, 12 June 1914, *H. R. E. Handel-Mazzetti 2890* (E E00612519, W 1930–4042); SE of Muli, 27.833333°N, 101°E, 3350 m, Sep. 1922, *G. Forrest 22402* (A 00279601, E E00612564, P P02313670, US 00946022); Zuosuo Qu, Meipeng Xiang, 2600 m, 12 June 1960, *Nanshuibeidiao Exp. 6011* (KUN 0178739, PE 01033139). **Zhaojue Xian:** Guangming Gongshe, 2470 m, 7 June 1979, *Zhaopu Team 0510* (SM SM704800151).

C & NE Yunnan. Plateau de Sen-kia-leang-tse, 2800 m, July [?1912], *E. E. Maire s.n.* (P P02682386). **Fumin Xian:** Yongding Gongshe, Maying Shan, 2600 m, 20 Oct. 1964, *B. Y. Qiu 596165* (KUN 0176361, 0178934, PE 01033192). **Kunming:** "Vicinity of Yunnan-sen," s.d., *E. E. Maire 2001* (E E00612537, K); Sanqing Ge, 2200 m, 25 June 1946, *S. E. Liu 16372* (IBSC 0091745, KUN 0178938–39, PE 01033206); Xishan, Sanqing Ge, 2300 m, 9 July 1958, *B. Y. Qiu 57089* (IBSC 0092356, KUN 176362–63). **Luquan Xian:** Malutang,

2730 m, 25 July 1965, *F. W. Zhu s.n.* (YUKU 02004113, 02004122). **Qiaojia Xian:** Dapingzi, 2300–2400 m, 13 Aug. 2003, *H. D. Wang 03-0556* (IBSC 0765249). **Songming Xian:** Hung-Shih-Yen, 2100 m, 14 Oct. 1950, *P. Y. Mao 222* (KUN 0176394, 0176397, PE 01033171, 01033173–74); Shaodian Qu, Guodong Cun, 2590 m, 21 Apr. 1953, *P. H. Yu 139* (IBSC 0091832–33, KUN 0176347–48, 0176357–58); Shaodian, Dazhuxiang, 23 Apr. 1956, *B. Y. Qiu 51666* (IBSC 0092353, KUN 0176354–56, PE 01033176).

NW Yunnan. Mekong–Salween divide, 28.17°N, 3350 m, Oct. 1914, *G. Forrest 13566* (E E00373489); Chungtien plateau, 27.92°N, 3350 m, July 1913, *G. Forrest 10633* (A 00279602, BM BM000566450, E E00373488, K K000395194). **Dêqên (Deqin) Xian:** Benzi Lan Gongshe, Dongzhulin, 3000 m, 8 July 1981, *Qinghai-Xizang Team 2198* (CDBI CDBI0027398, HITBC 003581, KUN 0176313–14, PE 01033143–44). **Eryuan Xian:** Dengchuan Qu, Jiaoshidong, 2750 m, 31 July 1963, *Northwest Yunnan Jinsha River Team 63-6351* (KUN 0176329, 0178935, PE 01033207); on hwy. S221, SE of Sanjia Cun, 26.105556°N, 100.170556°E, 2510 m, 14 Sep. 2013, *D. E. Boufford, Y. S. Chen & J. F. Harber 43530* (A 00914422, E E00770753, KUN 1278431, PE). **Gongshan Xian:** Si Qu, Gaoligongshan, 3200 m, 24 May 1960, *Nanshuibeidiao Exp 8529* (KUN 0176328, YUKU 02065671). **[Heqing Xian]:** "Montagnes de Hee-chan-men, 11 July 1883, *J. M. Delavay s.n.* (P P02313225); Hee Chan Men, Lan Kong, 2 May 1884, *J. M. Delavay 881* "C" (P P02682360); "Versant oriental de Heèchan men au village de Kiao-che ton," 2400 m, 2 May 1884, *J. M. Delavay s.n.* (P P02313089); "a Kiao-che tong, versant oriental du Hee chan men," 2500 m, 15 May 1884, *J. M. Delavay 881* "A" (P P06868284); "Au village de Kiao-che tong versant oriental de Mt Hee chan men," 9 Oct. 1884, *J. M. Delavay 881* "B" (P P02682359, P06868283); "Village de Kiao-che tong sur le Hee chan men," 2500 m, 21 Sep. 1885, *J. M. Delavay s.n.* (P P02313091, VNM). **Lijiang Shi:** W slope of Likiang Snow Range, Yangtze watershed, May 1923, *J. F. Rock 8521* (A 00279609, UC UC327332, US 00946018). **Weixi Xian:** Lidi Ping, 18 June 1940, *G. M. Feng 4855* (KUN 0176384–86); Badi Xiang, Najiluo He, 2800 m, 14 May 2004, *D. Z. Fu & H. Li 13* (TNM S137883). **Zhongdian (Xianggelila) Xian:** betw. Chiao-tou (Qiaotou) & Hsia-Chungtien (Xiao Zhongdian), 13 May 1939, *K. M. Feng 886* (A 00279597, KUN 0176378, 0176398).

SE Xizang (Tibet). Zayü (Chayu) Xian: Tsarong, Londre La, Mekong–Salwin Divide, 28.23333333°N, 98.66666667°E, 3050–3350 m, May 1922, *G. Forrest 21609* (A 00279604, E E00395942, P P02313672, UC UC253196, US 00946021); Me-kong, Tsa-wa-rung, 3000 m, Sep. 1935, *C. W. Wang 66136* (A 00279583, KUN 0176372, LBG 00064306, PE 01033215–17, TAI 047306, WUK 0047684); Chawalong, Songta Xueshan, 2700 m, 26 June 1982, *Qinghai-Xizang Team 7665* (KUN 0176390, PE 01033218–20).

171. Berberis jiulongensis T. S. Ying, Acta Phytotax. Sin. 37: 320. 1999; T. S. Ying, Cat. Type Spec. Herb. China, Suppl. 2: 54. 2007. TYPE: China. W Sichuan: Jiulong Xian, Naiqu Gongshe, 2500 m, 22 June 1974, *F. Y. Qiu 4660* (holotype, PE 00935146!; isotypes, CDBI [3 sheets] 0027532–34 images!).

Shrubs, deciduous, to 2 m tall; mature stems purplish or reddish brown, angled, sparsely black verruculose; spines solitary, concolorous, 0.5–1 cm, terete.

Petiole almost absent; leaf blade abaxially pale green, papillose, adaxially green, obovate-lanceolate or narrowly obovate, 1.5–3.5 × 0.5–0.8 cm, papery, midvein raised abaxially, slightly impressed adaxially, lateral veins and reticulation indistinct or obscure on both surfaces, base attenuate, margin entire, apex obtuse or acute, sometimes mucronate. Inflorescence a raceme, rarely partially paniculate, sometimes compound at base, 20- to 30-flowered, 4–10 cm; bracts leaflike, ca. 10 mm; pedicel ca. 2 mm. Sepals in 3 whorls; outer sepals triangular-ovate or ovate, 2–2.5 × 1.6–2 mm; median sepals broadly elliptic or suborbicular, ca. 3.2 × 2.8–3 mm; inner sepals broadly obovate, 4.5–5 × 4.1–4.5 mm; petals obovate or obovate-elliptic, 4–4.5 × 3–3.5 mm, base clawed, glands separate, apex emarginate. Stamens ca. 3.5 mm; anther connective extended, obtuse or truncate. Ovules 3 or 4, shortly stipitate. Berry red, globose, ca. 6 × 5 mm; style persistent.

Phenology. *Berberis jiulongensis* has been collected in flower in June and July and in fruit in July and August.

Distribution and habitat. *Berberis jiulongensis* is known from Jiulong and Muli Xian in west Sichuan. It has been collected from thickets, mountain slopes, and watersides at ca. 1900–2500 m.

Ying (in Jin & Chen, 2007: 54) validated *Berberis jiulongensis*, whose name was not validly published by Ying (1999: 320) because no type was indicated (Shenzhen Code, Art. 40.1 [Turland et al., 2018]).

The protologue describes *Berberis jiulongensis* as evergreen. However, since it cited no specimens from the winter months, this appears to be speculative and I have assumed it is actually deciduous. The inflorescence is also described as a corymbose panicle. This is not borne out either by the type or the other specimens listed above whose inflorescences are largely racemose. The evidence about the fruit is from *Q. Q. Wang 21428*.

On current evidence, *Berberis jiulongensis* appears to have a very restricted distribution, all of the collections being from the Naiqu area with the exception of *T. T. Yu 6707*, which was collected some 50 km westsouthwest of Naiqu. The fieldbook of T. T. Yu locates Maidilong in Jiulong Xian, but it is on the Yalong Jiang which currently, at least, is in Muli Xian, some 10 km from the Jiulong border.

Selected specimens.

W Sichuan. Jiulong Xian: Naiqu Xiang, 2350 m, 9 Aug. 1979, *Q. Q. Wang 21428* (CDBI CDBI0027572–73, IBSC 0091499); Naiqu Xiang, 2350 m, 12 Aug. 1979, *Q. Q. Wang 21568* (CDBI CDBI0027525, CDBI0027574, IBSC 0091500); 26 km S of Jiulong, 2450 m, 18 June 1984, *W. L. Chen & J. R. Chen 6436* (PE 00935149, 02044419, 02044453); [Naiqu

Xiang], betw. the city of Jiulong & the Yalong Jiang (River), right side of hwy. going toward Yalong Jiang, 28.74333333°N, 101.68083333°E, 2250–2500 m, 16 July 2005, *D. E. Boufford et al. 33011* (A 00279941, CAS 1078920). **Muli Xian:** Miu-ti-lung (Maidilong), 2300 m, 3 July 1937, *T. T. Yu 6706* (KUN 0177099–100, PE 00935147–48).

172. Berberis johannis Ahrendt, Gard. Chron., ser. 3, 109: 101. 1941. TYPE: Cultivated. Oct. 1939, from plant grown by W. J. Marchant, Stapehill, Wimborne, Dorset, U.K., from seeds reputedly from the same plant as *F. Kingdon-Ward 5936*, SE Xizang (Tibet), [Mainling (Milin) Xian], Pome Province, Tumbatse, 29.666667°N, 94.783333°E, 3350–3650 m, 13 July 1924 (lectotype, designated here, BM BM000554691!; isolectotype, WSY 0057573!).

Shrubs, deciduous, to 2 m tall; mature stems reddish or yellowish brown, sub-sulcate, scarcely black verruculose; spines 3-fid, yellow, 0.7–1.4 cm, slender, abaxially flat. Petiole almost absent; leaf blade abaxially pale gray-green, papillose, adaxially dull green, obovate or elliptic-obovate, 1.5–3 × 0.8–2.5 cm, papery, midvein and lateral veins raised abaxially, inconspicuous adaxially, reticulation inconspicuous abaxially, inconspicuous or indistinct adaxially, base cuneate, margin entire, rarely spinulose with 2 to 5 teeth on each side, apex acute or subacute, sometimes mucronate. Inflorescence a sub-raceme, sometimes a sub-fascicle or sub-umbel, 3- to 7-flowered, 2–3 cm overall; peduncle red, 3–10 mm, slender; bracts triangular-ovate, ca. 1 mm; pedicel red, (4–)6–8(–17) mm, slender. Sepals in 2 whorls; outer sepals oblong-ovate, ca. 4 × 2 mm, apex subobtuse; inner sepals elliptic or elliptic-orbicular, ca. 5 × 2.5–4 mm; petals elliptic-obovate or obovate, ca. 4.5 × 3 mm, base clawed, glands separate, elliptic, apex emarginate, lobes acute. Stamens ca. 2.5 mm; anther connective extended, obtuse. Ovules 3 to 5. Berry bright red, narrowly oblong-ellipsoid or oblong-ovoid, 11–13 × 3–4.5 mm, slightly contracted at middle, usually bent; style not persistent.

Phenology. *Berberis johannis* has been collected in flower between May and July and in fruit between July and September.

Distribution and habitat. *Berberis johannis* is known from Mainling (Milin) Xian and Bayi Qu (formerly Nyingchi (Linzhi) Xian in southeast Xizang (Tibet). It has been collected from thickets and mountainside thickets at ca. 3050–3650 m.

Berberis johannis is one of three species from Xizang whose types, designated by Ahrendt, are cultivated specimens grown from seeds collected by Kingdon-

Ward in the autumn of his 1924 expedition to the Tsangpo valley, the others being *B. temolaica* and *B. gyalaica*. In each of these cases, there are wild-collected specimens at K with the same number as those collected in the summer of 1924, the explanation being that it was Kingdon-Ward's practice on this expedition to collect herbarium specimens in the summer months, mark the plants, and return to collect seeds in the autumn.

As with *Berberis temolaica* and *B. gyalaica*, determining which specimen is the type of *B. johannis* presents problems. Ahrendt's protologue (which has a minimal description) gives the collector's number but cites no herbarium for any specimen, simply recording "I am grateful to Mr Marchant of Keeper's Hill Nursery, Stapehill Wimborne for bringing this unusual species to my attention." Subsequently (1941b: 67), Ahrendt cited the type as being cultivated material from Marchant in the "Herb Ahrendt." Later (1961: 142), Ahrendt inaccurately cited the type as being specimens at BM of 4 June 1942 and 4 May and 20 Oct. 1944 of cultivated material from his own living collection grown from seeds of *K. Kingdon-Ward 5936* (though this is more likely to refer to a plant grown from a cutting). Only two specimens whose identical labels state they are from Marchant have been found. These are at BM and WSY. That at BM has been chosen here as the lectotype because it also has a label in Ahrendt's hand identifying it as a type. None of the cultivated specimens Ahrendt cited in 1961 have been found at BM, nor are these specimens at K or WSY.

From the evidence of a line drawing by Schneider on the sheet, the flowers of the wild-collected specimen of *Kingdon-Ward 5936* at K largely correlate with the description of flowers given by Ahrendt (1941b: 66), though the maximum number of ovules is five versus the three or four given originally. The leaves are sometimes more broadly obovate than the lectotype but, overall, there appears to be enough evidence to suggest that the lectotype is a true species rather than a hybrid.

Berberis johannis is on the current IUCN Red List of Threatened Species (China Plant Specialist Group, 2004d).

Selected specimens.

SE Xizang (Tibet). Mainling (Milin) Xian: Pome Province, Tumbatse, 29.666667°N, 94.783333°E, 3350–3650 m, 13 July 1924, *F. Kingdon-Ward 5936* (K K000568151); Tsela Dzong, 29.433333°N, 94.366667°E, 3050 m, 31 May 1924, *F. Kingdon-Ward 5724* (E E00258231, K K000568150); betw. Pai & Duoxiongla, 3400 m, 13 Sep. 1974, *Qinghai-Xizang Team 74-4682* (KUN 0178755, PE 01840025). **Nyingchi (Linzhi) Xian (now Bayi Qu):** Lulang military depot, 3020 m, 27 July 1965, *Y. T. Zhang & K. Y. Lang 970* (KUN 0179164, PE 01037449–50); near Lulang, 3200 m, 4 Aug. 1975, *Qinghai-Xizang Supplementary Collections 751288* (HNWP 51901, KUN 0178650–51, PE 01037454–55).

173. Berberis kangdingensis T. S. Ying, Acta Phytotax. Sin. 37: 349. 1999. TYPE: China. W Sichuan [Kangding Xian]: "Kangting (Tachienlu); ad orientum urbis," ca. 2700 m, 4 Nov. 1934, *H. Smith 13039* (holotype, PE 00935150!; isotypes, MO [2 sheets] 4367261!, 6408630!, S S12–25248 image!, UPS BOT:V-040911 image!)

Shrubs, deciduous, to 30 cm tall; mature stems pale yellow-brown, terete or subangled; spines 3-fid, semi-concolorous, 0. 5–1 cm, terete, slender. Petiole almost absent; leaf blade abaxially yellow-green, papillose, adaxially green, narrowly elliptic or obovate-elliptic, 0.5–0.8 × 0.2–0.4 mm, papery, midvein raised abaxially, venation otherwise indistinct or obscure on both surfaces, base cuneate, margin entire or occasionally spinulose with 1 or 2 teeth on each side, apex acute. Inflorescence 1-flowered; pedicel 4–5 mm. Sepals in 2 whorls; outer sepals elliptic, ca. 4.5 × 3.5 mm; inner sepals obovate-oblong, ca. 5.5 × 4.1 mm; petals obovate, ca. 5 × 3.1 mm, base cuneate, glands separate, elliptic, apex emarginate. Stamens ca. 3 mm; anther connective slightly extended, rounded or obtuse. Ovules 5 to 7. Berry red, subglobose, ca. 9 × 8 mm; style persistent, short.

Phenology. *Berberis kangdingensis* has been collected in fruit between July and October. Its flowering season outside cultivation is unknown.

Distribution and habitat. *Berberis kangdingensis* is known from Kangding and Luding Xian in west Sichuan. It has been found in open areas, rock crevices on mountain slopes, and *Abies* forests at ca. 2700–3600 m.

The description of the flowers used above is that given in the protologue and is based on the cultivated specimen of *H. Smith 13039* at PE. The description in the protologue of the fruit as being densely pruinose is based on what appears to be dried mold on the holotype since none of the fruit of the other specimens cited above are pruinose.

The only collection located other than those of Harry Smith—*Z. G. Liu 20896*—is listed below.

Selected specimens.
W Sichuan. Kangding Xian: Liuba, Zimei Cun, 3375 m, 26 July 1979, *Z. G. Liu 20896* (CDBI CDBI0028264–65, IBSC 0092687). **[Luding Xian]:** Yülingkong, Yachiangan Mtns., ca. 3600 m, 18 Oct. 1934, *H. Smith 12795* (UPS BOT: V-040920); Yülingkong, Gomba la, ca. 3400 m, 19 Oct. 1934, *H. Smith 12893* (MO 4577172, PE 00935151, UPS BOT: V-040909).

Cultivated material:
Cultivated specimens.
Uppsala Botanical Garden, from *H. Smith 13039*, 20 June 1940 (UPS BOT: V-159619); 28 May 1943 (MO 5958155, UPS BOT: V-159620); 6 June 1946 (PE 01840099). From *H. Smith 12893*, 20 June 1940 (UPS: BOT V-159618); 13 Oct. 1945, (UPS: BOT: V-040910). From *H. Smith 12893*, 29 June 1944 (MO 94577172).

174. Berberis kangwuensis Harber, sp. nov. TYPE: China. SW Sichuan: Muli Xian, N of the city of Muli on S216 to Daocheng, at first pass N of the city of Muli, 28.124722°N, 101.160833°E, 3750 m, 7 Sep. 2013, *D. E. Boufford, J. F. Harber & X. H. Li 43494* (holotype, PE!; isotypes, A!, BM!, CAS!, E!, HAST!, IBSC!, K!, KUN!, MO!, PE!, TI!).

Diagnosis. *Berberis kangwuensis* has leaves similar to *B. pallens* of northwest Yunnan, but has fewer flowers and a different flower structure. It is found in the same area as *B. purpureocaulis*, but can be distinguished from it by color of its mature stems, the different shape of its leaves, and its flower structure.

Shrubs, deciduous, to 2.25 m tall; mature stems reddish brown, sulcate; spines mostly absent, when present 1- to 3-fid, pale brownish yellow, 0.8–3.2 cm, abaxially sulcate. Petiole almost absent or to 4 mm; leaf blade abaxially pale green, papillose, adaxially dark green, obovate-elliptic or obovate, (0.8–)2.5–4 × (0.6–)1–1.4(–1.8) cm, papery, midvein raised abaxially, impressed adaxially, lateral veins and reticulation indistinct abaxially, inconspicuous adaxially, margin entire or spinulose with 3 to 6 widely spaced teeth on each side, spines to 2 mm, apex subacute, obtuse, or rounded-mucronate, base attenuate. Inflorescence a fascicle, sub-fascicle, sub-umbel, or sub-raceme, sometimes fascicled at base, 2- to 8-flowered, to 4 cm overall; pedicel 6–12 mm, to 20 mm when from base; bracteoles lanceolate-ovate, 1.5 × 0.5 mm. Sepals in 2 whorls; outer sepals ovate, 2–2.5 × 1 mm; inner sepals obovate-elliptic, 3–4 × 2–2.5 mm; petals obovate, 3–4 × 2–2.5 mm, base distinctly clawed, glands widely separate, 0.6 mm, apex entire. Stamens ca. 2.5 mm; anther connective not or slightly extended, truncate. Ovules 2 to 4. Berry red, ellipsoid or oblong, 9–11 × 3–5 mm; style persistent, short.

Phenology. *Berberis kangwuensis* has been collected in flower in July and September and in fruit in September.

Distribution and habitat. *Berberis kangwuensis* is known from the area around Kangwu in Muli Xian in southwest Sichuan previously known as Kulu. It has been found on forest margins and stream and roadsides at 3200–4150 m.

IUCN Red List category. *Berberis kangwuensis* is assessed as DD or Data Deficient, according to IUCN (2001) criteria.

The description of flowers given above is from two very late flowers from the type collection.

The type and *D. E. Boufford, J. F. Harber & X. H. Li 43467* and *43497* have only immature fruit. The description of fruit given above is from *Yu 14298* (details below). This collection was mistakenly given as the type of *Berberis humidoumbrosa* var. *inornata* by Ahrendt (1961: 160). The specimen of *Yu 14298* at A has very different leaves from the specimens at E and KUN and is possibly *B. purpureocaulis* (the specimen at PE is too poor to make any judgment).

Selected specimens.
SW Sichuan. Muli Xian: "Mountains of Kulu," 4150 m, Sep. 1929, *J. F. Rock 18167* (A 00279719, US 00945964, 00945999); Kulu, 3200 m, 17 Sep. 1937, *T. T. Yu 14298* (A 01037200, E E00217945, KUN 0754341, PE 01037200); N of the city of Muli on S216 to Daocheng, then on Xian rd. Z025 toward Cunduochanghaizi, then E on Xian rd. Z020, 28.115°N, 101.177778°E, 3650 m, 6 Sep. 2013, *D. E. Boufford, J. F. Harber & X. H. Li 43467* (A, CAS, E, K, KUN, PE); on Xian rd. Z020 to near Kangwu Dasi SE of Cunduochanghaizi, 28.106944°N, 101.200278°E, 3710 m, 7 Sep. 2013, *D. E. Boufford, J. F. Harber & X. H. Li 43497* (A, CAS, E, K, KUN, PE, TI).

175. Berberis kartanica Ahrendt, J. Bot. 79(Suppl.): 68. 1941. TYPE: China. SE Xizang (Tibet): [Cona (Cuona) Xian], arid valley above Karta (Kadaxiang), ca. 28.076827°N, 92.362061°E, 4250 m, 15 June 1935, *F. Kingdon-Ward 11716* (holotype, BM BM000559596!).

Shrubs, deciduous, erect, height unrecorded; mature stems pale yellow, subterete, not verruculose; spines 3-fid, concolorous, 1.5–2.5 cm, terete, stout. Petiole almost absent; leaf blade abaxially pale green, papillose, adaxially green, slightly shiny, obovate, 1.3–1.9 × 0.6–1.2 cm, papery, midvein and lateral veins raised abaxially, venation inconspicuous or indistinct adaxially, base cuneate, margin entire, apex rounded, sometimes minutely mucronate. Inflorescence a sub-umbel or sub-raceme, 5- to 7-flowered, 1.8–3 cm overall; peduncle 0.7–1.4 cm; bracts ca. 2.5 mm; pedicel 5–10 mm. Flowers bright yellow, 10.5–11.5 mm diam. Sepals in 2 whorls; outer sepals lanceolate-ovate, 3.5–4 × 1.5–2 mm; inner sepals oblong-elliptic, 5.5–6 × ca. 3 mm; petals broadly obovate, ca. 3.7 × 3 mm, base scarcely clawed, glands separate, oblong-obovate, apex emarginate, lobes acute. Stamens ca. 2.8 mm; anther connective scarcely extended, truncate. Ovules 4, often stipitate. Berry unknown.

Phenology. *Berberis kartanica* has been collected in flower in June. Its fruiting season is unknown.

Distribution and habitat. *Berberis kartanica* is known from only the type collected in an arid valley in Cona (Cuona) Xian in southeast Xizang (Tibet) at 4250 m.

Berberis kartanica was included by Ying (1985: 132), but subsequently (2001) omitted without explanation.

176. Berberis koehneana C. K. Schneid., Bull. Herb. Boissier, sér. 2, 5: 814. 1905. TYPE: NW India. Kumaon, near Budhi Byans, 2440–2740 m, 17 July 1886, *J. F. Duthie 5309* (holotype, WU 0040033!; isotypes, CAL CAL0000004791 image!, DD [2 sheets] images!, K K000644937!).

Berberis koehneana var. *auramea* Ahrendt, J. Linn. Soc., Bot. 57: 210. 1961. TYPE: C Nepal. Langtang, 3500 m, 22 June 1949, *O. V. Polunin 506* (holotype, BM BM000884595!; isotypes, A 00038830!, E E00663671!).

Shrubs, deciduous, to 3 m tall; mature stems terete or slightly angled, dark reddish brown, verruculose; spines 3-fid, solitary toward apex of stems, semi-concolorous, 0.5–1.5 cm, terete or angular, stout. Petiole almost absent or 2–5 mm; leaf blade abaxially pale green, papillose, adaxially green, shiny, obovate, 2–5 × 0.7–1.5 cm, papery, midvein raised abaxially, lateral veins and reticulation indistinct abaxially, inconspicuous adaxially, base cuneate, apex obtuse, rarely acute, usually mucronate, margin usually entire, sometimes spinulose with 1 to 4 teeth on each side. Inflorescence a panicle, 15- to 70-flowered, 3–16 cm overall including peduncle 0.5–4 cm; bracts ovate-triangular, 1.5–3 mm; pedicel 5–10 mm, reddish brown. Flowers ca. 0.5–1 cm diam. Sepals in 3 whorls; outer sepals ovate-triangular, 1.5–2.5 × 1–1.5 mm; median sepals elliptic or elliptic-ovate, 2.5–5 × 2–3.5 mm; inner sepals obovate to broadly obovate, 3.5–6.5 × 3.5–4.5 mm; petals obovate, 3.5–6 × 2.5–4.5 mm, base cuneate, glands separate, 0.5–0.7 mm, apex incised. Stamens 2–3.5 mm; anther connective scarcely extended. Pistil 2–3.5 mm; ovules 2 or 3. Berry bright red, oblong, ovoid, or ellipsoid, 5–10 × 0.25–4 mm; style not persistent.

Phenology. In China *Berberis koehneana* has been collected in flower in June and July and in fruit in September.

Distribution and habitat. *Berberis koehneana* is known in China from Gyirong (Jilong) and Nyalam (Nielamu) Xian in south Xizang (Tibet). It has been found on forest margins at 2750–3150 m. It is also known from India and Nepal.

Adhikari et al. (2012: 486) synonymized *Berberis koehneana* var. *auramea*. *Berberis koehneana* was not included by Ying (1985, 2001).

Selected specimens.

S Xizang (Tibet). Gyirong (Jilong) Xian: Gyirong Qu, Buxing Logging Station, 3000 m, 14 June 1972, *Xizang Chinese Herbal Medicine Team 254* (HNWP 29983, PE 00049439, 01037561, 01037568); Nimu, 26 June 1975, *s. coll. 75-317* (PE 01037560); near Ruga, 3100 m, 22 July 1975, *Qinghai-Xizang Team Botany Group 5601* (PE 01037563, 01840033); near Gyirong Qu, Rema, 3150 m, 24 July 1975, *Qinghai-Xizang Team 7046* (KUN 0178769, 0179344, PE 01840034); near Gyirong Qu, 2750 m, 14 Sep. 1981, *Z. C. Ni et al. 2235* (PE 01037555); near Luga, 3000 m, 19 Sep. 1981, *Z. C. Ni et al. 2367* (PE 01037554) and *Z. C. Ni et al. 2374* (PE 01037553). **Nyalam (Nielamu) Xian:** Zhangmu, 3000 m, 3 Aug. 2012, *Phylogeny of Chinese Land Plants, National Taiwan University Expedition to Tibet 2012 747* (E E00662308).

177. Berberis kongboensis Ahrendt, J. Bot. 80 (Suppl.): 97. 1942, as "*kongroensis.*" TYPE: China. SE Xizang (Tibet): [Mainling (Milin) Xian], Kongbo, Lilung Chu, W bank of river, 29.066667°N, 93.933333°E, 3150–3200 m, 24 May 1938, *F. Ludlow, G. Sherriff & G. Taylor 4424* (holotype, BM [3 sheets] BM00559575–77!).

Shrubs, deciduous, to ca. 1.8 m tall; mature stems dark reddish brown or reddish purple, shiny, sulcate, verruculose; spines 3-fid, pale brownish yellow, 0.8–2.5 cm, abaxially sulcate. Petiole almost absent, rarely to 5 mm; leaf blade abaxially pale green, papillose, adaxially dark green, oblanceolate, 1.5–4.75 × 0.5–1 cm, papery, midvein, lateral veins, and reticulation conspicuous abaxially, inconspicuous adaxially, base narrowly attenuate, margin entire, apex acute, mucronate. Inflorescence a loose raceme, rarely compound or fascicled at base, sometimes sub-fascicled at the tips of stems, 7- to 25-flowered, (1–)3–8 cm including peduncle 1–2 cm; bracts 3–5 mm, apex acuminate; pedicels 10–20 mm, slender. Sepals in 2 whorls; outer sepals oblong-ovate, ca. 6.5 × 2.5 mm; inner sepals oblong-elliptic, ca. 8.5 × 4 mm; petals obovate, ca. 4.5 × 3 mm, base cuneate, glands separate, lanceolate, apex retuse with 2 rounded lobes. Stamens ca. 3 mm; anther connective slightly extended, truncate. Ovules 4. Berry unknown.

Phenology. *Berberis kongboensis* has been collected in flower in May; its fruiting season is unknown.

Distribution and habitat. *Berberis kongboensis* is known from only the type collected in a clearing in Quercus-Ilex forest at 3150–3200 m in Mainling (Milin) Xian in southeast Xizang (Tibet).

178. Berberis leboensis T. S. Ying, Acta Phytotax. Sin. 37: 328. 1999. TYPE: China. S Sichuan: Jinyang Xian, near Linchang, 2700 m, 12 Aug. 1964, *T. P. Zhu et al. 0205* (holotype, PE 00935152!; isotypes, CDBI CDBI0027473 image!, PE 00935153!).

Berberis pseudoamoena T. S. Ying, Acta Phytotax. Sin. 37: 331. 1999; T. S. Ying, Cat. Type Spec. Herb. China, Suppl. 2: 55. 2007, syn. nov. TYPE: China. S Sichuan: Leibo Xian, Huangmaogeng, 3500 m, 23 June 1959, *Liangshan Division Sichuan Economic Plant Exped. 0970* (holotype, PE! 00935190; isotype, PE 00935191!).

Shrubs, deciduous, to 2 m tall; mature stems dark purplish red, sulcate, black verruculose; spines 3-fid, yellow-brown, 0.5–1 cm, terete. Petiole almost absent; leaf blade abaxially pale gray, papillose, adaxially green, oblanceolate or narrowly elliptic, 1–2 × 0.3–0.4 cm, papery, midvein raised abaxially, lateral veins and reticulation indistinct abaxially, inconspicuous adaxially, base cuneate, margin entire, rarely spinose with 1 to 5 inconspicuous teeth on each side, apex acute or acuminate, occasionally rounded, obtuse. Inflorescence a sub-umbel or sub-raceme, rarely fascicled at base, 3- to 6-flowered, 2.5–5 cm overall; peduncle 1.5–3 cm; bracts leaflike, ca. 1.3 cm; pedicel 4–8 mm; bracteoles ovate-lanceolate, ca. 2.5 mm. Sepals in 2 whorls; outer sepals ovate-elliptic, ca. 2.8 × 1.5 mm; inner sepals broadly elliptic, 4–4.3 × ca. 3 mm; petals broadly elliptic, 4–4.5 × ca. 3 mm, base clawed, glands separate, elliptic, apex slightly incised. Stamens ca. 3 mm; anther connective extended, obtuse to rounded. Ovules 2 or 3. Immature berry obovate-oblong, 9–10 × 5–6 mm, contracted at lower part, style not persistent.

Phenology. *Berberis leboensis* has been collected in flower in June and July and with immature fruit in August.

Distribution and habitat. *Berberis leboensis* is known from south Sichuan and northeast Yunnan. It has been found in alpine thickets, open places, and trailsides on mountain slopes at 2700–3700 m.

The protologue stated that the type collection of *Berberis leboensis* was from Leibo Xian. However, the isotypes at CBDI, which have more detailed collection information, give this as Jinyang Xian.

Ying (in Jin & Chen, 2007: 55) validated *Berberis pseudoamoena* whose name was not validly published by Ying (1999: 331) because no type was indicated (Shenzhen Code, Art. 37.1 and 40.1 [Turland et al., 2018]). The protologue description differs little from that of *B. leboensis*.

Huangmaogeng is on the border between Leibo and Meigu Xian which presumably explains why *Liangshan Division Sichuan Economic Plant Exped. 0970* is stated to be collected from the former and *Botany Team 13277* from the latter.

Selected specimens.

S Sichuan. Dechang Xian: Yinlu Gongshe, Niujuan, 3500 m, 30 June 1976, *Dept. of Biology, Southwest Teachers' University 11998* (PE 00935160). **Huili Xian:** Hongqi Qu, Waibai Gongshe, Baima Dadui, 3000 m, 19 July 1976, *Southwest Normal University Department of Biology 12321* (CDBI CDBI0027471–2, 0027474, PE 00935158); Yimen, Longzhou Shan, 3580 m, 23 June 1978, *C. Y. Yin 0048* (SM SM704800139); same details, but *C. Y. Yin 0049* (SM SM704800140). **Jinyang Xian:** Boluo Xiang, San Cun, 3300 m, 20 May 1959, *Survey Team 3175* (PE 01036932, SM SM704800544). **Leibo Xian:** 2900 m, 16 Aug. 1934, *T. T. Yu 3797* (PE 00935154, 00935156, 00935161); Ahe Gou Linchang, 3000 m, 14 Aug. 1972, *236 Collection Team 0796* (PE 00935155, 00935157). **Meigu Xian:** Huangmaogeng, 3500 m, 29 July 1976, *Botany Team 13277* (CDBI CDBI0027470, CDBI0028002, PE 00935159). **Puge Xian:** betw. Jilixiong & Tuojue Qu, 3700 m, 7 Aug. 1960, *Survey Team 25071* (SM SM704800143). **Zhaojue Xian:** Qiliba, 3000 m, 1 June 1982, *Y. J. Li 781* (CDBI CDBI0027446–47).

NE Yunnan. Qiaojia Xian: Yaoshan, Xiaoqua Shan, 3500 m, 19 Aug. 1974, *Kunming Institute of Botany Phytogeography Group 19* (HITBC 074231, KUN 0175680).

179. Berberis lecomtei C. K. Schneid., Pl. Wilson. (Sargent) 1: 373–374. 1913. TYPE: China. NW Yunnan: [Heqing Xian], "Les bois au col de Kouala-po (Ho-kin)" (Heqing), 3000 m, 26 May 1884, *J. M. Delavay 1047* "B" (lectotype, designated here, P P00716521!; isolectotypes, KUN 1206554!, P P02682381!; possible isolectotypes A ex Herb. E. Cosson 00263205!, K K000077424!).

Berberis thunbergii DC. var. *glabra* Franch., Pl. Delavay. 35. 1889. TYPE: China. NW Yunnan [Eryuan Xian]: "Les bois de Kou Toui au dessus de Mo-so-yn," 3000 m, 17 May 1887, *J. M. Delavay s.n.* (lectotype, designated here, P P00835684!; isolectotype, P P00835685!; possible isolectotype K K000077425!).

Shrubs, deciduous, to 2 m tall; mature stems pale reddish brown, subterete or angled, sparsely black verruculose; spines 3-fid, pale yellowish brown, 0.4–1.5 cm, terete or abaxially sulcate, slender, solitary or absent at apex of stems. Petiole almost absent; leaf blade abaxially pale grayish green, papillose, adaxially green, narrowly obovate, 1.4–2 × 0.5–0.8 cm, midvein raised abaxially, lateral veins and reticulation inconspicuous on both surfaces, base attenuate, margin entire, apex rounded or subacute, sometimes minutely mucronate. Inflorescence a sub-fascicle, sub-umbel, or sub-raceme, 4- to 10-flowered, to 1.6 cm overall; pedicel 4–8 mm; bracteoles 1.5–1.7 × 0.8–1 mm. Sepals in 2 whorls; outer sepals narrowly elliptic, 3.5–4 × 1.8–2.2 mm, apex acute; inner sepals elliptic, 4–4.5 × 2.5–3 mm; petals obovate, 3.5–4 × 2.5–2.7 mm, base slightly clawed, glands separate, ca. 0.5 mm, oblong-obovate, apex rounded, emarginate. Stamens 3 mm; anther connective distinctly extended, apiculate. Ovules 2. Berry red, oblong-elliptic, ca. 8 × 5–6 mm; style persistent.

Phenology. *Berberis lecomtei* has been collected in flower in May and June and in fruit in October.

Distribution and habitat. *Berberis lecomtei* is known from Eryuan and Heqing Xian in northwest Yunnan. It has been found in woods and open ground at ca. 3000–3200 m. It is possibly known from Huize Xian in northeast Yunnan.

Schneider's protologue of *Berberis lecomtei* designated as type *Delavay 1047* "in silvis ad collum. Kouala-po (Ho-kein)," 3000 m, 26 May 1884, but did not cite a herbarium, nor did he subsequently (Schneider, 1918: 225). There are two specimens in the Paris herbarium corresponding to this description, while a third was sent to KUN following the herbarium's reorganization. The specimen at P with the most material has been chosen here as the lectotype. Besides these three specimens, there are two other dated Delavay collections numbered *Delavay 1047*: "Les bois, Mt. Hee chan men (Lankong)" of 23 August 1884, which is the type of *B. papillifera* and which is designated *Delavay 1047* "A" under that species, and a non-type specimen: "In monte Hee-chan-men sup. Lan Kong. Aug. 1887" (VNM) which is also *B. lecomtei*. The type collection of *B. lecomtei* of 26 August 1884 is designated here as *Delavay 1047* "B" and that of Aug. 1887 *Delavay 1047* "C." *Delavay 1047* "B" first appeared in Franchet (1887: 386) as "?*B. heteropoda* Schrenk" and there are undated specimens with floral material at A and K numbered *1047* with the same collection details as *Delavay 1047* "B;" these are designated here as probable isolectotypes. Just to complicate matters further, there is an undated specimen with fruit at P (P02465461) on a modern sheet with a much older label recording "*Berberis thunbergii* DC. var. *papillifera* Franch. Col de Hee cha-men" with the number *1047*, the latter possibly added later. This appears to be *B. lecomtei*, though its relationship to any other Delavay collection of the species with fruit is impossible to determine.

The protologue of *Berberis lecomtei* stated that *Delavay 1047* of 26 May 1884 and *Delavay 2447* "in collibus ad collum Yen-tze-hay (Lancong)" of "jun. 1886" "represent the type of Franchet's *B. sinensis typica*" and, in the case of *Delavay 1047*, this was repeated by Ahrendt (1961: 156). There is a double mistake here. First, Franchet's "*B. sinensis typica*" was, in fact, a reference to *B. sinensis* Desf. var. *typica* published by Regel (1873: 416), and the specimens Franchet listed were simply given as examples. Secondly, had this been a new variety, then *Delavay 1047* and *2447* would not have been the types since, though Franchet (1889: 35) did cite these as being this taxon, earlier (1887: 385–386) he had cited neither of these, but various other

Delavay specimens which are all, in fact, *B. amoena* (for further details, see under that species). It is also the case that not only is "*typica*" redundant, but *B. sinensis* is an illegitimate name (see under *B. chinensis*). Delavay *2447* (KUN 1206553, P P02682382, 06868288, VNM) has a much longer inflorescence than Delavay *1047* "B" and, as Schneider (1913: 374) noted, "has more elongated and umbellate racemes and non-apiculate anthers." Whether this is *B. lecomtei* or of an unrecognized species requires further investigation.

The protologue of *Berberis thunbergii* var. *glabra* cited only Delavay *s.n.* "in silvis Kou-toui supra Mo-so-yn" of 17 May 1887, but without citing a herbarium. Ahrendt (1961: 155–156) cited as type a Delavay specimen at K (K000077425) labelled as being from this location, but without date. There are, however, two specimens at P whose collection details correspond exactly to the protologue description. They are of equal quality. Since Schneider's protologue of *B. lecomtei* synonymized *B. thunbergii* var. *glabra*, the one that includes a label in Schneider's hand dated 16 June 1906 has been chosen here as the lectotype and the Kew specimen is treated as only a possible isolectotype.

On current evidence, *Berberis lecomtei* appears to be endemic to the border area of Eryuan and Heqing. But *NE Yunnan Team 282* from Huize Xian in northeast Yunnan (details below) looks remarkably like *B. lecomtei* and has fruit with two seeds. However, an account of flowers from a specimen from the same area is needed to prove this or otherwise. Beyond this, from the evidence of identifications of specimens in both Chinese and non-Chinese herbaria, the name has been used as a taxonomic "dustbin" for a variety of deciduous *Berberis* not only from other areas of Yunnan, but also from Sichuan and Xizang (Tibet). In this context, it is worth noting that all of the non-Delavay specimens cited by Ahrendt (1961) as *B. lecomtei* are, in fact, of other species. Some of all of this reflects a highly complex situation which includes a number of very similar species with either two or two or three ovules (see *B. bowashanensis*, *B. microtricha*, *B. papillifera*, *B. yingii*, *B. yulongshanensis*, and *B. zhongdianensis*) and which almost certainly include undescribed ones.

In September 2013, I made a brief visit to the area where Delavay collected *Berberis lecomtei* and found it to have been planted with dense forests of *Pinus yunnanensis* as a response to earlier forest clearances and logging and consequent soil erosion. Whether the species has survived in this area under such drastic changes is currently unknown.

Ying (2001: 195–196) treated the following as synonyms of *B. lecomtei*: *B. franchetiana* var. *macrobotrys*, *B. humidoumbrosa* var. *inornata*, and *B. tsarongensis* var. *megacarpa*. For the first, see under *B. polybotrys*;

for the second see under Taxa Incompletely Known; for the third, see under *B. moloensis*.

Selected specimens.
NW Yunnan. Heqing Xian: "In monte Hee-chan-men, sup. Lankong," Aug. 1887, *J. M. Delavay 1047* "C" (VNM); same details, but *J. M. Delavay 1047bis* (PE 01901815); "les bois de Fang yang Tchang au dessus de Mo-so-yn," 21 Oct. 1887, *J. M. Delavay s.n.* (K, P P00716518–20); "les bois de Fang yang Tchang, 24 May 1889, *J. M. Delavay s.n.* (KUN 1206510, P 02682372).

Possible specimens.
NE Yunnan. Huize Xian: Dahai, betw. Dishuiyan & Hongzi Kou, 3600 m, 21 July 1964, *NE Yunnan Team 282* (KUN 0176217, 0179351).

180. Berberis lepidifolia Ahrendt, Bull. Misc. Inform. Kew 1939: 269. 1939. TYPE: China. NW Yunnan: [Lanping Xian], Chien-Chuan–Mekong Divide, 26.33°N, 99.33°E, 3350 m, Aug. 1923, *G. Forrest 23614* (holotype, K K000077409!; isotypes, BM BM00056429!, E E00217954!, HBG HBG-506755 image!, IBSC 0091926 image!, PE 01032794!, S 08–783 image!).

Shrubs, deciduous, to 1.2 m tall; mature stems pale brownish yellow, sulcate, sparsely black verruculose; spines 1- to 3-fid, concolorous, 0.3–1 cm, weak, often absent, terete. Petiole almost absent; leaf blade abaxially gray-green, papillose, adaxially dull deep green, narrowly oblanceolate or obovate, 1.7–4.5 × 0.2–0.5 cm, papery, midvein obviously raised abaxially, slightly impressed adaxially, lateral veins and reticulation indistinct or inconspicuous on both surfaces, reticulate veins indistinct on both surfaces, base attenuate, margin entire, apex obtuse or acute. Inflorescence a sub-umbel or sub-raceme, 3- to 8-flowered, 1.5–2.5 cm overall; bracts lanceolate, ca. 1.5 mm; pedicel 5–10 mm, slender. Flowers ca. 4 mm diam. Sepals in 1 whorl, center red with a yellow margin, ovate, 2–2.5 × 1.3–1.5 mm, subacute; petals oblong-elliptic, ca. 2.5 × 1.5 mm, base cuneate, glands separate, ovate-elliptic, apex emarginate with acute lobes. Stamens ca. 2 mm. Ovules 2, shortly stipitate. Berry black-purple, ovoid-oblong, 8–11 × 5–7 mm; style persistent, short; seeds 2.

Phenology. *Berberis lepidifolia* has been collected in fruit in August. Its flowering period is unknown.

Distribution and habitat. *Berberis lepidifolia* is known from only two collections in Lanping Xian in northwest Yunnan. The collector's notes to the type records it as being found on the ledges of cliffs and stony slopes at 3350 m, that of *H. T. Tsai 53993* (see below) as being found in thickets at 3000 m.

The protologue of *Berberis lepidifolia* cited *G. Forrest 22051* "between Yung-peh and Yung-ning, S.W. Sichuan" (but actually collected in Ninglang Xian in northwest Yunnan) as a likely further example of the species. Later, Ahrendt (1961: 154) cited this as *B. lepidifolia* without qualification. This, however, is different and is it treated here as the type of a new species, *B. ninglangensis*.

The description of flowers used above is Ahrendt's from the protologue and is based on cultivated material. As such, it should be treated with caution.

X. H. Li (2010: 440) treated *Berberis heteropsis* from Guizhou as a synonym of *B. lepidifolia*. Given that the type gathering of *B. heteropsis* has dark reddish brown mature stems, abaxially pruinose leaves, and salmon-red globose fruit, this is unsustainable. *Berberis heteropsis* is treated here as a synonym of *B. wilsoniae*.

Selected specimens.
NW Yunnan. Lanping Xian: 3000 m, 17 Aug. 1933, *H. T. Tsai 53993* (A 00279652, IBSC 0091888, KUN 0176456, LBG 00064191, PE 01033917, SZ 00291393, WUK 0035741).

181. Berberis leptoclada Diels, Notes Roy. Bot. Gard. Edinburgh 5: 167. 1912. TYPE: China. NW Yunnan: [Zhongdian (Xianggelila) Xian], descent from the Chung-Tien Plateau to the Yangtze near Tang-Tui, 27.75°N, 3050 m, Sep. 1904, *G. Forrest 330* (holotype, E E00217955!).

Berberis weisiensis C. Y. Wu ex S. Y. Bao, Bull. Bot. Res., Harbin 5(3): 17. 1985, syn. nov. TYPE: China. NW Yunnan: Weixi Xian, Liu Qu, Xiaruo, Zhubaluo, 2000 m, 27 Oct. 1956, *P. Y. Mao 00864* (holotype, KUN 0250101!; isotype, PE 01032927!).

Shrubs, deciduous, to 1.8 m tall; mature stems purple, sulcate, thin, glaucous pruinose; spines 1- to 3-fid, pale brownish yellow, 0.2–1.3 cm, terete, weak. Petiole almost absent, rarely to 2 mm; leaf blade abaxially pale green, glaucous pruinose, adaxially green, obovate, rarely sub-spatulate, 0.5–1.5 × 0.4–0.7 cm, papery, midvein and lateral veins raised abaxially, venation indistinct or obscure adaxially, margin entire, rarely spinulose with 1 or 2 teeth on each side, base cuneate or attenuate, apex obtuse or subacute, mucronate. Inflorescence a raceme, sometimes a sub-raceme, sometimes fascicled at base, rarely a fascicle, sub-fascicle, or sub-umbel, 1- to 7-flowered, 2–4 cm overall including peduncle to ca. 1 cm; pedicel 4–7 mm (to 9 when from base), rachis and pedicels both partially glaucous pruinose; bracteoles lanceolate-ovate, ca. 2.5 × 1 mm. Sepals in 2 whorls; outer sepals elliptic, 3.5 × 1.75 mm; inner sepals elliptic-suborbicular, 5 × 3.75 mm; petals broadly obovate, ca. 4 × 2.5 mm, base slightly clawed or cuneate, glands separate, elliptic, apex emar-

ginate. Stamens ca. 3 mm; anther connective slightly extended, truncate. Ovules 3. Berry globose, red, 7 × 5 mm; style persistent; seeds 3.

Phenology. *Berberis leptoclada* has been collected in flower between May and September and in fruit in October.

Distribution and habitat. *Berberis leptoclada* is known from Zhongdian (Xianggelila) Xian and Weixi Xian in northwest Yunnan. It has been found on mountain slopes, dry rocky hillsides, and in dry *Pinus yunnanensis* scrub at ca. 2100–3050 m in Yulong and Zhongdian Xian and between a river and roadside in Weixi Xian.

Schneider (1918: 222) treated *Berberis leptoclada* as a possible synonym of *B. amoena*, though subsequently he made a dissection of a flower of the type (to be found on the type sheet and used in the description above) which showed it had a different structure from *B. amoena*. However, his original synonymy was followed by Ahrendt (1961: 154) and Ying (2001: 168). *Forrest 12887* was identified as *B. sichuanica* by Ying (1999: 329).

Berberis weisiensis was published solely on the basis of the holotype at KUN, though the isotype at PE is a better specimen. This collection is the source of fruit description given above.

Selected specimens.
NW Yunnan. Lijiang (now Yulong) Xian: "Li-kiang Hsien," 2700 m, July 1935, *C. W. Wang 71551* (A 00280041, IBSC 0091927, KUN 0176226, LBG 00064297, NAS NAS00313979, PE 01037335–36, WUK 0045506). **Zhongdian (Xianggelila) Xian:** "Yünnan bor.-occid.: In regionis subtropicae vallis fluvii Djinscha-djiang [Yangtse] ad boreo-occid. urbis Lidjiang [Lijiang], silvis siccisa vico Bölo ad Rouscha (ad affluentem)," 27.733333–27.766667°N, 2100–2200 m, 5 June 1916, *H. R. E. Handel-Mazzetti 8811* (A 00279984, E E00395985, W 1930–4039, WU 093258); Mekong–Yangtze divide, 27.666667°N, 3350–3660 m, July 1914, *G. Forrest 12887* (BM, E E00395981, E00395984, PE 00935200); Nixi, 42 km S of Benzilan, 2900 m, 27 May 1992, *Sino-Scottish Exped. to NW Yunnan SSY 123* (E E00073642).

182. Berberis lhunzensis Harber, sp. nov. TYPE: Cultivated. Sherwood Gardens, Exeter, Devon, U.K., 15 May 2013, *V. Gallavan s.n.* from seed of *A. Clark 4339*, China. SE Xizang (Tibet): Lhünzê (Longzi) Xian, Shopa, 4300 m, 20 Sep. 2000 (holotype, PE!; isotypes, A!, E!).

Diagnosis. *Berberis lhunzensis* has mature stems and leaves that are somewhat similar to *B. humidoumbrosa* (also from Lhünzê Xian, though collected at lower elevations) but with a fascicled rather than a racemose inflorescence and a different flower structure with a greater number of ovules.

Shrubs, deciduous, height outside cultivation unknown; mature stems pale reddish brown, sulcate, verruculose; spines 3-fid, pale reddish brown, 0.6–2.2 cm, abaxially sulcate. Petiole almost absent; leaf blade abaxially pale green, papillose, adaxially green, narrowly obovate or obovate-elliptic, (0.8–)1.2–2.5(–3) × 0.5–1.1 cm, papery, midvein raised abaxially, lateral veins and reticulation inconspicuous abaxially (conspicuous when dry), indistinct or obscure adaxially, base attenuate, margin entire, rarely spinulose with 1 to 4 teeth on each side, apex obtuse or subacute, mucronate. Inflorescence a fascicle, 4- to 6-flowered; pedicel 4–8 mm, slender; bracteoles triangular-ovate, 1–1.25 × 0.75–1 mm, apex acuminate. Sepals in 3 whorls; outer sepals ovate or lanceolate-ovate, 2.5–3 × 1–1.5 mm; median sepals oblong-ovate or ovate-elliptic, 3–3.5 × 2–2.5 mm; inner sepals broadly elliptic, 4–5 × 3–4 mm; petals broadly obovate, 3.75–4 × 3 mm, base slightly clawed, glands contiguous, apex entire. Stamens 3 mm; anther connective extended, truncate. Pistil 3 mm; ovules 5 to 7. Berry red, ovoid, 12–14 × 4–6 mm; style persistent.

Phenology. *Berberis lhunzensis* has been collected in fruit in September; its flowering season in the wild is unknown.

Distribution and habitat. *Berberis lhunzensis* is known from only cultivated plants grown from seed collected from a plant growing with *Rosa* in a yak meadow at 4300 m in Lhünzê (Longzi) Xian in southeast Xizang (Tibet).

IUCN Red List category. *Berberis lhunzensis* is assessed as DD or Data Deficient, according to IUCN (2001) criteria.

In cultivation in 2015, *Berberis lhunzensis* was some 2 m tall.

There are very few *Berberis* collections from Lhünzê Xian and it appears there is only one other from a comparable elevation. This is *Z. Y. Wu et al. 862* (HNWP 53885, KUN 0179328, 0179331) Longzi Gou, 47 km, 4530 m, 17 July 1975. This is fascicled and might be *B. lhunzensis*, though the mature stems appear to be darker than those of the type. From the evidence of images of the KUN specimens, this collection has floral material, though I have not had the opportunity to examine this.

Cultivated material:
Living cultivated plants.
Sherwood Gardens, Exeter, U.K. Plants from seed of *Clark 4339* (details above).

Cultivated specimens.
Sherwood Gardens, Exeter, Devon, U.K., 24 Sep. 2015, *V. Gallavan s.n.* from seed of *A. Clark 4339* (A, E, K, PE, WSY).

183. Berberis lhunzhubensis Harber, sp. nov. TYPE: China. C Xizang (Tibet): Lhünzhub (Linzhou) Xian, Reting, "60 miles [96.56064 km] N. of Lhasa," 30.366667°N, 91.466667°E, 4115 m, 12 July 1944, *F. Ludlow & G. Sherriff 9974* (holotype, BM BM000939687!; isotype, E E00351618!).

Diagnosis. *Berberis lhunzhubensis* is similar to *B. hemsleyana* but is 1(or 2)-flowered with a different flower structure including four or five ovules.

Shrubs, deciduous, to 2.5 m tall; mature stems purple or reddish purple, sulcate; spines 3-fid, pale yellowish brown, 0.6–1.5 cm, abaxially sulcate. Petiole almost absent; leaf blade narrowly obovate or elliptic-obovate, 1–1.5 × 0.4–0.6 cm, abaxially pale green, papillose, adaxially green, papery, midvein and lateral veins inconspicuous abaxially, indistinct adaxially, base attenuate, margin entire, rarely spinulose with 1 or 2 teeth on each side, apex subacute, sometimes minutely mucronate. Inflorescence 1-flowered, sometimes paired; pedicel 6–12 mm. Flowers pale yellow. Sepals in 2 whorls; outer sepals oblong-ovate to elliptic, 5–6 × 2–2.5 mm; inner sepals broadly obovate, 6.5–7.5 × 3.5–4.5 mm; petals obovate, 4–4.5 × 3–3.5 mm, base cuneate or clawed, glands obovoid, 1 mm, apex notched ca. 0.5 mm deep. Stamens 3–3.5 mm; anther connective not extended. Ovules 4 or 5. Berry red, ovoid, 10–12 × 4–6 mm; style persistent.

Phenology. *Berberis lhunzhubensis* has been collected in flower between May and July and in fruit in September and October.

Distribution and habitat. *Berberis lhunzhubensis* is known from Damxung (Dangxiong), Lhünzhub (Linzhou), and Maizhokunggar (Mozhugongka Xian) in central Xizang, and Biru Xian in northeast Xizang. It has been collected among *Abies* and from thickets on valley sides and mountain slopes at ca. 3800–4500 m.

IUCN Red List category. *Berberis lhunzhubensis* is assessed as DD or Data Deficient, according to IUCN (2001) criteria.

F. Ludlow & G. Sherriff 9974 was identified by Ahrendt (1961: 128) as an example of *Berberis dictyophylla* var. *epruinosa* whose type is from Kangding, treated here as a synonym of *B. ambrozyana*. Given that it was collected some 1000 km west of Kangding, this identification was inherently unlikely and, indeed, *B. ambrozyana* has a different flower structure and is always 1-flowered.

The description of the flowers given above is from a flower of the isotype. This had four ovules. Annotations on the sheets of *Qinghai–Xizang Botany Team 7411* and *G. D. Fu 203*, cited below, record four ovules and five ovules, respectively. *D. D. Tao 11343* from Biru Xian was collected a considerable distance to the northeast of the other collections, but appears to be of the same species. The specimens at KUN produced three to five seeds per fruit.

The collectors' notes to *F. Ludlow & G. Sherriff 9974* do not record the height of the plant while those of *Xizang Herbal Medicine Team 1971* and *D. D Tao* record 1.5–2 m and 2.5 m, respectively.

I have not located any other *Berberis* specimens of any description from Biru, Damxung, or Lhünzhub Xian.

Selected specimens.
C & NE Xizang (Tibet). Biru Xian: Baiga Qu, Houshan, 3800 m, 9 Sep. 1976, *Qinghai–Xizang Team, Nagqu Team, D. D Tao 11343* (HNWP 58694, KUN 0178802–03, PE 0137832–3). **Damxung (Dangxiong) Xian:** Lagende, 4500 m, 8 June 1960, *G. D. Fu 203* (PE 01037437); Xugen Si, Tanggulashan, 4400 m, 28 May 1975, *Qinghai–Xizang Botany Team 7411* (PE 01032833). **Lhünzhub (Linzhou) Xian:** Pangdu Qu, Dalongxiang, 4100 m, 9 Oct. 1972, *Xizang Herbal Medicine Team 1971* (HNWP 30647, 80291, PE 00049429, 01840052). **Maizhokunggar (Mozhugongka) Xian:** ca. 30 km E of Maizhokunggar on Rte. 318, 29.703333°N, 92.121389°E, 4256 m, 2 Sep. 2006, *Tibet McArthur 33* (US 00995016).

184. Berberis lixianensis Harber, sp. nov. TYPE: China. C Sichuan: Li Xian, Siboguo Shangzhai, on side rd. WSW off National Rd. G317 S of Dalangba Cun, 31.681944°N, 102.736111°E, 3040 m, 9 May 2014, *D. E. Boufford, S. Cristoph, C. Davidson, Y. D. Gao & Q. Y. Xiang 43624* (holotype, PE image!; isotypes, A image!, CDBI image!, E!, MO image!, TI image!).

Diagnosis. Berberis lixianensis has leaves that are very similar to those of *B. silva-taroucana*, except they appear to always have entire margins while those of *B. silva-taroucana* are sometimes spinose. Its mature stems are very pale brownish yellow versus the purple stems of *B. silva-taroucana*, its flowers are smaller, and its flower structure different.

Shrubs, deciduous, to 2 m tall; mature stems very pale brownish yellow, subterete; spines 1- to 3-fid, mostly solitary, mostly absent toward apex of stems, 3-fid to 3.2 cm at base of main branches, concolorous 0.5–2.2(–3.2) cm, abaxially sulcate. Petiole almost absent, sometimes to 13 mm; leaf blade abaxially pale gray-green, papillose, adaxially dull green, obovate or obovate-elliptic, 1.5–2.8(–3.5) × (0.4–)0.8–1.1(–1.6) cm, papery, midvein raised abaxially, impressed adaxially, lateral and reticulate veins indistinct abaxially, inconspicuous adaxially, base attenuate, often narrowly so, margin entire, apex obtuse or acute. Inflorescence a fascicle, sub-fascicle, umbel, or rarely a sub-raceme, 3- to 7(to 9)-flowered, to 2 cm overall; peduncle (when present) to 6 mm; pedicel 6–10 mm; bracteoles slightly reddish at base, triangular-ovate, 1 × 1 mm. Flowers pale yellow. Sepals in 2 whorls; outer sepals sometimes with partial pale reddish tinge, narrowly elliptic, 2.5–3.25 × 1.5–1.75 mm; inner sepals obovate, 3.5 × 2 mm; petals obovate-elliptic, ca. 2.5 × 1.75 mm, base clawed, glands separate, ca. 0.75 mm, apex entire. Stamens ca. 2.75 mm; anther connective not extended, truncate. Pistil ca. 2.25 mm; ovules 2. Berry red, ovoid or oblong, 9 × 6 mm; style not persistent.

Phenology. Berberis lixianensis has been collected in flower in May and in fruit between June and September.

Distribution and habitat. Berberis lixianensis is known from Li Xian in central Sichuan. It has been found on roadsides and in cut-over *Picea* forest with scrub vegetation at 2640–3200 m.

IUCN Red List category. Berberis lixianensis is assessed as DD or Data Deficient, according to IUCN (2001) criteria.

The type and all of the herbarium specimens cited below come from the same area of Li Xian. Whether it is found beyond this area requires further investigation. I do not know where in Li Xian *A. Clark 4894* was collected.

Selected specimens.
C Sichuan. Li Xian: Laisu Gou, Dalang Ba, 30 Gou, 3200 m, 3 Sep. 1956, *H. P. He 46156* (IBSC 0091603); Miyaluo, 2740 m, 22 June 1958, *S. Y. Chen 5289* (NAS NAS00314213, SM SM704800194); Lugan Qiao, 26 June 1958, *S. Y. Chen et al. 5358* (NAS NAS00314212, SM SM704800193); en route betw. Miyaluo Shixi Linchang & Lugan Qiao, 2640 m, 6 Aug. 1958, *Z. He et al. 5763* (NAS NAS00314211, SM SM704800192); Maxi Gou, 4 Aug. 1958, *Forestry Research Institute 33596* (PE 01033815, 01033818); Miyaluo, 3100 m, 11 May 2004, *P. D. Zuang 2004044* (WCSBG 001574–75); Miyaluo; 3100 m, 13 Sep. 2004, *Z. B. Feng 20041013* (WCSBG 001572–73).

Cultivated material:
Living cultivated plants.
Foster Clough, Mytholmroyd, West Yorkshire, U.K., from *A. Clark 4894*, Sichuan: Li Xian, 3000 m, 26 Sep. 2001.

Cultivated specimens.
From above plant, 30 Apr. 2015, *J. F. Harber 2015-26* (BM, E, K).

185. Berberis longipedicellata Harber, sp. nov. TYPE: China. NC Sichuan: Xiaojin Xian, SE of city of Xiaojin on rd. to pass over Jiajin Xian, 30.874444°N, 102.631389°E, 3350–3400 m, 28

July 2007, *D. E. Boufford, K. Fujikawa, S. L. Kelley, R. H. Ree, B. Xu, D. C. Zhang, J. W. Zhang, T. C. Zhang & W. D. Zhu 38505* (holotype, A!; isotypes, CAS!, E!, KUN not seen, MO!, TI!).

Diagnosis. Berberis longipedicellata is somewhat similar to *B. bawangshanensis* but has longer pedicels and fruit with two seeds versus the five ovules of *bawangshanensis*. It is somewhat similar to *B. diaphana*, particularly in relation to the bent tip of its berry, but is distinguished by being mostly single-flowered.

Shrubs, deciduous, to 2 m tall; mature stems pale yellowish gray, subterete to sulcate; spines 3-fid, solitary or absent toward apex of stems, pale brownish yellow, 0.8–2.8 cm, abaxially sulcate, weak. Petiole almost absent or to 5 mm; leaf blade abaxially pale green, adaxially mid-green, obovate, 2.5–3.5(–4) × 1.2–1.6 cm, papery, midvein raised abaxially, lateral veins and reticulation indistinct or inconspicuous abaxially, inconspicuous adaxially, base attenuate, margin minutely spinulose with 12 to 20 teeth on each side or entire, apex obtuse or subacute, sometimes minutely mucronate. Inflorescence 1(or 2)-flowered; pedicel (15–)25–30(–35) mm. Flowers unknown. Berry red, ovoid-oblong, 13–16 × 6 mm, apex markedly bent; style persistent; seeds 2.

Phenology. Berberis longipedicellata has been collected in fruit in July and September. Its flowering period is unknown.

Distribution and habitat. Berberis longipedicellata is known from Danba, Dawu (Daofu), Wenchuan, and Xiaojin Xian, and possibly Luding Xian (see below) in west and west-central Sichuan. The type was collected in the remnants of *Abies* forest at 3350–3400 m, *H. Smith 12765* on a forest margin at 3700 m, *S. Y. Hu 2604* among *Clematis*, *Rosa*, and *Spiraea* at 3350–3660 m.

IUCN Red List category. Berberis longipedicellata is assessed as DD or Data Deficient, according to IUCN (2001) criteria.

Berberis longipedicellata is one of a small number of species found in Sichuan with berries that are distinctly bent at the apex, the others being *B. aemulans*, *B. tachiensis*, *B. tischleri*, and *B. diaphana*, the last named being found mainly in Gansu and Qinghai. *Berberis longipedicellata* can be distinguished from all of these by having single (and occasionally paired) flowers with two ovules. However, all of these species sometimes are single-flowered toward the ends of branches, though all have more ovules. It is in this context that *Sichuan Botany Team 1108*, cited below, should be treated as only possibly *B. longipedicellata* in that,

though 0028234 has four very long pedicels, it has only extremely immature fruit (0028235 is sterile) and, only having seen an image, I have no information as to the number of seeds or ovules in these fruit. Nevertheless, its leaves are narrower than *B. tachiensis* (which is also recorded for Luding Xian) which suggests it is *B. longipedicellata*.

W. J. Zheng 10016 (KUN 0176513), an evergreen specimen from Emeishan Xian annotated by C. Y. Wu in 1963 as an unpublished *Berberis longipedicellata*, is probably *B. simulans*.

Selected specimens.
W Sichuan. Danba Xian: Yi Qu, Qianjin Gongshe, Shacong Gou, 3500 m, 23 Aug. 1974, *s. coll. 05438* (PE 01037761). [**Dawu (Daofu) Xian**]: "Between Taining (Ngata) and Taofu (Dawo), Tjedji La," 3700 m, 30 Sep. 1934, *H. Smith 12765* (MO 2388828, UPS BOT: V-078124). **Luding Xian:** Qianghuopeng, 3100 m, 28 June 1974, *Sichuan Botany Team 1108* (CDBI CDBI0028234–35). **Wenchuan Xian:** "Winchuan, Tsao-puh," 3350–3660 m, 6 Aug. 1942, *S. Y. Hu 2604* (ex Herb. Girling College, A 0028063). **Xiaojin Xian:** SE of the city of Xiaojin and S of Dawei-Xiang near Xiamucheng along provincial hwy. S210 (rd. to city of Baoxing), 30.88694444°N, 102.65333333°E, 3650 m, 31 July 2018, *D. E. Boufford, B. Bartholomew, J. F. Harber, Q. Li & J. P. Yue 44299* (A, CAS, E, KUN, PE).

186. Berberis ludlowii Ahrendt, J. Bot. 79(Suppl.): 43. 1941. TYPE: China. SE Xizang (Tibet): [Mainling (Milin) Xian], Kongbo, Singo Samba, Langong Chu, 28.866667°N, 93.866667°E, 3350 m, 10 Oct. 1938, *F. Ludlow, G. Sherriff & G. Taylor 6589* (holotype, [2 sheets] BM BM000559578–79!).

Shrubs, deciduous, to 4 m tall; mature stems dark purple, sulcate, shiny; spines 1- to 3-fid, sometimes absent, concolorous, 0.3–0.7(–1) cm, terete. Petiole almost absent, sometimes 2–5 mm; leaf blade abaxially grayish green, papillose, adaxially dark green, narrowly obovate or oblong-obovate, (2.5–)3.3–7 × 0.9–2 cm, papery, midvein raised abaxially, slightly impressed adaxially, lateral veins and reticulation inconspicuous on both surfaces, base attenuate, margin entire or rarely spinose-serrulate with 1 to 4 teeth on each side, apex obtuse or subacute, sometimes minutely mucronate. Inflorescence 1-flowered, otherwise unknown; fruiting pedicel 25–50(–55) mm. Berry red, ovoid or ellipsoid, 15–18 × 6–9 mm; style not persistent; seeds 5 to 7.

Phenology. Berberis ludlowii has been collected in flower in May and in fruit between August and October.

Distribution and habitat. Berberis ludlowii is known from Lhünzê (Longzi), Mainling (Milin), and Nang (Lang) Xian in southeast Xizang (Tibet). It has been collected from margins of *Abies* forests by riverbanks, among *Rhododendron*, mixed forests, and grassy valley sides at 3035–3950 m.

The protologue described the leaves of *Berberis ludlowii* as being up to 4 cm long and this was repeated by Ahrendt (1961: 114). In fact, they are often up to 6 cm and sometimes to 7 cm. The protologue also included an incomplete description of flowers. Only two specimens with flowers were cited: *F. Ludlow, G. Sherriff & G. Taylor 3851* (BM BM000939689, but also E E00395945) and *F. Ludlow & G. Sherriff 1626* (BM BM000939688). Both have pale yellowish brown mature stems and much smaller leaves than the type of *B. ludlowii*. The floral description must have come from the former since the flower structure of the latter (treated here as a new species, *B. cornuta*) is very different from that given by Ahrendt. It is likely that this former is of an undescribed species. I have therefore ignored Ahrendt's description.

Ludlow described *Berberis ludlowii* in the area where *F. Ludlow, G. Sherriff & G. Taylor 6337* (details below) was collected, growing "not in hundreds or thousands, but in tens and hundreds of thousands, whose fiery red leaves glowed so vividly that the whole hillside for mile upon mile seemed to be ablaze" (Fletcher, 1975: 221).

Berberis ludlowii var. *capillaris* and *B. ludlowii* var. *deleica*, both published by Ahrendt (1961: 115), would not appear to be related to *B. ludlowii*. The former is treated here as a synonym of *B. capillaris*, the latter is not recorded for China and there is no evidence to relate it to *B. ludlowii*.

Berberis ludlowii was treated as synonym of *B. muliensis* from southwest Sichuan by Ying (2001: 79). However, as well being found some 700 km to the west, *B. ludlowii* is a much taller species with much larger leaves and is always single-flowered and has five to seven seeds, whereas *B. muliensis* is 1- to 3-fascicled and has three to five seeds.

Selected specimens.
SE Xizang (Tibet). Lhünzê (Longzi) Xian: Zhari Xiang, 3035 m, 19 Sep. 2010, *Y. D. Tang & Q. W. Lin 2010-108* (PE 02031483); Zhari Xiang, Qusang Cun en route to Zhari, 6 Aug. 2013, *Y. S. Chen et al. 13-0574* (PE 01992036). **[Mainling Xian]:** Tsari Distr., Migyitun, 28.666667°N, 93.633333°E, 3050 m, 23 Oct. 1938, *F. Ludlow, G. Sherriff & G. Taylor 6623* (BM BM000939750). **[Nang (Lang) Xian]:** Bimbi La, 28.75°N, 93.466667°E, 3960 m, 14 Oct. 1938, *F. Ludlow, G. Sherriff & G. Taylor 6337* (BM BM000939749, E E00395944).

187. Berberis luhuoensis T. S. Ying, Acta Phytotax. Sin. 37(4): 323. 1999. TYPE: China. NW Sichuan: Luhuo Xian, Jiabangou, 3100 m, 11 Aug. 1960, *T. S. Ying et al. 4627* (holotype, PE 00935162!; isotype, KUN 0178452!).

Shrubs, deciduous, to 1.2 m tall; mature stems yellowish brown, angled; spines 1- to 3-fid, concolorous, 0.6–1.5 cm, absent toward end of stems, terete. Petiole almost absent or to 2 mm; leaf blade abaxially pale green, adaxially deep green, obovate or obovate-elliptic, 2.5–4 × 0.8–2 cm, papery, midvein raised abaxially, slightly impressed adaxially, lateral veins raised abaxially, inconspicuous adaxially, reticulation inconspicuous on both surfaces, base attenuate, margin spinose with 4 to 20 teeth on each side, apex obtuse or rounded. Inflorescence a narrow panicle, (20 to)30- to 50-flowered, 4–13 cm overall including peduncle (0.5–)3–4.5 cm; pedicel 2–7 mm. Flowers unknown. Immature berry oblong, 7–8 × 2–2.1 mm, slightly pruinose; style not persistent; number of seeds unrecorded.

Phenology. *Berberis luhuoensis* has been collected in fruit in August. Its flowering period is unknown.

Distribution and habitat. *Berberis luhuoensis* is known from only the type collection from Luhuo Xian, northwest Sichuan. This was found on a mountainside at 3100 m.

Besides the type, Ying cited *K. Y. Lang & L. Q. Li 539* (KUN 0178906, PE 00935163–64), Luding Xian, Xinxing, E flank of Gonga Shan, 2100 m, 7 July 1982, and the protologue is a composite description of both gatherings including the flowers of no. *539*. However, this collection differs from the type in having leaves whose margins are consistently entire and appears to be *Berberis prattii*.

188. Berberis mabiluoensis Harber, sp. nov. TYPE: Cultivated. Foster Clough, Mytholmroyd, West Yorkshire, U.K., 29 Apr. 2015, *J. F. Harber 2015-22*, from seeds of *Z. L. Dao, M. Lear, Y. H. Li & M. Wickenden (KWL) 304*, China. NW Yunnan: Gongshan Xian, Mabiluo River, 28.321083°N, 98.278033°E, 3130 m, 25 Sep. 2008 (holotype, PE!; isotypes, A!, BM!, CAS!, E!, IBSC!, K!, KUN!, MO!, TI!, WSY!).

Diagnosis. *Berberis mabiluoensis* is somewhat similar to *B. atroviridiana* (the type of which was collected some 25 km to the northeast of where *KWL 304* was collected) but has larger leaves and a more varied inflorescence with flowers with six or seven ovules versus the two seeds recorded for *B. atroviridiana*.

Shrubs, deciduous, height outside cultivation unknown; mature stems dark purplish red, sulcate, not verruculose; spines 3-fid, pale brown, 1–1.5 cm, abaxially sulcate, weak. Petiole almost absent or to 5 mm; leaf blade abaxially pale green, papillose, adaxially dull green, narrowly or broadly obovate, (2.5–)3–5.5 × 1.5–2.5(–3.3) cm, papery, midvein raised abaxially, slightly impressed adaxially, lateral veins and reticulation indistinct or obscure abaxially, indistinct or inconspicuous adaxially, base attenuate, margin entire or spinulose with 1 to 7(to 12) mostly inconspicuous teeth on each side, apex obtuse, sometimes minutely mucronate. Inflorescence a fascicle, sub-fascicle, sub-umbel, umbel, or sub-raceme, sometimes fascicled at base, 3- to 8-flowered, to 2.5 cm overall; peduncle (when present) to 1.2 cm; pedicel 8–18 mm; bracteoles triangular, 2.75 × 1.5 mm. Flowers yellow. Sepals in 2 whorls; outer sepals sometimes with reddish vertical stripe, obovate-elliptic or obovate-oblong, 6.5 × 4 mm; inner sepals obovate or obovate-elliptic, 7 × 4–4.5 mm; petals broadly obovate or obovate-orbicular, 4–6 × 4–5 mm, base cuneate or slightly clawed, glands separate, ovoid or oblong, ca. 0.7 mm, apex entire. Stamens ca. 4 mm; anther connective slightly extended, truncate. Ovules 6 or 7. Berry red, ellipsoid or oblong, 12 × 6 mm; style not persistent.

Phenology. Seeds of *Berberis mabiluoensis* have been collected in September. Its flowering season in the wild is unknown.

Distribution and habitat. *Berberis mabiluoensis* is known from only cultivated plants grown from seed of *KWL 304* collected in the Mabiluo River valley, a side valley of the northwest arm of the Dulong Jiang in Gongshan Xian in northwest Yunnan.

IUCN Red List category. *Berberis mabiluoensis* is assessed as DD or Data Deficient, according to IUCN (2001) criteria.

My plants of *Berberis mabiluoensis* were supplied by the late Michael Wickenden of Cally Gardens, Castle Douglas, Scotland. For an account of the expedition that collected the seeds of *KWL 304*, see Wickenden and Lear (2010).

The leaves of the type specimens are not fully formed. This is because the flowers and the leaves of the plant appeared almost simultaneously.

Though the height of *KWL 304* was unrecorded, from the evidence of my cultivated plants, it is likely to reach 2 m or more.

Cultivated material:
Living cultivated plants.
Foster Clough, Mytholmroyd, West Yorkshire, U.K., from *Z. L. Dao, M. Lear, Y. H. Li & M. Wickenden (KWL) 304.*

Cultivated specimens.
From above plant, *J. F. Harber 2015-25*, 5 Aug. 2015 (A, BM, CAS, E, K, KUN, PE, TI).

189. Berberis markamensis Harber, sp. nov. TYPE: China. E Xizang (Tibet): Markam (Mang-

kang) Xian, Mangcuo Hu, 4300 m, 15 June 1976, *Qinghai-Xizang Botany Team 11882* (holotype, KUN 0178848!; isotypes, HNWP 6560 not seen, KUN 0178849!, PE [2 sheets] 01037688!, 01037829!).

Diagnosis. *Berberis markamensis* is somewhat similar to *B. ambrozyana*, but with a different flower structure.

Shrubs, deciduous, to 2 m tall; mature stems purple, sulcate; spines 3-fid, pale orange, 0.5–1.7 cm, terete, internodes 8–18 mm. Petiole almost absent; leaf blade abaxially pale green, adaxially green, papillose, narrowly obovate or oblong-obovate, 0.6–0.8 × 0.2–0.35 cm, papery, midvein and reticulation conspicuous abaxially, inconspicuous adaxially, base attenuate, margin entire, apex obtuse or subacute, mucronate. Inflorescence 1-flowered; pedicel 2 mm. Sepals in 3 whorls; outer sepals oblong-ovate, 3 × 1.5 mm; median sepals oblong-elliptic, 4 × 2.5 mm; inner sepals broadly obovate, 4 × 3 mm, apex obtuse; petals obovate, 3 × 2.5 mm, base slightly clawed, glands approximate, apex slightly emarginate. Stamens ca. 2 mm; anther connective slightly extended, truncate. Ovules 1 or 2. Berry unknown.

Phenology. *Berberis markamensis* has been collected in flower in May and June. Its fruiting period is unknown.

Distribution and habitat. *Berberis markamensis* is known from Karuo Qu (formerly Qamdo [Changdu] Xian) and Markam (Mangkang) Xian in east Xizang (Tibet). It has been found on mountain slopes, thickets on grassland, and roadsides at 3900–4200 m.

IUCN Red List category. *Berberis markamensis* is assessed as DD or Data Deficient, according to IUCN (2001) criteria.

All the specimens cited here appear to have been taken from the tips of branches. It is possible that leaves further down may be somewhat larger than described above.

Selected specimens.
E Xizang (Tibet). Markam (Mangkang) Xian: Mangcuo Hu, 4200 m, 14 June 1976, *Qinghai-Xizang Botany Team 11842* (HNWP 62182, KUN 0179148, 0178860, PE 01037828). **Qamdo (Changdu) Xian (now Karub [Karuo] Qu):** Famuchang, 3900 m, 25 May 1976, *Qinghai-Xizang Botany Team 11607* (KUN 0179335).

190. Berberis mekongensis W. W. Sm., Notes Roy. Bot. Gard. Edinburgh 9: 82. 1916. TYPE: China. NW Yunnan: [Dêqên (Deqin) Xian], Bei Ma Shan, 28.33°N, 3655 m, Aug. 1914, *G. Forrest 13204* (lectotype, designated by Ahrendt [1961:

159] K missing, new lectotype, designated here, E E00217957!).

Berberis nutanticarpa C. Y. Wu ex S. Y. Bao, Bull. Bot. Res., Harbin 5(3): 15. 1985, syn. nov. TYPE: China. NW Yunnan: Weixi Xian, cattle farm betw. Yezhi & Shiba, 3000 m, 26 Oct. 1956, *P. Y. Mao 00856* (holotype, KUN 161490!; isotypes, KUN 161489!, PE 01037119!, SZ 00290653 image!).

Shrubs, deciduous, to 2 m tall; mature stems pale yellow, subterete to sulcate, stout, rigid, sparsely finely verruculose; spines 3-fid, slightly paler than stems, 1–2.5 cm, abaxially slightly sulcate. Petiole 3–10 mm; leaf blade abaxially pale green, papillose, adaxially deep yellow-green, shiny, obovate or broadly obovate, 2–4.5 × 1–2 cm, papery, midvein, lateral veins, and reticulation raised abaxially, midvein impressed adaxially, lateral veins and reticulation conspicuous adaxially, base cuneate, margin spinose with (3 to)10 to 15 teeth on each side, rarely entire, apex rounded or subacute. Inflorescence an umbellate raceme, rarely partially paniculate, sometimes with a few fascicled flowers at base, 6- to 20-flowered, 3–7 cm overall including peduncle 1–2.5 cm; bracts 1–1.5 mm; pedicel 4–15(–20) mm. Sepals in 3 whorls; outer sepals lanceolate, ca. 4 × 1.4 mm; median sepals oblong-elliptic, 5–5.5 × 2–3 mm; inner sepals obovate, 6–6.5 × 3.5–4 mm; petals obovate, 4–5 × 2.5–3.5 mm, base clawed, glands separate, apex acute, incised. Stamens ca. 3.5 mm; anther connective extended, truncate. Ovules 1 or 2. Berry red, oblong, 8–10 × 4–6 mm; style not persistent.

Phenology. *Berberis mekongensis* has been collected in flower in June and in fruit in August and September.

Distribution and habitat. *Berberis mekongensis* is known from Dêqên and Weixi Xian in northwest Yunnan and Zayü (Chayu) Xian in southeast Xizang (Tibet). It has been collected among scrub in open situations and mixed forests at ca. 3000–4100 m.

Wu differentiated *Berberis nutanticarpa* from *B. mekongensis* on the basis that its infructescence was nutant, it had longer pedicels, and two seeds. But the nutant infructescence appears to be a result of the way the holotype has been mounted since this characteristic is not to be found on the isotypes. Moreover, the pedicels of the holotype and isotypes are not up to 3.5 cm as Wu states, the longest pedicel being ca. 2 cm long, and the protologue of *B. mekongensis* gives the number of seeds as one or two. The protologue of *B. nutanticarpa* also cited *C. W. Wang 68731* (A 00279647, KUN 161491, PE 01037122–23) Wei-hsi Hsien (Weixi Xian), Yeh Chih (Yezhi), 3300 m, Aug. 1935. This has a semi-

fascicled inflorescence and mostly entire leaves and is of another species.

Selected specimens.
SE Xizang (Tibet). Zayü (Chayu) Xian: near Ridong Gongshe, 3500 m, 18 Sep. 1982, *Qinghai-Xizang Team 10466* (KUN 0178317, PE 01032809).
NW Yunnan. Dêqên (Deqin) Xian: E flank Bei-ma Shan, 28.2°N, 3660 m, June 1917, *G. Forrest 13816* (E E00681202, P P02313650); Mt. Peimashan (Beima Shan), Mekong–Yangtze divide betw. Atuntze & Pungtzera, (Benxilan), 1923, *J. F. Rock 9295* (A 00279978, UC UC328405, US 00945899); Beima Shan, 3000 m, Sep. 1935, *C. W. Wang 69781* (A 00280054, IBSC 0093739, KUN 0176566, LBG 00064354, NAS NAS00314078, WUK 0043294); near Lama Miao, 3500 m, 15 June 1960, *s. coll. 10063* (KUN 0176567; Xi Shan, W of Dêqên, 3845 m, 30 Sep. 1993, *Kunming-Gothenburg Bot Exped. NW Yunnan (KGB) 487* (GB); E slope of Baima Shan, along hwy. G214 at km marker 1918, 28.327222°N, 99.115556°E, 4020–4100 m, 19 Aug. 2013, *D. E. Boufford, J. F. Harber & Q. Wang 43131* (A 00914426, BM, CAS, E, IBSC, K, KUN 1278435, MO, PE, TI).

Cultivated material:
Living cultivated plants.
Howick Hall, Northumberland, U.K., from *KGB 487* (details above).

Cultivated specimens.
Royal Botanic Gardens, Kew, from *G. Forrest 13204*, 10 May 1921 (K K000568011–12).

191. Berberis mianningensis T. S. Ying, Acta Phytotax. Sin. 37: 347. 1999. TYPE: China. SC Sichuan: Mianning Xian, Tuowu, Yele Xiang, 2650 m, 19 July 1959, *S. G. Wu 2021* (holotype, PE 00935165!; isotypes, KUN 0177075!, PE 00935166!).

Shrubs, deciduous, ca. 50 cm tall; mature stems pale brownish yellow, subterete to sulcate; spines solitary or 3-fid, pale yellow, ca. 1 cm, slender, terete. Petiole almost absent; leaf blade abaxially pale yellow-green, adaxially green, elliptic or narrowly elliptic, occasionally oblanceolate, 1–2 × 0.2–0.4 cm, papery, midvein raised abaxially, lateral veins and reticulation inconspicuous on both surfaces, base cuneate, margin entire or spinose with 1 to 3 teeth on each side, apex acute or rounded. Inflorescence 1-flowered, otherwise unknown; fruiting pedicel 10–14 mm. Immature berry obovoid-ellipsoid, 8–9 × 4–5 mm; style persistent, short; seeds 1.

Phenology. *Berberis mianningensis* has been collected with immature fruit in July. Otherwise its phenology is unknown.

Distribution and habitat. *Berberis mianningensis* is known from only the type collection collected at 2650 m from a thicket in a mountainside meadow among *Potentilla* and *Hypericum*.

The isotype of *Berberis mianningensis* at KUN was annotated by C. Y. Wu as *B. shukungana*, an unpublished name. There appears to be few collections of *Berberis* from Mianning Xian.

192. Berberis microtricha C. K. Schneid., Oesterr. Bot. Z. 67: 223. 1918. TYPE: China. SW Sichuan: [Yanyuan Xian], river at Wo-lo-ho betw. Yenjuan Hsien (Yanyuan Xian) & Yung-ning (Yongning Xiang), ca. 2600 m, 23 June 1914, *C. K. Schneider 1543* (lectotype, designated by Ahrendt [1961: 159], K000077413!; isolectotypes, A 00038776!, E E00217958!).

Shrubs, deciduous, to 1.5 m tall; mature stems pale yellow, sub-sulcate; spines concolorous, 1- to 3-fid, 0.2–0.7 cm, terete, weak, mostly absent. Petiole almost absent; leaf blade abaxially pale gray-green, papillose, adaxially green, narrowly obovate or oblanceolate, 0.8–2.5 × 0.3–0.8 cm, papery, midvein slightly raised on both surfaces, both surfaces with inconspicuous 2 or 3 pairs of lateral veins and branched reticulation, base cuneate or attenuate, margin slightly revolute when dry, entire, apex rounded or subacute. Inflorescence a fascicle, sub-fascicle, or sub-raceme, 2- to 7(to 10)-flowered; pedicel 5–12 mm. Sepals in 2 whorls; outer sepals partially reddish, oblong-ovate, 1.5–2.25 × 1–1.5 mm, apex acute; inner sepals obovate, 2.5 × 2 mm; petals obovate, 2.5 × 2 mm, base clawed, glands widely separate, ca. 0.5 mm, apex entire. Stamens 1.25 mm; anther connective not or slightly extended, truncate. Pistil 1.75 mm; ovules 2. Berry red, ellipsoid, 6–7 × 4 mm; style not persistent; seeds 1 or 2.

Phenology. Berberis microtricha has been collected in fruit from between June and September. Its flowering period in the wild is unknown.

Distribution and habitat. Berberis microtricha is reliably known from only the type collection, one other collection from Yanyuan Xian, and one collection from Muli Xian in southwest Sichuan. The type was collected from woods on a riverside at ca. 2600 m, the second on a mountain slope at 3000 m, and the third in an open situation in a side valley.

The protologue of *Berberis microtricha* cited only the type, but without citing any herbarium. Ahrendt's (1961: 159) citation of the specimen at K as type was an effective lectotypification.

The description of flowers given above is from the cultivated specimen cited below grown from *Forrest 22395*. As such, it differs from that given by Ahrendt (1961) and followed by Ying (2001: 196–197). This appears to have been based on *T. T. Yu 15048* (BM, E

E00570049, KUN 0176590) Lichiang Snow Range (Yulongshan), 3000 m, 21 May 1937 (there is some doubt about this source since Ahrendt cites both the herbarium specimen at E and various living cultivated plants of this number, but the latter cannot be from *15048* since no seeds would be available in May). In any event, *Yu 15048* is treated here as an example of a new species, *Berberis yulongshanensis*.

Selected specimens.
SW Sichuan. [Muli Xian]: "Mountains SE of Muli," 27.833333°N, 101.3355°E, 3350–3660 m, Sep. 1922, *G. Forrest 22395* (A 00279662, CAL CAL0000027190, E E00612663, K, US 00945972). **Yanyuan Xian:** betw. Yanyuan & Bailing Gongshe, 3000 m, 6 Aug. 1983, *Qinghai-Xizang Team 12671* (KUN 0179004, PE 01033838).

Cultivated material:
Cultivated specimens.
Royal Botanic Gardens, Kew, *s. coll. s.n.*, from *Forrest 22395*, 26 June 1939 (BM).

193. Berberis minutiflora C. K. Schneid., Ill. Handb. Laubholzk. 2: 914. 1912. *Berberis angulosa* Wall. ex Hook. f. & Thomson var. *brevipes* Franch., Pl. Delavay. 39. 1889. *Berberis brevipes* (Franch.) C. K. Schneid., Bull. Herb. Boissier sér. 2: 8. 194. 1908, not Greene (1901). TYPE: China. NW Yunnan: "au col de Yen-tze-hay, Lankong [Eryuan]," 20 Oct. 1885, *J. M. Delavay 1046* "B" (lectotype, designated here, P P00716531!; isolectotypes, A fragm. 00107119!, LE fragm.!, P P00716532!).

Berberis graminea Ahrendt, J. Bot. 80(Suppl.): 110. 1944, syn. nov. TYPE: China. SW Sichuan: Muli [Xian], Wachin, near Lamasary, 3100 m, 4 Oct. 1937, *T. T. Yu 14429* (holotype, E E00217938!; isotypes, A 00038759!, BM BM001015564!, KUN 1204863!, PE 01030723!).
?*Berberis minutiflora* Ahrendt var. *yulungshanensis* S. Y. Bao, Bull. Bot. Res., Harbin 5(3): 3–4. 1985. TYPE: China. NW Yunnan: Lijiang Xian, Yulong Shan, Ganhaizi, 3200 m, 5 May 1962, *Coll. Bot. Gard. of Lijiang 100475* (holotype, KUN 0161462!; isotypes, KUN [3 sheets] 0161463!, 0176622–23!).

Shrubs, deciduous, to 40 cm tall; mature stems reddish brown or reddish purple, sulcate; spines 3-fid, pale yellowish brown, 0.4–1 cm, terete. Petiole almost absent; leaf blade abaxially pale green, papillose, adaxially green, narrowly obovate or oblanceolate, 1–2 × 0.25–0.4 cm, papery, midvein raised abaxially, slightly impressed adaxially, venation indistinct or inconspicuous on both surfaces, base cuneate, margin entire or with 1 to 3 spinulose teeth on each side, apex acute. Inflorescence 1-flowered, 5–6 mm diam.; pedicel 3–5 mm, slender; bracteoles red, ovate, ca. 1.4 cm. Sepals in 2 whorls; outer sepals narrowly elliptic, 5 × 1.8 mm; inner sepals obovate-elliptic, 5.5 × 3 mm; petals obovate, 4–6 × 2.5–4 mm, base cuneate, glands separate, ovate, apex emarginate. Stamens ca. 2.5 mm; anther connective extended, obtuse. Ovules 2. Berry red, sometimes slightly pruinose at first, ellipsoid or ovoid-ellipsoid, 8–9 × 5–6 mm; style persistent, short.

Phenology. *Berberis minutiflora* has been collected in flower in May and in fruit between August and October.

Distribution and habitat. *Berberis minutiflora* is known from northwest Yunnan and Muli Xian in southwest Sichuan. It has been found in thickets and on grassy and rocky slopes in open areas at ca. 2900–4000 m.

Specimens of *Berberis minutiflora* first appeared in Franchet (1887: 388) as examples of *B. angulosa*, and subsequently in Franchet (1889: 39) as examples of *B. angulosa* var. *brevipes*. Both articles cited *Delavay 1046* "ad montem Koua-la-po," 26 May 1884. Additionally, the former article also cited *Delavay s.n.* "ad collum Yen-tze-hay, Lankong," 20 October 1885, the latter *Delavay 1046* from Yen-tze-hay, 30 October 1885 (the earlier dating of this as 20 October presumably being an error). In renaming the taxon *B. brevipes*, Schneider cited no specimens but cited the type as being in "Herb. Paris." In renaming it *B. minutiflora*, Schneider (1912: 914) merely noted that the name *B. brevipes* had already been taken by Greene, and it was only later (Schneider, 1918: 221) that he designated *Delavay 1046* of 30 October 1885 as the type. In order to differentiate this from the *Delavay 1046* of 26 May 1884, the type is designated here as *Delavay 1046* "B" and the earlier collection *Delavay 1046* "A." Imchanitzkaja (2005: 269) cited *Delavay 1046* "B" at P as the lectotype, but was unaware there are two sheets there. Schneider's lectotypification is completed here by selecting the sheet with his annotations. Ahrendt (1961: 151) cited a *Delavay s.n.* from "keua le pe" of 26 May 1884 at K as a type, but this is missing.

There has been some confusion in relation to the stem color of *Berberis minutiflora*. Though *Delavay 1046* "B" does not have mature stems, *Delavay 1046* "A" does and these are reddish brown or reddish purple. However, the color was not noted either by Franchet (1886) or in any of Schneider's contributions cited above. Ahrendt (1944a: 110) mistakenly described the mature stems as brown and subsequently in a key (Ahrendt: 1961: 151) as yellow, this latter description being followed by Ying (2001: 86–87).

There is some uncertainty in relation to the flowers of *Berberis minutiflora*. Delavay gave no description of them beyond noting they were single. Schneider (1908: 194) reported he was unable to describe the flowers of the Delavay specimens in any detail, but those he had examined all had two ovules. And, indeed, the few flowers of *Delavay 1046* "B" are so frail and fragmentary that, even if any dissection was permitted, it is unlikely that any definitive account of them would result. Unfortunately, there appear to have been no subsequent collections from the type area or, indeed, from anywhere else in Eryuan Xian.

There are at least three sources for any description of flowers. The first is that provided by Ahrendt (1944a: 110). This he claimed was based on a cultivated plant that had been grown from seeds of *J. F. Rock 24392* from Siga Shan in Muli Xian in Sichuan, but which cannot have had this origin since *J. F. Rock 24392* was collected in May. The second is a brief description by Schneider (1918: 220) of the flowers of *Schneider 1271*, also collected in Muli Xian (details below). The third is a line drawing by Schneider attached to the specimen at US of *J. F. Rock 3327* from Yulongshan in northwest Yunnan. There is some variation among these accounts (Schneider's of 1918 being the most imprecise), but they all agree that the number of ovules is two. Given the uncertainty of the origin of Ahrendt's description, that used above is taken from Schneider's line drawing.

The protologue of *Berberis graminea*, whose type is another collection from Muli Xian, differentiated it from *B. minutiflora* by it having reddish purple rather than brown mature stems, pedicels of 2–3 mm versus 5–8 mm, and stylose rather than estylose berries. However, as noted above, the mature stems of the type of *B. minutiflora* are, in fact, not brown but reddish brown or reddish purple. Moreover, its pedicels are 3–4 mm long and its fruit are shortly stylose. Ahrendt's (incomplete) description of the flowers of *B. graminea* (apparently based on cultivated material grown from seed of the type) differs from Schneider's line drawing only by describing the anther connective as not extended and truncate. Again, the number of ovules is given as two. In these circumstances, *B. graminea* is treated here as a synonym of *B. minutiflora*.

The issue, however, still remains as to whether all of the specimens cited above have the same flower structure as the type and, indeed, whether all dwarf *Berberis* with reddish purple stems and flowers with two ovules from northwest Yunnan and neighboring areas of Sichuan are the same species. It should also be noted that there are also low-growing and dwarf *Berberis* species from northwest Yunnan with similar stem color but with three ovules (*B. exigua*), four ovules (*B. nanifolia*), or up to five seeds per fruit (*B. crassilimba*). Additionally, it should be noted that the first of these is also found in Yulongshan.

Given all this, the only way that individual dwarf specimens from this area of China with reddish brown or reddish purple mature stems can be properly identified is by dissecting their flowers, though dissections of fruit that produce more than two seeds should be taken as an indication that they are of species other than *Berberis minutiflora*.

Bao (1985: 3–4) published *Berberis minutiflora* var. *yulungshanensis* on the basis that it differed from *B. minutiflora* by having oblong-ovoid fruit. However, the protologue gave no description of the number of seeds. Variety *yulungshanensis* was treated as a synonym of *B. minutiflora* by Ying (2001: 87), but may actually be *B. exigua*.

Given all this, the list of specimens given below should be treated with caution, though it should be noted that the fruit from both *T. Y. A. Yang et al. 16920* from Zhongdian Xian and *D. E. Boufford, J. F. Harber & X. H. Li 43466* from Muli Xian have only two seeds. Because both *Berberis minutiflora* and *B. exigua* are found on Yulongshan, the specimens listed below from this area, apart from *J. F. Rock 3327*, should be treated with particular caution.

Ying (2001: 86) treated as a synonym of *Berberis minutiflora*, *B. minutiflora* var. *glabramea* Ahrendt (1961: 152), whose type is *F. Ludlow, G. Sherriff & H. H. Elliot 12437* from the Tzangbo gorge, Xizang (Tibet), which the collector's notes record as being up to 1.8 m tall. This is treated here as a separate species, *B. glabramea*. Ying's description of *B. minutiflora* is a conflation of both taxa and should therefore be discounted.

Selected specimens.

Syntypes. Berberis minutiflora. NW Yunnan, Ho-kin [Heqing Xian], "hautes montagnes de Koua-la-po," 26 May 1884, *J. M. Delavay 1046* "A" (A fragm. A 00038715, LE fragm., P [2 sheets] P00716533–34); same details, but *J. M. Delavay s.n.* (K missing).

Other specimens.
SW Sichuan. Muli Xian: "In montes Liuku-liangdse" 27.8°N, "inter oppidum Yenyüen et castellum Kwapi," 3450–3550 m, 17 May 1914, *H. R. E. Handel-Mazzetti 2285* (E E00623021, W, WU 0039261); same details, but *C. K. Schneider 1271* (A 00279700, E E00623009); Mt. Mutzuga, W of Muli Gomba, 3700 m, June 1928, *J. F. Rock 16086* (A 00279701, E E00623021, NY, US 00946004); Mt. Siga, W and overlooking the Yalang River, N of Karadi, 3350 m, May 1932, *J. F. Rock 23850* (A 00279549, E E00267780, NY); Mt. Siga, 3350 m, May 1932, *J. F. Rock 24392* (A 00279550, E E00267778, K, NA 0082381); Mt. Mi-tzu-ga, 3000 m, 22 May 1937, *T. T. Yu 5584* (A 00279544, KUN 0176245, PE 01030724, 01037692); near Kangwu, 3600 m, 23 Sep. 1959, *S. G. Wu 2751* (KUN 0176244, 0176246, PE 01037713); 913 Linchang, 3400 m, 9 May 1978, *Q. S. Zhao & G. Hu 4347* (CDBI CDBI0027403, CDBI0027405, SZ 00291022–24); N of the city of Muli on Xian Rd. Z020, 28.115°N, 101.177778°E, 3650 m, 6 Sep. 2013, *D. E. Boufford, J. F. Harber & X. H. Li 43466* (A 00914429, CAS, E E00833554, K, KUN, PE, TI).
NW Yunnan. Dali Shi: N end of Cangshan, Xiaohuadianba, 3100 m, 18 May 1981, *Sino-British Expedition to Cangshan 0735* (A 00279551, E E00623018, KUN 0176629–

30). **Dêqên (Deqin) Xian:** Baima Shan, 28.366667°N, 99.016667°E, 4160 m, 7 June 1993, *Kunming, Edinburgh, Gothenburg Exped. 919* (E E00623019). **Eryuan Xian:** "Les coteaux calcaires au Col du Lo-pin-chan (Lankong)," 3200 m, 25 May 1886, *J. M. Delavay s.n.* (K, KUN 1221655, P P06868326, P02682344). **Heqing Xian:** "Koua-la po (Hokin)," 3500 m, 5 Aug. 1885, *J. M. Delavay s.n.* (B); Machang, Hongshui Tang, 30 July 1929, *R. C. Qin 23433* (KUN 0176608–610, PE 01037709); **Lijiang Xian [now Yulong Xian]:** E slopes of Likiang Snow Range, 3050 m, May–Oct. 1922, *J. F. Rock 3327* (A 00279548, US 00946003); Yangtze watershed, W slopes of Likiang Snow Range, 30 May–6 June 1922, *J. F. Rock 4241* (A 00279399, US 00945931); Nguluko (Yuhu), 3000 m, 23 May 1937, *T. T. Yu 15061* (A 00279545, KUN 0176614, 0176626, PE 01037695–96); Xuesong Cun, 3300–3600 m, 12 Aug. 1942, *G. M. Feng 8963* (KUN 0176613, PE 01037694, 01037697–98); Xueshan, Liuou Gu, 3600 m, 15 Aug. 1942, *G. M. Feng 9043* (KUN 0176612, PE 01037693); Yi Qu, Yuhu Xiang en route to Shabai, 2900 m, 6 Sep. 1955, *G. M. Ma 21418* (KUN 0176611, 0176627, PE 01037699); Yulong Shan, Wenhai, 3200 m, 24 May 1962, *Coll. Bot. Gard. of Lijiang 100282* (KUN 0176596, 0176604–05); Yulong Shan, Ganhe Ba, 3250 m, 26 May 1985, *Kunming/Edinburgh Yulong Shan Expedition 214* (E E00623016, KUN 0176617); Yulong Shan, He Shui, Lou Shan, 3400 m, 31 May 1985, *Kunming-Edinburgh Yulong Shan Exped. 394* (E E00623011, KUN 0176618); Yulong Xue Shan, 3650–4000 m, 17–26 Sep. 1987, *S. G. Wu et al. 776* (KUN 0178734). **Zhongdian (Xianggelila) Xian:** rd., 18 km N of Zhongdian, 27.958056°N, 99.701944°E, 3390 m, 26 May 1993, *Kunming, Edinburgh, Gothenburg Expedition 254* (E E00623020); Geza, 28.003889°N, 99.704167°E, 3480 m, 24 Aug. 2004, *T. Y. A. Yang et al. 16920* (TNM S116793).

194. Berberis moloensis (Ahrendt) Harber, comb. et stat. nov. Basionym: *Berberis amoena* Dunn var. *moloensis* Ahrendt, J. Linn. Soc., Bot. 57: 154. 1961. TYPE: China. SE Xizang (Tibet): [Mainling (Milin) Xian], Kongbo Province, Molo, Langong Chu, 28.9°N, 93.9°E, 3200 m, 24 May 1938, *F. Ludlow, G. Sherriff & G. Taylor 3835* (holotype, BM BM000564306!; isotype, E E00327787!).

Berberis tsarongensis Stapf var. *megacarpa* Ahrendt, J. Linn. Soc., Bot. 57: 156. 1961, syn. nov. TYPE: China. SE Xizang: [Mainling Xian], Molo, Lilung Chu, 28.95°N, 93.883333°E, 3200 m, 30 Sep. 1938, *F. Ludlow, G. Sherriff & G. Taylor 6544* (holotype, BM 000794140!).

Shrubs, deciduous, to 2.4 m tall; mature stems dark purplish red, sulcate, sparsely verruculose; spines 1- to 3-fid, absent at the apex of stems, pale yellow, 0.5–1.7 cm, terete or abaxially sulcate. Petiole almost absent; leaf blade abaxially grayish green, slightly papillose, adaxially pale green, narrowly obovate, 1–2 × 0.4–0.5 cm, papery, midvein raised abaxially, lateral veins conspicuous on both surfaces, reticulation inconspicuous on both surfaces, base attenuate, margin entire, rarely spinulose with 1 to 5 teeth on each side, apex acute, mucronate. Inflorescence a fascicle, sub-fascicle, or umbellate sub-raceme, rarely an umbel, (3- to)5- to

10-flowered, 1.5–3 cm; pedicel 6–18 mm. Sepals in 2 whorls; outer sepals narrowly ovate, ca. 3.25 × 1.25 mm, apex subacute; inner sepals obovate-elliptic, ca. 4.5 × 7.75 mm; petals narrowly obovate, 4–4.5 × 2–2.25 mm, base clawed, glands ca. 0.9 × 0.4 mm. Ovules 1 or 2. Berry scarlet, obovoid or oblong, 12 × 4–5 mm; style not persistent.

Phenology. *Berberis moloensis* has been collected in flower in May and in fruit in September.

Distribution and habitat. *Berberis moloensis* is known from Mainling (Milin) Xian in southeast Xizang (Tibet). It has been found on river banks among *Rhododendron* at 3200 m.

"*Berberis moloensis* (Ahrendt) Harber ex J. Y. Fang, Zhi H. Wang & Z. Y. Tang" (J. Y. Fang et al., 2011) was invalidly published as a nomen nudum and without a basionym on the basis of my draft of the *Berberis* manuscript on the Flora of China website. The name, however, was not included in the published version (Ying, 2011).

The synonymy of *Berberis tsarongensis* var. *megacarpa* with *B. amoena* var. *moloensis* is confirmed by the cultivated specimens with floral material grown from seeds of the type cited below.

Berberis amoena var. *moloensis* was treated by Ying (2001: 177) as a synonym of *B. kongboensis*, but this latter has a paniculate inflorescence and four ovules. Ying also (2001: 174, 196) treated *B. tsarongensis* var. *megacarpa* as a synonym of both *B. tsarongensis* and *B. lecomtei.*

Selected specimens.
SE Xizang (Tibet). [Mainling (Milin) Xian]: Kongbo Province, Molo, Lilung Chu, E bank of river, 28.95°N, 93.883333°E, 3200 m, 23 May 1938; *F. Ludlow, G. Sherriff & G. Taylor 4394* (BM BM000939704).

Cultivated material:
Cultivated specimens.
Royal Botanic Garden Edinburgh, from seed of *F. Ludlow, G. Sherriff & G. Taylor 6544, s. coll. s.n.*, 31 May 1988 (E E00112411).

195. Berberis monticola Harber, sp. nov. TYPE: China. NW Yunnan: Zhongdian (Xianggelila) Xian, S side of Daxue Shan on hwy. S217 from Zhongdian (Xianggelila) to Xiangcheng, 28.573611°N, 99.834444°E, 4210 m, 29 Aug. 2013, *D. E. Boufford, J. F. Harber & Q. Wang 43304* (holotype, PE!; isotypes, A!, E!, KUN!).

Diagnosis. *Berberis monticola* is somewhat similar to *B. muliensis*, but has a different color of mature stem, is 1-flowered except for the apex of stems, and has five or six seeds per fruit versus the three or four recorded for *B. muliensis.*

Shrubs, deciduous, to 2 m tall; mature stems pale reddish brown, sulcate; spines 3-fid, pale yellowish brown, (0.6–)1.2–2.6 cm, abaxially sulcate. Petiole almost absent; leaf blade abaxially glaucous or graygreen, papillose, adaxially mid-green, narrowly obovate, but often more broadly obovate on plants in more exposed conditions, (1–5–)2–2.6(–5) × 0.8–1.3(–1.8) cm, papery, midvein raised abaxially, venation inconspicuous on both surfaces, base narrowly attenuate, margin entire or spinulose with 5 to 12 teeth on each side, apex subacute, acute, or obtuse, mucronate. Inflorescence 1-flowered, sometimes 2- or 3-flowered at apex of stems, otherwise unknown; fruiting pedicel (8–)13–30 mm. Berry orange, obovoid or oblong, 10–12 × 6–7 mm; style persistent, short; seeds 5 or 6.

Phenology. *Berberis monticola* has been collected in fruit from August to October. Its flowering period is unknown.

Distribution and habitat. *Berberis monticola* is known from both the Zhongdian (Xianggelila) and Xiangcheng Xian sides of Daxue Shan on the northwest Yunnan/west Sichuan border. It has been found on open forest areas, meadows, and steep slopes at ca. 3900–4300 m.

IUCN Red List category. *Berberis monticola* is assessed as DD or Data Deficient, according to IUCN (2001) criteria.

On current evidence, *Berberis monticola* appears to be endemic to Daxue Shan. From personal observation in September 2013, at the top of the pass taken by the S217, it is a sprawling scrub plant with small leaves and mostly short pedicels, whereas the colony on a more favored position on the south side of the pass is of taller plants with larger leaves and longer pedicels.

Selected specimens.
W Sichuan. Xiangcheng Xian: Daxue Shan, near Daoban, 4100 m, 3 Aug. 1981, *Qinghai-Xizang Team 3728* (CDBI CDBI0027516–17, HITBC 003577, KUN 0178924, 0178926, PE 01037479–80); Rewu Gongshe, betw. Balang & Qingda, 4000 m, 9 Aug. 1981, *Qinghai-Xizang Team 3933* (CDBI CDBI0027518, 0027520, HITBC 003606, KUN 0178326, 0178895, PE 01037787–88); Rewu Gongshe, Balang, 3900, 9 Aug. 1981, *Qinghai-Xizang Team 4010* (CDBI CDBI0027514–15, HITBC 003607, KUN 0178322, 0178925, PE 01037477–78); NE side of Daxue Shan, 28.585833°N, 99.837222°E, 4250–4600 m, 2 Sep. 2013, *D. E. Boufford, J. F. Harber & X. H. Li 43397* (A, CAS, E, K, KUN, PE).
NW Yunnan. Yunnan Zhongdian (Xianggelila) Xian: W side of Daxue Shan pass, 28.569722°N, 99.831944°E, 4200 m, 7 Oct. 1994, *Alpine Garden Exped. to China (ACE) 1845* (E E00039720, WSY WSY0057779); Sichuan–Yunnan border area, S side Daxue Shan, 28.573333°N, 99.811667°E, 4280–4600 m, 2 Sep. 2013, *D. E. Boufford, J. F. Harber & X. H. Li 43379* (A, CAS, E, K, KUN); same details, but

28.571667°N, 99.825833°E, 4170 m, 2 Sep. 2013, *D. E. Boufford, J. F. Harber & X. H. Li 43401* (A, E. KUN).

196. Berberis morrisonensis Hayata, J. Coll. Sci. Imp. Univ. Tokyo 30(1): 25. 1911. TYPE: China. Taiwan: Mt. Morrison (Yushan), Oct. 1906, *T. Kawakami & U. Mori 2289* (lectotype, designated by A. Shimizu et al. [2002: 24], TI 02626 image!).

Shrubs, deciduous, to 2 m tall; mature stems dark red or red-brown, sparsely verruculose, sulcate; spines 3-fid, concolorous, 1–2 cm, slender, terete. Petiole almost absent or to 8 mm; leaf blade abaxially pale green, sometimes gray-white or glaucous, papillose, adaxially dull green, obovate or obovate-lanceolate, 1.5–3(–4) × 0.5–1(–1.5) cm, papery, midvein slightly raised abaxially, slightly impressed adaxially, lateral veins and reticulation mostly indistinct on both surfaces (inconspicuous why dry), base attenuate, margin spinose with 4 to 7 teeth on each side or entire, apex obtuse, mucronate. Inflorescence a fascicle, (1- or)2- to 5(or 6)-flowered; pedicel 12–25(–30) mm, slender. Sepals in 3 whorls; outer sepals lanceolate or narrowly ovate, (3.5–)4–4.5 × 1.5–2.5 mm, apex acuminate; median sepals oblong-elliptic, 5.5–6.5 × 3–3.5 mm; inner sepals narrowly obovate, 6–7.5 × 3–4 mm; petals broadly elliptic, 5–6 × (3–)3.5–4 mm, apex rounded, emarginate. Anther connective obtuse or truncate. Ovules 4 to 7. Berry scarlet, subglobose, 8–9(–10) × 7–8 mm; style not persistent.

Phenology. *Berberis morrisonensis* has been collected in flower in May and June and in fruit between August and November.

Distribution and habitat. *Berberis morrisonensis* is known from the northern Central and Hsuehshan ranges and Yushan in Taiwan. It has been found in rocky areas among *Abies*, and in moist forest among *Yunshania* at ca. 3000–3900 m.

The protologue of *Berberis morrisonensis* cited both *Kawakami & U. Mori 2289* and *2297*, but without determining a type. A. Shimizu et al. (2002: 24), who lectotypified *T. Kawakami & U. Mori 2289*, stated that *2297* could not be found. In fact, it is at TAIF.

S. Y. Lu and Yang (1996: 579) erroneously cited a *Kawakami & Mori 2287* as the type.

Berberis morrisonensis is the only deciduous *Berberis* species found in Taiwan. In cultivation in the U.K. at least, its leaves appear exceptionally late in the year, sometimes not until late May.

Syntype. Taiwan: Mt. Morrison (Yushan), 3810 m, 18 Nov. 1906, *T. Kawakami & U. Mori 2297* (TAIF 9917).

Selected specimens.
Taiwan. Chiayi Hsien: Arisan to Mt. Morrison, 3500–5633 m, 25 Oct. 1918, *E. H. Wilson 10912* (A [3 sheets], US 01121972); from Paiyan lodge to Yushan peak, 3580–3900 m, 7 July 1995, *C. H. Chen & H. T. Hung 1095* (HAST 54300). **Hsinchu Hsien:** Wufeng Xiang, Sheipa Natl. Park, Tapachienshan, 24.463056°N, 121.258056°E, ca. 3400 m, 7 Sep. 1993, *C. L. Huang et al. 99* (HAST 40311). **Hualien Hsien:** Mt. Hohuan, 8 Aug. 1976, *M. T. Kao 8806* (TAI 167712–13). **Kaohsiung Hsien:** Taoyuan Hsiang, from campsite to Kuanshan, 23.229722°N, 120.5°E, 3300–3400 m, 16 May 1995, *K. Y. Wang 1107* (HAST 53102). **Miaoli Hsien:** Tsuichih, 3500 m, 12 Sep. 1986, *S. Y. Lu 19918* (TAIF 233627). **Nantou Hsien:** Taroko Natl. Park, summit of Mt. Hohuan N, 3400 m, 5 June 1996, *J. N. Panero 6522* (HAST 64324, TEX). **Taichung Hsien:** Mt. Tugitaka, Oct. 1925, *Y. Shimada 2567e* (NTUF); Hoping Hsiang, Taroko Natl. Park, Nanhutashan hiking trail, near Nanhupeishan, 3400 m, 17 July 1996, *T. Y. Liu 299* (HAST 64184). **Taitung Hsien:** Taoyuan Hsiang, betw. Chinching Bridge & Kuanshan, 23.255278°N, 120.910288°E, 2450–2820 m, 20 Oct. 1995, *L. Y. Lin 115* (HAST 59060, PE 01793789). **Yilan Hsien:** Tatung Hsiang, en route from Hsanga Camp to Pintienshan, 3250–3536 m, 14 Aug. 1993, *C. H. Chen 266* (HAST 27262).

197. Berberis mouillacana C. K. Schneid., Pl. Wilson. (Sargent) 1: 371–372. 1913. TYPE: China. W Sichuan: Tachien-lu (Kangding), 2400–2750 m, Sep. 1908, *E. H. Wilson 1039* (lectotype, designated by Ahrendt [1961: 166] K missing, new lectotype, designated here, A 00038777!; isolectotype, BM BM000564310!, US 00103891 image!).

Shrubs, deciduous, to 3 m tall; mature stems dark purple, reddish purple, or reddish brown, sub-sulcate; spines 3-fid, solitary or absent toward apex of stems, pale brownish yellow, (0.5–)1.2–2 cm, abaxially sulcate. Petiole almost absent or to 15 mm on leaves lower down the branches; leaf blade abaxially pale green, slightly shiny, papillose, sometimes partially lightly pruinose, adaxially green, obovate-elliptic to obovate, sometimes more broadly obovate toward base of branches, 1–2.5(–4) × 0.4–1.4(–2) cm, papery, midvein raised abaxially, lateral veins and reticulation conspicuous abaxially, inconspicuous adaxially, margin entire, very rarely spinulose with 1 to 8 inconspicuous teeth on each side, base attenuate, apex subacute or rounded, rarely minutely mucronate. Inflorescence a sub-raceme, sometimes fascicled at base, or a sub-fascicle toward apex of stems, 6- to 12-flowered, to 4.5 cm overall, otherwise unknown; fruiting pedicel 5–13 mm. Berry crimson red, slightly pruinose, ellipsoid, oblong, or ovoid, 8–10 × 5–7 mm; style not persistent; seeds 1 or 2.

Phenology. *Berberis mouillacana* has been collected in fruit from July to October. Its flowering season is unknown.

Distribution and habitat. *Berberis mouillacana* is known from Kangding Xian in west Sichuan. It has been found in thickets, forests, and mountainsides at ca. 2400–3000 m.

The protologue of *Berberis mouillacana* cited *Wilson 1039* as type, but without citing a herbarium. Ahrendt's (1961: 166) citing of a specimen at K as type was an effective lectotypification, but this is missing. The specimen at A has been chosen as a new lectotype because it has more material than the specimens at BM and US.

Besides the type, the protologue cited *Wilson 1041, 1283,* and *4123.* Apart from *X. L. Jiang 36424* and *36937* (details below), no definite specimens collected subsequently have been located.

The protologue described the mature stems of *Berberis mouillacana* as purplish and the evidence for this comes not from the type collection, but from all the other Wilson numbers cited.

Ahrendt (1961: 166) gave an abbreviated description of the flowers of *Berberis mouillacana*, but this was based on cultivated material grown at K from seed of *E. H. Wilson (Veitch) 1707* which, from the evidence of Wilson's fieldbook (Wilson MSS), was collected on Emei Shan. There are cultivated specimens with this number at K identified not as *B. mouillacana* but as *B. silva-taroucana*, though if their provenance is accurate, they are *B. emeishanensis*. Ahrendt's description has, therefore, been discounted as has the account of the flowers given by Ying (2001: 194), which included stating the number of ovules to be two to four versus the one or two seeds of the type of *B. mouillacana*. It is possible that *Z. Y. Chen & Z. X. Xiong 112125* (CDBI CDBI0028113, SZ 00291054) Kangding Xian, Zheduo Shan Guogu, 2770 m, 10 May 1980, is *B. mouillacana*. This has floral material, but I have not had the opportunity to investigate this.

There are various specimens at KUN, PE, and other Chinese herbaria from areas of Sichuan other than Kangding Xian that are identified as *Berberis mouillacana* (or more usually as "*mouilicana*"). These would seem to be of a variety of species. Whether any are *B. mouillacana* requires further research. Specimens collected by J. F. Rock in southwest Gansu and identified by Byhouwer (1928: 45–46) as *B. mouillacana* are identified here as *B. taoensis*.

Ying (2001: 194) treated as a synonym *Berberis boschanii* whose type is from Mao Xian. This is maintained here as a separate species.

Selected specimens.
WC Sichuan. Kangding Xian: Tachien lu, 2440–2740 m, Sep. 1908, *E. H. Wilson 1041* (A 000946157, BM BM000946157, E E00623027, K, US 0946002); Tachien lu, 2740–3750 m, Oct. 1908, *E. H. Wilson 1283* (A 00279704, BM BM000946156, K, US 00946001); Tachien lu, 2440–2740 m, Oct. 1910, *E. H. Wilson 4123* (A 00279703); Yulin Xiang, Simaqiao Linchang, 2900 m, 24 July 1953, *X. L. Jiang 36424* (HNWP 46481, IBK IBK00012820, IBSC 0092482, PE 01032984, SWCTU 00006300); Dapao Shan, Niuyuanping, en route to Zhonggu, 3350 m, 19 Sep. 1953, *X. L. Jiang 36937* (IBK IBK00012839, PE 01033841).

Cultivated material:
Cultivated specimens.
Arnold Arboretum, U.S., from accession no. 12739, grown from seeds of *E. H. Wilson 1039*, 15 Aug. 1916, *s. coll. s.n.* (A 00167877); from accession no. 7179, grown from seeds of *E. H. Wilson 1284*, 31 Aug. 1916, *s. coll. s.n.* (A 00167876).

198. Berberis muliensis Ahrendt, Bull. Misc. Inform. Kew 1939: 268. 1939. TYPE: China. SW Sichuan: [Muli Xian] Mu-li mtns., 28.2°N, 100.833333°E, 3350 m, Sep. 1921, *G. Forrest 20633* (holotype, K K000077422!; isotypes, A 00058267!, E E00217960!).

Shrubs, deciduous, to 1.5 m tall; mature stems dark reddish purple, sulcate; spines 3-fid, solitary or absent toward apex of stems, yellowish brown, (0.6–)1.5–2.5 cm, terete. Petiole almost absent or 2–7(–9) mm; leaf blade abaxially pale green, papillose, adaxially dark green, narrowly obovate or oblong-obovate, (2–)2.4–3 × 0.8–1.3 cm, papery, midvein raised abaxially and extending less than half-way up leaf before dividing, lateral veins and reticulation raised abaxially, conspicuous on both surfaces, base narrowly cuneate, margin entire, rarely spinulose with 1 to 3(to 5) teeth on each side, apex acute or subobtuse, mucronate. Inflorescence a fascicle or occasionally a sub-fascicle, 1- to 3-flowered, otherwise unknown; fruiting pedicel (12–)20–30 mm. Berry scarlet-crimson, not or very slightly pruinose, ovoid or oblong-ovoid, 12–16 × 8–10 mm, sometimes slightly bent at tip; style not persistent or persistent and short; seeds 3 or 4.

Phenology. *Berberis muliensis* has been collected in fruit in September. Its flowering period is unknown.

Distribution and habitat. *Berberis muliensis* is known from only the type collection and one other from Muli Xian in southwest Sichuan. Both were found in open alpine meadows, the type at 3350 m, the other at 4000 m.

There have been various confusions in relation to *Berberis muliensis*. The protologue described the inflorescence as consisting of single and occasionally paired, fascicled flowers. Later, Ahrendt (1961: 127), in a key, described the flowers simply as solitary, this latter description being repeated by Ying (2001: 79). But the infructescence of the holotype of *B. muliensis* has evidence of three solitary fruit and three paired fruit, two of which are fascicled and one sub-fascicled, while the isotype at A has two fascicled paired fruit, one paired fruit stalk with a fascicle at its base, and only one single fruit, while that at E is 1- to 3-fascicled. Besides the type, the protologue of *B. muliensis* cited *Forrest 20431* (K K000077421, but also A 00279706, E E00623031, K, UC UC253194, US 00945916) with exactly the same coordinates as the type, but collected at 3660 m in June 1921. This was the basis for the protologue description of flowers which was largely repeated by Ahrendt (1961: 129) and Ying (2001: 79). However, *Forrest 20431* is single-flowered with pedicels that do not exceed 6 mm versus the 20–30 mm of the type. Not only that, but from the evidence of a line drawing by Schneider of the specimen of *Forrest 20431* at E, the description of flowers by Ahrendt is inaccurate and this is actually *B. dictyophylla*.

A second confusion relates to Ying (2001), who treated as synonyms of *Berberis muliensis*, *B. capillaris* whose type is from Myanmar, *B. ludlowii* from Mainling (Milin) Xian in southeast Xizang (Tibet), and *B. tianbaoshanensis* from Zhongdian (Xianggelila) Xian in northwest Yunnan. If the first of these had been correct, then the synonymy would be the other way around since *B. capillaris* was published first, but it is not since *B. capillaris* has very different-shaped leaves and up to 10 ovules versus the three or four seeds of *B. muliensis*. Both *B. ludlowii* and *B. tianbaoshensis* are also different species from *B. muliensis*, the former having leaves up to 7 cm long and being 1-flowered with pedicels up to 55 mm long, the latter being 1(or 2)-flowered with four or five ovules.

I have found no specimens of *Berberis muliensis* beyond the type and *Zhao 6497* (details below). This latter is sterile, but its leaves and number and length of its pedicels suggest it is *B. muliensis*.

The report by X. H. Li et al. (2017) that *Berberis muliensis* is found in Baoxing Xian in west Sichuan is likely to be referable to *B. aemulans*.

The protologue of *Berberis muliensis* also included a variety *atuntzeana* and a variety *beimanica*. The former is treated here as a synonym of *B. capillaris*, the latter as a new species, *B. beimanica*.

Selected specimens.
SW Sichuan. Muli Xian: Basang Shan, 4000 m, 7 Aug. 1978, *Q. S. Zhao 6497* (CDBI CDBI0027510–11, CDBI CDBI0027524, SZ 00290501).

199. Berberis multiserrata T. S. Ying, Fl. Xizang. 2: 139. 1985. TYPE: China. SE Xizang (Tibet): Zayü (Chayu) Xian, Guyu Qu, 3100 m, 20 June 1973, *Qinghai-Xizang Team 73-247* (holotype, PE 00935229!; isotypes, KUN [2 sheets] 0177571–72!, PE 0093523!).

?*Berberis orthobotrys* "Bien. ex Aitch." var. *rupestris* Ahrendt, J. Linn. Soc., Bot. 57: 143. 1961, syn. nov. TYPE: India. Arunachal Pradesh, Delei Valley, 28.35°N, 96.616667°E, 3350–3660 m, 21 June 1928, *F. Kingdon-Ward 8350* "A" (lectotype, designated here, K K000568137!; isolectotype, K K000568138!).

Shrubs, deciduous, to 2 m tall; mature stems dark reddish or blackish purple, subterete, partially pruinose; spines solitary or 3-fid, orange-yellow, 0.5–0.7 cm, weak. Petiole almost absent; leaf blade abaxially grayish or whitish green, papillose, adaxially dark green, broadly obovate, occasionally orbicular, 1.7–5.5 × 1.5–3.5 cm, thickly papery, midvein raised abaxially, lateral veins and reticulation indistinct or inconspicuous abaxially, conspicuous adaxially, base cuneate, margin dentate with 2 to 7 coarse teeth on each side, rarely entire, apex rounded, sometimes retuse. Inflorescence a fascicle, sub-fascicle, or sub-umbel, 4- to 8-flowered, ca. 2 cm overall; peduncle 3–5 mm, purplish black; pedicel 10–17 mm, stout, slightly pruinose; bracteoles ovate, ca. 2 mm, pruinose, apex obtuse. Sepals in 2 whorls; outer sepals oblong-elliptic, ca. 8 × 4 mm; inner sepals broadly obovate, ca. 6 × 4.1 mm; petals obovate, ca. 6 × 3.2 mm, glands separate, apex emarginate. Stamens ca. 3.1 mm; anther connective extended, rounded. Ovules 4(to ?6). Berry deep purple-black, pruinose, oblong, ca. 14 × 5 mm; style not persistent.

Phenology. *Berberis multiserrata* has been collected in flower in June and July and in fruit in August and September.

Distribution and habitat. *Berberis multiserrata* is known from Zayü (Chayu) Xian and Bomê (Bomi) Xian in southeast Xizang (Tibet). It may also be found in that area of Arunachal Pradesh in India where there is a conflicting territorial claim by China. In Xizang, the species has been found on forest margins on mountain slopes at ca. 3100–3950 m.

The protologue of *Berberis orthobotrys* var. *rupestris* cited the type as being at BM. This appears to be an error. As can be seen above, there are two sheets of *Kingdon-Ward 8350* at K. Both also have specimens of another gathering with the same number. The ones that conform to the protologue description are designated here as *Kingdon-Ward 8350* "A." The other specimens, designated here as *Kingdon-Ward 8350* "B," (one of which is annotated as "*bis*" in an unknown hand) are poor ones with no mature stems, leaves that are somewhat similar to the description given in the protologue, but with a distinctly racemose semi-umbellate inflorescence and, as such, may be of another species. There are two dates and collectors' descriptions on sheet K000568137. The first description with the collection date of 21 June 1928 is the fullest and, following Ahrendt, I have assumed that this refers to *8350* "A." This is described as "A small shrub growing on gneiss cliffs in the more open parts of the *Abies* forest or along the ridge." The second description gives the collection date as 4 July 1928 and "In the alpine region, amongst dwarf *Rhododendron*, just coming into flower, stems red." This is given here in case this latter date and description actually applies to *8350* "A" rather than "B." *Kingdon-Ward* 8350 "A" on sheet K000568137 has been chosen as the lectotype because K000568138 has no collection information.

The type of *Berberis orthobotrys* var. *rupestris* has leaves very similar to the type of *B. multiserrata*, mature stems of the same color (including being partially pruinose), and a similar flower stalk. However, the flowers are smaller and the flower structure, from the evidence of a line drawing on the sheet (almost certainly by Schneider), is slightly different for that reported for *B. multiserrata*, being as follows: "Sepals in 2 whorls; outer sepals oblong-elliptic, 3.5 × 2 mm, apex subacute; inner sepals broadly elliptic, 4.25 × 3 mm, apex rounded; petals obovate, ca. 3.5 × 2.25 mm, apex slightly emarginate, glands separate, obovate, apex incised. Stamens 3.75 mm; anther connective conspicuously extended, slightly retuse. Ovules 4 to 6."

Given these differences, it is possible that this is a separate species, but in the absence of other specimens from the same area, it would seem unwise to recognize it as such, hence my tentative synonymy. In any event, Ahrendt's designation of it as a variety of *Berberis orthobotrys*, a species whose type is from the Iranian border with Turkmenistan and whose name was validly published not by Aitchison but by Schneider (see Browicz & Zieliński, 1975: 13–14), is not sustainable.

Selected specimens.
SE Xizang (Tibet). Bomê (Bomi) Xian: near Zhamu Linchang, 3600 m, 11 July 1979, *Xizang Team – G. F. Huang 686* (HNWP 85257); betw. Bomi & Gawalong Lake, 29.822967°N, 95.7091°E, 3453 m, 22 June 2009, *Tibet MacArthur 2644* (US 01061699). **Zayü (Chayu) Xian:** 10 km N of Baxue, 3950 m, 21 Aug. 1973, *J. W. Zhang 1117* (PE 00049422, 00935231); Guyu Qu, S of Demula, 3800–4500 m, 8 July 1979, *Xizang Team – G. F. Huang 569* (HNWP 85279); betw. Chayu & Ranwu on S201, 29.320556°N, 97.176111°E, 3640 m, 22 Sep. 2009, *H. Sun et al. Sun H-07ZX-2504* (KUN 1202194).

200. Berberis nambuensis (Ahrendt) Harber, comb. et stat. nov. Basionym: *Berberis everestiana* Ahrendt var. *nambuensis* Ahrendt, J. Linn. Soc., Bot. 57: 118. 1961. TYPE: China. SE Xizang (Tibet): [Nyingchi (Linzhi) Xian, now Bayi Qu] Kongbo, Nambu La, 29.983333°N, 94.316667°E, 4420 m, 7 July 1947, *F. Ludlow, G. Sherriff & H. H. Elliot 15385* (holotype, BM BM000559582!; isotypes, A 00038742!, E E00373492!).

Shrubs, deciduous, to 1.2 m tall; mature stems very pale yellowish brown, sulcate, verruculose; spines 3-fid, pale brown, 0.3–0.5 cm, terete. Petiole almost absent; leaf blade abaxially pale green, papillose, adaxially green, obovate, 2–2.2 × ca. 0.5–0.8 cm, papery, midvein, lateral veins, and reticulation conspicuous abaxially, inconspicuous adaxially, base attenuate, margin entire, apex subacute or obtuse. Inflorescence 1-flowered; pedicel 7–14 mm. Sepals in 2 whorls; outer sepals ovate, ca. 6 × 3 mm, apex acute; inner sepals ovate-elliptic, 7.5 × 4 mm; petals obovate, 4–5 × 3 mm, base cuneate, glands separate, apex entire. Stamens 3 mm; anther connective not extended, truncate. Ovules 5. Berry unknown.

Phenology. *Berberis nambuensis* has been collected in flower and with very immature fruit in July. Its phenology is otherwise unknown.

Distribution and habitat. *Berberis nambuensis* is known from only the type found on dry ground in Nambu La in Nyingchi (Linzhi) Xian (now Bayi Qu), Tibet, at 4220 m.

It is unclear why Ahrendt determined *F. Ludlow, G. Sherriff & H. H. Elliot 15385* as a variety of the dwarf species *Berberis everestiana*, found some 750 km away on the Xizang border with Nepal and with a different flower structure. The protologue also cited *F. Ludlow, G. Sherriff & H. H. Elliot 13884* (BM, E E00395934), Nambu La, Tongkyuk River, 3810 m, 12 June 1947. This appears to be of another species.

Berberis everestiana var. *nambuensis* was treated as a synonym of *B. parisepala* by T. S. Ying (2001: 76). See Excluded Taxa for this taxon.

201. Berberis nanifolia Harber, sp. nov. TYPE: China. NW Yunnan: Zhongdian (Xianggelila) Xian, SW side of Tianbaoshan, E of Xiao Zhongdian, 27.605278°N, 99.885833°E, 3680–3840 m, 4 Sep. 2013, *D. E. Boufford, J. F. Harber & X. H. Li 43437* (holotype, PE 01892852; isotypes, A 00914432!, BM BM001190939!, CAS!, E E00833562–63, HAST, IBSC!, K!, KUN!, MO 1204038!, TI!)

Diagnosis. *Berberis nanifolia* is similar to *B. exigua* and *B. minutiflora*. It has leaves similar to *B. minutiflora* but has four ovules rather than two. It is a smaller species than *B. exigua* and has a different flower structure. It has some similarities with another dwarf species, *B. crassilimba*, but this has up to five seeds per fruit and more obovate-elliptic leaves. Another dwarf species from northwest Yunnan, *B. scrithalis*, has four ovules but has pale brownish red mature stems versus the purple stems of *B. nanifolia* as well as different-colored leaves.

Shrubs, deciduous, to 40 cm tall; mature stems purple, sulcate; spines 3-fid, brown, 0.6–1.7 cm, abaxially sulcate. Petiole almost absent; leaf blade abaxially pale green, papillose, adaxially green, narrowly obovate or oblanceolate, 0.8–1.7 × 0.3–0.6 cm, papery, midvein raised abaxially, impressed adaxially, venation indistinct or inconspicuous on both surfaces, more distinct abaxially when dry, base attenuate, often narrowly so, margin entire, apex subacute or obtuse, mucronate. Inflorescence 1-flowered; pedicel 4–6 mm. Sepals in 2 whorls; outer sepals narrowly elliptic, 2 × 1.25 mm; inner sepals obovate, 6 × 4 mm; petals broadly obovate, 5.5–6.5 × 4.75 mm, base slightly clawed, glands separate, ca. 0.75 mm, apex entire. Stamens 3 mm; anther connective extended, rounded. Pistil 3 mm; ovules 4. Berry red, ovoid or subglobose, 8–9 × 5–6 mm; style persistent.

Phenology. *Berberis nanifolia* has been collected with immature fruit in early September and has been found with flowers in June and in early September. Its phenology is otherwise unknown.

Distribution and habitat. *Berberis nanifolia* is known from only the type and two other collections, all from Tianbao Shan in Zhongdian (Xianggelila) Xian in northwest Yunnan, the former on open valley side scree at 3680–3840 m, the latter in scrub at 3600 m.

The source of the flower structure given above is from one late flower on *D. E. Boufford, B. Bartholomew, J. L. Guo & J. F. Harber 45067* (details below). It should be noted that on this visit, we found the colony of *Berberis nanifolia* much diseased with most plants sterile. The source of the species flowering in June is from a visit to Tianbao Shan by Peter Edge, Christopher Parsons, and Edward Shaw of the U.K. (C. Parsons, pers. comm. and photograph 25 October 2016).

I have not had an opportunity to examine any of the specimens of *Qinghai-Xizang Team, Hengduan Shan 1112* cited below, but from the collection details and the fact that two other *Berberis* species (*B. calciprotorum* and *B. tianbaoshanensis*) we collected in the same valley were also collected by this team on the same day as no. *1112*, I deduce that these specimens were collected from the same area as *43437* and *45067* and hence are also *B. nanifolia*.

Selected specimens.
NW Yunnan. Zhongdian (Xianggelila) Xian: SW side of Tianbao Shan, 3600 m, 15 June 1981, *Qinghai-Xizang Team, Hengduan Shan 1112* (CDBI CDBI0027404, HITBC 003595, KUN 0176635, PE 01037702–03); same location as type collection (see above), 1 Sep. 2018, *D. E. Boufford, B. Bartholomew, J. L. Guo & J. F. Harber 45067* (A, CAS, E, K, KUN, P, PE, TI).

202. Berberis ngawaica Harber, sp. nov. TYPE: China. N Sichuan: Heishui Xian, W of the city of

Heishui on provincial rd. S302 near Shashiduo Xiang, 32.103056°N, 102.811111°E, 2675 m, 9 May 2014, *D. E. Boufford, S. Cristoph, C. Davidson, Y. D. Gao & Q. Y. Xiang 43640* (holotype, PE image!; isotypes, A image!, CDBI image!, E!, MO image!, TI image!).

Diagnosis. *Berberis ngawaica* has similar-shaped leaves to *B. silva-taroucana*, but these are smaller and almost always have entire margins. Though the flower stalks are similar to *B. silva-taroucana*, the pedicels are much shorter and the flower structure is different.

Shrubs, deciduous, to 3 m tall; mature stems pale reddish brown, subterete; spines 3-fid, solitary or absent toward apex of stems, pale brownish yellow, 0.6–2.2 cm, abaxially sulcate. Petiole to 9 mm, sometimes almost absent on leaves at apex of stems; leaf blade abaxially pale green, papillose, adaxially dull mid-green, obovate or obovate-elliptic, 1–2.2 × 0.6–1.2 cm, papery, midvein, lateral veins, and reticulation inconspicuous on both surfaces, base narrowly attenuate, margin entire, rarely spinulose with 1 to 5 indistinct teeth on each side, apex obtuse, rounded, or subacute, sometimes minutely mucronate. Inflorescence a fascicle, sub-fascicle, sub-umbel, or sub-raceme, 3- to 12-flowered, to 2.2 cm overall; pedicel 4–8 mm; bracteoles ovate-lanceolate, 2 × 0.7 mm. Sepals in 3 whorls; outer sepals ovate, 2 × 1.25 mm; median sepals oblong-obovate, 3–4 × 1.75–2 mm; inner sepals elliptic, 3–4 × 2.5 mm; petals elliptic-obovate, 2.75 × 1.25 mm, base clawed, glands separate, ca. 0.5 mm, apex entire. Stamens ca. 2.5 mm; anther connective slightly or not extended, obtuse or truncate. Pistil ca. 2.5 mm; ovules 2. Immature berry obovoid or ellipsoid, 7 × 4 mm; style not persistent.

Phenology. *Berberis ngawaica* has been collected in flower in May and with immature fruit in July and August.

Distribution and habitat. *Berberis ngawaica* is known from Heishui Xian in north Sichuan. It has been collected in an open cut-over area, a forest, and a mountain slope at 2675–2800 m.

Etymology. *Berberis ngawaica* is named after the Ngawa Tibetan and Qiang Autonomous Prefecture of Sichuan which includes Heishui Xian.

IUCN Red List category. *Berberis ngawaica* is assessed as DD or Data Deficient, according to IUCN (2001) criteria.

Selected specimens.
NC Sichuan. Heishui Xian: near Luhua, 2800 m, 31 July 1957, *X. Li 73931* (IBSC 0092607, PE 01033028, SZ 00291433); near Gua Gu, 2750 m, 14 Aug. 1957, *X. Li. 73976*

(IBSC 0092606, 0092715, MO 04124812, PE 01033029, SZ 00290554–55).

203. Berberis ninglangensis Harber, sp. nov. TYPE: China. NW Yunnan: [Ninglang Xian], "Mountains between Yuang-peh and Yangning," 27.466667°N, 100.8°E, 3050 m, July–Aug. 1922, *G. Forrest 22051* (holotype, E E00570041!; isotypes, A 00105609!, CAL CAL0000027192 image!, K!, US 009465992 image!).

Diagnosis. *Berberis ninglangensis* is somewhat similar to *B. lepidifolia* but has narrower and longer leaves often very narrowly elongated at the base and a longer racemose fruit stalk.

Shrubs, deciduous, to 1.5 m tall; mature stems pale yellowish brown, sulcate; spines solitary or absent, concolorous, 0.7–2 cm, terete. Petiole almost absent or to 6 mm; leaf blade abaxially pale green, papillose, adaxially green, very narrowly oblanceolate or linear-oblanceolate, (1.6–)2.4–6 × 0.2–0.8(–1.3) cm, papery, midvein obviously raised abaxially, slightly impressed adaxially, lateral veins indistinct or obscure on both surfaces, base attenuate, often elongated up to 2.2 cm, width 0.1–0.2 cm, margin entire, apex acute, sometimes minutely mucronate. Inflorescence a raceme, 4- to 8-flowered, 1.5–4.5 cm; pedicel 3–5 mm; bracteoles triangular, 1.5 × 1 mm. Sepals in 2 whorls; outer sepals ovate, 1.75 × 0.75 mm; inner sepals broadly obovate, 3 × 2 mm; petals obovate, 2 × 1 mm, base clawed, glands very close together, apex entire. Stamens ca. 1.75 mm; anther connective extended, apiculate. Ovules 2 or 3. Berry dark dull red, slightly pruinose, oblong or sub-globose, 7 × 4 mm, style persistent.

Phenology. The flowering and fruiting season of *Berberis ninglangensis* in the wild is unknown.

Distribution and habitat. *Berberis ninglangensis* is known from only the type collection from Ninglang Xian in northwest Yunnan. This was found growing among scrub on an open, rocky hillside at 3050 m.

G. Forrest 22051 was cited by Ahrendt (1939: 269, 1961: 154) first as a possible example and then a definite example of *Berberis lepidifolia* which is otherwise known from only Lanping Xian. However, the fruit stalks of *B. lepidifolia* are semi-fascicled, semi-umbellate, or sub-racemose, up to 2.5 cm long, with pedicels up to 10 mm long, whereas the fruit stalks of *B. ninglangensis* are uniformly racemose, up to twice as long, and with pedicels half the length. In addition, the leaves of *B. ninglangensis* are significantly longer and narrower, often with a very narrowly elongated leaf base, this last being a feature not found in any other *Berberis* species found in China.

All the wild-collected specimens of *Forrest 22051* are sterile. The description of the flowers and fruit given above is from the cultivated specimens cited below. According to the Kew database, the plant they were taken from was grown from a cutting from a plant originally grown from seed provided by Royal Botanic Garden Edinburgh, in 1923 of either *Forrest 22421* or *Forrest 22623*. *Forrest 22421* is *Berberis lepidifolia* while *Forrest 22623* is *B. tsarongensis*, so there appears to have been some confusion somewhere down the line.

It is worth noting that no other deciduous *Berberis* specimens from Ninglang Xian have been found at either KUN or PE or, indeed, at any other herbarium apart from *C. K. Schneider 3506* (A 00279731, E E00623039, K) which I am unable to identify.

Cultivated material:
Cultivated specimens.
Royal Botanic Gardens, Kew, from accession no. 1923–905, 30 May 1974, *s. coll. s.n.* (K K001273093); 18 May and 6 Sep. 1993, *s. coll. s.n.* (K K001273094).

204. Berberis nyingchiensis Harber, sp. nov. TYPE: China. SE Xizang (Tibet): [Nyingchi (Linzhi) Xian, now Bayi Qu], Pome, Sip Valley near Tongyuk Dzong, 29.966667°N, 94.833333°E, 2895 m, 21 May 1947, *F. Ludlow, G. Sherriff & H. H. Elliot 13731* (holotype, BM BM000794117!; isotypes, A 00279577!, E E00258233!).

Diagnosis. *Berberis nyingchiensis* is somewhat similar to *B. lecomtei* and *B. tsarongensis* but with a different flower structure.

Shrubs, deciduous, to 1.8 m tall; mature stems purplish red, sulcate, black verruculose; spines 3-fid, pale brownish yellow, 0.6–1.2 cm, terete or abaxially slightly sulcate, solitary or absent toward apex of stems. Petiole almost absent, sometimes to 5 mm; leaf blade abaxially pale green, papillose, adaxially green, obovate, 1.1–2 × 0.5–0.8 cm, papery, midvein, lateral veins, and reticulation conspicuous on both surfaces, base attenuate, margin entire, rarely spinulose with 1 to 5 teeth on each side, apex acute, subacute, or obtuse, mucronate. Inflorescence a fascicle, sub-fascicle, or sub-raceme, 4- to 7-flowered, to 2 cm overall; pedicel 4–10 mm; bracteoles lanceolate, 2 × 1 mm. Sepals in 2 whorls; outer sepals obovate, 2.5–3 × 2 mm; inner sepals obovate-orbicular or broadly obovate, 4.5–6 × 3–4.5 mm; petals obovate, 4.5 × 2.5 mm, base clawed, glands separate, oblong, ca. 0.7 mm, apex entire. Stamens ca. 2 mm; anther connective not or slightly extended, truncate or obtuse. Pistil 3 mm; ovules 3. Berry unknown.

Phenology. *Berberis nyingchiensis* has been collected in flower in May and at the very beginning of June. Its fruiting season is unknown.

Distribution and habitat. *Berberis nyingchiensis* is known from Bayi Qu (formerly Nyingchi [Linzhi] Xian) in southeast Xizang. It has been found in open jungle, deciduous forest, and open valleys at 2895–3200 m.

IUCN Red List category. *Berberis nyingchiensis* is assessed as DD or Data Deficient, according to IUCN (2001) criteria.

F. Ludlow, G. Sherriff & H. H. Elliot 13731 was identified as *Berberis johannis* by Ahrendt (1961: 42), while he identified (Ahrendt, 1961: 160) their nos. *13802, 13805,* and *13806* (details below) as *B. humidoumbrosa.* These appear to be indistinguishable from no. *13731.*

Selected specimens.
SE Xizang (Tibet). [**Nyingchi (Linzhi) Xian, now Bayi Qu**]: Pome, Lokmo, near Tongyuk Dzong, 30.016667°N, 94.75°E, 3200 m, 31 May 1947, *F. Ludlow, G. Sherriff & H. H. Elliot 13802* (BM BM000939716); same details, but 1 June 1947, *F. Ludlow, G. Sherriff & H. H. Elliot 13805* (A 00279578, BM BM00939717, E E00162253); *F. Ludlow, G. Sherriff & H. H. Elliot 13806* (A 00279576, BM BM000939718, P P02682340).

205. Berberis obovatifolia T. S. Ying, Fl. Xizang. 2: 146. 1985. TYPE: China. C Xizang (Tibet): Lhasa, No. 3 Guest House, 3900 m, 25 May 1975, *Qinghai-Xizang Complex Exp 75-55* (holotype, PE 00935179!).

Berberis sabulicola T. S. Ying, Fl. Xizang. 2: 133. 1985, syn. nov. TYPE: China. C Xizang (Tibet): Lhasa, No. 3 Guest House, 3810 m, 19 June 1975, *Z. Y. Wu et al. 75-342* (holotype, PE 00935194!; isotypes, HNWP 53355 image!, KUN [2 sheets] 0178677–78!, PE 01840006!).

Shrubs, deciduous, to 1 m tall; mature stems pale pink or yellowish pink, sub-sulcate to sulcate; spines 3-fid, sometimes solitary, semi-concolorous, 0.8–2 cm, terete or abaxially flat. Petiole almost absent; leaf blade abaxially green, shiny, adaxially dark green, obovate or obovate-elliptic, 0.8–2.3 × 0.3–1.4 cm, thickly papery, midvein and venation raised abaxially, inconspicuous adaxially, base cuneate, margin entire, rarely spinulose with 1 to 5 teeth on each side, apex rounded, mucronate. Inflorescence a fascicle, sub-fascicle, or sub-umbel, 2- to 5-flowered; pedicel 4–8 mm. Sepals in 2 whorls; outer sepals elliptic, ca. 3 × 1.5–2 mm; inner sepals oblong-elliptic or obovate-suborbicular, ca. 6.5 × 4 mm; petals broadly elliptic or obovate, ca. 3–4 × 2–3 mm, base clawed, glands separate, apex entire or emarginate. Stamens ca. 3 mm; anther connective truncate or apiculate. Ovules 2 or 3, shortly stipitate. Immature berry yellow turning red, 8–10 × 2–4 mm, oblong; style persistent.

Phenology. *Berberis obovatifolia* has been collected in flower in May and in fruit in June.

Distribution and habitat. *Berberis obovatifolia* is known from the Lhasa area of Xizang (Tibet). It has been found on rocky mountain and ravine sides and on the bed of a grassy valley at 3800–3900 m.

Given that the types of *Berberis obovatifolia* and *B. sabulicola* were collected at the same location within a few days of each other, it is unclear as to why they were published as separate species since the differences in the protologue descriptions are minor. The much fuller description of *B. sabulicola* subsequently given by Ying (2001: 180) included the statement that the inflorescence is racemose. This is not borne out by the type gathering which is fascicled, sub-fascicled, or sub-umbellate. Though both taxa are described as evergreen, the basis for this is unclear given the time of the year the specimens were collected and, from the appearance of the leaves, it would seem more likely that *B. obovatifolia* is deciduous.

Selected specimens.
Xizang (Tibet). Lhasa: 3595 m, 20 May 1942, *F. Ludlow & G. Sherriff 8597* (BM BM000939684); en route to Naikedong, 3800 m, 16 Aug. 1965, *Y. T. Zhang & K. Y. Lang 1548* (PE 01030831–32).

206. Berberis pallens Franch., Pl. Delavay. 36. 1889. TYPE: China. NW Yunnan: [Eryuan Xian], "ad collum Lo-pin-chan supra Lankong," 3200 m, 25 May 1886, *J. M. Delavay s.n.* (lectotype, designated here, P P00716579!; isolectotypes, B, K [2 sheets] K000077407–8!, KUN 1216485!, LE fragm.!, P [2 sheets] P00716578!, P02682016!).

Shrubs, deciduous, to 2 m tall; mature stems purple or shades of reddish purple, sulcate, verruculose; spines 3-fid, sometimes solitary toward apex of stems, pale yellowish or orange-brown, 1.2–2.5 cm, abaxially slightly sulcate, stout. Petiole almost absent; leaf blade abaxially pale green, papillose, adaxially dark green, oblong-obovate to obovate, 1.8–3(–4) × 0.5–1 cm, papery, midvein raised abaxially, lateral veins and dense reticulation indistinct or obscure on both surfaces (sometimes visible when dry), base attenuate, margin entire, apex obtuse or subacute, mucronate. Inflorescence a fascicle, sub-fascicle, or sub-raceme, 5- to 20-flowered, 1.5–4.5 cm overall; pedicel 7–10 mm, to 22 mm when from base. Sepals in 2 whorls; outer sepals ovate, 6–6.5 × 4–4.5 mm; inner sepals oblong-ovate, 6.5–7 × 6–6.5 mm; petals obovate, 5–5.5 × 3.5–4 mm, base very distinctly clawed, glands separate, ca. 1 mm, apex rounded, entire. Stamens 3.5–4 mm, extended, rounded. Pistil 3.5

mm; ovules 2. Berry red, oblong-ovoid or ellipsoid, 7–8 × 4–5 mm; style persistent.

Phenology. *Berberis pallens* has been collected in flower in May and in fruit in October.

Distribution. *Berberis pallens* is known from only the area of the type collection in Eryuan Xian in northwest Yunnan. It has been found at 3200 m in an unrecorded habitat.

The protologue of *Berberis pallens* cited only one gathering, that of 25 May 1886. There are three specimens of this at P (a fourth was sent to KUN as a result of the reorganization of the Paris herbarium in 2010–2012). P00716579 has been chosen here as the lectotype because it is the only specimen which has an original handwritten label. From Bretschneider (1898b), the place of collection appears to be a mountain range north and northwest of Eryuan town, marked as "Chang lo pin."

The description above is based solely on the specimens cited and, as such, it differs from that given by Ying (2001: 161). The protologue describes the flowers of *Berberis pallens* as having bracteoles. I have found no evidence for this.

Schneider (1918: 214–215), noting that he himself had not seen this species while in Yunnan and that Forrest had not collected any specimens in the mountains to the west of Eryuan, speculated that *Berberis pallens* might have a very local distribution. The apparent absence of any specimens since Delavay's collections suggests this may indeed be the case, though the number of *Berberis* collections of any description from Eryuan Xian since Delavay's time is extremely small in comparison with areas of Yunnan further north. Hopefully, the species is not extinct.

For a somewhat similar species from Muli Xian, Sichuan, see under *Berberis purpureocaulis*.

Selected specimens.
NW Yunnan. [**Eryuan Xian**]: "Au col du Lo pin chan, Lankong," 3200 m, 13 Oct. 1886, *J. M. Delavay s.n.* (KUN 1227365, P P02682369); "ad collum Lo-pin-chan supra Lankong," 1886, *J. M. Delavay s.n.* (P P02313226–7).

207. Berberis papillifera (Franch.) Koehne, Gartenflora 48: 21. 1899. *Berberis thunbergii* DC. var. *papillifera* Franch., Bull. Soc. Bot. France 33: 386. 1887; *Berberis finetii* C. K. Schneid., Bull. Herb. Boissier, sér. 2, 8: 203. 1908. TYPE: China. NW Yunnan: [Eryuan–Heqing Xian border], "Les bois, Mt. Hee chan men (Lankong)," 2800 m, 23 Aug. 1884, *J. M. Delavay s.n.* [*1047* "A"] (lectotype, designated here, P P00716515!; isolecto-

types, A fragm. A 00038749!, LE fragm.!, P P00716516!, PE 01901815!).

?*Berberis pruinocarpa* C. Y. Wu ex S. Y. Bao, Bull. Bot. Res., Harbin 5(3): 16. 1985, syn. nov. TYPE: China. NW Yunnan: Lijiang [now Yulong] Xian, Yulong Xue Shan, Dahuo Shan, 2700 m, 25 Aug. 1961, *R. L. Xiong & Y. F. Qi 610639* (holotype, KUN 1204047!; isotype, GXMI GXMI004393 image!).

Shrubs, deciduous, to 1 m tall; mature stems pale brownish yellow, subterete to angled; spines 3-fid, solitary toward apex of stems, concolorous, 0.5–1.5 cm, terete. Petiole almost absent; leaf blade abaxially pale green, slightly papillose, adaxially deep green, narrowly oblong-obovate or spatulate, 1–3 × 0.3–1 cm, thickly papery, midvein and lateral veins raised abaxially, inconspicuous adaxially, base attenuate, margin entire, apex rounded, often minutely mucronate. Inflorescence a fascicle, sub-umbel, or umbel, 2- to 4-flowered, 0.6–1 cm overall; pedicel 10–15 mm. Flowers largely unknown. Inner sepals ?obovate; petals obovate, apex emarginate. Ovules 2, sessile. Berry red, ellipsoid, 9 × 6 mm; style persistent.

Phenology. *Berberis papillifera* has been collected in flower in April and in fruit in July and August.

Distribution and habitat. *Berberis papillifera* appears to be reliably known only from the border of Eryuan and Heqing Xian and possibly from Yulong Xian in northwest Yunnan. It has been found in woods at 2300–2800 m and in a gorge at an unrecorded elevation.

There has been a lot of confusion in relation to *Berberis papillifera*. First is the name itself. Schneider, while accepting that Franchet's *B. thunbergii* var. *papillifera* was a quite separate species from *B. thunbergii* (native to Japan), originally did not accept Koehne's re-classification of it as *B. papillifera* and instead named it *B. finetii*. Schneider's reason for not accepting Koehne's name was because the accompanying description did not accord with that of Franchet's. But since Koehne had written that *B. papillifera* "= *Berberis thunbergii* var. *papillifera* Franchet," this made the differences between Koehne and Franchet's descriptions irrelevant, as Schneider himself later conceded (1918: 222).

Second is the type. Franchet's protologue of *Berberis thunbergii* var. *papillifera* cites only "Yun-nan, Hee-chan-men prope Lankong, altitude 2800 m; fr. 23 Aug, 1884 (Delv. n. 1047)." However, on the same page, he also lists a "? *B. Heteropoda* Schrenk" with only one example, "Yun-nan, in silvis ad collum Koua-la-po prope Ho-kin, alt. 3000 m, fl. 26 maj. 1884 (Delav. n.

1047)." This latter reappears in Franchet (1889: 35) as an example of "*B. sinensis* var. *typica*." In order to separate out the two gatherings, the first is designated here as *Delavay 1047* "A" and the second as *Delavay 1047* "B." It should be noted that neither of the two sheets of *Delavay 1047* "A" at P are specifically numbered *1047*, rather, on both, the collection details "Le bois; Mt. Hee Chan men, (Lankong), 2800 m, d'alt. le 23 aout. 1884" are followed by "(Fruit du No 1047?)," which suggests that the reason why both the gatherings of 26 May and 23 August have the same number is because Franchet thought they might be the same species. *Delavay 1047* "B" is treated here as *Berberis lecomtei*.

Third is Schneider's claim (1918: 222) that the collection details of *Delavay 1047* "A" really apply to a *Delavay 1087* of the same date and that *Delavay 1047* is a misprint. However, Franchet (1887: 385) cites *Delavay 1087* only as applying to gatherings at "Hee-chan-men' near Lankong" of a "*B. sinensis* var. *typica*" of 15 May and 7 October 1884 (for details of these, see under *Berberis amoena*). Nevertheless, the assertion that a *Delavay 1087* of 23 August 1884 is the type of *B. papillifera* was repeated by Ahrendt (1961: 159).

As noted above there are two specimens *of Delavay 1047* "A" at P. The one with the most material has been chosen here as the lectotype.

As noted elsewhere (see under *Berberis lecomtei*), there are numerous specimens of deciduous *Berberis* species from northwest Yunnan with very similar leaves and a variety of inflorescences, and assigning them to the right species is exceptionally difficult, especially if they have no floral material. The only other Delavay specimen that resembles the type of *B. papillifera* would seem to be *Delavay 827 bis*, cited below. This is probably the specimen cited (as *Delavay 827*) by Franchet (1889: 36) as an example of *B. thunbergii* var. *papillifera* and by Schneider (1908: 204) as *B. finetii*, though neither give any collection details. However, both Franchet and Schneider also cite *Delavay s.n.*, "Fang-yang Tchang," 21 October 1887. From the evidence of the various specimens of this at P, this appears to be *B. lecomtei*.

The height of the type of *Berberis papillifera* was not recorded. The height of *Delavay 827 bis* is recorded as 40 cm, but this appears to be a young plant. That of *Northwest Yunnan Jiansha Jiang Team 6227* is recorded as being 0.5–1 m.

Schneider (1908: 203) included a partial description of the flowers of *Berberis papillifera* presumably taken from *Delavay 827 bis*, and this is reproduced above (there are only fragments of flowers on this specimen). Ahrendt (1961: 158–159) included a full description of flowers (reproduced by Ying [2001: 169–170]). From Ahrendt's text, the only possible source of this appears to be cultivated material reputedly grown from *T. T. Yu*

10972 from Zhongdian (now Xianggelila) Xian but, if so, this differs substantially from the account of such flowers given by Ahrendt (1945a: 116) where *T. T. Yu 10972* is cited as the type of *B. humidoumbrosa* var. *inornata* (for further details, see under Taxa Incompletely Known below).

The protologue of *Berberis pruinocarpa* (which misreported the date of collection as 25 Oct. and omitted Y. F. Qi as one of the collectors) cited only the holotype, which has no mature stems and only a few fruit, described as being 1-seeded. The isotype is a better specimen and has mature stems the same color as those of the type specimens of *B. papillifera*. The berries of the isotype (though smaller and partially lightly pruinose) are the same shape of *B. papillifera* including having a distinct style. Importantly, a dissection of berries by Yunfeng Huang of GXMI produced one or two seeds per fruit (pers. comm. 16 Aug. 2019). The leaf shape and size of *B. pruinocarpa* and the number of fascicles per infructescence are the same as *B. papillifera*, though the pedicels are no longer than 10 mm. Given the absence of floral material, however, synonymy with *B. papillifera* cannot be conclusively proved. Information that *Xiong & Qi 610639* was collected in sparse woods on Dahuo Shan some 40 km west of Lijiang comes from the label on the isotype; the height of the plant was not recorded.

Apart from *Northwest Yunnan Jiansha Jiang Team 6227*, no *Berberis* specimens from either Eryuan or Heqing Xian have been found that resemble the type, and it is possible *B. papillifera* is endemic to a very small area there. Whether specimens from elsewhere identified as *B. papillifera* are such is exceptionally difficult to determine. In addition to the type of *B. pruinocarpa*, three of the most likely candidates are listed below (it should be noted that *Sino-American Bot. Exped. 1114* is recorded as being 1–2 m tall, while *S. Y. Bao 445*, the only collection of a deciduous *Berberis* from Jianchuan Xian found, is described as being 1 m tall). But whether they or any others are *B. papillifera* would seem unprovable without further investigation including obtaining additional specimens with floral material from the area where the type was collected. Only this can lay the basis for the proper mapping of the distribution of the species. However, as noted under *Berberis lecomtei*, much of the area in Eryuan and Heqing Xian where Delavay made many of his collections is now dominated by dense forests of *Pinus yunnanensis* planted as a response to earlier forest clearances and logging and consequent soil erosion. It is therefore possible *B. papillifera* is extinct in the type collection area.

Selected specimens.
NW Yunnan. [Eryuan Xian]: "Les gorges du Pe cha ho près de Mo-so-yin. (Lan Kong)," 23 Apr. 1884, *J. M. Delavay 827 bis* (P P00716517); "On the ascent of the Sung-Kwei pass

from the Lan-kong Valley," 100.2°E, 3050 m, s.d., *G. Forrest 2013* (E E00523044, K). **Eryuan Xian:** Sanying Qu, Mengbo, Yangyu Shan, 2300 m, 29 July 1983, *Northwest Yunnan Jiansha Jiang Team 6227* (KUN 0178973, PE 01032792–93).

Possible specimens.
NW Yunnan. Dali Shi: Diancang Shan range, vic. of Huadianba herbal medicine farm, 25.88333333°N, 100.01666667°E, 2900–3300 m, 18 July 1984, *Sino-American Bot. Exped. 1114* (A 00279727, B, E E00623047, KUN 0176684). **Jianchuan Xian:** Shizhong Shan, 2500 m, July 1987, *S. Y. Bao 445* (KUN 0178712–13). **[Ninglang Xian]:** betw. Yung-ning (Yongningxiang) & Yung-peh-ting (Yongsheng), betw. Piji & Mutichin, 2400 m, 24 June 1914, *C. K. Schneider 3506* (A 00279731, E E00623039, K).

208. Berberis pingbaensis M. T. An, Bull. Bot. Res., Harbin 28(6): 641. 2008. TYPE: China. C Guizhou: Pingba Xian (now Pingba Qu), Wuli, 1330–1350 m, 15 July 2007, *M. T. An & Q. L. Luo 2007136* (holotype, GZAC image!).

Shrubs, semi-evergreen, to 40 cm tall; mature stems pale brown, subterete; spines solitary or 3-fid, yellow-brown, 0.6–1 cm, terete. Petiole to 15 mm, sometimes almost absent; leaf blade abaxially grayish green, densely white pruinose, adaxially mid-green, obovate, narrowly ovate, or obovate-lanceolate, 0.8–3 × 0.4–1 cm, thinly leathery, midvein raised abaxially, venation indistinct or obscure on both surfaces, base narrowly attenuate, margin entire, apex rounded, mucronate. Inflorescence a panicle, 10- to 43-flowered, to 14 cm overall; peduncle 2.5–5 cm; bracts ovate, 1–3 mm, apex acuminate or 3-lobed-dentate; pedicel 5–8(–12) mm; bracteoles ovate, 2 × 1.5 mm. Flowers 5–7 mm diam. Sepals in 2 whorls; outer sepals ovate-elliptic or elliptic, 2.5–3.5 × 2–2.25 mm; inner sepals elliptic or obovate-elliptic, 4–5 × 2.5–4 mm; petals obovate-elliptic or elliptic, 3.5–4.5 × 2.5–3 mm, base clawed, glands separate, oblong, ca. 0.5 mm, apex entire or emarginate. Stamens ca. 2.8 mm; anther connective extended, rounded. Ovules 2, subsessile. Berry coral pink, ovoid-globose or ovate, 5–7 × 3–4 mm, white pruinose; style persistent.

Phenology. Berberis pingbaensis has been collected in flower in July and in fruit in September.

Distribution and habitat. Berberis pingbaensis is known from Pingba Xian, central Guizhou. It has been found on open hillsides on limestone at ca. 1330–1350 m.

The relationship between *Berberis pingbaensis* and *B. wilsoniae*, which is also found in Pingba Xian, needs further investigation, as does the relationship with *B. brevipaniculata*.

Selected specimens.
C Guizhou. Pingba Xian: Wuli, 1330–1350 m, 9 Sep. 2007, *M. T. An 2007136* (GZAC).

Photographs. Type plant taken by M. T. An in the field on the day of collection.

209. Berberis platyphylla (Ahrendt) Ahrendt, J. Linn. Soc., Bot. 57: 145. 1961. *Berberis yunnanensis* Franch. var. *platyphylla* Ahrendt, J. Bot. 79(Suppl.): 61. 1941. TYPE: China. NW Yunnan: [Gongshan Xian], Mekong–Salwin divide, Sewalongba (Nisai Benggu), 3500 m, 31 Aug. 1938, *T. T. Yu 22613* (holotype, E E00318263!; isotypes, A 00038780!, KUN 1204871!, PE 01037537!).

Shrubs, deciduous, to 1.25 m tall; mature stems dark purple, sulcate; spines 3-fid, semi-concolorous, ca. 1 cm, weak, abaxially sulcate. Petiole 2–5 mm, sometimes almost absent; leaf blade abaxially pale gray-green, slightly papillose, adaxially deep green, broadly obovate or elliptic, 2–5 × 1–1.4 cm, papery, midvein raised abaxially, flat adaxially, lateral veins and reticulation mostly inconspicuous on both surfaces, base cuneate, margin entire, very rarely spinose-serrulate with 1 to 6 teeth on each side, apex rounded or obtuse. Inflorescence a fascicle, (1- or)2- to 4-flowered; pedicel 13–25 mm. Sepals in 3 whorls, obovate, ca. 2.5–3 × 2 mm; petals obovate, 4–5 × 2.5–3.5 mm, base clawed. Stamens ca. 3 mm; anther connective extended, truncate. Ovules 4 to 6. Berry red, oblong, ca. 10 × 7 mm; style not persistent; seeds 3 to 6.

Phenology. Berberis platyphylla has been collected in flower in June and July and in fruit between August and October.

Distribution and habitat. Berberis platyphylla is known from Dêqên (Deqin), Gongshan, and Weixi Xian in northwest Yunnan, and Zayü (Chayu) Xian in southeast Xizang (Tibet). It has been found on mountain sides and under mixed forests at ca. 3060–4150 m.

Ahrendt (1961: 145) gave a description of the flowers of *Berberis platyphylla* apparently based on *J. F. Rock 23911* from the Yetsi Mountains in Muli Xian, Sichuan, and this was largely followed by T. S. Ying (2001: 163). This is identified here as *B. purpureocaulis* and Ahrendt's description should therefore be discounted. The description here is from *G. M. Feng 5186*, cited below (I am grateful to Z. W. Liu of KUN for this description).

Selected specimens.
SE Xizang (Tibet). Zayü (Chayu) Xian: near Ridong Gongshe, 3500 m, 18 Sep. 1982, *Qinghai-Xizang Team 10467* (KUN 0178318, PE 01032813).

NW Yunnan. Dêqên (Deqin) Xian: "Ad confines Tibet-icas sub jugo Dokerla," 28.25°N, 3800–4150 m, 17 Sep. 1915, *H. R. E. Handel-Mazzetti 8117* (A 00279754, WU 093240); Cizhong, 3650 m, 7 July 1940, *G. M. Feng 5186* (KUN 0176738–40); valley side of Canjiang, Yongzi, 3600 m, 10 Aug. 1940, *G. M. Feng 6503* (KUN 0177566–68, PE 01030764). **Wei-hsi Hsien (Weixi Xian):** 3500 m, June 1935, *C. W. Wang 63875* (A 00280065, IBSC 0092310, LBG 00064294, PE 01032822–23); Laboluo, Shibawodi, 3500 m, 31 Aug. 2003, *X. T. Chiou et al. 07729* (TNM S136238).

210. Berberis pluvisylvatica Harber, sp. nov. TYPE: China. SE Xizang (Tibet): [border betw. Mêdog (Motuo) Xian & Mainling (Milin) Xian], Kongbo, Deyang La, 29.366667°N, 94.866667°E, 3050 m, 30 May 1947, *F. Ludlow, G. Sherriff & H. H. Elliot 15077* (holotype, BM BM000939719!).

Diagnosis. *Berberis pluvisylvatica* has some similarities with *B. dispersa*, but has a different flower structure.

Shrubs, deciduous, to 1.5 m tall; mature stems dark red, sulcate, sparsely black verruculose; spines mostly absent, very rarely 1- or 2-fid, yellow, 0.3–0.7 cm, densely papillose, terete. Petiole almost absent, rarely to 4 mm; leaf blade abaxially pale gray-green, adaxially green, obovate, rarely oblanceolate, 1.3–2.6 × 0.6–1.2 cm, papery, midvein and lateral veins indistinct or inconspicuous on both surfaces, base attenuate, sometimes narrowly so, margin entire, sometimes spinose with 2 to 5 widely spaced teeth on each side, apex subacute to obtuse, rarely retuse, sometimes mucronate. Inflorescence a raceme or an umbel but 1-flowered at the apex of stems, 1- to 8-flowered, to 4 cm overall; peduncle 0.8–1.9 mm; pedicel 6–13 mm. Sepals in 3 whorls; outer sepals narrowly ovate, 3 × 1.25 mm; median sepals broadly elliptic, 3.75 × 3.5 mm; inner sepals obovate, 5.5 × 3.5–3.75 mm; petals obovate, 4–4.5 × 2.5 mm, base distinctly clawed, glands separate, oblong, ca. 1 mm, apex emarginate. Stamens ca. 3 mm; anther connective not extended, truncate. Pistil ca. 3 mm; ovules 2 or 3. Berry unknown.

Phenology. *Berberis pluvisylvatica* has been collected in flower and with very immature fruit in May. Its phenology is otherwise unknown.

Distribution and habitat. *Berberis pluvisylvatica* is known from only the type found in rain forest at 3050 m on the border between Mêdog (Motuo) Xian and Mainling (Milin) Xian in southeast Xizang (Tibet).

Etymology. *Berberis pluvisylvatica* is derived from the Latin pluvisylva, rain forest.

IUCN Red List category. *Berberis pluvisylvatica* is assessed as DD or Data Deficient, according to IUCN (2001) criteria.

F. Ludlow, G. Sherriff & H. H. Elliot 15077 was identified by Ahrendt (1961: 160) as an example of *Berberis humidoumbrosa* despite having a different mature stem color, a different leaf shape, and a different flower structure from that species.

211. Berberis polyantha Hemsl., J. Linn. Soc., Bot. 29: 302. 1892. TYPE: China. WC Sichuan: near Tachienlu (Kangding), 2750–4100 m, s.d. but 1890 or before, *A. E. Pratt 206* (lectotype, designated here, K K000077383!; isolectotypes, BM [2 sheets] 00056 5568!, BM000946155!, E E00373496!, G G00343521!, GH 00038782!, LE fragm.!, P [2 sheets] P P00716524–25!).

Berberis prattii C. K. Schneid. var. *laxipendula* Ahrendt, J. Roy. Hort. Soc. 79, 192. 1954, syn. nov. TYPE: Cultivated. Kew, 1921–1922, grown from seed of *E. H. Wilson (Veitch) 3152*, W China, July 1904, (lectotype, designated here, "251–437/055" K K000395226!).

Shrubs, deciduous, to 4.5 m tall; mature stems yellow-brown, sulcate; spines 3-fid, solitary toward apex of stems, concolorous, 1–3 cm, terete. Petiole almost absent; leaf blade abaxially pale green, adaxially bright mid-green, slightly shiny, oblong-obovate or obovate, 1.2–4.5 × 0.5–1.7 cm, leathery, abaxially gray pruinose, midvein raised abaxially, lateral veins and dense reticulation inconspicuous on both surfaces (conspicuous when dry), base cuneate, margin spinose with 5 to 11 teeth on each side, sometimes entire, apex rounded, sometimes retuse. Inflorescence a panicle, broad and much-branched, 30- to 100-flowered, ca. 5–15 cm overall; peduncle 0.3–2 cm; bracts ca. 1.5 × 2.5 mm; pedicel 2–4 mm; bracteoles triangular, 1–1.5 mm. Sepals in 2 whorls; outer sepals ovate, ca. 3.5 × 2 mm, apex acute; inner sepals obovate, 4.5–6.5 × 2.5–3 mm; petals obovate, 3.5–4 × ca. 2 mm, base clawed, glands separate, oblong, apex incised, lobes acute. Stamens ca. 3 mm; anther connective extended, obtuse. Ovules 2, subsessile. Berry red, narrowly ovoid, 7–8 × 3–4 mm; style persistent, conspicuous, to 1.5 mm; seeds purple.

Phenology. *Berberis polyantha* has been collected in flower between May and August and in fruit between July and October.

Distribution and habitat. *Berberis polyantha* is known from north, northwest, and west Sichuan, and northeast Xizang (Tibet) with one collection from Chongqing (northeast Sichuan). It has been collected from dry slopes, ravine sides, forest margins, and trail, road, and riversides at ca. 2600–3500 m.

In the protologue of *Berberis polyantha*, Hemsley gave a sparse description and cited three gatherings—

Pratt 80, 206, and *704*—but with no further details. Subsequently, Schneider (1918: 295) designated *Pratt 206* as the type, though without indicating a herbarium. His lectotypification has been completed by choosing the specimen at K because that it is where Hemsley was based, and the specimen concerned has an annotation by Schneider dated 16 April 1906. Earlier, Schneider (1913: 377) identified *Pratt 80* as not being *B. polyantha* but a new species, *B. prattii.*

In the protologue of *Berberis prattii* var. *laxipendula,* Ahrendt did not designate a type as such but stated, "This variety appears to have been collected also by *Wilson Veitch No. 3152* and grown from seed from it, e.g., Kew No. 351 (437/05)," but subsequently (1961: 203), he was more specific stating, "at Kew 351 (437/05) (*Type* K)." There is no specimen at K labelled as being from *351,* but there are two labelled as being from a cultivated plant *251,* both annotated by Ahrendt on 7 July 1941 as *B. prattii.* The one that is also labelled "437/05" and "B. polyantha 3152" has been chosen as the lectotype. *Wilson (Veitch) 3152* was identified by Schneider (1913: 376) as *B. polyantha* and this identification is confirmed by the wild-collected specimens at A, BM, and K. Ying (2001: 206) treated *B. prattii* var. *laxipendula* as a synonym of *B. prattii.*

The speculation by Ahrendt (1954: 189–193, 1961: 201–202) that *Berberis polyantha* is possibly a natural hybrid with a small geographical range around Kangding was misplaced. In fact, as the specimens cited below show, it is found across a wide area of Sichuan, an area which extends into northeast Xizang.

The collection from Chengkou Shi in Chongqing (northeast Sichuan), cited below, is interesting in that it is a considerable distance away from any other recorded collection. Perhaps significantly, there is a similar disjunct in relation to *B. aggregata.*

Ying (2001: 210) described *Berberis polyantha* as semi-evergreen. No evidence has been found to support this.

Syntypes. Berberis polyantha. China. W Sichuan: near Tachienlu (Kangding), 2750–4100 m, s.d. but 1890 or before, *A. E. Pratt 704* (BM BM000565568, G 00343523, P P00716526); same details, *A. E. Pratt 80* (A 00279769, BM BM000810309, G, K K000742001, fragm. LE, P P00716523).

Selected specimens.
Chongqing (Sichuan). Chengkou Shi: Jiuchongshan, Zhouxi Xiang, 31.835775°N, 108.535583°E, 1929 m, 14 July 2008, *Bashan Collection Team 1690* (PE 01861151).
Sichuan. W China, [Sichuan], s. loc., July 1904, *E. H. Wilson 3152* (A 00279759, BM, K). **Baiyu Xian:** W of Baiyu, 31.269722°N, 98.793611°E, 2965 m, 21 Aug. 2006, *D. E. Boufford et al. 36971* (A). **Barkam (Ma'erkang) Xian:** Dalangzu Gou, 2700 m, 26 Oct. 1957, *J. X. Zou & X. Li 72733* (PE 010375300); Hexi, 2700 m, 11 July 1984, *W. L. Li et al. 7586* (PE 01037516, 02044450-51). **Dawu (Daofu) Xian:** Qiangning Qu, 3000 m, 7 Aug. 1974, *Sichuan Botany Team*

5663 (CDBI CDBI002755, CDBI0027564, CDBI0027582, PE 01037502). **Dêgê Xian:** 5 Aug. 1951, *Y. W. Cui 5155* (PE 01037501); rd. from Dêgê to Baiyu, Jinsha Jiang, 3040–3100 m, 14 Aug. 2006, *D. E. Boufford et al. 36403, 36427* and *36434* (A). **Heishui Xian:** Lama Si, 2800 m, 7 May 1959, *s. coll. 1069* (PE 01037509–10). **Jinchuan Xian:** ca. 3 km SW of E'ri Zhai, 2920 m, 31 May 1958, *Sichuan 8th Forest Brigade 5248* (IBSC 0091936, 0091942). **Kangding Xian:** Ta-tsien-lou 1893, *J. A. Soulié 191, 493* (G, P); N of Tachien-lu, 2440–3050 m, July and Sep. 1908, *E. H. Wilson 1048* "A" and "B" (A 00279755–56, US 00945957, W 1914/4799); Yala Xiang, Sandao Qiao, Hanjia Gou, 3100 m, 8 Oct. 1953, *X. L. Jiang 37095* (IBK IBK00012835, PE 01037520). **Li Xian:** Dabanzhao Lama Si, 2600 m, 19 July 1957, *D. Z. Deng 1772* (SZ 00289739). **Luding Xian:** Bazifang, 2300 m, 15 Sep. 1983, *Botany Team 31529* (CDBI CDBI0027575–76). **Mao Xian:** Min Valley N of Mao-chou, 2130–2590 m, Aug. 1910, *E. H. Wilson 4634* (A 00279757); Donghua Gou, 25 June 1952, *Z. He 13000* (IBK IBK00012850, PE 01037507, WH 06008730). **Ngawa (Aba) Xian:** Shayang Xiang, Make He, 2950 m, 29 Sep. 1971, *Guoluo Team 790* (HNWP 25860). **Tianquan Xian:** Erlang Shan, 5 Oct. 1953, *X. L. Jiang 37057* (IBK IBK00012856, PE 01037505, 01037518). **Xiaojin Xian:** Munanba Gou, 2700 m, s.d., *X. S. Zhang & Y. X. Ren 06690* (PE 01037499). **Xinlong Xian:** Zuoge, 3200 m, 26 Jun 1974, *S. A. Lu 06362* (CDBI CDBI0027585–87, PE 01037497). **Zamtang (Rangtang) Xian:** Shangzhai Qu, Kalong Shan, above rd., 2860 m, 18 July 1975, *Botany Team 9437* (CDBI CDBI0027546–49, IBSC 009194, PE 01037504, SWCTU 00006287); Dangmuda, N of Dongwu Cun along Aliang Rd. betw. the cities of Rangtang & Aba, 32.54305556°N, 101.06777778°E, 3230 m, 9 Aug. 2018, *D. E. Boufford, B. Bartholomew, J. L. Guo, J. F. Harber, Q. Li & J. P. Yue 44530* (A, CAS, E, KUN). **Zoigê (Ruo'ergai) Xian:** near Baxi Xiang, 33.649444°N, 103.338889°E, 2750 m, 22 Aug. 2007, *D. E. Boufford et al. 40231* (A, E).

NE Xizang (Tibet). Jomda (Jiangda Xian): near Jomda, 3500 m, 24 July 1976, *Qinghai-Xizang Exped. 12472* (HNWP 61503, KUN 0179145–46, PE 01037299–300).

212. Berberis polybotrys Harber, sp. nov. TYPE: Cultivated. Royal Botanic Garden Edinburgh, from 19943553*B, 28 May 2016, *P. Brownless 1048,* grown from *Alpine Garden Society China Exped. (ACE) 1921,* China. NW Yunnan: Zhongdian (Xianggelila) Xian, Zhongdian to Sanba, 27.733889°N, 99.952750°E, 3700 m, 11 Oct. 1994 (holotype, E E00831859!; isotypes, A!, PE!).

Berberis franchetiana C. K. Schneid. var. *macrobotrys* Ahrendt, J. Bot. 80(Suppl.): 114. 1945. TYPE: China. NW Yunnan: Chungtien (Zhongdian, now Xianggelila Xian), Pica (Pika), 3200 m, 30 Oct. 1937, *T. T. Yu 13903* (holotype, E E00217934!; isotypes, A 00038752!, BM BM001015562!, KUN 0754713!, PE 01037538!).

Diagnosis. Berberis polybotrys has leaves that are somewhat similar in shape to *B. pseudotibetica,* but are adaxially not as dark green. Its inflorescence is longer and distinctly racemose versus the semi-racemose and sometimes partially paniculate inflorescence of *B. pseudotibetica.* Unlike *B. pseudotibetica* it is sometimes fascicled at its base.

Shrubs, deciduous, to 1.5 m tall; mature stems pale reddish brown, sulcate; spines 3-fid, solitary or absent toward apex of stems, brownish yellow, 0.5–0.9 cm, terete. Petiole almost absent or to 3(–8) mm; leaf blade abaxially pale green, papillose, adaxially mid-green, narrowly obovate, oblong-obovate, or oblanceolate, 2.2–3.5 × (0.6–)1–1.5 cm, papery, midvein slightly raised abaxially, venation and reticulation inconspicuous on both surfaces, base narrowly attenuate, margin entire, rarely spinose with 1 to 6(to 8) teeth on each side, apex obtuse or subacute, rarely rounded. Inflorescence a loose raceme, sometimes fascicled at base, (4- to)8- to 12-flowered, 2.5–6 cm overall; peduncle 1.5–2.5 cm; pedicels 6–10 mm, to 16 mm when from base. Flowers lemon yellow. Sepals in 3 whorls; outer sepals narrowly ovate, 2.75 × 1.25 mm; median sepals oblong-ovate, 3.75–4.25 × 2.5 mm; inner sepals obovate, 4.5 × 3.25 mm; petals elliptic, 4 × 2.5–2.75 mm, base clawed, glands separate, ca. 0.7 mm, apex subacuminate, entire. Stamens ca. 2 mm; anther connective slightly extended, obtuse. Pistil 2.25 mm; ovules 2. Berry deep red, obovoid or oblong, 7–8 × 4 mm; style persistent.

Phenology. Berberis polybotrys has been collected in fruit in October. Its flowering season in the wild is unknown.

Distribution and habitat. Berberis polybotrys is known only from Zhongdian (Xianggelila) Xian in northwest Yunnan. Seeds of the type were collected on dry ground among *Rhododendron hippophaeoides* at 3700 m, seeds of *Yu 13903* on a grassy slope at 3200 m.

IUCN Red List category. Berberis polybotrys is assessed as DD or Data Deficient, according to IUCN (2001) criteria.

With the exceptions of the very distinctive *Berberis jamesiana* and *B. ninglangensis*, *Berberis* species from northwest Yunnan with racemose inflorescences with two or three ovules are very difficult to identify. Besides *B. polybotrys*, four such species are included in this treatment: *B. biguensis*, *B. forrestii*, *B. zhaoi*, and *B. pseudotibetica*. In the first three cases, there is an account of their flower structure (though see the discussion under *B. forrestii* regarding its accuracy). *Berberis polybotrys* differs in its flower structure from all three (and from *B. forrestii* in other ways). Though the flowers of *B. pseudotibetica* are unknown, other of its characteristics are different from those of *B. polybotrys*.

The collection information for *ACE 1921* (details below) include describing the fruit as being "in short panicles or racemes." Neither the specimen itself nor the inflorescences of the plants grown from seed of *ACE 1921* provide any evidence of panicles.

Berberis franchetiana var. *macrobotrys* was published on the basis of the holotype. This has abundant fruit

stalks, but these are on an old twig with a tiny number of leaves, only one of which is complete. The protologue conceded that the association with *B. franchetiana* (treated here as a synonym of *B. yunnanensis*) was largely speculative. The only conclusion one can draw from all this is that the taxon should never have been published.

Unknown to Ahrendt, there are four isotypes of *Yu 13903* (details above), but these provide no further taxonomic evidence in that they have no mature stems either and only one leaf among them. Given all this, I would have assigned *Berberis franchetiana* var. *macrobotrys* to my "taxa incompletely known" section until I discovered there was a living plant in the Royal Botanic Garden Edinburgh (RBGE), originating from wild-collected *Yu 13903* seed. The origin of this is that the RBGE together with the Arnold Arboretum partly funded Yu's collecting expedition in Yunnan in 1937. From the accession records of RBGE kindly provided for me by Peter Brownless, the seeds of *Yu 13903* were received on 25 May 1938. The current living plant dates from 1964 and appears to be from a cutting from the original planting. The label on the cultivated specimens listed below record the original accession number as 380229. The structure of the flowers of this plant is little different from that of *B. polybotrys*.

It has proved impossible to locate exactly where the type of *Berberis franchetiana* var. *macrobotrys* was collected. But from the evidence of T. T. Yu's fieldbook, it appears to be somewhere between Zhongdian City and Haba Shan and to be much nearer the former than the latter. Yu (or his collection team) collected there between 30 October and 4 November, some of the entries recording it as Pika Xueshan. Given this, there appears to be only two candidates—Shika Shan and Tianbao Shan—both of which are sometimes referred to as being Xue (snow) Shan. The similarity of names suggests the former is more likely. If so, this is some 35 km to the west of where the seeds of *ACE 1921* were collected.

Besides the holotype, the protologue of *Berberis franchetiana* var. *macrobotrys* cited *T. T. Yu 15053*, Lichiang, 27 May 1937 (E E00612483), though it gave no account of its flowers. This is treated here as being of a new species, *B. zhaoi*.

Selected specimens.
NW Yunnan. Zhongdian (Xianggelila) Xian: Zhongdian to Sanba, 27.733889°N, 99.952750°E, 3700 m, 11 Oct. 1994, *Alpine Garden Society China Exped. (ACE) 1921* (E E00039715).

Cultivated material:
Living cultivated plants.
Royal Botanic Garden Edinburgh, from *ACE 1921* 19943553*A and 19943553*B; from *Yu 13903*, 19644005B.

Cultivated specimens.
Royal Botanic Garden Edinburgh, from 19644005, 1 June 1989, *s. coll. s.n.* (E E00112214); 16 May 1977, *s. coll. C31789* (E E00112215).

213. Berberis potaninii Maxim., Trudy Imp. S.-Peterburgsk. Bot. Sada 11: 41. 1891, as "*potanini.*" TYPE: China. S Gansu: "Tshung-dsha-wan," 11 Sep. 1885, *G. N. Potanin s.n.* (lectotype, designated by Imchanitzkaja [2005: 276], LE!).

Berberis sphalera Fedde, Bot. Jahrb. Syst. 36 (Beibl. 82): 44. 1905. TYPE: China. Shaanxi: "Monti del Lung-san huo," May 1895, *G. Giraldi 62* (lectotype, designated here, FI image!; ?isolectotype, B†).
Berberis liechtensteinii C. K. Schneid., Pl. Wilson. (Sargent) 1: 377–378. 1913. TYPE: China. NC Sichuan: near Maochow (Mao Xian), Min Valley, 1200–1700 m, 26 May 1908, *E. H. Wilson 2871* (lectotype, designated by Ahrendt [1961: 92–93], A 0038775!; isolectotypes, BM BM00059574!, E E00217965!, HBG HBG-506694 image!, K K000644912!, LE!, US 00103887 image!, W 1914–4802!).

Shrubs, evergreen, to 1.75 m tall; mature stems dark purplish red, terete or angled; spines 3-fid, concolorous, (1.5–)2–6 cm, often longer than distance between internodes, stout, abaxially slightly sulcate. Petiole almost absent; leaf blade abaxially pale yellow-green, sometimes pruinose at first, adaxially shiny, deep green, lanceolate, narrowly obovate, or obovate, 1–3(–4) × (0.25–)0.5–1.2 cm, thickly leathery, midvein raised abaxially, venation indistinct on both surfaces, base attenuate or cuneate, margin often thickened, spinose or dentate with 1 to 4(to 6) teeth on each side, rarely entire, apex acute, mucronate. Inflorescence a raceme or an umbellate sub-raceme, 4- to 12-flowered, 2–4 cm; pedicel 5–10 mm. Sepals in 2 whorls; outer sepals elliptic to obovate, 4–5 × 3–4 mm, apex obtuse; inner sepals obovate, 5–7 × ca. 4.25 mm; petals obovate, 4.25–5 × 3–3.5 mm, base truncate, glands separate, lanceolate, apex entire. Stamens 4–5 mm; anther connective obtuse. Ovules 1 or 2, sessile. Berry red, not pruinose or sometimes slightly so, oblong or ellipsoid, 6–9 × 4–6 mm; style persistent, conspicuous.

Phenology. *Berberis potaninii* has been collected in flower in May and in fruit between May and November.

Distribution and habitat. *Berberis potaninii* is known from south Gansu, east, south, and southwest Shaanxi, and north-central and northern Sichuan. It has been collected on sunny slopes and trail and streamsides at 200–1700 m.

The protologue of *Berberis potaninii* cited "ad fl. Pei-shui, 22 Junii" and "Tshung-dsha-wan ad rivulum 11 Septbr. '85," but without citing any herbarium. A rea-sonable deduction is that it was LE and, indeed, specimens of both are there. Ahrendt (1961: 92) cited "Hei-shu, fl and immature fr 2 June 1885" as the type (again without citing any herbarium). This was in line with his policy (noted in the introduction) of if no type was indicated in a protologue, then simply choosing the first listed as type. As such, even ignoring the mistaken date, this must be regarded as a largely mechanical method of selection as defined by Art. 10.5 of the Shenzhen Code (Turland et al., 2018) and discounted. Imchanitzkaja (2005: 276), ignoring Ahrendt, lectotypified *Potanin s.n.*, 11 September 1885, on the basis that there is an annotation on the sheet by V. I. Grubov of March 1957 identifying this as the lectotype, and this was repeated by Borodina-Grabovskaya (2010: 101). Despite Grubov's lectotypification not being published, Imchanitzkaja's lectotypification stands.

The protologue of *Berberis sphalera* cited only *Giradi 62* but did not cite any herbarium. It is likely that its author, Fedde, based his description partly or wholly on a specimen of this at Berlin. However, any such specimen was destroyed in WWII. The specimen at FI has, therefore, been designated as the lectotype. I have been unable to identify where in Shaanxi *Giraldi 62* was collected.

The protologue of *Berberis liechtensteinii* cited *Wilson 2871* as type but without citing a herbarium. Ahrendt's citing (1961: 92–93) of the specimen at A as type, albeit without having seen it, was an effective lectotypification.

Schneider (1906: 199) synonymized *Berberis sphalera* and (1918: 214) *B. liechtensteinii.* Ahrendt (1961) continued to recognize *B. liechtensteinii* as a separate species without any commentary on Schneider's synonymy. *Berberis sphalera* was not noticed by Ying (2001).

The unusually long spines with the internodes often very close together make *Berberis potaninii* a very distinctive species. The leaf shape, however, can vary considerably, those of *Giraldi 62* and *J. X. Yang 2468* from Shaanxi are noticeably consistently narrower than specimens from elsewhere. No wild-collected specimens of *B. potaninii* with floral material have been located. The description of flowers used here is that of Ahrendt (1961: 93), which is based on cultivated plants grown from seeds of *E. H. Wilson 4154*.

K. S. Hao 414 from Wen Xian, Gansu, cited below, was mistakenly identified by its collector, K. S. Hao (1938), as *Berberis lycium*, a species of the west Himalayas.

No evidence has been found to support the report of *Berberis potaninii* in Henan by B. Z. Ding et al. (1981: 489).

Selected specimens.
Syntype. Berberis potaninii. S Gansu: "ad fl. Pei-shui," 22 June 1885, *G. N. Potanin s.n.* (LE).

Other specimens.

S Gansu. Hui Xian: Jialingxiang, 750 m, 19 Sep. 1958, *Z. Y. Zhang 952* (IBK IBK00322522, KUN 0176756, NAS NAS00314171). **Kang Xian:** Zhongzai, 23 Sep. 1951, *Z. B. Wang 14959* (PE 01030738, WUK 0020776). **Liangdang Xian:** Zhangjiazhuang, 27 Aug. 1965, *Z. X. Peng s.n.* (LZU 00009244). **Min Xian:** Kangduo, 6 June 1951, 1950 m, *Z. B. Wang 14094* (KUN 0178354, WUK 0068505). **Têwo (Diebu) Xian:** on the way to Kang Xian, 1800 m, *B. Z. Guo 5637* (WUK 00231236). **Wenhsian (Wen Xian):** 910 m, 15 June 1930, *K. S. Hao 414* (PE 01033866); Chengguan Gongshe, 870 m, 15 Oct. 1973, *J. X. Yang & Z. S. Hu 3458* (PE 01030737). **Wudu Xian:** Cike, 2000 m, 6 June 1959, *Z. Y. Zhang 3661* (WUK 00153788). **Zhugqu (Zhouqu) Xian:** Gongba, 1700 m, 28 May 1999, *Bailong Jiang Investigation Team 1467* (PE 01556134).

E & S Shaanxi. Feng Xian: near Tangzang, 1200 m, 2 July 1960, *K. J. Fu 12927* (WUK 00162534). **Shangnan Xian:** Xianghe Gongshe, Shuigou Xiang, 26 June 1960, *Northwest University Dept. of Biology 565* (WUK 00184770); Xianghe Zhen, Shuxilou Cun, 33.284350°N, 110.949617°E, 200 m, 6 Oct. 2013, *S. F. Li et al. 18080* (XBGH XBGH009217). **Shanyang Xian:** Manshuanguan Gongshe, 340 m, 1 May 1964, *J. X. Yang 2468* (PE 01037026–27). **Xixiang Xian:** Chazhen, Cengji He, 450 m, 22 Sep. 1952, *B. Z. Guo 2124* (CDBI CDBI0027490–93, IBSC 0092777, PE 01037025).

SW Shaanxi. Lueyang Xian: Baishuijiang, 700 m, 5 Apr. 1963, *Z. B. Wang 18759* (KUN 0178257, WUK 00213793, 0048347).

NC & N Sichuan. Li Xian: 1952, *Z. He & Z. L. Zhou 13934* (IBSC 0092557, PE 01033763–64, WH 06008714). **Mao-chow (Mao Xian):** Min valley, 1525–2135 m, Nov. 1910, *E. H. Wilson 4154* (A 00279767); Gokou Gongshe, 1800 m, 2 Oct. 1983, *Y. L. Cao et al. 012* (CDBI CDBI0172384–85). **Nanping (now Jiuzhaigou) Xian:** Beihe Qu, 1500 m, 25 Oct. 1983, *Y. L. Cao et al. 246* (CDBI CDBI0027483–84, CDBI027488). **Songpan Xian:** Baohua Dadui, 1300 m, 31 Aug. 1961, *X. N. Tang et al. 00622* (CDBI CDBI0027592). **Wenchuan Xian:** 13 June 1958, *Z. L. Guan 31832* (PE 01033763).

214. Berberis pratensis Harber, sp. nov. TYPE: China. W Sichuan: Daocheng Xian, S of Bowa Cun, on hwy. S216 toward Bowa Shan, 28.961389°N, 100.276389°E, 3900 m, 28 Aug. 2013, *D. E. Boufford, J. F. Harber & Q. Wang 43293* (holotype, PE!; isotypes, A!, E!, KUN!).

Diagnosis. *Berberis pratensis* has leaves similar to *B. reticulinervis* which is also consistently fascicled but with longer pedicels. The flower structure of *B. pratensis*, however, is different, including two whorls of sepals and three or four ovules versus the three whorls and two or three ovules of *B. reticulinervis*.

Shrubs, deciduous, to 2 m tall; mature stems reddish brown, sulcate; spines 3-fid, solitary or 2-fid toward apex of stems, pale brownish yellow, 0.5–1.5 cm, abaxially not sulcate. Petiole almost absent; leaf blade abaxially pale green or glaucous, papillose, adaxially dull mid-green or grayish green, narrowly obovate or narrowly elliptic, (0.8–)1.2–1.6 × 0.4–0.6 cm, papery, midvein raised abaxially, lateral veins and reticulation con-

spicuous on both surfaces, base attenuate or cuneate, margin entire, apex subacute or obtuse, mucronate. Inflorescence a fascicle, 1- to 3-flowered; bracts lanceolate, 1.5 × 0.5 mm, pinkish red; pedicel 5–6 mm. Sepals in 2 whorls; outer sepals ovate, 4.5 × 3 mm; inner sepals broadly elliptic or orbicular-elliptic, 4.5–5.5 × 4–4.5 mm; petals obovate, 4–6 × 2.5–4 mm, base clawed, glands approximate, elliptic, 0.7 mm, apex entire. Stamens 2.5–3 mm; anther connective slightly extended, obtuse or truncate. Ovules 3 or 4. Immature berry yellow turning red, some tinged purple, ellipsoid or oblong, 8–10 × 4–5 mm; style persistent.

Phenology. *Berberis pratensis* has been collected in flower and with immature fruit in August. Its phenology is otherwise unknown.

Distribution and habitat. *Berberis pratensis* is currently reliably known from only one collection in Daocheng Xian in west Sichuan. This was found in a roadside meadow at 3900 m.

IUCN Red List category. *Berberis pratensis* is assessed as DD or Data Deficient, according to IUCN (2001) criteria.

The type was collected from a large colony, many of which were sterile. The description of flowers comes from the one very late flower found.

215. Berberis prattii C. K. Schneid., Pl. Wilson. (Sargent) 1: 376. 1913. *Berberis aggregata*, C. K. Schneid. var. *prattii* (C. K. Schneid.) C. K. Schneid., Pl. Wilson. (Sargent) 3: 443. 1917. TYPE: China. W Sichuan: Tachien-lu (Kangding), 2130–2590 m, June 1908, *E. H. Wilson 1261* "A" (lectotype, designated here, specimen with flowers A 00038784!; isolectotypes, BM BM001040391!, E E00259001!, HBG HBG-506690 image!, K K000077382!, LE!, US 00103899 image!, W 1914–0004801!).

Berberis polyantha Hemsl. var. *oblanceolata* C. K. Schneid., Pl. Wilson. (Sargent) 1: 376. 1913. *Berberis oblanceolata* (C. K. Schneid.) Ahrendt, Bull. Misc. Inform. Kew 1939, 275. 1939. TYPE: China. WC Sichuan: Hsaochin-ho valley, near Monkong Ting (Xiaojin Xian), 2150–2750 m, June 1908, *E. H. Wilson 2868* (lectotype, designated by Ahrendt [1939: 275], K K000395200!; isolectotypes, A 00038783!, BM BM001015556!, E E00217964!, HBG HBG-506693 image!, LE!, US 00103898 image!, W 1914–4799!).

Berberis prattii var. *recurvata* C. K. Schneid., Pl. Wilson. (Sargent) 1: 377. 1913. *Berberis aggregata* var. *recurvata* (C. K. Schneid.) C. K. Schneid., Pl. Wilson. (Sargent) 3: 443. 1917. TYPE: China. WC Sichuan: Mupin (Baoxing), 1830–2300 m, June 1908, *E. H. Wilson 1073* "A" (lectotype, designated here, specimen with flowers A 00038786!; isolectotypes, A 00038785!, BM

BM001015579!, E E00259002!, HBG HBG-506691 image!, K K000077381, US 00103900 image!, W1914–0004797!).

Berberis aggregata C. K Schneid. var. *integrifolia* Ahrendt, J. Linn. Soc., Bot. 57: 203. 1961, syn. nov. TYPE: China. WC Sichuan: Pao-hsing-hsien (Baoxing Xian), 2500 m, 25 June 1936, *K. L. Chu 2939* (holotype, E E00217968!; isotypes, BM BM001015570!, IBSC [2 sheets] 0000582 image!, 0092572 image!, P P02313368!, PE [3 sheets] 01037397–99!).

Shrubs, deciduous, to 3 m tall; mature stems pale reddish brown, sulcate, sparsely verruculose; spines solitary or 3-fid, pale brownish yellow, 0.5–1.5 cm, terete, weak. Petiole 1–3 mm; leaf blade abaxially grayish green, papillose, adaxially slightly shiny, yellow-green, obovate-elliptic or obovate, 1–3(–4) × 0.5–1.5 cm, papery, midvein raised abaxially, slightly raised or flat adaxially, lateral veins and dense reticulation indistinct or inconspicuous abaxially, conspicuous adaxially especially when dry, base cuneate or attenuate, margin entire or spinulose with 3 to 8 teeth on each side, apex rounded, sometimes minutely mucronate. Inflorescence a narrow rigid panicle, 15- to 80-flowered, 5–15 cm, glabrous or puberulent; bracts 1–2.5 mm; pedicel 2.5–4 mm. Sepals in 3 whorls; outer sepals lanceolate, 4–5 × 1.5–2 mm, apex acuminate; median sepals ovate-lanceolate, 4–5 × 1.5–2 mm, apex subacuminate; inner sepals obovate, ca. 5.5 × 4 mm; petals narrowly obovate, ca. 4.5 × 2.3 mm, base clawed, glands separate, oblong, apex emarginate, lobes acute. Stamens ca. 2.5 mm; anther connective extended, apiculate. Ovules 2, sessile. Berry salmon pink, turning red, ovoid, ca. 6.5 × 4.5 mm; style persistent.

Phenology. *Berberis prattii* has been collected in flower in June and July and in fruit between September and November.

Distribution and habitat. *Berberis prattii* is known from Baoxing, Kangding, Luding, and Xiaojin Xian in Sichuan. It has been collected from thickets, forests, and riversides at ca. 1670–3000 m.

The type of *Berberis prattii*, *Wilson 1261*, consists of two gatherings, that of June 1908 is designated here as *Wilson 1261* "A" and that of October 1908 as *Wilson 1261* "B." *Wilson 1261* "A" at A has been chosen as the lectotype because the protologue was based on a specimen sent by Harvard to Schneider for identification and flowers are more useful for diagnosis than fruit.

The protologue of *Berberis prattii* var. *recurvata* cited only "June and October 1908 (No 1073)." The gatherings are designated here as *Wilson 1073* "A" and "B," respectively. A specimen of *Wilson 1073* "A" at A with flowers has been designated as the lectotype for the

same reasons as for *B. prattii*. In this case, there are two specimens at A with flowers, and the one on the sheet which also includes *Wilson 1073* "B" has been chosen because this seems to be the sheet which most faithfully reflects the protologue citation.

The protologue of *Berberis polyantha* var. *oblanceolata* cited only *Wilson 2868* but did not designate a herbarium. Ahrendt's designation (1939: 275) of the specimen at K as the type of *B. oblanceolata* is treated here as an obligate lectotypification.

Schneider published *Berberis prattii* var. *recurvata* solely on the basis that it had a narrower inflorescence than *B. prattii* with evidence of curved pedicels on *E. H. Wilson 1073* "B," while Ahrendt (1961: 202) stated, apparently on the basis of a cultivated plant grown from seeds of *Wilson 1073* "B," that the variety was "For cultural purposes scarcely distinguishable from *B. prattii*." The synonymy with *B. prattii* was made by Ying (2001: 206).

As can be seen above, at various times *Berberis prattii* has been associated with both *B. polyantha* and *B. aggregata*. Hemsley (1892) listed *Pratt 80*, cited below, as an example of *B. polyantha*, and it was Schneider (1913: 377) who identified it as *B. prattii*. In comparing the two species, Schneider noted, "*B. polyantha* has much thicker leaves with a very fine reticulation; the inflorescences are broader and looser, the bracts shorter and the elliptical dried fruits are swollen and bear a rather long style." However, at the same time, Schneider recognized a new *B. polyantha* var. *oblanceolata*, even though the type (which has flowers rather than fruit) has a narrow inflorescence and leaves that are thinner than *B. polyantha*. The synonymy with *B. prattii* was made by Ying (2001: 206).

The association of *Berberis prattii* with *B. aggregata* is more complex. As can be seen above, Schneider (1918: 296) decided that his *B. prattii* was not a separate species but a variety of *B. aggregata*, which is also found in Sichuan, but farther north. Both taxa have certain morphological features in common, including a narrowly paniculate inflorescence, two ovules, and fruit that turn salmon pink before turning red. However, both the leaf morphology and the flower structure of the taxa are different and I have followed Ahrendt (1961: 202) and Ying (2001: 206) in maintaining *B. prattii* as a separate species.

But a further complexity is provided by *Berberis aggregata* var. *integrifolia*. Besides the type, the protologue cited *K. L. Chu 3863* (for details, see below), *E. H. Wilson (Veitch) 3155a* (K K000077380), western China, and *E. E. Maire 1997* (E, K) Yunnan-sen (Kunming). However, only the first of these is the same as the type; *Wilson (Veitch) 3155a* is *B. aggregata* and *Maire 1997* is *B. wilsoniae*.

Ahrendt describes the inflorescence of variety *integrifolia* as up to 4 cm long. However, on one of the isotypes (PE 01037399) they are up to 6.5 cm long.

The report by Bao (1997: 81) of *Berberis aggregata* var. *integrifolia* being found in Yunnan appears to be based on misidentification of various specimens of *B. wilsoniae*.

Syntype. *Berberis prattii.* WC Sichuan: Tachien-lu (Kangding), 2130–2590 m, Oct. 1908, *E. H. Wilson 1261* "B" (A 0038784, BM BM001040391, E E00259002, HBG HBG-506690, K K000077383, LE, US00103899, W 1914–0004801).

Syntype. *Berberis prattii* var. *recurvata.* Mupin (Baoxing), 1830–2300 m, Oct. 1908, *E. H. Wilson 1073* "B" (A 00038786, BM BM001050024, E E00259002, HBG HBG-506691 image, K K000644944, US 00103900, W1914–0004797).

Selected specimens.
W, WC & NC Sichuan. Baoxing Xian: Mupin (Baoxing), 1830 m, Oct. 1908, *E. H. Wilson 1050a* (A 00279325, BM, E E00395965, HBG HBG-506692, K, US 0945889); Mupin, 2440–2745 m, Oct. 1910, *E. H. Wilson 4286* (A 00279324); Dengchi Gou, 1670 m, 2 Nov. 1933, *T. H. Tu 4901* (IBSC 0092530, PE 01037395); Pao-hsing-hsien (Baoxing Xian), 2000 m, 17 Sep. 1936, *K. L. Chu 3863* (BM, E E00267371, IBSC 0091506, 0092640, K, P P02565473, PE 01037394, 01037396); Jianfashan Gou, 2300 m, 10 June 1958, *X. S. Zhang & Y. X. Ren 5270* (GF 09001046, PE 01037393); Denglong Gou, Jiandao Shan, 2500 m, 29 Sep. 1958, *Sichuan Agricultural College 7339* (CCNU 09001050, CDBI CDBI0027609–10, GF 09001050, IBSC 0091507). **Kangding Xian:** near Tachien-lu (Kangding), 2750–4100 m, s.d. but 1890 or before, *A. E. Pratt 80* (A 00279769, BM, BM000810309, G, K K00742001, P P00716523); SE of Tachien-lu, June 1908, *E. H. Wilson 1300* "A" (A 00279772, K, US 0945888); same details, but Oct. 1908, *E. H. Wilson 1300* "B" (A 00279771); Tachien-lu, uplands, 2130–2590 m, Oct. 1910, *E. H. Wilson 4173* (A 00279770). **Luding Xian:** Reshui Gou, 1800 m, 28 Sep. 1980, *Botany Group 23723* (CDBI CDBI0027526–27); Reshui Gou, 2800 m, 7 Oct. 1982, *G. H. Xu 41154* (CDBI CDBI0027577–79); Xinxing, Gongga Shan, 2100 m, 7 July 1982, *K. Y. Lang et al. 539* (PE 00935163–4).

216. Berberis pruinosifolia Harber, sp. nov. TYPE: China. W Sichuan: [Kangding Xian],. "grasslands of Yulong-Hsi, Minya Country, south of Tatsienlu," 4000 m, June 1929, *J. F. Rock 17775* (holotype, E E00392889!; isotypes, A [2 sheets] 00279469!, 00279472!, IBSC 0091670 twig from specimen at A image!, US 00945922 image!).

Diagnosis. *Berberis pruinosifolia* is somewhat similar to *B. dictyophylla*, but has smaller leaves, an inflorescence that is fascicled rather than single-flowered, and a different flower structure.

Shrubs, deciduous, to 2.5 m tall; mature stems reddish purple, sulcate, partially glaucous pruinose; young shoots densely pruinose; spines 3-fid, orange-brown,

0.6–1.3 cm, slender, abaxially sulcate. Petiole almost absent; leaf blade concolorous green on both surfaces, abaxially often glaucous pruinose especially on young shoots, narrowly elliptic-obovate, 0.8–1.5 × 0.4–0.5 cm, papery, midvein inconspicuous on both surfaces, dense reticulation inconspicuous abaxially, conspicuous adaxially, base cuneate or attenuate, margin entire or spinulose with 1 to 3 teeth on each side, apex acute, sometimes mucronate. Inflorescence a fascicle, 1- to 3-flowered; pedicel 5–7 mm. Sepals in 3 whorls; outer sepals oblong-ovate, ca. 2.5 × 1.5 mm; median sepals oblong-elliptic, ca. 4 × 3.25 mm; inner sepals obovate, 4 × 3 mm; petals obovate-elliptic, ca. 3.25 × 2.5 mm, base slightly clawed, glands widely separate, ca. 0.6 mm, apex obtuse. Stamens ca. 2.25 mm; anther connective slightly extended. Pistil 2.5 mm; ovules 3 or 4. Berry red, pruinose, oblong or oblong-ovate, 10 × 5 mm; style persistent.

Phenology. *Berberis pruinosifolia* has been collected in flower in June and in fruit in August and September.

Distribution and habitat. *Berberis pruinosifolia* is known from only the type and two other collections from south Kangding Xian in west Sichuan, the type in grasslands at 4000 m, the other collections on a mountainside at 3600 m and a yak-grazed area at 3237 m.

IUCN Red List category. *Berberis pruinosifolia* is assessed as DD or Data Deficient, according to IUCN (2001) criteria.

Berberis pruinosifolia is a very distinctive species. From the map in Rock (1930: 388), Yulong Hsi can be confirmed as the present-day Yulongxi situated to the west of Minya Konka (the present day Gongga Shan) in a valley south-southwest of the Djesi La pass. It appears that there have been no collections of *Berberis* in this valley since Rock. I am told by the non-botanist Michael Woodhead of Sydney, Australia (pers. comm. 14 May 2013), that when he visited Yulongxi in 1995, he found much of grasslands barren and overgrazed, though there were still areas of bush. *K. Y. Lang et al. 997* and *SICH 2224* were collected ca. 15–20 km from the type in a parallel valley to the west.

Selected specimens.
W Sichuan. Kangding Xian: Shade Qu, Suoji en route to Liuba Gongshe, 3600 m, 4 Aug. 1982, *K. Y. Lang et al. 997* (KUN 0178923, PE 0103762–63); 70 km SW of Zheduo pass, by Niqiu River, 29.69375°N, 101.479528°E, 3237 m, 25 Sep. 2001, *A. Kirkham et al. SICH 2224* (CAS 1024135, K).

217. Berberis pseudothunbergii P. Y. Li, Acta Phytotax. Sin. 10: 211. 1965. TYPE: China. SW Shaanxi: Lueyang Xian, Baishuijiang, Maliutang,

1300 m, 2 Apr. 1963, *T. P. Wang 18735* (lectotype, designated here, WUK 213771 image!; isolectotypes, HNWP 55916 image!, KUN 0178256!, WUK 00131521 image!).

Shrubs, deciduous, to 1.5 m tall; mature stems dark red, sulcate; spines solitary, concolorous, 0.8–1.5 cm, terete, slender. Petiole 1–3 mm, sometimes almost absent; leaf blade abaxially grayish green, papillose, adaxially pale green, rhombic-obovate, 1–2 × 0.3–1 cm, papery, venation largely obscure on both surfaces, base cuneate or attenuate, margin entire, apex obtuse, rarely subacute. Inflorescence 1-flowered or a loose raceme or sub-umbel, 2- to 6-flowered, 1.5–3 cm including peduncle 1–2.5 cm; pedicel 13–25 mm; bracteoles oblong-obovate, 1.5–2.5 × 1–1.5 mm. Sepals in 2 whorls; outer sepals oblong-obovate, ca. 4 × 2 mm; inner sepals suborbicular, ca. 4 × 2 mm, apex rounded; petals broadly obovate, ca. 4 × 2 mm, base clawed, glands slightly separate, orange, 0.6–0.8 × ca. 0.5 mm, apex rounded. Stamens ca. 2 mm; anther connective rounded. Ovules 2 or 3. Berry ?black, oblong to obovate-oblong, 6–7 × 3–3.5 mm; style persistent.

Phenology. *Berberis pseudothunbergii* has been collected in flower in April and May and in fruit between July and October.

Distribution and habitat. *Berberis pseudothunbergii* is known from southwest Shaanxi and southeast Gansu. It has been found at the base of cliffs and on mountain slopes at ca. 1000–1400 m.

The protologue of *Berberis pseudothunbergii* cited *Wang 18735* as "Typus" but did not cite a herbarium. The specimen at WUK with the sheet number 213771, which is annotated "Typus," has been selected as the lectotype because the author was based there. The isolectotype at WUK is identified on the label as "*B. reticulata.*"

Besides the type, *Fenzel 510*, and *Wei 2070*, the protologue cited three specimens from Xiaolong Shan in Tianshui Xian, Gansu: *W. Y. Hsia 5932*, 1400 m, 28 Aug. 1939; *Dept. of Biology, Univ. of Lanchou 714*, 1400 m, 17 July 1959; and *K. T. Fu 19386*, 29 Aug. 1964, all apparently at WUK. These were not found. The protologue describes the number of ovules as two, but G. *Fenzel 510* at W has two or three.

Berberis pseudothunbergii is unusual in that some specimens (including the holotype) have evidence of only single flowers while others (including the isotype at KUN) have a more varied inflorescence.

The Piasetzki specimen at LE, cited below, is likely to be the Piatsetzki specimen of "*B. thunbergii*" from "Tsingling shan" referred to by Diels (1901: 341).

Berberis pseudothunbergii was not included in Ying (2001).

Selected specimens.
S Gansu. Hui Xian: 1000 m, 26 Aug. 1958, *C. P. Wei 2070* (WNU 00007032, WUK 109195); Jialing Gongshe, Yuya Dadui, 20 July 1964, *R. N. Zhao 646082* (LZU 00117831–33); Zhangjiadian, 1280 m, 20 Aug. 1965, *Z. X. Peng 5706* (LZU 00009223–4). **[Min Xian]:** Min Hsien Ho, 19 Oct. 1930, *G. Hummel 5288* (S GH-966). **Têwo (Diebu) Xian:** Duoji Gou, Taili'ao, 33.8785°N, 103.7775°E, 30 July 1998, *Bailongjiang Exped. 1020* (PE 01556144).
SW Shaanxi. *P. J. Piasetzki s.n.,* 7 June 1875 (LE); Tuti-ling, 13 May 1934, *G. Fenzel 510* (W 0032272, WUK 72652).

218. Berberis pseudotibetica C. Y. Wu, Acta Phytotax. Sin. 25(2): 159. 1987, as "*pseudo-tibetica.*" TYPE: China. NW Yunnan: Dêqên (Deqin) Xian, Benzilan Gongshe, Chezhu Lin, 2800–3200 m, 8 July 1981, *Qinghai-Xizang Exped. 2191* (holotype, KUN 1204049!; isotypes, CDBI [2 sheets] 0028132 image!, 0028172 image!, HITBC 003584 image!, KUN 0107914!, PE [2 sheets] 01033065–66!).

Shrubs, deciduous, to 2 m tall; mature stems reddish brown, terete, sparsely black verruculose; spines 3-fid, sometimes solitary or absent toward apex of stems, semi-concolorous, 0.8–1.7 cm, terete. Petiole almost absent or to 3 mm; leaf blade abaxially pale green, adaxially deep green, narrowly oblong-obovate or oblanceolate, 2.5–5 × 0.6–1.5 cm, thinly leathery, midvein raised abaxially, lateral veins and reticulation partially raised abaxially, midvein and lateral veins inconspicuous adaxially, reticulation indistinct adaxially, base narrowly attenuate, margin entire, sometimes spinose with 1 to 5 often coarse teeth on each side, apex obtuse or subacute, sometimes minutely mucronate. Inflorescence a raceme or sub-umbellate raceme, rarely partially paniculate, 3- to 11-fruited, 1.5–4 cm overall including peduncle 0.5–2 cm; pedicel 5–8 mm. Flowers unknown. Berry red, oblong or ellipsoid, 8 × 4 mm; style not persistent or persistent and short; seeds 2.

Phenology. *Berberis pseudotibetica* has been collected in fruit from July and August. Its flowering season is unknown.

Distribution and habitat. *Berberis pseudotibetica* is known from only the type and one other collection from Dêqên (Deqin) Xian. The first was found under *Pinus* forests at 2800–3400 m, the second on a roadside edge of a mixed forest at 3230 m.

The protologue stated that *Berberis pseudotibetica* is similar to *B. dawoensis*. However, this was clearly based on the mistaken description of *B. dawoensis* given by Ahrendt (1961: 161) for which, see under that species. The protologue also cited as *B. pseudotibetica* G. M. *Feng 20981*, Zhongdian (Xianggelila) Xian, Sanbaxiang, Haba Xueshan, 2500–3000 m, 2 Oct. 1955 (KUN 0107913, 0177016, PE 01037204), but this has leaves that are more broadly obovate and a shorter infructescence and appears to be of another species.

Evidence for the color of the mature stems is from PE 01033066 which was not cited in the protologue.

As noted in the section Taxa Incompletely Known below, reporting on *Berberis franchetiana* vars. *gombalana* and *glabripes*, the issue of semi-racemose *Berberis* from northwest Yunnan is complex and will need more research with more specimens before it can be satisfactorily resolved. It appears, however, that *B. pseudotibetica* is not the same as either of these taxa.

Selected specimens.
NW Yunnan. Dêqên (Deqin) Xian: NE slope of Baima Xueshan, along hwy. G214 NW of Benzilan, 28.293611°N, 99.160556°E, 3230 m, 19 Aug. 2013, *D. E. Boufford, J. F. Harber & Q. Wang 43134* (A 00914427, CAS, E E00833556, K, KUN 1278433, PE). **Zhongdian (Xianggelila) Xian:** Sanbaxiang, Haba Xueshan, 2500–3000 m, 2 Oct. 1955, *G. M. Feng 20981* (KUN 0107913, 0177016, PE 01037204).

219. Berberis purdomii C. K. Schneid., Pl. Wilson. (Sargent) 1: 372. 1913. TYPE: China. N Shaanxi: S Yenan Fu (Yan'an), 1910, *W. Purdom 3* (lectotype, designated by Ahrendt [1961: 175], K missing, new lectotype, designated here, A 0038790!).

Shrubs, deciduous, to 1.5 m tall; mature stems purplish red, sub-sulcate to subterete; spines solitary, rarely 3-fid, pale yellow, 1–3 cm, terete. Petiole almost absent; leaf blade abaxially pale green, adaxially green, lanceolate or obovate-lanceolate, but often narrowly elliptic on young shoots, 1–4 × 0.4–0.8 cm, papery, midvein raised abaxially, slightly impressed abaxially, lateral veins and reticulation inconspicuous on both surfaces, sometimes conspicuous when dry, base attenuate, margin spinose with 3 to 9(to 15) often widely spaced teeth on each side, sometimes entire especially on young shoots, apex acute, sometimes minutely mucronate. Inflorescence a spikelike raceme, 15- to 25-flowered, 2.5–4.5 cm overall; peduncle 1.5–3 cm; pedicel 1–3 mm; bracteoles reddish, subulate-lanceolate, apex caudate. Sepals in 2 whorls; outer sepals obovate-orbicular or ovate-elliptic, ca. 2.2 × 1.5 mm; inner sepals oblong, 3.2–4 × ca. 2.6 mm; petals obovate-oblong, 3–3.2 × 1.6–2 mm, base clawed, glands separate, apex emarginate. Stamens ca. 2.2 mm; anther connective not extended, truncate. Ovules 2. Berry black, oblong, 5–6 × 3–4 mm, slightly pruinose; style not persistent.

Phenology. *Berberis purdomii* has been collected in flower in May and June and in fruit between July and September.

Distribution and habitat. *Berberis purdomii* is known from north Shaanxi and west Shanxi. It has been found on loess hills at 1025–1400 m.

The protologue of *Berberis purdomii* cited *Purdom 3* and *Purdom 345* without citing any herbarium. Ahrendt (1961: 175) lectotypified *Purdom 3* at K, but this is one of a number of type specimens that are missing (see introduction for further details). The specimen at A has, therefore, been designated the new lectotype.

It should be noted that the middle specimen on the sheet *Purdom 345* at A has leaves that are distinctly different both from the other twigs on the sheet and the specimens of *Purdom 345* at K and US, and from those of *Purdom 3* and the other specimens cited below.

As Schneider noted in the protologue, the flowers of *Purdom 3* and *Purdom 345* are not fully developed. The account of their structure and of fruit given above is taken from Ying (2001: 159). It would be possible to verify this (e.g., for flowers from the PE specimen of *Shaanxi-Gansu Botany Team 10451* and for fruit from *K. J. Fu 7524*, cited below), but I have not had an opportunity to do this.

Berberis purdomii has been confused with both *B. chinensis* and *B. caroli*. It can be distinguished from *B. chinensis* by its predominately single spines and by having two ovules, and from *B. caroli* by having spinose leaves. This latter characteristic is very evident in the type specimen and some of the specimens cited below, including all those from west Shanxi. It is less evident on other specimens, particularly on the leaves of young shoots which can be entire; for a particularly good example of this, see *Huang He Team 980*, cited below. If the description of the fruit being black is correct, this would further distinguish *B. purdomii* from both *B. chinensis* and *B. caroli*.

Ying (2001: 159) stated that *Berberis purdomii* is found in Gansu and Qinghai. No evidence in support of the former has been found; the only possible candidate from Qinghai is *Z. Y. Zhang 19152*, Xunhua Xian, Mengda Gongshe, Suotong Cun, 1800 m, 24 May 1983 (WUK 0439488–89) which has a few spinose leaves. If this is *B. purdomii* rather than a particular form of *B. caroli*, this would represent a very unusual disjunct.

P. Y. Li (1965: 212) cited *Tao River Exped. 3004* (WUK) from Hwei-Chan Hsien (Huichan Xian), Gansu, as *Berberis purdomii* (there are further specimens at IBK, KUN, MO, and PE). This is *B. aggregata*.

Selected specimens.
Syntype. Shaanxi: Yenan Fu, 1910, *W. Purdom 345* (A 00038789, K K000077400, US 00945949).

Other specimens.

N Shaanxi. Ansai Xian: upper Qingjian He, 1380 m, 3 Sep. 1953, *K. J. Fu 7524* (IBSC 0091981, PE 01032569, WUK 0069618). **Fu Xian:** Niuwuzhen, 1250 m, 3 May 1954, *K. J. Fu 8041* (WUK 0069605). **Hengshan Xian:** betw. Gaozhen & Qilin Gou, 24 July 1956, *Huang He Team 7545* (PE 01032760, 02087755, WUK 0088229). **Jingbian Xian:** Dashi Ling, 1050 m, 14 Aug. 1953, *Shaanxi-Gansu Botany Team 10451* (PE 01032558, SHM 0014327); Chaxujia Gou, 1150 m, 1 Sep. 1953, *K. J. Fu 7484* (IBSC 0091982). **Mizhi Xian:** Qi Qu, Guandao, 1025 m, 12 July 1953, *K. J. Fu 6806* (PE 01032561, WUK 0069619). [**Qingjian Xian:**] Cuijiata, 1400 m, 21 June 1955, *Huang He Investigation Team 1530* (PE 01032555, 01609084, WUK 0077376). **Yan'an Xian:** "Hochster Berg bei Yennan," May 1938, *H. Jettmar 77* (W 1938–7658); Ninwan, Jiulongquan, 1260 m, 31 Aug. 1954, *Huang He Team 980* (PE 01036986, WUK 0069625). **Zichang Xian:** near Liuquxing, 1200 m, 6 Sep. 1953, *K. J. Fu 7651* (PE 01032759, WUK 00069620). **Zizhou Xian:** Sanhuangjie, 29 May 1956, *Huang He Team 6991* (PE 01032761, WUK 00086619).

W Shanxi. Lishi Xian: Dujiashan Cun, Anguosi, 30 June 1955, *Huang Exped. 2nd Team 1651* (PE 01032745, 01609085, WUK 00077374). **Zhongyan Xian:** Wannianbao Cun, Muhuhe Shan, 5 June 1955, *Huang He Team 1489* (KUN 0178371, PE 01032556–7, 01609083, WUK 0077311); SE of Wannianbao, Zhongshu Gou, 1420 m, 7 June 1955, *Huang He Exped., A Team 24* (PE 01032558, 01609082).

220. Berberis purpureocaulis Harber, sp. nov. TYPE: China. SW Sichuan: Muli Xian, "Mountains of Kulu," 4150 m, June 1929, *J. F. Rock 17957* (holotype, A 00279718!; isotypes, K!, P P02682339!, US [2 sheets] 00945962, 00946000 images!).

Diagnosis. *Berberis purpureocaulis* is similar to *B. pallens* but has narrower leaves and a different flower structure. It is found in the same area as *B. kangwuensis* but can be distinguished from it by the color of its mature stems, the different shape of its leaves, and its flower structure.

Shrubs, deciduous, to 2 m tall; mature stems dark purple or blackish purple, becoming dark brown with aging, sulcate; spines 3-fid, solitary or absent toward apex of stems, pale brownish yellow, (0.6–)1–2.2 cm, abaxially sulcate, stout. Petiole almost absent; leaf blade abaxially pale green, papillose, adaxially mid-green, narrowly obovate or oblanceolate, 1.4–4 × (0.4–)0.6–1 cm, papery, midvein raised abaxially, lateral veins and dense reticulation inconspicuous on both surfaces, margin entire, apex acute, subacute, or obtuse, mucronate, base attenuate. Inflorescence a sub-umbel or sub-raceme, sometimes fascicled at base, 4- to 10-flowered, to 3 cm overall; bracts triangular, 2 × 1 mm; pedicel 3–12 mm. Sepals in 2 whorls; outer sepals oblong-ovate, 5–6 × 1.8 mm, apex acute; inner sepals narrowly elliptic, 7 × 3.5 mm, apex obtuse; petals orbicular-obovate, 6 × 4 mm, base distinctly clawed, glands separate, ovoid, apex rounded, incised. Stamens ca. 4.5 mm; anther connective extended, obtuse. Ovules 2. Berry red, oblong or oblong-ovate, 10–12 × 4 mm; style persistent.

Phenology. *Berberis purpureocaulis* has been collected in flower from May to September and in fruit in September.

Distribution and habitat. *Berberis purpureocaulis* is known from Muli Xian in southwest Sichuan. It has been found in alpine regions on dry slopes and streamsides at 3710–4150 m.

The purple or blackish purple mature stems and narrow leaves make *Berberis purpureocaulis* a very distinctive species.

Ahrendt (1961: 145) identified *J. F. Rock 23911* (details below) as *Berberis platyphylla*, though it is unclear why since that species has quite different leaves and a very different inflorescence. Earlier (11 March 1941), he annotated the specimen at Kew as the type of an unpublished *B. yunnanensis* var. *yetsiana*. Though no. *23911* has smaller leaves and a denser inflorescence than the type of *B. purpureocaulis*, a line drawing by Schneider attached to the specimen at A (but actually of the specimen at E) shows the structure of the flowers to be the same.

Selected specimens.

SW Sichuan. Muli Xian: "Mountains of Yetsi, north of Kulu, Muli Territory," 3350 m, May 1932, *J. F. Rock 23911* (A 00279521, BM, E E00623070, K K000644926); "Western slopes of Mt. Mitzuga," 4115 m, May–June 1932, *J. F. Rock 24025* (A 00279656, BM, E E00392899, UC UC516215, US 00945971); same details, *J. F. Rock 24513* (A 00279655, E E00392898, UC UC516217); Yaju, Luchang Gou, 3400 m, 4 July 1978, *Y. B. Yang 7226* (CDBI CDBI0028005, CDBI0028190–92); N of the city of Muli on S216 to Daocheng, at first pass N of the city of Muli, 28.124722°N, 101.160833°E, 3750 m, 7 Sep. 2013, *D. E. Boufford, J. F. Harber & X. H. Li 43495* (A, E, K, KUN, MO, PE, TI); on Xian rd. Z020 to near Kangwu Dasi SE of Cunduochanghaizi, 28.106944°N, 101.200278°E, 3710 m, 7 Sep. 2013, *D. E. Boufford, J. F. Harber & X. H. Li 43498* (A, CAS, E, K, KUN, MO, PE, TI); same details, *D. E. Boufford, J. F. Harber & X. H. Li 43500* (A, BM, CAS, E, K, KUN, MO, PE, TI).

221. Berberis qamdoensis Harber, sp. nov. TYPE: China. NE Xizang: Qamdo (Changdu) Prefecture, Riwoqê (Leiwuqi) Xian, E of the city of Riwoqe on hwy. 317 + 214 from Riwoqê to Changdu (Chamdo) along Gei Qu (river), 31.184167°N, 96.605278°E, 4300 m, 12 Aug. 2004, *D. E. Boufford, J. H. Chen, S. L. Kelley, J. Li, R. H. Ree, H. Sun, J. P. Yue & Y. H. Zhan 32224* (holotype, A 00279934; isotype, KUN not seen).

Diagnosis. *Berberis qamdoensis* has similarities to *B. campylotropa*, whose type is also from Riwoqê Xian, but has a different-color mature stem, longer spines, and smaller and different-shaped berries on much shorter pedicels.

Shrubs, deciduous, to 1.5 m tall; mature stems reddish brown, sulcate; spines 3-fid or solitary, pale yellow, 0.7–1.4 cm, terete, weak. Petiole almost absent; leaf blade abaxially pale green, papillose, adaxially midgreen, obovate or obovate-elliptic, 0.9–1.2 × 0.3–0.6 cm, papery, midvein raised abaxially, lateral veins and reticulation inconspicuous on both surfaces, conspicuous when dry, base cuneate, margin entire, rarely spinulose with 1 to 5 teeth on each side, apex obtuse, sometimes subacute, sometimes minutely mucronate. Inflorescence a fascicle, 1- to 5-flowered, otherwise unknown; fruiting pedicel 3–4 mm. Immature berry turning red, ellipsoid, 6–7 × 5–6 mm; style persistent; seeds 3.

Phenology. *Berberis qamdoensis* has been collected in fruit in July and August. Its flowering season is unknown.

Distribution and habitat. *Berberis qamdoensis* is known from Gonjo (Gongjue), Jombda (Jiangda), and Riwoqê (Leiwuqi) Xian, and Karub (Karuo) Qu in northeastern Xiang (Tibet). It has been collected on open slopes and in meadows at 3600–4300 m.

IUCN Red List category. *Berberis qamdoensis* is assessed as DD or Data Deficient, according to IUCN (2001) criteria.

There is an image taken in the field of the plant that produced the type specimen at http://hengduan.huh.harvard.edu/fieldnotes/.

Selected specimens.
NE Xizang (Tibet). Gonjo (Gongjue) Xian: near Gonjo, 3600 m, 3 Aug. 1976, *Qinghai-Xizang Botany Team 12587* (HNWP 61965, KUN 0178843, 0178852); Waba Gongshe, 3700 m, 20 Aug. 1976, *Xizang-Qinghai Team, Botany Team 9658* (PE 01037685). **Jombda (Jiangda) Xian:** on S side of Du Qu (Du River) ca. 0.5 km W of center of city of Jiangda, 31.488056°N, 98.39°E, 3675–3900 m, 29 July 2004, *D. E. Boufford et al. 31275* (A 00279933). **Qamdo (Changdu) Xian [now Karub (Karuo) Qu]:** Langcula Shan, 31 Aug. 1952, *B. Q. Zhong 5285* (PE 01037714–15).

222. Berberis qiaojiaensis S. Y. Bao, Bull. Bot. Res., Harbin 5(3): 1. 1985. TYPE: China. NE Yunnan: Qiaojia Xian, Yaoshan Gongshe, 3300 m, 24 Oct. 1964, *Expedition of NE Yunnan 1395* (holotype, KUN 1204051!; isotypes, KUN 1204054!, LBG 00064252 not seen).

Shrubs, deciduous, to 1 m tall; mature stems dark brown, angled; spines 3-fid, concolorous, 1–1.2 cm, slender, abaxially slightly sulcate. Petiole almost ab-

sent; leaf blade abaxially pale green, papillose, adaxially deep green, shiny, oblong-oblanceolate, 1–1.3 × 0.4–0.5 cm, papery, midvein raised abaxially, slightly impressed adaxially, venation indistinct or obscure on both surfaces, base cuneate, margin entire, apex rounded, mucronate. Inflorescence 1-flowered, very rarely paired; pedicel 4–5 mm. Sepals in 2 whorls; outer sepals oblong-ovate, ca. 5.5–7 × 3.5–4 mm; inner sepals oblong-obovate or obovate-elliptic, 7.5 × 5.5–6.5 mm; petals broadly obovate, ca. 5 × 5 mm, base slightly clawed, glands widely separate, ca. 0.8 mm, oblong, apex rounded, entire. Stamens ca. 4 mm; anther connective slightly extended, truncate. Ovules 4. Berry red, ovoid or ellipsoid, 10–12 × 5–7 mm; style not persistent or persistent and short.

Phenology. *Berberis qiaojiaensis* has been collected in flower in April and in fruit between July and October.

Distribution and habitat. *Berberis qiaojiaensis* is known from Huili Xian in south Sichuan and Huize, Leibo, and Qiaojia Xian in northeast Yunnan. It has been found in scrub meadows on mountain slopes, and on exposed ridges at 3260–3650 m.

The protologue of *Berberis qiaojiaensis* cited only the type which has fruit with four seeds. *D. E. Boufford, J. F. Harber & X. H. Li 43528* (details below) also has four seeds per fruit. Flowers from my plant of *B. & S. Wynn-Jones 7881* from seed collected in the same area as *Y. Y. Geng et al. 20070155*, cited below, have four ovules and I have, therefore, deduced that there is only one dwarf, one-flowered *Berberis* species from Qiaojia Xian and neighboring areas of northeast Yunnan and Huili Xian. The description of flowers given above comes from my plants. Floral material from collections in Qiaojia and Huize Xian is needed to definitely confirm whether my deduction is correct.

It should be noted that plants I found of *Berberis qiaojiaensis* on the Daihai grasslands of Huize Xian in September 2013 varied considerably in height, depending on whether they were growing on dry turf or on streamsides.

Selected specimens.
S Sichuan. Huili Xian: Longzhou Shan, 3400 m, 24 Apr. 2007, *Y. Y. Geng et al. 20070155* (WCSBG 015096–97). **Leibo Xian:** Ahe Gou Linchang, 14 Aug. 1972, *236 Task Force 0806* (PE 01037716–17).
NE Yunnan. Huize Xian: Dahai, Hengshan, 3360 m, 27 Sep. 1963, *Dongchuan Team 93-90* (KUN 0175976, 0176040, LBG 00064069); Dahai, Dasong, Naoke, 3650 m, 12 Oct. 1963, *Dongchuan Team 218* (KUN 0175977, 0176625); Dahai, Dishuiyan, 3600 m, 28 July 1964, *NE Yunnan Team 427* (KUN 0175725, 0177510); Dahai, 3674 m, 2 June 1995, *H. Peng & L. G. Lei 2265* (KUN 0179284); Dahai Xiang, near

Dishuiyan and S of Mopan Shan, 26.190278°N, 103.269167°E, 3600 m, 10 Sep. 2013, *D. E. Boufford, J. F. Harber & X. H. Li 43528* (A 00914595, CAS, E E0077075, K, KUN 1278436, PE 01892853-54, TI). **Qiaojia Xian:** Yaoshan, 3400 m, 16 July 1973, *B. X. Sun 1001* (IBSC 0092286, KUN 0176595, PE 01037710); Yaoshan, 3800 m, 12 Aug. 1974, *Kunming Institute of Botany; Geobotany Section 2520144* (KUN 0176594, 0176600).

Cultivated material:
Living cultivated plants.
Foster Clough, Mytholmroyd, West Yorkshire, U.K., and Sherwood Gardens, Devon, U.K., both from *B. & S. Wynn-Jones 7881*, Sichuan, Huili Xian, Longzhou Shan, 3260 m, 9 Oct. 2000.

Cultivated specimens.
From Foster Clough plant cited above, 12 May 2015, *J. F. Harber 2015-18* (A, E, KUN).

223. Berberis qinghaiensis Harber, sp. nov. TYPE: China. SE Qinghai: Jiuzhi (Jigzhi) Xian, NW of the city of Aba on hwy. S302 (Sichuan) then ca. 10 km N on hwy. 101 (Qinghai) toward Kangsai, 33.3025°N, 101.527222°E, 3720 m, 9 Aug. 2007, *D. E. Boufford, K. Fujikawa, S. L. Kelley, R. H. Ree, B. Xu, D. C. Zhang, J. W. Zhang, T. C. Zhang, W. D. Zhu 39340* (holotype, A!; isotypes, CAS!, E!, F!, KUN [2 sheets] not seen, PE!, TI!).

Diagnosis. Berberis qinghaiensis is similar to *B. ambrozyana* but differs in its mature stem color, leaves that are usually spinose rather than entire, and different flower structure.

Shrubs, deciduous, to 2 m tall; mature stems pale yellowish brown, sulcate; spines 3-fid, sometimes solitary toward apex of stems, concolorous, 0.5–1 cm, terete. Petiole almost absent; leaf blade abaxially green, papillose, adaxially pale green, narrowly obovate or very narrowly elliptic, 1.3–1.7 × 0.4–0.6 cm, papery, midvein raised abaxially, flat adaxially, lateral veins and reticulation partially raised abaxially, inconspicuous adaxially, base attenuate, margin spinose with 4 to 7 teeth on each side, sometimes entire, apex obtuse or subacute, sometimes minutely mucronate. Inflorescence 1-flowered; pedicel 5–10 mm, reddish brown. Flowers bright yellow, ca. 0.5–1 cm diam. sepals in 2 whorls; outer sepals obovate to broadly obovate, 3–4 × 2–2.5 mm; inner sepals elliptic, 4–5(–6.5) × 3.5–4 mm; petals obovate, 3.5–5 × 2.5–3.5 mm, base cuneate. Stamens 3–4 mm; anther connective scarcely extended. Pistil 3–4 mm; ovules 3 or 4. Immature berry turning red, ellipsoid, 8–11 × 4 mm; style persistent; seeds 2 to 4, brown.

Phenology. Berberis qinghaiensis has been collected in flower in June and fruit in August.

Distribution and habitat. Berberis qinghaiensis is known from Jiuzhi (Jigzhi) Xian in southeast Qinghai.

It has been collected from thickets in meadows and sunny mountain slopes at ca. 3700–4000 m.

The description of flowers used here is from *Guoluo Team 83*, kindly provided by W. J. Li of HNWP.

Selected specimens.
SE Qinghai. Jiuzhi Xian: Hagaya He, 3920 m, 13 June 1971, *Guoluo Team 83* (HNWP 25165); near Mentang Gongshe, 18 June 1971, *Xizang Medicine Team 277* (HNWP 23948); by Longka Hu, 3980 m, 18 Aug. 1971, *Guoluo Team 566* (HNWP 25640).

224. Berberis reticulata Bijh., J. Arnold Arbor. 9: 132. 1928. TYPE: Cultivated. Arnold Arboretum, Jamaica Plains, Boston, Massachusetts, U.S., 11 & 21 May 1927, from seeds of *W. Purdom 644*, N China, without precise locality (lectotype, designated here, A 00038681 second twig on left!).

Shrubs, deciduous, to 2 m tall; mature stems dark reddish purple, angled; spines solitary or 3-fid, to 1 cm, concolorous, terete, weak. Petiole 2–12 mm; leaf blade abaxially gray-green, papillose, adaxially green, obovate or spatulate-obovate, 2–4.5 × 0.8–2 cm, papery, midvein raised abaxially, impressed adaxially, lateral veins and reticulation indistinct or inconspicuous abaxially, conspicuous adaxially, base attenuate, sometimes narrowly so, margin spinose with 10 to 24 teeth on each side, apex obtuse to rounded. Inflorescence an often dense, sub-umbellate or sub-corymbose raceme, sometimes a fascicle or sub-umbel, 1- to 8-flowered, 1.4–2 cm; pedicel 4–7 mm; bracteoles ca. 1 mm. Flowers ca. 11 mm diam. Sepals in 2 whorls; outer sepals ovate, 4–5 × 3 mm; inner sepals obovate, ca. 7 mm; petals obovate, ca. 7 mm, apex emarginate; ovules 2, sessile. Berry red, ovoid, 8–11 × ca. 7 mm, slightly pruinose; style not persistent.

Phenology. Berberis reticulata has been collected in flower in June and in fruit in August and September.

Distribution and habitat. Berberis reticulata is known from Baoji Shi, Long, Taibai, and Yang Xian in Shaanxi. It has been found on sunny mountain slopes, by roadsides, and in forests at ca. 2600–3000 m.

The protologue of *Berberis reticulata* cited as type material from a cultivated plant from the Arnold Arboretum made on 11 and 21 May 1927. There is a sheet at A exactly matching the protologue description and marked as type. However, there is no indication which of the four twigs on the sheet was collected on which date. The twig with the most material has been chosen as the lectotype.

Only one wild-collected specimen from Taibai Shan with floral material has been located, *Shaanxi Herbal Survey Team 284* (details below), and I have seen only an image of this. I have, therefore, relied on the description from the type given by Byhouwer supplemented by additional information from Michael Dosmann of the Arnold Arboretum (pers. comm. 18 Oct. 2012). Ying (2001: 161) gave a more detailed description including giving the number of ovules as five or six. This should be discounted.

Interestingly, *Berberis circumserrata* is found on Taibai Shan at the same elevations as *B. reticulata*. This has somewhat similar leaves to *B. reticulata* but is 1- to 3-fascicled and has four ovules.

It is possible that *Berberis shensiana* (Ahrendt, Gard. Chron., ser. 3, 112: 155. 1942) is a synonym of *B. reticulata* (see Excluded Taxa below).

Selected specimens.
SW & W Shaanxi. Baoji Shi: betw. Baoji & Huang Shan, 2840 m, 23 June 1950, *J. M. Liu 10937* (PE 01032843, WUK 0020767). **Lung Hsien (Long Xian):** Kwanshan (Kuanshan), 2200 m, 2 June 1936, *T. P. Wang 4132* (CDBI CDBI0028203, PE 01033704); same details, but 2000 m, 6 June 1936, *T. P. Wang 4260* (PE 01033695, WUK 0066936). **Taibai Xian:** Taibai Shan, s.d., *C. Wang 673* (PE 01033700); vic. of Tatien, 29 Aug. 1937, *P. C. Tsoong 371* (PE 01033701); Toumukung, 3000 m, 29 Sep. 1937, *T. N. Liou & P. C. Tsoong 1330* (PE 01033699, KUN 0178812, WUK 004108); on way from Pinganszu (Ping'an Si) to Mingsingszu, 24 July 1938, *T. N. Liou & P. C. Tsoong 2664* (PE 01032846, WUK 00134878); near Ping'an Si, 2600 m, Sep. 1955, *F. Z. Wang et al. 718* (PE 01032845); Doutian Gong, 2800 m, 29 Sep. 1955, *F. Z. Wang et al. 10153* (PE 01032848); near Ping'an Si, 2975 m, 29 Sep. 1955, *D. Qin & W. Tian 19* (PE 01032847); 2800 m, 30 Sep. 1955, *F. Z. Wang et al. 0153* (PE 01032848); Doumu Gong, 19 June 1970, *Shaanxi Herbal Survey Team 284* (WUK 0286981, XBGH 000978). **Yang Xian:** Canger Ya, 8 Sep. 1959, *59 Nianqin Ling Team 10903* (PE 01387599).

Cultivated material:
Cultivated specimens.
Arnold Arboretum, Jamaica Plains, Boston, Massachusetts, U.S., from *W. Purdom 644* (accession no. 200881); 15 May 1918 (A 00168013); May 1919 (A 00168016); 15 May 1922 (A 00168012); 14 Sep. 1922 (A 00168011); 22 Sep. 1927 (A 00168010).

225. Berberis reticulinervis T. S. Ying, Acta Phytotax. Sin. 37: 305. 1999. TYPE: China. E Xizang (Tibet): Markam (Mangkang) Xian, Dongshan, 3800 m, 22 June 1976, *Qinghai-Xizang Exped. 11937* (holotype, PE 00935239!; isotypes, HNWP 61428 image!, KUN [2 sheets] 0178846–47!).

Shrubs, deciduous, to 1.5 m tall; mature stems pale orange-brown, sulcate; spines 3-fid, pale yellow, 1–2.5 cm, terete or abaxially sulcate. Petiole almost absent; leaf blade abaxially pale green, papillose, adaxially dark green, oblanceolate or narrowly elliptic, 0.7–2.1 × 0.3–

0.8 cm, papery, midvein and lateral veins raised abaxially, flat adaxially, venation indistinct or inconspicuous adaxially, base attenuate, margin entire, rarely spinulose with 1 to 4 teeth on each side, apex acute, mucronate. Inflorescence a fascicle, (1- to)3- or 4-flowered; pedicel 10–15 mm, slender. Sepals in 3 whorls; outer sepals lanceolate-oblong, ca. 3 × 1 mm; median sepals elliptic, ca. 4 × 2 mm; inner sepals oblong-elliptic, 5.5–6 × 2.5–2.7 mm; petals elliptic, base cuneate, glands separate, apex acute, incised with acute lobes. Stamens ca. 2.5 mm; anther connective extended, rounded. Ovules 2 or 3, very shortly stipitate. Berry unknown.

Phenology. *Berberis reticulinervis* has been collected in flower in June and July. Its fruiting season is unknown.

Distribution and habitat. *Berberis reticulinervis* is known from Markam (Mangkam) Xian in east Xizang. It has been found on mountain slopes and in and on the edge of forested areas at ca. 3800–4145 m.

The isotypes of *Berberis reticulinervis* are much better specimens than the holotype. The protologue also cited *S. Li 71410* (PE) from Barkam (Ma'erkang) Xian in Sichuan as this species. This specimen has not been located at PE, but duplicates at IBSC (0092709, 0092453), KUN (0175845), and NAS (NAS NAS00313881–82) appear to be of a different species.

The protologue also published a *Berberis reticulinervis* var. *brevipedicellata* from Gansu. There appears to be no evidence to connect this taxon with *B. reticulinervis*. For further details, see under Taxa Incompletely Known.

S. G. Wu 2720 (KUN 1204884), annotated by S. Y. Wu as an unpublished "*Berberis reticulinervis,*" is *B. yui*.

Selected specimens.
E Xizang (Tibet). Markam (Mangkang) Xian: Yanjing, Hong La Shan, 3800 m, 7 June 1976, *Qinghai-Xizang Botany Group 8265* (PE 01840058); Lawu La, 3800 m, 22 June 1976, *Qinghai-Xizang Exped. 11963* (HNWP 60923, KUN 0179143–44); Hong La, just on N side of pass, S of Markam, 29.273889°N, 98.672778°E, 4083–4145 m, 12 July 2000, *D. E. Boufford et al. 29395* (A 00279939, HAST 115037, P P02797876).

226. Berberis retusa T. S. Ying, Acta Phytotax. Sin. 37: 338. 1999. TYPE: China. W Sichuan: Daocheng Xian, Mengzi Xiang, Esha She, 3000 m, 20 Aug. 1973, *Sichuan Botany Exped. 2596* (holotype, PE 00935240!; isotypes, CDBI [2 sheets] CDBI0027630–31 images!, KUN 0178738!, SITC 00012138 image!, SWCTU 00006189 image!).

Shrubs, deciduous, to 3 m tall; mature stems dark reddish brown, sparsely verruculose, angled; spines solitary or absent, semi-concolorous, ca. 1 cm, terete. Petiole almost absent; leaf blade abaxially pale green, adaxially green, obcordate or subcuneate, 0.8–1.4 × 0.6–0.9 cm, papery, midvein conspicuously raised abaxially, venation indistinct or obscure on both surfaces, base cuneate, margin entire, apex slightly retuse or truncate, sometimes obtuse or subacute. Inflorescence a raceme, 6- to 12-flowered, sometimes sub-verticillate on upper rachis, 1.6–2.2 cm overall including peduncle 0.5–1 cm; bracts lanceolate, ca. 1.5 mm. Flowers unknown; fruiting pedicel ca. 6 mm. Berry red, ellipsoid, 8–9 × 5–6 mm, pruinose; style not persistent; seeds 2.

Phenology. *Berberis retusa* has been collected in flower in July and in fruit in August and September.

Distribution and habitat. *Berberis retusa* is known from Daocheng Xian in southwest Sichuan. It has been found in sunny, dry valleys and in the understory of a *Quercus* forest at 2850–3000 m.

The type and all the specimens cited below come from a very small area. The KUN and PE specimens of *Sichuan Botany Exped. 2090* have floral material, but I have only seen images and so am unable to give a description of its flowers.

Selected specimens.
W Sichuan. Daocheng Xian: betw. Mengzi & Riwa Xiang, 3000 m, 1 Nov. 1959, *S. K. Wu 2917* (KUN 0176568–69); Julong Xiang, Walong She, 2840 m, 11 July 1973, *Sichuan Botany Exped. 2090* (CDBI CDBI0028057–58, KUN 0179295, PE 01033136, SITC 00012154, SWCTU 00006190); Riwa Gongshe, Kanggu, 2850 m, 14 Aug. 1982, *Z. G. Liu 29379* (CDBI CDBI0027503–04, SWCTU 0009338).

227. Berberis salicaria Fedde, Bot. Jahrb. Syst. 36 (Beibl. 82): 42. 1905. *Berberis brachypoda* Maxim. var. *salicaria* (Fedde) C. K. Schneid., Bull. Herb. Boissier, sér. 2, 8: 262. 1908. TYPE: China. Shaanxi: "In kia po, sopra la metà del del monte Si-ku-tzui san," May 1900, *G. Giraldi 7019* (lectotype, designated by Ahrendt [1961: 197], K!; isolectotypes, B†, photograph of specimen at B and fragm. A 00058265!, FI image!).

?*Berberis giraldii* Hesse, Gard. Chron., ser. 3: 52. 321. 1912. TYPE: Cultivated. Neotype designated here, photograph in H. A. Hesse, Mitt. Deutsch. Dendrol. Ges. 1913, 267.

Shrubs, deciduous, to 2 m tall; mature stems yellowish brown, sulcate, pubescent; spines solitary, rarely 2- or 3-fid, 0.6–2.5(–4) cm, concolorous, terete. Petiole to 25 mm; leaf blade abaxially pale green, slightly shiny, adaxially dull green, lanceolate, 3–14 × 1–3 cm, papery, midvein raised abaxially, impressed adaxially,

lateral veins and reticulation partially raised abaxially, inconspicuous adaxially, base attenuate, margin spinose with 15 to 40 teeth on each side, apex subacuminate. Inflorescence a spikelike raceme, ca. 25-flowered, 4–7 cm overall including peduncle ca. 1.5 cm; pedicel 2–3 mm; bracteoles lanceolate-triangular, 1.75 × 0.75 mm, apex narrowly acuminate. Sepals in 2 whorls; outer sepals oblong to oblong-ovate, ca. 2 × 1 mm, inner sepals oblong-elliptic–obovate, 3–3.5 × 2 mm; petals obovate, ca. 3 × 2 mm, base distinctly clawed, glands separate, ovoid, ca. 0.5 mm, apex rounded. Stamens ca. 3 mm; anther connective not or slightly extended, obtuse or slightly apiculate. Ovules 2. Berry red, oblong or ellipsoid, 10–11 × 5–7 mm; style not persistent.

Phenology. *Berberis salicaria* has been collected in flower between April and June and in fruit from July to October.

Distribution and habitat. *Berberis salicaria* is known from east Gansu, south Ningxia, and central and west Shaanxi. It has been found in thickets, on mountain slopes, and road and streamsides at ca. 1200–1950 m.

The protologue of *Berberis salicaria* cited only *Giraldi 7019*, but cited no herbarium. As noted under *B. gilgiana*, Giraldi specimens of many genera were sent from FI to Berlin for identification. That a specimen of *Giraldi 7019* was retained in Berlin is clear both from the photograph at A and from the specimen at K, which is annotated as being ex-B. The Berlin specimen, however, was destroyed in WWII. Ahrendt's citation (1961: 197) of the K specimen as type was an effective lectotypification. This specimen, however, is little more than a fragment, having only one leaf and a small portion of inflorescence with only five flowers. The isolectotype at FI is a much better one.

The protologue of *Berberis giraldii* referred simply to a Chinese plant newly introduced in the U.K. by the horticultural firm James Veitch and Sons, and contained a description only 11 words long. Subsequently, Hesse (1913: 267, 272) gave a brief description in German together with a photograph, without giving any origin of the taxon except that it was from China. Schneider (1918: 286) reported a plant with this name supplied by Hesse's nursery in Weener am Ems in Germany, then growing in the Arnold Arboretum, which Schneider thought (presumably because of its name) was probably grown from seed collected by Giraldi, possibly in Shaanxi. No actual type was cited and there is no specimen from the plant among the extensive cultivated collection of *Berberis* at the Jamaica Plain Herbarium at Harvard. Subsequently, Ahrendt (1961: 196) reported a plant in his own living collection "raised by the late W. J. Marchant from seed sent by Hesse," specimens of which "may be regarded as a substitute for the type." There are specimens from this plant made in 1940 and 1941 at BM (00054785–97, 000896000) and at OXF, but given that Ahrendt's description of his plant differs from that of Schneider's description of the Arnold Arboretum plant, the provenance of Ahrendt's plant would seem doubtful (the seed Hesse referred to may have been from a cultivated plant, rather than wild-collected). Hence the neotypification here of Hesse's photograph.

I have been unable to discover exactly where the type of *Berberis salicaria* was collected (see the introduction for the difficulties with the transcription of Chinese placenames of Giraldi collections), but it is likely to have been in the Taibai Shan area.

The protologue description of the flowers of *Berberis salicaria* was imprecise. The more detailed description given here is from *s. coll. 5* from Chang'an Xian, cited below. It differs somewhat from Schneider's description of *B. giraldii*, who described the inner sepals as being 6 mm long, but differs substantially from Ahrendt, who reported two series of bracteoles and three of sepals (Ahrendt also reported the color of the mature stems as yellow). The synonymy of *B. giraldii* made by Ying (2001: 157) is, therefore, unprovable. Ying's use of Ahrendt's description of the flowers of *B. giraldii* for his entry for *B. salicaria* should, therefore, be discounted.

Selected specimens.

E Gansu. Heshui Xian: Jiajia, near Gouquan, 1400 m, 20 Oct. 1953, *Z. B. Wang 17392* (IBSC 0093686, PE 01037014, WUK 0069622); Taibai Zhen, Guanshang Gou, 1335 m, 17 July 1954, *Huang He Team 566* (IBSC 0069607, PE 01037013, WUK 0069607). **Hui Xian:** Jialing Xiang, summit of Tieli Shan, 1800 m, 6 Sep. 1958, *Z. Y. Zhang 574* (WUK 00102871). **Kang Xian:** Anmenkou, 4 July 1959, *Kang Xian Investigation Team s.n.* (LZU 00063869). **Pingliang Xian:** Kongtong Shan, 8 Aug. 1956, *Chinese Academy of Science, Huang He Gansu 1st Team 2035* (PE 01037015, WUK 0082067), *2105* (PE 01032771, WUK 0083319), and *2124* (KUN 0178343, PE 01032772, WUK 0083337). **Qingshui Xian:** Shanmen Xiang, Dabao, 1750 m, 24 June 1986, *J. X. Yang 6874* (MO 04468406, WUK 00469349–50). **Zhangjiachuan Xian:** Malu Xiang, Siwan Cun, 3 June 2015, *Y. He & J. C. Hao GSL2015051101* (BNU 0020656).

S Ningxia. Jingyuan Xian: Xinmin Linchang, 6 June 1984, *X. Liu s.n.* (LZU 0009196); Longtan Linchang, 1950 m, 6 July 1984, *X. Liu s.n.* (LZU 0009197).

C & W Shaanxi. Chang'an Xian: Ziping Gongshe, Guanyin Shan, 1600 m, 4 June 1960, *s. coll. 5* (KUN 0176241). **Chencang Qu:** 1800 m, 11 May 2014, *Y. He & J. C. Hao GSL2014050110* (BNU 000971). **Feng Xian:** Xinjia Shan, 1600 m, 4 July 1960, *K. J. Fu 12999* (WUK 00161813). **Hu Xian (now Huyi Qu):** Shiyi Gou, Liusi Gou, 10 July 1951, *B. Z. Guo 264* (PE 01032778, WUK 0020785); Taiping Yu, Xisi Gou, Guoyuanzi, 1800 m, 6 June 1959, *Medicinal Plant Exploration Team 2064* (PE 01032777); Laoyu, Duijiaocha, 1200 m, 20 Sep. 1962, *K. J. Fu 14830* (KUN 0178233). **Long Xian:** Guanshan Linchang, Chunshutan, 1400 m, 17 May 1983, *J. X. Yang et al. 4060* (WUK 00438820–21). **Mei**

Xian: Liujiaya, near Haoping Si, 1200 m, 9 Oct. 1935, *F. Z. Wang & K. J. Fu 283* (PE 01036981); Taibaishan, Liouchiayai, 1400 m, 25 Apr. 1938, *T. N. Liou & P. C. Tsoong 77* (PE 01036990, WUK 004112); Haoping Si, NW of the temple, 26 Sep. 1956, *Seed Collection Team 271* (PE 01811889, WUK 0080824, 0099530); Haoping Si, 1200 m, 1 May 1959, *J. X. Yang 120* (PE 01036983, WUK 00132623), and *121* (PE 01036989, WUK 00132637); same details, but 1400 m, 2 May 1959, *J. X. Yang 179* (PE 01036997, WUK 0133014); same details, but 1200 m, 4 May 1959, *J. X. Yang 219* (PE 01036989, 01036996, WUK 00132987). **Taibai Xian:** Taibai Shan, Taiwan, 1250 m, 29 Apr. 1959, *J. X. Yang 81*, (KUN 0176242, PE 01036987, WUK 0131783); same details, but 30 Apr. 1959, *J. X. Yang 80* (PE 01033277, WUK 00131782). **Zhouzhi Xian:** Chenhe Xiang, Qinggangbian, 1560 m, 14 Aug. 1958, *X. M. Zhang 81* (KUN 0178762, WUK 00103635).

228. Berberis saltuensis Harber, sp. nov. TYPE: China. SW Sichuan: Muli Xian, N of the city of Muli on S216 to Daocheng, at first pass N of the city of Muli, 28.124722°N, 101.160833°E, 3750 m, *D. E. Boufford, J. F. Harber & X. H. Li 43496* (holotype, PE!; isotypes, A!, BM!, CAS!, E!, K!, KUN!, MO!, TI!).

Diagnosis. Berberis saltuensis has some similarities with *B. muliensis*, but is always single-flowered except rarely at or toward the tips of branches. The pedicels are much shorter and the fruit have four to six seeds versus the three or four of *B. muliensis*.

Shrubs, deciduous, to 3.2 m tall; mature stems reddish brown or reddish purple, subterete; spines mostly absent, but when present 1- to 3-fid, pale brownish yellow, 0.4–1.6 cm, abaxially sulcate. Petiole almost absent, sometimes to 3(–6) mm; leaf blade abaxially pale gray-green or glaucous, papillose, adaxially dull green, narrowly elliptic, obovate, or oblanceolate, (2–)3–4(–5.5) × 0.8–1.4(–1.7) cm, papery, midvein slightly raised abaxially, slightly impressed adaxially, lateral veins indistinct or inconspicuous on both surfaces, base narrowly attenuate, margin entire, sometimes spinose-serrulate with 4 to 8 widely spaced teeth on each side, apex acute or subacute. Inflorescence 1-flowered, very rarely 2- or 3-flowered toward or at the apex of stems, otherwise unknown; fruiting pedicel 3–5(–8) mm, but to 12 mm at apex of stems. Immature berry turning red, ovoid, 12–15 × 6–10 mm, apex sometimes bent; style persistent; seeds 4 to 6.

Phenology. Berberis saltuensis has been collected with immature fruit in September, otherwise its phenology is unknown.

Distribution and habitat. Berberis saltuensis is known from only the type collection made on a trail through a forested area fringed with dense thickets at 3750 m in Muli Xian in southwest Sichuan.

Etymology. *Berberis saltuensis* is derived from the Latin saltuensis meaning "of forests."

IUCN Red List category. *Berberis saltuensis* is assessed as DD or Data Deficient, according to IUCN (2001) criteria.

I found only one plant of *Berberis saltuensis*. This was overwhelmingly single-fruited, but I collected the only examples of 2- or 3-fascicled fruit from the ends of branches of the plant.

229. Berberis saxatilis Harber, sp. nov. TYPE: China. NW Yunnan: Dêqên (Deqin) Xian, E entrance of Baima Shan Natl. Park, near Yueliang Wan, 28.259194°N, 99.273944°E, 2541 m, 25 July 2007, *J. Wen, Z. Nie, R. J. Soreng, K. B. Ramkin, L. L. Yue, M. Wang & X. K. Yue: Tibet-MacArthur 1282* (holotype, US 00971881!; isotype, KUN image!).

Diagnosis. *Berberis saxatilis* has a mature stem color somewhat similar to *B. leptoclada*, but its stems are much stouter and its spines much longer. Its infructescence has some similarities with the inflorescence of *B. leptoclada*, though that of the latter is sometimes more varied. The leaves of *B. saxatilis* have mostly spinose margins while the margins of *B. leptoclada* are mostly entire. Fruit of *B. saxatilis* produced one or two seeds versus the three ovules recorded for the flowers of *B. leptoclada*.

Shrubs, deciduous, to 1.2 m tall; mature stems dark reddish brown, partially pruinose, subterete; spines 3-fid, sometimes solitary toward apex of stems, pale brownish yellow, 1–2.8 cm, terete or abaxially sulcate; internodes 10–20 mm. Petiole almost absent; leaf blade adaxially green, abaxially pale green, papillose, partially pruinose, obovate, 1.2–1.5 × 0.5–0.6 cm, papery, midvein, lateral veins, and reticulation raised abaxially, conspicuous adaxially, margin spinose with 1 or 2(to 4) widely spaced coarse teeth on either side, sometimes entire on smaller leaves, base cuneate or attenuate, apex subacute or obtuse, sometimes mucronate. Inflorescence a sub-raceme, rarely a fascicle, 2- to 4(to 6)-flowered, 1.8–4.5 cm overall including peduncle to 1.8 cm; pedicel 2–6 mm. Flowers unknown. Fruit red, subglobose or ellipsoid, 5–7 × 3–5 mm; style conspicuous to 2 mm; seeds 1 or 2.

Phenology. *Berberis saxatilis* has been collected in fruit in July and August. Its flowering season is unknown.

Distribution and habitat. *Berberis saxatilis* is known from three collections in Dêqên (Deqin) Xian in northwest Yunnan. The type was found in a dry, rocky area at 2541 m, and *Forrest 13208* in open scrub among rocks at 3660 m. The habitat of *Z. Y. Wu 4436* was unrecorded.

Etymology. Saxatilis is Latin for dwelling or found among rocks.

IUCN Red List category. *Berberis saxatilis* is assessed as DD or Data Deficient, according to IUCN (2001) criteria.

This is a very distinctive and rather ferocious plant with very large spines in relation to its size and with internodes close together, characteristics that are likely to have evolved as a deterrent to animals in an area of scarce vegetation.

Forrest 13208 has leaves and an infructescence the same as the type, but mature stems that are much more distinctly pruinose. It was mistakenly cited as an example of *Berberis sichuanica* by Ying (1999: 329).

Selected specimens.
NW Yunnan. Dêqên (Deqin) Xian: Bei Ma Shan, Mekong–Yangtze divide, 28.333333°N, 3660 m, Aug. 1914, *G. Forrest 13208* (E E00395980, E00395983, PE 00935199); Hongla Shan pass, 3600 m, 10 Aug. 1976, *Z. Y. Wu 4436* (KUN 0178334–35).

230. Berberis scrithalis Harber, sp. nov. TYPE: China. NW Yunnan: Zongdian (Xianggelila) Xian, W side of Daxue Shan pass, 28.5785°N, 99.825833°E, 4200 m, 7 Oct. 1994, *Alpine Garden Society China Exped. (ACE) 1847* (holotype, E E00039648!; isotypes, LIV LIV.2005.15.1184!, WSY WSY0057777!).

Diagnosis. *Berberis scrithalis* is somewhat similar to *B. mianningensis*, but has shorter pedicels, pale reddish brown mature stems, and four ovules (vs. the pale yellowish brown mature stems and the one seed per fruit recorded for the latter). The leaves have some similarities with those of *B. minutiflora*, but this has reddish purple or dark reddish brown mature stems and flowers with only two ovules. *Berberis scrithalis* has the same number of ovules as *B. nanifolia* has seeds per fruit, but the latter has purple mature stems and leaves that are not bluish green.

Shrubs, deciduous, to 30 cm tall; mature stems pale reddish brown, sulcate; spines 3-fid, semi-concolorous, 0.6–1 cm, terete. Petiole almost absent; leaf blade abaxially pale green, papillose, adaxially bluish green, narrowly obovate, obovate-elliptic, or oblanceolate, 0.8–1.5 × 0.3–0.4 cm, papery, midvein slightly raised abaxially, slightly impressed adaxially, venation indistinct or obscure on both surfaces, base attenuate, margin entire, rarely spinulose with 1 tooth on each side, apex acute, mucronate. Inflorescence 1-flowered; pedicel 10–12 mm. Sepals in 2 whorls; outer sepals obovate, 5–6.5 × 3–4.5 mm, apex slightly retuse; inner sepals obovate, 4.5–6 × 3.75 mm; petals orbicular-elliptic, 4–5 ×

3.75–4 mm, base clawed, glands separate, elliptic, ca. 0.75 mm, apex entire. Stamens ca. 2.75 mm; anther connective slightly or not extended, truncate or obtuse. Ovules 4. Berry red, globose, 8–9 × 6–7.5 mm; style persistent, short.

Phenology. Seeds of *Berberis scrithalis* have been collected in October. Its flowering season in the wild is unknown.

Distribution and habitat. *Berberis scrithalis* is known from only one collection on the Zhongdian (Xianggelila) Xian side of the Daxueshan pass. This was found on a slope of limestone scree and rocks at 4200 m.

Etymology. *Berberis scrithalis* is derived from scritha, a Latin adaption of the Norse word for scree.

IUCN Red List category. *Berberis scrithalis* is assessed as DD or Data Deficient, according to IUCN (2001) criteria.

I looked for this species on the Daxue Shan pass in September 2013, but there was dense cloud and I failed to find it. The description of flowers comes from my plant grown from *ACE 1847* (details below). I am grateful to Robert Potterton of Pottertons Nursery, Caistor, Lincolnshire, U.K., who supplied it, Robert being a member of the Alpine Garden Society Expedition that collected the seed.

Cultivated material:
Cultivated plants.
Foster Clough, Mytholmroyd, West Yorkshire, U.K., from seeds of *ACE 1847*.

Cultivated specimens.
From above plant; *J. F. Harber 2015-23*, 20 Apr. 2015 (E).

231. Berberis sherriffii Ahrendt, J. Bot. 79(Suppl.): 77. 1941. TYPE: China. SE Xizang (Tibet): [Lhünzê (Longzi) Xian], Charme Distr., betw. Charme & Kyimpu, 28.416667°N, 93.016667°E, 3350–3650 m, 2 Nov. 1938, *F. Ludlow, G. Sherriff & G. Taylor 6653* (holotype, BM BM000573791!).

Shrubs, deciduous, to 3 m tall; mature stems dark reddish brown, subterete to sub-sulcate, slightly verruculose; spines solitary, yellow, 0.4–1 cm, but 3-fid to 1.4 cm at base of stems, terete, weak. Petiole 1–5 mm; leaf blade abaxially pale gray-green, papillose, adaxially green, obovate, 1.6–4 × 0.5–1.3 cm, papery, midvein raised abaxially, flat adaxially, lateral veins and reticulation indistinct or inconspicuous on both surfaces, base cuneate, margin entire, occasionally spinu-lose with 1 tooth on 1 or both sides, apex obtuse, mucronate. Inflorescence a panicle, 10- to 20-flowered, (2.5–)3.5–5 cm overall including peduncle 0.5–1.2 cm; bracts 1.5–2.5 mm; pedicel 3–4(–5) mm; bracteoles oblong, ca. 2 mm, apex acute. Sepals in 3 whorls; outer sepals oblong, ca. 3.5 × 2.5 mm, apex acute; median sepals oblong-elliptic, ca. 4.3 × 3.5 mm; inner sepals obovate, ca. 6 × 5 mm; petals obovate, ca. 5 × 3.8 mm, base clawed, glands widely separate, apex slightly emarginate, lobes acute. Stamens ca. 3 mm; anther connective slightly extended, shortly apiculate. Ovules 1 or 2 (or 3). Berry purplish red when mature, slightly blue pruinose, ovoid-oblong, 6–7 × 3–4 mm; style not persistent.

Phenology. *Berberis sherriffii* has been collected in flower in July and in fruit between October and November.

Distribution and habitat. *Berberis sherriffii* is known from Lhünzê (Longzi) Xian and south Mainling (Milin) Xian in southeast Xizang (Tibet). It has been collected from thickets and forest edges on dry mountainsides at ca. 2740–3350 m.

The protologue of *Berberis sherriffii* cited *F. Ludlow, G. Sherriff & G. Taylor 6653* as the type. Subsequently, Ahrendt (1961: 219) mistakenly cited the type to be *F. Ludlow, G. Sherriff & G. Taylor 6629*, while omitting any mention of *F. Ludlow, G. Sherriff & G. Taylor 6653*.

The above description of flowers is Ahrendt's of a cultivated plant grown from seed from *F. Ludlow, G. Sherriff & G. Taylor 6629*. *Qinghai-Xizang Supplementary Collections 750546* has floral material, but I have not investigated their structure.

Berberis sherriffii is very similar to *B. gyalaica*, but has a different flower structure with fewer ovules. There is also a distinct geographical separation.

Selected specimens.
SE Xizang (Tibet). Lhünzê (Longzi) Xian: Sananqulin, Long Zhan, 2740 m, 14 July 1975, *Qinghai-Xizang Supplementary Collections 750546* (HNWP 51180, KUN 0179336–37, PE 01037284); Zhari Xiang, 28.660183°N, 93.372933°E, 2925 m, 19 Sep. 2010, *Y. D. Tang & Q. W. Lin 2010-100* (PE 02014925). **[Mainling (Milin) Xian]:** Tsari Distr., Migyitun, 28.666667°N, 93.633333°E, 3050–3350 m, 24 Oct. 1938, *F. Ludlow, G. Sherriff & G. Taylor 6629* (BM BM000939708).

232. Berberis shunningensis Harber, nom. et stat. nov. Replaced synonym: *Berberis sikkimensis* (C. K. Schneid.) Ahrendt var. *glabramea* Ahrendt, J. Bot. 80: 87. 1942. TYPE: China. WC Yunnan: Shunning, [Fengqing Xian], Wumulung (Wumulong), 2300 m, 8 July 1938, *T. T. Yu 16607* (holotype, E E00438551!; isotypes, A 00038793!,

KUN [2 sheets] 0177076–77, PE [2 sheets] 01037339–40!).

Shrubs, evergreen, to 2.5 m tall; mature stems ?pale yellowish brown, sulcate; spines 3-fid, shiny, pale yellow-brown, 0.5–2 cm, abaxially slightly sulcate. Petiole short; leaf blade abaxially pale green, sometimes pruinose, adaxially deep green, shiny, elliptic or obovate-elliptic, 1.5–2.7 × 0.5–1 cm, leathery, midvein raised abaxially, impressed adaxially, lateral veins and reticulation mostly conspicuous on both surfaces, base cuneate, margin spinose with 1 to 5 teeth on each side, sometimes entire, apex acute or rounded, mucronate. Inflorescence a sub-umbel, sub-raceme, or sub-panicle, 6- to 12-flowered, 3–5 cm; fruiting pedicel 4–8 mm. Immature berry narrowly ovoid, ca. 1.3 cm; style persistent, distinct, ca. 1.5 mm; seeds 4.

Phenology. *Berberis shunningensis* has been collected in fruit in July. Its flowering season is unknown.

Distribution and habitat. *Berberis shunningensis* is known from only the type. This was collected in thickets at 2300 m.

Ahrendt's determination (1942: 87, 1961: 99) that *T. T. Yu 16607* was a variety of *Berberis sikkimensis* was misplaced. This latter taxon (which is a synonym of *B. aristata*), known from India (Sikkim), Bhutan, Myanmar (Burma), Nepal, and south Xizang (Tibet) is deciduous, whereas *B. shunningensis*, which has leaves that are much broader and more leathery than *B. aristata* and that are adaxially shiny, appears to be evergreen.

I have renamed the taxon *Berberis shunningensis* because the epithet *glabramea* is unavailable.

Though the collector's notes on the specimen sheet describe it as "common," *Yu 16607* is the only known collection of this species.

Berberis sikkimensis var. *glabramea* was mistakenly treated as a synonym of *B. sikkimensis* by Ying (2001: 210).

233. Berberis sibirica Pall., Reise Russ. Reich. 2(2): 737–738. 1773. TYPE: Russia. S.d., *P. S. Pallas s.n.* (neotype, designated here, specimen ex Herb. Pallas, ex. Herb Fisher, with 3 fruit on top right hand side of sheet annotated "Typus," LE!).

Berberis altaica Pall., Fl. Ross. (Pallas) 1(2): t. 67 (p. 42). 1789. TYPE: Russia. [Altai] (neotype designated here, illustration).
Berberis borealisinensis Nakai, J. Jap. Bot, 15. 528. 1939. TYPE: China. Hebei, [Yu Xian], Hsiao-wu-tai shan (Xiao Wutai Shan), Aug. 1935, *Yö Takenaka 38* (holotype, TI image!).

Berberis sibirica f. *subintegerrima* Krasnob., Sist. Zametki Mater. Gerb. Krylova Tomsk. Gosud. Univ. Kuybysheva 85(27): 2. 1974 [publ. 1975], syn. nov. TYPE: Russia. W Sayan Mtns., upper Ustju-Eldyg-Chem, 2050 m, 5 Aug. 1968, *I. Krasnoborov & G. Czajko 6179* (NS image!).
Berberis xinganensis G. H. Liu & S. Q. Zhou, Fl. Intramongolica, ed. 2, 2: 712, 579. 1991, syn. nov. TYPE: China. Nei Mongol, Hulunbuirmeng, Erguzuo Qi (Genhe), 1350 m, 27 June 1974, *Wastelands Team 105* (holotype, HIMC 0010777 image!).

Shrubs, deciduous, to 1 m tall; mature stems yellowish brown, very sulcate; young shoots puberulent; spines 3–9-fid, very pale brown, 0.3–1.1(–1.5) cm, slender, spreading at base to 2 mm wide, or partly foliaceous. Petiole 3–5 mm; leaf blade abaxially shiny, yellow-green, adaxially deep green, obovate, oblanceolate, or obovate-oblong, 1–2.5(–3.5) × 0.5–0.8 cm, papery, midvein raised abaxially, flat adaxially, lateral veins and reticulation inconspicuous on both surfaces, sometimes conspicuous when dry, base cuneate, margin sometimes repand, coarsely dentate with 5 to 12 teeth on each side, sometimes entire, apex subacute or obtuse, mucronate. Inflorescence 1-flowered, rarely paired at the apex of branches; pedicel 5–12 mm. Sepals in 2 whorls; outer sepals oblong-ovate or elliptic, 4–7 × ca. 2 mm; inner sepals obovate, 4.5–6.5 × 2.5–3 mm; petals obovate, ca. 4.5 × 2.5 mm, glands separate, apex shortly emarginate. Stamens 2.5–3 mm; anther connective truncate. Ovules 5 to 8. Berry bright red, obovoid, 7–9 × 6–7 mm; style not persistent.

Phenology. In China *Berberis sibirica* has been collected in flower in June and in fruit between July and September.

Distribution and habitat. In China, *Berberis sibirica* is known from Beijing, Hebei, Heilongjiang, Liaoning, Nei Mongol, Ningxia, Qinghai, Shanxi, and Xinjiang. It is also known from Kazakhstan, Mongolia, and Russia (Altai and Siberia). In China, it has been collected in alpine pastures, in crevices in rocky areas, by streamsides, and in desert regions at ca. 1000–3900 m.

Determining an appropriate type of *Berberis sibirica* is difficult. The protologue of *B. sibirica* was contained in the report of an expedition by Pallas in 1770–1771 and described the species as being found in the Altai mountains and Siberia, but cited no specimens. It included a description of fruit but noted "Flores non vidi." These specimens were presumably in Pallas's personal herbarium, together with others from a subsequent expedition described in Pallas (1799–1801), and possibly specimens Pallas acquired from other sources. According to Miller (1970: 534–535), Pallas's herbarium was first purchased by a Mr. Cripps, then bought by Lambert in 1808, but subsequently widely dispersed following the breakup of the Herb. Lambert. It is this history that explains why Pallas specimens of *B. sibirica* are scattered among various herbaria.

Some of these Pallas specimens have floral material and, therefore, cannot have been those referred to in the protologue. Fedtschenko (1937: 555) stated the type to be at LE, but gave no details. There are various specimens on three sheets at LE identified as *Berberis sibirica* in Pallas's own hand and subsequently labelled as "Typus." These are all identified as ex. Herb. Fischer, one being additionally identified as ex Herb. Pallas. A number of the specimens on these sheets have floral material and, as such, must post-date the protologue description. There are three specimens with fruit, but since the sheets are undated, it appears impossible to prove that any these were those referred to in the protologue; hence, lectotypification is not possible. Nevertheless, because it might date from Pallas's first expedition, the specimen with fruit on the sheet marked ex Herb. Pallas has been chosen here as the neotype. Other apparent Pallas specimens of *B. sibirica* in various herbaria are listed below.

Berberis altaica was published solely as an illustration and, from its name, presumably depicted a plant from Altai. The first reference I have found synonymizing the taxon is Schneider (1905: 396), but the synonymy may have been made earlier by another author.

The protologue of *Berberis borealisinensis* mistakenly stated that Hsiao-wu-tai shan is in Hubei rather than Hebei. The synonymy was made by Ying (2001: 90).

The shape and size of the leaves of *Berberis sibirica* can vary substantially, even on the same plant; for example, those of *Z. D. Wang 1862* (PE 01037855) from Nei Mongol, cited below, include leaves that are obovate-elliptic and 2 × 1 cm, narrowly obovate and 3 × 0.5 cm, and elliptic and 0.8 × 0.5 cm, with margins ranging from pronounced-spinose to entire. It is within this perspective that *B. sibirica* f. *subintegerrima* and *B. xinganensis* are treated here as synonyms. The rationale for the former, given in the protologue, was that it has oblong-obovate to obcuneiform leaves, 3.5 × 1 cm, with margins that are entire or 1- to 3-spinose, while *B. xinganensis* was published largely on the basis that, while otherwise being similar to *B. sibirica*, it has oblanceolate or obovate-lanceolate leaves. Ying (2001: 90) noted the supposed differences between *B. xinganensis* and *B. sibirica* reported by G. H. Liu and S. Q. Zhou, but was uncertain about the relationship between the two taxa. Subsequently, Ying (2010) omitted any reference to *B. xinganensis*.

The wide and sometimes foliaceous spines of *Berberis sibirica* make it easy to distinguish from other low-growing and dwarf species. Foliaceous spines are otherwise found in Asian *Berberis* only in the much taller

B. koreana (Palibin), native of Korea, and which has much larger leaves and a racemose inflorescence.

Berberis sibirica is one of hardiest species in the genus, surviving in exceptionally low temperatures.

Selected specimens.
Specimens that appear to be ex Herb. Pallas. S.d., *P. S. Pallas s.n.* (specimens ex-Herb. Fisher on same sheet as neotype [LE]); s.d., *P. S. Pallas s.n.* (specimens ex. Herb Fisher on 2 sheets marked "Typus" [LE]); Siberia, 1781, *P. S. Pallas s.n.* (LE ex Herb. Pott, S 08–762); s.d., *P. S. Pallas s.n.* (BM BM00559448); "? Itinera sibirica," s.d., "*?Pallas*" *s.n.* (HBG HBG-506696); *P. S. Pallas s.n.*, ex Herb Fisher, 1819 (G-DC G00201911); "*Berberis altaica*," Siberia, s.d., *P. S. Pallas s.n.* (W-Jacq. 0015476–1).

Other specimens.
Beijing. Beijing mun., Hsi-ling (Xiling) Shan, Sep. 1933, *C. W. Wang 61028* (A 00279850, PE 01037847); Mentougou, Donglingshan, 1700–2000 m, 18 July 1996, *D. E. Boufford et al. 27141* (A 0027985, CAS 943625, E E00074698).

Hebei. Yu Xian: Xiaowutai Shan, 2400 m, 20 June 1959, *s. coll. 686* (HNWP 114137, PE 01037846). **Zhuolu Xian:** Xilingshan, 30 June 1952, *C. G. Yang 717* (PE 01037852).

Heilongjiang. Fuyu Xian: Haxiong Gou, 18 Aug. 1956, *C. C. Ren 2079* (CDBI CDBI0027685).

Liaoning. Chaoyang Shi: 27 Sep. 1951, *C. W. Kuang 176* (IFP 05201001x0002, 05201002x0002).

Nei Mongol. "Mongolia borealis," 1897, *E. Klementz 85* (LE, US 01121989). **Arun (Arong) Qi:** 27 May 2014, *Z. L. Zhang 0208* (YAK 0000699). **Arxan (A'ershan) Shi:** Yi'ershi, 1197 m, 1 Aug. 2006, *L. Yan. s.n.* (NMAC 0027100). **E'erguna Zuo Qi (now E'erguna Shi):** Dahei Shan, 24 June 1984, *Y. J. Li 624* (HIMC 0010778–79). **Genhe Shi:** Haolibao Station en route to Kengai Li, 9 Sep. 1956, *S. E. Liu et al. 8930* (IBK IBK00013141, IBSC 0092025, PE 1037839). **Horqin (Ke'erqin) Yuoyi Qian Qi:** Bailang Shan, 1450 m, 27 July 1973, *Mongolian Medicinal Plant Survey Team 341* (HIMC 0010771). **Jungar (Zhunge'er) Qi:** Wuziwan, Wagui Miao, 24 July 1975, *Mongolian Medicinal Plant Survey Team 25* (HIMC 0010770). **Oroqen (Elunchun) Qi:** 15 June 2015, *Z. L. Zhang 0244* (YAK 0000108). **Otog (Ertuoke) Qi:** Zhuozi Shan, 23 May 2007, *Y. Zhou s.n.* (NMAC 0046644). **Ulanhot (Wulanhaote) Shi:** Wucha Gou, Balang, 9 June 1983, *Medicine Investigation Team 1246* (NMAC 005189). **Wuchuan Xian:** Aug. 1982, *s. coll. s.n.* (NMAC 005187). **Yakeshi Shi:** 18 June 2015, *Z. L. Zhang 0244* (YAK 0002459). **Zhalantun Shi:** Cgai He, Ji'erguo Shan, 30 July 1984, *Y. J. Li s.n.* (NMAC 005190). **Zhenglan Qi:** Sanggendalai Gongshe, 8 Sep. 1960, *Zhaotin Detachment-Nei Mongol 84* (HIMC 0010768).

Ningxia. Pingluo Xian: Chonggang, Dashui Gou, 1350 m, 11 Sep. 1980, *Z. Y. Yu 758* (MO 04738624, WUK 0449421).

Qinghai. Wulan Xian: Xili Gou, 3400 m, 17 July 1975, *B. Z. Guo & W. Y. Wang 11503* (HNWP 0221510); Chanong, 3900 m, 16 July 1981, *Q. Du 172* (HNWP 99937).

Shanxi. Liulin Xian: Anguozi, 28 June 1955, *Huang He 2nd Team 1634* (PE 01037856).

N Xinjiang. Altay (Aletai) Shi: Haxiong Gou, 1800 m, 8 Aug. 1965, *T. Y. Zhou et al. 652079* (PE 01037862). **Burqin (Bu'erjin) Xian:** en route to Kana Sihu, 1340 m, 16 Aug. 1987, *F. Konta et al. 0079* (PE 01196976–77);. Kanas Nature Preserve, 48.701667°N, 87.0275°E, 1350 m, 9 July 2004, *B. Bartholomew et al. 9230* (CAS 10644). **Fuyun Xian:** Haxiong Gou, 2600 m, 18 Aug. 1956, *R. C. Qin 2079* (CDBI CDBI0027685, IBK IBK00357895, IBSC 0092026, PE

01037858–59, XJBI 00036302). **Habahe Xian:** Baihaba, near Hala Sihu, 1500 m, *Y. R. Lin 74-1190* (IBSC 0092029, PE 01037861). **Hoboksar (Hebukesai'er) Xian:** Songshu Gou, Damu Qia, 11 June 1980, *L. D. Yin 80080* (XJA 000604072). **Kargilik (Yecheng) Xian:** Sukepiya, 2900 m. 14 Aug. 1987, *Qingzang Team, Y. H. Wu 1029* (HNWP 145288). **Qinggil (Qinghe) Xian:** Wulate Linchang, 2000 m, 29 Aug. 1980, *G. J. Liu et al. A925* (XJBI 00036299). **Tacheng Shi:** Beishan, 7 Aug. 1976, *C. Y. Yang 760521* (XJA 00064063–66). **Toli (Tuoli) Xian:** Hongtu, 1100 m, *L. M. Ke 084* (XJBI 00036296).

KAZAKHSTAN. Markakolskyi Rayon, S slope of Sarimsakti Range, 4–5 km W of the Karakaba River, 49.06°N, 85.97°E, 6 Aug. 1995, *J. Solomon 20526* (MO 1084946).

MONGOLIA. By side of Lake Kossogol (Khövsgöl), 2 Aug. 1897, *P. Mikhno s.n.* (G); Baga, Bogdo, Altai Mtns., 2130–2440 m, 1925, *R. W. Chaney 257* (UC UC295250, US 01121988); Arkhangai, 17 May 1999, *M. L. Zhang et al. 99X031* (PE 01807927); Khövsgöl, 1 July 1996, *L. R. Caddick 72* (E E00196359).

RUSSIA. Altay Republic, Tschuja Valley, 50.346111°N, 87.409444°E, 31 July 2002, *M. Staudinger 5430* (W 2003–12657); Altay Republic, Kosh-Agachskiy Rayon, Kokorya Valley, 2060 m, 9 Aug. 2008, *L. Martins & M. Schnittler 2469* (B B10 0274657).

234. Berberis sichuanica T. S. Ying, Acta Phytotax. Sin. 37: 329. 1999. TYPE: China. W Sichuan: Batang Xian, 2600 m, 1 July 1972, *236 Coll. Team 0028* (holotype, PE 00935195!; isotype, PE 00935196!).

Berberis batangensis T. S. Ying, Acta Phytotax. Sin. 37: 344. 1999. TYPE: China. W Sichuan: Batang Xian, 3000 m, 2 July 1972, *236 Coll. Team 0051* (holotype, PE 00935123!; isotype, PE 00935124!).

Shrubs, deciduous, to 2 m tall; mature stems reddish purple, angled; spines solitary or 3-fid, pale yellow, 1–1.5 cm, terete. Petiole almost absent; leaf blade abaxially gray-green, papillose, adaxially slightly shiny green, obovate or obovate-elliptic, 0.5–1.5 × 0.3–0.8 cm, thickly papery, midvein, lateral veins, and reticulation inconspicuous abaxially, conspicuous adaxially, base attenuate, margin entire or spinose with 1 to 5 teeth on each side, apex subacute, obtuse, or rounded. Inflorescence a raceme, 6- to 15-flowered, 1.5–3.5 cm overall; peduncle 0.4–1.2 cm; pedicel 3–6 mm bracts leaflike, obovate, apex aristate. Sepals in 3 whorls; outer sepals ovate-lanceolate, 1.6–2 × ca. 1 mm; median sepals obovate-elliptic, 2.8–3.1 × 1.5–1.7 mm; inner sepals elliptic, ca. 4 × 3 mm; petals obovate-elliptic, ca. 4 × 2 mm, base clawed, glands separate, apex rounded, entire. Stamens ca. 3 mm; anther connective rounded. Ovules 3 or 4. Berry red, ellipsoid, 7–8 × 4–5 mm; style persistent.

Phenology. *Berberis sichuanica* has been collected in flower in July and August and in fruit in August and September.

Distribution and habitat. *Berberis sichuanica* is known from Batang in west Sichuan. It has been found in thickets and on dry slopes at ca. 2600–3340 m.

Besides the type and one other collection from Batang Xian, *236 Coll. Team 052*, the protologue of *Berberis sichuanica* cited *Forrest 12887* (BM, E, PE) and *13208* (E, PE) from northwest Yunnan, and these were the source of the protologue's description of fruit. However, *Forrest 13208* is *B. saxatilis* and *Forrest 12887* is probably *B. leptoclada*. The protologue of *B. sichuanica* described the species as evergreen. However, from the evidence of plants in the field (pers. obs. 24 August 2018), it appears to be deciduous.

Besides the type, the protologue of *Berberis batangensis* (describing it as deciduous) cited *236 Coll. Team 0031* which, as can be seen from the details below, was collected on the same day as the type of *B. sichuanica*, the type of *B. batangensis* being collected the following day. The leaves of *236 Coll. Team 028, 031, 051*, and *052* are all the same shape and size, and the protologue account of the inflorescence of *B. batangensis* is little different from that of *B. sichuanica*, both being racemose with flowers having three whorls of sepals. The only major difference is that the racemes of the type of *B. sichuanica* are only 1.5 cm long.

There is a photograph of the fruit of *D. E. Boufford et al. 35455* (details below) on http://hengduan.huh.harvard.edu/fieldnotes.

Though the collection details of the type of *Berberis sichuanica* and of *236 Coll. Team 031, 051*, and *052* do not indicate where they were collected beyond Batang Xian, the other four collections listed below are all from a very small area some 20 km across, on, or near the G318 highway between Litang and Batang.

Selected specimens.

W Sichuan. Batang Xian: 2600 m, 1 July 1972, *236 Coll. Team 0031* (PE 00935125–26); 3000 m, 2 July 1972, *236 Coll. Team 0052* (PE 00935197–98); near Yawa Qu, 3000 m, 18 Sep. 1973, *Sichuan Botanical Survey Team 3941* (CDBI CDBI0027206–0, KUN 0178736, SWCTU 00006348); Yidun, 17 Sep. 1984, *F. D. Pu 352* (CDBI CDBI0028162–63); NE of Batang on rd. to Litang, S of first pass N of Batang, 30.285278°N, 99.212778°E, 3340 m, 30 July 2006, *D. E. Boufford et al. 35455* (A, E); Sungduo Xiang, Sungduo Cun, along the Da Qu, tributary of the Ba Qu, N of state hwy. G318, 30.23722222°N, 99.25305556°E, 3040 m, 24 Aug. 2014, *D. E. Boufford, B. Bartholomew, J. L. Guo & J. F. Harber 44902* (A, BM, CAS, E, K, KUN, MO, PE, TI).

235. Berberis silva-taroucana C. K. Schneid., Pl. Wilson. (Sargent) 1: 370–371. 1913. TYPE: China. NC Sichuan: [Mao Xian], Chiu-ting Shan (Jiuding Shan), 1830 m, 23 May 1908, *E. H. Wilson 2860* (lectotype, designated by Ahrendt [1961: 169], K K000077410!; isolectotypes, A 00038794!,

BM BM001015559!, E E00259005!, US image 00103906!).

Shrubs, deciduous, to 3 m tall; mature stems dark reddish purple or reddish brown, subterete; spines solitary or absent, rarely 2-fid, pale brownish yellow, 0.3–1.2 cm, terete. Petiole to 12(–20) mm, sometimes almost absent; leaf blade abaxially pale green, papillose, adaxially green, obovate or obovate-elliptic, 1.5–4.5 × 0.6–1.8 cm, papery, midvein raised abaxially, flat adaxially, lateral veins inconspicuous on both surfaces, base narrowly attenuate; margin entire or spinulose with 1 to 7 teeth on each side, apex obtuse, rounded, or subacute, sometimes mucronate. Inflorescence a fascicle, sub-fascicle, umbel, sub-umbel, or sub-raceme, 3- to 8-flowered, to 2.5 cm overall; peduncle (when present) to 0.4 cm; pedicel 10–18 mm; bracteoles oblong-lanceolate, 2 × 0.75 mm. Sepals in 2 whorls; outer sepals ovate-elliptic, 3.25 × 2.25 mm; inner sepals broadly obovate, 5 × 4 mm; petals obovate, ca. 4.5 × 3.5 mm, base distinctly clawed, glands widely separate, ca. 0.75 mm, apex entire. Stamens ca. 2.5 mm; anther connective slightly or not extended, rounded or truncate. Pistil ca. 3 mm. Ovules 2. Berry unknown.

Phenology. *Berberis silva-taroucana* has been collected in flower in May. Its fruiting season is unknown.

Distribution and habitat. *Berberis silva-taroucana* is currently reliably known from only the type from Mao Xian in north-central Sichuan and a collection from Wen Xian in south Gansu. The former was found in thickets at 1830 m, the latter in a rocky forest above a stream at 2400–2700 m.

The protologue cited *Wilson 2860* as type, but did not indicate any herbarium. Ahrendt's citing (1961: 169) of the specimen at K as type was an effective lectotypification.

The protologue of *Berberis silva-taroucana* cited the type from Jiuding Shan and 12 other Wilson collections across a wide area of Sichuan. None of these other collections appear to be the same as the type. *Wilson (Veitch) 955* and *4726* and *Wilson 1012a* and *2861* are treated here as a new species, *B. emeishanensis*, and *Wilson 1012* and *2857* as another new species, *B. wenchuanensis* (for details, see under those species). I have been unable to identify *Wilson (Veitch) 3151a* (A 00279859, 00279862), *Wilson 1059* (A 00279858, BM BM000810085, E E00612582, LE, US 00945941), *Wilson 4288* (A 00279870), or *Wilson 2867* (A 00279853); the last two seem to be the same species, but none of them appear to be the same as the type.

The protologue of *Berberis silva-taroucana* described its inflorescence as a loose raceme. In fact, *Wilson 2860*

is largely fascicled, sub-fascicled, or sub-umbellate. The protologue also gave an imprecise account of the flowers of *B. silva-taroucana* and gave no indication as to which specimen or specimens it was based on. The account given above is based on my dissection of a flower from the isolectotype at A. As such, it largely agrees with that given by Ahrendt (1961).

Jiuding Shan is currently the name of a specific mountain in Mao Xian, but from the evidence of the map at the beginning of his diary (Wilson MSS 1908) and Wilson (1913: 119), "Chiu-ting Shan" was the name Wilson gave for the whole range that includes this mountain. From his diary, it appears the collection was made at or near a pass to the northeast of the current Jiuding Shan. No other *Berberis* specimen of any description from this particular area has been located.

I have found no other specimens beyond the type that are definitely *Berberis silva-taroucana* apart from *Boufford & Jia 37792* (details below), which has the same leaves and flower structure as the type, though reddish brown mature stems versus the reddish purple stems of the type. On current evidence, the species therefore appears to have a very limited distribution. However, perhaps because of the confusions in the protologue, it has become a classic "dustbin" species as illustrated by the report by Ying (2001: 173) of it being found in Fujian, Gansu, Yunnan, and Xizang (Tibet). In the case of Fujian, this was likely to have been based on Jiang (1985: 45), and in the case of Yunnan, on Bao (1997: 80). It was also reported from Hunan by L. H. Liu (2000: 712). All of these reports appear to be based on misidentifications.

Selected specimens.
S Gansu. Wen Xian: Motianling Shan, Baishui Jiang Nature Reserve, W of the city of Wenxian, Qiujiaba, along SW branch of Baima He, 32.916667°N, 104.310833°E, 2400–2700 m, 20 May 2007, *D. E. Boufford & Y. Jia 37792* (A, CAS, E, MO, PE, TI).

236. Berberis subsessiliflora Pamp., Nuovo. Giorn. Bot. Ital., n. s. 22: 293. 1915. TYPE: China. NW Hubei: [Zhushan Xian], Zan-lan-scian (Canglang Shan), Lao-sciu-ze-ze, Nov. 1913, *C. Silvestri 4099* (lectotype, designated here, image FI!, photograph and fragments of FI specimen A 00038801!).

Berberis mitifolia Stapf, Bot. Mag. 154, t. 9236. 1931, syn. nov. TYPE: China. W Hubei: Paokang (Baokang Xian), May 1901, *E. H. Wilson (Veitch) 1915* (lectotype, designated here, K K000395198!; isolectotypes, A 00280252!, DBN image!, E E00327786!, HBG HBG-506731 image!, K K000395199!, LE!, NY 1365272!, P P00835676!, US 00945874 image!, W 1907–8584!).

Shrubs, deciduous, to 2.4 m tall; mature stems yellow, sub-sulcate, puberulous, sparsely black verruculose; spines 3-fid, rarely solitary, concolorous, 1–3 cm, abaxially sulcate. Petiole 10–15 mm; leaf blade abaxially pale grayish green, pubescent, adaxially green, elliptic to broadly lanceolate or oblong-obovate, 3.5–8 × 1.5–4(–5) cm, thickly papery, midvein, lateral veins, and reticulation raised abaxially, slightly impressed adaxially, base cuneate, margin flat, spinose with 15 to 25(to 40) teeth on each side, sometimes entire at the apex of stems, apex acute or obtuse, sometimes minutely mucronate. Inflorescence a raceme, 20- to 30-flowered, 5–9 cm overall including peduncle 1.5–2.5 cm; pedicel 3–4 mm, puberulous; bracteoles red, triangular, 1.8 × 1.2 mm. Sepals in 2 whorls; outer sepals oblong to lanceolate-oblong, 2.5–3 × 1.5–2 mm, apex acute; inner sepals obovate, 4.5–5.75 × 3.75–3.9 mm, apex rounded; petals oblong-elliptic to obovate, 4.5–5 mm, base distinctly clawed, glands separate, oblong, ca. 1.5 mm, apex rounded, incised. Stamens ca. 2.75 mm; anther connective slightly extended, truncate at perimeter, centrally apiculate or subapiculate. Ovules 2. Berry red, oblong to ellipsoid-subglobose, 8–10 × 6–8 mm; style not persistent.

Phenology. Berberis subsessiliflora has been collected in flower from April to June and in fruit from July to November.

Distribution and habitat. Berberis subsessiliflora is known from west Henan, Hubei, and southeast Shaanxi. It has been found in upland thickets and forested areas on valley and mountain sides at ca. 700–2740 m.

The protologue of *Berberis subsessiliflora* cited *Silvestri 4064* and *4099* but did not designate a type. *Silvestri 4099* has been chosen as the lectotype because it has the most fruiting material. The "U-kia-pi" reported for *Silvestri 4064* is likely to be "Wujia pi," a vernacular name for *Berberis*.

Stapf's protologue of *Berberis mitifolia* is somewhat difficult to decipher in relation to typification. It cited three collections from Hubei—*Wilson (Veitch) 1915*, *Wilson 554*, and *Wilson 4416*—as being examples of the new species, noting that the first had been listed (erroneously as *Henry 1915*) as *B. brachypoda* by Schneider (1908: 262), and all three as *B. brachypoda* by Schneider (1913: 375). It further noted that "*B. mitifolia* has under the name *B. brachypoda*, been in cultivation for some time, the plants having been partly raised from Wilson's no. 554 and partly from no. 4416, both distributed by the Arnold Arboretum." The protologue gave a detailed description of the taxon but cited no herbarium specimens of the Wilson numbers. It included a plate, the source of which was given as "prepared from specimens kindly communicated by Sir Oscar Warburgh, who grows the plant in his garden at Headly,

Epsom." Warburgh's specimens presumably were from either or both *Wilson 554* and *4416*. Curtis's Botanical Magazine was published from Kew, but there are no Warburgh specimens of the taxon in the Kew herbarium.

Ahrendt (1961: 197) cited as type of *Berberis mitifolia* "cultivated at Kew, from *Wilson 4416*, for plate of Bot. Mag. T. 9326 (Type, K)." This was at variance with the information given in the protologue, and there appears to be no such specimen at K. There are cultivated specimens of a *Wilson 441/b* (a likely misreading of *4416*), but both are on a sheet where one (dated 1939) is annotated as being from a plant grown at the Royal Horticultural Gardens at Wisley, Surrey, and the other (dated 1940) from Ahrendt's living collection at Watlington near Oxford (confusingly both specimens are also annotated by Ahrendt as "*B. mitifolia* var. *brevibracteata*," a name that was never published).

There is, however, a specimen of *Wilson (Veitch) 1915* at K annotated by Stapf as *Berberis mitifolia* on 19 February 1931. Though unmentioned in the protologue, the dating of Stapf's annotation suggests it is likely he drew upon it and can thus be regarded as original material. It has, therefore, been lectotypified here.

There is a second specimen at K of *Wilson (Veitch) 1915*, acquired in the 1950s, which is ex HK. The information that the type collection was made in Baokang Xian comes from this specimen and the isolectotype at NY.

Collection details of *Wilson 554* and *4416* are given below. *Wilson 554* consists of two gatherings, 5 June 1907 and November 1907, the former is designated here as *Wilson 554* "A" and the latter *Wilson 554* "B."

Berberis subsessiliflora has had a shadowy existence. The protologue account was rudimentary and failed to give any description of the shape of the leaves. The species was not noticed by Schneider in any of his publications, nor by Fu (2001) or Ying (2001, 2011), while Ahrendt (1961: 196) simply reported the syntypes without any description. The two syntypes have only fruit, but on the same page, Pampanini listed a number of Silvestri specimens, also from Zan-lan-cian, some of which he identified as *B. amurensis* (a species not found in Hubei) and some as *B. henryana*. I have not seen any of these except *Silvestri 4072* and *4073*, listed as *B. amurensis*. These, together with specimens not cited by Pampanini, *Silvestri 4071* and *4071a*, have leaves that are indistinguishable from those of the syntypes of *B. subsessiliflora*, but also have floral material. An examination of the flowers of the specimen of *4071a* at US showed the structure to be more or less the same as that reported by Stapf in the protologue of *B. mitifolia*; hence my synonymy. It should be noted that this flower structure is somewhat different from the account given for *B. mitifolia* by Ahrendt (1961: 198), who reported three whorls of sepals.

There are a number of *Farges 550* specimens from Chengkou in the Paris herbarium, only some of which are *Berberis subsessiliflora* (others appear to be *B. gilgiana*). Unfortunately, at the time of writing, none of these have barcodes.

Selected specimens.

Syntype. Berberis subsessiliflora. [Zhushan Xian], Zanlan-scian (Canglang Shan), "U-kia-pi," Oct. 1913, *C. Silvestri 4064* (FI, photograph and fragments of FI specimen A 00038802!).

Other specimens.

Chongqing (NE Sichuan). Chengkou Xian: "District de Tchen-kéou tin," s.d., *P. G. Farges 550* (P).

W Henan. Neixiang Xian: Baotianman Nature Reserve, betw. Pingfan & Saozhouchang, on border with Nanzhao Xian, 1650 m, 2 June 1994. *D. E. Boufford et al. 26414* (A 00280257). **Xixia Xian:** Huangsi'an, 990 m, 4 Oct. 1956, *Henan Forestry Department 1312* (PE 01032443–45); Taipingzhen, 820 m, 10 May 1959, *Henan Team 438* (PE 01032438–39).

Hubei. Lung men Ho, 1524 m, 23 Aug. 1922, *W. Y. Chun 4014* (A 00280258, NAU 00024186, PE 01032504). **Baokang Xian:** Chenkong, 1800 m, 25 May 1974, *R. H. Huang 2737* (HIB 0129268–9); Maoqiao, Lidian, Wangshi Laolin, 1600 m, 24 May 1975, *S. Y. Wang 382* (HIB 0129277–78). **Danjiangkou Shi:** Ping Gongshe, Beishan, 700 m, 4 July 1973, *Y. J. Ma 3105* (HIB 0129322). **Fang Xian:** (part now in Shennongjia Lin Qu), "Fang Hsien," 2440–2740 m, Nov. 1910, *E. H. Wilson 4416* (A 00280253); Shennongjia, Jiuhuping, 12 Oct. 1957, *Q. M. Hu 00829* (LBG 00064059–60). **Luotian Xian:** Tiantangzhai Linchang, 1600 m, *R. H. Huang 3272* (HIB 0129390). **Shennongjia Lin Qu:** Dawanyu, 1700 m, 4 July 1976, *Shennongjia Team 21283* (HIB 0129259–60, PE 01032508–09); Zhushanyazi pass on the W side of the Dajiuhu basin, 31.5°N, 110.5°E, 1780 m, 13 Sep. 1980, *Sino-Amer. Bot. Exped. 1204* (A 00280256, E E00612472, HIB 129478, KUN 0178360, NA 0017392, NAS NAS00314100–01, PE 00996108, UC UC1491497, WH 06008558); 1700 m, 30 May 1982, *Z. D. Jiang & G. F. Tao 350* (E E00612474, HIB 0129247). **Shiyan:** [Maojian Qu], Xiaochuan, Qingyan Shan, 1000 m, 20 May 1989, *Z. X. Chen 3451* (HIB 0129271). **Xingshan Xian:** (part now in Shennongjia Lin Qu), Jiuchong Xiang, Laojunshan, 26 May 1957, *H. J. Li 418* (HIB 0129239, LBG 00064064, PE 01032774, 01033364, 01036977). **Xingshan Xian:** "Hsing shan hsien," 1280–1580 m, 5 June 1907, *E. H. Wilson 554* "A" (A 00280254–55, BM BM001015557, E E00327785, HBG HBG-506732, LE, P P02482732, US 00945877, W 1914–4810); same details, but Nov. 1907, *E. H. Wilson 554* "B" (A 00423655, HBG HBG-506732, US 00945877); Muzhuzia, 23 July 1956, *H. J. Li 46* (HIB 0129273–74, PE 01032496); Sanyikou, 15 July 1957, *H. J. Li 1083* (HIB 129175–76, PE 01032497). **[Zhushan Xian]:** Zan-lan-scian, Apr. 1912, *C. Silvestri 4071* (FI, US 0945884); Apr. 1912, *C. Silvestri 4071a* (FI); Apr. 1912, *C. Silvestri 4073* (A 00280172, FI); Zan-lan-scian, June 1913, *C. Silvestri 4072* (FI).

E & SE Shaanxi. Pingli Xian: Longmen Xiang, Nanmu Gou, 1500 m, 17 Aug. 1992, *Z. H. Wu 1218* (WUK 0481499). **Shanyang Xian:** Tianzhu Shan, near Tiezhongping, 1900 m, 3 Sep. 1952, *Z. B. Wang 16448* (PE 01032779, WUK 0062118).

237. Berberis tachiensis (Ahrendt) Harber, comb. et stat. nov. Basionym: *Berberis diaphana* Maxim.

var. *tachiensis* Ahrendt, J. Linn. Soc., Bot. 57: 123. 1961. TYPE: China. W Sichuan: Kangtin Hsien (Kangding Xian), Tachienlu (Kangding), 2590–2750 m, 28 Sep. 1928, *W. P. Fang 3655* (holotype, K K000567991!; isotypes, A 00038736!, E [2 sheets] 00217990–91, IBSC 0091657 image!, NAU 00024176 image!, NY 01104641!, P P02313617!, PE 01037541!).

Shrubs, deciduous, to 4 m tall, mature stems pale yellowish brown, sulcate; spines 3-fid, solitary or absent toward apex of stems, concolorous, (0.5–)1.2–3 cm, abaxially sulcate. Petiole almost absent, rarely to 3 mm; leaf blade abaxially pale green, papillose, adaxially dull gray-green, obovate or obovate-elliptic, 1.5–2.4(–4.7) × 1–2.3 cm, papery, midvein slightly raised abaxially, lateral veins and reticulation conspicuous abaxially, inconspicuous adaxially, base attenuate, margin with 3 to 10(to 15) spinulose teeth on each side or entire, apex obtuse, sometimes subacute. Inflorescence a fascicle, sub-fascicle, or sub-raceme, sometimes fascicled at base, 2- to 9(to 15)-flowered, 2–4.5(–5.5) cm overall including peduncle (when present) to 3.5 cm; pedicel 8–18 mm (but to 28 mm at base); bracteoles triangular, 3 mm. Sepals in 2 whorls; outer sepals broadly elliptic or elliptic-ovate, 3–4 × 3.5–4 mm; inner sepals broadly elliptic or elliptic-ovate, 4.5–5 × 3.5 mm; petals obovate, 3.25 × 2.5 mm, base clawed, glands separate, 0.7 mm, apex rounded, entire. Stamens 3.5 mm; anther connective extended, apiculate. Ovules 3 or 4. Berry red, narrowly ovoid-oblong, 8–10 × 3–4 mm, apex attenuate, often slightly bent; style persistent.

Phenology. Berberis tachiensis has been collected in flower in June and July and in fruit between July and October.

Distribution and habitat. Berberis tachiensis is known from Kangding, Luding, and Yajiang Xian in west-central Sichuan. It has been found in thickets on rocky slopes, on open hillsides, and forest understories at ca. 2745–3800 m.

Berberis diaphana var. *tachiensis* was published on the basis of the type *W. P. Fang 3655* from Kangding Xian and *W. P. Fang 7711* (K but also IBSC 0091658) from Emei Shan. This latter, however, is *B. aemulans. Fang 3655* is clearly of the same species as *Wilson 2853* and *4134*, also collected in Kangding Xian, which Schneider (1913: 355) identified as *B. tischleri* whose type is from Songpan in north Sichuan. This identification was understandable since herbarium specimens with fruit from the two areas are more or less impossible to distinguish, and this no doubt why Ying (2001 173) treated *B. diaphana* var. *tachiensis* as a synonym of *B. tischleri*. However, comparing the flowers of the living plant at Howick Hall of *SICH 375* from Kangding with those from a living plant of *B. tischleri* from Songpan Xian at the Royal Botanic Garden Edinburgh, at Dawyck, it is clear that their flower structures are not the same. And while the leaves of the Howick Hall plant are adaxially dull grayish green, those of the Dawyck plant are dark green, a difference that is more or less impossible to discern from herbarium specimens.

Selected specimens.
W Sichuan. Jiulong Xian: Hongba Xiang, 3413 m, 22 June 2006, *Z. B. Liu et al.: Jiulong 4* (Ganzi Institute of Forestry Herbarium 22384). **Kangding Xian:** NE of Tachien lu,. 2745–3050 m, 6 July 1908, *E. H. Wilson 2855* (A 00279930, P P02482726, US 00945989, W 1914–4787); N of Tachienlu, 2745–3050 m, 9 July 1908, *E. H. Wilson 2853* (A 00279926, HBG HBG-506757, K, US 0946056); Tachienlu, 2440–2745 m, Oct. 1910, *E. H. Wilson 4134* (A 00279927); Kangting (Tachienlu), Distr. Cheto La (Zheduo La), 3800 m, 3 Aug. 1934, *H. Smith 11013* (MO 4367278, PE 01032913, S 12–25446, UPS: BOT V-040865); Kangting (Tachienlu), Distr. Yülingkong, Yachiakong Mtns., 3600 m, 18 Oct. 1934, *H. Smith 12873* (MO 4367280, PE 01032914, S 12–25441, UPS: BOT V-040868); Yulingong, Yingba, 6 Sep. 1951, *W. G. Hu & Z. He 11138* (HGAS 013612, PE 01032803); Yulin Xiang, Laoyulin, Liuhuang Gou, 3150 m, 16 June 1963, *L. X. Jiang 36021* (IBK IBK00012743, IBSC 0092412, PE 01033825); Gongga Si, 3600 m, 20 July 1979, *X. H. Hu 20674* (CDBI CDBI0027198–99, IBSC 0091493); same details, but 3700 m, *X. H. Hu 20677* (CDBI CDBI0027196–97, IBSC 0091492); Dagai Gou, 3200 m, 21 Sep. 1981, *Z. J. Zhao & J. B. Shi 115519* (CDBI CDBI0028067, CDBI0028076); Shade Qu, Gonga Si en route to Zimei Cun, 3500 m, 3 Aug. 1982, *K. Y. Lang et al. 938* (KUN 0178904, PE 01037324–25); E side of Zheduo pass, 3740 m, 21 Sep. 1991, *SICH 566* (K); Ertaizi, 30.04027778°N, 101.83833333°E, 3700–3750 m, 25 Aug. 1997, *D. E. Boufford et al. 27517* (A 00279949, CAS 1051364, E E00324026, MO 5771205); Jintang Xiang, Heika Liangzi, 3800 m, 1 July 2007, *C. Zhang 20071103* (WCSBG 015844–45). **Luding Xian:** Bazifang, Niba Gou, 2400 m, 22 June 1983, *Botany Group 31142* (CDBI CDBI0027194–95). **Yajiang Xian:** Jianzi Wanshan, 3700–3950 m, 6 Aug. 1983, *K. Y. Lang et al. 2892* (KUN 0178886, PE 01037327–28).

Cultivated material:
Living cultivated plants.
Howick Hall, Northumberland, U.K., from *SICH 375*, Kangding Xian, betw. pass on Zhidoa (Zheduo) Shan & Kangding. 3690 m, 3 Oct. 1988.

238. Berberis taoensis Harber, sp. nov. TYPE: China. Gansu Jonê (Zhuoni) Xian: "T'ao River basin; beyond Tatsuto," 2620 m, June 1925, *J. F. Rock 12443* (holotype, A 00279330!; isotypes, E E00623028!, P P02313649!)

Diagnosis. *Berberis taoensis* has similar-shaped leaves to *B. tianshuiensis* and *B. silva-taroucana*, though smaller than either. The flower stalk is also similar to these two species, but its flower structure is different.

Shrubs, deciduous, to 1.8 m tall; mature stems pale yellowish brown, sulcate; spines 1- to 3-fid, semi-

concolorous, 1–3 cm, abaxially sulcate, slender. Petiole almost absent, rarely to 10 mm; leaf blade abaxially pale green, papillose, adaxially green, elliptic to obovate-elliptic, occasionally obovate-orbicular, (0.8–)1.6–3 × (0.3–)0.6–1.2 cm, papery, midvein raised abaxially, flat adaxially, lateral veins and reticulation inconspicuous on both surfaces, base attenuate, margin spinulose with 6 to 15 teeth on each side, sometimes entire especially toward the tip of stems, apex subacute to obtuse. Inflorescence a sub-fascicle, sub-umbel, or sub-raceme, 3- to 5-flowered; bracts lanceolate, ca. 2.5 mm; pedicel 4–6 mm; bracteoles oblong-lanceolate, 2.5 × 0.75 mm. Sepals in 2 whorls; outer sepals oblong-elliptic, 4 × 2.5 mm; inner sepals broadly obovate, 4 × 3.5 mm; petals narrowly elliptic, 4.5 × 2.25 mm, base slightly clawed, glands separate, apex obtuse, incised. Anther connective not extended, truncate. Ovules 3 or 4. Berry scarlet, ellipsoid or oblong, 8–9 × 5–6 mm; style not persistent or persistent and short.

Phenology. *Berberis taoensis* has been collected in flower in June and in fruit from August to October.

Distribution and habitat. *Berberis taoensis* is known from Jonê (Zhuoni) Xian in south Gansu. It has been found in scrub on grassy slopes, on loess hills, in *Picea* forests, and on a roadside at 2530–3050 m.

IUCN Red List category. *Berberis taoensis* is assessed as DD or Data Deficient, according to IUCN (2001) criteria.

Byhouwer (1928: 45) identified *J. F. Rock 12443, 12303, 13202,* and *14898* as *Berberis mouillacana*, and *12429* and *12458* as probably so, while identifying *J. F. Rock 13522* and *14913* as *B. silva-taroucana*, this despite the fact that they all appear to be of the same species and both *B. mouillacana* and *B. silva-taroucana* have dark purple mature stems.

As can be seen below, only two collections of *Berberis taoensis* subsequent to Rock's expedition of 1925–1926 have been located. *SQAE 191* has pedicels up to 16 mm long but appears to be *B. taoensis*.

Selected specimens.
S Gansu. [Jonê (Zhuoni) Xian]: "T'ao River basin," 2680 m, June 1925, *J. F. Rock 12303* (A 00279976, AU 038857, NAS NAS00314097, NAS00314099); "On loess hills, back of Choni, en route to Taochow," 2530 m, Aug. 1925, *J. F. Rock 13202* (A 00279975, E E00623029, P P02313646); "Minshan range; valley of Kwadjaku, beyond Tatsuto," 3050 m, June 1925, *J. F. Rock 12429* (A 00279321, NAS NAS00314098); "N bank between Choni & Kwadjaku," 2620 m, June 1925, *J. F. Rock 12458* (A 00279976, E E00623026, IBSC 0091905, NTUF); "Adjuan, Taku, East Tebbu country," 3050 m, Oct. 1925, *J. F. Rock 13522* (A 00279337); "slopes of Laliku," 2745 m, Oct. 1925, *J. F. Rock 14913* (A 00279335, E E00612587, P P02313291); "Maerkhu, Minshan," 2745 m,

Aug.–Sep. 1926, *J. F. Rock 14898* (A 00279338, NAS NAS00314042, NAS0031496); Lali Gou, 3000 m, 30 June 1987, *S. Fang s.n.* (LZU 0063731–32); Dayu Valley, 34.465278°N, 103.586389°E, 2540 m, 17 June 2000, *Sino-British Qinghai Alpine Garden Society Expedition (SQAE) 191* (CAS 1087375, E E00274440, WSY WSY0105669).

239. Berberis temolaica Ahrendt, Gard. Chron. ser. 3, 109: 101. 1941, as "*telomaica.*" TYPE: Cultivated. Messrs. Hillier, Winchester, U.K., "10 Nov. 1938," reputedly grown from seed from same plant as *F. Kingdon-Ward 5733*, China. SE Xizang (Tibet): [Nyingchi (Linzhi) Xian, now Bayi Qu], Kongbo Province, Temo La, 3950–4350 m, 13 June 1924 (lectotype, designated here, specimen with flowers, annotated by Ahrendt "Type cult. Hillers" (BM BM000939735!).

Berberis temolaica var. *artisepala* Ahrendt, J. Bot. 79(Suppl.): 55. 1941. TYPE: China. SE Xizang (Tibet): [Mainling (Milin) Xian], Takpo Province, Langong, 28.85°N, 93.783333°E, 3650 m, 28 May 1938, *F. Ludlow, G. Sherriff & G. Taylor 3911* (holotype, BM [2 sheets] BM000559585–86!).
Berberis erythroclada Ahrendt var. *trulungensis* Ahrendt, J. Linn. Soc., Bot. 57: 118. 1961, syn. nov. TYPE: China. SE Xizang (Tibet): [Bomê (Bomi) Xian], Pome Province, above Trulung, 3800 m, 26 June 1947, *F. Ludlow, G. Sherriff & H. H. Elliot 13219* (holotype, BM BM000559584!).

Shrubs, deciduous, to 2 m tall; mature stems dark reddish purple, sulcate, not or scarcely verruculose, young shoots white pruinose, often densely so; spines 3-fid, pale purplish brown, 0.5–1.5 cm, abaxially slightly sulcate. Petiole 1–4 mm or almost absent; leaf blade abaxially densely white or glaucous pruinose, adaxially blue-green, oblong-obovate, 2–7 × 1.2–3 cm, thickly papery, midvein raised abaxially, slightly impressed adaxially, lateral veins and reticulation partially conspicuous abaxially, inconspicuous or indistinct adaxially, conspicuous when dry, base cuneate, margin dentate with 3 to 9 teeth on each side, sometimes entire, apex rounded. Inflorescence 1(rarely 2- or 3)-flowered; pedicel 8–15 mm, pruinose. Sepals in 3 whorls; outer sepals ovate, ca. 5 × 3.2 mm, pruinose, apex acute; median sepals obovate, ca. 7 × 5 mm; inner sepals obovate-orbicular, ca. 8 × 7.5 mm; petals broadly obovate, ca. 7 × 6 mm, base clawed, glands separate, apex emarginate to incised. Stamens ca. 4 mm; anther connective slightly extended, truncate or obtuse. Ovules 6 to 11. Berry red, pruinose, oblong-ovoid, 11–14 × 6–7 mm, apex bent; style persistent, short.

Phenology. *Berberis temolaica* has been collected in flower from May to July and in fruit from August to October.

Distribution and habitat. *Berberis temolaica* is known from Bomê (Bomi), Mainling (Milin), Mêdog (Motuo), and Nyingchi (Linxhi) Xian in southeast Xizang (Tibet). It has been found in open, grassy meadows and among *Rhododendron*, and on the margins of *Abies* forests at ca. 3600–4350 m.

Berberis temolaica is one of three species from Xizang whose types, designated by Ahrendt, are cultivated specimens grown from seeds collected by Kingdon-Ward in his expedition to the Tsangpo valley in 1924, the others being *B. johannis* and *B. gyalaica*. In each case, Ahrendt cited, in addition to the collector's number, a 1924 summer collection date; i.e., a time when no seeds would have been available. The solution to this seeming paradox is that it was Kingdon-Ward's frequent practice to collect floral material in the summer months, mark their source, and return in the autumn to collect seeds and, indeed, Kingdon-Ward (1926b: 66) specifically mentions no. *5773* in this respect.

In each of the three cases, Ahrendt justified his designation of cultivated material as the types by stating that he was unable to locate a corresponding wild-collected herbarium specimen, though in fact, there are such specimens of all three and this had already been made clear by Marquand (1929: 152, 159) who included them in his listing of material at Kew from Kingdon-Ward's 1924 expedition.

However, in each of the three cases, determining exactly which cultivated specimen is the type is somewhat complicated. In the case of *Berberis temolaica*, the protologue simply cited "*Kingdon-Ward 5778* (also *10008*)" and stated, "I acknowledge my material of this species from the nurseries of Messrs. Hillier, Winchester." Subsequently, Ahrendt (1941b: 55) cited a cultivated *Kingdon-Ward 5772* as the type with the location being "Herb. Ahrendt" (from the evidence of many other cultivated specimens of other taxa annotated by Ahrendt as "type," this reference is not necessarily to a single specimen). The only other specimen cited then was a wild-collected *Kingdon-Ward 10008* (BM), Burma–Tibet Frontier, 28.416667°N, 97.916667°E, 3560 m, 2 September 1931. There are three cultivated specimens of *Kingdon-Ward 5778*, and one cultivated and one wild-collected specimen of *10008* at BM. The four cultivated specimens are all dated 10 November 1938, two of *5778* (one with flowers and one with fruit) are annotated no. 640 and as "cult. Hilliers," the third as "cult. Wisley no. 641," that of *10008* appears to be from Ahrendt's living collection (the common date of 10 November clearly does not represent when all or perhaps any of these specimens were collected). All three of no. *5778* were annotated by Ahrendt on 31 December 1941 as "Type." An annotation by Ahrendt to the cultivated *Kingdon-Ward 10008* dated 21 January

1942 simply identified this as "*Berberis telomaica*" with no mention of type, while an annotation by Ahrendt to the wild-collected *10008* dated 22 September 1939 stated "*Berberis telomaica* (Type is *KW 5773*)."

On the basis of the above evidence, I think that the three cultivated specimens of *Kingdon-Ward 5773* at BM are probably the ones referred to by Ahrendt (1941b: 54) and the two ex-Hilliers were the ones referred to in the protologue (and are, therefore, syntypes), and these were subsequently transferred from Ahrendt's personal herbarium to BM. Therefore, Ahrendt's lectotypification has been completed here by selecting the ex-Hillier specimen with floral material. These cultivated specimens are no different from the wild-collected *Kingdon-Ward 5778* at Kew and Edinburgh. However, Ahrendt's subsequent statement (1961: 121) that the wild-collected specimen at Kew is the type is incompatible with the Shenzhen Code, Art. 9.3 and 9.4 (Turland et al., 2018).

It should also be noted that *5778* and *10008* are not the same taxon. This is apparent from both the cultivated and wild-collected specimens of *10008*. For while *5778* is 1-flowered (and, as such, corresponds to the description given in the protologue), the wild-collected *10008* (with very immature fruit) is 4- or 5-fascicled or sub-umbellate while the cultivated *10008* has a 3-fascicled fruit stalk with one fruit. The leaves of the latter are mostly ovate versus the oblong-obovate leaves of *Kingdon-Ward 5772*. It is possible that *10008* is *Berberis multiserrata*. It is impossible to be completely sure whether Ahrendt's reference to *Kingdon-Ward 10008* in the protologue of *B. temolaica* referred to the cultivated specimen or the wild-collected specimen at BM or, indeed, to either. Given it is anyhow arguable whether the reference fulfills the requirements of Art. 40.3 Note 2 of the Shenzhen Code (Turland et al., 2018), I think it unhelpful to regard either specimen as a syntype.

Ying (2001: 73) synonymized *Berberis temolaica* var. *artisepala*. The protologue of *B. erythroclada* var. *trulungensis* cited only the type (mistakenly reported in the protologue as *F. Ludlow, G. Sherriff & H. H. Elliot 1319* rather than *13219*). This is described in the collectors' notes as being 4–6 ft. (1.2–1.8 m) tall. Its association by Ahrendt with the dwarf species *B. erythroclada* is difficult to understand, especially as *13219* is very similar to *F. Ludlow, G. Sherriff & H. H. Elliot 13219a*, cited below, collected the day after of the collection of *13219* at the same location, the BM specimen of which was identified by Ahrendt on the sheet as *B. temolaica* (the notes on this sheet state "collected mixed with no *13219*").

Selected specimens.

Syntype. Cultivated. Messrs. Hillier, Winchester, U.K., "10 Nov. 1938," reputedly grown from seed from same plant as *F. Kingdon-Ward 5733*, specimen with fruit annotated by Ahrendt "Type cult. Hillers" (BM BM000939738).

SE Xizang (Tibet). Bomê (Bomi) Xian: Pome, Nambu La, Tongyuk River, 29.983333°N, 94.31667°E, 3660 m, 3 June 1947, *F. Ludlow, G. Sherriff & H. H. Elliot 13831* (A 00279914, BM BM000939744, E E00395950); Kongbo, Ba La, Pasum Chu, 30.366667°N, 94.15°E, 4115 m, 22 June 1947, *F. Ludlow, G. Sherriff & H. H. Elliot 13965* (A 00279911, BM BM000939743, E E00395946); Pome Province, above Trulung, 3810 m, 27 June 1947, *F. Ludlow, G. Sherriff & H. H. Elliot 13219a* (BM BM000939745, E E00395947); Kongbo, Nambu La, 3810 m, 10 July 1947, *F. Ludlow, G. Sherriff & H. H. Elliot 15373* (A 00279910, BM BM000939740, E E00395951); same location, 4420 m, 14 July 1947, *F. Ludlow, G. Sherriff & H. H. Elliot 15449* (A 00279913, BM BM000939740, E E00395948); same location, 3810 m, 27 Sep. 1947, *F. Ludlow, G. Sherriff & H. H. Elliot 15793* (BM BM000939741, E E00395949); Guxiang, 4000 m, 3 Aug. 1965, *T. S. Ying & D. Y. Hong 650882* (PE 01037456–58). **[Mainling (Milin) Xian]:** Tsari, Bimbi La, 28.833333°N, 93.466667°E, 4115 m, 7 June 1936, *F. Ludlow & G. Sherriff 1785* (BM BM000939747, MO 1620266); Lilung Valley, near Pamchi, ca. 28.95°N, 93.766667°E, 3960 m, 21 June 1936, *F. Ludlow & G. Sherriff 1850* (BM BM000939746). **[Mêdog (Motuo) Xian]:** Kongbo Province, Lusha Chu, 29.333333°N, 94.583333°E, 4115 m, 14 June 1938, *F. Ludlow, G. Sherriff & G. Taylor 4812* (BM BM000939721). **Nyingchi (Linzhi) Xian (now Bayi Qu):** Kongbo Province, Temo La, 3950–4350 m, 13 June 1924, *F. Kingdon-Ward 5733* (E, E00259019, K K000077359); [Bayi], near Hougou Bridge behind Agricultural College, 4000 m, 8 Aug. 1983, *B. S. Li et al. 06212* (PE 01037488–89); Doshong La, 22 Oct. 1995, *K. D. Rushforth 3496* (E E00073529).

240. Berberis tengchongensis Harber, sp. nov. TYPE: China. W Yunnan: Tengchong Xian, vic. of Gaoshidong in Guyong Linchang, ca. 11.4 km direct E of Houqiao (Guyong), 25.36533333°N, 98.32416667°E, 2060 m, 27 May 2006, *Gaoligong Biodiversity Survey 2006, 30682* (holotype, CAS 00120135!; isotypes, GH 00285287!, HAST 124191 image!, KUN 1409997 image!, MO 6057223!).

Diagnosis. Berberis tengchongensis is somewhat similar to *B. atroviridiana*, but is a much taller shrub with narrower leaves, a peduncle which is often shorter, and longer pedicels.

Shrubs, deciduous, to 3 m tall; mature stems dark reddish brown, semi-terete; spines solitary, absent toward apex of stems, yellowish brown, 0.5–3.2 cm, terete. Petiole almost absent; leaf blade abaxially gray-green, papillose, adaxially yellowish green, narrowly obovate or oblanceolate, 1–2.8 × 0.2–0.6 cm, papery, veins semi-camptodromous, inconspicuous particularly abaxially, base cuneate or attenuate, margin entire, apex semi-acute or rounded, sometimes minutely mucronate. Inflorescence a sub-umbel or semi-umbellate raceme, 4- to 8-flowered, 2–5 cm overall including peduncle 0.6–2.6 cm; pedicel 6–9 mm, brownish yellow. Sepals in 3 whorls; outer sepals ovate, 1–1.5 × 0.8–1.2 mm, apex acute; median sepals ovate-elliptic, 3–3.5 × 2–2.5 mm, apex rounded; inner sepals elliptic-obovate, 4–4.5

× 3–3.5 mm, apex rounded; petals obovate-elliptic, 3–4 × 2–2.5 mm, base slightly clawed, glands separate, elliptic, 0.8–1 mm, apex rounded, entire. Stamens 2.5–3 mm; anther connective slightly extended, rounded. Pistil 2.5–3 mm; ovules 3. Immature berry turning red, ellipsoid, 5 × 3 mm; style persistent.

Phenology. *Berberis tengchongensis* has been collected in flower from March to May and with immature fruit in May.

Distribution and habitat. *Berberis tengchongensis* is known from a very small area of Tengchong Xian in west Yunnan, no more than 25 km across. It has been found on forest margins and subtropical broadleaf forest disturbed by agriculture and felling at ca. 1800–2060 m.

Berberis tengchongensis is the only deciduous species recorded for Tengchong Xian. No deciduous species are recorded for any of the adjoining Xian, though given that *s. coll. 1051* and *Xiangliao Investigation Team 85-280* (details below) were collected very near the border with Myanmar, it is possible it is found there. Though the collection details for *s. coll. 1051* record the elevation as 1050 m and that of *Xiangliao Investigation Team 85-280* as 1600 m, from map evidence the actual elevations appear to be ca. 2200 m and ca. 1800 m, respectively. No other deciduous *Berberis* species are recorded in Yunnan as being found at below 2000 m.

Selected specimens.
W Yunnan. Tengchong Xian: Guyong Qu, Houqiao Gongshe, Heini Tang, 1950 m, 19 May 1964, *S. G. Wu 6696* (KUN 0175918–19); Banwa, 25.583333°N, 98.383333°E, 1050 m, 30 Mar. 1980, *s. coll. 1051* (HITBC 074439); Houqiao, Heini Tang, 1600 m, 17 Apr. 1985, *Xiangliao Investigation Team 85-280* (KUN 0176449–53).

241. Berberis tenuipedicellata T. S. Ying, Acta Phytotax. Sin. 37: 343. 1999. TYPE: China. WC Sichuan: Kangding, near Jiaoba, 2450 m, 22 May 1960, *T. S. Ying et al. 3542* (holotype, PE 00935201!; isotypes, SZ [2 sheets] 00289795–96 images!).

Shrubs, deciduous, to 1.5 tall; mature stems dark purplish red, angled; spines solitary or 3-fid, pale brownish yellow, 0.6–1.8(–2.5) cm. Petiole 2–3 mm, sometimes almost absent; leaf blade abaxially pale green, adaxially bright pale green, obovate or obovate-elliptic, 0.8–2(–3.5) × 0.5–1.2 cm, thinly leathery, midvein, lateral veins, and reticulation raised abaxially, conspicuous adaxially, base attenuate, margin spinose with 10 to 25 teeth on each side, sometimes entire, apex rounded, obtuse, or acute. Inflorescence a raceme, 3- to 10-flowered, often with 1- to 3-fascicled flowers at base, 1–3.5 cm;

pedicel 5–12 mm. Sepals in 2 whorls; outer sepals elliptic-ovate, ca. 2 × 1 mm, apex rounded; inner sepals elliptic-obovate, ca. 4 × 2.75 mm, apex rounded, base slightly clawed; petals elliptic, ca. 3 × 1.8 mm, base slightly clawed, glands separate, apex rounded, incised. Stamens ca. 2 mm; anther connective not extended, rounded. Ovules 2 or 3(or 4). Berry red, subglobose, 9–10 × 8–9 mm; style not persistent.

Phenology. *Berberis tenuipedicellata* has been collected in flower in May and in fruit from May to July.

Distribution and habitat. *Berberis tenuipedicellata* is known from Jiulong, Kangding, Muli, and Yajiang Xian in Sichuan. It has been collected on river, stream, and trailsides and field margins at ca. 2300–3500 m.

The protologue of *Berberis tenuipedicellata* describes the species as evergreen, but it is unclear what the evidence is for this since all the gatherings it cites were made between May and July and none are known outside the summer months. In the absence of this evidence, it appears that it is much more likely to be deciduous. The protologue did not include a description of flowers and the description above comes from *J. F. Rock 17447* (details below).

T. T. Yu's fieldbook and the collection details of *T. T. Yu 6708* (see below) describe Maidilong as being in Jiulong. Currently at least, it is in Muli Xian on the Yalong Jiang, some 10 km from the border with Jiulong. The specimen sheets of *T. T. Yu 6708* at KUN are annotated by C. Y. Wu as an unpublished "*Berberis depressocarpa.*"

Selected specimens.
W & SW Sichuan. Jiulong Xian: "Chiu-lung Hsien Territory, east of the Yalung River," 2775 m, May 1929, *J. F. Rock 17447* (A 00279994–95, IBSC 0092824, K, US 0945960, 0945983). **Kangding Xian:** by riverside, Songyu, 2500–3100 m, 8 May 1960, *T. S. Ying et al. 3353* (PE 00935202); Shade Qu, near Guanji Gou, 25 May 1961, *S. Jiang et al. 02998* (KUN 0178460, PE 00935203). [**Muli Xian**]: Kiulung (Jiulong), Mo-ti-lung (Maidilong), 2300 m, 3 July 1937, *T. T. Yu 6708* (A 00280073, KUN 0175946–47, PE 00935204–05). **Yajiang Xian:** 4 July 1930, *Z. P. Huang & S. P. Zhu 00554* (PE 00935207); Zangge, 2700 m, 12 May 1964, *Water Resources Exped. 3026* (PE 01037334); by Yalong Jiang, 2740 m, 20 June 1984, *W. L. Chen et al. 6600* (PE 00935206, 02044411, 02044463); suburbs of Yajiang, Yalong, 27 May 1991, *J. S. Yang 168* (KUN 0178577) and *169* (KUN 0178576); beside Yalong Jiang, 2700 m, 27 May 1991, *J. S. Yang 170* (IBSC 0092545, KUN 0178570, PE 01037167).

242. Berberis tenuispina Harber, sp. nov. TYPE: China. NW Yunnan: Dêqên (Deqin) Xian, near Dêqên, [Gushui], 28.610056°N, 98.755472°E, 2121 m, 27 July 2007, *Tibet-MacArthur 1373* (holotype, US 00972148 image!; isotype, KUN not seen).

Diagnosis. *Berberis tenuispina* has leaves that are very similar to *B. yulongshanensis* and *B. microtricha*, but has a different mature stem color and a different flower structure. Its spines are much longer and it appears to be only fascicled versus the more varied inflorescence of these two species.

Shrubs, deciduous, to 1 m tall; mature stems reddish purple, subterete or sub-sulcate; spines 2- or 3-fid, very pale yellow, 1.2–2.3 cm, very slender, abaxially sulcate. Petiole almost absent, rarely to 4 mm; leaf blade abaxially pale green, papillose, adaxially mid-green, oblanceolate or narrowly obovate, (0.9–)1.2–2.5 × 0.3–0.6 cm, papery, midvein raised abaxially, venation and reticulation inconspicuous abaxially, conspicuous adaxially, base cuneate or attenuate, margin entire, apex subacute, sometimes obtuse, sometimes mucronate. Inflorescence a fascicle, 2- to 7-flowered; pedicel 4–6 mm. Sepals in 2 whorls; outer sepals ovate, 3 × 1.8 mm; inner sepals obovate, 4 × 3 mm; petals broadly elliptic, 2.8 × 1.5 mm, glands 0.3 mm, apex incised. Stamens 2 mm; anther connective extended, truncate. Pistil 2.2 mm; ovules 2. Berry unknown.

Phenology. *Berberis tenuispina* has been collected in flower and very immature fruit in July. Its fruiting season is otherwise unknown.

Distribution and habitat. *Berberis tenuispina* is known from only the type collection from Dêqên (Deqin) Xian in northwest Yunnan. This was collected at 2121 m, its habitat being unrecorded.

IUCN Red List category. *Berberis tenuispina* is assessed as DD or Data Deficient, according to IUCN (2001) criteria.

The very long, weak spines of *Berberis tenuispina* (mostly less than a millimeter in diameter) are both distinctive and very unusual for such a low-growing *Berberis*. From the evidence of the co-ordinates given, *Tibet-MacArthur 1373* was collected adjacent to the G214 road on the section that runs from Dêqên to the Yunnan–Tibet border, following the upper Lancong River. There appear to be very few *Berberis* collections from this arid area, but they include the types of the very different *B. deqenensis* and *B. saxatilis*.

I am grateful to C. C. Yu of NTU for dissecting flowers of the holotype on my behalf while staying in the United States and thus avoiding the necessity for the specimen being sent on loan.

243. Berberis thomsoniana C. K. Schneid., Bull. Herb. Boissier, sér. 2, 5: 454. 1905. TYPE: India.

Sikkim, Lachen Valley, "8–10,000 ped.," s.d., *J. D. Hooker s.n.* (holotype, W 0024722!; isotypes, BM BM001010932!, C 10008312 specimen "a" image!, K K000644907!, M M0164276 image!).

Shrubs, deciduous, to 3 m tall; mature stems pale grayish yellow or brown, sulcate; spines 3-fid, absent toward apex of stems, concolorous, 0.5–2 cm, terete or abaxially flat. Petiole almost absent, sometimes to 6 mm; leaf blade abaxially pale green, adaxially green, obovate, 2–4.5 × 1–2 cm, thickly papery, midvein, lateral veins, and reticulation raised abaxially, conspicuous adaxially, margin entire, rarely spinulose with 2 to 6 teeth on each side, apex obtuse, mostly mucronate, base cuneate or attenuate. Inflorescence a sub-umbel or sub-raceme, sometimes fascicled at base, 4- to 10-flowered, 2–4.5 cm overall including peduncle to 1 cm, bracts narrowly ovate or triangular, 1.5–2.5 mm; pedicel 10–20 mm. Sepals in 4 whorls; outer sepals oblong-ovate, 4–6 × 1–2 mm; outer median sepals narrowly ovate, 4.5–7 × 2–4 mm; inner median sepals broadly obovate-elliptic, 6–8 × 4–6 mm; inner sepals broadly obovate-elliptic, 5–8 × 3–5 mm; petals obovate, 4.5–6.5 × 3.5–5 mm, base cuneate, glands obovoid, 0.8–1.3 mm, apex obtuse or emarginate. Stamens 3–4 mm; anther connective scarcely extended or not. Ovules 2 to 5. Berry red, oblong-obovoid, 9–10 × 4 mm; style persistent and short or absent.

Phenology. In China *Berberis thomsoniana* has been collected in flower in July and in fruit in August.

Distribution and habitat. The only specimens of *Berberis thomsoniana* currently known from China were found in Gyirong (Jilong) Xian in south Xizang (Tibet) at 2900–3700 m and (where recorded) on mountain slopes. The species is also known from Bhutan, India (Sikkim), and Nepal.

The protologue of *Berberis thomsoniana* cited only "Sikkim: c 3000 m (Hooker), typus in Herb. Hofm. Wien." Subsequently, Schneider (1908: 201) cited as *B. thomsoniana* an additional specimen "Sikkim: lg Hooker: Nr 39 Tonglo, 10000" without citing any herbarium. Ahrendt (1961: 128) cited a specimen with these latter details at K (000644909) as "type." However, this K specimen has different collection details from the specimen at W, whereas another specimen at K (000644907) has the same collection details as the holotype except it includes the additional information that it was collected in the Lachan Valley. Therefore, 000644907 seems a better candidate for an isotype. The specimens at BM, C, and M have the same collection details as the one at W. The latter two have annotations by Schneider identifying them as *B. thomsoniana*.

Schneider's annotation dated 10 February 1916 to the specimen at C includes identifying the material (b) on the sheet as "Type of B. thomsoniana." Adhikari et al. (2012: 484) simply cited *Hooker s.n.* from Sikkim at K as an isotype, but did not indicate the specimen they were referring to.

Selected specimens.

S Xizang (Tibet). Gyirong (Jilong) Xian: Jilong Gou, 2900 m, Aug. 2012, *Phylogeny of Chinese Land Plants, Academia Sinica Xizang Exped. 1021* (E E00662305); Jilong Zhen, pass betw. Zha Cun & Salei Xiang, "3700–3600" m, 19 July 2016, *L. Wei & Y. He BNUXZ2016222* (BNU 0028395).

BHUTAN. [1838], *W. Griffith 1745* (K K000644909, M).

INDIA. **Sikkim:** Lachen, "9000 ft," 4 June 1849, *J. D. Hooker s.n.* (K K000644908); T'h'lo rungong & Michum schzen, May–June s. anno, *J. D. Hooker 39* (K K000644909); Tonglo, "10 000 ped.," s.d., *J. D. Hooker s.n.* (K K000644906); "10 000 ped.," s.d., *J. D. Hooker s.n.* (K K000644904); Sandakphu; 27.15°N, 88.083333°E, 3350 m, 29 July 1914, *G. H. Cave s.n.* (E E00623256).

NEPAL. Rukum, near Seng Khola, 3350 m, 6 Oct. 1954, *J. D. A. Stainton, V. R. Sykes & L. Williams 4706* (BM); Sindhupalchok Distr., Phunboche Danda, S of Bhairab Kund, 27.968611°N, 85.886111°E, 3808 m, 16 Sep. 2011, *EKSIN 214* (E E00576209).

244. Berberis tianbaoshanensis S. Y. Bao, Acta Phytotax. Sin. 25(2): 158. 1987. TYPE: China. NW Yunnan: Zhongdian (Xianggelila) Xian, Tianbao Shan, Hongsha Lin Chang, 3600 m, June 15 1981, *Qinghai-Xizang Team 1510* (holotype, KUN 0161680!; isotypes, CBDI 0027512 image!, HITBC 003624 image!, KUN 0161681!, PE [2 sheets] 01037459–60!).

Shrubs, deciduous, to 2.5 m tall; mature stems purplish brown, terete; spines largely absent, when present 1- to 3-fid, pale yellowish brown, 0.5–2 cm terete, weak. Petiole almost absent or to 4 mm; leaf blade abaxially grayish green, papillose, adaxially dull green, narrowly oblong-obovate or narrowly elliptic, 2–3 × 0.7–1 cm, papery, midvein and lateral veins raised abaxially, inconspicuous adaxially, reticulation largely indistinct or obscure on both surfaces, base cuneate, margin entire or rarely spinulose with 1 to 4 teeth on each side, apex obtuse or acute, mucronate. Inflorescence 1(or 2)-flowered; pedicel 15–25 mm. Sepals in 2 whorls; outer sepals elliptic, ca. 11 × 7 mm; inner sepals oblong, ca. 10 × 5 mm; petals obovate, ca. 10 × 5.5 mm, apex rounded, incised, base cuneate, glands separate. Stamens ca. 6 mm; anther connective extended, obtuse. Ovules 4 or 5. Immature berry turning red, ovoid or oblong-ovoid, 10–17 × 6–9 mm; style persistent.

Phenology. Berberis tianbaoshanensis has been collected in flower in June and with immature fruit in September.

Distribution and habitat. Berberis tianbaoshanensis is known from only two collections from Tianbaoshan in Zhongdian (Xianggelila) Xian in northwest Yunnan. The type was found under *Abies* forest at 3600 m, *D. E. Boufford, J. F. Harber & X. H. Li 43436* in a degraded *Abies* forest area at 3680–3840 m.

The protologue which cited only the holotype at KUN stated that *Berberis tianbaoshanensis* was single-flowered; however, the isotypes at CDBI and KUN also have flowers in pairs. All the type specimens are poor ones, the best being that at CDBI. The protologue also mistakenly gave the elevation of collection as 2600 m.

On a brief visit to Tianbao Shan in September 2013, I could find only one plant of the species. This (which was largely 1-fruited but with some fruit in pairs) is the source of the description of fruit used here.

Ying (2001: 79) treated *Berberis tianbaoshensis* as a synonym of *B. muliensis*, but that species is 1- to 3-fascicled or sub-fascicled with three or four seeds per fruit.

Selected specimens.

NW Yunnan. Zhongdian (Xianggelila) Xian: Xiao Zhongdian, Tianbao Shan, 27.605278°N, 99.885833°E, 3680–3840 m, 4 Sep. 2013, *D. E. Boufford, J. F. Harber & X. H. Li 43436* (A 00914429, BM BM001190943, CAS, E E00770758, K, KUN 1278352, MO, PE, TI).

245. Berberis tianchiensis Harber, sp. nov. TYPE: China. NW Yunnan: Zhongdian (Xianggelila) Xian, Bigu, Tianchi, 27.525556°N, 99.6425°E, 3890 m, 31 Aug. 2013, *D. E. Boufford, J. F. Harber & X. H. Li 43334* (holotype, PE!; isotypes, A!, CAS!, E!, K!, KUN!).

Diagnosis. Berberis tianchiensis is somewhat similar to *B. dictyophylla*, but has much longer pedicels, five to eight seeds per fruit, and young shoots which are not densely pruinose.

Shrubs, deciduous, to 2 m tall; mature stems dark reddish purple, sulcate, epruinose, young shoots bright pink, slightly pruinose; spines 3-fid, sometimes solitary toward apex of stems, yellowish brown, 0.8–1.8 cm, terete. Petiole almost absent; leaf blade abaxially gray-green, adaxially mid-green, blue-green on young shoots, obovate or obovate-elliptic, 1.8–3(–3.5) × 0.6–0.8(–1) cm, papery, abaxially densely white pruinose, midvein raised abaxially, lateral veins and reticulation inconspicuous on both surfaces, base attenuate, margin entire, apex obtuse or subacute, sometimes mucronate. Inflorescence 1-flowered, otherwise unknown; fruiting pedicel 15–25 mm. Immature berry turning red, ovoid, 15 × 8 mm; style persistent, short; seeds 5 to 8.

Phenology. Berberis tianchiensis has been collected with immature fruit in August. Its flowering period is unknown.

Distribution and habitat. *Berberis tianchiensis* is known from only Bigu Tianchi (lake) in Zhongdian (Xianggelila) Xian in northwest Yunnan, where I found it to be the commonest *Berberis* in the open areas surrounding the lake at 3825–3890 m.

IUCN Red List category. *Berberis tianchiensis* is assessed as DD or Data Deficient, according to IUCN (2001) criteria.

The young shoots of *Berberis tianchiensis* make this a particularly attractive species.

Selected specimens.
NW Yunnan. Zhongdian (Xianggelila) Xian: Xiao Zhongdian, Tianchi Lake, 27.631667°N, 99.648611°E, 3825 m, 23 Sep. 1994, *Alpine Garden Society China Exped. (ACE) 1209* (E E00039600, LIV).

246. Berberis tianshuiensis T. S. Ying, Acta Phytotax. Sin. 37: 341. 1999. TYPE: China. SE Gansu: Tianshui Shi, Maiji Shan, 1800 m, 5 May 1984, *Q. R. Wang 11375* (holotype, PE 00935241!).

Shrubs, deciduous, to 2 m tall; mature stems pale yellowish brown, angled, not verruculose; spines solitary or 3-fid, concolorous, 1–3 cm, abaxially sulcate. Petiole 4–10 mm, sometimes almost absent; leaf blade abaxially pale green or gray, adaxially green, elliptic or obovate, occasionally oblong-elliptic, (0.7–)2–6 × (0.4–)1–2.7 cm, thinly papery, midvein, lateral veins, and reticulation raised abaxially, inconspicuous adaxially, base broadly or narrowly attenuate, margin mostly entire, sometimes spinulose with 12 to 25 inconspicuous teeth on each side, apex subacute to rounded. Inflorescence a sub-umbel, fascicle, sub-fascicle, or raceme, sometimes compound at base, (2- to)5- to 8-flowered, 3–4 cm; bracts lanceolate, ca. 1.5 mm; pedicel ca. 10 mm. Sepals in 2 whorls; outer sepals suborbicular or ovate-orbicular, ca. 3.5 × 3.1 mm; inner sepals suborbicular, ca. 4.8 × 4.6 mm; petals obovate-elliptic, ca. 5 × 3.5 mm, base clawed, glands separate, apex emarginate, lobes acute. Stamens ca. 2.5 mm; anther connective not extended, truncate. Ovules 3, sessile. Berry red, oblong, 10–12 × 5–6 mm; style persistent, ca. 1 mm.

Phenology. *Berberis tianshuiensis* has been collected in flower in May and June and in fruit from May to October.

Distribution and habitat. *Berberis tianshuiensis* is known from Tianshui Shi in south Gansu. It has been found in mixed forests, on sunny valley side slopes, river banks, and roadsides at 1750–1930 m.

Selected specimens.
S Gansu. Tianshui Shi: Maiji Shan, 11 Sep. 1950, *Z. G. Wu 20188* (PE 00935245); same location, 12 Sep. 1950, *Z. G. Wu 20231* (CAF 00009658, PE 01037181); Dangchuan, Baojia Gou, 2050 m, 25 June 1963, *Q. X. Li 277* (PE 00935244); same location, 1800 m, 26 June 1963, *Q. X. Li 345* (PE 00935242); same location, 1750 m, 26 June 1963, *Q. X. Li 352* (PE 00935243); Jiaochuan, Baojia Gou, Yindong Gou, 1850 m, 17 May 1963, *Q. X. Li 163* (PE 01032895); Jiaochuan, Dong Gou, 1930 m, 24 May 1963, *Q. X. Li 218* (PE 01033743); Dangchuan, Houyan Shan, 1800 m, 22 June 1964, *K. J. Fu 15411* (WUK 0230866); Dangchuan, Baojia Gou, Dong Gou, 1850 m, 26 June 1964, *K. J. Fu 15679* (WUK 0230435); Maiji Shan, Houya Gou, 1750 m, 2 July 1964, *K. J. Fu 15784* (WUK 0230493).

247. Berberis tischleri C. K. Schneid., Bull. Herb. Boissier, sér. 2, 8: 201. 1908. TYPE: China. N Sichuan: near Nereku River, 26 July 1885, *G. N. Potanin s.n.* (holotype, LE missing; neotype, designated here, China. N Sichuan: Songpan Xian, Huanglong Reserve, 3000 m, 9 Sep. 1986, *P. Cox et al. 2537* [E E00395660!]).

Shrubs, deciduous, to 3 m tall; mature stems pale yellowish brown to pale brown, sulcate; spines 3-fid, solitary or absent toward apex of stems, concolorous, (0.6–)2–3.5 cm, abaxially sulcate. Petiole mostly almost absent but sometimes to 12 mm; leaf blade abaxially pale green, papillose, adaxially dark green, obovate, elliptic-obovate, or oblong-obovate, 2.2–3.6 × 1.1–1.7 cm, papery, midvein, lateral veins, and reticulation raised abaxially, lateral veins inconspicuous adaxially, base attenuate, margin with 6 to 12 spinulose teeth on each side, sometimes entire, apex obtuse, sometimes subacute. Inflorescence a fascicle, sub-fascicle, sub-umbel, or sub-raceme, sometimes fascicled at base, 2- to 7(to 10)-flowered, 2–4.5 cm overall including peduncle (when present) to 2 cm; pedicel 10–17 mm (but to 36 mm at base). Sepals in 2 whorls; outer sepals narrowly obovate, 3.5–4 × 2–3 mm; inner sepals broadly obovate or elliptic-obovate, 4–6 × 3–4.5 mm; petals obovate, 3–4.5 × 2–2.5 mm, base clawed, glands widely separate, elliptic, ca. 0.5 mm, apex distinctly incised. Stamens ca. 3 mm; anther connective not extended, truncate. Ovules 3 to 5. Berry red, narrowly ovoid-oblong or ovoid, 10–12 × 3–7 mm, apex attenuate, often slightly bent; style persistent.

Phenology. *Berberis tischleri* has been collected in flower between June and August and in fruit from July to September.

Distribution and habitat. *Berberis tischleri* is known from Hanyuan, Nanping (now Jiuzhaigou), Pingwu, and Songpan Xian in north Sichuan. It has been found in meadows, thickets on mountain slopes, and forest understories at ca. 1525–3400 m.

The protologue of *Berberis tischleri* cited only *Potanin s.n.* "ad fl. Nereku" in north Sichuan, 26 July 1885, at LE (this specimen had previously been identified as *B. heteropoda* var. *oblonga* by Maximowicz [(1891: 41)]. From Bretschneider (1898a: 1014), the place of collection can be identified as Djan-la (Zangla) in Songpan Xian. I could not find this specimen when I visited LE in 2009 and, from subsequent correspondence with LE, it appears to be missing.

Schneider's description of *Berberis tischleri* was brief, but included stating that the inflorescence was "fasciculao-racemosae" with evidence of up to eight flowers, and that the immature fruit had three or four ovules. The specimens cited below from north Sichuan appear to be consistent with the protologue description. I have, therefore, designated as a neotype the only one of the specimens from Songpan Xian I have had an opportunity to examine in person, *P. Cox et al. 2537*. This was collected at Huanglong, which is only some 25 km from Zangla. The description of flowers given above is from living plants at the Royal Botanic Garden Edinburgh, at Dawyck, grown from seeds of *2537*. Some of these had up to five ovules.

Subsequently, Schneider (1913: 355–356) decided that specimens from other areas in Sichuan were *Berberis tischleri*. However, examination of the various specimens he cited reveals that they are not all of the same species and, indeed, only one, *Wilson 1177* from Mao Xian (details below), appears to be *B. tischleri*. Of the six other collections cited, three—*Wilson 2853, 2854*, and *4134*—are from Kangding Xian, and three—*Wilson 2856, 2859*, and *4307*—are from the Xiaojin–Li Xian border area. One of the Kangding collections—*Wilson 2854*—was designated by Ahrendt (1961: 125) as the type of a *B. tischleri* var. *abbreviata*. This has different leaves from *B. tischleri* and only two seeds per fruit and is treated here as a new species, *B. abbreviata*. The other two Kangding collections have leaves and fruit that are very similar to *B. tischleri*, but from the evidence from other collections from the same area, have a different flower structure and are treated here as being from another new species, *B. tachiensis*. I have been unable to identify the Wilson specimens from Xiaojin–Li Xian border area.

Ying (2001: 172–173) treated *Berberis tischleri* var. *abbreviata*, *B. diaphana* var. *tachiensis*, and *B. ellioti* as synonyms of *B. tischleri*. For the second, see under *B. tachiensis*; for the third, see under Taxa Incompletely Known.

Selected specimens.
N Sichuan. Min Valley, 1525 m, Aug. 1910, *E. H. Wilson 1177* (A 00279928, BM BM001053865, K, US 00946054). **Hongyuan Xian:** Shuajingsi Qu, near Kangle Gongshe, 26 June 1975, 3200 m, *Sichuan Botany Team 9040* (CDBI CDBI0027980, CDBI0028205, PE 01033920, SWCTU

00006209). **Nanping (now Jiuzhaigou) Xian:** Baihe Qu, Canjin Xiang, Ganhaizi, 3000 m, 5 Sep. 1964, *J. S. Li & X. C. Zhao 3128* (PE 01030897). **Pingwu Xian:** Dujian Shan, Yakou, 3250 m, 12 Aug 2007, *D. H. Zhu et al. 4688* (WCSBG 017779–80). **Songpan Xian:** valley W of Sung-pan, 3200 m, 14 July 1922, *H. Smith 2749* (A 00279923, LD 1969089, MO 4367262, PE 01032918, UPS: BOT V-04095); "Vicinity of Sungpan," 3200 m, 20 Sep. 1937, *K. T. Fu 1883* (WUK 004103); Huanglong Si, 3100 m, 9 July 1959, *Songpan Group 1603* (CDBI CDBI0027822, PE 01033827, SM SM704800108); Zhangla Qu, Zhangjin Xiang, 25 Sep. 1961, *X. N. Tan et al. 00931* (CDBI CDBI0027189); Huanglong Si, 3200 m, 21 June 1982, *W. H. Li et al. H82-377* (PE 01033924); Huanglong Si, 3400 m, 13 June 1983, *K. Y. Lang et al. 1785* (KUN 0179137, PE 01032961–62); same details, but 3200 m, *K. Y. Lang et al. 1792* (KUN 0179138, PE 01032957, 01032960).

Cultivated material:
Living cultivated plants.
Royal Botanic Garden Edinburgh, Dawyck Garden, Stobo, Scotland, from seeds of *P. Cox et al. 2537* (collection details above).

248. Berberis tomentulosa Ahrendt, J. Bot. 80(Suppl.): 112. 1944. TYPE: China. NW Yunnan: [Gongshan Xian], Upper Kiukiang (Dulong Jiang) valley, Sangolila, 2500 m, 5 Aug. 1938, *T. T. Yu 19640* (holotype, E E00117392!; isotypes, A 00038805!, KUN 1204057!, PE [2 sheets] 01037539–40!).

Shrubs, deciduous, to 1 m tall; mature stems pale yellow-brown, very sulcate; spines 3-fid, solitary toward apex of stems, concolorous, 0.3–1 cm, terete. Petiole almost absent; leaf blade abaxially pale yellow-green, adaxially bright deep green, obovate, 0.5–1.4 × 0.2–0.5 cm, papery, midvein, lateral veins, and reticulation raised abaxially, inconspicuous adaxially, base cuneate, margin spinose with 5 to 8 teeth on each side, but mostly entire on young shoots, apex rounded. Inflorescence a sub-umbellate raceme, 10-flowered, 1.5–2 cm overall including peduncle ca. 0.5 cm; pedicel 2–4 mm, slightly puberulous, becoming subglabrous; bracts ovate, 1–1.5 mm, apex acuminate; bracteoles oblong, ca. 2 × 1 mm. Flowers pale yellow, 5–6 mm diam. Sepals in 2 whorls; outer sepals broadly oblong, ca. 2.5 × 1.5 mm; inner sepals obovate-oblong, ca. 4 × 2.1 mm; petals obovate, ca. 3 × 1.5 mm, base clawed, glands separate, apex emarginate. Stamens ca. 2 mm; anther connective slightly extended, obtuse to rounded. Ovules 2. Berry unknown.

Phenology.   *Berberis tomentulosa* has been collected in flower in May. Its fruiting period is unknown.

Distribution and habitat.   *Berberis tomentulosa* is known from only the type from Gongshan Xian, northwest Yunnan. This was found on an open, rocky mountain slope at 2500 m.

As noted under *Berberis bracteata* (whose type was collected by T. T. Yu in the same area the following day), I have been unable to locate exactly where the type was collected. The Dulong (Drung) Jiang rises in Zayü (Chayu) Xian in southeast Xizang, then flows into Gongshan Xian in Yunnan and thence into Myanmar. Though the protologue states *T. T. Yu 19640* was collected in Zayü Xian, the collection details on the sheet give Yunnan. The coordinates given in the protologue appear to be speculation by Ahrendt since the collector's notes do not record this.

249. Berberis trichohaematoides Ahrendt, J. Bot. 79(Suppl.): 62. 1941. TYPE: China. SE Xizang (Tibet): [Mêdog (Motuo) Xian], Kongbo Province, near Paka, Kulu Phu Chu, 29.266667°N, 94.433333°E, 3650 m, 22 Sep. 1938, *F. Ludlow, G. Sherriff & G. Taylor 6513* (holotype, BM BM000559599!).

Shrubs, deciduous, to 2.4 m tall; mature stems dark purple, shiny, subterete or sub-sulcate, verruculose; spines solitary or absent, rarely 3-fid, pale yellow, 0.5–1.3 cm, terete. Petiole almost absent, sometimes to 3 mm; leaf blade abaxially pale green, papillose, adaxially green, slightly shiny, obovate, (1.4–)2–3(–4) × 0.6–1.5 cm, papery, midvein raised abaxially, impressed adaxially, lateral veins and reticulation inconspicuous on both surfaces, base narrowly attenuate, margin entire, apex obtuse, mucronate. Inflorescence a fascicle, subfascicle, or short sub-raceme, 3- to 11-flowered, to 3 cm overall; peduncle (when present) to 0.8 cm; pedicel 13–17 mm. Flowers unknown. Berry bright scarlet, oblong or ovoid, 12–13 × 4–4.5 mm; style not persistent or persistent and short; seeds 4 or 5, pale brown.

Phenology. *Berberis trichohaematoides* has been collected in fruit in September. Its flowering season is unknown.

Distribution and habitat. *Berberis trichohaematoides* is known from only the type collection from Mêdog (Motuo) Xian in southeast Xizang (Tibet). This was collected in an *Abies* forest at 3650 m.

Berberis trichohaematoides was not noticed by Ying (1985, 2001).

The cultivated specimen cited below is sterile but has the same fruit stalk and leaves as the type. A second cultivated specimen at E (E00112407), also annotated as being from seed of the type, is racemose with leaves with spinose margins and, as such, is almost certainly from another source.

Cultivated material:
Cultivated specimens.
Royal Botanic Garden Edinburgh, from seed of type, s.d., *s. coll. s.n.* (E E00038753).

250. Berberis tsangpoensis Ahrendt, Gard. Chron., ser. 3, 109: 101. 1941. TYPE: China. SE Xizang (Tibet): [Bomê (Bomi) Xian], Tsangpo (Zangbo) Gorge, near Pemakochung, 2133 m, 29 Nov. 1924, *F. Kingdon-Ward 6326* (holotype, K K000077357!).

Berberis medogensis T. S. Ying, Acta Phytotax. Sin. 37(4): 350. 1999, syn. nov. TYPE: China. SE Xizang (Tibet): Mêdog (Motuo) Xian, btw. Lugu Baguopuba & Bayu Jiaqiangbangma, 3300 m, 20 Nov. 1982, *B. S. Li & S. Z. Cheng 01875* (lectotype, designated here, PE 02073015).

Shrubs, semi-evergreen, subprostrate, to 20 cm tall; mature stems pale yellow, shiny, sulcate; spines 3-fid, concolorous, (0.3–)1–4 cm. Petiole almost absent; leaf blade adaxially dull green, obovate, (0.7–)1.3–2.1 × (0.3–)0.6–0.9 cm, papery, abaxially thickly pruinose, midvein and dense reticulation raised abaxially, conspicuous adaxially, base cuneate, margin thickened but not revolute, spinose with (1 or)2 to 5 teeth on each side, apex aristate-cuspidate or rounded, mucronate. Inflorescence 1-flowered; pedicel 25–30 mm, slender; bracteoles absent or 1 or 2, oblong, ca. 4.5 × 2.5 mm, apex acute. Sepals in 3 whorls; outer sepals elliptic, 7–9 × 5–5.5 mm, apex obtuse; median and inner sepals obovate, 9–10 × 6–7 mm; petals obovate, 6–7 × 4.5–5.3 mm, base scarcely clawed, glands approximate, apex entire or very slightly incised. Stamens ca. 4.5 mm; anther connective not or scarcely extended, truncate. Ovules 12 to 15. Berry red, slightly pruinose at first, subglobose, 12–13 × 9–11 mm; style not persistent.

Phenology. *Berberis tsangpoensis* has been collected in fruit in November. Its flowering period outside cultivation is unknown.

Distribution and habitat. *Berberis tsangpoensis* is known from Bomê (Bomi) and Mêdog (Motuo) Xian in southeast Xizang (Tibet). It has been collected in alpine thickets on steep slopes at ca. 2100–3600 m.

Berberis tsangpoensis and *B. medogensis* are known only from their type collections and, in the case of the former, from cultivated plants grown from *K. Rushforth 5623* (details below). I could not find the type of *B. medogensis* at PE while there in May 2007 and subsequent inquiries failed to locate it. But in 2019, an image of *B. S. Li & S. Z. Cheng 01875* appeared on the Chinese Virtual Herbarium and other Chinese websites.

The specimen is an exceptionally poor one and has no species identification. This suggests that there is (or was) another specimen identified as the type. For this reason, I have lectotypified 02073015 rather than treated it as a holotype. From the shape of its leaves and the length of its pedicels (as depicted in the line drawing accompanying the protologue), it would appear to be *B. tsangpoensis*. I have been unable to locate the exact place where the collection was made, but Bayu is in the Zhangbo valley some 10 km distant from where *Kingdon-Ward 6236* was collected.

The description of the flowers of *Berberis tsangpoensis* used here is that of Ahrendt (1941b: 50), which is based on cultivated material grown from seeds of the type. This description is confirmed by my plant of *K. Rushforth 5623*.

Berberis tsangpoensis was included by Ying (1985: 150), but subsequently (Ying: 2001) omitted without explanation.

Cultivated material:
Living cultivated plants.
Foster Clough, Mytholmroyd, West Yorkshire, U.K., from *K. Rushforth 5623*, SE Xizang (Tibet), Bomê (Bomi) Xian, Showa La, 29.883889°N, 95.399806°E, ca. 3600 m, 17 Oct. 1997.

251. Berberis tsarica Ahrendt, J. Bot. 79(Suppl.): 48. 1941. TYPE: China. SE Xizang (Tibet): [Mainling (Milin) Xian], Tsari Distr., near Langong, Chinang, 28.8°N, 94.7°E, 4400 m, 5 June 1938, *F. Ludlow, G. Sherriff & G. Taylor 3961* (holotype, BM [2 sheets] BM000559591–92!).

Berberis longispina T. S. Ying, Fl. Xizang. 2: 148. 1985, syn. nov. TYPE: China. S Xizang (Tibet): Tingri (Dingri) Xian, Kada He Gu, 4050 m, 27 May 1959, *X. G. Wang 76* (holotype, PE 00935228!; isotype, N 093057135 image!).

Shrubs, deciduous, to 50 cm tall; mature stems dark red or dark purple, sulcate; shoots pubescent; spines (3- to)5-fid, concolorous, 0.3–1.1(–2) cm, abaxially sulcate, weak. Petiole absent; leaf blade abaxially grayish or grayish white or glaucous, papillose, adaxially dark green, obovate, 0.5–1.2 × 0.2–0.5 cm, midvein and lateral veins inconspicuous abaxially, indistinct adaxially, base cuneate, margin entire, apex obtuse. Inflorescence 1-flowered; pedicel 4–7 mm, glabrous or sparsely puberulent; bracteoles red, ovate, ca. 2.3 × 1 mm, apex acute. Flowers yellow with red tips. Sepals in 2 whorls; outer sepals ovate, 3.5–5 × 2.5–3 mm, apex acute; inner sepals oblong-obovate, 5–6.5 × 3.5–4.5 mm; petals oblong-obovate, 3.5–4 × 2–2.3 mm, base cuneate, glands oblong-elliptic, apex emarginate with 2 acute lobes. Stamens 2.5–3 mm; anther connective slightly extended, truncate or rounded. Ovules 3 to 5, shortly

stipitate. Berry obovoid, 8–9 × ca. 6 mm; style persistent, short.

Phenology. In China *Berberis tsarica* has been collected in flower in June and in fruit in August.

Distribution and habitat. In China *Berberis tsarica* is known from southeast and south Xizang (Tibet). It has been found in thickets and among dwarf *Juniperus* and in meadows on open mountainsides at ca. 3900–4700 m. It is also known from Bhutan and Nepal and, according to Grierson (1984: 324), from India (Sikkim).

The type of *Berberis longispina* has unusually long spines, but otherwise appears to be a typical *B. tsarica*. Information on the name of the collector comes from the isotype at N.

Though *F. Ludlow, G. Sherriff & J. H. Hicks 20718*, cited below, is recorded as being from Bhutan, from the co-ordinates (taken from Stearn [1976: 266]), it appears it was collected on the Xizang side of the border in Cona (Cuona) Xian.

Selected specimens.
S Xizang (Tibet). Dinggyê (Dingjie) Xian: Riwuxiang, Kelabei Shan, 4450 m, 15 Aug. 1990, *B. S. Li et al. 13003* (PE 01487670–71). **[Yadong Xian]:** Khamba, 4370 m, 12 July 1939, *B. J. Gould 2368* (K); Dotha, 3960 m, 20 June 1945, *B. & K. Ram 20530* (K).
SE Xizang (Tibet). [Bomê Xian]: Kongbo, Nambu La, 29.983333°N, 94.316667°E, 4420 m, 11 July 1947, *F. Ludlow, G. Sherriff & H. H. Elliot 15384* (A 00280078, BM, E E00395955). **Cona (Cuona) Xian:** 1 km E of Boshan Kou, 4500 m, 6 Aug. 1974, *Qinghai-Xizang Team, Botany Group 2314* (PE 00049421). **[Mainling (Milin) Xian]:** Pome, Nyuma La, 4370–4570 m, 20 June 1924, *F. Kingdon-Ward 5811* (K); Tsari, Bimbi La, 28.833333°N, 93.466667°E, 4420 m, 3 June 1936, *F. Ludlow & G. Sherriff 1763* (BM, MO 1618550); Kongbo, Sang La, 29.583333°N, 94.716667°E, 4270 m, 29 June 1938, *F. Ludlow, G. Sherriff & G. Taylor 5050* (BM). **[Mêdog (Motuo) Xian]:** Kongbo, Deyang La, 29.366667°N, 94.866667°E, 4115 m, 5 June 1947, *F. Ludlow, G. Sherriff & H. H. Elliot 15154* (BM, E E00395953). **[Nang (Lang) Xian]:** Tsari, Chosam, 28.733333°N, 93.166667°E, 4270 m, 17 Oct. 1938, *F. Ludlow, G. Sherriff & G. Taylor 6376* (BM, E E00395956). **Nyingchi (Linzhi) Xian (now Bayi Qu):** Seqingla Shan, 4720 m, 4 Aug. 1983, *Q. Z. Yu 83845* (IBSC 0091834).
NE BHUTAN. Shingbe, Me La, 27.966667°N, 91.61667°E, 4270 m, 9 June 1949, *F. Ludlow, G. Sherriff & J. H. Hicks 20718* (BM, E E00168980).
NW BHUTAN. Lingshi Dzong, 27.916667°N, 89.45°E, 3960 m, 22 May 1949, *F. Ludlow, G. Sherriff & J. H. Hicks 16303* (BM, E E00168979, US 01122019).
C NEPAL. Dolkha, Rolwaling, 4570 m, 30 June 1964, *J. D. A. Stainton 4718* (BM).
E NEPAL. Solukhumbu, Khumbu, Tsolu Khola, 4550 m, 25 June 1964, *S. A. Bowes-Lyon 2103* (BM); Solukhumbu, Dole-Luza, 4300 m, 15 May 2004, *DNEP1 153* (E E00667701); Solukhumbu, Bhote Koshi, 4700 m, 21 Sep. 2005, *DNEP3 BY134* (E E00665479); Solukhumbu, Langmuche Valley, 4400 m, 24 Sep. 2005, *DNEP3 BY188* (E E00667696).

252. Berberis tsarongensis Stapf, Bot. Mag. 156, t. 9332. 1933. TYPE: China. NW Yunnan: [Dêqên (Deqin)–Gongshan Xian border], Mekong-Salween (Nu Jiang-Lancang Jiang) divide, 28°N, 3650–3950 m, Sep. 1917, *G. Forrest 14920* (lectotype, designated by Ahrendt [1961: 156], K K000077414!; isotype, E E00259009!).

Berberis stearnii Ahrendt, Gard. Chron., ser. 3, 109: 101. 1941, syn. nov. TYPE: Cultivated. Stonefield, Watlington, Oxford, U.K., 7 Oct. 1942, *L. W. A. Ahrendt s.n.* from *G. Forrest 29042*, 7 Oct. 1942, (neotype, designated here, OXF!).

Shrubs, deciduous, to 4 m tall; mature stems dark reddish brown, angled; spines solitary or 3-fid, yellowish brown, 1–2 cm, weak. Petiole almost absent; leaf blade abaxially pale green, papillose, adaxially bright green, obovate or oblong-elliptic, 1–1.8 × 0.5–0.8 cm, papery, midvein slightly raised abaxially, flat adaxially, lateral veins and reticulation indistinct on both surfaces, base cuneate, margin spinose with 1 to 4 teeth on each side, sometimes entire, apex rounded. Inflorescence a fascicle, sub-umbel, or sub-raceme, 4- to 8-flowered, 1.5–2 cm; pedicel 8–15(–20) mm; bracteoles ca. 2 mm, apex acute. Sepals in 2 whorls; outer sepals oblong-elliptic, ca. 3.75 × 2 mm; inner sepals obovate, ca. 5 × 4 mm; petals oblong-obovate, ca. 5.5 × 3.5 mm, base cuneate, glands separate, ovate, apex emarginate, lobes rounded. Stamens ca. 3.5 mm; anther connective rounded, apiculate. Ovules 2 or 3. Berry red, oblong-ellipsoid, 8–15 × 4–7 mm; style not persistent.

Phenology. *Berberis tsarongensis* has been collected in flower from May to July and in fruit in September and October.

Distribution and habitat. *Berberis tsarongensis* is known from Dêqên (Deqin) and Weixi Xian in northwest Yunnan, and the Chawalong area of Zayü (Chayu) Xian in southeastern Xizang (Tibet). It has been found on roadsides, *Pinus* forest margins, and among scrub on open, bouldery slopes and alpine meadows at ca. 3100–3950 m.

The protologue designated *Forrest 14920* as the type. The lectotypification of the specimen at K was by Ahrendt (1961: 156), though he mistakenly cited it as *Forrest 14290*.

The protologue of *Berberis stearnii* simply cited cultivated material from *Forrest 29042* without any information as to where this might be found. Subsequently, Ahrendt (1961: 155) listed a number of specimens of various dates taken from a plant or plants in his living collection grown from *Forrest 29042*, these dates being followed by "(Type. O)," O being his acronym for OXF.

There are no cultivated specimens with this number at OXF dated before the publication of the protologue, but there is one dated 7 October 1942 with collection details in Ahrendt's hand, and this has been chosen as the neotype. This is a typical example of *B. tsarongensis* as is the wild-collected specimen of the same number at E (details below).

Ying (2001: 196) treated *Berberis tsarongensis* var. *megacarpa*, published by Ahrendt (1961: 146), as a synonym of *B. tsarongensis*. This variety from Mainling Xian, Xizang, is treated here as a synonym of *B. moloensis*.

Selected specimens.
SE Xizang (Tibet). Zayü (Chayu) Xian: Salwin Kiu–Chiang divide, 28.666667°N, 98.25°E, 15 July 1919, *G. Forrest 18992* (A 00280007, E E00117389, K); Londre pass, Mekong–Salwin divide, 28.3°N, 98.666667°E, 3960 m, Sep. 1921, *G. Forrest 20802* (A 00280080, CAL, E E00612620, K, P P02682389, US 00945958); Salwin–Kiu Chiang divide, W of Chamtong, 28.3°N, 98.45°E, 3660 m, Oct. 1922, *G. Forrest 22623* (A 00280079, E E00117390, K, US 00946053); Dzernar, Tsa-wa-rung (Chawalong), 3400 m, Sep. 1935, *C. W. Wang 66431* (A 00279955, KUN 0177018, LBG 00064312, NAS NAS00314084, PE 01033905–06, WUK 0035424); Chawalong Qu, betw. Wabu & Jialang, 3800 m, 5 Oct. 1982, *Qinghai-Xizang Team 10974* (KUN 0178159, 01030755, PE 01030755).

NW Yunnan. Dêqên (Deqin) Xian: Doker-la, Mekong–Salwin divide, 28.416667°N, 3350 m, June 1918, *G. Forrest 16554* (A 00280006, E E00612617, K). **Weixi Xian:** Mekong–Salwin divide, 27.6°N, 98.933333°E, 3660–3960 m, Oct. 1921, *G. Forrest 20621* (A 00280081, CAL, E E00612619, K, P P02482737, US 00945959, W 1925–62676); Nujiang valley, 3300 m, 27 May 1940, *G. M. Feng 4221* (KUN 0177256–58, PE 01033049); Bading, 3100 m, 10 May 1982, *Qinghai-Xizang Team 6422* (KUN 0177259–60, PE 01032930–31); s. loc., s.d., *G. Forrest 29042* (E E00612591).

253. Berberis tsienii T. S. Ying, Acta Phytotax. Sin. 37: 307. 1999. TYPE: China. W Guizhou: Pan Xian, Badashan, 2100 m, 23 Aug. 1959, *Anshun Team 962* (holotype, PE 00935248!; isotypes, GF 09001096 image!, HGAS 013664!, KUN 0177021!, PE 00935249!).

Shrubs, deciduous, to 1.5 m tall; mature stems dark brown, very conspicuously sulcate; spines 3-fid, pale yellow, 0.2–2.3 cm, terete, weak. Petiole almost absent; leaf blade abaxially yellow-green, adaxially shiny dark green, elliptic or obovate-elliptic, sometimes obovate, 0.7–1.5 × 0.3–0.6 cm, papery, midvein raised abaxially, flat or slightly impressed adaxially, venation indistinct or obscure on both surfaces, base cuneate, margin entire, rarely spinulose with 1 to 4 teeth on each side, apex acute, sometimes mucronate. Inflorescence a fascicle, 3- to 6-flowered, otherwise unknown; pedicel purplish red, 3–4 mm. Berry purplish, ellipsoid, 6–8 × ca. 3 mm; style not persistent; seeds 1.

Phenology. *Berberis tsienii* has been collected in fruit from June to August. Its flowering season is unknown.

Distribution and habitat. *Berberis tsienii* is known only from Hezhang, Pan, and Weining Xian in west Guizhou, and a collection from the neighboring Fuyuan Xian in east Yunnan. The type was found in thickets in a mountain valley at 2100 m, and the Fuyuan specimen among *Azalea* and *Quercus* in a limestone area at 2400 m. The habitat of *K. M. Lan 00335* and *C. H. Yang 4640* was not recorded.

The isotype of *Berberis tsienii* is at KUN and was annotated by C. Y. Wu as an unpublished "*Berberis pterocaulis.*"

Hongshui He Expedition 2202 at PE has much longer spines than those recorded in the protologue of *Berberis tsienii.*

Selected specimens.
W Guizhou. Hezhang Xian: Liuqu He, 9 Nov. 1973, *K. M. Lan 00335* (GZAC GZAC14011887). **Pan Xian:** Xichong Qu, Lianghe Xiang, Wang Haidi, 3 June 1991, *T. L. Zhang & Y. Q. Huang 1* (GZTM 0011503–07). **Weining Xian:** Lao Hu Shan, 8 June 2005, *C. H. Yang 4640* (GF 9003499–500).
E Yunnan. Fuyuan Xian: Housuo, Laohei Shan, 2400 m, 17 June 1989, *Hongshui He Expedition 2202* (KUN 0179207–08, PE 01986307).

254. Berberis ulicina Hook. f. & Thomson, Fl. Ind. 1: 227. 1855. TYPE: India. Kashmir, Ladakh, Nubra Valley, Sasser, 4250 m, s.d., *T. Thomson s.n.* (lectotype, designated here, second specimen on right with flowers K K000395222!).

Berberis kaschgarica Rupr., Sert. Tianchan. 38. 1869, syn. nov. TYPE: China. [W Xinjiang]: Tian Shan, Thal Suukty, 31 July 1867, *F. P. Osten-Sacken s.n.* (lectotype, designated here, LE!).

Shrubs, deciduous, to 1 m tall; mature stems reddish purple, subterete; spines 3(to 5)-fid, pale yellow, 0.8–1.7 cm, stout, abaxially flat or slightly sulcate; internodes 5–12 mm. Petiole almost absent; leaf blade concolorous, slightly bluish green on both surfaces, linear-oblanceolate, 0.6–1.3 × 0.1–0.4 cm, papery, midvein raised abaxially, flat adaxially, venation indistinct or obscure on both surfaces, base attenuate, margin spinose with 3 teeth on each side, occasionally entire, apex aristate. Inflorescence a fascicle, sometimes subracemose or sub-umbellate, 3- to 6-flowered; pedicel 2–5 mm. Sepals in 2 whorls; outer sepals lanceolate or elliptic, 3–4.5 × 0.8–1 mm; inner sepals obovate-oblong, (4–)6–6.5 × 3–3.5 mm; petals oblanceolate or oblong, 4–5.1 × 2–2.2 mm, base clawed, glands widely separate, apex incised or emarginate with acute lobes.

Stamens 2.5–3.5 mm; anther connective not extended, shortly apiculate. Ovules 3 to 5. Berry black, globose, 3–3.5 × ca. 3 mm; style persistent, ca. 0.8 mm.

Phenology. In China *Berberis ulicina* has been collected in flower in May and June and in fruit from July to September.

Distribution and habitat. *Berberis ulicina* is known in China from west and northwest Xinjiang and west Xizang (Tibet). It has been found on stony mountainsides, alpine grasslands, and river beaches at 1960–4300 m. It is also known from northwest India (Kashmir), Kyrgyzstan, Pakistan, and Tajikistan.

The protologue of *Berberis ulicina* cited "In Tibetica occidentali; Nubra . . . 14–16,000 ped." and commented "all our specimens are very uniform in appearance" without giving any further details. Later, Schneider (1905: 400) cited Thomson specimens from "Kashmir: Ladakh" and "W Thibet, Nubra" apparently without being aware that, for the part of Kashmir that Hooker and Thompson were referring to, the names Ladakh and West Tibet are interchangeable. Schneider, however, cited no herbarium. Ahrendt (1961: 228) designated *Hooker s.n.* "W Tibet adjacent to Nubra 14–16000 ft. 1849 (K)" as the type; however, no such specimen could be found, suggesting this description was simply a paraphrase of the protologue description. But sheet K K000395222, which is labelled in Hooker's hand as "Berb. ulicina, Hab. Sassur, 14000 ft," was annotated by Ahrendt on 1 October 1946 as "Type specimen." The sheet, however, clearly has more than one and possibly as many of five different gatherings on it. The specimen with the best floral material has been chosen as the lectotype.

The protologue of *Berberis kaschgarica* (sometimes erroneously spelled "*kashgarica*" or "*kasgarica*") cited *Osten-Sacken*, 31 July 1867, but cited no herbarium. Maximowicz (1889b: 31) cited the same specimen (as well as *Krasnow s.n. 1886*), but again, without any herbarium. Fedtschenko (1937: 555) and Borodina-Grabovskaya et al. (2007: 165) stated the type was at LE, but neither cited an actual specimen; hence my lectotypification.

Jafri (1975: 28) stated that *Berberis kaschgarica* "is hardly different" from *B. ulicina*, but did not formally describe it as a synonym.

The density and length of the spines of *Berberis ulicina*, which presumably have evolved as a deterrent to animals in terrain where vegetation is sparse, make the species one of the most ferocious in the whole genus.

Berberis kaschgarica in Kyrgyzstan and China is on the current IUCN Red List of Threatened Species (Participants of the FFI/IUCN SSC Central Asian regional tree Red Listing workshop, 2007b).

Syntypes. *Berberis ulicina.* India, Kashmir. *T. Thomson s.n.*, specimens other than lectotype on sheet K K000395222; Ladakh "14000 ft," s.d., *T. Thomson s.n.* (BM BM001015548, BR BR0000013206192, C, CAL CAL0000006780, CGE 32531–3, CGE Herb. Bunbury 32521, E E00623268, G G00343564–66, GH 00872740, GH 969153, M M0164278, MANCH, NY 00000045, P [3 sheets] P02797909, P06868537, P00580431, S S12–25336, TCD TCD0017859, VT UVMVT13982, W [2 sheets] Rchb-1889–8366, 0032263).

Selected specimens.
W & NW Xinjiang: Tian Shan, Ili River, 1886, *A. N. Krasnow s.n.* (G, K K000395197, P). **Akqi (Aheqi) Xian:** Ta'erdibulake, 6 Sep. 1992, *B. Wang 92-1934* (XJA 00063940). **Aksu (Akesu) Shi:** en route from Akesu to Talake, 1960 m, 18 June 1978, *Xinjiang Integrated Investigation Team 517* (PE 01037029). **Akto (Aketao) Xian:** Tuohai, 2580 m, 10 June 1974, *Institute of Biology, Xizang Exped. 3076* (HNWP 42283, PE 01037034). **Awat (Awati) Xian:** Talakule, 1960 m, 18 June 1978, *Xinjiang Integrated Investigation Team 517* (PE 01037029, XJBI 00036269). **Baicheng Xian:** 15 km N of Kugan, 2000 m, 6 Sep. 1958, *Xinjiang Integrated Investigation Team 8217* (KUN 0178305, PE 01037028). **Hotan (Hetian) Shi:** Ate, 3600 m, 21 July 1987, *B. S. Li et al. 10737* (PE 01352423–24). **Kapgal (Chabucha'er) Xian:** 15 June 1959, *Xinjiang Integrated Investigation Team K 352* (KUN 0178304, PE 01037035). **Kargilik (Yecheng) Xian:** Mazha, Mazha Dala, 4100 m, 24 Aug. 1986, *R. F. Huang 63* (HNWP 134966). **Kuchar (Kuche) Xian:** betw. Yiqike Oilfield & Layishu Linchang, 2120 m, 15 June 1978, *Xinjiang Integrated Investigation Team 475* (PE 01037030, XJBI 00036260). **Luntai Xian:** Baxi Daban, 1 Sep. 1983, *s. coll. 06-100* (XJA 00063926). **Makit (Maigaiti) Xian:** Xincan Gongshe. 3700 m, 3 June 1959, *Xinjiang Integrated Investigation Team K 478* (PE 01037980). **Minfeng Xian:** 14 Aug. 2007, *H. Y. Ren 106* (XJA 00063958). **Pishan Xian:** Kalaqi, 3600 m, 9 July 1987, *Qinghai-Xizang Team – Y. H. Wu 545* (HNWP 144380). **Qarqan (Qiemo) Xian:** Wuyilake, 24 July 1984, *s. coll. 317* (XJA 0063927). **Qira (Cele) Xian:** Yekeda'er Xiang, 20 May 1959, *Xinjiang Integrated Investigation Team 00142* (PE 01037986–87, XJBI 00036271). **Taxkorgan (Tashiku'ergan) Xian:** Xinliu Gou, 3800 m, 8 July 1987, *B. S. Li et al. 10463* (PE 01352405–07). **Ulugqat (Wuqia) Xian:** Positidie Like, 2600–2900 m, 25 June 1987, *B. S. Li et al. 10233* (PE 01352408–10). **Uqturpan (Wushi) Xian:** Mengkuzi, 4 Aug. 1983, *B. Wang 93-587* (XJA 00063931). **Yarkant (Shache) Xian:** Kalatuzikuang Qu, 2700 m, 22 July 1987, *Qinghai-Xizang Team – W. Y. Hu 700* (HNWP 144443). **Yutian Xian:** Pulu Dadui, 3300 m, 7 July 1988, *B. S. Li 11860* (PE 02046402–03, 02046407).
W Xizang (Tibet). Rutog (Ritu) Xian: Sada, 4300 m, 17 Aug. 1987, *S. B. Li et al. 11045* (PE 01352403–04).
INDIA. Kashmir, [W] Tibet, Kuenlun Plains, 3505–5180 m, 1892, *Capt. H. P. Picot s.n.* (K K000395196); Ladakh, below Chaluk, Markha valley, 3750 m, s.d., *L. Hartmann 3054* (G); above Shigar, Baltistan, 8000 ft., 21 Aug. 1936, *W. Koelz 9687* (NA 0082443, NY).
KYRGYZSTAN. E Tian Shan, Sarydzhaz Range, right bank Sarydzhaz River, near mouth of Bolshoi Taldysu River, 19 July 1969, *5177- V. Vassiljeva s.n.* (BM BM001010959, G, LE, NY, P P02313668, UC UC1559290, US 01121980, W 1974/10302).

PAKISTAN. Karakorum, Chupersan valley, Yishkuk, 36.8°N, 74.35°E, 3330 m, 19 June 2000, *E. Eberhart 7012* (MSB MSB-144503); Karakorum, Misgar valley, Ramuhar valley, upward of Ramuhar village, 36.75°N, 74.783333°E, 16 July 2000, *E. Eberhart 8163* (MSB MSB-144505).
TAJIKISTAN. Pamir Mtns., Dzamental, ca. 3800 m, 10 Sep. 1933, *6462-Júlris s.n.* (BM BM001010931, BR, G, LE, P P02313668, NY, UC UC1559323, W 1988–5751).

255. Berberis umbratica T. S. Ying, Fl. Xizang. 2: 135. 1985. TYPE: China. SE Xizang (Tibet): Bomê (Bomi) Xian, Guxiang, 29.916667°N, 95.5°E, 3340 m, 19 June 1965, *T. S. Ying & D. Y. Hong 650285* (holotype, PE 00935250!; isotypes, PE [3 sheets] 00935251–52!, 01840004!).

Shrubs, deciduous, to 1.2 m tall; mature stems reddish brown, shiny, sub-sulcate, scarcely black verruculose; spines 3-fid, solitary or absent toward apex of stems, concolorous, 0.6–0.9 cm, terete, weak. Petiole almost absent or to 5 mm; leaf blade abaxially pale green, papillose, adaxially deep green, obovate or obovate-lanceolate, 1.5–5 × 0.6–2.1 cm, papery, midvein and lateral veins slightly raised abaxially, slightly impressed adaxially, margin entire, base narrowly attenuate, apex rounded or acute. Inflorescence a sub-umbel, 3- to 5-flowered, 2–3 cm overall; peduncle 0.6–1.2 cm; pedicel 5–11 mm, slender; bracteoles ovate, ca. 1.8 × 1 mm. Sepals in 2 whorls; outer sepals ovate-lanceolate, ca. 5.5 × 3 mm; inner sepals elliptic, ca. 6.2 × 4 mm; petals obovate, ca. 5 × 3 mm, base slightly clawed, glands separate, apex slightly emarginate or entire. Stamens ca. 3.2 mm; anther connective extended, rounded. Ovules 4, shortly stipitate. Fruit black, sometimes lightly pruinose, oblong, 8–9 × 5 mm; style not persistent or persistent and short.

Phenology. *Berberis umbratica* has been collected in flower in June. Its fruiting season is unknown outside cultivation.

Distribution and habitat. *Berberis umbratica* is known from only the type gathering from Guxiang Bomê (Bomi) Xian, southeast Xizang, and a cultivated plant grown from seed collected in the same valley (details below). The former was collected from an *Abies* forest at 3340 m, the latter from an unrecorded habitat at 3300 m.

The information about the fruit above is from my plant of *Clark 3657*.

Cultivated material:
Living cultivated plants.
Foster Clough, Mytholmroyd, West Yorkshire, U.K., from *A. Clark 3657*, Bomê (Bomi) Xian, Po Tsangpo Valley, 3300 m, autumn 1997.

Cultivated specimens.
From above plant, 5 Sep. 2017, *J. F. Harber 2017-1* (A, E, PE, KUN).

256. Berberis virescens Hook. f., Bot. Mag. 116: t. 7116. 1890. TYPE: India. Sikkim, Lachen, 2745 m, 28 May 1849, *J. D. Hooker s.n.* (lectotype, designated by Adhikari et al. [2012: 503], K K000340167!).

Berberis ignorata C. K. Schneid., Bull. Herb. Boissier, sér. 2, 5. 661. 1905. *Berberis virescens* var. *ignorata* (C. K. Schneid.) Ahrendt, J. Bot. 79(Suppl.): 60. 1941. *Berberis vulgaris* L. var. δ *brachybotrys* Hook. f. & Thomson, Fl. Brit. India 1: 19. 1872, pro parte. TYPE: India. Sikkim, Lachen, "9–10,000 ped.," 6 July 1849, *J. D. Hooker s.n.* (lectotype, designated here, W 0029519!; isolectotypes, B†, CGE 32530!, GH 00872737!, K K000340169!, P P02327418!, TCD TCD0017857 image!).

Shrubs, deciduous, to 3 m tall; mature stems reddish brown, terete or slightly angled; spines 3-fid, concolorous, 0.8–1 cm, terete or abaxially sulcate, stout. Petiole almost absent; leaf blade abaxially slightly glaucous, adaxially dark green, obovate or obovate-elliptic, 0.8–2 × 0.3–1 cm, papery, midvein, lateral veins, and reticulation raised abaxially, inconspicuous adaxially, base cuneate, margin entire or spinulose with 3 to 5 teeth on each side, apex obtuse, mucronate. Inflorescence a short, condensed raceme or sub-umbellate raceme, rarely a fascicle, 1–3 cm, 2- to 8-flowered; peduncle (if present) 0.2–0.5 cm; pedicel 3–10 mm; bracts ovate-triangular, 2–2.5 × 1–1.5 mm. Flowers yellow, ca. 0.8 cm diam. Sepals in 3 whorls; outer sepals ovate, 2.5–3.5 × 1–2 mm; median sepals ovate-elliptic, 3.5–5 × 2–3 mm; inner sepals broadly obovate, 5–8 × 4–7 mm; petals obovate-elliptic, 4–6 × 2–4 mm, base clawed, apex notched 0.5 mm deep, margin entire, venation distinct with 1 central vein and 1 pair of lateral veins, glands ovoid, ca. 1 mm. Stamens 3–4 mm; anther connective extended, obtuse or slightly retuse. Pistil 2.5–4 mm; ovules 3 or 4. Berries red, oblong-ellipsoid, 8–12 × 5 mm; style not persistent or persistent and short.

Phenology. In China *Berberis virescens* has been collected in flower in May and June and in fruit in July.

Distribution and habitat. In China *Berberis virescens* is known from Nyalam (Nielamu) and Yadong Xian in south Xizang (Tibet). It has been found in thickets on mountain slopes at 2900–4100 m. It is also known from India (Sikkim) and Nepal.

Though the protologue of *Berberis virescens* contained a reference to Griffith specimens from Bhutan, cultivated material apparently from seeds of Sikkim origin and "a similar plant from the N.W. Himalaya," it

was very specific that the species was being published largely on the basis of Hooker specimens collected in the Lachen valley, Sikkim, on 28 May 1849; however, no herbarium was cited. The lectotypification of K000340167 was by Adhikari et al. (2012: 503).

Importantly, the protologue of *Berberis virescens* also referred to specimens at Kew from Sikkim which Hooker and Thomson had earlier considered to be a variety of *B. vulgaris*, but "which I am now disposed to think should be referred to *B. virescens*." This reference was undoubtedly to *B. vulgaris* var. δ *brachybotrys* (Hooker and Thomson, 1855: 220–221) which cites "Sikkim . . . 9–11,500 ped." The only Hooker specimen located at Kew that corresponds to this is K K000340169 which has two labels, one with "Lachen, 9–10,000 ft, July 6 1849," and another identifying it as "*Berberis vulgaris* L var. δ *brachybotrys* . . . Sikkim, 9–10,000 ped."

The protologue of *Berberis ignorata* stated "Typus in Herb. Hofm Wien; Herb Berlin etc." This was the format used by Schneider to describe specimens that were likely to be duplicates from an originating herbarium (in this case, clearly Kew). The specimen at W has been chosen as the lectotype, any specimen at Berlin having been destroyed in World War II. The GCE and W specimens are also annotated as "*Berberis vulgaris* L. var. δ *brachybotrys*" in Hooker's hand and are duplicates of K K000340169. Three years after *B. ignorata* was published, Schneider (1908: 259) stated he now regarded it as a synonym of *B. virescens*, having on 2 June 1906 annotated K K000340169 as "That is my *ignorata* which now I think is only the same as *B. virescens*."

From then, the situation was muddied by Ahrendt who (1941b: 60) published *Berberis virescens* var. *ignorata* as a stat. nov. for Schneider's *B. ignorata*. This made no reference to the specimens at B and W cited in Schneider's protologue and designated K K000340169 as the type (given the existence of the W specimen, this cannot be treated as a lectotypification). Ahrendt's protologue also made no reference to Schneider's synonymy of 1908 nor to his comments on the sheet of K K000340169. It gave only the briefest of descriptions and, in addition to Hooker's Sikkim specimen, cited three collections from southeast Xizang. Subsequently, Ahrendt (1961: 145) restored *B. ignorata* as a separate species with a full description of flowers, citing a further three specimens from southeast Xizang. These eight specimens are of a variety of species, none of them *B. virescens*. It is unclear which of these were the basis for Ahrendt's description of flowers of *B. ignorata* (reproduced in Ying [2001: 202] which also recognized this taxon), but it can be discounted.

Ahrendt (1961: 125) also cited four specimens from southeast Xizang as *Berberis virescens*: *F. Ludlow &*

G. Sherriff 1758 and *F. Ludlow, G. Sherriff & G. Taylor 3831, 3834*, and *4394* (all BM). None of these are *B. virescens*.

Though the description of the stems and leaves in the protologue of *Berberis virescens* was fairly full, that of the flowers was cursory. The above description of flowers is based on the type. It differs somewhat from that given by Ahrendt (1961: 125) and reproduced by Ying (2001: 170), which may have been based on the cultivated plant Ahrendt cites in his own living collection, reputedly grown from seed of Sikkim origin.

Berberis spraguei var. *pedunculata*, which was synonymized as *B. virescens* by Ying (2001: 170), is treated here as a synonym of *B. concolor*.

The report of *Berberis virescens* in Qinghai by L. H. Zhou (1997: 374) is mistaken.

Selected specimens.
S Xizang (Tibet). Nyalam (Nielamu) Xian: Fuqu He, near Bululin, 4100 m, 25 June 1975, *Qinghai-Xizang Botany Team 4332* (PE 01840035, 01032938–39); near Nyalam, 4000 m, 25 June 1975, *Xizang Team 5894* (HNWP 48936, KUN 0178768, 0179345, PE 01032937, 01032942); 3700 m, 22 July 1975, *s. coll. 755* (PE 01032936). **Yadong Xian:** W side of Pi He, 2900 m, 29 May 1975, *Qinghai-Xizang Supplementary Team 750063* (HNWP 50696, 96906, KUN 0179317–18, PE 01037093, 01840027).
NEPAL. Kyängshar Khola, rive droite, 2800 m, 10 May 1952, *A. Zimmerman 480* (G G00302061).

257. Berberis virgetorum C. K. Schneid., Pl. Wilson. (Sargent) 3: 440. 1917. TYPE: China. NW Jiangxi: Kuling, 1370 m, 29 July 1909, *E. H. Wilson 1517* (lectotype, designated here, A 00038814!; isolectotypes, A 00038815!, BM BM000564311!, E E00259011!, GH 00038816!, K K000644924!, MO 2270576!).

Berberis chekiangensis Ahrendt, J. Linn. Soc., Bot. 57: 185. 1961. TYPE: China. E Zhejiang: Tiantai Shan, 2000 m, 1889, *E. Faber 260* (holotype, K K000077394!).
Berberis pingjiangensis Q. L. Chen & B. M. Yang, Acta Phytotax. Sin. 20(4): 483. 1982. TYPE: China. NE Hunan: Pingjiang Xian, Shifeng, 6 Apr. 1981, *R. M. Yang 02266* (holotype, HNNU 00002906 image!; isotypes, HNNU (5) 00002908–11 images! 00002913 image!, NAS NAS00315806 image!).

Shrubs, deciduous, to 2 m tall; mature stems pale grayish yellow or pale grayish brown, angled or sulcate, not verruculose; spines solitary, occasionally 3-fid, concolorous, 1–2.5 cm, abaxially sulcate. Petiole 4–15 mm, sometimes almost absent; leaf blade abaxially yellow-green, lightly papillose, adaxially dull deep yellow-green, oblong-rhombic, 3.5–10 × 1.5–3.5(–4) cm, thinly papery, midvein raised abaxially, flat or slightly impressed adaxially, lateral veins and reticulation largely raised abaxially, inconspicuous or indistinct adaxially, base narrowly attenuate, margin entire, apex acute. Inflorescence a raceme, rarely a sub-umbel, 3- to 15-flowered, 2–5 cm overall; peduncle 1–2 cm; bracts lanceolate, 1–1.5 mm, apex acuminate; pedicel 4–8 mm, slender. Sepals in 2 whorls; outer sepals oblong-ovate, 1.5–2 × 1–1.2 mm; inner sepals oblong-obovate, ca. 4 × 1–1.8 mm, apex obtuse; petals elliptic-obovate, 3–3.5 × 1–1.8(–2.5) mm, base clawed, glands separate, oblong, apex obtuse, entire. Stamens ca. 3 mm; anther connective not extended, obtuse. Ovules 1, sessile. Berry red, oblong-ellipsoid, 8–12 × 3–4.5 mm; style not persistent.

Phenology. *Berberis virgetorum* has been collected in flower between March and May and in fruit between June and November.

Distribution and habitat. *Berberis virgetorum* is known from Anhui, Fujian, Guangdong, north Guangxi, northeastern Guizhou, east and southwest Hubei, northeastern, southwestern, and south Hunan, Jiangsu, Jiangxi, Shandong, and Zhejiang. It has been found in montane thickets, forests, and on road and riversides at ca. 100–1500 m.

The protologue of *Berberis virgetorum* cited *E. H. Wilson 1517* as the type, but without indicating any herbarium. Ahrendt (1961: 166) gave this as A; however, there are two specimens of this there. The lectotypification has been completed by selecting the specimen that has "Type" in Schneider's hand on the label.

C. M. Hu (1986: 11) synonymized *Berberis chekiangensis* and *B. pingjiangensis*.

F. X. Liu et al. 2220, listed below, is the only specimen located from Jiangsu. The specimen sheet notes it as a first collection from that province. Though undated, from specimens of other genera in the same sequence, it appears to have been collected in June 1956. Other specimens from Jiangsu in Chinese herbaria are from cultivated plants; e.g., *F. Liu 894* (HHBG HZ009014, IBSC 0092257, PE 01036974–76), Nanjing, Zhongshan (Sun Yat-sen) Mausoleum, 24 April 1951. Whether it is still found in the wild is unknown.

There are a number of specimens at IBK from Guangxi which are identified as *Berberis virgetorum*. These, however, have leaves that are orbicular-elliptic or orbicular-ovate. As such, they are different from *S. L. Yu 900189* cited below from farther north in Guangxi, which has the typical oblong-rhombic leaves of *B. virgetorum*. Unfortunately, these other Guangxi specimens (which are listed separately below) are all sterile, so their inflorescence is unknown. It is possible they are actually *B. honanensis* (for which, see under Taxa Incompletely Known). Clearly, further investigation is needed here.

Selected specimens.

Anhui. Huoshan Xian: Mozitan Fish Farm, 120 m, 22 Apr. 1983, *M. B. Deng & G. Yao 81465* (NAS NAS00314275–77). **Jing Xian:** Suhong Gongshe, 29 Oct. 1959, *s. coll. 677* (NAS NAS00314272). **Jinzhai Xian:** Baima Zhai Linchang, 800 m, 20 May 1984, *M. B. Deng 81768* (NAS NAS00315812–13). **Qianshan Xian:** Shuihou Ling, Sanli Xiang, Heping Cun, Quanxing Shan, 750 m, 15 Oct. 1953, *Huadong Work Team 7215* (NAS NAS00314274, PE 01033069). **She Xian:** Sanyang, Daxiyuan, 1200 m, 15 May 1959, *s. coll. 1808* (NAS NAS00314271). **Xuanzhou Qu:** betw. Xikou & Kengxuan, 7 Nov. 1959, *s. coll. 484* (NAS NAS00198107). **Yuexi Xian:** 1000 m, 14 Aug. 1997, *Z. W. Xie et al. 97029* (A 00280091, CAS 940798, PE 01604650); Yaoluoping Natl. Nature Reserve, 900 m, 7 May 2006, *R. Zhang & Q. F. Gao 06187* (Yaoluoping Natl. Nature Reserve Herbarium NAU0508470).

Fujian. Changting Xian: s.d., *Ling 4934* (PE 01033116). **Chong'an Xian (now Wuyishan Shi):** en route to Huanggang Shan, 1300 m, 2 May 1981, *Wukao Team 2424* (IBSC 0092153). **Hua'an Xian:** Gao'an Gongshe, Futian Dadui, 5 June 1959, *K. M. Wu 60069* (IBSC 0092151). **Jianyang Xian (now Shi):** Huangkeng Gongshe, Guilin, 400 m, 6 Sep. 1979, *Wukao Team 1290* (IBSC 0092152). **Pinghe Xian:** Luxi, 21 Aug. 1958, *C. J. Huang 12228* (AU 007433, FUS 00014278). **Shaowu Shi:** Shuibei Gongshe, 23 Mar. 1975, *s. coll. 75057* (PE 01033115). **Xianyou Xian:** Mapu, 24 June 1931, *Y. Lin 415* (IBSC 0092150, PE 01033118). **Yongding Xian:** Feb. 1946, *J. X. Chen s.n.* (PE 01033117).

Guangdong. Liannan Xian: Beimang Xiang, 11 Aug. 1958, *P. X. Tan 58934* (IBSC 0092135, PE 01033121); Xinyi Shi, Huochang Ping, 25 Nov. 1934, *Z. Huang 38027* (IBK IBK00012788, 00012819, IBSC 0092137). **Renhua Xian:** Tsengshing (Zhencheng Xian), To Wonglin, Naamkwan Shan, 4 Apr. 1932, *W. T. Tsang 20104* (A 00280092, IBSC 0092143, K, KUN 0177286, MO 1261420, N 093057102, NA 0082448, NAS NAS00314049, NY, P P06868333, PE 01033119, 01033122, UC UC 612146, US 00946041).

Guangxi. Quanzhou Xian: Daxijiang Xiang, 400 m, 2 June 2015, *Z. Y. Zhang et al. 2015-246* (PE 02062372). **Xing'an Xian:** Jiu Qu, Tangdong Xiang, Huachou Bridge, 570 m, 14 Nov. 1956, *S. L. Yu 900189* (IBK IBK00012787, IBSC 0092145).

NE Guizhou. Jiangkou Xian: Fanjing Shan, 900 m, 14 May 1981, *Y. D. Yang 1651* (FAN); SE side of Fanjing Shan, Yuao, along Heiwan River, 900–950 m, 28 Aug. 1986, *Sino-American Guizhou Bot. Exped. 529* (A 00280089, BR BR000000522909, 000000522924, CAS 774367, HGAS 013668, PE 01033114, NY). **Songtao Xian:** Lengiaba, 820–1120 m, 5–9 Oct. 1986, *Sino-American Guizhou Bot. Exped. 2107* (A 00280088, BM, CAS 801304, HGAS 013669, PE 01033113). **Yanhe Xian:** Mayanghe Nature Reserve, 700 m, 15 Apr. 2008, *H. H. Zhang 1157* (Mayanghe Herbarium 951).

E & SW Hubei. Jianshi Xian: Longtanping, 7 Aug. 1957, *Hubei Institute of Forestry 412* (HIB 0129617). **Jingmen Shi:** Dongbao Qu, Lixi, Qianqi Liuzu, 326 m, 4 Nov. 2005, *J. Q. Wu & X. Y. Ma 828* (HIB 0134394, 0142926). **Lichuan Xian:** Yujiawan, E of Xiahohe, 1500 m, 8 Oct. 1980, *SW Hubei Sino-American Exped. 2079* (A 00280090, CM, E E00612632, HIB 0129611–12, NA 0016564, NAS NAS00314371, NY, PE 00996099, UC).

S & SW Hunan. Xinning Xian: Ziyunshan, 1100 m, 4 Sep. 1984, *Ziyunshan Team 635* (PE 01033107). **Yizhang Xian:** Mangshan, Qingguai Keng, 430 m, 16 Jan. 1942, *Q. C. Shao 2499* (AU 038835, IBK IBK00012786, IBSC 0092158, KUN 0177287, MO 04120389, 4178262, NAS NAS00314372).

Jiangsu. Xing Shi, s.d., *F. X. Liu et al. 2220* (NAS NAS00110834).

Jiangxi. Boyang Xian: Qianqiu He, 6 Nov. 1955, *J. S Yue 1989* (NAS NAS00314392, 00314396–97); Shangrao Shi, Wufushan, 1300 m, 11 Sep. 1958, *M. X. Nie 4944* (FUS 00014275, KUN 0179079, LBG 00013860, PE 01033095, SHM 0014335). **Dayu Xian:** Zuoba Gongshe, 14 July 1962, *J. S. Yue et al. 1642* (NAS NAS00314380, NAS00314390). **Guixi Xian:** Tianhua Shan, 400 m, 23 July 1958, *M. X. Nie 3723* (IBSC 0092243, LBG 00013857, PE 01033094, SHM 0014336). **Lichuan Xian:** Wuyishan, 700 m, 18 Oct. 1957, *M. J. Wang et al. 2368* (HHBG HZ009019). **Longnan Xian:** Chenshuixi Cun, 410 m, 13 May 1958, *J. Xiong 00807* (LBG 00013861, PE 01033100). **Lushan Qu:** 1929, *W. Y. Hsia 2062* (PE 01033090). **Nanfeng Xian:** Yi Qu, Junfeng Xiang, 5 May 1958, *M. X. Nie 2432* (KUN 0178652, 0179077, LBG 00013863–64, SHM 0014337-38). **Ningdu Xian:** opposite Xiobu Linchang, 300 m, 4 Aug. 1979, *D. C. Wu 79-0127* (PE 01358342, 01358350). **Suichuan Xian:** Dafen Qu, Linyang, 673 m, 17 Sep. 1963, *J. S Yue et al. 3823* (KUN 0179080, NAS NAS00314389). **Tonggu Xian:** Sandu, 650 m, 26 May 1959, *J. Xiong 03934* (LBG 00013858). **Wuning Xian:** Shimenlou, Yinlu, 1939, *Y. G. Xiong 1462* (LBG 00013855). **Wuyuan Xian:** betw. Sikou & Wuyuan, 4 Apr. 1959, *Q. H. Li & C. Chen 00271* (LBG 00013862, PE 01033093). **Xiushui Xian:** Huanglong, Dapo Ling, 900 m, 12 Aug. 1959, *J. Xiong 05898* (LBG 00013856). **Yuanzhou Qu:** Feijiantan Xiang, Feijiantan reservoir, 180 m, 4 Aug. 2014, *H. G. Ye & F. Y. Zheng LXP10-918* (IBSC 0771542). **Zixi Xian:** Shixiaxiang, 450 m, 19 May 1951, *M. J. Wang 0468* (LBG 00013846, NAS NAS00314385).

Shandong. 1935, *Y. C. Wang s.n.* (PE 01036973, WUK 004161).

Zhejiang. Chun'an Xian: Linqi Gongshe, 260 m, 26 May 1959, *M. L. She et al. 27510* (NAS NAS00314370). **Lin'an Shi:** Damingshan, 730 m, 12 Oct. 1957, *X. Y. He 26768* (HHBG HZ009027, NAS NAS00314280). **Longquan Shi:** Badu, 21 Oct. 1958, *s. coll. 22151* (HHBG HZ009020, HZ009022, NAS NAS00314289). **Taishun Xian:** near Siqian, 4 Dec. 1958, 100 m, *s. coll. 24067* (HHBG HZ008989, NAS NAS00314288). **Tiantai Xian:** Huading Linchang, 1000 m, 16 May 1993, *Z. X. Jin 93256* (HTC HTC0006090).

Specimens from Guangxi whose identification as Berberis virgetorum is doubtful:
Beiliu Shi: near Beiliu, 4 Oct. 1958, *Y. C. Chen 0442* (IBK IBK00012807). **Cangwu Xian:** Dapo, 4 Nov. 1958, *S. Q. Zhong 302061* (IBK IBK00012808, IBK00012810). **Luchuan Xian:** Liupan, 4 Oct. 1958, *Y. C. Chen 571* (IBK IBK00012806). **Pingnan Xian:** Dapo, 10 June 1959, *K. L. Yin 404611* (IBK IBK00013102). **Xingye Xian:** Xiaopingshan Gongshe, 23 May 1959, *K. Y. Yin 404254* (IBK IBK00012805).

258. Berberis wanhuashanensis Y. J. Zhang, Acta Bot. Boreal. Occid. Sin. 11: 258. 1991. TYPE: China. N Shaanxi: Yan'an Shi, Hua-yuan-tou, Wanhuashan, 1100 m, 20 May 1982, *Z. Y. Zhang 18532* (lectotype, designated here, WUK 433921 image!; isolectotype, WUK 433920 image!).

Shrubs, deciduous, to 70 cm tall; mature stems purple, terete; spines 3-fid, solitary or absent toward apex of stems, concolorous, 0.5–1.5 cm, terete. Petiole almost absent, occasionally to 3 mm; leaf blade yellowish green on both surfaces, obovate or obovate-elliptic,

1–4 × 0.8–2 cm, midvein raised abaxially, impressed adaxially, venation indistinct on both surfaces, base cuneate, margin entire or spinulose with 1 to 6 teeth on each side, apex obtuse, sometimes subacute. Inflorescence a spikelike raceme, 12- to 30-flowered, 3–5 cm overall; peduncle 0.5–1 cm; pedicel 2–4 mm. Sepals in 3 whorls; outer sepals suborbicular, 1.8 × 1.5 mm; median sepals oblong, 2–2.5 × 1.8 mm; inner sepals oblong, 3 × 2.5 mm; petals spatulate, 2.8 × 1.5 mm, glands separate, oblong, apex emarginate. Stamens 1.8 mm. Ovules 2 or 3. Berry red, ellipsoid, 5 × 3 mm; style not persistent; seeds 2 or 3.

Phenology. Berberis wanhuashanensis has been collected in flower in May and in fruit in August.

Distribution and habitat. Berberis wanhuashanensis is known from Wanhuashan, southwest of Yan'an Shi in Shaanxi. It has been collected at 1100 m. Its habitat is not recorded.

The protologue cited *Z. Y. Zhang 18532* at WUG (now WUK) as the type of *Berberis wanhuashanensis*. There are two specimens of this at WUK. The specimen with the most floral material has been chosen as the lectotype. The protologue described the number of ovules as two, but *Biological Collection Team 5070* is annotated as having two or three seeds.

Berberis wanhuashanensis was not noticed by T. S. Ying (2001).

Selected specimens.
Shaanxi. Yan'an Shi: Wanhuashan, Aug. 1941, *Biological Collection Team 5070* (PE 01033050).

259. Berberis weiningensis T. S. Ying, Acta Phytotax. Sin. 37: 326. 1999. TYPE: China. W Guizhou: Weining Xian, Yanjiaping, 19 Sep. 1929, *F. C. Tsoong 1817* (holotype, PE 00935208!).

Shrubs, evergreen, to 1 m tall; mature stem pale yellow, angled; spines 3-fid, concolorous, 0.5–1 cm, terete, slender. Petiole almost absent; leaf blade abaxially grayish green, adaxially deep green, narrowly obovate-elliptic, narrowly elliptic, or obovate, 0.4–2 × 0.2–0.5 cm, thinly leathery, midvein and lateral veins raised, reticulate veins inconspicuous, base cuneate, margin thickened, slightly revolute, entire or spinulose with 1 to 6 teeth on each side, apex acute or obtuse, aristate. Inflorescence a sub-umbel or sub-raceme, 3- to 6-flowered, 2–3 cm overall including peduncle 1.3–2.3 cm; bracts ovate, apex acuminate; pedicel 3–4 mm, slender; bracteoles triangular-ovate, 1–1.7 × 0.8–1.1 mm. Flowers golden-yellow, 4–5 mm diam. Sepals in 2 whorls; outer sepals elliptic, ca. 3.2 × 2.5 mm; inner sepals broadly obovate, ca. 3.5 × 3.2 mm; petals obovate, ca. 3.1 × 2

mm, base attenuate, glands separate, lanceolate, apex incised, lobes acute. Stamens ca. 2.5 mm; anther connective extended, rounded. Ovules 3, subsessile. Berry reddish turning purplish blue, ovate-oblong, 7–9 × 4–6 mm; style not persistent.

Phenology. Berberis weiningensis has been collected in flower between April and June and in fruit between July and October.

Distribution and habitat. Berberis weiningensis is known from Hezhang and Weining Xian in west Guizhou. It has been found in thickets and meadows on mountain slopes and summits at 1950–2500 m.

This species is different from the unpublished *Berberis weiningensis* from the *Berberis* sect. *Wallichianae* described in He et al. (1995: 646). For further details, see under *B. zhaotongensis* above.

Selected specimens.
W Guizhou. Hezhang Xian: Shilin, s.d., *S. H. Wei 0408085* (GZTM 0011411–12); Jiucaiping Shilin, 2580 m, 11 June 2011, *T. Feng 01024* (BJ); same details, but 1 July 2012, *T. Feng H 2003* (BJ). **Weining Xian:** Gali Xiang, Mabai Huoshan, 2500 m, 10 July 1959, *Bijie Team 139* (GF 09016508, HGAS 013680, PE 00935209–11); Zhong Yangchang, 2400 m, 28 Apr. 1963, *Qianxi Team 119* (HGAS 013677); Zhong Yangchang, 1950 m, 7 Sep. 1963, *Qianxi Team 63* (HGAS 013679); Yizhongshui Qu, 2200 m, 18 Mar. 1983, *Y. K. Li 11272* (HGAS 013681); Caohaiyang, Guanshan, 2150 m, 7 Oct. 1984, *Weining Team 508* (HGAS 013682); Xianglishan, 2100 m, 16 May 1985, *S. Z. He et al. s.n.* (GZTM, PE 01037168); Yancang, 8 June 2005, *C. H. Yang 4636* (GF 09003484); Mazha Xiang, 22 June 2009, *s. coll. w0910003* (GUES 001398). **S. loc.:** betw. 1 May & 31 July 1934, *H. C. Zhou 996* (HIB 0129620).

260. Berberis wenchuanensis Harber, sp. nov. TYPE: China, NC Sichuan: [Wenchuan Xian], summit Niu-tou shan, W of Kuan Hsien (Guan Xian, now Dujiangyan Shi), 3050 m, 20 June 1908, *E. H. Wilson 2857* (holotype, A 00279860; isotypes, BM BM000810083!, E E006125831!, K!, LE!, US 00945943 image!).

Diagnosis. Berberis wenchuanensis has leaves similar to *B. silva-taroucana*, but has a different inflorescence and flower structure.

Shrubs, deciduous, to 3 m tall; mature stems dark reddish or purplish brown, sulcate; spines solitary or absent, pale brownish yellow, 0.2–0.3 cm, abaxially sulcate. Petiole to 10 mm, sometimes almost absent; leaf blade abaxially pale green, papillose, adaxially green, obovate, sometimes obovate-elliptic 1.3–3.5 × 0.6–1.8 cm, papery, midvein, lateral veins, and reticulation raised abaxially, lateral veins inconspicuous adaxially, base narrowly attenuate, margin entire, rarely spinulose with 4 to 16 teeth on each side, apex obtuse or rounded, sometimes mucronate. Inflorescence a loose raceme, 8- to 13-flowered, to 5.5 cm overall; peduncle to 1 cm; pedicel (5–)12–22 mm. Sepals in 2 whorls; outer sepals oblong-lanceolate, 3.75 × 2.2 mm; inner sepals oblong-obovate, 5 × 3 mm; petals obovate, 3 × 2 mm, base clawed, glands widely separate, ca. 0.5 mm, apex entire. Stamens ca. 2.5 mm; anther connective not extended, truncate. Pistil ca. 3 mm; ovules 2, sessile. Berry red, ellipsoid, 8–10 × 6 mm; style not persistent or persistent and short.

Phenology. *Berberis wenchuanensis* has been collected in flower in June and in fruit from July to September.

Distribution and habitat. *Berberis wenchuanensis* is known from Danba, Wenchuan, and Xiaojin Xian in west and north-central Sichuan. The type was collected on a mountain summit at 3050 m, *Wang 21161* was collected from a thicket at 2200 m, and *Boufford et al. 38018* in an open, disturbed area at 2800–2900 m.

IUCN Red List category. *Berberis wenchuanensis* is assessed as DD or Data Deficient, according to IUCN (2001) criteria.

In the protologue of *Berberis silva-taroucana*, Schneider (1913: 370–371) listed both *Wilson 2857* and *Wilson 1012* as examples of that species and, in the case of *Wilson 2857*, this was repeated by Ahrendt (1961: 169). However, these have a different inflorescence and flower structure from the type of *B. silva-taroucana*, *Wilson 2860*. This was somewhat obscured by the protologue which described the inflorescence of *B. silva-taroucana* as a lax raceme which, while it is an accurate description of *Wilson 2857* and of *Wilson 1012*, it is not of *2860* which is largely fascicled, sub-fascicled, or sub-umbellate. *Wilson 1012* consists of two gatherings. That of July 1908 is designated here as *Wilson 1012* "A," and that of September 1908 as *Wilson 1012* "B."

It should noted that there are specimens from "West of Wenchuan Hsien" with pale yellowish brown mature stems and somewhat shorter inflorescences which are mostly sub-umbellate rather than racemose, e.g., *F. T. Wang 20987*, 2900 m, May 1930 (A 00279865, KUN 0177094); *F. T. Wang 21130*, 2600 m, 2 June 1930 (A 00279866, KUN 0177095, NAS NAS00314207, PE 01033017), and *R. C. Ching 3370*, 2500 m, September 1931 (LBG 00064348, NAS NAS00314208, 00314214, PE 01033021). All three of these collections have been identified as *Berberis silva-taroucana*, but are not this species. The color of their mature stems suggests they are not *B. wenchuanensis* either. *Wang 21130* has floral

material, but I have not had an opportunity to investigate this. *E. H. Wilson 4288* (A 00279870), "Pan-lan-shan, W of Kuan Hsien," 2440–2745 m, October 1910, also cited by Schneider (1913) as an example of *B. silva-taroucana*, may belong here. Further investigation of all of these is clearly needed.

Selected specimens.
W & NC Sichuan. "West of Wen-chuan Hsien," 2200 m, 5 June 1930, *F. T. Wang 21161* (A 00279971, KUN 0177096, PE 01032964, 01033019–20, WUK 0045140). **Danba Xian:** on hwy. 303, 30.611111°N, 101.756111°E, 2800–2900 m, 24 July 2007, *D. E. Boufford et al. 38018* (A, E). **Wenchuan Xian:** "Wa-ssu Country," 2745 m, July 1908, *E. H. Wilson 1012* "A" (A 00279855, US 00945940); same details, but Sep. 1908, *E. H. Wilson 1012* "B" (A 00279855, E E00612581, HBG HBG-506686, LE, US 00945940). **Xiaojin Xian:** Er Qu, Huaniu Gou, 3200 m, 29 Aug. 1958, *X. S. Zhang & Y. X. Ren 6923* (CBDI 0027997, 0028179–80, PE 01033040).

261. Berberis wilsoniae Hemsl., Bull. Misc. Inform. Kew 1906: 151. 1906, as "*wilsonae.*" TYPE: China. "Western China," [Sichuan]: 600–1800 m, July 1903, *E. H. Wilson (Veitch) 3147* (lectotype, designated here, A 00038823!; isolectotypes, BM BM01015578!, HBG HBG-506682 image!, K [3 sheets] 000742002–4!, P P00716545!).

Berberis parvifolia Sprague, Bull. Misc. Inform. Kew 1908, 445. 1908, not Lindl. (1847). *Berberis wilsoniae* var. *parvifolia* (Sprague) Ahrendt, J. Linn. Soc., Bot. 57: 215. 1961. TYPE: China. [Sichuan, 1903 or 1904], *E. H. Wilson (Veitch) 4154a* "A" (lectotype, designated here, K K000077370!; isolectotype, K K000077371!).
Berberis subcaulialata C. K. Schneid., Repert. Spec. Nov. Regni Veg. 6: 267. 1909. *Berberis wilsoniae* var. *subcaulialata* (C. K. Schneid.) C. K. Schneid., Oesterr. Bot. Z. 67: 298. 1918. TYPE: Cultivated. [Czech Republic]. Pruhonitz, "e horto Les Barres, M. de Vilmorin," Sep. 1913, *s. coll. 292* (neotype, designated here, W 1913–14692).
Berberis bodinieri H. Lév., Repert. Spec. Nov. Regni Veg. 9: 454. 1911. TYPE: China. C Yunnan: [Kunming], "environs de Yun-nan-sen," 17 Nov. 1896, *E. M. Bodinier 4* (holotype, E E00259012!; isotypes, A, KUN, P [3 sheets] P06865535–7!, PE [2 sheets] 01969514–15!).
Berberis coryi Veitch, Gard. Chron., ser. 3, 52: 321. 1912, syn. nov. TYPE: Cultivated. From seed collected by E. H. Wilson in Sichuan, no original material found.
Berberis stapfiana C. K. Schneid., Bull. Misc. Inform. Kew 1912, 35. 1912. *Berberis wilsoniae* var. *stapfiana* (C. K. Schneid.) C. K. Schneid., Oesterr. Bot. Z. 57: 298. 1918, syn. nov. TYPE: Cultivated. Royal Botanic Gardens, Kew, U.K., 16 Oct. 1911 from 510–1905 m, *Vilmorin 4039* received from Les Barres Botanic Garden, France, grown from seeds sent from W China (neotype, designated here, K K000568007!).
Berberis favosa W. W. Sm., Notes Roy. Bot. Gard. Edinburgh 11: 200. 1920. *Berberis wilsoniae* var. *favosa* (W. W. Sm.) Ahrendt, J. Linn. Soc., Bot. 57: 215. 1961, syn. nov. TYPE: Myanmar (Burma). Border with Yunnan, Hpimaw, 2100 m, 2 Aug. 1914, *F. Kingdon-Ward 1852* (holotype, E E00267779!).

Berberis subcaulialata C. K. Schneid. var. *guhtzunica* Ahrendt, J. Bot. 79(Suppl.): 76. 1942. *Berberis wilsoniae* var. *guhtzunica* (Ahrendt) Ahrendt, J. Linn. Soc., Bot. 57: 216. 1961, syn. nov. TYPE: China. SW Sichuan: Muli Xian, Guhtzun, 3100 m, 5 Dec. 1937, *T. T. Yu 14840* (holotype, E E00259013!; isotypes, A 00274774!, BM BM001015560!, KUN 1204872!).

Berberis heteropsis Ahrendt, J. Linn. Soc., Bot. 57: 213. 1961, syn. nov. TYPE: China. SC Guizhou: Longli, 15 Nov. 1907, *J. Cavalerie 3042* (holotype, K K000077375!; isotypes, E [2 sheets] E00217942–3!).

Shrubs, evergreen, to 1.5 m tall; mature stems reddish brown, sulcate, scarcely black verruculose; spines 3-fid, concolorous or slightly lighter, 1–2 cm, terete, slender. Petiole almost absent; leaf blade abaxially pale green or bluish green, papillose, sometimes glaucous or white pruinose, adaxially dull gray-green or bluish green, obovate, obovate-spatulate, or oblanceolate, 0.6–2.5 × 0.2–0.6 cm, thinly leathery, midvein and dense reticulation inconspicuous on both sides, conspicuous when dry, margin entire, occasionally spinulose with 1 or 2 teeth on each side, apex obtuse or subacute. Inflorescence a fascicle, sub-fascicle, stalked corymb, sub-raceme, raceme, umbel, sub-umbel, or very rarely a panicle, 4- to 7(to 15)-flowered, 0.3–1.5(4) cm overall; pedicel brownish, 3–7(–11) mm; bracteoles ovate. Sepals in 2 whorls; outer sepals ovate, (2–)3–4 × (1.3–)2–3 mm; inner sepals obovate-orbicular or obovate, (3.5–)5–5.5 × (2.5–)3.5–4 mm; petals obovate, 3–4 × ca. 2 mm, apex emarginate with acute lobes. Stamens 2–3 mm; anther connective slightly extended, obtuse. Ovules 3 to 5. Berry salmon-red, translucent, slightly pruinose, globose or subglobose, 6–7 × 4–5 mm; style persistent.

Phenology. *Berberis wilsoniae* has been collected in flower from June to September, and in fruit from October to February of the following year.

Distribution and habitat. *Berberis wilsoniae* is known from Guizhou, Sichuan, southeast Xizang, and Yunnan, and Myanmar (Burma). It has been found in thickets, on river banks and beaches, stream and trailsides, dry slopes, rock ledges, and in crevices particularly in limestone areas at ca. 1000–3730 m.

The protologue of *Berberis wilsoniae* cited *Wilson 3154* and *3147* but without type designation or herbarium. Schneider (1918: 297) designated *Wilson 3147* as the type, but without citing a herbarium. Schneider's lectotypification has been completed by choosing the specimen at A because that is where Schneider was then based. This specimen is dated July 1903, as are all the duplicates except K K000742002, which as well as this date, is also annotated in Wilson's hand as 26/6/03. This specimen is ex HK and it is possible this

was acquired by HK at the end of Wilson's 1903–1905 expedition. The lectotype at A has only floral material whereas the sheets of the isolectotype have fruiting as well as floral material. In the case of K K000742002, these are only immature, but otherwise they are mature, suggesting these fruit specimens are actually from a later gathering. Ahrendt (1961: 214–215) cited only *Wilson (Veitch) 3154* (K) as the type of *B. wilsoniae*. There are three specimens of this at K which should be regarded as syntypes.

The protologue of *Berberis coryi* gave the following description of a plant submitted in 1912 to the U.K.'s Royal Horticultural Society, Floral Committee, by Messrs. James Veitch & Sons: "has round, glaucous currant-red berries, very densely borne round the stem and appears to be evergreen, bearing clusters of spoon shaped leaves, which are glaucous below." The taxon reappeared in a subsequent catalogue of Veitch and Sons (1913: 7) entitled "New Hardy Plants from Western China (Introduced through Mr. E. H. Wilson)" which would indicate that the seeds of *B. coryi* had been collected by Wilson. No material from such a plant has been found. Schneider (1918: 298) treated the taxon as a synonym of *B. wilsoniae* var. *subcaulialata*, but Ahrendt (1961: 207–208) maintained it as a separate species, citing *Forrest 18516* at K (details below) as the type. At most, *Forrest 18516* could only be regarded as a neotype, but given that it was collected in 1919 from northwest Yunnan, an area never visited by Wilson, it would seem unproductive to give it such a status. The protologue description of *B. coryi* would seem sufficient to identify it as *B. wilsoniae*.

Schneider's protologue of *Berberis subcaulialata* cited only cultivated material sent to him by Vilmorin of Les Barres, France, the material having been grown from seeds collected in Tibet (which then was held to include parts of what are now northwest Yunnan and western Sichuan). Later, Schneider (1918: 298), while renaming the taxon a variety of *B. wilsoniae*, stated that the seed of the Vilmorin plant was collected by Soulié, probably in 1894 on a journey between Tachien lu (Kangding) and Tseku (Dêqên Xian). Schneider cited no herbarium for the taxon, and the specimen from a plant at Pruhonitz has been designated as the neotype, since not only do the collection details identify it as *B. subcaulialata* and record it as being from a Vilmorin plant, but in the period before he left for China in 1913, Schneider, as General-Secretary of the Austro-Hungarian Dendrological Society, was a frequent visitor to Pruhonitz, the seat of its President, Count Silva-Tarouca (see Kriechbaum, 1951).

The protologue of *Berberis stapfiana* cited three cultivated plants grown at Kew: one, *Vilmorin 4039* from Les Barres, France; one from seeds of *Wilson 1284* supplied by the Arnold Arboretum (this being a mix-up of some sort since the wild-collected *Wilson 1284* is the type of the section *Wallichianae Berberis atrocarpa*); and one supplied by Veitch & Sons grown from seeds of *Wilson 1560* (from the evidence of the Wilson manuscripts, these seeds were collected at Tachien Lu [Kangding]). It is possible that *Vilmorin 4039* was grown from the same seeds as the plant which was the source of the material that Vilmorin sent to Schneider and which he had named as *B. subcauliata*, but this cannot be proved. It is more than possible that the Veitch plant was the one advertised by them as *B. coryi*, but again this is unprovable. Ahrendt (1961: 216) cited *Vilmorin 4039* at K as the type. Since the protologue of *B. stapfiana* states "the above description was drawn up by Dr Schneider from a specimen of *Vilmorin 4039*, sent to him from Kew" and there is no evidence that Schneider returned it, any specimen at K could only be regarded as a neotype. There are, in fact, two of these, both labelled as "co-types." Ahrendt's neotypification has been completed by selecting the one with a specific date of collection.

The protologue of *Berberis parvifolia* cited *Wilson 3154a* at K. There are two specimens of this at K. The specimen that has both flowers and immature fruit has been chosen as the lectotype. Though neither specimen is dated, from the numbering it appears the collection was made in Sichuan in 1903 or 1904.

As can be seen from the synonyms listed above and their various manifestations, there have been significant taxonomic disagreements in relation to *Berberis wilsoniae*. There appear to be three interrelated reasons for this. First, *B. wilsoniae* has the widest distribution of any *Berberis* species in southwestern China, and early specimens and seeds were collected from different areas at different times and were named by different authors. Secondly, the attractiveness of the plant (at least outside China) as an ornamental led those with a horticultural interest, including Schneider and Ahrendt, to pay particular attention to cultivated plants. And thirdly is the considerable variation, particularly in flower stalk, but also in height, leaf shape, size, and color.

Because there are many specimens of *Berberis wilsoniae* available (A, CDBI, KUN, PE, and SZ alone have well over 500 between them), it might be expected that it would be possible to correlate these variations with particular geographical areas, but I have been unable to do this except in the broadest of terms. This would seem to show that stalked corymbose, fascicled, or sub-fascicled specimens (including the type and syntype of *B. wilsoniae*) predominate in Sichuan, northwest and western Yunnan, and southeast Xizang, whereas more diverse inflorescences (racemose, sub-racemose, semi-umbellate, or umbellate) predominate in central and northeast Yunnan, Guizhou, and Nanchuan in Chong-

qing (southeast Sichuan). Smaller-leaved plants are mostly found in the first of these areas, while leaves from the second area are sometimes wider than those in the first and are often paler and sometimes abaxially pruinose. This latter characteristic is particularly evident in Guizhou where leaves can be abaxially densely chalky-white. In all of this, however, there are exceptions with racemose and sub-racemose inflorescences sometimes being found in the first area and fascicled and sub-fascicled inflorescences sometimes being found in the second.

It is in this light that all the various taxa from China above are cited as synonyms. Their types differ from the types of *Berberis wilsoniae* in some small variation in flower and/or fruit size and, in some cases, their leaf size. With the types of variety *guhtzunica* and with *B. bodinieri* and *B. heteropsis*, there also is some variation in the flower or fruit stalks which, as well as being sub-fascicled, are also partially shortly sub-racemose, and in the case of variety *guhtzunica*, also partially sub-umbellate, but these particular variations only partially reflect the full range of inflorescence evidenced by the totality of the specimens cited below. Just to give two examples, *Boufford et al. 34872* from Anning Xian in central Yunnan is partially paniculate while *Wang 77236* from Pingba Xian in Guizhou is semi-umbellate with peduncles up to 18 mm long.

The type of variety *favosa* from the Myanmar border with Yunnan is described in the collector's notes as semi-prostrate and less than 30 cm high. *Kingdon-Ward 3661* and *Farrer 1738* from the same area are reported as being of similar height. Population studies are needed to establish whether this is sustainable as a separate variety. But it should be noted that dwarf forms are also recorded from northeast Yunnan; for example, *Southeast Yunnan Team 1099* from Zhenxiong is recorded as being only 10–20 cm high. And this height is confirmed from a plant in my own collection grown from *A. Clark 1014* from northeast Yunnan (details below). Both, however, differ from the Myanmar specimens, which have narrowly oblanceolate, abaxially epruinose leaves, and in having leaves that are obovate and abaxially partially pruinose.

Ying (2001: 105–106) synonymized varieties *parvifolia*, *stapfiana*, and *subcaulialata* while maintaining variety *guhtzunica* as a separate taxon on the basis that it was racemose, whereas *B. wilsoniae* var. *wilsoniae* was described as being fascicled. As can be seen from the discussion above, this simple subdivision is unsustainable.

Schneider (1908: 202) identified *Bodinier 4* as *Berberis wilsoniae*, but Léveillé appears not to have been aware of this. Later, (1918: 297) Schneider treated *B. bodinieri* as a possible synonym of *B. wilsoniae*. *Berberis bodinieri* was not noticed by Ahrendt (1961) or

Ying (2001), and the latter also did not notice *B. het-eropsis*. X. H. Li (2017), who did not notice Schneider's treatment of the taxon, synonymized *B. bodinieri* as *B. wilsoniae* var. *guhtzunica*.

Ying (2001: 211) maintained *Berberis coryi* as a separate species from Yunnan and followed Ahrendt (1961: 207–208) in describing the inflorescence as "a panicle, 7- to 20-flowered, 3–5 cm, sometimes reduced to a short raceme or sub-fascicle." But this description (which appears to be based on cultivated plants of un-specified origin in Ahrendt's living collection) is not only different from that of the protologue of *B. coryi*, but is not even evidenced by specimens of *Forrest 18516* claimed by Ahrendt as the type (see above) which are fascicled.

X. H. Li (2010: 440) treated *Berberis heteropsis* from Guizhou as a synonym of *B. lepidifolia*, whose type is from northwest Yunnan. Given that the latter has abaxially epruinose leaves and black fruit whereas the leaves of the type gathering of *B. heteropsis* are abaxi-ally pruinose and have salmon-red fruit, this synonymy is unsustainable.

The predominance of racemose inflorescences in Guizhou and the Kunming area of Yunnan has led to many specimens of *Berberis wilsoniae* from there being mistakenly identified as *B. aggregata*, a deciduous species found in north Sichuan, Gansu, Qinghai, and Shaanxi, which has similar fruit to *B. wilsoniae* or *B. aggregata* var. *integrifolia* (treated here as a synonym of *B. prattii*). This latter misidentification appears to have originated with Ahrendt (1961: 204), who so mis-identified *Maire 1997* from Kunming cited below (a further specimen from Kunming, *Schoch 8*, also cited below, was misidentified by Ahrendt [1961: 154] as *B. amoena*).

Many of the specimens cited below were collected in dry limestone areas. This does not seem to be because *Berberis wilsoniae* is an exclusively limestone-loving species, since in cultivation it grows well on any soil that is not water-logged. Rather it would seem likely that, in the wild, it thrives in dry limestone areas be-cause here it has fewer competitors than elsewhere.

The reports of *Berberis wilsoniae* in Shaanxi by Insti-tuto Botanico Boreali-Occidentali Academiae Sinica (1974: 312), in Qinghai by Zhou (1997: 378), and in Gansu by Ying (2001: 105–106), are likely to be mostly referable to *B. aggregata*. The reports of *B. wilsoniae* in Hubei by Fu (2001: 384) and in Hunan by L. H. Liu (2000: 712) are likely to be referable to *B. brevi-paniculata*.

It should be noted in relation to *Wilson 1356* from Washan, Ebian Xian, Sichuan, cited below, that I have designated it *Wilson 1356* "A." This is because, as noted by Schneider (1913: 368–369), there are four other Wilson collections of *Berberis wilsoniae* numbered *1356*

from other areas of Sichuan: two from Mongkong Ting in Baoxing Xian, and two from Tachienlu (Kangding).

For *Berberis wilsoniae* var. *latior* Ahrendt (1961: 216), see under Excluded Taxa.

Berberis wilsoniae, which was named after Wilson's wife, is frequently misspelled as *B. wilsonae* and, more frequently, as *B. wilsonii*.

Selected specimens.

Syntype. *B. wilsoniae*. "Western China," 1645 m, Aug. 1903, *E. H. Wilson (Veitch) 3154* (A 00038824, BM BM01015577, HBG HBG-506681, IBSC 0000592, K [3 sheets] 000077372–4, P P00716546).

Other specimens.

Chongqing (SE Sichuan). Nanchuan Xian: Xiaohe Xiang, Sancha Cun, Chengqiang Yan, 1700 m, 21 July 1957, *J. H. Xiong & Z. L. Zhou 92189* (PE 01036925, SZ 00289542); Jinfoshan, Leidaping, 1700 m, 12 Aug. 1978, *Liu 784175* (PE 01036924); Jinfoshan, 1750 m, 13 Sep. 1978, *Liu 784192* (PE 01036923).

Guizhou. Dafang Xian: Lihua Xiang, 1480 m, 29 July 2012, *DF 003-04* (BJ). **Guiyang Shi:** Huaxi, 1150 m, 2 Nov. 1956, *Sichuan-Guizhou Team 2319* (PE 01036905, SZ 00289540); Huishi, Gaopo, 28 Sep. 2000, *S. Z. He 20242* (GZTM 0011511–12, 0011514); Wudang, 22 June 2004, *S. H. Wei 2142* (GZTM). **Hezhang Xian:** Shuitang Linchang, 2150 m, 10 May 1984, *R. B. Jiang 537* (IBSC 0092196). **Jinsha Xian:** 17 Aug. 1985, *C. B. Pu 6-91* (GZTM). **Leishan Xian:** Leigongshan, 31 May 1987, *K. Y. Lan 870405* (GZAC GZAC0029994). **Longli Xian:** 10 June 2003, *S. C. 105091* (GZTM 0011508–9). **Pan Xian:** 28 Aug. 1929, *B. Q. Zhong 1725* (PE 01036902). **Pingba Xian:** Longjing Fruit Farm, 14 July 1978, *Y. Wang 77236* (PE 01036906). **Qianxi Xian:** 1410 m, 27 Sep. 1986, *Z. W. Wang 396* (GZTM 0011480–81). **Shuicheng Shi:** outside city, 1500 m, 20 July 1964, *Shuicheng Team 507* (HGAS 013601). **Weining Xian:** out-side city, 2160 m, 26 Apr. 1964, *Weining Team 94* (HGAS 013671). **Zhenning Xian:** 1370 m, 21 Jan. 1987, *C. B. Pu 214* (GZTM).

Sichuan. Baoxing Xian: Shaoqi, 2500 m, 13 July 1933, *T. H. Tu 4411* (PE 01030823). **Barkam (Ma'erkang) Xian:** Lieshi Mu, 2600 m, 14 Oct. 1957, *X. Li & J. X. Zhou 72509* (KUN 0177526, PE 01030807, SZ 00287693); A'mu, Jiao Gou, 2600 m, 18 July 1957, *Z. R. Zhang & H. F. Zhou 23064* (KUN 0177117, PE 01036919, SZ 00289558). **Beichuan Xian:** Guanrong Dadui, 1200 m, 28 Aug. 1984, *C. L. Tang et al. 503* (CDBI CDBI0027915–16). **Danba Xian:** 1 km W of Maoniu, 2855 m, 20 July 1959, *Nanshuibeidiao Exp.- S. Jiang et al. 02180* (PE 01030850). **Daocheng Xian:** Riwa Gong-she, Gayong, 3200 m, 24 Aug. 1981, *Qinghai-Xizang Team 4227* (CDBI CDBI0027958, HITBC 003608, KUN 0175662–63, PE 01036917). **Dawu (Daofu) Xian:** Leishi, Lingyuan, 3300 m, 29 June 1974, *Y. T. Wu & B. C. Gao 111651* (CDBI CDBI0027925, PE 01030860, SZ 00289521). **Dechang Xian:** Luoji Shan, Xumuchang, Ningmu Gou, 2650 m, 3 July 1976, *Southwest Normal University Department of Biology 11865* (CDBI CDBI0027890,. CDBI0027935, CDBI0027945, PE 01036930). **Dêrong Xian:** Dingqu He 2600 m, 3 Aug. 1981, *Qinghai-Xizang Team 3160* (CDBI CDBI0027909, CDBI0027911, HITBC 003604, KUN 0175660–61, PE 01030809–10). **[Ebian Xian]:** Wa-shan, June 1908, 1830 m, *E. H. Wilson 1356* "A" (A 00280034, E E00612639, HBG-506750–51, US 00946028, 00946031). **Emeishan Shi:** Omei Hsien (Emei Xian), Mt. Omei, 2 July 1940, *W. P. Fang 15160*

(IBSC 0092542, PE 01036915); betw. Jinjing & Longchi, near Sanpan, 8 July 1961, *X. L. Jiang & X. S. Zhang 31573* (IBK IBK00012790, PE 01036916). **Ganluo Xian:** Haitang, 2000 m, 19 July 1959, *s. coll. 4016* (PE 01031012). **Hanyuan Xian:** 19 Apr. 1930, *W. C. Cheng 684* (LBG 00064200, 00064204, N 093057106, NAS NAS00314407, P P02313565, PE 01030842). **Heishui Xian:** WNW of the city of Heishui and NW of Shashiduo-Xiang on side rd. off hwy. S302 near Sibugong, 32.10472222°N, 102.8025°E, 2800 m, 5 Aug. 2018, *D. E. Boufford, B. Bartholomew, J. F. Harber, Q. Li & J. P Yue 44437* (A, CAS, E, KUN, PE). **Hongya Xian:** betw. Zhangcun & Gaomiao, 8 Aug. 1938, *Z. W. Yao 2702* (KUN 0177532, PE 01030857). **Hsi-chang Hsien (Xichang Xian):** 2200 m, 1 Aug. 1932, *T. T. Yu 1218* (A 00280164, CQNM 0005184, PE 01030844–46). **Huidong Xian:** Zizhi Qu, Lingpangqing, 14 June 1959, 3000 m, *S. G. Wu 655* (CDBI CDBI0027972, KUN 0177427, SZ 00287720). **Huili Xian:** Bazi 1800 m, 25 June 1965, *Woody Oil Plant Resource Exp. 226* (KUN 0175638, 0177446). **Jinchuan Xian:** near Kasa Gou, 1920 m, 29 Sep. 1957, *X. Li 76534* (IBSC 0092199, KUN 0177426, PE 01030803–4, SZ 00289575). **Kangding Xian:** Yala Xiang, Jiajin Gou, 3050 m, 3 Oct. 1953, *X. L. Jiang* [or *W. G. Hu*] *37035* (IBK IBK00012796, PE 01031025). **Leibo Xian:** Huangrang Qu, Duimen Shan, 2000 m, 9 Aug. 1976, *Sichuan Botany Team 13430* (CDBI CDBI0027884, 0027905, 0027924, PE 01030878). **Li Xian:** E of Lifan Hsien, 1900 m, 10 July 1930, *F. T. Wang 21702* (A 00280131, KUN 0177431, LBG 00064053, NAS NAS00314406, PE 01031017, 01031019, WUK 0045139). **Luding Xian:** Moxi, Gongga Shan, E side, downstream from Hailuo Guo glacier, 29.6033333°N, 102.0783333°E, 1780–2035 m, 17 Aug. 1997, *D. E. Boufford et al. 27246* (A 00280015, E E00324027, MO 5771206). **Lu-shan-hsien (Lushan Xian):** Chinglongchang, 1000 m, 16 Oct. 1936, *K. L. Chu (G. L. Qu) 4006* (A 00280044, BM, E E0021794, IBSC 0092442, 0092492, 0092645, P P02465474, PE 01036878–79). **Mao Xian:** Ciliu Gou, 1600 m, 8 June 1959, *Nanshuibeidiao Exp. (S. Jiang et al.) 00593* (KUN 0178463, 0179253, PE 01030872). **Meigu Xian:** outskirts of Meigu, 2000 m, 16 July 1976, *Sichuan Botany Team 12968* (CDBI CDBI0027859, CDBI0027957–58, PE 01036933). **Mianning Xian:** Tuowo Xiang, 2220 m, 15 Aug. 2007, *H. Y. He & X. H. Miao 070815037-02* (Southwest Jiaotong University Herbarium). **Miyi Xian:** Puwei Xiang, 1700 m, 24 July 1986, *Y. J. Li 062* (CDBI CDBI0027864, CDBI0027944). **Muli Xian:** 2440–2745 m, 21 Aug. 1911, *F. Kingdon-Ward 4767* (E E00612645); Zhang'aobaidong Xiang, 2150–2800 m, 24 Sep. 1959, *S. G. Wu 3216* (KUN 0177438–39); Er Qu, Xiaojin He, 2100 m, 15 Aug. 1983, *Qinghai-Xizang Team 12923* (KUN 0179002, 0179042, PE 01036926–27). **Ningnan Xian:** Paoma, San Dadui, 2500 m, 12 July 1978, *Ningnan Team 0327* (SM SM704800443). **Puge Xian:** near Tuomu Gou, 1800 m, 27 Aug. 1959, *Puge Xian Survey Team 5559* (KUN 0177415, PE 01030875–76, SM SM704800462). **Shimian Xian:** Jinwo, 2200 m, 11 Aug. 1981, *G. H. Xu 25564* (CDBI CDBI0027830–31). **Tianquan Xian:** Huang Cun, Erlang Shan, 1900 m, 28 July 1953, *X. L. Jiang 35092* (IBK IBK00012793, IBSC 0092487, PE 01030815). **Wenchuan Xian:** near Wenchuan City, 1800 m, 28 July 1975, *Sichuan Botany Team 8486* (CDBI CDBI0027899–900, 0027937). **Xiangcheng Xian:** San Qu, Dongsong Xiang, 2700 m, 18 July 1973, *Sichuan Plant Investigation Team 2953* (CDBI CDBI0027839–40, PE 01036912, SWCTU 00006332). **Xiaojin Xian:** near Sha'nong Gongshe, 2400 m, 12 Aug. 1975, *Sichuan Botany Team 9706* (CDBI CDBI0027895, 0027936, IBSC 0092202, PE 01036929, SWCTU 00006330). **Yanbian Xian:** on the way from Yanyuan to Panzhihua Shi, 6.5 km S of the border of Yanyuan Xian and Yanbian Xian, near km marker

598, 27.159778°N, 101.272861°E, 2756 m, 6 Aug. 2007, *Tibet-McArthur 1986* (US 00972176). **Yanyuan Xian:** Daoguabie Qu, Huolushan, 2600 m, 27 July 1983, *Qinghai-Xizang Team 12528* (KUN 0179000, 0179034–35, 0179152, PE 01037063–64). **Yingjing Xian:** Sanhe Gongshe, Lianshan Cave, 2000–2500 m, 7 July 1978, *Yingjing Team 78-0192* (SM SM704800486). **Yuexi Xian:** Luoji Shan, Xumuchang, Ningmu Gou, 2650 m, 3 July 1976, *s. coll. 11865* (PE 01036930). **Zhaojue Xian:** "In montium Daliang-schan (territorii Lolo) ad orientem urbis Ningyüen . . . prope vicum Lemoka," 1900–2270 m, 23 Apr. 1914, *H. R. E. Handel-Mazzetti 1569* (E E00612640, GZU 000294024, P P0231567, WU 0039273).

SE Xizang (Tibet). Markam Xian: Zhubalong, 2540 m, 31 Aug. 1981, *Qinghai-Xizang Team 005231* (CDBI CDBI0027854, 0027866, HITBC 003578, KUN 0177469–70, PE 01031071–72). **Zayü (Chayu) Xian:** Dzer-nar, Tsawa-rung, Aug. 1935, *C. W. Wang 65429* (A 00280152, IBSC 0092214, KUN 0177531, LBG 00064298, NAS NAS00314439, PE 01036910).

C, E & NE Yunnan. Anning Xian: vic. of Shinan Forestry College, 1850–2000 m, 24.933333°N, 102.483333°E, 29 July 1984, *Sino-Amer. Bot. Exped. 1428* (A 00280134, B, KUN 0175632–33, L L.0831889, US 00945890); S side Bijia Shan, 24.993611°N, 102.458889°E, 2070–2150 m, 20 Aug. 2005, *D. E. Boufford et al. 15134872* (A 00268147, HAST 115068, P P02797885). **Chengjiang Xian:** Lianwang Shan, 2800 m, 27 Mar. 1991, *s. coll. 4461* (KUN 0177508–09). **Fumin Xian:** Yongding Gongshe, Laoqing Shan, 2200–2500 m, 19 Oct. 1964, *B. Y. Qiu 596108* (KUN 0177390, 0177506, PE 01036940). **Fuyuan Xian:** Houlang Qu, Heishan, 2300 m, 26 June 1987, *Spice Plant Investigation Team 870164* (KUN 0175664–65). **Kunming:** "Yunnan-sen," s.d., *E. E. Maire 1997* (E E00270088, K); "Yunnan-fu," 18 May 1916, *O. Schoch 8* (A 00280026, K, US 00946030, WU 0039275); Xishan, Sanqing Ge, 2100–2200 m, 28 Aug. 1953, *B. Y. Qiu 60091* (KUN 0175649–52). **Luoping Xian:** Dabailashan, 1900–2000 m, 3 June 1989, *Hongshui He Exped. 1994* (KUN 0179217–18). **Luquan Xian:** Er Qu, E'mao Xiang, E'mao Cun, Jiashu Shan, 2500 m, 29 Oct. 1952, *P. Y. Mao 1493* (HIB 0129632, IBK IBK00357953, IBSC 0092191, KUN 0177396–97, PE 01031041). **Qiaojia Xian:** Yanmai Qu, betw. Dacun Gongshe & Xiaohe, Lujia Gou, 1700 m, 27 Oct. 1974, *Southeast Yunnan Team 1450* (KUN 0177493). **Qujing Xian:** 24.828889°N, 104.012222°E, 2040 m, 15 Aug. 2006, *C. H. Chen et al. 07469* (TNM S122087). **Songming Xian:** Longtannao, near Dajian Shan, 2500–2700 m, 22 Mar. 1964, *B. Y. Qiu 58756* (KUN 0177391, PE 01036936). **Xundian Xian:** Le-lang, 2700 m, 19 Nov. 1940, *Y. P. Chang 1000* (IBSC 0092195, KUN 0177434–35). **Zhanyi Qu:** Zhujiangyuan Fengjingqu, 2003 m, 15 Aug. 2015, *E. D. Liu et al. 4532* (KUN 1263480). **Zhaotong Diqu:** Yandui Shan, 5 Oct. 1958, *X. Y. Li 253* (KUN 0175636, 0175659, LBG 00064195). **Zhenxiong Xian:** Niuchang, Huashan, 1870 m, 28 Sep. 1972, *Southeast Yunnan Team 1099* (KUN 0175648, 0177492, YUKU 02065656).

NW & W Yunnan. N'Marka–Salween divide, 26.333°N, 2750–3050 m, Sep. 1919, *G. Forrest 18516* (A 00280143, E E00117394, K K000644945, WSY 0057682). **Dêqên (Deqin) Xian:** Bei-ma Shan, 28.3°N, 99.166667°E, 3350–3660 m, Oct. 1921, *G. Forrest 20929* (E E00612657); A-tun-tze, 2700 m, Sep. 1935, *C. W. Wang 70367* (A 00280157, IBSC 0092215, KUN 0177496, LBG 00064303, PE 01031048–49, WUK 0045814); Benzelan Gongshe, Yin Gu, 2100 m, 1 July 1981, *Qinghai Xizang Team 2121* (CDBI CDBI0027886–87, HITBC 003591, KUN 0177467–68, PE 01037044, 01037048); rd. from Benzelan to Dêqên, 28.260278°N, 99.241944°E, 2800 m, 24 Sep. 1994, *Alpine Garden Society China Exped. (ACE) 1247* (E E00039601, LIV 2005.15.1076, WSY 0057767). **Gongshan Xian:** Der-la, Cham-pu-tung, 2300 m, Oct. 1935, *C. W. Wang 66825* (A 00280151, IBSC 0091634, KUN 0177494, LBG 00064293, NAS NAS00313883, PE 01036945–46, WUK 0036496); Kiukiang valley (Taron), Mt. Chingutin-glaka, 1900 m, 28 July 1938, *T. T. Yu 19499* (E E00117393, KUN 0176762, PE 01036939, 010369341); Ne Wa Long River valley, 28.04°N, 98.579167°E, 1800–2000 m, 13 Sep. 1997, *Gaoligongshan Exped. 1997, 9026* (E E00121061, MO 5160253). **Lijiang [Yulong Xian]:** Yulongshan, above Bai-shui He, 3000 m, 13 July 1962, *S. W. Yu & A. L. Zhang 100994* (KUN 0177516); Yulongshan, Baishui, 27.116667°N, 100.246111°E, 2920 m, 23 Oct. 1994, *Alpine Garden Society China Exped. (ACE) 2462* (E E00039697). **[Lushui Xian]:** Shweli–Salwin divide, 25.916667°N, 98.75°E, 3050 m, July 1925, *G. Forrest 27044* (A 00280147, E E00117396, US 00946035). **[Tengchong Xian]:** upper Mingkwang valley, 25.59°N, 98.75°E, 2135 m, Sep. 1924, *G. Forrest 25085* (E E00117984, US 00946036). **Weihsi (Weixi) Xian:** Mekong–Yangtze divide, 3660 m, 27.6°N, 99.166667°E, Sep. 1921, *G. Forrest 20373* (A 00280146, E E00612658, US 00946033); W of Tungchuling (Dongshuling), 3000 m, 14 Nov. 1937, *T. T. Yu 10708* (A 00280160, E E00612649, KUN 0177442, PE 01031050). **Zhongdian (Xianggelila) Xian:** Sanba Xiang, Habaxueshan, Haba Cun, 2500 m, 16 Oct. 1955, *G. M. Feng 21058* (KUN 0177413–14, PE 01031061, SZ 00289511); on rd. from Gongxia (on hwy. G214) to Bigu Tianchi Hu, 27.630278°N, 99.662222°E, 3730 m, 31 Aug, 2013, *D. E. Boufford, J. F. Harber & X. H. Li 43338* (A 00914776, E E00833567, KUN 20130831).

MYANMAR (BURMA). Luksang and Hpimaw, 2440 m, 5 Oct. 1919, *R. Farrer 1378* (E E00117397); border with Yun-nan, Hpimaw, 2440 m, 19 Sep. 1923, *F. Kingdon-Ward 3661* (E E00117395).

Cultivated material:
Living cultivated plants.
Foster Clough, Mytholmroyd, U.K., from *A. Clark 1014*, NE Yunnan, Zhaotong, "Maushan, Lahhua," 2399 m, 1995.

Cultivated specimens.
Royal Botanic Gardens, Kew, U.K., s.d., from *Vilmorin 4039* (K K000568006).

262. **Berberis woomungensis** C. Y. Wu ex S. Y. Bao, Bull. Bot. Res., Harbin 5(3): 4. 1985. TYPE: China. N Yunnan: Luquan Xian, Diwu Qu, Wu-meng Shan, 3400 m, 10 May 1957, *P. Y. Mao 670* (holotype, KUN 1204063!; isotypes, HITBC 003654 image!, IBK IBK00357954 image!, IBSC 0000593 image!, KUN 0177538!).

Shrubs, deciduous, to 50 cm tall; mature stems pur-plish red, angled; spines 3-fid, pale brownish yellow, 1–1.3 cm, slender, abaxially slightly sulcate. Petiole almost absent; leaf blade abaxially pale green, adaxi-ally green, slightly papillose, oblong-obovate or oblan-ceolate, 1–2 × 0.5–0.7 cm, papery, midvein and lateral veins slightly raised abaxially, inconspicuous or in-distinct adaxially, base cuneate, entire, apex rounded, mucronate. Inflorescence a fascicle, 1- to 4-flowered; pedicel 5–10 mm. Sepals in 2 whorls; outer sepals ob-

long, ca. 7 × 3–4 mm; inner sepals oblong-elliptic, ca. 6 × 4 mm; petals oblong-obovate, ca. 4 × 2 mm, base clawed, glands separate, oblanceolate, apex emarginate, with acute lobes. Stamens ca. 4 mm; anther connective not extended, truncate. Ovules 3, stipitate. Berry red, oblong, ca. 10 × 5–6 mm, slightly pruinose; style not persistent.

Phenology. *Berberis woomungensis* has been col-lected in flower in May and fruit in November.

Distribution and habitat. *Berberis woomungensis* is known from Luquan Xian in north Yunnan. It has been found among bushes and in scrub on dry mountain slopes at ca. 3400–4400 m.

The protologue of *Berberis woomungensis* described the species as being single-flowered and this was re-peated by Ying (2001: 78). However, both the holotype and the isotype at HITBC are distinctly fascicled. The prologue also stated the species to be 0.5–1 m tall, but no height is recorded on the type specimens, nor on *Zhang 601*, while *Mao 1040* records 0.5 m.

Berberis woomungensis is on the current IUCN Red List of Threatened Species (China Plant Specialist Group, 2004g).

Selected specimens.
N Yunnan. Luquan Xian: Wumeng Shan, 1 Nov. 1940, 4400 m, *Y. B. Zhang 601* (1204064, IBSC 0092357, KUN 0177535, PE 01037435); Wumeng Xiang, Wumeng Shan, 3700 m, 26 May 1952, *P. Y. Mao 1040* (IBK IBK00357956, KUN 0177539–40).

263. **Berberis xanthophlaea** Ahrendt, J. Bot. 79(Suppl.): 73. 1942. TYPE: China. SE Xizang (Tibet): [Lhünzê (Longzi) Xian], Charme Distr., Chayul Chu, 28.3°N, 92.8°E, 3650 m, 25 Oct. 1938, *F. Ludlow, G. Sherriff & G. Taylor 6406* (holotype, BM [2 sheets] BM000565569–70!).

Shrubs, deciduous, to 2 m tall; mature stems very pale yellow, sulcate, not verruculose; spines 3-fid, some-times solitary, concolorous, 1.2–2.4 cm, stout, terete. Petiole almost absent; leaf blade abaxially pale green, papillose, adaxially green, obovate, 1–2.6 × 0.8–1.3 cm, thickly papery, midvein raised abaxially, slightly im-pressed adaxially, veining inconspicuous or indistinct on both surfaces, base cuneate, margin entire, rarely spinose with 1 to 4 teeth on each side, apex subacute or obtuse, sometimes minutely mucronate. Inflorescence a panicle, 15- to 30-flowered, sometimes with a few ad-ditional flowers at base, 3–6(–9) cm overall; peduncle 1–3 cm; pedicel 4–7 mm; bracteoles ovate, ca. 1.5 × 1 mm, apex acute. Sepals in 2 whorls; outer sepals ellip-tic, 3–3.5 × 2.5–3 mm; inner sepals elliptic-obovate,

5–6.3 × 2.5–3.5 mm; petals obovate, 3.5–4.5 × 2.5–3 mm, base clawed, widely separate, submarginal, apex deeply incised. Stamens ca. 3 mm; anther connective not extended, truncate. Ovules (1 or)2 or 3, shortly stipitate. Berry red, blue pruinose, oblong-obovoid, 7–9 × 5–6 mm; style persistent.

Phenology. Berberis xanthophlaea has been collected in flower in October and November and with immature fruit in November.

Distribution and habitat. Berberis xanthophlaea is known from Lhünzê (Longzi) and Nang (Lang) Xian in southeastern Xizang (Tibet). It has been collected from stony river banks and gravel terraces at 3050–3659 m.

Berberis xanthophlaea is known from only the type collection and *F. Ludlow, G. Sherriff & G. Taylor 6651*. What is most unusual is that both have floral material despite being collected so late in the year, though *6651* also has immature fruit. I have checked with the fieldbooks of the collectors at BM and these confirm these dates. Neither collection, however, is recorded in the diaries of Ludlow and Sherriff, held respectively by the British Library and the Edinburgh Herbarium Library (these show that Ludlow and Taylor were at Chayul Dzong on 24 October, while Sherriff was in the Charme area on 1 November), but the diaries record only plants that they found of particular interest.

In the protologue, Ahrendt did not state whether *Berberis xanthophlaea* was deciduous or evergreen, though he did say it had an affinity with *B. polyantha*, which is deciduous. Subsequently, Ahrendt (1961: 201) implied it to be deciduous and this is followed here.

Specimens from south Xizang at PE identified as *Berberis xanthophlaea* are not this species and are mostly *B. koehneana*.

Berberis xanthophlaea is on the current IUCN Red List of Threatened Species (China Plant Specialist Group, 2004h).

Selected specimens.
SE Xizang (Tibet). [**Nang (Lang) Xian border with Lhünzê (Longzi) Xian**]: Charme, Char Chu, 28.433333°N, 93.083333°E, 3050 m, 1 Nov. 1938, *F. Ludlow, G. Sherriff & G. Taylor 6651* (BM BM000939754–55).

264. Berberis xiangchengensis Harber, sp. nov. TYPE: W Sichuan: Xiangcheng Xian, W of Reda Xiang along rd. Z005 to Muyucun, 29.145°N, 99.671667°E, 3475 m, 24 Aug. 2013, *D. E. Boufford, J. F. Harber & Q. Wang 43202* (holotype, PE!; isotypes, A!, CAS!, E!, K!, KUN!, TI!).

Diagnosis. Berberis xiangchengensis has leaves somewhat similar to *B. yui*, though more varied and often much larger, but has an inflorescence that is racemose or partially panicu-

late rather than mainly fascicled with much shorter pedicels. The inflorescence is somewhat similar to *B. mekongensis*, which has much broader leaves.

Shrubs, deciduous, to 2 m tall; mature stems pale brownish yellow, sulcate; spines 3-fid, solitary or absent toward apex of stems, pale yellow, (0.6–)1–2.6 cm, abaxially sulcate. Petiole almost absent; leaf blade abaxially pale green, adaxially mid- to dark green, obovate, mostly narrowly so, sometimes oblanceolate, sometimes obovate-elliptic toward apex of stems, (1–)3.5–5(–6) × (0.8–)1.6–2.5 cm, papery, abaxially lightly papillose, midvein raised abaxially, impressed adaxially, lateral veins and reticulation conspicuous on both surfaces, base attenuate, often very narrowly so, margin spinose with (2 to)8 to 15 teeth on each side, sometimes entire toward apex of stems and on young shoots, apex obtuse or subacute. Inflorescence a raceme or panicle, sometimes fascicled at base, 2.5–4 cm overall, 12- to 25-flowered; pedicel 3–6 mm. Flowers unknown. Immature berry red, oblong, 6–9 × 2–4 mm; style persistent, short; seeds 1 or 2.

Phenology. Berberis xiangchengensis has been collected with the remnants of flowers in June and with immature fruit in July and August; its flowering and fruiting periods are otherwise unknown.

Distribution and habitat. Berberis xiangchengensis is known only from Xiangcheng Xian in west Sichuan. It has been found in riverside meadows and adjacent slopes at 3475–3850 m.

IUCN Red List category. Berberis xiangchengensis is assessed as DD or Data Deficient, according to IUCN (2001) criteria.

The leaves of *Berberis xiangchengensis* are exceptionally varied in shape and size, even on the same plant and, as such, provide a case study of the importance of attempting to evidence the full range of leaves when making *Berberis* collections.

Selected specimens.
W Sichuan. Xiangcheng Xian: Si Qu, Hongqi Xiang, Lama Si, 3850 m, 30 June 1972, *Sichuan Economic Plant Exped. 1000* (CDBI CDBI0027183, CDBI0027190); Reda Lincha upstream on Mayi River from the town of Reda, 29.14194444°N, 99.64972222°E, 3450–3500 m, 17 July 1998, *D. E. Boufford et al. 28815* (A 00279588, CAS 1011357, E E00281180, HAST 89736, MO 5315748); W of Reda Xiang along rd. Z005 to Muyucun, 29.145°N, 99.671667°E, 24 Aug. 2013, *D. E. Boufford, J. F. Harber & Q. Wang 43206* (A, CAS, E, K, KUN, PE) and *43207* (A, CAS, E, K, KUN).

265. Berberis xiaozhongdianensis Harber & Xin Hui Li, sp. nov. TYPE: China. NW Yunnan: Zhongdian (Xianggelila) Xian, Xiao Zhongdian, Tianbao

Shan, 27.605278°N, 99.885833°E, 3680–3840 m, 4 Sep. 2013, *D. E. Boufford, J. F. Harber & X. H. Li 43440* (holotype, PE!; isotypes, A!, E!, KUN!).

Diagnosis. *Berberis xiaozhongdianensis* has leaves that are somewhat similar to *B. yunnanensis*, but has mature stems that are a different color and are largely spineless. It has fruit with three to five seeds versus the four ovules of *B. yunnanensis*.

Shrubs, deciduous, to 1.3 m tall; mature stems brown, sulcate; spines absent or rarely 1- to 3-fid, semi-concolorous, 0.2–0.5 cm, terete, weak. Petiole almost absent, sometimes to 3 mm; leaf blade abaxially pale green, papillose, adaxially dark green, narrowly obovate, 1.6–2.8 × 0.7–1.1 cm, papery, midvein raised abaxially, flat adaxially, lateral venation inconspicuous on both surfaces, base narrowly attenuate, margin entire, rarely spinulose with 2 to 4 teeth on each side, apex subacute or obtuse, mucronate. Inflorescence a sub-raceme or sub-umbel, sometimes fascicled at base, or a sub-fascicle, 5- to 10-flowered, otherwise unknown, 1.2–3 cm overall; fruiting pedicel pinkish purple, 3–8 mm, to 20 mm when from base. Immature berry turning white then pink or purple, oblong or oblong-ovate, 8–9 × 5–6 mm; style persistent, short; seeds 3 to 5, pale brown.

Phenology. *Berberis xiaozhongdianensis* has been collected with immature fruit in September. Its flowering period is unknown.

Distribution and habitat. *Berberis xiaozhongdianensis* is known from only the type collection from Tianbaoshan in Zhongdian (Xianggelila) Xian in northwest Yunnan. This was found on a slope in a degraded *Abies* forest at 3680–3840 m.

IUCN Red List category. *Berberis xiaozhongdianensis* is assessed as DD or Data Deficient, according to IUCN (2001) criteria.

266. Berberis yaanica Harber, sp. nov. TYPE: China. WC Sichuan: Ya'an Shi, Baoxing Xian, N of the city of Baoxing on the Baoxing-Xiaojin Rd. (state rd. G210), Xinzhazi Gou (known locally as Mahuang Gou) just N of Jiajin Shan Natl. Forest Park, 30.808889°N, 102.724444°E, 2415 m, 14 May 2014, *D. E. Boufford, S. Cristoph, C. Davidson & Y. D. Gao 43738* (holotype, CDBI!; isotypes, A !, CAS!, E!, MO!, P!, PE!, SRP!, TI!).

Diagnosis. *Berberis yaanica* has a similar inflorescence (though shorter and with fewer flowers) and a similar color of mature stems to *B. mouillacana*, but a different leaf shape that is not abaxially pruinose and which has spinose margins.

Shrubs, deciduous, to ca. 2.2 m tall; mature stems dark reddish brown, subterete, densely verruculose; spines solitary, absent toward apex of stems, pale brown, (0.6–)1–2.6 cm, terete. Petiole 3–8 mm, sometimes almost absent; leaf blade abaxially pale green, papillose, adaxially green, obovate or obovate-elliptic, (1–)1.7–3 × 0.9–1.3 cm, papery, midvein raised abaxially, impressed adaxially, lateral veins and reticulation inconspicuous on both surfaces, margin spinulose with 4 to 8(to 10) inconspicuous, widely spaced teeth on each side, rarely entire, base attenuate, often narrowly so, apex subacute, sometimes minutely mucronate. Inflorescence a sub-fascicle, sub-raceme, or raceme, sometimes fascicled at base, 2- to 5(to 7)-flowered, to 2.2 cm overall; pedicel 8–16 mm; bracteoles reddish, triangular, 1.5 × 0.7 mm, apex acute. Sepals in 2 whorls; outer sepals elliptic-ovate or elliptic, 2.5–4 × 2–3 mm; inner sepals broadly elliptic or elliptic-orbicular, 4–6 × 3.5–4 mm; petals elliptic, 4–5 × 2.5–3.5 mm, base cuneate, glands oblong, widely separate, ca. 1 mm, apex obtuse, crenate or slightly notched. Stamens ca. 3 mm; anther connective not or slightly extended, truncate or slightly obtuse. Pistil 3 mm; ovules 2. Berry unknown.

Phenology. *Berberis yaanica* has been collected in flower in May. Its fruiting season is unknown.

Distribution and habitat. *Berberis yaanica* is known from only the type collection from north Baoxing Xian in west-central Sichuan. This was found on a steep, open slope at 2415 m.

IUCN Red List category. *Berberis yaanica* is assessed as DD or Data Deficient, according to IUCN (2001) criteria.

It would be interesting to compare the flower structure of *Berberis yaanica* with that of *B. mouillacana* but, unfortunately, the flower structure of the latter is unknown.

267. Berberis yalongensis Harber, sp. nov. TYPE: China. NW Sichuan: Garzê (Ganzi) Xian, along tributary of Yalong Jiang, 31.813056°N, 99.717778°E, 3520–3550 m, 2 Aug. 2005, *D. E. Boufford, J. H. Chen, K. Fujikawa, S. L. Kelley, R. H. Ree, H. Sun, J. P. Yue, D. C. Zhang & Y. H. Zhang 33983* (holotype, A 00279952!; isotypes, CAS 1078422 image!, KUN not seen, TI image!).

Diagnosis. *Berberis yalongensis* has leaves that are somewhat similar to those of *B. aemulans*, but has much shorter pedicels and only one or two seeds per fruit versus the seven to 11 ovules of *B. aemulans*. On current evidence, it also appears to be a much smaller species.

Shrubs, deciduous, to 60 cm tall; mature stems color uncertain (purple?), sulcate; spines 3-fid, solitary or absent toward apex of stems, very pale brownish yellow, 0.7–1.6 cm, terete, weak. Petiole almost absent, rarely to 2 mm; leaf blade abaxially pale green, papillose, adaxially dark green, obovate, (1–)1.8–2.4(–3.2) × (0.5–)0.8–1.3 cm, papery, midvein, lateral veins, and reticulation raised abaxially, conspicuous adaxially, base attenuate, margin spinulose with 5 to 14 prominent teeth on each side, sometimes entire, apex obtuse, sometimes subacute. Inflorescence 1- or 2-flowered, otherwise unknown; pedicel of infructescence 7–10 mm. Immature berry turning dark purple, narrowly oblong or obovoid, 6–8 × 3–4 mm; seeds 1 or 2; style persistent.

Phenology. *Berberis yalongensis* has been collected with immature fruit in August; its flowering period is unknown.

Distribution and habitat. *Berberis yalongensis* is known from only the type and one other collection, both made in a heavily disturbed area (cutting, grazing) with alluvial thickets on a slope below a cliff in a tributary of the Yalong Jiang at 3520–3550 m in Garzê (Ganzi) Xian in northwest Sichuan.

IUCN Red List category. *Berberis yalongensis* is assessed as DD or Data Deficient, according to IUCN (2001) criteria.

The leaves of the type have very prominent spinulose margins, whereas those of *Boufford et al. 33982*, collected in the same area, are largely entire. Nevertheless, they appear to be of the same species.

There are very few *Berberis* collections of any description from Garzê Xian.

Selected specimens.
NW Sichuan. Garzê (Ganzi) Xian: along tributary of Yalong Jiang, 31.813056°N, 99.717778°E, 3520–3550 m, 2 Aug. 2005, *D. E. Boufford et al. 33982* (A 00279944).

268. Berberis yanyuanensis Harber, sp. nov. TYPE: China. SW Sichuan: Yanyuan Xian, "inter Tuyungpu et Yanyuanhsien," ca. 3000–3500 m, 12 May 1914, *C. K. Schneider 1196* (holotype, A 00279733!; isotype, K!).

Diagnosis. *Berberis yanyuanensis* has mature stems of a similar color and somewhat similarly shaped leaves to two fascicled or sub-fascicled species from southwest Sichuan and which also have two ovules: *B. bowashanensis* and *B. difficilis*. *Berberis yanyuanensis* differs from both these species in having leaves that are abaxially pruinose and a different flower structure.

Shrubs, deciduous, to 2.5 m tall; mature stems pale reddish brown, terete; spines 3-fid, solitary toward apex of stems, pale brownish yellow, (0.4–)1–2.1 cm, terete. Petiole almost absent; leaf blade abaxially pale green, pruinose, adaxially green, narrowly obovate, (0.9–)1.7–3.2 × 0.5–0.8 cm, midvein slightly raised abaxially, slightly impressed adaxially, venation inconspicuous on both surfaces, base attenuate, margin entire, apex subacute, rarely obtuse, sometimes minutely mucronate. Inflorescence an umbel, 3- to 5-flowered; peduncle ca. 4 mm; pedicel 8–10 mm. Sepals in 2 whorls; outer sepals narrowly oblong-elliptic, 3.5 × 1.25 mm; inner sepals obovate-elliptic, 5 × 3 mm; petals obovate-elliptic, 4 × 2.5 mm, base cuneate or slightly clawed, glands obovoid, ca. 0.5 mm, separate, apex emarginate. Stamens ca. 3 mm; anther connective extended, obtuse. Pistil ca. 3 mm; ovules 2. Berry unknown.

Phenology. *Berberis yanyuanensis* has been collected in flower in May. Its fruiting season is unknown.

Distribution and habitat. *Berberis yanyuanensis* is known from only the type collection from Yanyuan Xian in southwest Sichuan. This was collected from a thicket in open ground at ca. 3000–3500 m.

IUCN Red List category. *Berberis yanyuanensis* is assessed as DD or Data Deficient, according to IUCN (2001) criteria.

The account of the flower structure given above is based on an undated line drawing by Schneider attached to the specimen sheet of the isotype at Kew.

Tuyungpu referred to in the collection details appears as Tukungpu on the map depicting part of the route of Schneider and Handel-Mazzetti's joint expedition to Yunnan and Sichuan in 1914 (Handel-Mazetti, 1925: 12–13). This is some 15 km northeast of Yanyuan town. I have been unable to find its current name.

The only subsequent specimens I have found from near where *Schneider 1196* was collected are of *M. Y. He & Q. S. Zhao 116831* (CDBI CDBI0028083, SZ 00294056–57), Xiaogao Shan, 3100 m, 19 May 1983. These may be *Berberis yanyuanensis*, but the images I have seen of them are too poor to make any judgement.

269. Berberis yarigongensis Harber, sp. nov. TYPE: China. W Sichuan: Batang Xian, "Yargong, forêts des montagnes," July–Aug. 1904, *J. A. Soulié 3057* (holotype, P P02682030!; isotypes, K K001273511!, MO 6318150!, P P02682028!, PE [2 sheets] 01969512–13 images!, SING 0155672 image!).

Diagnosis. *Berberis yarigongensis* has leaves and a flower stalk which are similar to those of *B. abbreviata*, which is found to the west of Batang in Kangding Xian. However, *B. abbreviata* has very pale reddish brown mature stems and a different flower structure.

Shrubs, deciduous, height unknown; mature stems dark reddish purple or dark reddish brown, sulcate; spines 3-fid, pale orangish or yellowish brown, 0.7–0.9 cm, terete. Petiole almost absent; leaf blade abaxially pale green, papillose, adaxially green, obovate, often narrowly so, 1.5–2.6 × 0.4–0.7 cm, papery, midvein flat abaxially and adaxially, lateral venation and reticulation conspicuous abaxially, inconspicuous adaxially, base attenuate, margin entire, rarely spinulose with 1 to 3 teeth on each side, apex subacute or obtuse, sometimes mucronate. Inflorescence an often dense fascicle, sub-fascicle, or sub-raceme, 3- to 10-flowered, to 1.7 cm overall; pedicel 4–8 mm; bracteoles triangular, ca. 1.5 mm. Sepals in 2 whorls; outer sepals narrowly oblong-ovate, 5 × 2.5 mm; inner sepals obovate-elliptic, 4.5–5 × 2.5 mm; petals obovate, 3–3.5 × 2–2.5 mm, base clawed, glands narrowly separate, apex obtuse. Stamens 2.5 mm; anther connective conspicuously or slightly extended, obtuse. Ovules 2. Berry unknown.

Phenology. *Berberis yarigongensis* has been collected in flower in June and July. Its phenology is otherwise unknown.

Distribution and habitat. *Berberis yarigongensis* is known from only three collections, all from the Yarigong area of south Batang in west Sichuan. All are recorded as being collected in mountain forests.

IUCN Red List category. *Berberis yarigongensis* is assessed as DD or Data Deficient, according to IUCN (2001) criteria.

Soulié 3055, 3056, and *3057* were all collected by Soulié during his tenure at the Catholic missionary station in Yaregong (now Yarigong). For his own account of this locality, see Soulié (1904). There appear to have been no subsequent *Berberis* collections from what currently remains a remote area (though one that will be more easily accessible once the new road from Batang to Derong, parallel to the Jinsha Jiang, under construction at the time of writing, is completed).

Selected specimens.
W Sichuan. Batang Xian: Yargong, June–July 1903, *J. A. Soulié 3055* (P P02682033); Yargong, July 1903, *J. A. Soulié 3056* (P P02682022).

270. Berberis yingii Harber, sp. nov. TYPE: Cultivated. Royal Botanic Garden Edinburgh, 20 May 2014, *P. Brownless 492* from seed of *Alpine Garden Society China Expedition (ACE) 1829*, China. W Sichuan: Xiangcheng Xian, Wengshui to Xiang-

cheng, N side of Daxue Shan pass, 28.6115°N, 99.847167°E, 4100 m, 7 Oct. 1994 (holotype, E E00705631!).

Diagnosis. *Berberis yingii* is similar to various mainly fascicled and sub-fascicled species with two ovules from northwest Yunnan and west and southwest Sichuan, but with a different flower structure; for further details, see the multi-access key (Key 9).

Shrubs, deciduous, to 1.5 m tall; mature stems pale yellowish brown, sulcate, verruculose; spines 3-fid, solitary or absent toward apex of stems, pale yellowish brown, 0.6–2 cm, abaxially sulcate. Petiole almost absent, rarely to 3 mm; leaf blade abaxially pale green, slightly papillose, adaxially dark green, obovate, mostly narrowly so, or oblanceolate, 1.2–2.8 × 0.5–1 cm, papery, midvein raised abaxially, impressed adaxially, lateral veins and reticulation inconspicuous on both surfaces, more conspicuous when dry, base attenuate, margin entire, rarely spinulose with 1 to 5 teeth on each side, apex subacute or obtuse, sometimes minutely mucronate. Inflorescence a fascicle, sub-fascicle, or sub-raceme, sometimes fascicled at base, 4- to 9-flowered, to 3 cm overall; pedicel 7–16 mm. Sepals in 2 whorls; outer sepals broadly ovate, 3 × 2 mm, apex shortly acuminate; inner sepals broadly obovate or elliptic-obovate, 3.5 × 2.5 mm; petals obovate, 2–2.5 × 1.5 mm, base slightly clawed, glands separate, apex entire. Stamens 2 mm; anther connective not or slightly extended, apex obtuse. Ovules 2. Berry red, oblong or ellipsoid, ca. 8 × 4–5 mm; style persistent; seeds red.

Phenology. *Berberis yingii* has been collected in fruit in October, and in cultivation has flowered in May.

Distribution and habitat. *Berberis yingii* is reliably known from only the type collection made on a roadside on Daxue Shan in Xiangcheng Xian in west Sichuan at 4100 m.

Etymology. *Berberis yingii* is named after Ying Junsheng (应俊生) in recognition of his contribution to the study of Chinese *Berberis*.

IUCN Red List category. *Berberis yingii* is assessed as DD or Data Deficient, according to IUCN (2001) criteria.

Berberis yingii is one of a number of species with similar leaves and two ovules, but which can be distinguished from each other mainly by their different flower structures. It is for this reason that a cultivated specimen of *ACE 1829* has been chosen as the type rather than one of the wild-collected specimens listed below.

It is possible that *Boufford et al. 43305*, detailed below, from very near where the type was collected is *Berberis yingii*. This had very similar leaves and fruiting stalks and fruit with two seeds but reddish mature stems. There were remnants of late flowers on the shrub concerned, but unfortunately, it was impossible to draw any conclusions from them.

Selected specimens.
W Sichuan. Xiangcheng Xian: Wengshui to Xiangcheng, N side of Daxue Shan pass, 28.6115°N, 99.84716°E, 4100 m, 7 Oct. 1994, *Alpine Garden Society China Expedition (ACE) 1829* (E E00045897, LIV, WSY WSY0057780); N side of Daxue Shan on hwy. S217, S of Ranwu Xiang, 28.621667°N, 99.835833°E, 4100 m, 29 Aug. 2013, *D. E. Boufford, J. F. Harber & Q. Wang 43305* (A, BM, CAS, E, K, KUN, PE, TI).

Cultivated material:
Living cultivated plants.
Royal Botanic Garden Edinburgh, Inverleith and Dawyck Gardens, *19943511A-C* from *ACE 1829*.

271. Berberis yui T. S. Ying, Acta Phytotax. Sin. 37: 309. 1999, as "*yuii*." TYPE: China. SW Sichuan: Muli, Wachin, Ching-chang, 3800–3900 m, 21 June 1937, *T. T. Yu 6530* (holotype, PE 00935255!; isotypes, A 00279403!, KUN 0177029!, PE 00935256!).

Shrubs, deciduous, to 1.5 m tall; mature stems pale grayish or brownish yellow, sulcate; spines 3-fid, pale yellow, 0.5–1.6 cm, terete, slender. Petiole almost absent; leaf blade abaxially pale green, adaxially dark green, obovate or obovate-elliptic, rarely obovate-oblanceolate, 1–2.5 × 0.4–0.9 cm, papery, midvein, lateral veins, and reticulation conspicuous abaxially, inconspicuous or indistinct adaxially, base attenuate, margin spinulose with 9 to 16, often conspicuous teeth close together on each side, apex rounded, occasionally acute, sometimes minutely mucronate. Inflorescence a fascicle, rarely a sub-fascicle, sub-umbel, or short sub-raceme, 4- to 8-flowered; pedicel 12–17 mm; bracteoles lanceolate, ca. 2.5 × 0.6 mm. Sepals in 3 whorls; outer sepals lanceolate, ca. 4.2 × 1.5 mm; median sepals elliptic, ca. 4.5 × 2 mm; inner sepals elliptic, ca. 4 × 2 mm; petals obovate, ca. 3.5 × 2 mm, base clawed, glands separate, apex incised. Stamens ca. 2.5 mm; anther connective extended, rounded. Ovules 2, stipitate. Berry red, ovoid, 9 × 5–6 mm; style not persistent.

Phenology. *Berberis yui* has been collected in flower in June and July and in fruit in September.

Distribution and habitat. *Berberis yui* is known from Daocheng, Jiulong, and Muli Xian in west and southwest Sichuan. It has been found in thickets, field sides,

forest understories, and on open slopes at ca. 3600–4600 m.

Berberis yui appears to be known from only the type and the specimens cited below.

Selected specimens.
SW & W Sichuan. Daocheng Xian: SSW of the city of Daocheng along provincial rd. S216 betw. Sela-Xiang & Chitu-Xiang, 28.81666667°N, 100.155°E, 4050 m, 29 Aug. 2018, *D. E. Boufford, B. Bartholomew, J. L. Guo & J. F. Harber 45041* (A, CAS, E, KUN, PE). **Jiulong Xian:** Mengdong Shan, 4200 m, 20 June 1960, *T. S. Ying 4066* (PE 00935258); same details, but 4100 m, *T. S. Ying 4067* (PE 00935257); Jidan Shan, 3600–3900 m, 1 July 1979, *Q. Q. Wang 20302* (CDBI CDBI0027148, IBSC 0091624). **Muli Xian:** Wachin, Ching-chang, 3600 m, 31 June 1937, *T. T. Yu 6532* (A 00279684, PE 01037762); Baidiao to Kangwu, 3800 m, 21 Sep. 1959, *S. G. Wu 2720* (KUN, 0177028, 1204884, PE 01037116); Mairi Muchang, 4600 m, 30 July 1978, *Q. S. Zhao 6329* (CDBI CDBI0028052–53, CDBI0028120).

272. Berberis yulongshanensis Harber, sp. nov. TYPE: China. NW Yunnan: Lijiang (now Yulong) Xian, Yulong Shan, Wenhai, 3200 m, 5 June 1985, *Kunming-Edinburgh Yulong-Shan Exped. (KEY) 538* (holotype, E E00612570!; isotype, KUN 0176571!).

Diagnosis. *Berberis yulongshanensis* is similar to various mainly fascicled and sub-fascicled species with two ovules from northwest Yunnan and west and southwest Sichuan, but with a different flower structure; for further details, see the multi-access key (Key 9).

Shrubs, deciduous, to 2 m tall; mature stems reddish brown, sulcate; spines 3-fid, solitary or absent toward apex of stems, pale brownish yellow, 0.5–1.6 cm, terete. Petiole almost absent or to 2 mm; leaf blade abaxially pale green, papillose, adaxially green, narrowly obovate or oblanceolate, 1.4–3 × 0.4–0.7 cm, papery, midvein raised abaxially, lateral venation and reticulation inconspicuous or indistinct on both surfaces, base narrowly attenuate, margin entire, apex obtuse or subacute, mucronate. Inflorescence a fascicle, sub-fascicle, sub-umbel, or sub-raceme, 3- to 6-flowered; pedicel 4–10 mm. Sepals in 3 whorls; outer sepals narrowly elliptic, 3–3.5 × 1.5–2 mm; median sepals elliptic-obovate, 4.5–5 × 3–3.5 mm; inner sepals elliptic-orbicular, 5–5.5 × 3.5–4 mm; petals obovate, 4–4.5 × 2.3–2.5 mm, base slightly clawed, glands separate, ca. 1 mm, apex crenate or slightly notched. Stamens ca. 3 mm; anther connective not or slightly extended, truncate. Pistil 2.5–3 mm; ovules 2. Berry red, oblong or ellipsoid, 7 × 3 mm; style persistent.

Phenology. *Berberis yulongshanensis* has been collected in flower in May and June and in fruit in September.

Distribution and habitat. *Berberis yulongshanensis* is known from Yulong Shan in Yulong Xian in northwest Yunnan. It has been found in forest understories and clearings, among *Pinus*, and on stream and road sides at 3000–3600 m.

IUCN Red List category. *Berberis yulongshanensis* is assessed as DD or Data Deficient, according to IUCN (2001) criteria.

Kunming-Edinburgh Yulong Shan Expedition 538 was first identified as *Berberis microtricha* and most of the specimens listed below were identified either as this or as *B. lecomtei*. *Berberis lecomtei* also has two ovules, but has more flowers on its flower stalks and a different flower structure. *Berberis microtricha* also has a different flower structure as well as distinctly pale yellow stems versus the reddish brown stems of *B. yulongshanensis*.

It should be noted that there are other specimens at KUN and PE from Yulongshan with very similar leaves and flower stalks to *Berberis yulongshanensis*, but with dark reddish purple or very dark reddish brown mature stems. I have not had an opportunity to examine the flower structure of these. There are also specimens without evidence of the color of mature stems which may or may not be *B. yulongshanensis*, e.g., *R. C. Ching 30645*, Xuesong Cun, 18 September 1939 (A 00279640, KUN 0176580–81, 0176587).

There are also specimens with flowers from both Weixi and Zhongdian (Xianggelila) Xian that are very similar to *Berberis yulongshanensis*. Determining whether they are or not in each case requires examining their flower structures, something I have not attempted.

Selected specimens.
NW Yunnan. Lijiang (now) Yulong Xian: Yulongshan, "Lichiang Snow Range," 3000 m, 21 May 1937, *T. T. Yu 15048* (BM, E E00570049, KUN 0176590); Wenhai, 3600 m, 6 June 1964, *S. W. Yu 63-118* (KUN 0178974); Yunshanping, 3100 m, 26 May 1981, *Y. L. Hu & H. Wen H81-33* (PE 01840108–09); Ganhe Ba, 3250 m, 26 May 1985, *Kunming-Edinburgh Yulong Shan Expedition 215* (E E00612576, KUN 0176574); betw. Haligu & Pianbaihua, 3400 m, 29 May 1985, *Kunming-Edinburgh Yulong Shan Expedition 319* (E E00612572, KUN 0176573); Luomei, Luo Gou, 3200 m, 30 May 1985, *Kunming-Edinburgh Yulong Shan Expedition 382* (E E00612569, KUN 0176572); E flank of Lijiang Range, 3000 m, 26 Sep. 1986, *P. Cox et al. 2643* (E E00612568).

273. Berberis yunnanensis Franch., Bull. Soc. Bot. France 33: 388. 1887. TYPE: China. NW Yunnan: [Heqing Xian], "au col de Yen-tze hey (Lankong)," 3200 m, 18 Sep. 1885, *J. M. Delavay 1660 bis* "A" (lectotype, designated here, P P00835680!; isolectotypes, fragm. A 00038826!, KUN 1207481!, P P02682376!, VNM image!).

Berberis franchetiana C. K. Schneid., Oesterr. Bot. Z. 67: 223. 1918, syn. nov. TYPE: China. Yunnan: s. loc., s.d., *J. M. Delavay s.n.* (holotype, K K000077417!).

Shrubs, deciduous, to 2 m tall; mature stems reddish or brownish purple, sulcate, verruculose; spines 3-fid, sometimes solitary toward apex of stems, pale brownish yellow, stout, 0.5–2.5 cm, abaxially sulcate. Petiole almost absent; leaf blade abaxially pale green, papillose, adaxially green, obovate to oblong-obovate, 1.5–3(–4) × 0.8–1.5(–2) cm, papery, midvein raised abaxially, venation inconspicuous abaxially, largely indistinct adaxially, base narrowly attenuate, margin entire, rarely spinulose with 1 to 3 teeth on each side, apex rounded, obtuse or subacute, sometimes mucronate. Inflorescence a fascicle or sub-fascicle, 3- to 5(to 7)-flowered; pedicel (10–)15–25(–40) mm. Sepals in 3 whorls; outer sepals narrowly ovate, 5.5–6 × 2–2.5 mm; median sepals elliptic, 5.5–6.5 × 3.5–4.5 mm, apex semi-acute; inner sepals elliptic, 5.5–6.5 × 4–5 mm, apex semi-acute; petals obovate-elliptic, 5–5.5 × 3–3.5 mm, based clawed, glands separate, emarginate, apex obtuse, entire. Stamens 4–4.5 mm; anther connective conspicuously apiculate, filaments sometimes dentate. Ovules 4. Berry red, oblong-ovoid, 10–14 × 7–8 mm; style not persistent.

Phenology. *Berberis yunnanensis* has been collected in flower in May and June and in fruit between July and September.

Distribution and habitat. *Berberis yunnanensis* is known from Heqing Xian on the border with Eryuan Xian and Yangbi Xian in northwest Yunnan. It has been found in open, stony situations among scrub and near streamsides at 2745–ca. 3800 m.

The protologue of *Berberis yunnanensis* cited only *Delavay 1660 bis* of 18 September 1885. There are two specimens of this at P, while a third was sent to KUN as a result of the reorganization of the Paris herbarium in 2010–2012; a fourth is at VNM. The best of these has been chosen as the lectotype. These specimens have fruit and are designated here as *Delavay 1660 bis* "A." Later, Franchet (1889: 38–39) included a brief description of flowers while still citing only *Delavay 1660 bis*, though now without date. This description appears to be derived from one or more of three undated Delavay specimens with flowers at P with no collection details but numbered *1660 bis*. These are designated here as *1660 bis* "B," though given that, according to the sheets, 00716574 and 06868295 were received from Delavay on 18 June 1887, while 00716573 was received on 10 April 1889, it is possible these are actually from two different gatherings. There is, in addition, a further specimen with fruit numbered *1666 bis* from the same collection area as the type, but undated. The fruit of this specimen appears to be immature in comparison with those of the type collection, so it is designated here as *Delavay 1660 bis* "C."

Ahrendt (1961: 161) described a specimen received at Kew on 31 May 1888 as "type" of *Berberis yunnanensis* and which he stated was "*Delavay 1660.*" This was a reference to K K000077411 which has both flowers and fruit on the sheet. Contrary to Ahrendt's report, it is unnumbered and, since the fruit resemble those of *Delavay 1660 bis* "C" rather than those of *Delavay 1660 bis* "B," is unlikely to be an isolectotype.

The placing of "Yen-tze hey" in Heqing Xian is from Bretschneider (1898b).

It is not completely clear why Schneider published *Berberis franchetiana* since the specimen he cited (which has fruit) is indistinguishable from the specimens of *Delavay 1660 bis* "C" and those on K K000077411. This was noticed by Stapf who, on 1 July 1933, commented on the sheet of the type "This is evidently part of Delavay's Yen-tze hey plant *B. yunnanensis* and misplaced in distribution or mounting and renamed by C. K. Schneid." (see also his comments in Stapf [1933]).

In the protologue of *Berberis franchetiana*, Schneider also noted that the holotype was on the same sheet as a specimen of *B. thunbergii* var. *glabra*, and it was questionable whether the collection details "in silvis ad Kou tui" given on the sheet specifically for that taxon also applied to *B. franchetiana*. This caution was ignored by Ahrendt (1961: 153), who reported the type of *B. franchetiana* as being collected at "woods at Kou Tui." The two specimens have since been mounted on separate sheets.

The description of the flowers of *Berberis yunnanensis* given above is from *Delavay 1660 bis* "B" (P P00716573). This differs from that of Ahrendt (1961: 16) which was largely followed by Ying (2001: 94), and which appears to have been based on a specimen from a cultivated plant of unknown origin that once grew in the Royal Botanic Gardens, Kew.

As noted under *Berberis lecomtei*, when I made a brief visit in September 2013 to the area where Franchet made his collections of *B. yunnanensis*, I found it to have been planted with dense forests of *Pinus yunnanensis* as a response to earlier forest clearances and logging and consequent soil erosion. Whether *B. yunnanensis* has survived such drastic changes there is currently unknown.

The report of *Berberis yunnanensis* in Sichuan and Xizang by Ying (2001) appears to be based on various specimens at PE from Leibo, Muli, Yanyuan, and Xiangcheng Xian in Sichuan and Zayü (Chayu) Xian in southeast Xizang (Tibet), which are annotated as such. These are of a variety of species, none of which are *B. yunnanensis*.

Ahrendt (1945a: 114–115) published a *Berberis franchetiana* var. *glabripes* with the type from Dokerla on the Zhongdian (Xianggelila) Xian–Xizang border and a *B. franchetiana* var. *macrobotrys*, whose type is from Pica, also in Xianggelila Xian. In both cases, he admitted that connecting them with *B. franchetiana* was somewhat speculative since "the precise character of this species [*B. franchetiana*] . . . is rather imperfectly known." Variety *macrobotrys* is treated here as a synonym of separate species *B. polybotrys*. For variety *glabripes*, see under Taxa Incompletely Known.

Bao (1985: 15) published a *Berberis franchetiana* var. *gombalana*, whose type is from Sungta in Zayü Xian on the border with Gongshan Xian in Yunnan. For details of this, see also under Taxa Incompletely Known.

Were varieties *glabripes* and *gombalana* to be sustainable as varieties, they both would have to be renamed as varieties of *Berberis yunnanensis*.

Selected specimens.
NW Yunnan. S. loc.: specimens with flowers, *J. M. Delavay 1660 bis* "B," (P P00716573–4, P06868295). [**Heqing Xian**]: "ad collum Yen-tze hey, prope Lankong," s.d., specimen with fruit, *J. M. Delavay 1660 bis* "C," (P P06868525); "au col de Yen-tze hey," Lankong, 3200 m, 1 June 1886, *J. M. Delavay s.n.* (K, KUN 1206512, P P02682375); "ad collum Yen-tze hey prope Lankong," Sep. 1887, *J. M. Delavay s.n.* (K ex Herb, Casson 000077412, PE 01901509); "Col de Yen ze hey au dessus Mo-se-yn," 3200 m, 17 July 1889, *J. M. Delavay s.n.* (P P02465463–4, P06868524, P06868294, PE 01921429, SING); "ad collum Yen-tze hey prope Lankong," s.d., specimens with fruit and flowers, *J. M. Delavay s.n.* (K K000077411); "jugi Sanshishao pr. urb. Hodjing," 3250 m, 19 May 1916, *H. R. E. Handel-Mazzetti 8736* (A 00279724). **Yangbi Xian:** W side of Diancang Shan, Baiyunfeng Peak above Malutang, 25.76666667°N, 100.01666667°E, 3500–3800 m, 26 June 1984, *Sino-Amer. Bot Exped. 567* (A 00280067, CAS 723900, KUN 0176733–34, US 00945995).

274. Berberis yushuensis Harber, sp. nov. TYPE: China. S Qinghai: Yushu Xian, SE of Baiting He basin in side valley of Tongtian He (upper Chang Jiang), 32.75°N, 97.35°E, 3700 m, 18 Aug. 1996, *T. N. Ho, B. Bartholomew, M. F. Watson & M. G. Gilbert 2028* (holotype, GH 00279965!; isotypes, BM BM001010958!, CAS 935986!, E E00065379!, HNWP 0286562 image!, MO 5329351 image!, PE 01840129!).

Diagnosis. Berberis yushuensis is similar to *B. reticulinervis*, but the color of its mature stems and its flower structure are different.

Shrubs, deciduous, to 1.5 m tall; mature stems very pale grayish brown, sulcate, verruculose; young shoots bright purple; spines 3-fid, pale brown, 0.4–0.9 cm, terete or abaxially sulcate. Petiole almost absent; leaf blade abaxially pale gray-green, adaxially mid-green, oblanceolate or obovate, 0.8–1.5 × 0.3–0.5 cm, papery, midvein raised abaxially, flat or slightly impressed adaxially, lateral veins and reticulation conspicuous on both surfaces, base attenuate, margin entire, sometimes spinose with 1 to 4 teeth on each side, apex acute or rounded, mucronate. Inflorescence a fascicle, sub-umbel, or short sub-raceme, sometimes fascicled at base, 1- to 4(to 6)-flowered; peduncle 0.3–0.7 cm; pedicel 5–8 mm; bracts triangular, pink, 1.5 mm. Sepals in 2 whorls; outer sepals oblong-ovate, ca. 4–4.5 × 3 mm; inner sepals broadly elliptic, 4–5.25 × 2.75–3.25 mm; petals obovate, 3.5 × 2.25–2.5 mm, base clawed, glands separate, ca. 1 mm, apex rounded. Stamens ca. 2.5 mm; anther connective truncate. Ovules 2. Immature berry turning red, oblong-ellipsoid, 8 × 3 mm; style persistent.

Phenology. *Berberis yushuensis* has been collected in flower and with immature fruit in August.

Distribution and habitat. *Berberis yushuensis* is known from only the type from Yushu Xian and one collection from Chindu Xian in south Qinghai. The former was found on a steep, rocky slope in a *Juniperus–Picea–Betula* community, the latter growing on a field side and on slopes, both at 3700 m.

Selected specimens.
S Qinghai. Chindu (Chenduo) Xian: Xiwu Xiang, Shang Saiba, E of Chumda, 32.983333°N, 97.35°E, 3700 m, 15 Aug. 1996, *T. N. Ho et al. 1834* (BM 201010938, CAS 940714, E 3059996, GH 00280127, HNWP 00001390, MO 5190726, PE 01840140).

275. Berberis zayulana Ahrendt, J. Bot. 79(Suppl.): 64. 1941. TYPE: China. SE Xizang (Tibet): Zayü (Chayu) Xian, Rong Tö–Dibang divide [side valley of Rongo Tö Chu, SW of Mugu], 3650 m, 24 Nov. 1933, *F. Kingdon Ward 11017* (holotype, BM BM000559595!).

Shrubs, deciduous, height unrecorded; mature stems brownish red, terete or sub-sulcate, black verruculose; spines 1- to 3-fid, pale brownish yellow, 0.5–1(–1.2) cm, terete or abaxially slightly sulcate, weak. Petiole 1–2 mm or almost absent; leaf blade abaxially pale green, adaxially green, obovate or oblanceolate, 1–3 × 0.6–1.1 cm, papery, midvein obviously raised abaxially, slightly raised adaxially, lateral and reticulate veins conspicuous abaxially, lateral veins in 3 or 4 pairs adaxially, reticulate veins inconspicuous adaxially, base cuneate, margin entire or spinulose with 1 to 4 teeth on each side, apex rounded, mucronate. Inflorescence a sub-umbel or sub-raceme, 2- to 7-flowered, 2.5–3.5 cm overall including peduncle 1.2–1.5 cm; pedicel 8–13 mm; bracteoles triangular, acuminate, whitish, 1.5 × 1.5 mm. Flowers 11–14 mm diam. Sepals in 2 whorls;

outer sepals oblong-elliptic or oblong-ovate, 6 × 3–3.5 mm; inner sepals obovate-elliptic, 7.6 × 5 mm; petals 4.5 × 3.5 mm, base scarcely clawed, glands separate, oblong-obovate, 1 × 0.4 mm, apex entire or sub-emarginate. Stamens 3 mm; anther connective extended, truncate at the edge, but shortly apiculate in the center. Ovules 6 to 9, often stipitate. Berry scarlet, not or slightly pruinose, oblong-ellipsoid, 7–10 × 3–5 mm; style not persistent; seeds 2.

Phenology. *Berberis zayulana* has been collected in fruit in November. It possibly flowers in June.

Distribution and habitat. *Berberis zayulana* is known from the type and possibly one other collection, both from Zayü (Chayu) Xian in southeast Xizang (Tibet). The former was collected from thickets in open ground at 3650 m, the latter from thickets at 3660–4265 m.

There are unresolved issues in relation to *Berberis zayulana.* On 28 September 1939, Ahrendt annotated both *Kingdon-Ward 11017* (details above) and *Kingdon Ward 10548* (details below) as "Type specimen." It is not clear why he chose to publish the former, which has fruit and little other material, as the type versus the latter which has abundant flowers and leaves. The description of flowers given above is from the protologue and must have been based on no. *10548.* However, a dissection of a fruit of the type produced only two seeds and the discrepancy between this and the six to nine ovules of the flowers suggests the two specimens may be of different species.

Clearly more research is needed here (there appear to have been no collections of *Berberis* in this area since Kingdon-Ward).

Simultaneously with *Berberis zayulana,* Ahrendt published a variety *dolichocentra,* the type being *F. Ludlow & G. Sherrif 1072* (BM BM000573952) from Sakden in east Bhutan. This was omitted from Ahrendt (1961). There would seem to be no evidence to associate this taxon with *B. zayulana.*

Berberis zayulana was not noticed by Ying (1985, 2001).

Selected specimens.
SE Xizang (Tibet). Zayü (Chayu) Xian: Chutong Camp, Atakang La, ca. 29.069374°N, 96.813583°E, 3660–4265 m, 30 June 1933, *F. Kingdon-Ward 10548* (BM BM000939692, MO 1618680).

276. Berberis zhaoi Harber, sp. nov. TYPE: China. NW Yunnan: [Yulong Xian], "Eastern flank of the Lichiang Range," 27.25°N, 3050–3650 m, May 1910, *G. Forrest 5553* (holotype, E E00612485!;

isotypes, BM BM000810140!, IBSC 0091908 image!, K K001273250!, N 093057077 image!, P P02313228!, PE [2 sheets] 01033864–65!, !, UC UC224593!).

Diagnosis. *Berberis zhaoi* has mature stems with a similar color to those of *B. pallens* and similar-shaped leaves, but the inflorescence of *B. pallens* is fascicled, sub-fascicled, or sub-racemose with larger flowers with a different structure. *Berberis zhaoi* has similar leaves to *B. pseudotibetica*, but its inflorescence has more flowers and is sometimes fascicled from the base. The inflorescence of *B. zhaoi* is similar to *B. forrestii*, but is shorter. The color of its mature stems is also different from *B. forrestii*, whose flower structure (in as far as it is known) is different and whose fruit have two or three seeds versus the two of *B. zhaoi*.

Shrubs, deciduous, to 4.5 m tall; mature stems reddish purple or reddish brown, turning pale reddish brown, sulcate; spines 3-fid, solitary or absent toward apex of stems, pale brownish yellow, (0.5–)1.4–2.4 cm, abaxially sulcate. Petiole almost absent, rarely to 3 mm; leaf blade abaxially pale green, papillose, adaxially green, narrowly obovate to obovate-oblanceolate, (1.5–)2.2–4 × (0.4–)0.8–1.3 cm, papery, midvein slightly raised abaxially, slightly impressed or flat adaxially, lateral veins and reticulation conspicuous abaxially, inconspicuous adaxially, base narrowly attenuate, margin entire, rarely spinose-serrulate with 1 to 4 teeth on each side, apex subacute or obtuse, sometimes minutely mucronate. Inflorescence a loose raceme, sometimes fascicled at base, 5- to 18-flowered, 2–4.5 cm overall including peduncle to 1.2 cm; pedicel 4–8 mm, but to 16 mm from base. Sepals in 2 whorls; outer sepals oblong-ovate, 3 × 1.75 mm; inner sepals elliptic-ovate, 5 × 3.5 mm; petals elliptic, 5 × 2.75–3.5 mm, base cuneate, glands separate, apex incised. Stamens 3 mm; anther connective extended, obtuse. Ovules 2. Berry red, oblong, 9–11 × 5–6 mm; style not persistent.

Phenology. *Berberis zhaoi* has been collected in flower in May and June and with fruit in August and October.

Distribution and habitat. *Berberis zhaoi* is known from Yulong Shan in Yulong Xian in northwest Yunnan. It has been collected in open areas and sparse *Pinus* forests at 2950–3550 m.

Etymology. *Berberis zhaoi* is named in honor of Zhao Cheng-Zhang (赵 成章) of the Naxi village of Nvlvk'ö (Xuesong Cun) on Yulong Shan, who played an indispensable role between 1906 and 1932 in organizing and leading George Forrest's local collectors and preparing his specimens for dispatch to the U.K.; for details, see Mueggler (2011).

IUCN Red List category. *Berberis zhaoi* is assessed as DD or Data Deficient, according to IUCN (2001) criteria.

Specimens of the type collection were originally identified as *Berberis lecomtei* or *B. pallens*, and subsequently by Ahrendt (1961: 172), along with *Forrest 2271* and *5554*, as *B. forrestii*. Various of the specimens cited below have been identified as *B. concolor*, *B. francheti-ana* var. *macrobotrys*, *B. jamesiana*, and *B. mekongen-sis*, none of which, except for *B. jamesiana*, are found on Yulong Shan.

Selected specimens.
NW Yunnan. Lijiang (now Yulong) Xian: "Eastern flank of the Lichiang Range," 27.2°N, 3050–3350 m, June 1906, *G. Forrest 2271* (E E00612484, K); same details, but 27.166667°N, June 1906, *G. Forrest 2343* (E E006124790); same details, but 27.25°N, 3350 m, May 1910, *G. Forrest 5554* (BM BM000810139, E E00612486, K); "Likiang-Hsien," 2700 m, July 1935, *C. W. Wang 70959* (A 00279648, IBSC 0091719, KUN 0176221, LBG 00064305, NAS NAS00313982, PE 01033062, 01033859, WUK 0066479); "Lichiang Snow Range," 3000 m, 23 May 1937, *T. T. Yu 15053* (A 00279650, E E00612483, KUN 00279650, PE 01033892); Xue Shan, 15 Oct. 1939, *G. M. Feng 2970* (KUN 0176188–89); Ma Huang Ba, 3500 m, 13 Aug. 1942, *R. C. Chin 8979* (KUN 0176199, PE 01033061, 01033063–64); Yi Qu, Yuhu Xiang, Xuesong Cun, 3400 m, 30 Aug. 1955, *G. M. Feng 21224* (KUN 0176213–4, PE 01032807); Yulong Shan, near Ma Huang Ba, 3300 m, 8 Aug. 1959, *G. M. Feng 22572* (KUN 0176207–08, 0176317, PE 01032806); Yulong Shan, 3550 m, 20 June 1962, *Lijiang Botanical Garden 100311* (KUN 0176218, 0178421, 0178937); Yulong Shan, Xianyiyan, 3400 m, 1 June 1964, *S. W. Yu 70* (KUN 0176209); Yulong Shan, below Ma Huang Ba, 3200 m, 27 May 1985, *Kunming-Edinburgh Yulong Shan Exped. 254* (E E00612480, KUN 0177264); Yulong Shan, Liyugao to Xianjinyuan, 2950 m, 3 June 1985, *Kunming-Edinburgh Yulong Shan Exped. 480* (E E00612428, KUN 0175841).

277. Berberis zhongdianensis Harber, sp. nov. TYPE: China. NW Yunnan: Zhongdian (now Xianggelila) Xian, Xiaoxue Shan, summit of pass, 28.323333°N, 99.75222°E, 3830 m, 21 June 1994, *Alpine Garden Society China Exped. (ACE) 507* (holotype, E E00045830!; isotype, K!).

Diagnosis. *Berberis zhongdianensis* is similar to various mainly fascicled and sub-fascicled species with two ovules from northwest Yunnan and west and southwest Sichuan, but with a different flower structure; for further details, see the multi-access key (Key 9). *Berberis zhongdianensis* also differs from these in having two or three ovules. It has somewhat similar leaves to *B. bowashanensis*, *B. lecomtei*, *B. yingii*, and *B. yulongshanensis*, which all also have similarly varied flower stalks, but has a different flower structure from all of these including two or three rather than two ovules.

Shrubs, deciduous, to 3 m tall; mature stems pale yellowish or orangish brown, sulcate; spines 3-fid, solitary or absent toward apex of stems, pale yellowish brown, 0.4–1.5 cm, terete. Petiole almost absent, rarely to 4 mm; leaf blade abaxially pale green, papillose, adaxially dark green, narrowly obovate or obovate-oblanceolate, (0.6–)1.6–2.5 × (0.4–)0.8–1(–1.2) cm, midvein raised abaxially, flat adaxially, lateral veins and reticulation inconspicuous on both surfaces, but more conspicuous on dried specimens, margin entire or spinulose with 1 to 15 teeth on each side, apex obtuse or subacute, mucronate. Inflorescence a fascicle, sub-fascicle, sub-umbel, or sub-raceme, (2- to)4- to 10(to 18)-flowered; pedicel 5–12 mm. Flowers pale lemon yellow. Sepals in 3 whorls; outer sepals narrowly oblong-ovate, 3.5 × 1.5 mm; median sepals narrowly oblong-ovate, 5–6 × 2 mm; inner sepals obovate, 4 × 3 mm; petals obovate, 3 × 2.5 mm, base cuneate, glands approximate, ca. 0.5 mm, apex entire. Stamens 2 mm; anther connective slightly extended, obtuse. Pistil 2 mm; ovules 2 or 3. Immature berry yellow, turning red, oblong or ovoid, ca. 6–8 × 3–4 mm; style persistent.

Phenology. Berberis zhongdianensis has been collected in flower in June and with immature fruit in September.

Distribution and habitat. Berberis zhongdianensis is known from the type collection, one other collection from the top of the Xiaoxue Shan pass, and one collection made some 65 km to the south-southeast of this, both areas being in Zhongdian (Xianggelila) Xian in northwest Yunnan. They were made at 3742–3940 m, the first two in open areas, in the latter case in a regenerating area recovering from fire, and in the third from roadside scrub.

IUCN Red List category. Berberis zhongdianensis is assessed as DD or Data Deficient, according to IUCN (2001) criteria.

When I visited the Xiaoxue Shan pass in September 2013, there were various *Berberis* species on the north side immediately below the pass, but *B. zhongdianensis* was the dominant one.

Selected specimens.
NW Yunnan. Zhongdian (Xianggelila) Xian: rd. to Haba Shan, 27.734722°N, 99.969167°E, 3742 m, 14 June 1994, *Alpine Garden Exped. to China 1994 (ACE) 227* (E E00194914); Xiaoxueshan, N side of pass and ridge on W side of pass off hwy. S217, 28.32°N, 99.757222°E, 3900–3940 m, 3 Sep. 2013, *D. E. Boufford, J. F. Harber & X. H. Li 43420* (A, CAS, E, KUN, PE).

Taxa Incompletely Known

Berberis elliotii Ahrendt, J. Linn. Soc., Bot. 57: 126. 1961. TYPE: China. SE Xizang (Tibet): [Mêdog (Motuo) Xian], Kongbo, Lucha La, 29.3°N, 94.616667°E, 3650 m, 1 Oct. 1947, *F. Ludlow, G. Sherriff & H. H. Elliot 13293* (holotype, BM BM000559589!; isotype, E E00259017!).

Both the holotype and isotype of *Berberis elliotii* are poor specimens and it is impossible to write a description that would definitively differentiate it from other species. Though the collection details describe seeds as being collected, no cultivated material has been found.

Berberis franchetiana C. K. Schneid. var. **glabripes** Ahrendt, J. Bot. 80(Suppl.): 114. 1945. TYPE: China. NW Yunnan: Atunze (Dêqên [Deqin] Xian), Doker-la pass, 3200 m, 4 Nov. 1937, *T. T. Yu 7864* (holotype, E E00217996!; isotypes, A 00038751!, BM BM001015563!, KUN 0754714, PE 01033852!).

The protologue cited the holotype and two other specimens: *T. T. Yu 10713* (E E00612476), Weihsi (Weixi Xian), W of Tungchuling, 15 November 1937, and *T. T. Yu 10819* (E E00612477), Atunze, Baima Shan, Sancha He, 3600 m, 28 November 1937. The holotype is an exceptionally poor specimen with only four complete leaves and only a few fruit. The other two specimens, as Ahrendt admitted, were "devoid of leaves," and determining that the three represented a variety of *Berberis franchetiana* was, in effect, largely speculative since "the precise character of this species . . . is rather imperfectly known."

Later, Ahrendt (1961: 154) gave a description of the flowers apparently based on cultivated material grown from *Yu 7864*, while discounting any cultivated material from either of the other two Yu collections, since "plants distributed" as *Yu 10819* were *Berberis forrestii*, and those of *Yu 10713* were a variety of *B. spraguei*. He also now cited *F. Ludlow, G. Sherriff & H. H. Elliot 13993* (A 00279526, BM, E E00612478), Ba La, Basum Chu, Kongbo, Tibet, as *B. franchetiana* var. *glabripes*.

All of this needs to be unpicked. As can be seen above, there are three isotypes of the taxon. There are also duplicates of the two other collections cited by Ahrendt: *T. T. Yu 10713* (A 00279522, KUN 0176194) and *T. T. Yu 10819* (KUN 0176195, PE 01033851). These are all poor specimens; in the case of *10713* and *10819*, none have leaves. The best specimen of the type collection is the one at KUN, but there is not enough material to identify it as a published species or publish it as a new one (for why it is not the same as *Berberis dokerlaica* whose type is also from Doker-la, see under that species). I have been unable to identify *F. Ludlow, G. Sherriff & H. H. Elliot 13993*, but it is not the same as *Yu 10713*.

Berberis franchetiana is treated here as a synonym of *B. yunnanensis*; but there is no evidence to support *Yu 7864* being a variety of this species.

Berberis franchetiana C. K. Schneid. var. **gombalana** C. Y. Wu & S. Y. Bao, Bull. Bot. Res., Harbin 5(3): 15. 1985. TYPE: China. NW Yunnan: Gongshan Xian, Sungta, Xue Shan, 2900 m, 17 June 1960, *Nanshuibeidiao Exped. 8995* (holotype, KUN 0160899!; isotype, PE 01037202!).

The protologue of *Berberis franchetiana* var. *gombalana* gave the most minimal of description, restricting itself to the statement that it differed from *B. franchetiana* by having glabrous stems, larger leaves, three ovules, and epruinose fruit. However, there appears to be no reason to associate it with *B. franchetiana* (as noted above, treated here as a synonym of *B. yunnanensis*), though this also has glabrous stems and epruinose fruit. The type of variety *gombalana* has floral material. Were a dissection of these to be possible, it is likely that *8995* would turn out to be of an undescribed species. The protologue also cited *K. M. Feng 24697*, Gongshan Xian, Yi Qu, Baihanluo, 23 November 1959 (KUN 0160900, 0176206); but these specimens have no leaf material, and a further specimen at PE (01037203) has only three complete leaves, so whether they are the same as no. *8995* is impossible to determine.

Berberis honanensis Ahrendt, Gard. Ill. 64: 426. 1944. TYPE: Cultivated. Stonefield, Watlington, Oxford, U.K., from seed reputedly collected by an unknown collector at an unknown date in "Lushan, Honan," 19 May 1943, *L. W. A. Ahrendt s.n.* (lectotype, designated here, OXF!; isolectotype, BM).

The protologue stated the type to be at OXF but gave no date of collection. Subsequently, Ahrendt (1961: 185–186) cited as "Types" *L. A. 395* of 19 May 1943, 24 September 1943, and 25 September 1944 at OXF and BM. There are specimens of the first two at OXF, and of the first at BM, though none are numbered *395*. That of 19 May 1943 has been chosen as the lectotype because it has floral material.

In the protologue, Ahrendt stated his plant was one of those distributed as *Berberis virgetorum* by "Mr Marchant and Mr Hillier" (the U.K.'s leading nurserymen in the interwar period) and were from seeds sent from "Lu Shan in Honan." In his 1961 entry, Ahrendt stated the seeds were from "Honan: Lu shan Xsien." However, it appears *Berberis* does not occur in Lushan Xian in Henan and it is possible that the actual origin of the seed was Lushan Botanic Garden in Jiangsu, founded in 1934. Interestingly, the leaves of Ahrendt's plants are very similar to a number of specimens from

Guangxi at IBK identified as *B. virgetorum*, but as noted in the discussion under that species, these are all sterile. Clearly more research is needed here.

Berberis honanensis was included in Ying (2001: 183) though with a slightly different description from that of Ahrendt.

Berberis humidoumbrosa var. **inornata** Ahrendt, J. Bot. 80(Suppl.): 116. 1945. TYPE: China. NW Yunnan: Chungtian (Zhongdian now Xianggelila Xian), Lichiashica, 3450 m, 16 Nov. 1937, *T. T. Yu 10972* (holotype, E E00295773!; isotypes, BM BM001015561!, GH 00274575!, KUN 0176686!).

Berberis humidoumbrosa var. *inornata* has a curious history. The protologue cited only the holotype at E. This is a poor specimen with some fruit but only a few leaves. The protologue, however, gave a full description of flowers without citing any source. Subsequently, Ahrendt (1961: 160), quite illegitimately, cited the type as being *T. T. Yu 14298* from Muli Xian in Sichuan (treated here as an example of a new species, *B. purpureocaulis*) while describing the flowers of variety *inornata* as being the same as *B. humidoumbrosa* var. *humidoumbrosa*, whose type is from Lhünzê (Longzi) Xian in Xizang, nearly 1000 km from where *T. T. Yu 10972* was collected. This description of flowers is not the same as given in the protologue description of variety *inornata*. Even more confusingly, *T. T. Yu 10972* was now cited by Ahrendt (1961: 158) not under variety *inornata* but as an example of *B. papillifera*, the type of which is from the Eryuan Xian–Heqing Xian border.

All of the isotypes of *Berberis humidoumbrosa* var. *inornata* are poor ones with the exception of the specimen at KUN. But even the KUN specimen does not have enough evidence to associate it with a published species or to recognize it as new. There is no reason, however, to associate it with *B. humidoumbrosa* var. *humidoumbrosa*.

Berberis jaeschkeana C. K. Schneid. var. **bimbilaica** Ahrendt, J. Bot. 79(Suppl.): 65. 1941. TYPE: China. SE Xizang (Tibet): [Nang (Lang) Xian], Tsari Distr., Bimbi La, 28.833333°N, 93.466667°E, 3350–4100 m, 13 Oct. 1938, *F. Ludlow, G. Sherriff & G. Taylor 6305* (holotype, [2 sheets] BM 00559597–98!).

There seems to be no reason to associate this collection with *Berberis jaeschkeana*. The protologue describes the mature stems of variety *bimbilaica* as pale yellow. This is inaccurate; the type clearly has dark purple stems. This is almost certainly an undescribed species, but there is not enough evidence to publish it as such.

Rao et al. (1998b: 122) cite as an example of *Berberis jaeschkeana* var. *bimbilaica*, *B. S. Aswal 10064* (LWG), India. Himachal Pradesh, Lahul valley, Tandi-Gonda, 3400 m, 23 Sep. 1978. I have not seen this, but the chances of it being that same as *F. Ludlow, G. Sherriff & G. Taylor 6305* would seem remote.

Berberis reticulinervis var. **brevipedicellata** T. S. Ying, Acta Phytotax. Sin. 37: 307. 1999. TYPE: China. S Gansu: Min Xian, E of Dangchang, 1600 m, 23 May 1959, *S. Jiang & T. L. Chin 443* (holotype, PE 00935238!; isotype, KUN 0179252!).

The protologue of *Berberis reticulinervis* var. *brevipedicellata* indicated that it differed from *B. reticulinervis* var. *reticulinervis* by having much shorter pedicels and obovate-oblong inner sepals. Given that *B. reticulinervis* appears to be endemic to Markam Xian in northeast Tibet some 700 km distant from where the only cited specimen of variety *brevipedicellata* was collected, it seems quite unlikely that they are the same species. A full dissection of the flowers of *Chin 443* would probably show it to be of an undescribed species. It is possible that *Bailong Jiang Survey Team 1020*, Gansu, Têwo (Diébù) Xian, Duoji Gou, Taili'ao, 30 July 1998 (PE 01556144) is of the same species.

Excluded Taxa

Berberis ambigua Ahrendt, J. Bot. 79(Suppl.): 60. 1941. TYPE: Cultivated. S. loc. (but probably from a plant in Ahrendt's living collection), s.d., *L. W. A. Ahrendt 765* (lectotype, designated here, BM BM000794132!).

Neither in the protologue nor subsequently (1961: 144–145) did Ahrendt cite any herbarium for the type. The only specimen I have found and which is annotated by Ahrendt as type has been chosen as the lectotype.

According to the protologue, the plant concerned came from W. J. Marchant of Keeper's Hill Nursery, Wimborne, Dorset, U.K., with its reputed origin being from among mixed seed collected by Forrest in the Dali Range in northwest Yunnan between 1917 and 1919. In all probability, this was a cultivated hybrid.

Berberis approximata Sprague, Bull. Misc. Inform. Kew 1909: 256. 1909. *Berberis dictyophylla* Franch. var. *approximata* (Sprague) Rehder, Mitt. Dendrol. Ges. 21. 183. 1912. TYPE: Cultivated. Arboretum nursery, Royal Botanic Gardens, Kew, U.K., "May 13 1901," *572-1897 Vilmorin* (lectotype, designated here, K K000395236!).

The protologue stated "Introduced by Vilmorin, Andrieux & Co., and described from specimens cultivated at Kew." Vilmorin and Bois (1904: 19) advertised *Berberis dictyophylla* grown from seeds of *Farges 1080* collected in Sichuan. There are four specimens at K, all labelled as being from *572-1897 Vilmorin*. The only specimen that includes specifying it is from no. *1080* and has been chosen as the lectotype. These four specimens were the source of *B. dictyophylla* in Bot. Mag. 228, t. 7833, 1903. They differ little from *B. dictyophylla*, except they sometimes have flowers or fruit in pairs, a characteristic that is rare in the wild, but not unknown. The claim that they were grown from Farges seed is highly dubious since it seems he collected exclusively in the Chengkou area of Chongqing and there are no wild-collected specimens known from this area which remotely resemble these ones. Whether the Vilmorin plants were a form of *B. dictyophylla* or a hybrid grown from cultivated seed is impossible to determine. The taxon was listed as a species found in China by Ying (2001: 75).

Berberis approximata Sprague var. **campylogyna** Ahrendt, J. Bot. 79(Suppl.): 53. 1941. *Berberis dictyophylla* Franch. var. *campylogyna* (Ahrendt) Ahrendt, J. Linn. Soc., Bot. 57: 128. 1961. TYPE: Cultivated. Stonefield, Watlington, Oxford, U.K., reputedly from *G. Forrest 13224*, 13 May 1942, *L. W. A. Ahrendt s.n.* (neotype, designated here, OXF!).

The protologue cited cultivated material from Kew from *J. F. Rock 24276* and "also from the Sunningdale Nursery grown from the seed of Forrest 13224," but cited no herbarium specimens. In renaming the taxon as a variety of *Berberis dictyophylla*, Ahrendt cited a number of cultivated specimens from *Forrest 13224* at OXF as type. There are three such specimens at OXF, none with the dates given by Ahrendt and all dated after the publication of the protologue. The specimen with floral material has been chosen as the neotype. This is not the same as a wild-collected specimen of *G. Forrest 13224* (E E00375453), which I have been unable to identify. The OXF specimens have many similarities with *B. dictyophylla* but apparently have flowers with three rather than two whorls of sepals and, as such, are likely to be of hybrid origin.

Berberis aristata DC. var. **sinensis** K. Koch, Hort. Dendrol. 18. 1853.

A cultivated plant of unknown origin.

Berberis atroprasina Ahrendt, Gard. Chron., ser. 3, 112: 155. 1942. TYPE: Cultivated. Stonefield, Watlington, Oxford, U.K., 31 Oct. 1941, *L. W. A. Ahrendt s.n.* (lectotype, designated here, OXF!).

The protologue simply cited as type cultivated material "from seeds collected on Forrest's last journey" at OXF. There are three specimens there. Two with fruit are dated before the date of the protologue. The one with the most material has been chosen as the lectotype. The third specimen has floral material, as does a specimen dated 13 May 1943 at Kew (K K000395229). No wild-collected specimens corresponding to these have been found.

Berberis beaniana C. K. Schneid. Pl. Wilson. (Sargent) 3: 439. 1917. TYPE: Cultivated. Royal Botanic Gardens, Kew, Oct. 1916, from 1930W, 52–13 Veitch, no. 30, *s. coll. s.n.* (lectotype, designated here, A 00038730).

The protologue cited cultivated plants at Kew which flowered on 18 June 1914 and 3 October 1914, but did not cite a herbarium. A specimen at A dated October 1916, which was annotated by Schneider as type, has been chosen as the lectotype. The protologue stated the 1930W (see above) was a reference to Wilson "Veitch Exped. Seed No. 1930." However, from the evidence of "Numerical List of seeds collected 1899–1901" (Arnold Arboretum Wilson Papers Series WII Box 4 Folder 3), no. 1930 for his first Veitch Expedition (1899–1902) was not *Berberis* (though I have been unable to decipher what it is recorded as) and seed numbers recorded for his second Veitch Expedition (1903–1905) only go up to no. 1910 with herbarium specimens starting at no. 3000. No wild-collected specimens resembling this taxon have been found. The taxon was listed as a species found in China by Ying (2001: 207).

Berberis bretschneideri Rehder, Trees & Shrubs 2: 21. 1907.

This was originally published on the basis of cultivated material grown from seeds collected near Beijing, but subsequently corrected by Schneider (1918: 291) as being from seeds from Japan. This correction was not noticed by Ahrendt (1961: 19). For further information about *Berberis bretschneideri* in Japan, see under *B. amurensis* above.

Berberis caraganifolia Banks ex DC., Syst. Nat. 2: 18. 1821. TYPE: China. Shandong, s.d., *G. Staunton s.n.* (BM BM000997386!).

This is not *Berberis* but *Caragana frutex* (L.) K. Koch.

Berberis circumserrata C. K. Schneid. var. **subarmata** Ahrendt, J. Bot. 79(Suppl.): 56. 1941. TYPE: Cultivated. Royal Horticultural Society Garden, Wisley, Surrey, U.K., reputedly from

R. Farrer 238, 1940, *s. coll. s.n.* (lectotype, designated here, BM BM000554675!).

The protologue cited cultivated material grown from seed of *Farrer 238*, collected in Gansu at an unspecified date, but with no reference to any herbarium specimen. Later, Ahrendt (1961: 123) cited as type specimens with various dates at BM. None with these dates have been found. The only specimen found at BM has been chosen as the lectotype. This is labelled as "Type collection." Number *238* is not included in the list of *Berberis* in Farrer (1916: 47–114, 1917, 2: 319), while the wild-collected *Farrer & Purdom 238* (E E00386339) is not *Berberis* and has been identified as *Rhodiola dumulosa* (Franch.) S. H. Fu. The cultivated material was probably a hybrid of *B. diaphana*. The taxon was treated as a synonym of *B. circumserrata* by Ying (2001: 97).

Berberis concinna Hook. f., Bot. Mag. 79, t. 4744. 1853.

The species was described as being found in China by Ying (2001: 88); however, based on present evidence, it appears to be endemic to Nepal and India (Sikkim).

Berberis consimilis C. K. Schneid., Oesterr. Bot. Z. 66: 324. 1916. TYPE: Cultivated. Arnold Arboretum, Jamaica Plains, Boston, Mass., U.S., 1 June 1915, from No. 187 Hort. Vilmorin (accession no. 7565), *s. coll. s.n.* (lectotype, designated here, A 00168856 image!).

The protologue cited the type as being at A and in the Herb. Schneider. There are various specimens at A (and at BKL and US). The only one at A with a date preceding the date of the protologue has been designated as the lectotype. This includes a line drawing by Schneider which includes the word "Type." Any specimen in the Herb. Schneider was destroyed in WWII. The protologue stated the plant to be of Sichuan origin, but this information is not to be found on the Arnold Arboretum's accession card for 7576, which simply records it being grown from seed sent by Vilmorin in 1904.

Berberis farreri Ahrendt, J. Linn. Soc., Bot. 57: 192. 1961.

The protologue cited as type a specimen at OXF from a cultivated plant grown from *R. Farrer 318* apparently collected in Gansu at an unknown date. This could not be found. Number *318* is not included in the lists of *Berberis* in Farrer (1916: 47–114, 1917, 2: 319). The taxon was included as a species found in China by Ying (2001: 183).

Berberis faxoniana C. K. Schneid., Oesterr. Bot. Z. 66: 325. 1916. TYPE: Cultivated. Arnold Arboretum, Jamaica Plains, Boston, Mass., U.S., from Vilmorin 1189 (accession no. 4768), 17 June 1912, *A. Rehder s.n.* (lectotype, designated here, A 00167684 image!).

The protologue cited as type cultivated material from the Arnold Arboretum at A and in the Herb. Schneider, reputedly originating from Vilmorin. There are various specimens corresponding to this description at A. One that includes a sheet of line drawings by Schneider with the word "Type" has been chosen as the lectotype. Any specimen in the Herb. Schneider was destroyed in WWII. The Arboretum accession card for no. 4768 gives its origin as a plant received from Vilmorin in 1902. Schneider stated the plant was evidently of Chinese origin and speculated it was from either Sichuan or Yunnan, but I have found no evidence to support this.

Berberis kerriana Ahrendt, J. Linn. Soc., Bot. 57: 91. 1961. TYPE: Cultivated. Royal Horticultural Society Garden, Wisley, Surrey, U.K., *s. coll. s.n.*, 11 Oct. 1943 (lectotype, designated here, BM BM000554636!).

The protologue cited as type two specimens at BM from a plant at Wisley grown from seeds of *Kerr 39* collected in some unspecified part of China in 1935. The only specimen at BM has been designated the lectotype. This was annotated as type by Ahrendt on 20 October 1943. I have been unable to identify who the Kerr referred to was. Ying (2001: 209) included *Berberis kerriana* as a species found in China. It is possibly a hybrid of *B. potaninii*.

Berberis macrosepala Hook. f. & Thomson, Fl. Ind. 1: 228. 1855.

The syntypes of this species come from Sikkim. Ahrendt (1961: 116) cited a number of specimens from southeast Xizang at BM as *Berberis macrosepala*. These are of other species.

Berberis magnifolia Ahrendt, J. Linn. Soc., Bot. 57: 198. 1961.

In the protologue, Ahrendt cited only a cultivated plant in his own collection from seed which he described as probably from "the northern Chinese provinces." No herbarium was given for any specimen and none has been found.

Berberis metapolyantha Ahrendt, J. Bot. 79(Suppl.): 75. 1942. TYPE: Cultivated. Royal Botanic Gardens, Kew, 25 Oct. 1939, *s. coll. s.n.* (lectotype, designated here, BM BM000554813!).

The protologue simply cited cultivated material sent from Kew "recorded as growing from *Forrest 27773*." Later, Ahrendt (1961: 207) cited as type two specimens at OXF. Neither have been found. A specimen at BM with one of the dates specified in the protologue has been designated the lectotype. This specimen does not resemble the wild-collected specimen of *Forrest 27773* (E E00612643) which is *Berberis wilsoniae*.

Berberis miqueliana Ahrendt, J. Linn. Soc., Bot. 57: 168–169. 1961.

This was described by Ahrendt as a cultivated plant, distributed as *Berberis silva-taroucana* by W. J. Marchant of Wimborne, Dorset, U.K. The type was described as being at OXF but could not be found.

Berberis nummularia Bunge, Index Seminum [Tartu], 6. 1843.

This was reported from Xinjiang by An (1995: 4), but is likely to be based on misidentification of *Berberis integerrima*. For further details, see the discussion under that species.

Berberis orthobotrys "Bien. ex Aitch." var. **conwayi** Ahrendt, J. Linn. Soc., Bot. 57: 143. 1961.

This was published solely on the basis of *W. M. Conway 146* (K K000567984) which Ahrendt reported was collected in Xinjiang on the Kashmir border. However, it is clear from Conway (1894: 318), which specifically mentions the collection of this particular specimen (as *Berberis vulgaris*), that it was collected in what is now Pakistan (Northern Areas) rather than Xinjiang. For why "Bienert ex Aitchison" is not the author of the name *B. orthobotrys*, see the discussion under *B. multiserrata* above.

Berberis parisepala Ahrendt, Gard. Chron., ser. 3, 109: 100. 1941. TYPE: Cultivated. Royal Horticultural Society Gardens, Wisley, Surrey, U.K., 28 June 1938, *L. W. A. Ahrendt 578*, reputedly grown from seed of *F. Kingdon-Ward 8350* from the Delei Valley on the Assam–Tibet frontier (lectotype, designated here, BM specimen with flower on sheet BM00554651!).

The type of *Berberis parisepala* is different from the wild-collected specimen of *Kingdon-Ward 8350* cited by Ahrendt (1961 143–144) as the type of *B. orthobotrys* var. *rupestris* and treated here as a possible synonym of *B. multiserrata*.

Berberis parisepala was listed as a species found in China by Ying (2001: 76).

Berberis rockii Ahrendt, J. Bot. 79(Suppl.): 72. 1942. TYPE: Cultivated. Royal Botanic Gardens, Kew, U.K., 13 Oct. 1939, *s. coll. s.n.* (lectotype, designated here, BM BM000794131; isolectotypes, BM [2 sheets] BM0007941105!, 000794130!).

The protologue cited as the type "cultivated material at Kew from seed collected by Rock under the number 23339," but without citing any herbarium. There are three specimens with this origin at BM, all annotated by Ahrendt as type. The best of these has been chosen as the lectotype. These specimens, however, do not resemble the wild-collected specimen of *J. F. Rock 23339* at A (00279646), which is identified as *Berberis lecomtei*, but probably is not.

Berberis shensiana Ahrendt, Gard. Chron., ser. 3, 112: 155. 1942. TYPE: Cultivated. Royal Horticultural Society Garden, Wisley, U.K., reputedly from seed of *W. Purdom 543*, 5 May 1939, *L. W. A. 1460* (lectotype, designated here, specimen with floral material, BM BM000554717!).

The protologue cited cultivated material from seed of *W. Purdom 543*, with the type at OXF. Later, Ahrendt (1961: 148–149) cited two differently dated specimens at OXF. Neither of these could be found. A specimen at BM whose date corresponds to one of those given by Ahrendt has been chosen as the lectotype. There is a much better specimen at K (000395239) from a plant in Ahrendt's living collection dated 20 May 1943. Though Ahrendt stated the seed was from Shaanxi, he provided no evidence for this. No wild-collected specimens corresponding to the taxon have been found (all specimens seen in Chinese herbaria identified as *Berberis shensiana* are other species).

Ying (2001: 190) included *Berberis shensiana* as a species found in China with a description significantly different from that given by Ahrendt.

Berberis sinensis (Desf.) var. **crataegina** (DC.) Regel, Trudy Imp. S.-Peterburgsk. Bot. Sada 2: 417. 1873.

No specimens were cited in the protologue of this taxon. When little was known about *Berberis* in China, the taxon name was applied to a range of deciduous specimens that could not otherwise be identified. For example, *A. David 2619*, "Mongolie, Ta-tsing-chan," s.d. (P P02313045) cited by Franchet (1883: 178) as this taxon consists of three specimens, two of which are *B. caroli*, the other being *B. dubia*.

Berberis spraguei Ahrendt, Gard. Chron., ser. 3, 109: 101. 1941. TYPE: Cultivated. Stonefield, Watlington, Oxford, U.K., (from plant received November 1936 from Knap Hill Nursery, Surrey, U.K., reputedly from *G. Forrest 30513*), 12 Oct. 1938, *L. A. 75* (lectotype, designated here, BM BM000794129!).

The protologue cited plants grown from *G. Forrest 30513* and *30717*, but without citing any herbarium. Later, Ahrendt (1961: 162) published a *Berberis spraguei* var. *pedunculata* with the type being from plants of *30717*, while the type of *B. spraguei* was cited as being various cultivated specimens at BM and OXF from his own collection of *30513*. No specimens from any of the dates cited were found at either location. A specimen at BM which gives details of its origin has been chosen as the lectotype. This specimen (and others at BM and OXF) which may be hybrids are not the same as wild-collected specimens of *G. Forrest 30613* (E E00612563, HBG HBG-506741) which are *B. jamesiana*. *Berberis spraguei* var. *pedunculata* is treated here as a synonym of *B. concolor*.

Berberis suberecta Ahrendt, Bull. Misc. Inform. Kew 1939: 271. 1939. TYPE: Cultivated. 23 June 1939, *L. W. A. Ahrendt s.n.* (lectotype, designated here, K K000395231!).

The protologue described a plant of unknown origin with the type at K. There are two specimens at K with different dates. The one with floral material has been chosen as the lectotype. Later, Ahrendt (1941b: 57) stated "The origin of this species has now been partially revealed," in that he believed that this plant was the same as material that had been sent to him by a Mr. Arnold Vivien via two nurseries and had apparently had its origin in seed labelled "*B. dictyophylla* from Tali Range," in the possession of J. C. Williams of Caerhays Castle, Cornwall, and recorded as being collected by Forrest.

Berberis thibetica C. K. Schneid., Repert. Spec. Nov. Regni Veg. 6: 268. 1909.

Schneider cited as "type" material sent to him for identification by Maurice Vilmorin from a cultivated plant grown at Les Barres, France, probably from seed collected by Soulié in northwest Yunnan near Tse-Kou (Lijiang) at an unspecified date. Subsequently (1918: 224), Schneider stated the Soulié seed number to be *3448* and the date of collection to be 1903 while speculating that the seed might have actually been collected near Kangding in west Sichuan. In neither of Schneider's articles was any herbarium cited for a specimen from the type plant and none has been found (if Schneider retained the material sent to him by Vilmorin, it will have been lost in the destruction of the Berlin-Dahlem herbarium in 1943). It is possible that the Vilmorin Herbarium may shed further light on this. Currently inaccessible, images of its specimens are scheduled to be eventually available on the Paris Herbarium website.

Berberis thunbergii DC., Syst. Nat. 2: 9. 1821.

This is native to Japan, but is included by Ying (2001: 155) on the grounds that it is commonly grown as an ornamental throughout China, especially in large cities. It is also listed for the same reason in various provincial floras.

Berberis umbellata Lindl., Edwards's Bot. Reg. 30: t. 44. 1844.

Hao (1938: 599) identified three specimens from Qinghai—*K. S. Hao 824* (PE 01037947, S 12–25473, WUK 0004089), *1296* (NKU NK003824, P P04447318, PE 01030892, 01037951–54, S 12–25475), and *1261*—as this species, otherwise found in Nepal and northwest India. Borodina-Grabovskaya et al. (2007: 168) reported the same specimens as *Berberis umbellata* var. *brianii* (Ahrendt), otherwise reported only from Sikkim. *K. S. Hao 824* and *1296* are *B. diaphana*. *K. S. Hao 1261* could not be found, but it is safe to assume it is not *B. umbellata*.

Berberis validisepala Ahrendt, Bull. Misc. Inform. Kew 1939: 270. 1939. TYPE: Cultivated. Royal Botanic Gardens, Kew, U.K., from "No. 152," 26 Oct. 1938 (lectotype, designated here, BM BM000554677!).

Berberis validisepala f. **primoglauca** Ahrendt, J. Bot. 79(Suppl.): 62. 1941. *Berberis validisepala* var. *primoglauca* Ahrendt, J. Linn. Soc., Bot. 57: 130. 1961. TYPE: Royal Botanic Gardens, Kew, U.K., 5 Oct. 1939, *L. W. A. Ahrendt 579* (neotype, designated here, BM BM000554681!).

The protologue of *Berberis validisepala* cited only the type, "No. 152" at K. No specimen dated before the

protologue was found at K. A specimen at BM dated before the protologue has been chosen as the lectotype. Given that there is no information at all about the provenance of the cultivated plant, I would not have referred to it in this treatment except for Ahrendt's variety *primoglauca*. The protologue of forma *primoglauca* cites only material sent to him by Major Stern of Goring, Sussex, U.K., reputedly "grown from seed sent to England by Forrest's native collectors after his death in China." No herbarium was cited for this material but later, Ahrendt (1961: 130) cited as the type of variety *primoglauca* a specimen at BM from a Stern plant. However, the only specimens at BM annotated as types of this taxon are cultivated ones from Kew whose origin is not stated. The best of these has been chosen as the lectotype. It should be noted that the description of the flowers of the taxon given by Ahrendt in 1961 differs from the description he gave in 1939.

Berberis vulgaris L., Sp. Pl. 1: 330. 1753. TYPE: *Herb. Clifford 122* (lectotype, designated by Browicz & Zieliński [1975: fasc. 111/16. 10. 14], BM BM000558505!).

This was cited as "*Berberis vulgaris* var. *normalis*" by Maximowicz (1889a: 29) as being found in Gansu, and in Qinghai by Zhou (1997: 377) but, in fact, is a native of Europe and possibly southwest Asia.

Berberis wilsoniae Hemsl. var. **latior** Ahrendt, J. Linn. Soc., Bot. 57: 216. 1961. TYPE: Cultivated. Stonefield, Watlington, Oxford, U.K., reputedly from seeds of *G. Forrest 30459*, 12 Oct. 1945, *Ahrendt s.n.* (neotype, designated here, OXF!).

The protologue cited as type three specimens at OXF from Ahrendt's living collection. None with the dates cited are at OXF. The only specimen there (annotated by Ahrendt as type) has been chosen as a neotype. This specimen is not *Berberis wilsoniae* and, despite Ahrendt's assertion, cannot have been grown from *G. Forrest 30459* since a wild-collected specimen of this number at E (E00351896) with no collection details other than Yunnan has only flowers and very immature fruit which could not have produced viable seed. This is not *B. wilsoniae* either. It is treated as a synonym of *B. wilsoniae* var. *guhtzunica* by Ying (2001: 106).

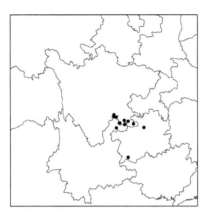

Map 1. *Berberis acuminata*

Map 2. *Berberis alpicola*

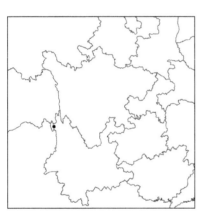

Map 3. *Berberis amabilis*

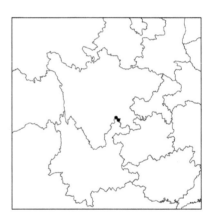

Map 4. *Berberis arguta*

Map 5. *Berberis aristatoserrulata*

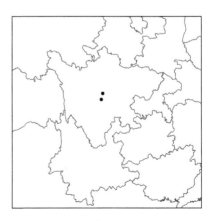

Map 6. *Berberis asmyana*

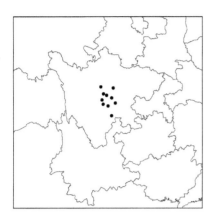

Map 7. *Berberis atrocarpa*

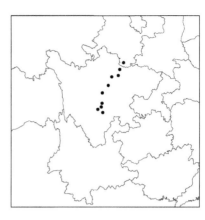

Map 8. *Berberis bergmanniae*

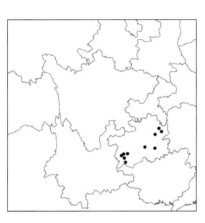

Map 9. *Berberis bicolor*

Map 10. *Berberis brevisepala*

Map 11. *Berberis calliantha*

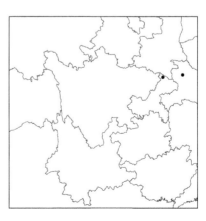

Map 12. *Berberis candidula*

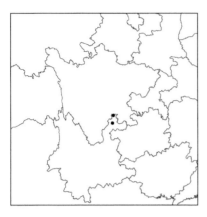

Map 13. *Berberis caudatifolia*

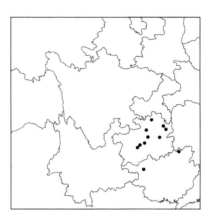

Map 14. *Berberis cavaleriei*

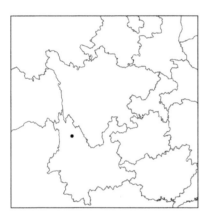

Map 15. *Berberis centiflora*

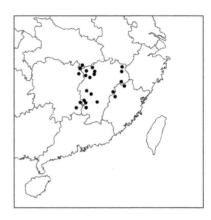

Map 16a. *Berberis chingii* subsp. *chingii*

Map 16b. *Berberis chingii* subsp. *wulingensis*

Map 17. *Berberis chingshuiensis*

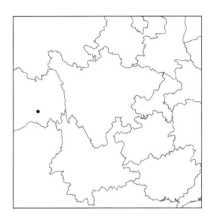

Map 18. *Berberis chrysophaera*

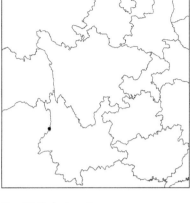

Map 19. *Berberis coxii*

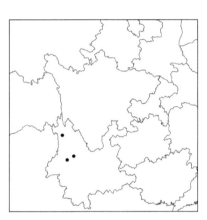

Map 20. *Berberis davidii*

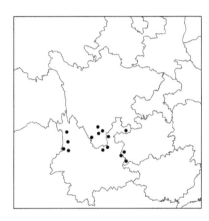

Map 21. *Berberis deinacantha*

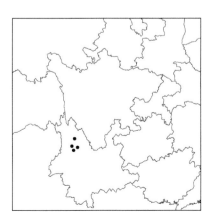

Map 22. *Berberis delavayi*

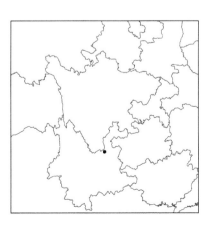

Map 23. *Berberis dongchuanensis*

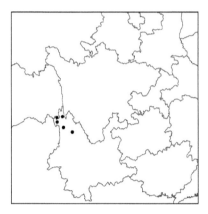

Map 24. *Berberis dumicola*

Map 25. *Berberis ebianensis*

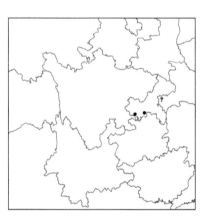

Map 26. *Berberis fallaciosa*

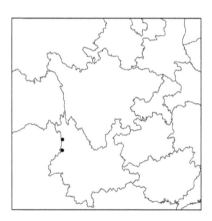

Map 27. *Berberis fallax*

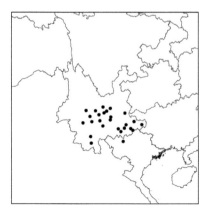

Map 28. *Berberis ferdinandi-coburgii*

Map 29. *Berberis fujianensis*

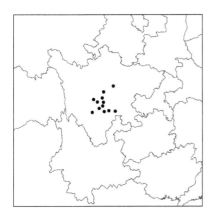

Map 30. *Berberis gagnepainii*

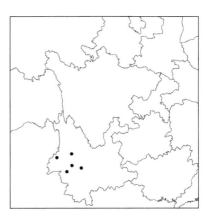

Map 31. *Berberis griffithiana*

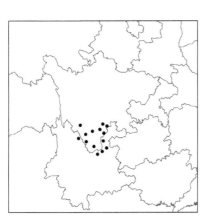

Map 32. *Berberis grodtmanniana*

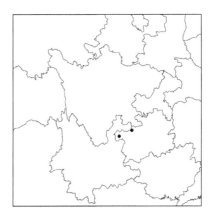

Map 33. *Berberis guizhouensis*

Map 34. *Berberis hayatana*

Map 35. *Berberis holocraspedon*

Map 36a. *Berberis hookeri* subsp. *hookeri*

Map 36b. *Berberis hookeri* subsp. *longipes*

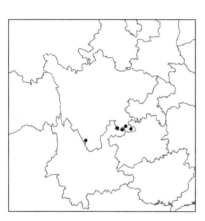

Map 37. *Berberis hsuyunensis*

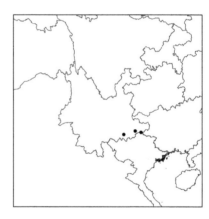

Map 38. *Berberis hypoxantha*

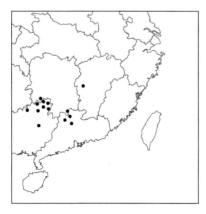

Map 39. *Berberis impedita*

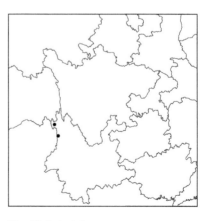

Map 40. *Berberis incrassata*

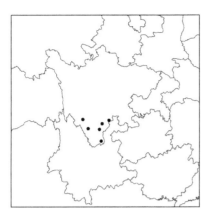

Map 41. *Berberis insolita*

Map 42. *Berberis jiangxiensis*

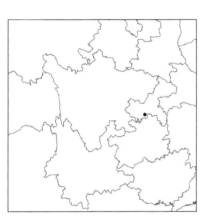

Map 43. *Berberis jinfoshanensis*

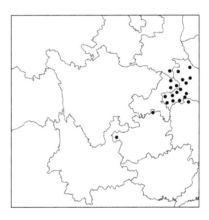

Map 44. *Berberis julianae*

Map 45. *Berberis kawakamii*

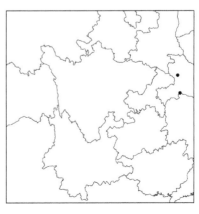

Map 46. *Berberis laojunshanensis*

Map 47. *Berberis lempergiana*

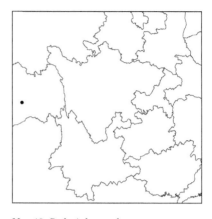

Map 48. *Berberis leptopoda*

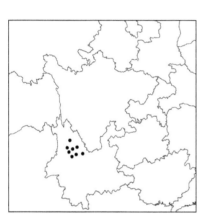

Map 49. *Berberis levis*

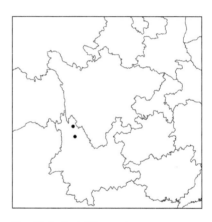

Map 50. *Berberis lijiangensis*

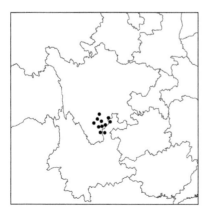

Map 51. *Berberis liophylla*

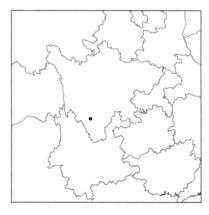

Map 52. *Berberis lubrica*

Map 53. *Berberis mingetsensis*

Map 54. *Berberis morii*

Map 55. *Berberis nantoensis*

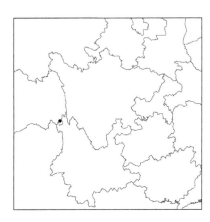

Map 56. *Berberis nujiangensis*

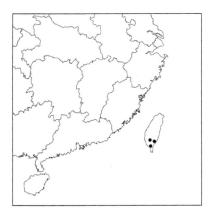

Map 57. *Berberis pengii*

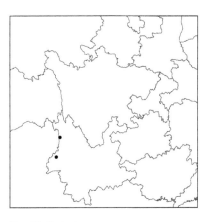

Map 58. *Berberis petrogena*

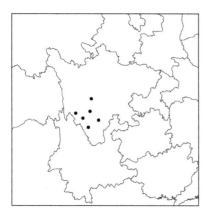

Map 59. *Berberis phanera*

Map 60. *Berberis photiniifolia*

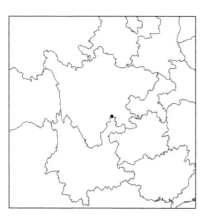

Map 61. *Berberis pingshanensis*

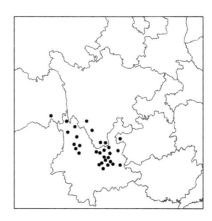

Map 62. *Berberis pruinosa*

Map 63. *Berberis ravenii*

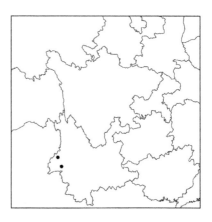

Map 64. *Berberis replicata*

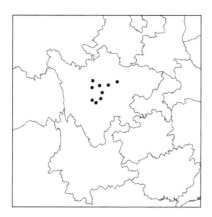

Map 65. *Berberis sanguinea*

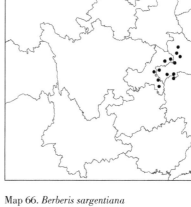

Map 66. *Berberis sargentiana*

Map 67. *Berberis schaaliae*

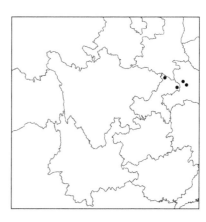

Map 68. *Berberis silvicola*

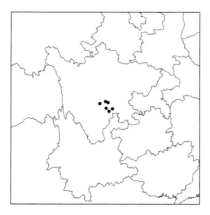

Map 69. *Berberis simulans*

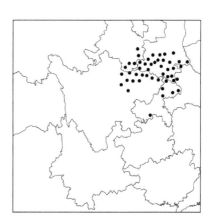

Map 70. *Berberis soulieana*

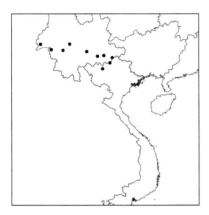

Map 71. *Berberis subacuminata*

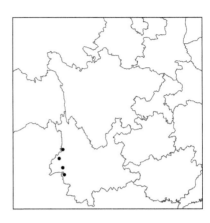

Map 72. *Berberis sublevis*

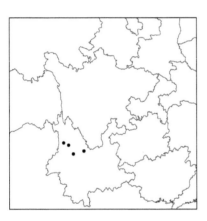

Map 73. *Berberis taliensis*

Map 74. *Berberis tarokoensis*

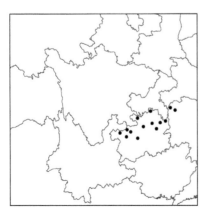

Map 75. *Berberis tengii*

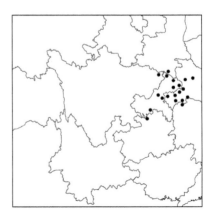

Map 76. *Berberis triacanthophora*

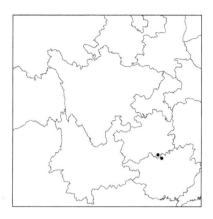

Map 77. *Berberis uniflora*

Map 78. *Berberis veitchii*

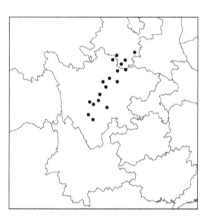

Map 79. *Berberis verruculosa*

Map 80. *Berberis vietnamensis*

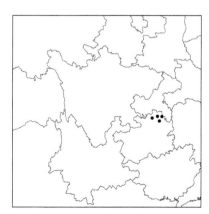

Map 81. *Berberis wuchuanensis*

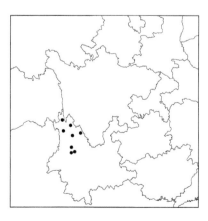

Map 82. *Berberis wui*

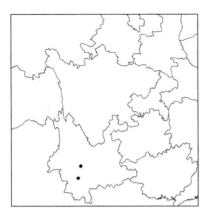

Map 83. *Berberis wuliangshanensis*

Map 84. *Berberis wuyiensis*

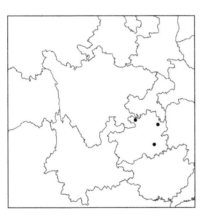

Map 85. *Berberis xanthoclada*

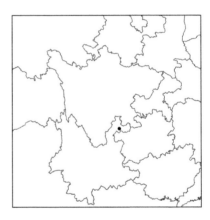

Map 86. *Berberis yiliangensis*

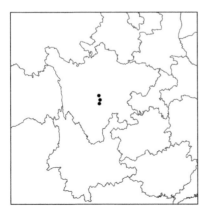

Map 87. *Berberis yingjingensis*

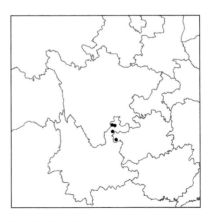

Map 88. *Berberis zhaotongensis*

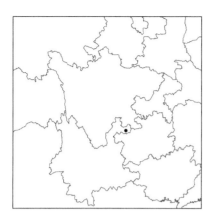

Map 89. *Berberis zhengxiongensis*

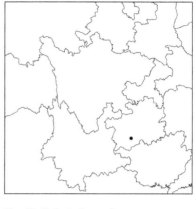

Map 90. *Berberis ziyunensis*

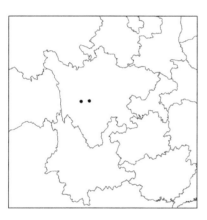

Map 91. *Berberis abbreviata*

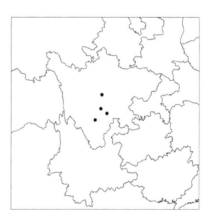

Map 92. *Berberis aemulans*

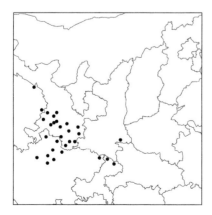

Map 93. *Berberis aggregata*

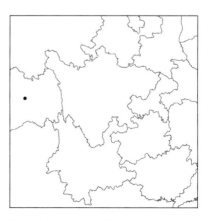

Map 94. *Berberis agricola*

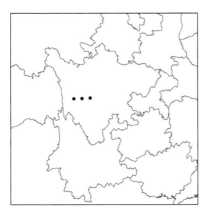

Map 95. *Berberis ambrozyana*

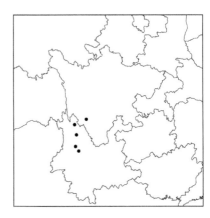

Map 96. *Berberis amoena*

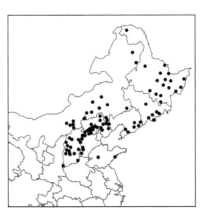

Map 97. *Berberis amurensis*

Map 98. *Berberis angulosa*

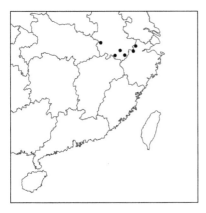

Map 99. *Berberis anhweiensis*

Map 100. *Berberis aristata*

Map 101. *Berberis asiatica*

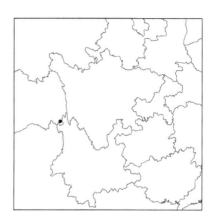

Map 102. *Berberis atroviridiana*

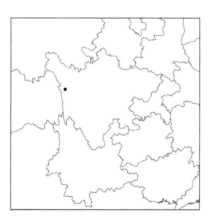

Map 103. *Berberis baiyuensis*

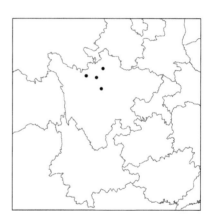

Map 104. *Berberis barkamensis*

Map 105. *Berberis basumchuensis*

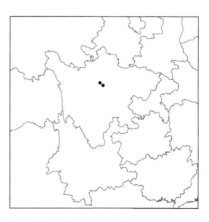

Map 106. *Berberis bawangshanensis*

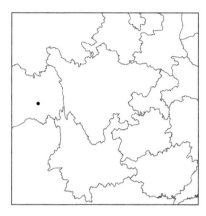

Map 107. *Berberis baxoiensis*

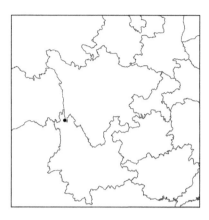

Map 108. *Berberis beimanica*

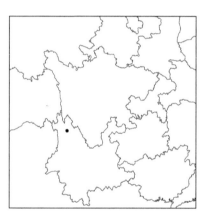

Map 109. *Berberis biguensis*

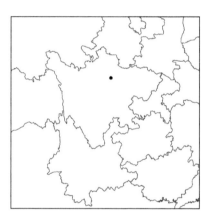

Map 110. *Berberis boschanii*

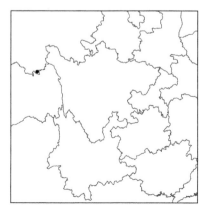

Map 111. *Berberis bouffordii*

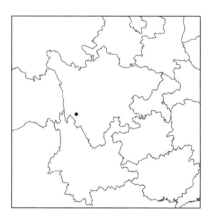

Map 112. *Berberis bowashanensis*

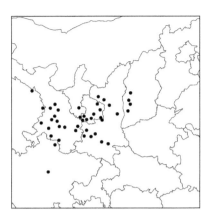

Map 113. *Berberis brachypoda*

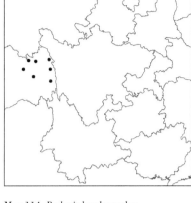

Map 114. *Berberis brachystachys*

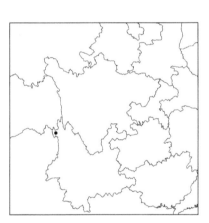

Map 115. *Berberis bracteata*

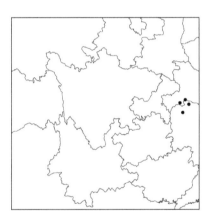

Map 116. *Berberis brevipaniculata*

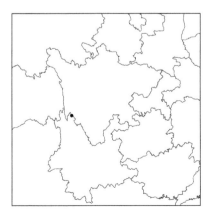

Map 117. *Berberis brevipedicellata*

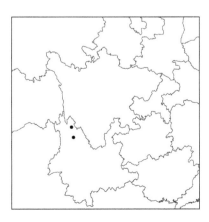

Map 118. *Berberis calcipratorum*

Map 119. *Berberis calliobotrys*

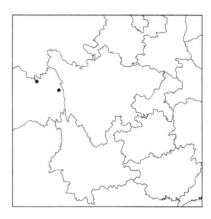

Map 120. *Berberis campylotropa*

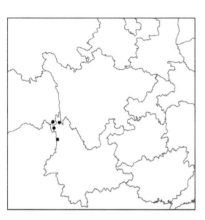

Map 121. *Berberis capillaris*

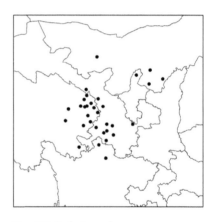

Map 122. *Berberis caroli*

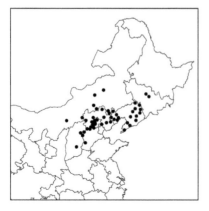

Map 123. *Berberis chinensis*

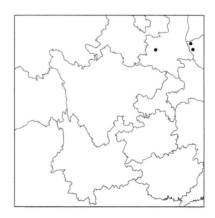

Map 124. *Berberis circumserrata*

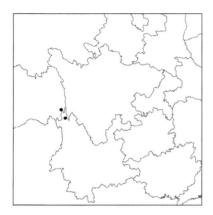

Map 125. *Berberis concolor*

Map 126. *Berberis cooperi*

Map 127. *Berberis cornuta*

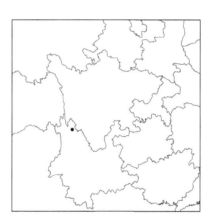

Map 128. *Berberis crassilimba*

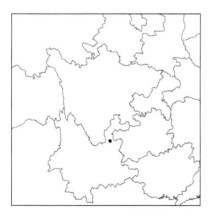

Map 129. *Berberis dahaiensis*

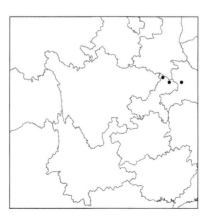

Map 130. *Berberis daiana*

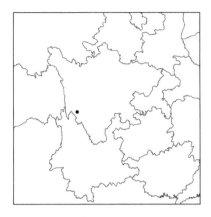

Map 131. *Berberis daochengensis*

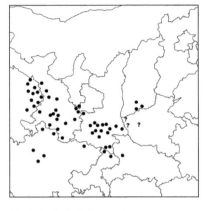

Map 132. *Berberis dasystachya*

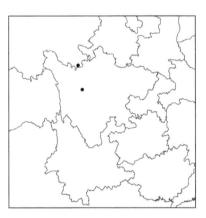

Map 133. *Berberis dawoensis*

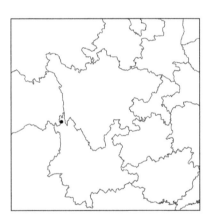

Map 134. *Berberis deqenensis*

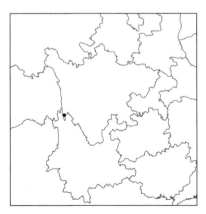

Map 135. *Berberis derongensis*

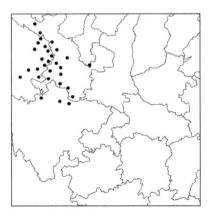

Map 136. *Berberis diaphana*

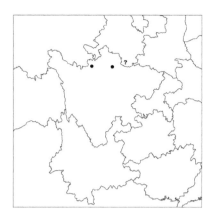

Map 137. *Berberis dictyoneura*

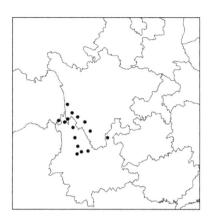

Map 138. *Berberis dictyophylla*

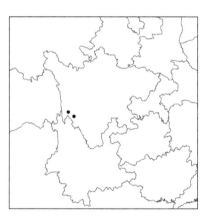

Map 139. *Berberis difficilis*

Map 140. *Berberis dispersa*

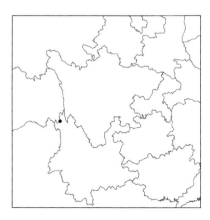

Map 141. *Berberis dokerlaica*

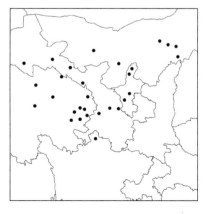

Map 142. *Berberis dubia*

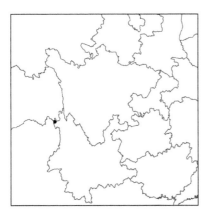

Map 143. *Berberis dulongjiangensis*

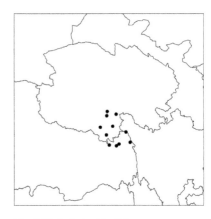

Map 144. *Berberis elliptifolia*

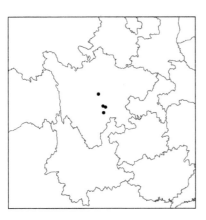

Map 145. *Berberis emeishanensis*

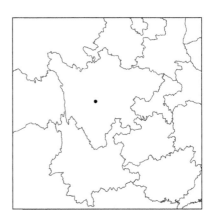

Map 146. *Berberis epedicellata*

Map 147. *Berberis erythroclada*

Map 148. *Berberis everestiana*

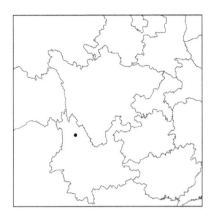

Map 149. *Berberis exigua*

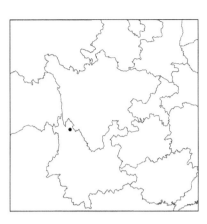

Map 150. *Berberis fengii*

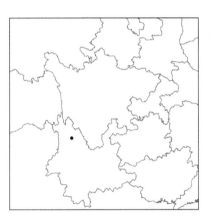

Map 151. *Berberis forrestii*

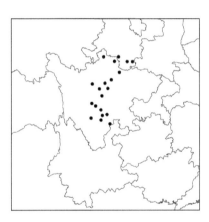

Map 152. *Berberis francisci-ferdinandi*

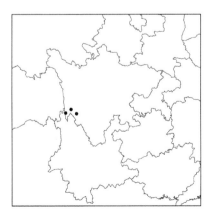

Map 153. *Berberis gaoshanensis*

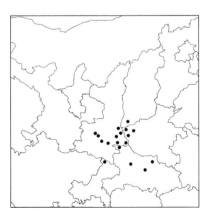

Map 154. *Berberis gilgiana*

Map 155. *Berberis gilungensis*

Map 156. *Berberis glabramea*

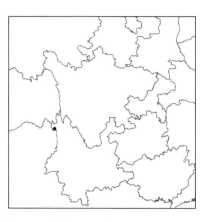

Map 157. *Berberis gongshanensis*

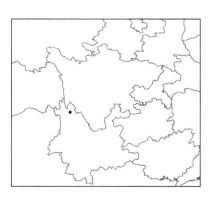

Map 158. *Berberis gyaitangensis*

Map 159. *Berberis gyalaica*

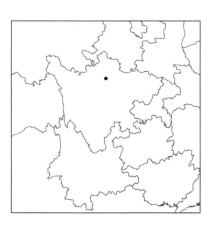

Map 160. *Berberis heishuiensis*

Map 161. *Berberis hemsleyana*

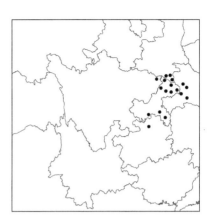

Map 162. *Berberis henryana*

Map 163. *Berberis heteropoda*

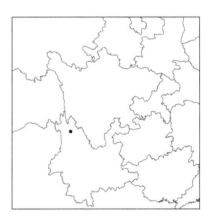

Map 164. *Berberis hubianensis*

Map 165. *Berberis humidoumbrosa*

Map 166. *Berberis hypericifolia*

Map 167. *Berberis integerrima*

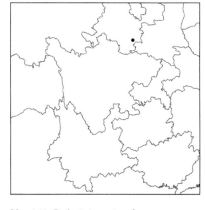

Map 168. *Berberis integripetala*

Map 169. *Berberis jaeschkeana* var. *usteriana*

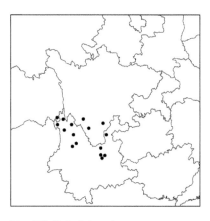

Map 170. *Berberis jamesiana*

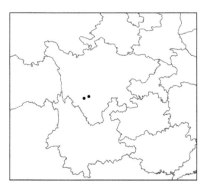

Map 171. *Berberis jiulongensis*

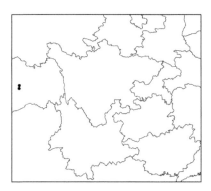

Map 172. *Berberis johannis*

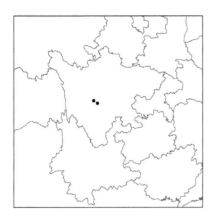

Map 173. *Berberis kangdingensis*

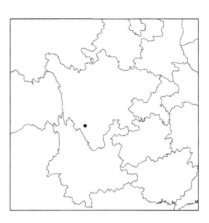

Map 174. *Berberis kangwuensis*

Map 175. *Berberis kartanica*

Map 176. *Berberis koehneana*

Map 177. *Berberis kongboensis*

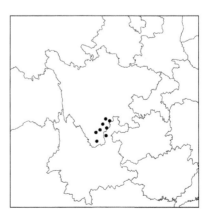

Map 178. *Berberis leboensis*

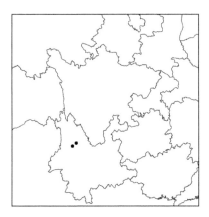

Map 179. *Berberis lecomtei*

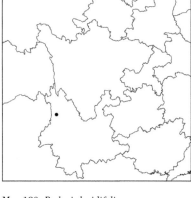

Map 180. *Berberis lepidifolia*

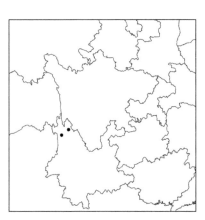

Map 181. *Berberis leptoclada*

Map 182. *Berberis lhunzensis*

Map 183. *Berberis lhunzhubensis*

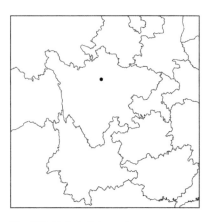

Map 184. *Berberis lixianensis*

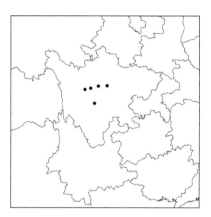

Map 185. *Berberis longipedicellata*

Map 186. *Berberis ludlowii*

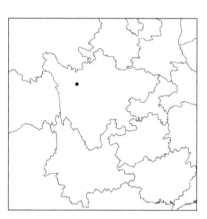

Map 187. *Berberis luhuoensis*

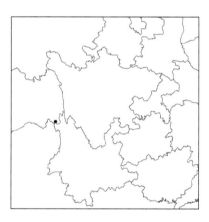

Map 188. *Berberis mabiluoensis*

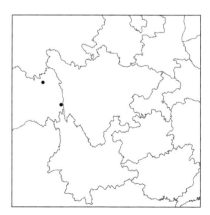

Map 189. *Berberis markamensis*

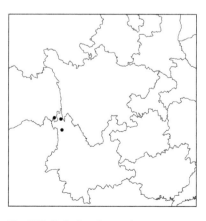

Map 190. *Berberis mekongensis*

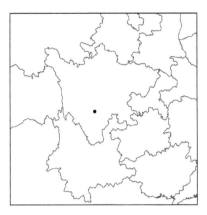

Map 191. *Berberis mianningensis*

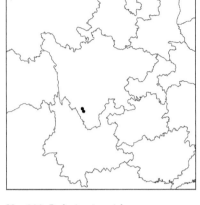

Map 192. *Berberis microtricha*

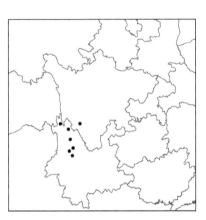

Map 193. *Berberis minutiflora*

Map 194. *Berberis moloensis*

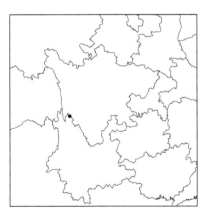

Map 195. *Berberis monticola*

Map 196. *Berberis morrisonensis*

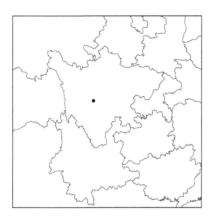

Map 197. *Berberis mouillacana*

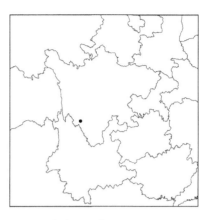

Map 198. *Berberis muliensis*

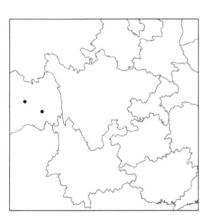

Map 199. *Berberis multiserrata*

Map 200. *Berberis nambuensis*

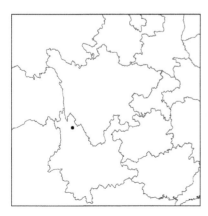

Map 201. *Berberis nanifolia*

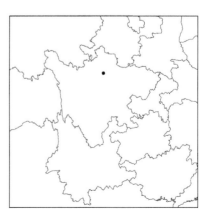

Map 202. *Berberis ngawaica*

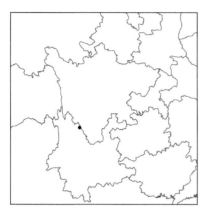

Map 203. *Berberis ninglangensis*

Map 204. *Berberis nyingchiensis*

Map 205. *Berberis obovatifolia*

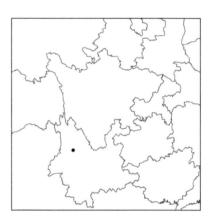

Map 206. *Berberis pallens*

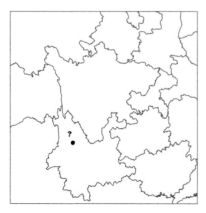

Map 207. *Berberis papillifera*

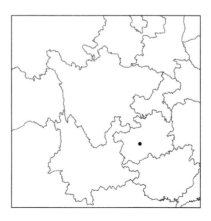

Map 208. *Berberis pingbaensis*

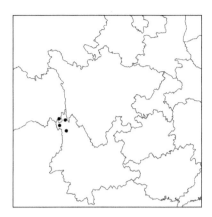

Map 209. *Berberis platyphylla*

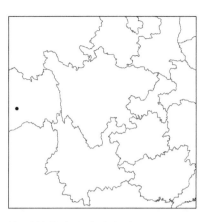

Map 210. *Berberis pluvisylvatica*

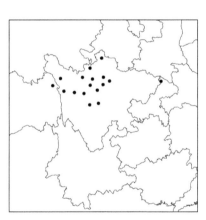

Map 211. *Berberis polyantha*

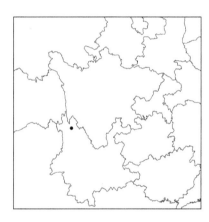

Map 212. *Berberis polybotrys*

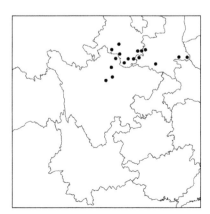

Map 213. *Berberis potaninii*

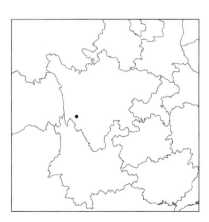

Map 214. *Berberis pratensis*

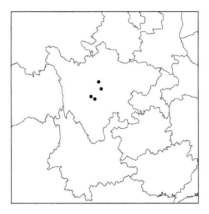

Map 215. *Berberis prattii*

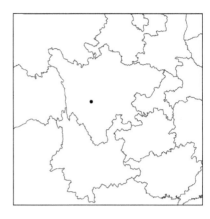

Map 216. *Berberis pruinosifolia*

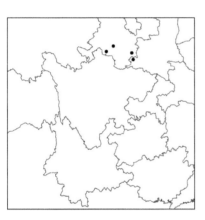

Map 217. *Berberis pseudothunbergii*

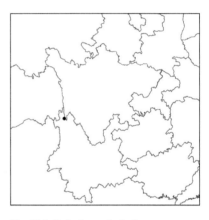

Map 218. *Berberis pseudotibetica*

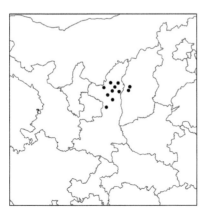

Map 219. *Berberis purdomii*

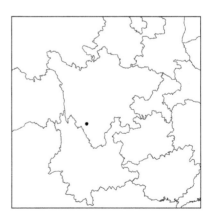

Map 220. *Berberis purpureocaulis*

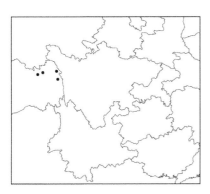

Map 221. *Berberis qamdoensis*

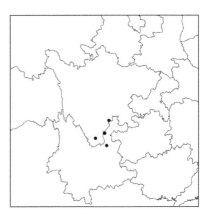

Map 222. *Berberis qiaojiaensis*

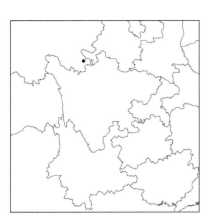

Map 223. *Berberis qinghaiensis*

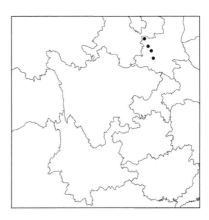

Map 224. *Berberis reticulata*

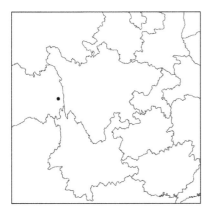

Map 225. *Berberis reticulinervis*

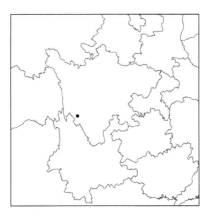

Map 226. *Berberis retusa*

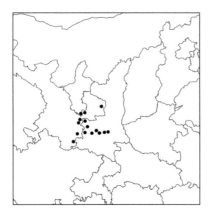

Map 227. *Berberis salicaria*

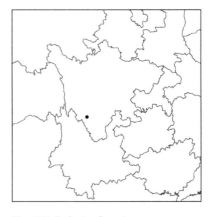

Map 228. *Berberis saltuensis*

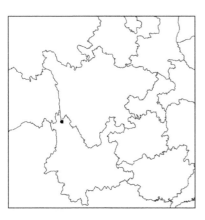

Map 229. *Berberis saxatilis*

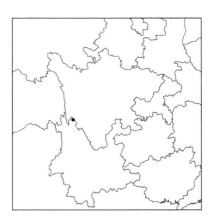

Map 230. *Berberis scrithalis*

Map 231. *Berberis sherriffii*

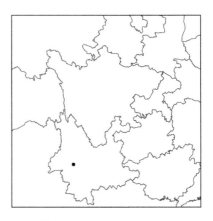

Map 232. *Berberis shunningensis*

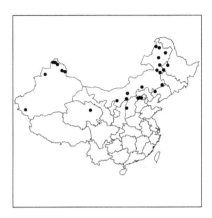

Map 233. *Berberis sibirica*

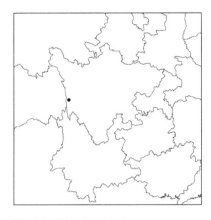

Map 234. *Berberis sichuanica*

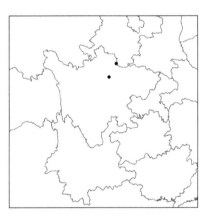

Map 235. *Berberis silva-taroucana*

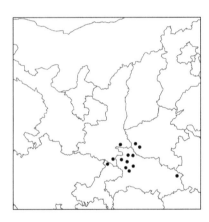

Map 236. *Berberis subsessiliflora*

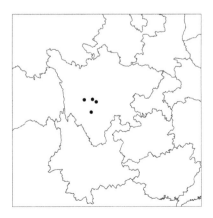

Map 237. *Berberis tachiensis*

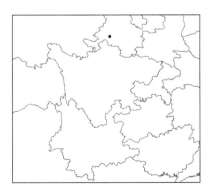

Map 238. *Berberis taoensis*

Map 239. *Berberis temolaica*

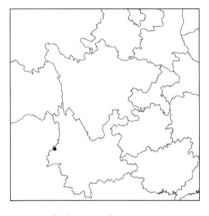

Map 240. *Berberis tengchongensis*

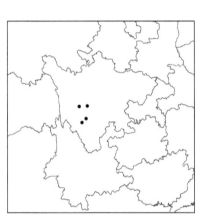

Map 241. *Berberis tenuipedicellata*

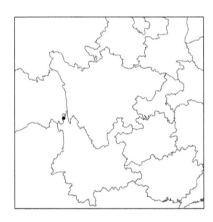

Map 242. *Berberis tenuispina*

Map 243. *Berberis thomsoniana*

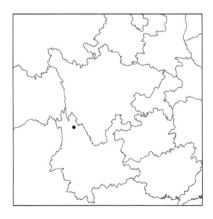

Map 244. *Berberis tianbaoshanensis*

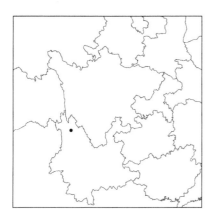

Map 245. *Berberis tianchiensis*

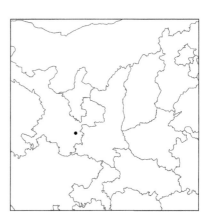

Map 246. *Berberis tianshuiensis*

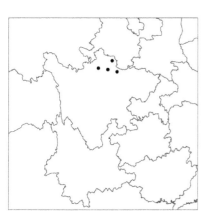

Map 247. *Berberis tischleri*

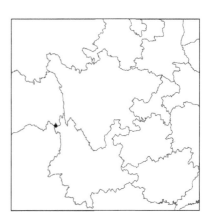

Map 248. *Berberis tomentulosa*

Map 249. *Berberis trichohaematoides*

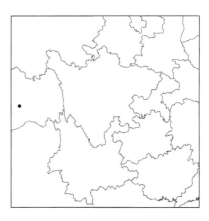

Map 250. *Berberis tsangpoensis*

Map 251. *Berberis tsarica*

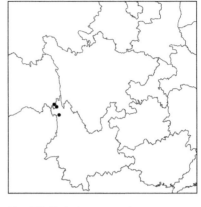

Map 252. *Berberis tsarongensis*

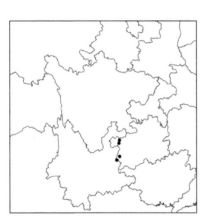

Map 253. *Berberis tsienii*

Map 254. *Berberis ulicina*

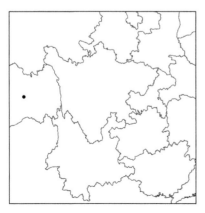

Map 255. *Berberis umbratica*

Map 256. *Berberis virescens*

Map 257. *Berberis virgetorum*

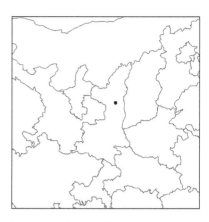

Map 258. *Berberis wanhuashanensis*

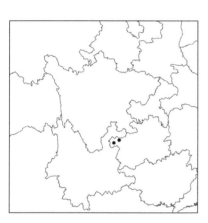

Map 259. *Berberis weiningensis*

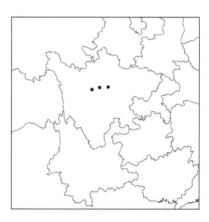

Map 260. *Berberis wenchuanensis*

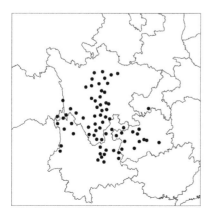

Map 261. *Berberis wilsoniae*

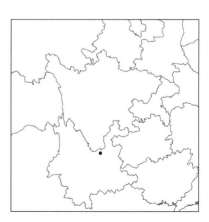

Map 262. *Berberis woomungensis*

Map 263. *Berberis xanthophlaea*

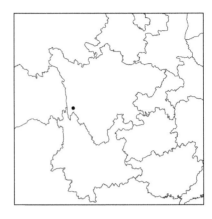

Map 264. *Berberis xiangchengensis*

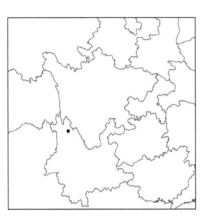

Map 265. *Berberis xiaozhongdianensis*

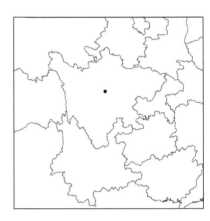

Map 266. *Berberis yaanica*

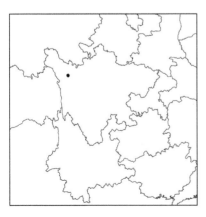

Map 267. *Berberis yalongensis*

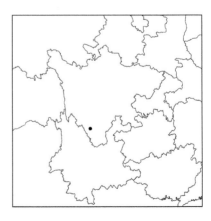

Map 268. *Berberis yanyuanensis*

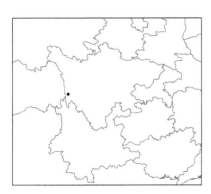

Map 269. *Berberis yarigongensis*

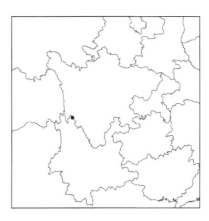

Map 270. *Berberis yingii*

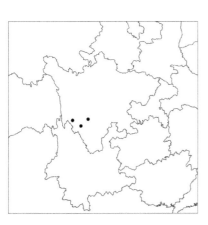

Map 271. *Berberis yui*

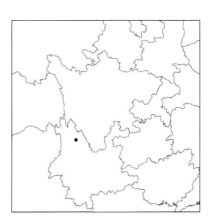

Map 272. *Berberis yulongshanensis*

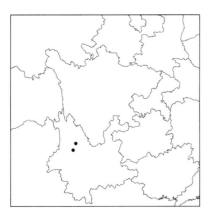

Map 273. *Berberis yunnanensis*

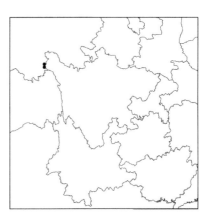

Map 274. *Berberis yushuensis*

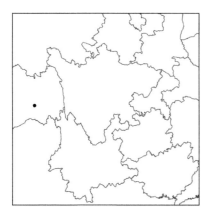

Map 275. *Berberis zayulana*

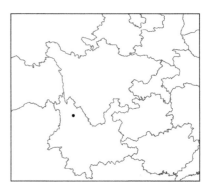

Map 276. *Berberis zhaoi*

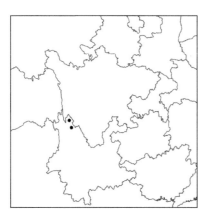

Map 277. *Berberis zhongdianensis*

Addenda. *Berberis concinna*

BIBLIOGRAPHY

Manuscript Material

Ludlow, F. Diaries, 1914–1949, Mss. Eur. D979, British Library, London.

Ludlow, F., G. Sherriff & G. Taylor. 1938. Collection Book, Botany Library, Natural History Museum, London.

Purdom, W. Papers, 1909–1912. Arnold Arboretum, Boston.

Royal Botanic Gardens, Kew, Herbarium Loans Books.

Royal Horticultural Society, London. Lindley, Library, Minutes of Committee and Publications Committee meetings, 1923–1947.

Sherriff, G. Diary 1938. Library, Royal Botanic Garden Edinburgh.

Wilson, E. H. Papers, 1896–1952. Arnold Arboretum, Boston.

Yu, T. T. Field Notes 1932–1938 (originals destroyed but imaged on http://rheum.huh.harvard.edu/ttyu/default).

Literature Cited

Adhikari, B. 2010. Systematics and phylogeographic studies of *Berberis* L. (Berberidaceae) in the Nepal Himalaya. Ph.D. Thesis, University of Edinburgh, Edinburgh.

Adhikari, B., C. A. Pendry, R. T. Pennington & R. Milne. 2012. A revision of *Berberis* s.s. (Berberidaceae) in Nepal. Edinburgh J. Bot. 69: 447–522.

Adhikari B., R. Milne, R. T. Pennington, T. Särkinen & C. A. Pendry. 2015. Systematics and biogeography of *Berberis* s.l. inferred from nuclear ITS and chloroplast ndhF gene sequences. Taxon 64(1): 39–48.

Ahrendt, L. W. A. 1939. Some new Asiatic barberries in cultivation. Kew Bull. 1939: 261–275.

Ahrendt, L. W. A. 1941a. New deciduous berberises. Gard. Chron., ser. 3, 109: 100–101.

Ahrendt, L. W. A. 1941b. A survey of the genus *Berberis* in Asia. J. Bot. 79(Suppl.): 1–80.

Ahrendt, L. W. A. 1942. A survey of the genus *Berberis* in Asia. J. Bot. 80(Suppl.): 81–104.

Ahrendt, L. W. A. 1944a. A survey of the genus *Berberis* in Asia. J. Bot. 80(Suppl.): 105–112.

Ahrendt, L. W. A. 1944b. Newcomers to the Barberries—II. Gard. Ill. 64: 425–426.

Ahrendt, L. W. A. 1945a. A survey of the genus *Berberis* in Asia. J. Bot. 80(Suppl.): 113–116.

Ahrendt, L. W. A. 1945b. Some new and little known *Berberis* from India. J. Roy. Asiat. Soc. Bengal 11: 1–5.

Ahrendt, L. W. A. 1954. On the relationship of *Berberis prattii* and *B. polyantha*. J. Roy. Hort. Soc. 79: 189–193.

Ahrendt, L. W. A. 1956. Berberidaceae in résultats des expéditions scientifiques genevoises au Népal en 1952 et 1954 (partie botanique). Candollea 15: 153–155.

Ahrendt, L. W. A. 1961. *Berberis* and *Mahonia*: A taxonomic revision. J. Linn. Soc., Bot. 57: 1–410.

Airy Shaw, H. K. 1965. Diagnoses of new families, new names, etc., for the seventh edition of Willis's Dictionary. Kew Bull. 18: 249–273.

Aitchison, J. E. T. 1881. On the flora of Kuram Valley, etc., Afghanistan (Part I). J. Linn. Soc., 18: 1–113.

Aitchison, J. E. T. 1882. On the flora of Kuram Valley, etc., Afghanistan (Part II). J. Linn. Soc., 19: 139–200.

An, M. T. 2008. *Berberis pingbaensis*, a new species of *Berberis* Linn (Berberidaceae) from Guizhou, China. Bull. Bot. Res., Harbin 28(6): 641.

An, Z. X. 1995. *Berberis*. Pp. 1–5 *in* Z. M. Mao (editor), Flora Xinjiangensis, Vol. 2(2). Xinjiang Science & Technology & Hygiene Publishing House, Xinjiang.

Anonymous. 1982. Abbreviations of English names of institutions cited in Acta Phytotaxonomica Sinica. Acta Phytotax. Sin. 20(5): 245–256.

APG II (Angiosperm Phylogeny Group). 2003. An update of the Angiosperm Phylogeny Group classification for the orders and families of flowering plants: APG II. Bot. J. Linn. Soc. 141(4): 339–436.

APG III (Angiosperm Phylogeny Group). 2009. An update of the Angiosperm Phylogeny Group classification for the orders and families of flowering plants: APG III. Bot. J. Linn. Soc. 161(2): 105–121.

Averyanov, L. V., Nguyen Tien Hiep, D. K. Harder & Phan Ke Loc. 2002. The history of discovery and natural habitats of *Xanthocyparis vietnamensis* (Cupressaceae). Turczaninowia 5(4): 31–39.

Bao, S. Y. 1985. New taxa of *Berberis* Linn. from Yunnan. Bull. Bot. Res., Harbin 5(3): 1–35.

Bao, S. Y. 1997. *Berberis*. Pp. 27–89 *in* C. Y. Wu, Flora Yunnanica, Vol. 7. Science Press, Beijing.

Borodina-Grabovskaya, A. E., V. I. Grubov & M. A. Mikhailova. 2007. *Berberis*. Pp. 159–168 *in* Plants of Central Asia, Vol. 12. Science Publications, Enfield, New Hampshire (trans. of Rasteniya Tsentral'noi Asii, Vol. 12. 141, Izd-vo Sankt-Peterburgskoi goc. Khimiko-farmatzevticheskoi akademii, St. Petersburg, 2001).

Borodina-Grabovskaya, A. E. (editor). 2010. Catalogue of the Type Specimens of East-Asian Vascular Plants in the Herbarium of the V. L. Komarov Botanical Institute (LE) Part 2 (China). KMK Scientific Press, Moscow, St. Petersburg.

Boufford, D. E. 2013. *Mahonia* (Berberidaceae) in Asia: Typification, synonymy and notes. Mem. New York Bot. Gard. 108: 251–283.

Boufford, D. E. & P. P. van Dijk. 2000. South-central China. Pp. 338–351 *in* R. A. Mittermeier, N. Myers & C. G. Mittermeier (editors), Hotspots: Earth's Biologically Richest and Most Endangered Terrestrial Ecoregions. Cemex/Agrupación Sierra Madre, Mexico City.

Bretschneider, E. 1898a. History of European Botanical Discoveries in China. Sampson Low, Marston and Co., London.

Bretschneider, E. 1898b. Croquis des Regions du Yunnan, ou des Collections de Plantes. History of European botanical discoveries in China, Supplement V. A. Iliin, St. Petersburg.

Briggs, R. 1993. 'Chinese' Wilson, a Life of Ernest H. Wilson 1876–1930. HMSO, London.

Britton, N. L. & A. Brown. 1913. An Illustrated Flora of the Northern United States, Canada and the British Possessions. 2nd ed. Charles Scribner's Sons, New York.

Browicz, K. 1988. Chorology of Trees and Shrubs in Southwest Asia and Adjacent Regions, Vol. 6. Institute of Dendrology, Kórnik, Poland.

Browicz, K. & J. Zieliński. 1975. *Berberis*. Pp. 1–24 *in* K. H. Rechinger, Flora Iranica: Flora des Iranischen Hochlandes und der Umrahmenden Girbinge, Persien, Afghanistan. Teile von West-Pakistan, Nord-Iraq, Azerbaidian, Turkmenistan. Fasc. 111/16, Vol. 10. Akademische Druck und Verlagsanstalt, Graz.

Bunge, A. A. 1843. *Berberis integerrima, Berberis nummularia*. Index Seminum [Tartu], 6.

Bunge, A. A. 1847. Alexandri Lehman, Reliquiae Botanicae, Dorpat.

Byhouwer, J. T. P. 1928. Berberidaceae. *In* A. Rehder & E. H. Wilson, Enumeration of the Ligneous Plants Collected by J. F. Rock on the Arnold Arboretum Expedition to Northwestern China and Northeastern Tibet. J. Arnold Arbor. 10: 43–48.

Callaghan, C. 2011. Wa Shan–Emei Shan, a further comparison. Int. Dendrol. Soc. Year Book 2010: 72–88.

de Candolle, A. P. 1821. Regni Vegetabilis Systema Naturale, Vol. 2. Treuttel et Würtz, Paris.

Chamberlain, D. F. 1982. A revision of *Rhododendron*. II. Subgenus *Hymenanthes*. Notes Roy. Bot. Gard. Edinburgh 39: 209–486.

Chamberlain, D. F. & C. M. Hu. 1985. A synopsis of *Berberis* section *Wallichianae*. Notes Roy. Bot. Gard. Edinburgh 42. (3): 529–557.

Chamberlain, D. F. & S. J. Rae. 1990. A revision of *Rhododendron*. IV. Subgenus *Tsutsusi*. Edinburgh J. Bot. 47(2): 89–200.

Chen, F. L, R. C. Peng & Y. F. Deng. 2012. Identity of *Berberis ziyunensis* P. G. Hsiao & Z. Y. Li (Berberidaceae). J. Trop. Subtrop. Bot. 20(6): 605–607.

Cheng, W. C. 1934. *Berberis chingii*. Contr. Biol. Lab. Sci. Soc. China, Bot. Ser. 9: 191.

Chung, K. F. 2009. Unearthing a forgotten legacy of 20th century floristics: The collection of Taiwanese plant specimens in the herbarium of the Academy of Natural Sciences (PH). Taiwania 54(2): 159–167.

China Plant Specialist Group. 2004a. *Berberis bicolor*. The IUCN Red List of Threatened Species 2004: e. T46555A11067019. <http://dx.doi.org/10.2305/IUCN.UK .2004.RLTS.T46555A11067019.en>, accessed 10 June 2019.

China Plant Specialist Group. 2004b. *Berberis candidula*. The IUCN Red List of Threatened Species 2004: e. T46556A11067057. <http://dx.doi.org/10.2305/IUCN.UK .2004.RLTS.T46556A11067057.en>, accessed 10 June 2019.

China Plant Specialist Group. 2004c. *Berberis iteophylla*. The IUCN Red List of Threatened Species 2004: e. T46557A11067100. <http://dx.doi.org/10.2305/IUCN.UK .2004.RLTS.T46557A11067100.en>, accessed 10 June 2019.

China Plant Specialist Group. 2004d. *Berberis johannis*. The IUCN Red List of Threatened Species 2004: e. T46558A11067138. <http://dx.doi.org/10.2305/IUCN.UK .2004.RLTS.T46558A11067138.en>, accessed 10 June 2019.

China Plant Specialist Group. 2004e. *Berberis silvicola*. The IUCN Red List of Threatened Species 2004: e. T46559A11058895. <http://dx.doi.org/10.2305/IUCN.UK .2004.RLTS.T46559A11058895.en>, accessed 10 June 2019.

China Plant Specialist Group. 2004f. *Berberis taronensis*. The IUCN Red List of Threatened Species 2004. e. T46560A11059389. <http://dx.doi.org/10.2305/IUCN.UK .2004.RLTS.T46560A11059389.en>, accessed 10 June 2019.

China Plant Specialist Group. 2004g. *Berberis woomungensis*. The IUCN Red List of Threatened Species 2004: e. T46561A11059242. <http://dx.doi.org/10.2305/IUCN.UK .2004.RLTS.T46561A11059242.en>, accessed 10 June 2019.

China Plant Specialist Group. 2004h. *Berberis xanthophloea*. The IUCN Red List of Threatened Species 2004: e. T46562A11059551. <http://dx.doi.org/10.2305/IUCN.UK

.2004.RLTS.T46562A11059551.en>, accessed 10 June 2019.

Conway, W. M. 1894. Climbing and Exploration in the Karakoram Himalayas. T. Fisher Unwin, London.

Creasey, L. B. 1935. *Berberis sanguinea*. Gard. Chron., ser. 3, 98: 11.

Cullen, J. 1980. Revision of *Rhododendron*. I. Subgenus *Rhododendron* sections *Rhododendron* and *Pogonanthum*. Notes Roy. Bot. Gard. Edinburgh 39: 1–207.

Desfontaines, R. L. 1804. Tableau de l'école de botanique du Muséum d'histoire naturelle. J. A. Brosson, Paris.

Desfontaines, R. L. 1809. Histoire des Arbres et Arbrisseaux qui peuvent être cultivés en pleine terre sur le sol de la France, II. J. A. Brosson, Paris.

Diels, F. L. E. 1901. *Berberis*. *In* Die Flora von Central-China. Bot. Jahrb. 29: 340–341.

Ding, B. Z., S. Y. Wang & Z. Y. Gao. 1981. *Berberis*. Pp. 488–494 *in* B. Z. Ding (editor), Flora of Henan, Vol. 1. Hunan Science and Technology Press, Zhanlanguanlu.

Ding, C. S. 1986. *Berberis*. Pp. 312–315 *in* J. X. Wang (editor), Flora of Zhejiang, Vol. 2. Zhejiang Science and Technology Publishing House, Zhejiang.

Dinh, V. M. 1999. Biodiversity of medicinal plants in Hoang Lien Mountains. Pp. 7–30 *in* R. T. Sobey (editor), Biodiversity Value of Hoang Lien Mountains and Strategies for Conservation. Proceedings of Seminar and Workshop, 7–9th December, 1998, Sa Pa, Lao Cai Province, Vietnam. Society for Environmental Exploration, London.

Dunn, S. T. & W. J. Tutcher. 1912. Flora of Kwantung and Hongkong (China). Bull. Misc. Inform. Kew, addl. ser. 10: 32–33.

Fang, J. Y., Z. H. Wang & Z. Y. Tang. 2011. Atlas of Woody Plants in China: Distribution and Climate. Springer, Berlin, Heidelberg.

Farrer, R. 1916. Report of work in 1914 in Kansu and Tibet. J. Roy. Hort. Soc. 42: 47–114.

Farrer, R. 1917. On The Eaves of The World, Vols. 1 & 2. Edward Arnold, London.

Fedde, F. K. G. 1901. Versuch einer Monographie der Gattung Mahonia. Bot. Jahrb. Syst. 31: 30–131.

Fedtschenko, B. A. 1937. *Berberis*. Pp. 553–560 *in* L. Komarov, Fl. USSR 7. Akademiya Nauk SSSR, Leningrad (English trans., Fl. USSR 7: 422–427, 1970, Israel Program for Scientific Translations, Jerusalem).

Fletcher, H. R. 1975. A Quest of Flowers: The Plant Expeditions of Frank Ludlow and George Sherriff. Edinburgh University Press, Edinburgh.

Forbes, F. B. & W. B. Hemsley. 1886–1888. *Berberideae*. *In* An Enumeration of All the Plants Known from China Proper, Formosa, Hainan, Corea, the Luchu Archipelago, and the Island of Hong Kong. J. Linn. Soc., Bot. 23: 31–32.

Franchet, A. R. 1883. *Berberis*. *In* Plantae Davidianae. Nouv. Arch. Mus. Hist. Nat., sér. 2, 5: 177–178.

Franchet, A. R. 1885. *Berberis sanguinea*. *In* Plantae Davidianae. Nouv. Arch. Mus. Hist. Nat., sér. 2, 8: 194.

Franchet, A. R. 1887. *Berberis*. *In* Plantae Yunnanensis. Bull. Soc. Bot. France 33: 385–388.

Franchet, A. R. 1889. Plantae Delavayanae. P. Klincksieck, Paris.

Fu, S. X. 2001. *Berberis*. Pp. 382–394 *in* Flora Hubeiensis. Vol. 1. Hubei Science & Technology Publishing House, Hubei.

Gagnepain, F. 1938. *Berberis wallichiana*. P. 144 *in* J. H. Humbert, Supplement a la Flore Générale de L'Indo-Chine, Vol. 1. Muséum National d'Histoire Naturelle, Paris.

Government of the Socialist Republic of Vietnam, Ministry of Science, Technology and Environment. 1996. Sách Đỏ Việt

Nam (Red Data Book of Vietnam), Vol. 2. Thực vật (Plants). Science and Technics Publishing House, Hanoi.

Government of the Socialist Republic of Vietnam. 2006. Decree no. 32/2006/nd-cp of March 30, 2006. On Management of Endangered, Precious and Rare Forest Plants and Animals. Official Gazette, Hanoi.

Grierson, A. J. C. 1984. *Berberis*. Pp. 322–327 *in* A. J. C. Grierson & D. G. Long, Flora of Bhutan, Vol. 1(2). Royal Botanic Garden, Edinburgh.

Griffith, W. 1847. Posthumous Papers: Journals of Travels in Assam, Burma, Bhootan, Afghanistan and the Neighbouring Countries. Bishop's College Press, Calcutta.

Griffith, W. 1848. Posthumous Papers, Vol. 2: Itinerary Notes of Plants Collected in the Khasyah and Bootan Mountains, 1837–38, in Afghanistan and Neighbouring Countries, 1839 to 1841. J. F. Bellamy, Calcutta.

Griffith, W. 1849. Icones Plantarum Asiaticarum, Part II, on the Higher Crytogamous Plants. C. A. Serrao, Calcutta.

Grossheim, A. A. 1930. Flora Kavcaza, Vol. 2. Tiflis [Tbilisi], Georgia.

Gubanov, I. A., T. V. Bagdasarova & T. P. Balandina. 1998. Nauchnoe nasledie vydayushchikhsya russikh florister (Scientific heritage of the outstanding Russian experts in floristics). G. S. Karelin & I. P. Kirilov, Moscow State University, Moscow.

Handel-Mazzetti, H. 1925. Naturbilder aus Südwest China, Österreichischer Bundesverlag für Unterricht, Wissenschaft und Kunst, Vienna. (English trans., A Botanical Pioneer in South West China. David Winstanley, Brentwood, 1996.)

Handel-Mazzetti, H. 1931. *Berberis pruinosa* Franchet var. *centiflora* (Diels). Symbolae Sinicae, Vol. 7. Verlag von Julius Springer, Vienna.

Hao, K. S. 1938. Berberidaceae. *In* Pflanzengeographische Studien über den Kokonor-See und über das angrenzende Gebiet. Bot. Jahrb. Syst. 68: 599.

Harber, J. 2010. Getting to grips with *Berberis*. Plantsman, n.s., 9(2): 106–133.

Harber, J. 2012. Two new *Berberis* section *Wallichianae* from Western China. Curtis's Bot. Mag. 29(2): 115–121.

Harber, J. 2015. *Berberis* in Taiwan. Int. Dendrol. Soc. Year Book 2014: 49–57.

Harber, J. 2017. *Berberis bowashanensis*. Curtis's Bot. Mag. 34(2): 105–110.

Hayata, B. 1911. *Berberis kawakamii* and *Berberis morrisonensis*. *In* Materials for a flora of Formosa. J. Coll. Sci. Imp. Univ. Tokyo 30(1): 24–25.

Hayata, B. 1913. *Berberis aristato-serrulata* and *Berberis brevisepala*. Icon. Pl. Formosan. 3: 13–14.

Hayata, B. 1915. *Berberis mingetsensis*. Icon. Pl. Formosan. 5: 4–5, pl. II.

He, S. Y. 1992. *Berberis*. Pp. 260–262 *in* S. Y. He (editor), Flora of Beijing, 2nd ed. Beijing Publishing House, Beijing.

He, S. Z. & Q. H. Chen. 2004. *Berberis*. Pp. 11–38 *in* Q. H. Chen, Flora Guizhouensis. Vol. 10. Guizhou Science and Technology Publishing House, Guiyang Hien.

He, S. Z., T. Zhang, Y. Huang & W. Xu. 1995. Medicinal plant resources of *Berberis* L. in Guizhou province. China J. Chin. Mater. Med. 20(11): 646–649.

Helmerson, G. von. 1852. Alexander Lemann's Reise nach Buchara und Samarkand in den Jahren 1841 und 1842. Buchdrukerei für Keiselichen Akademie der Wissenschaften, St. Petersburg.

Hemsley, W. B. 1892. *Berberis polyantha*. *In* New Chinese Plants. J. Linn. Soc., Bot. 29: 302.

Herner, G. 1988. Harry Smith in China: Routes of his botanical travels. Taxon 37(2): 299–308.

Hien, L. T. T., N. N. Linh, P. L. B. Hang, N. P. Mai, H. T. T. Hue & N. V. Huan. 2018. Developing DNA barcodes for species identification of *Berberis* and *Dysosma* genera in Vietnam. Int. J. Agric. Biol. 20: 1097–1106.

Hillier Nurseries. 2014. The Hillier Manual of Trees and Shrubs, 8th ed. Royal Horticultural Society, London.

Hitchcock, A. S. 1929. *Berberis vulgaris*. International Botanical Congress. Cambridge (England), 1930. P. 147 *in* Nomenclature. Proposals by British Botanists. HMSO, London.

Hooker, J. D. 1855. Himalayan Journals; Notes of a Naturalist in Bengal, The Sikkim, Nepal, Himalayas, the Khasia Mountains etc, Vol. 1., John Murray, London.

Hooker, J. D. 1865. Catalogue of the plants distributed at the Royal Gardens, Kew, (under the sanction of the Secretary of State for India) from the herbaria of Griffith, Falconer, and Helfer. John E. Taylor, London.

Hooker, J. D. & T. Thomson. 1855. Flora Indica. W. Pamplin, London.

Hooker, J. D. & T. Thomson. 1875. Flora of British India 1. L. Reeve, London.

Hsiao, P. K. & W. C. Sung. 1974. Study on the medicinal plant resources of *Berberis* in China. Acta Phytotax. Sin. 12: 383–404, pl. 77 & 78.

Hu, C. M. 1986. A study of the genus *Berberis* L. from east and south China. Bull. Bot. Res., Harbin 6(2): 1–19.

Hu, C. M. 1995. *Berberis*. Pp. 11–13 *in* F. H. Chen & T. L. Wu, Flora of Guangdong, Vol. 2. Guangdong Science & Technology Press, Guangzhou.

Hu, C. M. & M. F. Watson. 2015. Plant Exploration in China. Pp. 212–235 *in* D. Y. Hong & S. Blackmore (editors), Plants of China: A Companion to the Flora of China. Science Press, Beijing; Cambridge University Press, Cambridge.

Hu, H. H. & H. Y. Chien. 1927–1937. Icones Plantarum Sinicarum, 5 vols. Commercial Press, Ltd., Shanghai.

Hu, Z. G., H. Y. Ma, J. S. Ma & D. Y. Hong. 2015. History of Chinese botanical institutions. Pp. 237–255 *in* D. Y. Hong & S. Blackmore (editors), Plants of China: A Companion to the Flora of China, Science Press, Beijing; Cambridge University Press, Cambridge.

Husain, T., B. Datt & R. R. Rao. 1994. On the identity of two taxa of *Berberis* (Berberidaceae) from Tibet. Sida 16(1): 17–21.

Husain, T., B. Datt & R. R. Rao. 1997. Palynological evidence supporting the identity of two taxa of *Berberis* (Berberidaceae) from Tibet. Sida 17(3): 575–578.

Hutchinson, J. 1959. The Families of Flowering Plants. Clarendon Press, Oxford.

Hyun, C. W. & Y. D. Kim. 2008. Morphological variation of *Berberis amurensis* complex. Taxon 38(2): 98–109.

Imchanitzkaja, N. 2005. Specimina Typica Taxorum Nonnullorum e Fasmilia Berberidaceae Florae Chinae in Herbario Instituti Botanici nomine V. L. Komarovii (LE) Conservata. Novosti Sist. Vyssh. Rast. 37: 268–283.

Institute of Botany in Jiangsu Province & Chinese Academy of Sciences (editors). 1982. Berberidaceae. Pp. 186–187 *in* Flora of Jiangsu, Vol. 2. Jiangsu Science and Technology Press, Nanjing.

Instituto Botanico Boreali-Occidentali Academiae Sinica. 1974. *Berberis*. Pp. 307–324 *in* Flora Tsinlingensis, Vol. 1, Spermatophyta. Part 2. Science Press, Beijing.

International Union for Conservation of Nature and Natural Resources (IUCN). [Unión Internacional para la Conservación de la Naturaleza y Recursos Naturales (UICN).] 2001. IUCN Red List Categories and Criteria, 2nd ed., vers. 3.1. IUCN, Gland, Switzerland; Cambridge, United Kingdom.

Jafri, S. M. H. 1975. Berberidaceae. Flora of West Pakistan, Vol. 87. Fakhiri Press, Karachi.

Jiang, Y. Z. 1985. *Berberis* Pp. 44–46 *in* L. K. Ling, Flora Fujianica, Vol. 2. Fujian Science and Technology Publishing House, Fuzhow.

Jin, S. Y. & Y. L. Chen. 1994. A Catalogue of Type Specimens in the Herbaria of China. Science Press, Beijing.

Jin, S. Y. & Y. L. Chen. 1999. A Catalogue of Type Specimens in the Herbaria of China (Suppl.). China, Forestry Publishing House, Beijing.

Jin, S. Y. & Y. L. Chen. 2007. A Catalogue of Type Specimens in the Herbaria of China, Suppl. 2. China, Forestry Publishing House, Beijing.

Jin, X. F., B. Y. Ding & H. Z. Wang. 2008. Validation of *Berberis chunanenis* (Berberidaceae), the name of a species endemic to Zhejiang, eastern China. Novon 18: 494.

Judd, W. S. & W. S. Kron. 1995. A revision of *Rhododendron* sections *Sciadorhodion, Rhodora* and *Viscidula.* Edinburgh J. Bot. 52: 1–54.

Karelin, G. S. & I. P. Kirilov. 1842. Enumeratio plantarum in desertis Songoriae orientalis et in jugo summarum alpium Alatau anno 1841 collectarum. Bull. Soc. Imp. Naturalistes Moscou 15.

Kilpatrick, J. 2014. Fathers of Botany: The Discovery of Chinese Plants by European Missionaries. Kew Publishing, Royal Botanic Gardens Kew; The University of Chicago Press, Chicago and London.

Kim, Y. D., S. H. Kim & L. R. Landrum. 2004. Taxonomic and phytogeographic implications from ITS phylogeny in *Berberis* (Berberidaceae). J. Pl. Res. 117: 175–182.

Kingdon-Ward, F. 1923. Mystery Rivers of Tibet. Seeley Service & Co. Ltd., London.

Kingdon-Ward, F. 1926a. Explorations in south-eastern Tibet. Geogr. J. (London) 67(2): 97–119,192.

Kingdon-Ward, F. 1926b. The Riddle of the Tsangpo Gorges. Edward Arnold, London.

Kingdon-Ward, F. 1934. A Plant Hunter in Tibet. Jonathan Cape Ltd., London.

Kingdon-Ward, F. 1936. Botanical and geographical explorations in Tibet, 1935. Geogr. J. (London) 88(5): 385–410.

Koehne, B. A. E. 1899. Gartenflora; Monatschrift für Deutsche und Schweizerische Garten-und Blumenkunde. Erlangen, Stuttgart, Berlin.

Kouznetsov, A. N. & Phan Luong. 1999. Flora of Fansipan Mountain. Pp. 11–18 *in* R. T. Sobey (editor), Biodiversity Value of Hoang Lien Mountains and Strategies for Conservation, Proceedings of Seminar and Workshop, 7–9th December 1998, Sa Pa, Lao Cai Province, Vietnam. Society for Environmental Exploration, London.

Kreichbaum, [W.] 1951. Die Tragödie eines deutchen Gartenbauschriftstellers. (Zum Tode Camillo Schneiders), Gart.-Z. Gärtn. Ill. Fl. 74(3): 27–29.

Kress, W. J., R. A. DeFilipps, E. Farr & R. Y. Y. Kyi. 2003. A Checklist of Trees, Shrubs, Herbs, and Climbers of Myanmar. Department of Systematic Biology-Botany, National Museum of Natural History, Washington, D.C.

Kron, K. A. 1993. A revision of *Rhododendron* section *Pentanthera.* Edinburgh J. Bot. 50: 249–364.

Laferrière, J. E. 1997. Transfer of specific and infraspecific taxa from *Mahonia* to *Berberis.* Bot. Zhurn. 82(9): 96–99.

Lamond, J. M. 1970. The Afghanistan collections of William Griffith. Notes Roy. Bot. Gard. Edinburgh 30: 159–175.

Landrum, L. R. 1999. Revision of *Berberis* (Berberidaceae) in Chile and adjacent southern Argentina. Ann. Missouri Bot. Gard. 86: 793–834.

Lee, W. T. 1996. Lineamenta Florae Koreae. Academy Press, Seoul.

Lemaire, C. A. 1859. L'Illustration Horticole, Vol. 6. Imprimerie et lithographie de F. et E. Gyselnyck, Gand, Belgium.

Léveillé, H. 1915. Flore du Kouy-Tchéou. Le Mans.

Léveillé, H. 1916. *Gymnosporia esquirolii.* China Rev. [1]: 18.

Li, H. L. 1952. Berberidaceae. *In* Notes on some families of Formosan Phanerogams. J. Wash. Acad. Sci. 42: 41–42.

Li, H. L. 1963. Berberidaceae. Pp. 168–172 *in* Woody Flora of Taiwan. Livingstone Publishing Company, Narberth, Pennsylvania.

Li, H. N., A. Q. Sun, J. H. Zuo & C. S. Lin. 2011. Genetic relationship between three *Berberis* species. J. Guizhou Agric. Sci. 39(10): 11–16.

Li, P. Y. 1965. Additional notes on the *Berberis* species of provinces Shensi, Kansu and north-eastern Chinghai. Acta Phytotax. Sin. 10: 210–214.

Li, S. C. 1987. *Berberis.* Pp. 347–349 *in* Editorial Committee, Flora of Anhui, Vol. 2. Chinese Prospect Publishing House, Beijing.

Li, X. H. 2008. The identity of *Berberis wuliangshanensis* C. Y. Wu ex S. Y. Bao and *B. jingguensis* G. S. Fan & X. W. Li. J. Trop. Subtrop. Bot. 16(2): 176–178.

Li, X. H. 2010. Two new synonyms of *Berberis* L. from China. Guihaia 30(4): 440–442.

Li, X. H. 2017. Identity of *Berberis wilsoniae* and *B. bodinieri* (Berberidaceae). Guihaia 37(10): 1339–1341.

Li, X. H. & S. J. Lu. 2013. Variation patterns of the characters of fruits and spines of *Berberis amurensis* Rupr. (Berberidaceae) in Taishan Mountain. Acta Bot. Boreal.-Occid. Sin. 33(3): 478–482.

Li, X. H. & L. C. Zhang. 2014. Variation patterns and characters of flowers and fruits of *Berberis replicata* W. W. Smith (Berberidaceae) from Yunnan Province. Acta Bot. Boreal.-Occid. Sin. 34(4): 720–726.

Li, X. H., W. H. Li, L. C. Zhang & X. M. Yin. 2015a. *Berberis* × *baoxingensis* (Berberidaceae), a new putative hybrid from western Sichuan, China. Phytotaxa 227(1): 25–34.

Li, X. H., S. Yuan, X. P. Shi & L. C. Zhang. 2015b. Variation patterns of *Berberis paraspecta* Ahrendt (Berberidaceae) from Yunnan Province. J. Trop. Subtrop. Bot. 23(5): 495–500.

Li, X. H., L. C. Zhang, W. H. Li, X. M. Yin & S. Yuan. 2017. New taxa of *Berberis* (Berberidaceae) with greenish flowers from a biodiversity hotspot in Sichuan Province, China. Pl. Diversity 39(2): 94–103.

Limpricht, W. 1922. Botanische Reisen in den Hochgebirgen Chinas und Ost-Tibets. Repert. Spec. Nov. Regni Veg. Beih. 12: 1–515, maps 1–9.

Liu, G. H. & S. Q. Zhou. 1991. *Berberis.* Pp. 578–583 *in* Y. C. Ma, Flora Intramongolica, 2nd ed., Vol. 2. Inner Mongolia Publishing House, Huhhot.

Liu, L. H. 2000. *Berberis.* Pp. 706–714 *in* K. M. Liu, Flora of Hunan, Vol. 2. Hunan Science & Technology Press, Hunan.

Liu, S. Z. 1988. *Berberis.* Pp. 543–547 *in* S. X. Lee, Flora Liaoningica, Vol. 1. Liaoning Science and Technology Press, Liaoning.

Liu, T. S. 1976. *Berberis.* Pp. 14–17 *in* H. L. Li et al. (editors), Flora of Taiwan, 1st ed., Vol. 2. Epoch Publishing Co., Taipei.

Liu, T. W. & the Editorial Committee of Shanxi Flora. 1998. *Berberis.* Pp. 1–2 *in* T. W. Liu & the Editorial Committee of Shanxi Flora, Flora Shanxiensis, Vol. 2. China Science and Technology Publishing House, Beijing.

Long, D. G. 1979. The Bhutanese itineraries of William Griffith and R. E. Cooper. Notes Roy. Bot. Gard. Edinburgh 37(2): 355–368.

Lu, S. Y. & Y. P. Yang. 1996. *Berberis.* Pp. 575–581 *in* T. C. Huang (editor-in-chief), Flora of Taiwan, 2nd ed., Vol. 2. Editorial Committee of the Flora of Taiwan, Taipei.

Lu, Z., S. L. Liu & Z. H. Lu. 1992. *Berberis*. Pp. 120–125 *in* Y. L. Chou, Flora Heilongjiangensis, Vol. 5. Northeast Forestry University Press, Harbin.

Ma, D. Z. 2007. *Berberis*. Pp. 249–257 *in* H. L. Liu & F. X. Hu, Flora Ningxiaensis, Vol. 1. Ningxia People's Press, Yingchuan.

Ma, J. 1986. *Berberis*. Pp. 480–484 *in* S. Y. He, Flora Hebeiensis, Vol. 1. Hebei Science of Technology Publishing House, Hebei.

Mabberley, D. J. 2008. The Plant-Book. Cambridge University Press, Cambridge.

Marchant, W. J. 1937. Catalogue of Choice Trees, Shrubs, Wall Plants and Climbers. Wimborne, Dorset.

Marquand, C. V. H. 1929. The botanical collection made by Captain F. Kingdon-Ward in the eastern Himalaya and Tibet in 1924–25. J. Linn. Soc., Bot. 48: 149–229.

Maung, A. M. 2011. In the Name of Pauk-Phaw: Myanmar's China Policy since 1948. Institute of Southeast Asian Studies, Singapore.

Maximowicz, C. J. 1877. *Berberis*. *In* Diagnosis plantarum novarum asiaticarum. Bull. Acad. Petersburgsk., sér. 3, 23: 307–310.

Maximowicz, C. J. 1889a. *Berberis*. Pp. 29–32 *in* Flora Tangutica. Academiae Imperialis Scientiarum Petropolitanae, Petropoli [St. Petersburg].

Maximowicz, C. J. 1889b. *Berberis*. Pp. 32–33 *in* Enumeratio Plantarum hucusque in Mongolia. Academiae Imperialis Scientiarum Petropolitanae, Petropoli [St. Petersburg].

Maximowicz, C. J. 1891. *Berberis*. Trudy Imp. S.-Peterburgsk. Bot. Sada 11: 40–42.

Mikhailova, M. A. 2000. Berberidaceae. Pp. 60–61 *in* V. I. Grubov (editor), Catalogue of the Type Specimens of Central Asian Vascular Plants in the Herbarium of the V. L. Komarov Botanical Institute (LE). St. Petersburg University Press, St. Petersburg.

Miller, H. S. 1970. The herbarium of Aylmer Bourke Lambert: Notes on its acquisition, dispersal, and present whereabouts. Taxon 19(4): 489–553.

Mizushima, M. 1954. Questions on Formosan barberries. Misc. Rep. Res. Inst. Nat. Resources 35: 28–32.

Moran, R. 1982. *Berberis claireae*, a new species from Baja California; and why not *Mahonia*? Phytologia 52: 221–226.

Mori, U. 2000. Exploring Aborigines—Mori Ushinosuke's Adventures in Taiwan. Translated and commented by Nan-Chun Yang. Yuan-Liou Publishing Co., Ltd, Taipei.

Morley, B. D. 1979. Augustine Henry: His botanical activities in China, 1882–1890. Glasra 3: 21–81.

Mueggler, E. 2011. The Paper Road. Archive and Experience in the Botanical Exploration of West China and Tibet. University of California Press, Berkeley.

Nakai, T. 1909. Flora Koreana. J. Coll. Sci. Imp. Univ. Tokyo 26(1): [1]–304.

Nakai, T. 1936. Flora Sylvatica Koreana, Vol. 21. Government of Chosen, Seoul.

Nguyên, N. T. 1998. Ða dang thuc vât có mach vùng núi cao Sa Pa-Phan Si Pan (Diversity of Vascular Plants of High Mountain Area Sa Pa, Phan Si Pan). Vietnam National University Publishing House, Hanoi.

Nuttall, T. 1818. The Genera of North American Plants and a Catalogue of the Species, to the Year 1817, Vol. 1. D. Heartt, Philadelphia.

Osborn, T. & E. Fanning (editors). 2003. Sa Pa Integrated Environmental Education Project. An End of Project Report. Frontier Vietnam, Hanoi.

Palibin, I. V. 1899. *Berberis koreana*. Trudy Imp. S.-Peterburgsk. Bot. Sada 17(1): 22.

Pallas, P. S. 1773. Reise durch verschiedene Provinzen des Russischen Reichs. Kayserlichen Academie der Wissenschaften, St. Petersburg.

Pallas, P. S. 1789. Flora Rossica seu stirpium Imperii Rossici per Europam et Asiam indigenarum descriptiones et icones. St. Petersburg.

Pallas, P. S. 1799–1801. Bemerkungen auf einer Reise in die Südlichen Statthalterschaften des Russischen Reichs in den Jahren 1793 und 1794. Gottfried Martini, Liepzig.

Pampanini, R. 1910. Le piante vascolari raccolte dal. Rev. P. C. Silvestri della Hu-peh durante gli anni 1904–1907 (e negli anni 1909, 1910). Nuovo Giorn. Bot. Ital. 17: 223–215.

Pampanini, R. 1915. Le piante vascolari raccolte dal. Rev. P. C. Silvestri della Hu-peh durante gli anni 1910–13. Nuovo Giorn. Bot. Ital., n. s., 22: 293–294.

Participants of the FFI/IUCN SSC Central Asian regional tree Red Listing workshop (Bishkek, Kyrgyzstan, 11–13 July 2006). 2007a. *Berberis iliensis*. The IUCN Red List of Threatened Species 2007: e.T63407A12666327. <http://dx.doi.org/10.2305/IUCN.UK.2007.RLTS.T63407A 12666327.en>, accessed 10 June 2019.

Participants of the FFI/IUCN SSC Central Asian regional tree Red Listing workshop (Bishkek, Kyrgyzstan, 11–13 July 2006). 2007b. *Berberis kaschgarica*. The IUCN Red List of Threatened Species 2007: e.T63493A12669507. <http://dx.doi.org/10.2305/IUCN.UK.2007.RLTS.T63493A 12669507.en>, accessed 10 June 2019.

Petélot, A. 1952. Les Plantes Médicinales du Cambodge, du Laos et du Viêtnam, Tom. 1. Archives des Recherches Agronomiques au Cambodge, au Laos et au Viêtnam, Saigon.

Pham, H. H. 1999. Cây có Viêt Nam (An Illustrated Flora of Vietnam), Vol. 1. Nhà Xuất Bản Trẻ, Ho Chi Minh City.

Philipson, W. R. & M. N. Philipson. 1986. A revision of *Rhododendron*. III. Subgenera *Azaleastrum*, *Mumeazalea*, *Candidastrum* and *Therorhodion*. Notes Roy. Bot. Gard. Edinburgh. 44: 1–23.

Poiret, J. L. M. 1808. Encyclopédie Méthodique, Botanique, Vol. 8. H. Agasse, Paris.

Rao, R., R. T. Hussain, B. Dutt & A. Garg. 1998a. Revision of the family Berberidaceae of India I. Rheedea 8(1): 1–66.

Rao, R., R. T. Hussain, B. Dutt & A. Garg. 1998b. Revision of the family Berberidaceae of India II. Rheedea 8(2): 109–143.

Regel, E. A. 1873. Synopsis Berberidis specierum varietatumque sectionis follies simplicibus cadusis Europam, Asiam Mediam, Japaniam et Americam Borealam incolenteum in Descriptiones; Plantarum Novarum in Regionibus Turkestanicus a CL. Viris Fedjenko, Korolkow, Kuschakewicz et Krause Collectis cum adnotationibus ad plantas vivas in Horto Imperiali Botanico Petropolitano cultas. Trudy Imp. St.-Peterburgsk. Bot. Sada 2: 407–421.

Regel, E. A. 1877. Berberideae. *In* Plantae Regiones Turkestanicus Incolentes, Secundum Specimina Secca. Trudy Imp. St.-Peterburgsk. Bot. 5(1): 226–228.

Rehder, A. 1936. Berberidaceae. *In* Notes on the ligneous plants described by H. Léveillé from eastern Asia. J. Arnold Arbor. 17(4): 321–323.

Rock, J. F. 1930. The glories of the Minya Konka. Natl. Geogr. Mag. 58: 4.

Roy, S., A. Tyagi, V. Shukla, A. Kumar, U. M. Singh, L. B. Chaudhary, B. Bag, et al. 2010. Universal plant DNA barcode loci may not work in complex groups: A case study with Indian *Berberis* species. PLOS ONE 5(10): e 13674.

Ruprecht, F. J. I. 1857. *Berberis amurensis*. Bull. Phys.-Math. Acad. St. Petersburgsk., sér. 2, 15: 260.

Ruprecht, F. J. I. 1869. *Berberis kaschgarica*. P. 38 *in* F. P. v. d. Osten-Sacken & F. J. Ruprecht, Sertum Tianchanicum.

Commissionnaires de l'Académie Impériale des sciences, St. Petersburg.

Schneider, C. K. 1904. Die Gattung *Berberis* (*Euberberis*). Bull. Herb. Boissier, sér. 2, 5: 33–48.

Schneider, C. K. 1905. Die Gattung *Berberis* (*Euberberis*). Bull. Herb. Boissier, sér. 2, 5: 133–148, 391–403, 449–464, 655–679, 800–831.

Schneider, C. K. 1906. Bemerkungen über die *Berberis* des Herbar Schrader. Mitt. Deutsch. Dendrol. Ges. 15: 175–181.

Schneider, C. K. 1908. Weitere Beiträge zur Kenntnis der Gattung *Berberis* (*Euberberis*). Bull. Herb. Boissier, sér. 2, 8: 192–204, 258–266.

Schneider, C. K. 1912. *Berberis minutiflora*. P. 914 *in* Illustriertes Handbuch der Laubholzkunde, Vol. 2. G. Fischer, Jena.

Schneider, C. K. 1913. Berberidaceae. Pl. Wilson. (Sargent) 1(3): 353–378.

Schneider, C. K. 1916. Weitere Beiträge zur Kenntnis der chinesischen Arten der Gattung *Berberis* (*Euberberis*). Oesterr. Bot. Z. 66: 313–326.

Schneider, C. K. 1917. *Berberis*. Pl. Wilson. (Sargent) 3(3): 434–443.

Schneider, C. K. 1918. Weitere Beiträge zur Kenntnis der chinesischen Arten der Gattung *Berberis* (*Euberberis*). Oesterr. Bot. Z. 67: 15–32, 135–146, 213–228, 285–300.

Schneider, C. K. 1923. Notes on hybrid *Berberis* and some other garden forms. J. Arnold Arbor. 4: 193–231.

Schneider, C. K. 1939. Neue *Berberis* der Sect. *Wallichianae*. Spec. Nov. Regni Veg. 46: 245–267.

Schneider, C. K. 1942. Die *Berberis* der Section *Wallichianae*. Mitt. Deutch. Dendr. Ges. 55: 1–60.

Schrader, H. A. 1838. Berberidaceae. *In* Reliquiae Schraderianae. Linnaea 12: 360–387.

Schrenk, A. G. 1841. *Berberis heteropoda*. Pp. 102–103 *in* F. E. L. Fischer & C. A. Meyer, Enumeratio Plantarum Novarum, Vol. 1. Typis G. Fischeri, Petropoli [St. Petersburg].

Shi, X. P., S. Yuan, X. H. Li & G. P. Xing. 2016. Leaf characteristics and geographical distribution of *Berberis amurensis* and *B. anhweiensis* of Berberidaceae. J. Zhejiang Univ. Sci. 42(6): 687–693.

Shimizu, A., H. Ohba & S. Akiyama. 2002. *Berberis*. Pp. 23–25 *in* Catalogue of the Type Specimens Preserved in the Herbarium. Department of Botany, the University Museum, the University of Tokyo Material Reports, No. 48(2), Tokyo.

Shimizu, T. 1963. *Berberis chingshuiensis*. J. Fac. Textile Sci. Technol. Shinsu Univ., A 36(12): 29.

Shimizu, T. 1964. *Berberis chingshuiensis*. Acta Phytotax. Geobot. 21: 26.

Smith, W. W. 1916. Diagnoses: Specierum novarum in herbario Horti Regii Botanici Edinburgensis cognitarum; species Chinensis. Notes Roy. Bot. Gard. Edinburgh 9: 81–83.

Soulié, J. A. 1904. Géographie de la principauté de Bathang. Géographie 9: 87–104.

Stapf, O. 1908. *Berberis acuminata*. Bot. Mag. 134, t. 8185.

Stapf, O. 1926. *Berberis lyciodes*. Bot. Mag. 151, t. 9102.

Stapf, O. 1928. *Berberis prainiana*. Bot. Mag. 152, sub t. 9153.

Stapf, O. 1931. *Berberis mitifolia*. Bot. Mag. 154, t. 9236.

Stapf, O. 1933. *Berberis tsarongensis*. Bot. Mag. 156, t. 9332.

Staunton, G. 1799. An Authentic Account of an Embassy from the King of Great Britain to the Emperor of China. Robert Campbell, Philadelphia.

Stearn, W. T. 1951. Camillo Karl Schneider 1976–1951. Gard. Chron. 129: 32.

Stearn, W. T. 1976. Frank Ludlow (1885–1972) and the Ludlow-Sherriff Expeditions to Bhutan and South-eastern Tibet of 1933–1950. Bull. Brit. Mus. Nat. Hist. Bot. 5(5): 242–308.

Steven, C. V. 1827. *Berberis iberica*. Syst. Veg. (ed. 16) [Sprengel] 4(2, Cur. Post.): 138.

Sun, H. & C. Y. Wu. 2015. Phytogeographical Regions of China. Pp. 176–204 *in* D. Y. Hong & S. Blackmore (editors), Plants of China: A Companion to the Flora of China. Science Press, Beijing; Cambridge University Press, Cambridge.

Sweet, R. 1826. *Berberis iberica*. P. 13 *in* Hortus Britanicus. James Ridgway, London.

Takhtajan, A. 1969. Flowering Plants: Origin and Dispersal. Smithsonian Institute Press, Washington, D.C.

Terabayashi, S. 2006. Berberidaceae. Pp. 342–343 *in* K. Iwatsuki, D. E. Boufford & H. Ohba (editors), Flora of Japan, Vol. IIa. Kodansha, Tokyo.

Tiwari, L. & B. S. Adhikari. 2011. *Berberis rawatii* sp. nov. (Berberidaceae) from India. Nordic J. Bot. 29: 184–188.

Tordoff, A., S. Swan, M. Grindley & H. Siurua (editors). 1999. Hoang Lien Nature Reserve. Conservation Evaluation 1997/98. Frontier Vietnam Environmental Research Report 17. Society for Environmental Exploration, U.K., and Institute of Ecology and Biological Resources, Hanoi.

Turland, N. J., J. H. Wiersema, F. R. Barrie, W. Greuter, D. L. Hawksworth, P. S. Herendeen, S. Knapp, et al. 2018. International Code of Nomenclature for algae, fungi, and plants (Shenzhen Code). Regnum Veg. 159.

Veitch, J. & Sons. 1913. New Hardy Plants from Western China (Introduced through Mr. E. H. Wilson). Autumn, Chelsea.

Veitch, J. H. 1906. Hortus Veitchii: A History of the Rise and Progress of the Nurseries of Messrs. James Veitch and Sons, Together with an Account of the Botanical Collectors and Hybridists Employed by Them and a List of the Most Remarkable of Their Introductions. J. Veitch & Sons, London.

Vierle, C. 1998. Camillo Schneider, Dendrologe und Gartenbauschriftsteller: Eine Studie zu seinem Leben und Werk. Technische Universität, Berlin.

Vilmorin, M. & D. G. J. M. Bois. 1904. Fruticetum Vilmorinianum. O. Doin, Paris.

Võ, V. C. 2007. Sách tra cứu tên cây cỏ Việt Nam (A Portable Dictionary of Generic Names of Vietnamese Higher Vascular Plant Families). Giao Duc, Hanoi.

Wang, Y. & S. Z. He. 2015. Leaf venation of *Berberis* (Berberidaceae) in Guizhou. Guihaia 35(4): 476–486.

Wang, Y. S. 1991. *Berberis*. Pp. 311–313 *in* S. K. Lee (editor), Flora of Guangxi, Vol. 1. Guangxi Science & Technology Publishing House, Nanning.

Wickenden, M. & M. Lear. 2010. Exploring the Upper Dulong River: The KWL Expedition to North-West Yunnan, September-October 2008. Alba Printers, Dumfries.

Wilson, E. H. 1913. A Naturalist in Western China, 2 vols. Methuen, London.

Woodman, D. 1962. The Making of Burma. Cresset Press, London.

World Health Organization. 1990. Medicinal Plants in Viet Nam. World Health Organization Regional Office for the Western Pacific, Manila; Institute of Materia Medica, Hanoi. Originally published as Cây Thuoc Việt Nam (Science and Technics Publishing House, Hanoi, 1990).

Ying, T. S. 1985. *Berberis*. Pp. 121–153 *in* Z. Y. Wu (editor), Flora Xizangica, Vol. 2. Science Press, Beijing.

Ying, T. S. 1993. *Berberis*. Pp. 549–557 *in* W. T. Wang (editor), Vascular Plants of the Hengduan Mountains, Vol. 1. Science Press, Beijing.

Ying, T. S. 1999. New taxa of *Berberis* Linn. (Berberidaceae) from China. Acta Phytotax. Sin. 37(4): 305–350.

Ying, T. S. 2001. *Berberis*. Pp. 54–214 *in* Flora Reipublicae Popularis Sinicae, Vol. 29. Science Press, Beijing.

Ying, T. S. 2011. *Berberis*. Pp. 715–771 *in* Z. Y. Wu, P. H. Raven & D. Y. Hong (editors), Flora of China, Vol. 19. Science Press, Beijing; Missouri Botanical Garden Press, St. Louis.

Ying, T. S., D. E. Boufford & A. R. Brach. 2011. Berberidaceae. Pp. 714–800 *in* Z. Y. Wu, P. H. Raven & D. Y. Hong (editors), Flora of China, Vol. 19. Science Press, Beijing; Missouri Botanical Garden Press, St. Louis.

Yu, C. C. & K. F. Chung. 2014. Systematics of *Berberis* sect. *Wallichianae* (Berberidaceae) of Taiwan and Luzon with description of three new species, *B. schaaliae*, *B. ravenii*, and *B. pengii*. Phytotaxa 184(2): 61–99.

Yu, C. C. & K. F. Chung. 2016. The rediscovery of *B. aristatoserrulata* and *B. brevisepala*. Sci. Amer. 178: 104–106.

Yu, C. C. & K. F. Chung. 2017. Why *Mahonia*? Recircumscribing *Berberis* s.l., with the description of two new genera, *Alloberberis* and *Moranothamnus*. Taxon 66(6): 1371–1392.

Zhang, Y. J. 1991. A new species of *Berberis* from Shaanxi. Acta Bot. Boreal. Occid. Sin. 11: 258–259.

Zhou, L. H. 1997. *Berberis*. Pp. 372–378 *in* S. W. Liu (editor), Flora Qinghaiica, Vol. 1. Qinghai People's Publishing House, Xining.

Zhu, G. F. 2004. *Berberis*. Pp. 194–201 *in* Y. Lin (editor), Flora of Jiangxi, Vol. 2. China Science and Technology Press, Beijing.

Zhu, Y. M. 1997. *Berberis*. Pp. 52–56 *in* S. B. Chen, Y. J. Zheng & F. Z. Li (editors), Flora of Shandong, Vol. 22. Qingdao Press, Shandong.

Website Names

Since over time many institutions and organizations change their website addresses, rather than give ones that may become obsolete, the following are the names of the main websites drawn on.

CHINA (INCLUDING TAIWAN)

Chinese Virtual Herbarium (CVH)

National Specimen Information Structure (NSII)–Botany Section

Chinese National Herbarium, Institute of Botany, Chinese Academy of Sciences, Beijing (PE)

Kunming Institute of Botany, Chinese Academy of Sciences (KUN)

Academica Sinica, Taiwan (HAST)

National Taiwan University (TAI)

Taiwan Forestry Research Institute (TAIF)

NON-CHINESE WEBSITES

Botanischer Garten und Botanisches Museum Berlin-Dahlem (B)

Conservatoire et Jardin botaniques de la Ville de Genève (G)

Harvard University Herbarium (A, GH)

Index Herbariorum (Barbara Thiers, editor)

JSTOR Global Plants

Muséum National d'Histoire Naturelle, Paris (P)

National Herbarium of the Netherlands (L, U)

Natural History Museum, London (BM)

Royal Botanic Garden Edinburgh (E)

Royal Botanic Gardens, Kew (K)

Royal Horticultural Society Herbarium, Wisley (WSY)

Smithsonian Institution, Washington (US)

Sweden's Virtual Herbarium (GB, S, UPS)

The New York Botanical Garden (NY)

Tropicos, Missouri Botanical Garden (MO)

Virtual Herbaria JACQ (maintained by WU)

Index to Accepted Species

Numbers in parentheses correspond to the species number in this treatment. The first page number indicates the main entry, the second the distribution map.

Index to All Taxa of *Berberis*

Names of taxa found in China are in Roman, while those not found in China (including synonyms of accepted taxa) are in *italic*. Accepted taxa are in **boldface**. For those names occurring more than once in the treatment, **bold** page numbers indicate the key reference where the taxon is described, synonymized, or informally treated. Page numbers for distribution maps are provided in the Index to Accepted Species. Keys 1–11 are located on pages 22–37.

ADDENDA

Berberis concinna Hook. f., Bot. Mag. 79: t. 4744. 1853. TYPE: India. Sikkim, Lachen, 3650 m, 31 July 1849, *J. D. Hooker s.n.* (lectotype, designated by Ahrendt [1961: 119], K K000077361!).

Berberis concinna var. *brevior* Ahrendt, J. Roy. Asiat. Soc. Bengal 11: 3. 1945, as "*breviora*." TYPE Nepal. Namlang, 2745 m, 28 Oct. 1931, *K. N. Sharma E269* (holotype, BM BM000573943!; isotype, E E00465284!).

Berberis concinna was listed above in the Excluded Taxa section on the grounds that though it was included by Ying (2001: 88) as being found in China I could find no evidence for this. However, as this monograph was at the proof page stage, I discovered the specimen listed below. Identified only as *Berberis sp.* it appears to have been a very recent addition to the PE herbarium database. Though the specimen is sterile it is recognizably *B. concinna* (an identification confirmed by Bhaskar Adhikari, pers. comm. 18 Nov. 2019). It was collected some 7 km north of the border with Nepal.

Selected specimens.
S Xizang (Tibet). Gyirong (Jilong) Xian: Jilong Zhen, Zha Cun, Laduo mtn. pass, 28.39733°N, 85.41017°E, 4165 m, 11 Dec. 2011, *Y. S. Chen et al. 402* (PE 01992035).

Julian Harber, Xiaojin Xian, Sichuan (1 August 2018). Photo by Bruce Bartholomew.